AF521852

INTRODUCTION TO
COMPLEX VARIABLES

FREDERICK P. GREENLEAF
Associate Professor of Mathematics
Department of Mathematics, New York University

1972
W. B. SAUNDERS COMPANY PHILADELPHIA • LONDON • TORONTO

W. B. Saunders Company: West Washington Square
Philadelphia, Pa. 19105

12 Dyott Street
London, WC1A 1DB

1835 Yonge Street
Toronto 7, Ontario

Introduction to Complex Variables ISBN 0-7216-4260-8

Library of Congress catalog card number 70–180179.

Print No.: 9 8 7 6 5 4 3 2 1

To Sandra McKnight

PREFACE

In applying complex variable theory there is no substitute for a firm grasp of the basic mathematical facts. For mathematics majors this is obvious; however, students inclined toward applications face constantly evolving demands, and are badly served by a hasty outline or compendium of useful formulas. There are excellent books on complex analysis but, unfortunately, most are directed toward students with considerable mathematical maturity. The present book strikes a balance between such intensely mathematical treatment and the opposite extreme, an applications-oriented formulary.

The manuscript has evolved through several revisions during the past six years, as a result of my experience at Berkeley and New York University, teaching complex variable theory to mathematicians, engineers, and physicists. Every effort has been made to give a careful, self-contained account of the mathematical side of the theory, that does justice to basic applications such as conformal mapping, potential theory, and boundary value problems in physics.

The book is designed to allow a smooth transition from a background of basic calculus. Courses in advanced calculus or modern analysis are not presumed; the necessary mathematical techniques are introduced within the text, with many interpolated examples and illustrations. There is constant reference to the geometrical and intuitive meaning of the concepts discussed and, wherever possible, their practical significance. Some advanced topics are outlined without proofs but, in these cases, many references for further reading are provided. Any reasonably competent student should be able to read most of the book on his own, without having to turn to his instructor for interpretations. Thus, instructors will be able to cover the main points and be confident that the student has a detailed account of the subject, with many worked examples, to fall back on. In order to ease the student's entry into the subject, power series and other topics familiar from calculus are treated as early as possible; moreover, the relevant facts from calculus are reviewed in the early chapters in a way that emphasizes their role in complex variable theory.

Chapter 1 summarizes elementary facts about the complex number system. For most students a superficial reading should suffice to re-acquaint them with this material and introduce them to the notation used throughout the book. Section 1.3 deserves special attention: it introduces the triangle inequality, and explains how sets may be specified by inequalities. Chapter 2 deals with limits of sequences and of functions of a complex variable; most students will be able to read the early sections on their own. Complex variable theory begins in earnest in Section 2.6, where elementary functions such as e^z and $\log(z)$ are introduced. Sequences and series of functions are discussed in Chapter 3; uniform convergence, power series, analytic functions, and the maximum modulus theorem are treated at this stage. Chapter 4 gives a detailed account of holomorphic mappings, with strong emphasis on global mapping properties and their geometric interpretation. Fractional linear transformations are studied, the complex sphere is introduced, and specific mappings such as $w = 1/z$ and $w = \sin(z)$ are examined in detail. Integration theory is handled in Chapter 5. The connection between invariance of contour integrals and antiderivatives is emphasized in the early sections, and Cauchy's Theorem is interpreted from this point of view. The chapter concludes with a modern and eminently practical version of the Cauchy theorems based on the notion of winding number.

Chapters 2 through 5 contain the heart of the mathematical development; the remaining chapters are relatively independent, and deal with various applications. Chapter 6 is devoted to residues and singularities (discussed from several points of view), and the applications of the residue calculus. Boundary value problems for Laplace's equation are the central topic in Chapter 7; the conformal mapping principle is studied, with many worked examples of its use in solving boundary value problems; finally, the Poisson integral formula is discussed. Chapter 8 provides a unique, self-contained account of the connection between complex variable theory and its physical applications. Electrostatics, gravitation, heat propagation, fluid flow, and the mathematical similarities between these subjects are discussed. Finally, Chapter 9 deals with advanced topics such as the Schwarz reflection principle, the Riemann mapping theorem, and the Schwarz-Christoffel formula.

Almost every section ends with an extensive list of exercises, many with answers, interpretive commentary, and hints. Generally, the problems increase in difficulty from beginning to the end of each set of exercises. The more difficult problems often introduce important new concepts, or extend the basic results of the section. Students should be encouraged to read and understand the significance of these problems, even if they do not try to solve them in detail. Certain sections, marked with an asterisk, can be skipped or left for the student to read on his own.

I am deeply indebted to Mr. George Fleming, mathematical editor for the W. B. Saunders Company and personal friend, for encouraging me to begin this book, offering considerable moral support along the way, and making many useful suggestions about content and style. I am equally indebted to Professor Gerald MacLane, of Purdue, who devoted many hours of his time to reviewing the final manuscript; his enthusiastic, painstaking, and urbane commentary has had a substantial influence on the form and spirit of the finished book. I am also grateful to Professor Glenn Schober, of Indiana University, and Mr. Carlos Puig, editorial assistant at Saunders, for their efforts in preparing independent critical reviews of the manuscript. The editorial and production staffs at Saunders, especially Herb Powell and Lorraine Battista, have my hearty appreciation for their efforts in preparing the book for production. In addition, the preparation of the manuscript was aided, in its early stages, by Professors Bernard Epstein and James Osborn, who reviewed parts of early drafts, and by Miss Maie Croner, who painstakingly typed the exercises.

Last, but by no means least, my appreciation to my wife for her encouragement during the whole project.

FREDERICK P. GREENLEAF

New York, New York

CONTENTS

Chapter 3

Power Series and Analytic Functions 108

Chapter 4

Mapping Properties of Holomorphic Functions 173

Chapter 5

Integration in the Complex Plane 250

Chapter 9

1 BASIC FACTS ABOUT THE COMPLEX NUMBER SYSTEM

This chapter provides a self-contained description of the complex number system that particularly emphasizes its geometric aspects. Its content will be more or less familiar to anyone who has had a previous exposure to complex numbers as part of an Algebra or Calculus course, and these readers may wish to turn to Chapter 2 immediately. However, we encourage the reader to examine Sections 1.3 to 1.5 to make sure he understands the terminology used throughout this book and the basic facts about absolute values of complex numbers.

Let us consider some of the steps in the development of the complex number system. The first hint of inadequacies in the real number system comes when we realize that certain algebraic operations are not always meaningful—particularly the operation of taking square roots. A square root of a number a is any number x such that $x^2 = x \cdot x = a$. As is well known, the number $a = 0$ has $x = 0$ for its only square root, and a positive number $a > 0$ has two square roots, $x = +\sqrt{a}$ and $x = -\sqrt{a}$; but if a is *negative* it has no real square roots at all. For example, if $a = -1$, there are no real numbers x that satisfy the equation $x^2 = -1$, because the square of any real number x is non-negative and cannot equal -1. Thus the operation $\sqrt{a}$ is not meaningful for negative real numbers.

Another difficulty arises when we use the quadratic formula, which gives us the solutions of a general quadratic equation

$$ax^2 + bx + c = 0; \tag{1}$$

the solutions explicitly involve the square root operation, and are given by

(2) $$x_+ = \frac{-b + \sqrt{b^2 - 4ac}}{2a} \qquad x_- = \frac{-b - \sqrt{b^2 - 4ac}}{2a}.$$

However, if $b^2 - 4ac$ is *negative*, the square root in these formulas is meaningless: equation (1) has no solutions in the real number system. For example, consider the equation $x^2 + x + 1 = 0$, which has no real solutions. Similar difficulties are also encountered in trying to solve equations of higher degree.

The German mathematician C. F. Gauss (1777–1855) showed how to construct a new system of numbers that

(i) includes all the real numbers, so that it is a *larger* number system than the real numbers, and
(ii) includes a number i (*not* a real number) with the property $i^2 = -1$.

This is the system of complex numbers; the extra number i plays the role of $\sqrt{-1}$. The complex number system is actually not difficult to construct; our first order of business, in Section 1.1 will be to display this construction. Complex numbers will be interpreted as points in a plane, just as real numbers are thought of as points on a line.

The presence of the number i in the complex number system allows us to interpret square roots of negative real numbers as complex numbers: if a is positive real, the square roots $\sqrt{-a}$ will be given by the complex numbers $i\sqrt{a}$ and $-i\sqrt{a}$.† In fact, we will see that every non-zero real number has two square roots. Moreover, if we allow complex numbers x as solutions of the quadratic equation (1) we find that it always has solutions, given by equation (2), because the radicals involved make sense. Even more is true: Gauss proved that every polynomial equation of degree N ($N = 1, 2, \ldots$)

$$a_N x^N + \cdots + a_1 x + a_0 = 0$$

has exactly N solutions in the complex number system, provided multiple roots are counted appropriately. This result, known as the Fundamental Theorem of Algebra, is discussed in Chapter 5.

The moral of these observations is that one is using the wrong number system when one tries to solve general (or even quadratic) polynomial equations with real numbers—sometimes there are no solutions, sometimes too few (fewer than the degree N of the equation); even if some of the solutions are real, others might appear only in the larger system of complex numbers. This lesson from the theory of equations should not be lost on those who are using real numbers in other, perhaps very different, ways. The Calculus, the theory of differential equations, and many other subjects of interest to scientists all rest

† Since $i^2 = -1$, the number $i\sqrt{a}$ that we get by multiplying i and $\sqrt{a}$ has for its square the number $(i \cdot \sqrt{a})^2 = i \cdot i \cdot \sqrt{a} \cdot \sqrt{a} = -1 \cdot a = -a$, so that $i\sqrt{a}$ is a square root of $-a$. Similar reasoning applies to $-i \cdot \sqrt{a}$.

on the use of the real number system; perhaps there is something new and important to be learned by considering complex numbers in these situations.

And indeed there is! The introduction of complex numbers has been fruitful in ways surpassing the dreams of its creators. These discoveries are the subject of this book.

1.1 THE DEFINITION OF THE COMPLEX NUMBER SYSTEM

To define complex numbers we start with a plane equipped with Cartesian coordinates, so that a point P is located by specifying a pair (a, b) of real numbers (Figure 1.1). Two pairs of coordinates (a, b) and (a', b') specify the same point in the plane if and only if $a = a'$ and $b = b'$; thus the coordinates $(0, 1)$ and $(1, 0)$ specify quite different points in the plane, even though the same real numbers 0 and 1 enter into each pair. We shall use the symbol $\mathbf{R}^2$ to indicate the **Cartesian plane** (the plane equipped with Cartesian coordinates); this notation properly suggests that the Cartesian plane is obtained by putting together two copies of the real number system, which is indicated by the symbol $\mathbf{R}$.

The complex number system consists of the Cartesian plane $\mathbf{R}^2$, together with two algebraic operations $(+)$ and $(\cdot)$, which allow us to "add" and "multiply" points in the plane. In terms of coordinates, the **addition operation** is defined by

$$(x, y) + (u, v) = (x + u, y + v), \tag{3}$$

and the **multiplication operation** by

$$(x, y) \cdot (u, v) = (xu - yv, xv + yu). \tag{4}$$

The auxiliary operations of subtraction and division will be derived from addition and multiplication later on. Let us immediately point out that these operations *commute;* if $z = (x, y)$ and $w = (u, v)$ are points in the plane, then

$$z + w = w + z$$
$$z \cdot w = w \cdot z.$$

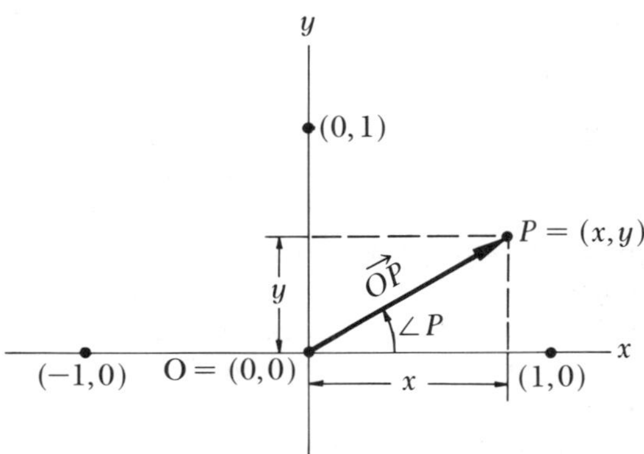

Figure 1.1 The point P is located by its Cartesian coordinates (x, y) as shown.

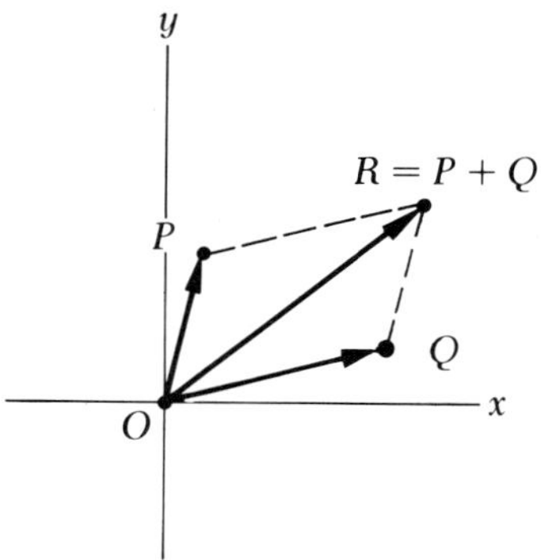

Figure 1.2 The parallelogram law for addition.

These algebraic properties can be verified from the definitions by direct calculations. We could give a long list of other algebraic laws, such as the *distributive law*,

$$w \cdot (z_1 + z_2) = w \cdot z_1 + w \cdot z_2.$$

and we will list some of these laws in the exercises. We will not take the time to present them here since they will be familiar to many readers.

The definitions of $(+)$ and $(\cdot)$ are chosen so these operations will have certain natural geometric properties. Let $P = (x, y)$ and $Q = (u, v)$ be points in the plane and let $O = (0, 0)$ be the origin of coordinates. If we consider the directed line segments $\overrightarrow{OP}$ and $\overrightarrow{OQ}$, as shown in Figure 1.2, we can verify that the point $R = P + Q$ given by formula (3) is located so that the directed segment $\overrightarrow{OR}$ is the diagonal of the parallelogram that has $\overrightarrow{OP}$ and $\overrightarrow{OQ}$ as its sides. Conversely, we would be forced to accept definition (3) if we wanted addition to satisfy this geometric "*parallelogram law*."

> Such a parallelogram law arises naturally in many physical situations; it is reflected in the addition laws of mechanical forces, electromagnetic fields, and other natural phenomena. Thus we may expect a number system whose addition operation has similar properties to be useful in studying physical problems. More will be said about this in Chapters 7 and 8.

A geometric interpretation of multiplication will be given later on.

Hereafter, when we speak of the Cartesian plane equipped with the addition and multiplication operations just defined, we shall call it the **complex plane** (or **complex number system**) and use the symbol **C** for it. Several remarks should be made about the algebraic features of this system. First, the real numbers **R** can be identified with a subset of **C**; this is done by letting the real number x in **R** correspond to the complex number $(x, 0)$ in **C**. This correspondence between x and $(x, 0)$ is one-to-one in the sense that $(x, 0) \neq (y, 0)$ in **C** if $x \neq y$ in **R**, and it identifies **R** with the set of points on the x-axis in the plane. Furthermore, the correspondence $x \to (x, 0)$ respects the algebraic operations in both **R** and **C** since

$$(x, 0) + (u, 0) = (x + u, 0)$$
$$(x, 0) \cdot (u, 0) = (x \cdot u, 0)$$

for all x and u in **R**. Thus, when we identify **R** with the x-axis in **C**, the algebraic operations that this subset inherits from **C** agree precisely with the original operations in **R**. Hereafter we shall feel justified in regarding **R** either by itself or as a subset of **C**, as most suits our needs. In accordance with this, if we have some real number a in mind we will often write a for the corresponding complex number, which should properly be written $(a, 0)$. For example, when we speak of -1, 0, or $+1$ as complex numbers we really mean the complex numbers $(-1, 0)$, $(0, 0)$, and $(1, 0)$ respectively. Once we have identified **R** as a subset of the complex number system **C**, we can make sense of the operation of multiplying a complex number $z = (x, y)$ by a real number a; thus

(5) $$a \cdot z = a \cdot (x, y) \quad \text{means} \quad (a, 0) \cdot (x, y) = (ax, ay).$$

If $a = -1$, we will always write $-z$ for the number $(-1) \cdot z$; the number $-z$ is referred to as the **negative of z**. Differences $w - z$ (and the operation of subtraction) are defined by taking

$$w - z = w + (-z).$$

The "real" numbers 0 and 1 in **C** have special algebraic significance; they are the only elements of **C** that have the properties:

$$\left.\begin{aligned} 0 + z &= z \\ 1 \cdot z &= z \end{aligned}\right\} \text{ for all } z \text{ in } \mathbf{C}.$$

Furthermore, it is easy to check that

$$\left.\begin{aligned} &0 \cdot z = 0 \\ &z + (-z) = z - z = 0 \end{aligned}\right\} \text{ for all } z \text{ in } \mathbf{C}.$$

There is another special number in **C**, the number $i = (0, 1)$, which is a square root of -1; thus

$$i^2 = -1$$

since $i^2 = i \cdot i = (0, 1) \cdot (0, 1) = (-1, 0) = -1$. The negative $-i$ of i is also a square root of -1:

$$(-i)^2 = -1$$

since $(-i)^2 = (-1) \cdot i \cdot (-1) \cdot i = (-1)^2 \cdot i^2 = i^2 = -1$. It is not very hard to see that the numbers $+i$ and $-i$ are the *only* numbers z in **C** such that $z^2 = -1$. The existence of even one solution of this equation is a distinguishing

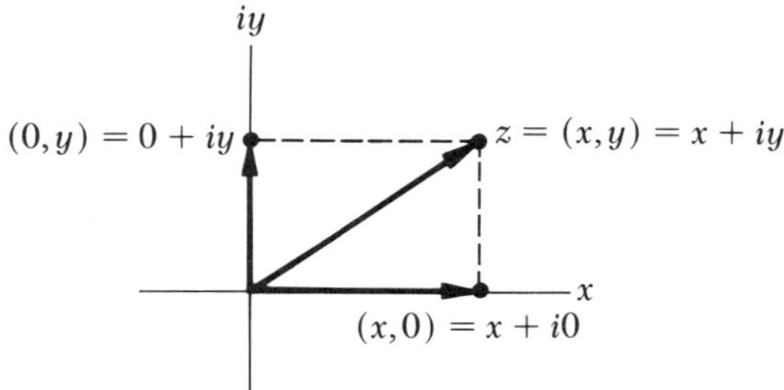

Figure 1.3 The representation of $z = (x, y)$ as the sum of the real number $x = x + i0$ and the pure imaginary number $iy = 0 + iy$.

property of the complex number system; in the system of real numbers **R** there are no solutions of this equation.

We say that a complex number is **real** if it has the special form $z = (x, 0)$. On the other hand, we say that z is **pure imaginary** if it has the form $a \cdot i = (0, a)$, where a is some real number. When **C** is represented as the Cartesian plane, the real numbers in **C** fill up the x-axis while the pure imaginary numbers fill up the y-axis. Every complex number z has a unique representation as the sum of a real number and a pure imaginary number. In fact, if $z = (x, y)$ then x and y are real numbers and

$$\begin{aligned} z = (x, y) &= (x, 0) + (0, y) \\ &= x \cdot (1, 0) + (0, 1) \cdot y \\ &= x \cdot 1 + i \cdot y = x + iy. \end{aligned} \tag{6}$$

This decomposition is illustrated in Figure 1.3. Clearly, if we take the system of real numbers and adjoin a single new (complex) number i, we get the whole complex number system by forming (real) linear combinations of the two numbers 1 and i. If z is a complex number having the form $z = x + iy$ when represented as in (6), we shall refer to x as the **real part of z** and write $x = \mathrm{Re}(z)$; likewise we shall call y the **imaginary part of z** and denote it by $y = \mathrm{Im}(z)$. Thus,

$$z = \mathrm{Re}(z) + i\,\mathrm{Im}(z) \tag{7}$$

if z is a complex number. Notice that both the numbers $\mathrm{Re}(z)$ and $\mathrm{Im}(z)$ are *real numbers.*

Besides $(+)$ and $(\cdot)$, there is also a division operation in **C**; if z and w are points in **C** such that $w \neq 0$, we may define a quotient z/w. As with real numbers, z/w is defined to be the unique complex number p such that

$$z = p \cdot w. \tag{8}$$

Obviously, the quotient is being defined so that $z = (z/w) \cdot w$. One can easily figure out how to compute z/w; by using the defining equations for the operation $(\cdot)$ we get the following explicit formula.

$$\frac{z}{w} = \frac{x + iy}{u + iv} = \left(\frac{xu + yv}{u^2 + v^2}\right) + i\left(\frac{yu - xv}{u^2 + v^2}\right). \tag{9}$$

If we let $w = 0$ in equation (8), the equation will have either no solutions at all or infinitely many solutions. In either case there is no well defined candidate for the role of $z/0$. This is why division by zero makes no sense. Of special interest is the **reciprocal** of a non-zero complex number $w = u + iv$:

$$\frac{1}{w} = \frac{1}{u + iv} = \left(\frac{u}{u^2 + v^2}\right) + i\left(\frac{-v}{u^2 + v^2}\right). \tag{10}$$

It is not at all hard to see that

$$\frac{z}{w} = z \cdot \left(\frac{1}{w}\right), \tag{11}$$

so that computing quotients can be reduced to the computation of reciprocals. Throughout this book we will use the symbol z^{-n} ($n = 1, 2, \ldots$ an integer) for the reciprocal $1/z^n$ of z^n. For various algebraic properties of quotients, we refer the reader to Exercise 9.

EXERCISES

1. Plot the location in the plane of the following complex numbers. Calculate the lengths of the segments $\overrightarrow{OP}$ and determine the angle from the positive real axis to this segment, whenever this is possible without resorting to trigonometric tables.

(i) $-i$ (iv) $i + 1$

(ii) $\frac{3}{2}$ (v) $\dfrac{1}{2} + i\dfrac{\sqrt{3}}{2}$

(iii) -1 (vi) $2 + i$

2. Carry out the following calculations.

(i) $\dfrac{1}{1 - i} - \dfrac{1}{1 + i} = i$ (iv) $\dfrac{-2}{1 + i\sqrt{3}} = -\dfrac{1}{2} + i\dfrac{\sqrt{3}}{2}$

(ii) $(i - 1)^{-4} = -\frac{1}{4}$ (v) $\dfrac{26}{i(1 - i)(3 - 2i)} = 5 - i$

(iii) $(1 + i)^4 = -4$

3. By direct calculations, verify that $+i$ and $-i$ are the *only* solutions of the equation $z^2 = -1$ in the complex number system. Find all solutions of the equation $z^4 = 1$ in the complex number system.
Answers: $+1, -1, +i, -i$ are the solutions of $z^4 = 1$.

4. If $z = x + iy$, express the numbers

(i) z^3 (ii) $1/z$ (assuming $z \neq 0$)

in the form $u + iv = U(x,y) + iV(x,y)$.

Answers: (i) $(x^3 - xy^2) + i(x^2y - y^3)$; (ii) $\left(\frac{x}{x^2 + y^2}\right) + i\left(\frac{-y}{x^2 + y^2}\right)$

5. Verify the parallelogram law for addition, using analytic geometry.

6. Verify the following properties of addition and multiplication.

(i) $z + w = w + z$

(ii) $z \cdot w = w \cdot z$

(iii) $1 \cdot z = z$ (all z)

(iv) $p \cdot (q \cdot r) = (p \cdot q) \cdot r$

(v) $p \cdot (z + w) = (p \cdot z) + (p \cdot w)$

7. Given complex numbers $z = x + iy$ and $w = u + iv$ (such that $w \neq 0$), solve the equation $z = p \cdot w$ to establish formula (9) directly.

8. Prove that the product $z \cdot w$ of two complex numbers is zero if and only if at least one of the numbers z and w is zero. Use this to prove the following "cancellation law": if $p \neq 0$ and if $p \cdot z = p \cdot w$, then $z = w$.

9. Verify the following properties of reciprocals.

(i) $1 \Big/ \left(\frac{1}{z}\right) = z, \quad \text{if } z \neq 0$

(ii) $\frac{1}{z \cdot w} = \left(\frac{1}{z}\right) \cdot \left(\frac{1}{w}\right), \quad \text{if } z \neq 0 \text{ and } w \neq 0$

(iii) $\left(\frac{z}{w}\right) + \left(\frac{p}{q}\right) = \frac{zq + wp}{wq}, \quad \text{if } w \neq 0 \text{ and } q \neq 0.$

Hint: Use the results of Exercise 6. These formulas can be proved by appealing to the algebraic definition of reciprocal, rather than computational formulas (9) and (10).

1.2 THE CONJUGATE $\overline{Z}$ OF A COMPLEX NUMBER Z

The conjugation operation in the complex number system has no counterpart in the real number system. If $z = x + iy$, where x and y are real, we define complex conjugate of z to be the number

$$\bar{z} = x - iy. \tag{12}$$

Geometrically, this operation reflects z across the real axis, as shown in Figure 1.4. This makes it clear that every real number $z = x + i0$ is left fixed by conjugation; $\bar{z} = \overline{(x + i0)} = (x - i0) = z$; however, if z is pure imaginary,

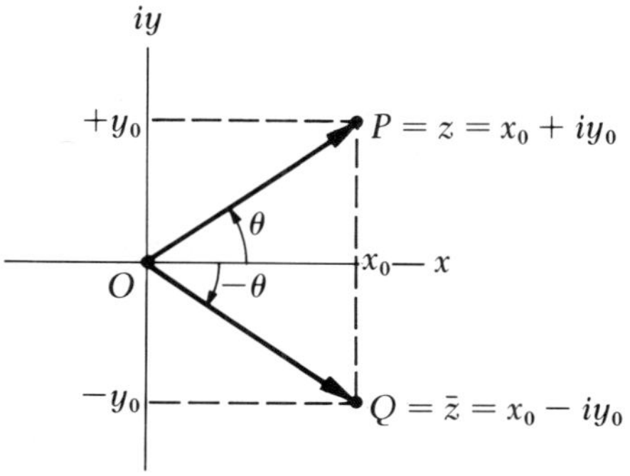

Figure 1.4 The geometric interpretation of the conjugate $\bar{z}$ as z reflected through the x-axis.

$z = 0 + iy$, then $\bar{z} = \overline{(0 + iy)} = 0 - iy = -z$. The conjugation operation interacts nicely with the other operations $(+)$ and $(\cdot)$; one can easily verify that

(13)
$$\text{(A)}\ \overline{z \cdot w} = \bar{z} \cdot \bar{w}$$
$$\text{(B)}\ \overline{z + w} = \bar{z} + \bar{w}$$
$$\text{(C)}\ \overline{(z/w)} = \bar{z}/\bar{w}$$
(D) $\bar{z} = z$ if and only if z is *real.*

It is also interesting to notice that application to z of the conjugation operation twice in succession yields z again:

$$\overline{(\bar{z})} = z$$

for all z in **C**, and that complex conjugation is closely related to the real and imaginary parts of z by the formula

(14)
$$\operatorname{Re} z = \frac{1}{2}(z + \bar{z}) \qquad \operatorname{Im} z = \frac{1}{2i}(z - \bar{z}),$$

as the reader can easily check. For other properties of complex conjugation, see Exercises 1 to 3.

EXERCISES

1. Let z be any non-zero complex number, and consider its powers z^n $(n = 0, \pm 1, \pm 2, \ldots)$, taking $z^0 = 1$. Verify that $\overline{(z^n)} = (\bar{z})^n$.

2. Describe the relative positions of z, $-z$, $\bar{z}$, $-\bar{z}$, $1/z$, $1/\bar{z}$, and $-1/\bar{z}$ if $z \neq 0$. Sketch these numbers taking $z = 2 + i$.

3. Verify the following identities.

(i) $\overline{z \cdot w} = \bar{z} \cdot \bar{w}$

(ii) $\overline{1/z} = 1/\bar{z}$, provided $z \neq 0$

(iii) $\overline{z + w} = \bar{z} + \bar{w}$

(iv) $\text{Re}(z) = \frac{1}{2}(z + \bar{z})$

(v) $\text{Im}(z) = \frac{1}{2i}(z - \bar{z})$

(vi) $\overline{iz} = -i\bar{z}$

1.3 GEOMETRY OF THE COMPLEX NUMBERS

We must introduce a **distance function** if we wish to discuss limit processes in a number system. Such a function assigns a distance $d(z, w)$, a non-negative real number, to any pair of points z and w. In the real number system **R**, there is a natural distance

(15) $$d_{\mathbf{R}}(x, y) = |x - y|$$

for all x and y in **R**. In **C** there is also a natural distance, whose definition is strongly suggested by our use of the Cartesian plane as a model for **C**; we simply take the usual **Euclidean distance** between points in the plane. Thus, the distance from a point $z = x + iy$ to another point $w = u + iv$ in **C** is defined to be

(16) $$d_{\mathbf{C}}(z, w) = \sqrt{|x - u|^2 + |y - v|^2} = \sqrt{(x - u)^2 + (y - v)^2}.$$

(See Figure 1.5.) Distance functions may be introduced not only in **C** and in **R**, but also in other geometric systems, as we will indicate.

It is evident that when we consider the real numbers as a subset of **C**, the distance this subset "inherits" from **C** agrees with the original distance in **R** given by (15). In fact,

$$d_{\mathbf{C}}((a, 0), (b, 0)) = \sqrt{|a - b|^2 + 0^2} = \sqrt{|a - b|^2} = |a - b| = d_{\mathbf{R}}(a, b)$$

for real numbers a and b.

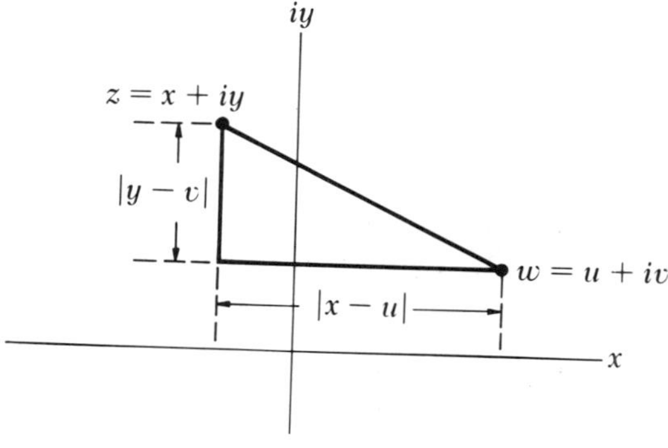

Figure 1.5 The distance $d(z, w)$ between points $z = x + iy$ and $w = u + iv$ is given by $d(z, w) = \sqrt{|x - u|^2 + |y - v|^2}$.

In discussing distances in **C**, it is very helpful to define the **absolute value** $|z|$ of a complex number z:

(17) $$|z| = \sqrt{|x|^2 + |y|^2} = \sqrt{x^2 + y^2}$$

if $z = x + iy$. From definitions (16) and (17) we see that

(18) $$|z - w| = d_{\mathbf{C}}(z, w)$$

for any pair of complex numbers z and w; this formula for distances in **C** is similar to formula (15), which defines distances in **R** in terms of absolute values. The absolute value $|z|$ should be interpreted as giving the distance from the point z to the origin, since

(19) $$|z| = |z - 0| = d_{\mathbf{C}}(z, 0).$$

It is also identified with the length of the line segment from 0 to z in the plane.

In the real number system there is a notion of *order*—i.e., a statement such as "$x > y$" has a natural meaning. The order relation in **R** plays an important role in finding limits, derivatives, and other functions. Furthermore, the order relation is directly involved in the definition of the absolute value $|x|$ for real numbers, and thus in the definition of distances in **R**, which is basic to all limit problems. There is no reasonable way to put an order relation "$>$" into the complex number system; thus, arguments that involve an order relation are simply not possible in the system **C**. A good illustration of this difference between **R** and **C** is that there are only two ways in which we may approach a point in **R**: from the right, or from the left. There are infinitely many directions from which we may approach a point in **C**.

The absolute values $|z|, |w|, \ldots$ of complex numbers are all *real* (and in fact $|z| \geq 0$); thus it is possible to discuss order relations between *absolute values* of two complex numbers, even though it makes no sense to speak of order relations between the numbers themselves. This does not look like much of a grip on the situation, but it turns out to be enough to understand the geometry and limit problems associated with the complex number system. Such inequalities are often used to specify sets in the plane.

Example 1.1 Let $p = x_0 + iy_0$ be a point in the complex plane and let $r > 0$ be a fixed real number. The set consisting of all points z that satisfy $|z - p| = r$ is just the set of points whose distance from p is equal to r, since $d_{\mathbf{C}}(z, p) = |z - p|$. This can also be seen by noting that $|z - p| = r$ if and only if $|z - p|^2 = r^2$, which implies that

$$(x - x_0)^2 + (y - y_0)^2 = |z - p|^2 = r^2.$$

This is the equation of a circle about p with radius r. Similarly, the set defined by taking points z such that

$$|z - p| < r$$

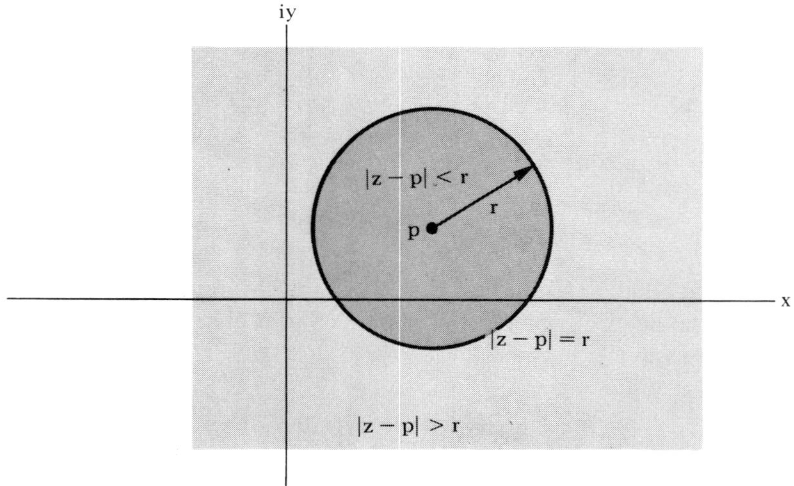

Figure 1.6 The set determined by $|z - p| < r$ is the disc about p; the set determined by $|z - p| > r$ is the region exterior to the circle $|z - p| = r$.

consists of those points whose distance from p, $d_{\mathbf{C}}(z, p) = |z - p|$, is less than r; this is just a disc of radius r about p, not including the boundary circle $|z - p| = r$. On the other hand, the inequality

$$|z - p| > r$$

determines the set of points that lie outside of the circle $|z - p| = r$, excluding this circle. These sets are shown in Figure 1.6.

Example 1.2 Consider the set E determined by the inequality $\mathrm{Im}(z) > 0$. If $z = x + iy$, this inequality means that we must have $y > 0$, while x is unrestricted. These points form the **upper half plane,** consisting of all points that lie above the real axis, as shown in Figure 1.7. The real axis itself is not part of the set E; in fact, the real axis consists of the points z such that $\mathrm{Im}(z) = 0$. Several other examples are presented in Exercises 3 and 6.

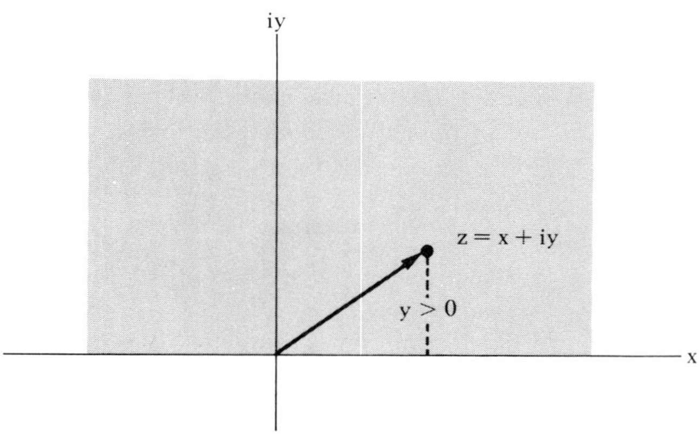

Figure 1.7 The upper half plane determined by the inequality $\mathrm{Im}(z) > 0$. The real axis is not part of this set.

Next, we examine the basic properties of $|z|$. There are several elementary algebraic properties, whose straightforward proofs will not be given here:

(20)

(A) $|z \cdot w| = |z| \cdot |w|$

(B) $|\bar{z}| = |z| = |-z|$

(C) $\left|\dfrac{z}{w}\right| = \dfrac{|z|}{|w|}$ and $\left|\dfrac{1}{w}\right| = \dfrac{1}{|w|}$, provided $w \neq 0$.

(D) $|z| = 0$ if and only if $z = 0$

(E) $|z|^2 = (\mathrm{Re}(z))^2 + (\mathrm{Im}(z))^2 = |\mathrm{Re}(z)|^2 + |\mathrm{Im}(z)|^2$

It is a useful consequence of (20E) that

(21)
$$|z|^2 = z \cdot \bar{z}$$

for all z in **C**. There are several elementary order relations between absolute values:

(22)

(A) $|z| \geq 0$

(B) $|z| \geq |\mathrm{Re}(z)|$

(C) $|z| \geq |\mathrm{Im}(z)|$

for all z in **C**. The last two inequalities follow by omitting either $|\mathrm{Re}(z)|^2$ or $|\mathrm{Im}(z)|^2$ in the identity (20E) above. The most important property of absolute values is the "triangle inequality" (whose name is explained below).

Theorem 1.1 (Triangle inequalities for absolute values) *If z and w are any complex numbers, then*

(23)
$$\big||z| - |w|\big| \leq |z \pm w| \leq |z| + |w|.$$

PROOF: First we prove that

(24)
$$|z + w| \leq |z| + |w|$$

for all complex numbers z and w; if we then replace w with $-w$, we get the companion inequality $|z - w| \leq |z| + |w|$. Now $0 \leq |z|^2 = z \cdot \bar{z}$ for every complex number z; therefore,

$$\begin{aligned} 0 \leq |z + w|^2 = (z + w)(\bar{z} + \bar{w}) &= z\bar{z} + w\bar{w} + z\bar{w} + w\bar{z} \\ &= |z|^2 + |w|^2 + z\bar{w} + \overline{(z\bar{w})} \\ &= |z|^2 + |w|^2 + 2\,\mathrm{Re}(z\bar{w}). \end{aligned}$$

In view of formula (22), $\text{Re}(z\bar{w}) \le |\text{Re}(z\bar{w})| \le |z\bar{w}| = |z| \cdot |\bar{w}| = |z| \cdot |w|$; thus, by replacing $\text{Re}(z\bar{w})$ with $|z| \cdot |w|$ above, we get

$$0 \le |z + w|^2 \le |z|^2 + |w|^2 + 2\,|z|\,|w| = (|z| + |w|)^2.$$

The desired formula (24) now follows by taking square roots.

To obtain the other half of the triangle inequality we first write $z = (z - w) + w$ and apply (24) to get $|z| \le |z - w| + |w|$, or $|z| - |w| \le |z - w|$; we then write $w = (w - z) + z$ and apply (24) to get $|w| \le |z - w| + |w|$, or $|w| - |z| \le |z - w|$. Since both $|z| - |w|$ and its negative $|w| - |z|$ are less than or equal to $|z - w|$, we conclude that

$$\big|\,|z| - |w|\,\big| \le |z - w|, \tag{25}$$

for all z and w in **C**. Using $-w$ in place of w here, we get the companion inequality $\big|\,|z| - |w|\,\big| \le |z + w|$. We have proved the full set of inequalities in (23). ▮

From the inequalities (23) we can easily deduce three facts about the distance function in **C**.

(26)

(i) $d(z, w) \ge 0$; $d(z, w) = 0$ if and only if $z = w$.

(ii) $d(z, w) = d(w, z)$.

(iii) If z and w are given, and if p is any other point, then $d(z, w) \le d(z, p) + d(p, w)$.

The first two are trivial, and the third follows from the triangle inequality for absolute values by means of the following calculation.

$$d_{\mathbf{C}}(z, w) = |z - w| = |(z - p) - (w - p)| \le |z - p| + |w - p|$$
$$= d_{\mathbf{C}}(z, p) + d_{\mathbf{C}}(p, w).$$

It is equation (iii) for distances in **C** which gives rise to the name "triangle inequality." If we draw a triangle connecting the points z, w, and p in the plane, then the distance directly from z to w is always less than or equal to the distance traveled in going from z to p and then from p to w (this is true no matter where the third point p is located). The other triangle inequality, formula (25), indicates that the difference in length of two sides can not exceed the length of the third side:

$$d_{\mathbf{C}}(z, w) \ge |d_{\mathbf{C}}(z, p) - d_{\mathbf{C}}(w, p)|.$$

(See Figure 1.8.) The triangle inequalities give upper and lower bounds for $|z - w|$, but cannot tell us exactly how large the distance $d_{\mathbf{C}}(z, w) = |z - w|$ is unless we know more about the particular numbers z and w (see Exercise 8).

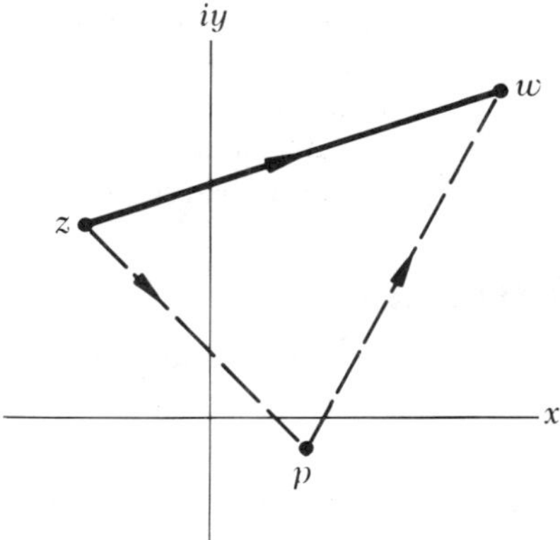

Figure 1.8 The meaning of the triangle inequality.

The distance function in any reasonable geometric system must satisfy the relationships (26). There is another geometric system equipped with a distance function that has these properties. This system is important when one considers functions of *several* independent variables.

The space $\mathbf{R}^n$ consists of all ordered n-tuples $\mathbf{x} = (x_1, \ldots, x_n)$ in which each x_k is a real number. If $n > 3$, this space has too many degrees of freedom to allow a graphical representation; if $n = 2$ it may be identified with the Cartesian plane.

There is a certain amount of algebraic structure in $\mathbf{R}^n$. We may add elements $\mathbf{x} = (x_1, \ldots, x_n)$ and $\mathbf{y} = (y_1, \ldots, y_n)$ to get

$$(x_1, \ldots, x_n) + (y_1, \ldots, y_n) = (x_1 + y_1, \ldots, x_n + y_n),$$

and we may multiply elements $\mathbf{x} = (x_1, \ldots, x_n)$ by any real number a:

$$a \cdot (x_1, \ldots, x_n) = (ax_1, \ldots, ax_n).$$

If $n \geq 3$, there is no reasonable multiplication operation (one multiplying n-tuples by n-tuples to give us new n-tuples).

So much for the algebraic structure in $\mathbf{R}^n$. There is a natural distance in $\mathbf{R}^n$, which is obtained by taking elements $\mathbf{x} = (x_1, \ldots, x_n)$ and $\mathbf{y} = (y_1, \ldots, y_n)$ and then defining the distance from $\mathbf{x}$ to $\mathbf{y}$ to be the **Euclidean distance**:

$$d_{\mathbf{R}^n}(\mathbf{x}, \mathbf{y}) = \sqrt{|x_1 - y_1|^2 + \cdots + |x_n - y_n|^2}.$$

One can prove that this distance function has the properties listed in (26). One can also define a **length** or **absolute value** $\|\mathbf{x}\|$ for elements of $\mathbf{R}^n$:

$$\|\mathbf{x}\| = \sqrt{|x_1|^2 + \cdots + |x_n|^2} = d_{\mathbf{R}^n}(\mathbf{x}, \mathbf{0}),$$

where $\mathbf{0} = (0, \ldots, 0)$. This definition of $\|\mathbf{x}\|$ leads to the familiar equation

$$d_{\mathbf{R}^n}(\mathbf{x}, \mathbf{y}) = \|\mathbf{x} - \mathbf{y}\|.$$

Note: If we similarly define $\mathbf{C}^n$ to be the set of all n-tuples $\mathbf{z} = (z_1, \ldots, z_n)$ with *complex* entries, we may give this space an algebraic structure and distance function exactly as above. This geometric structure is the complex variable analog of the n-dimensional Euclidean space $\mathbf{R}^n$, and it turns up whenever one considers functions of *several* complex variables.

EXERCISES

1. Verify that $|z|^2 = z \cdot \bar{z} = |\text{Re } z|^2 + |\text{Im } z|^2$ for every complex number z.

2. Verify the properties of absolute value listed in equation (20).

3. Sketch the subsets of $\mathbf{C}$ determined by the following inequalities and equalities.

(i) $\text{Im}(z) \neq 0$

(ii) $|z + i| < 1$

(iii) $|z| - 1 < 2$

(iv) $\text{Re}(z - 1) > 0$

(v) $|z - i| = |z + i|$

(vi) $|z - i| < |z + i|$

(vii) $|\text{Im}(z)| < \pi$

(viii) $|\text{Im}(z)| \leq 1$ and $|\text{Re}(z)| \leq 1$.

Answers: (i) $\mathbf{C}$ with real axis removed; (ii) open disc about $-i$ with $r = 1$; (iii) same as $|z| < 3$, a disc; (iv) right half plane bounded by vertical line $x = 1$; (v) points equidistant from $+i$, $-i$ (the real axis); (vi) all points *above* the real axis; (vii) horizontal strip of width 2π (points $z = x + iy$ such that $-\pi < y < +\pi$); (viii) a square centered at 0, including the boundary segments.

4. A common way to calculate real and imaginary parts of a quotient z/w is to multiply and divide by $\bar{w}$; thus $z/w = z\bar{w}/w\bar{w} = z\bar{w}/|w|^2$, and the new denominator has no imaginary part. Express the following quotients in the form $U(x, y) + iV(x, y)$, if $z = x + iy$.

(i) $1/z^2$

(ii) z^4

(iii) $z + \dfrac{1}{z}$

(iv) $\dfrac{1 + z}{1 - z}$

(v) $\dfrac{1}{z^2 + 1}$

5. Prove the general triangle inequality: $\left|\sum_{k=1}^{m} z_k\right| \leq \sum_{k=1}^{m} |z_k|$. Prove that $|z^n| = |z|^n$ for $n = \pm 1, \pm 2, \ldots$ (assume $z \neq 0$ if n is negative).

6. Sketch the following loci in the plane.

(i) $\text{Re}(1/z) < 1$

(ii) $|z - 1| < |z|$

(iii) $|4z + i| > 2$

(iv) $\text{Re}(z^2) = \text{const}$

(v) $|z|^2 = 2\,\text{Re}(z)$.

7. Use the triangle inequality to prove that the equation $z^4 + z + 4 = 0$ has all its roots outside of the unit disc $|z| \leq 1$.

Hint: How large can $|z^4 + z|$ be if $|z| \leq 1$?

8. Exhibit a pair of non-zero complex numbers z and w such that $|z + w| = |z| + |w|$. Exhibit another pair such that $\big||z| - |w|\big| = |z + w|$. If $w = 2i$ and if z lies on the circle $|z| = 1$, what is the range of values for $|z - w| = |z - 2i|$ as z varies?

Answer: (i) Take $z = i$, $w = i$; (ii) Take $z = +i$, $w = -i$; (iii) $1 \leq |z - 2i| \leq 3$.

9. Prove that the locus $|z - p| = \alpha\,|z - q|$ $(0 < \alpha < +\infty)$ is a circle that separates p from q, except when $\alpha = 1$. What is the locus if $\alpha = 1$?

1.4 NOTATION AND TERMINOLOGY FOR SETS IN THE PLANE

It will be convenient to use the following notation to describe sets in the complex plane. If $P(z)$ is some statement that makes sense for points z in **C**, we shall write $\{z : P(z)\}$ for the subset consisting of all points z for which $P(z)$ is *true*. For example, if $P(z)$ is the statement "$\text{Im}(z) = 0$," then $\{z : \text{Im}(z) = 0\}$ is just the real axis in the complex plane. Other examples are

(i) $\{z : \text{Im}(z) > 0\}$, the upper half plane.

(ii) $\{z : z \neq 0\}$, the complex plane with the single point $z = 0$ removed.

(iii) $\{z : |z| = 1\}$, the circle of radius one centered at $z = 0$.

In the last example, notice that $|z| = |z - 0|$.

Next we define two special types of sets in **C**. If p is a point in the plane and r is a positive number, the disc $\{z : |z - p| < r\}$ is called the **open disc** of radius r about p, while the disc $\{z : |z - p| \leq r\}$ (which includes the boundary circle $|z - p| = r$) is called the **closed disc** of radius r about p. We have discussed these sets in Example 1.1.

A subset E in **C** is called an **open set** if it has the following property: for each point p in E there is some open disc with positive radius about p which lies *entirely within* E; of course, the radius r will depend on which point p we are looking at. Some of the possibilities are illustrated in Figure 1.9. The half

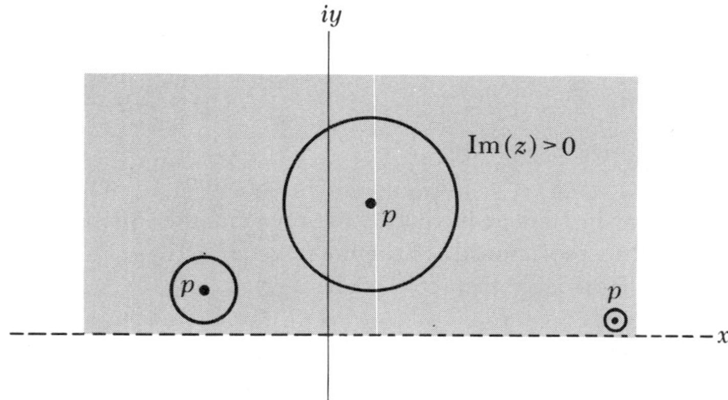

Figure 1.9 The set $E = \{z:\text{Im}(z) > 0\}$ as an example of an *open* set. About various points p we show the discs of positive radius that lie entirely within E.

plane $\{z:\text{Re}(z) > 0\}$, the open unit disc about the origin $\{z:|z| = |z - 0| < 1\}$, and the ring-shaped set centered at the origin $\{z:\frac{1}{2} < |z| < 1\}$ are all open sets in the plane. But the closed unit disc $\{z:|z| \leq 1\}$ or the unit circle $\{z:|z| = 1\}$ are *not* open sets (try to find a suitable disc about the point $p = 1 + i0$ that lies entirely within these sets).

A subset E in the plane is **bounded** if it lies within some disc of finite radius about $z = 0$ (i.e., the absolute value $|z|$ is bounded as z varies through points in E); otherwise, we say that E is an **unbounded** set. As an example, consider the half plane $E = \{z:\text{Re}(z) > 0\}$, which is unbounded.

We will repeatedly use a few standard symbols from set theory as a kind of shorthand. We will write $A \subseteq B$ if the set A is contained within B (the possibility $A = B$ is allowed!). If A and B are sets, we may combine them in various ways, as shown in Figure 1.10.

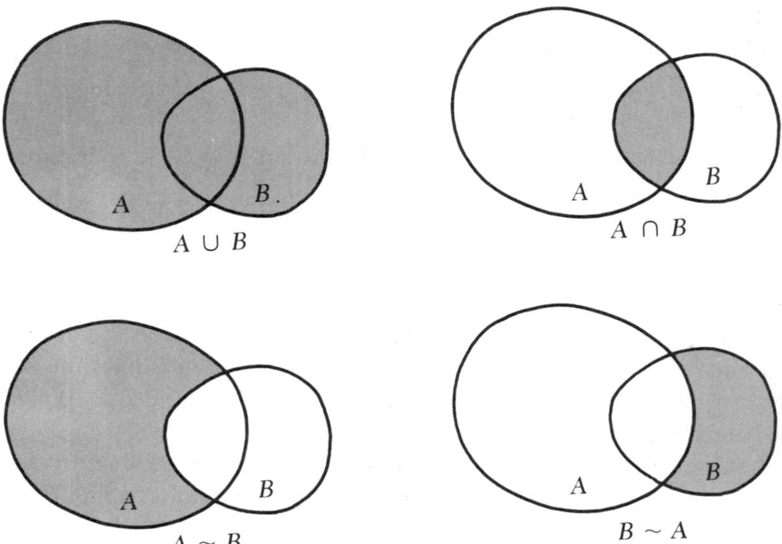

Figure 1.10 The various combinations $A \cup B, \ldots$ of two sets A and B are shaded in each diagram.

(i) $A \cup B$, the **union** of A and B, is formed by taking the points in A and the points in B together.
(ii) $A \cap B$, the **intersection** of A and B, consists of those points which belong to both A and B.
(iii) $A \sim B$, the **difference** of A and B, is obtained by deleting from A all points (if any) which also appear in B. There is also a difference set $B \sim A$, which might not be the same set as $A \sim B$; this possibility is illustrated in Figure 1.10.

If A is a subset of the complex plane, the difference set $\mathbf{C} \sim A$ consists of all points z in $\mathbf{C}$ which do not appear in A. This set, often written $A^{\sim} = \mathbf{C} \sim A$, is called the **complement of A**.

1.5 THE POLAR FORM OF COMPLEX NUMBERS

The location of a complex number z can be described by specifying the length of the line segment from 0 to z, which is precisely the absolute value $r = |z|$, and the angle θ from the positive real axis to this segment. Angles are measured in radians, and counterclockwise angles (such as the one in Figure 1.11) are considered to be positive. As illustrated in Figure 1.11, the Cartesian coordinates (x, y) of z can be obtained from the "polar coordinates" r, θ by means of the equations

(27) $$x = r(\cos \theta) \qquad y = r(\sin \theta),$$

so that $z = r(\cos \theta) + ir(\sin \theta) = r(\cos \theta + i \sin \theta)$.

Polar coordinates in the complex plane are best handled by introducing the **complex exponential function** e^z, which is defined for all z in the complex plane by the formula:†

(28) $$e^{x+iy} = e^x(\cos y + i \sin y) = (e^x \cos y) + i(e^x \sin y).$$

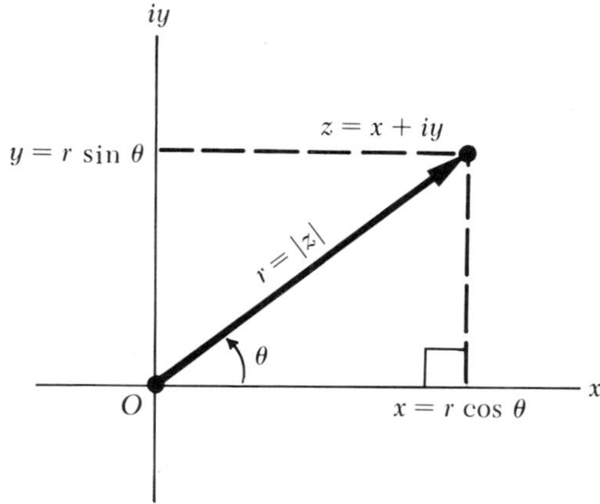

Figure 1.11 The polar coordinates r, θ of a point in the plane.

† When we encounter the exponential of a complicated expression, we will sometimes write $\exp(z)$ in place of e^z.

When $y = 0$, so that $z = x + i0$ is *real*, this exponential function gives us the expected value $e^{x+i0} = e^x$. Other reasons for choosing this particular formula for the exponential of a complex number will be explained in Section 2.6. Using this definition, we can verify by direct computations (left as Exercise 4) that the complex exponential function satisfies the identity

(29) $$e^{z+w} = e^z \cdot e^w$$

for all z, w in **C**. From (29) and the definition of reciprocals, it follows that

(30) $$e^{-z} = 1/e^z,$$

because $e^z \cdot e^{-z} = e^{z-z} = e^0 = 1$ for any complex number. Similarly,

(31) $$e^{w-z} = e^w/e^z$$

for any complex numbers w and z.

If θ is any real number, the complex number $e^{i\theta} = (\cos\theta) + i(\sin\theta)$ has absolute value one, since $|e^{i\theta}| = \sqrt{\cos^2\theta + \sin^2\theta} = 1$; thus, these points lie on the circle of radius one about the origin defined by the condition $|z| = 1$. Furthermore, the angle determined by the positive real axis and $e^{i\theta}$ is just θ, because $e^{i\theta}$ has Cartesian coordinates $x = \cos\theta$ and $y = \sin\theta$ (see Figure 1.11). As particular examples of numbers with the form $z = e^{i\theta}$, let us consider the values θ shown in Table 1.1.

TABLE 1.1 VALUES OF $e^{i\theta}$ FOR A FEW CHOICES OF θ (REAL).

	$\theta = 0$	$\theta = \pi/4$	$\theta = \pi/2$	$\theta = \pi$	$\theta = 3\pi/2$
$e^{i\theta} =$	$+1$	$\frac{1}{\sqrt{2}}(1 + i)$	$+i$	-1	$-i$

Notice that

(32) $$1/e^{i\theta} = e^{-i\theta}$$

for all real θ.

For any non-zero complex number z we can obtain a "polar decomposition"

(33) $$z = re^{i\theta}$$

in which θ is real and $r \geq 0$.† Obviously $r = |z|$, since $|e^{i\theta}| = 1$, so that the

† If $z = 0$, there is no well defined angle θ, and this is why we insist that z be non-zero in defining the polar decomposition (33). Actually, if $r = 0$ in equation (33), we get $z = 0 \cdot e^{i\theta} = 0$ for all choices of θ.

number r in (33) is uniquely determined. On the other hand, since θ is real, the number $z = re^{i\theta}$ has Cartesian form $(r\cos\theta) + i(r\sin\theta)$, and it should be apparent from Figure 1.11 that θ is the angle determined by the positive real axis and z. Clearly, θ is defined only up to an added multiple of 2π; in fact, from equation (28) it is a simple matter to show that

(34) $e^{i\phi} = 1$ if and only if $\phi = 2\pi k$ for some integer $k = 0, \pm 1, \pm 2, \ldots$

(details in Exercise 6). Therefore, $e^{i\theta'} = e^{i\theta''}$ if and only if $1 = e^{i\theta'}/e^{i\theta''} = e^{i(\theta' - \theta'')}$, and we conclude that

(35) $e^{i\theta'} = e^{i\theta''}$ if and only if $\theta' - \theta''$ is an integral multiple of 2π.

From (35) the degree of indeterminacy of θ in the polar decomposition is evident. Combining all these observations, we see that two complex numbers represented in polar form $z = re^{i\theta}$ and $w = \rho e^{i\phi}$ are *equal* if and only if

(36)
(i) $r = \rho$
(ii) $\theta - \phi = 2\pi k$ for some integer $k = 0, \pm 1, \pm 2, \ldots$.

The polar form and formula (29) provide a natural geometric interpretation of multiplication. If $Z = re^{i\theta}$ and $W = \rho e^{i\phi}$, then

$$Z \cdot W = (re^{i\theta}) \cdot (\rho e^{i\phi}) = r\rho e^{i\theta} \cdot e^{i\phi} = (r\rho)e^{i(\theta + \phi)}.$$

Thus, the length of the segment from the origin to the product $Z \cdot W$ is the *product* of the lengths $|Z|$ and $|W|$, since $|ZW| = r\rho = |Z| \cdot |W|$. Similarly, the angle associated with ZW is the *sum* of the angles for Z and W respectively:

$$\angle ZW = \theta + \phi = \angle Z + \angle W.$$

This is illustrated in Figure 1.12. There is also an interpretation of division

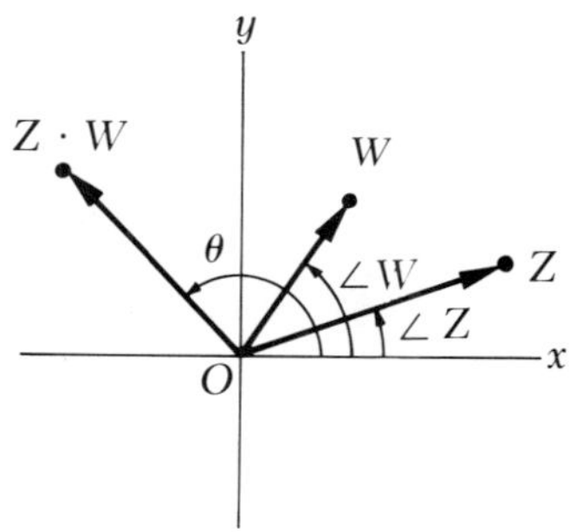

Figure 1.12 The geometric meaning of multiplication. Here $\theta = \angle Z + \angle W$.

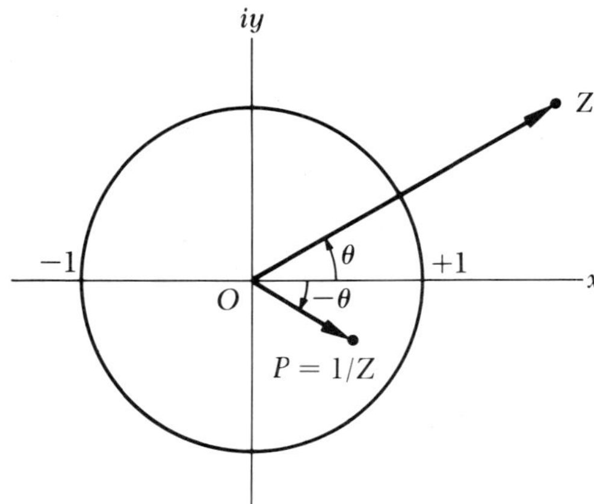

Figure 1.13 The geometric relation between Z and its reciprocal $1/Z$; $\angle(1/Z) = -\angle Z$ and the length of $\overrightarrow{OP}$ is the reciprocal of the length of $\overrightarrow{OZ}$. The circle shown consists of points unit distance from the origin O.

based on formula (31): if Z is non-zero, then

$$|W/Z| = |W|/|Z|$$
$$\angle(W/Z) = \phi - \theta = \angle W - \angle Z.$$

The relation between Z and its reciprocal $1/Z = 1/re^{i\theta} = (1/r)(1/e^{i\theta}) = (1/r)e^{-i\theta}$ is a special case of this, in which

$$|1/Z| = 1/|Z| \quad \text{and} \quad \angle(1/Z) = -\angle Z,$$

as indicated in Figure 1.13.

EXERCISES

1. Express the following complex numbers in polar form.

(i) $1 + i$ (iv) $1 - i\sqrt{3}$

(ii) $-i$ (v) $-\pi - i\pi$.

(iii) -1

Answers: (i) $\sqrt{2}e^{i\pi/4}$; (ii) $e^{-i\pi/2}$; (iii) $e^{i\pi} = e^{-i\pi}$; (iv) $2e^{-i\pi/3}$; (v) $\pi\sqrt{2}e^{-3\pi i/4}$

2. Show that $e^{z+2\pi ni} = e^z$ for every integer n and every complex number z.

3. Show that e^z is never zero. Show that $|e^z| = e^{\text{Re}(z)}$ for any z.

4. Using standard trigonometric identities, prove that $e^{z+w} = e^z \cdot e^w$ for all complex z and w. Start with definition (28).

5. Prove *De Moivre's Theorem:*

$$(\cos\theta + i\sin\theta)^n = \cos n\theta + i\sin n\theta$$

for real θ and any integer $n = 0, \pm 1, \pm 2, \ldots.$

6. If z is *real*, show that $e^{iz} = 1$ if and only if z is an integral multiple of 2π. Show that no new solutions appear if we allow z to be complex (thus *all* solutions are real, and have the form $2\pi k + i0$).

7. Find all complex solutions of the following equations.

(i) $e^z = 1$

(ii) $\frac{1}{2}(e^z + e^{-z}) = 0$

(iii) $e^z = 2$

Answers: (i) $z = 2\pi ni$ $(n = 0, \pm 1, \pm 2, \ldots)$; (ii) $z = i\left(\frac{\pi}{2} + n\pi\right)$ $(n = 0, \pm 1, \pm 2, \ldots)$; (iii) $z = (\log 2) + i(2\pi n)$ for $n = 0, \pm 1, \pm 2, \ldots$.

1.6 THE EQUATION $z^n = 1$ AND THE nth ROOTS OF A COMPLEX NUMBER

The polar decomposition is useful in solving certain kinds of equations and we will use it now to determine the nth roots of a complex number w. These complex numbers are the solutions of the following equation (the **$\mathbf{n}^{\text{th}}$ root equation**):

$$z^n = w. \tag{37}$$

First notice that when $w = 0$ there is just one solution to this equation, namely $z = 0$. We will show that every other complex number $w \neq 0$ has exactly n distinct n^{th} roots.

It is easy to determine the **$\mathbf{n}^{\text{th}}$ roots of unity** (the roots of $w = +1$). Simply write z in polar form $z = re^{i\theta}$. By applying equation (29) several times, we see that

$$1 = z^n = re^{i\theta} \cdot \ldots \cdot re^{i\theta} = r^n(e^{i\theta})^n = r^n e^{in\theta}; \tag{38}$$

therefore, r and θ must satisfy the following conditions, in view of formula (36):

(i) $r^n = 1$

(ii) $n\theta = 2\pi k$ for some $k = 0, \pm 1, \pm 2, \ldots$.

Obviously, r must equal 1, since r must be a positive real number; the allowable values of θ are

$$\theta = 2\pi\left(\frac{k}{n}\right) \qquad k = 0, \pm 1, \pm 2, \ldots.$$

Since $e^{i\theta'} = e^{i\theta''}$ if θ' and θ'' differ by an integer multiple of 2π, these values of θ give us only n distinct numbers of the form $e^{i\theta}$, so that equation (38) has

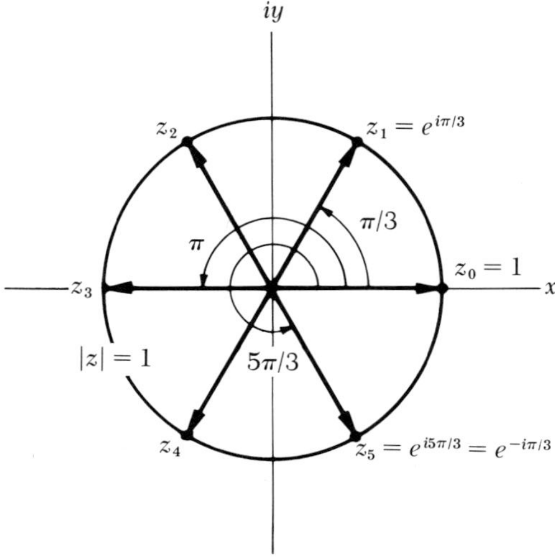

Figure 1.14 The successive 6th roots of unity $z_0 = 1, z_1, z_2, \ldots, z_5$ are separated by the angle $\theta = 2\pi/6 = \pi/3$.

exactly n distinct solutions:

(39)
$$\begin{aligned} z_0 &= e^{2\pi i(0/n)} = 1 \\ z_1 &= e^{2\pi i(1/n)} \\ z_2 &= e^{2\pi i(2/n)} \\ &\cdot \\ &\cdot \\ &\cdot \\ z_{n-1} &= e^{2\pi i(n-1/n)}. \end{aligned}$$

The angle determined by the positive real axis and a typical root $z_k = e^{2\pi i(k/n)}$ is just $\theta = 2\pi(k/n)$, so that these roots of unity are equally spaced around the circle $|z| = 1$. As an example, Figure 1.14 shows the positions of the roots if $n = 6$. All roots of unity have absolute value one, since $|z^n| = |z|^n = 1$ implies that $|z| = 1$.

A great simplification in solving the equation $z^n = 1$ has been achieved by expressing z in polar form and making systematic use of the exponential formula (29). The reader might try his hand at solving the equation $z^3 = 1$ by writing z as $z = x + iy$; if we then multiply out $z^3 = (x + iy)^3$ and separate it into real and imaginary parts, we get

$$(x^3 - 3xy^2) + i(3x^2y - y^3) = 1 + i0.$$

This is equivalent to a rather formidable system of cubic equations in x and y:

(40)
$$\begin{aligned} x^3 - 3xy^2 &= 1 \\ 3x^2y - y^3 &= 0. \end{aligned}$$

The solutions (x, y) of this system agree with the ones derived above, but greater computational difficulties are involved in using (40).

The general n^{th} root problem can be handled by similar methods. Let $w = \rho e^{i\phi}$ be any non-zero complex number; if $z = re^{i\theta}$ is any n^{th} root of w, then

$$z^n = r^n(e^{i\theta})^n = r^n e^{in\theta} = w = \rho e^{i\phi}.$$

By formula (36) of the last section, this implies that

(i) $r^n = \rho$

(ii) $n\theta = \phi + 2\pi k$ for some integer $k = 0, \pm 1, \pm 2, \ldots$.

The first condition means that $r = \rho^{1/n} = |w|^{1/n}$ (the ordinary n^{th} root of a positive real number), so that

$$|z| = |w|^{1/n} \tag{41}$$

for any n^{th} root of w. Meanwhile, the angle variable θ must assume one of the values

$$\theta = \frac{\phi}{n} + 2\pi\left(\frac{k}{n}\right) \qquad k = 0, \pm 1, \pm 2, \ldots. \tag{42}$$

These values of θ give us exactly n distinct numbers of the form $e^{i\theta}$, so that the nth roots of w have the form

$$\begin{aligned} z_0 &= |w|^{1/n} \exp\left[i\left(\frac{\phi}{n} + 2\pi\left(\frac{0}{n}\right)\right)\right] = |w|^{1/n}\, e^{i\phi/n} \\ z_1 &= |w|^{1/n} \exp\left[i\left(\frac{\phi}{n} + 2\pi\left(\frac{1}{n}\right)\right)\right] \\ &\ \ \vdots \\ z_{n-1} &= |w|^{1/n} \exp\left[i\left(\frac{\phi}{n} + 2\pi\left(\frac{n-1}{n}\right)\right)\right]. \end{aligned} \tag{43}$$

Notice that the various n^{th} roots of w can be obtained from the first one, $z_0 = |w|^{1/n}e^{i\phi/n}$, by multiplying it by the various n^{th} roots of unity, $1, e^{2\pi i/n}, \ldots, e^{2\pi i(n-1)/n}$:

$$z_k = |w|^{1/n} \exp\left[i\left(\frac{\phi}{n} + 2\pi\left(\frac{k}{n}\right)\right)\right] = |w|^{1/n}\, e^{i\phi/n} e^{2\pi i k/n} = z_0 \cdot e^{2\pi i k/n}$$

for $k = 0, 1, \ldots, n-1$. Indeed, from *any* one of the nth roots of w we can obtain all the others by multiplying it by the various n^{th} roots of unity.

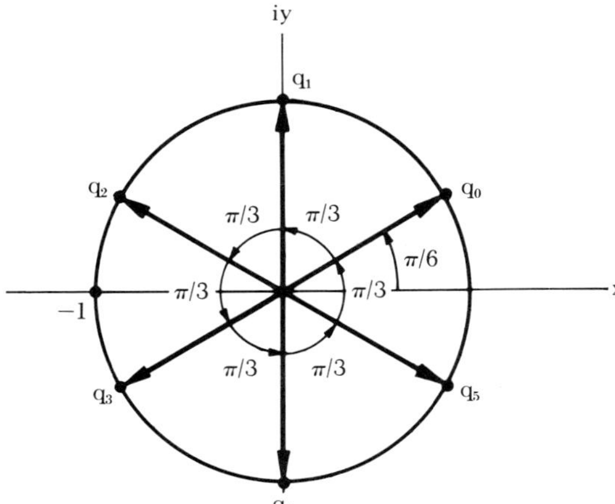

Figure 1.15 The 6th roots of -1.

Example 1.3 Let us determine the nth roots of $+1$ and -1 for $n = 6$. The roots of $+1$ are obtained by dividing $\theta = 2\pi$ into six equally spaced angles: $\theta_k = 2\pi k/6$ for $0 \leq k \leq 5$. Thus the roots are

$$p_0 = 1; \quad p_1 = e^{i\pi/3}; \quad p_2 = e^{2\pi i/3}; \quad p_3 = e^{i\pi}; \quad p_4 = e^{4\pi i/3}; \quad p_5 = e^{5\pi i/3}.$$

Their positions on the unit circle $|z| = 1$ are shown in Figure 1.14. To determine the roots of -1, it is sufficient to find a single one of them and multiply it by the 6th roots of unity listed above. The number $-1 = e^{i\pi}$ makes an angle of π radians with the positive real axis, and we get a 6th root by taking $e^{i\theta}$ where $6\theta = \pi$; thus $e^{i\pi/6} = q_0$. The other roots are†

$$q_0 = e^{i\pi/6} \cdot p_0 = e^{i\pi/6}$$

$$q_1 = e^{i\pi/6} \cdot p_1 = e^{i\pi/6} \cdot e^{i\pi/3} = \exp[i\pi(\tfrac{1}{6} + \tfrac{1}{3})] = e^{i3\pi/6} = e^{i\pi/2}$$

$$q_2 = e^{i\pi/6} \cdot p_2 = e^{i\pi/6} \cdot e^{i2\pi/3} = \exp[i\pi(\tfrac{1}{6} + \tfrac{2}{3})] = e^{i5\pi/6}$$

$$q_3 = e^{i\pi/6} \cdot p_3 = e^{i\pi/6} \cdot e^{i\pi} = \exp[i\pi(\tfrac{1}{6} + 1)] = e^{i7\pi/6} = e^{-i5\pi/6}$$

$$q_4 = e^{i\pi/6} \cdot p_4 = e^{i\pi/6} \cdot e^{i4\pi/3} = \exp[i\pi(\tfrac{1}{6} + \tfrac{4}{3})] = e^{i9\pi/6} = e^{-i\pi/2}$$

$$q_5 = e^{i\pi/6} \cdot p_5 = e^{i\pi/6} \cdot e^{i5\pi/3} = \exp[i\pi(\tfrac{1}{6} + \tfrac{5}{3})] = e^{i11\pi/6} = e^{-i\pi/6}$$

The locations of these numbers on the circle $|z| = 1$ (equally spaced with respect to the initial point q_0) are shown in Figure 1.15.

In Chapter 2, after we have studied the exponential function more carefully, we shall have more to say about the roots of complex numbers.

EXERCISES

1. Determine the cube roots of $+i$ and plot their locations in the plane.

Answers: $z_1 = e^{i\pi/6}$; $z_2 = e^{5\pi i/6}$; $z_3 = e^{-i\pi/2} = -i$.

† Since $\theta' = 7\pi/6$ and $\theta'' = -5\pi/6$ differ by 2π, we have $e^{i\theta'} = e^{i\theta''}$; likewise, the remaining entries in our list have been converted to values of θ such that $-\pi < \theta < \pi$.

2. Prove that the quadratic formula

$$z_{\pm} = \frac{-b \pm \sqrt{b^2 - 4ac}}{2a}$$

yields the roots of any quadratic equation $az^2 + bz + c = 0$ $(a \neq 0)$, even if the coefficients a, b, and c are *complex*.

Hint: Recall how the quadratic formula is proved for real polynomials by "completing the square." The same idea works for complex numbers.

3. Determine all fourth roots of $z = +1$ and of $z = -1$, and plot their locations in the plane.

Answer: Roots are $+1$, -1, $+i$, $-i$ in both cases.

4. Use the quadratic formula (2) to determine the roots of the equation $z^2 + z + 1 = 0$. The square roots appearing in (2) are to be determined as explained in this section.

Answer: $z_1 = e^{2\pi i/3}$, $z_2 = e^{-2\pi i/3} = e^{i4\pi i/3}$.

2 LIMITS AND FUNCTIONS OF A COMPLEX VARIABLE

Most of calculus is devoted to the study of limits in the space **R** or, when functions of several real variables are studied, the space $\mathbf{R}^n$. The complex number system **C** has an addition operation, multiplication operation, and a natural distance function, so that notions such as continuity and differentiability of functions have exact analogs for functions of a complex variable. In the first two sections of this chapter we will review the notions of "limits of sequences" and "limits (or sums) of infinite series"; however, we are going to consider sequences and infinite series of *complex* numbers. Then we will use these concepts to study functions of a complex variable. Readers familiar with sequences and series of complex numbers may wish to begin with Section 2.3.

2.1 LIMITS OF SEQUENCES OF COMPLEX NUMBERS

A **sequence of complex numbers** is just a correspondence between the integers $n \geq 1$ and certain points s_n in **C**; each integer has assigned to it a complex number s_n. The sequence is indicated by the symbol $\{s_n\}$ or $\{s_n\colon n = 1, 2, \ldots\}$. Sometimes, if it is convenient, we will allow sequences to be indexed on the integers starting with $n = 0$, or with $n = -1$, or with $n = 2$, or any other integer.

Example 2.1 (A constant sequence) Let w be a fixed complex number, and define $\{s_n\}$ so that $s_n = w$ for all $n = 1, 2, \ldots$.

Example 2.2 Let i be the complex number $(0, 1)$, as usual, and define a sequence $\{s_n\}$ by assigning values

$$s_n = i^n = \underbrace{i \cdot i \cdot \ldots \cdot i}_{\text{n times}} \quad \text{for } n = 1, 2, \cdots.$$

Now $i^2 = -1$; $i^3 = -i$; and $i^4 = +1$; thus, if we are given any power i^n, we may split off successive factors $i^4 = 1$ to get

$$i^n = i^{n-4} = i^{n-8} = \cdots.$$

This shows that i^n is always one of the four numbers $i, -1, -i, +1$. Thus the sequence $\{s_n\}$ takes on only *four* distinct values as n increases.

Example 2.3 Let z be a fixed complex number and let $s_n = z^n$ (the nth power of z):

$$s_n = z^n = \underbrace{z \cdot \ldots \cdot z}_{\text{n times}} \quad \text{for } n = 1, 2, \ldots.$$

The behavior of s_n as n increases depends on the location of the point $z = s_1$ that we start with; this behavior will be discussed below.

One of the most important features of a sequence $\{s_n\}$ is the behavior of its terms s_n as n increases. The notion of convergence, and of the limit $\lim_{n\to\infty} s_n$, should be familiar in the special case when we consider sequences of *real* numbers.

Definition 2.1 *A sequence $\{s_n\}$ of complex numbers* **converges** *to the complex number s if the distances $d(s_n, s) = |s_n - s|$ converge to zero as $n \to \infty$. Since $|s_n - s|$ is a real number for $n = 1, 2, \ldots$, the meaning of the statement that $\lim_{n\to\infty} |s_n - s| = 0$ is clear.*

This definition can be expressed in another way. A sequence $\{s_n\}$ of complex (or real) numbers converges to a complex (or real) number s if and only if:

For every positive number ε, we have

$$(1) \qquad d(s_n, s) = |s_n - s| < \varepsilon$$

for all sufficiently large indices n, say $n \geq N(\varepsilon)$.

Thus, for any number $\varepsilon > 0$, the distance from s_n to s is "eventually" less than ε. How far out in the set of integers n we must go to achieve (1) will depend on the size of ε; we may have to take $N(\varepsilon)$ very large to get (1) if ε is small.

If $\{s_n\}$ is a sequence that converges to the number s, we will write

$$s = \lim_{n\to\infty} s_n = \lim_{n\to\infty} \{s_n\},$$

or

$$s_n \to s \quad \text{as} \quad n \to \infty,$$

as an abbreviation for the phrase "s_n converges to s." If there is no number s to which a given sequence $\{s_n\}$ converges, we say that s_n **diverges**.

We may often simplify limit problems in **C** by reducing them to questions about *pairs* of limits in **R**, as follows.

Theorem 2.1 *Let $\{z_n\}$ be a sequence in **C** and write $z_n = x_n + iy_n$; thus, there are two sequences $\{x_n\}$ and $\{y_n\}$ of real numbers associated with $\{z_n\}$. Then*

$$z_n \to w = x + iy \quad \text{as} \quad n \to \infty \text{ (in } \mathbf{C})$$

if and only if

$$\left.\begin{array}{l} x_n \to x \\ y_n \to y \end{array}\right\} \quad \text{as} \quad n \to \infty \text{ (in } \mathbf{R}).$$

PROOF: Clearly $z_n - w = (x_n - x) + i(y_n - y)$. If $z_n \to w$ as $n \to \infty$ in **C**, this means that

$$|z_n - w| \to 0 \quad \text{as} \quad n \to \infty;$$

when this is combined with the fact that $|z_n - w| \geq |x_n - x| \geq 0$ for all n (see equation (22), Chapter 1) we get

$$|x_n - x| \to 0 \quad \text{as} \quad n \to \infty,$$

so that $x_n \to x$ in **R**. Likewise, we see that $y_n \to y$ in **R**, and half of the theorem is proved.

On the other hand, if $x_n \to x$ and $y_n \to y$ as $n \to \infty$ (in **R**), and if ε is any positive number, then the distances $|x_n - x|$ and $|y_n - y|$ are less than $\varepsilon/2$ for all large n. This implies that

$$|z_n - z| = \sqrt{|x_n - x|^2 + |y_n - y|^2} \leq \sqrt{\varepsilon^2/4 + \varepsilon^2/4} < \varepsilon$$

for all large n. Since this conclusion is valid for any choice of $\varepsilon > 0$, we conclude that $z_n \to z$ as $n \to \infty$. ▌

It is not always advisable to resort to Theorem 2.1 in studying limit problems; sometimes it is better to work directly with the definition of complex limits and the triangle inequality for absolute values. In Example 2.5 below, our analysis would only be confused by considering real and imaginary parts of the sequence separately.

Example 2.4 The sequence $s_n = i^n$ from Example 2.2 is divergent (does not converge to any complex number). Indeed, we may write i in polar form as $i = e^{i\pi/2} = \cos(\pi/2) + i\sin(\pi/2)$, and thus $i^n = (e^{i\pi/2})^n = e^{in\pi/2} = \cos(n\pi/2) + i\sin(n\pi/2) = x_n + iy_n$ for $n = 1, 2, \ldots$. Neither of the sequences $\{x_n\}$ or $\{y_n\}$

converges in **R**. Therefore, by Theorem 2.1, the sequence of complex numbers $s_n = i^n = x_n + iy_n$ diverges.

Example 2.5 Fix a complex number z and form the sequence of complex numbers $s_n = z^n$ for $n = 1, 2, \ldots$. Let us investigate which choices of z lead to a convergent sequence of complex numbers. Our analysis depends on two well known facts about sequences of real numbers.

(i) If r is a real number such that $0 \leq r < 1$, then $\lim_{n\to\infty} r^n = 0$.

(ii) If r is a real number such that $r > 1$, then the sequence $\{r^n\}$ diverges (its terms increase without bound).

There are three cases to consider:

Case 1. If $|z| < 1$, then the distance $|z^n - 0| = |z^n| = |z|^n$ converges to zero as $n \to \infty$, in view of (i); thus,

$$\lim_{n\to\infty} z^n = 0 \quad \text{if} \quad |z| < 1.$$

Case 2. If $|z| > 1$ the sequence of absolute values $|s_n| = |z|^n$ increases without bound; that is, given any positive number K, the values $|z^n|$ are eventually larger than K:

$$|z^n| = |z|^n > K \quad \text{for all large } n.$$

Thus $\{z^n\}$ cannot converge to any complex number α because $|z^n - \alpha| \geq |z^n| - |\alpha|$ does not converge to zero as n increases.

Case 3. If $z = 1 + i0 = 1$ we get $z^n = 1^n = 1$, so that $\lim_{n\to\infty} 1^n = 1$. One can prove that the sequence $\{z^n\}$ diverges for any other choice of z such that $|z| = 1$; the proof is outlined in Exercise 8.

The behavior of $\{z^n\}$ as n increases is shown in Figure 2.1 for typical points z. This behavior is easily deduced by writing z in polar form $z = re^{i\theta}$ and noticing that $z^n = r^n(e^{i\theta})^n = r^n e^{in\theta}$.

Various combinations of convergent sequences of complex numbers lead to new convergent sequences. Analogous results hold for sequences of *real* numbers; the proofs for complex sequences are similar, and we leave the details as Exercise 4.

Theorem 2.2 *Let $\{z_n\}$ and $\{w_n\}$ be convergent sequences of complex numbers.*

(i) *The sequence $\{z_n \pm w_n\}$ converges and*

$$\lim_{n\to\infty} \{z_n \pm w_n\} = \left(\lim_{n\to\infty} z_n\right) \pm \left(\lim_{n\to\infty} w_n\right).$$

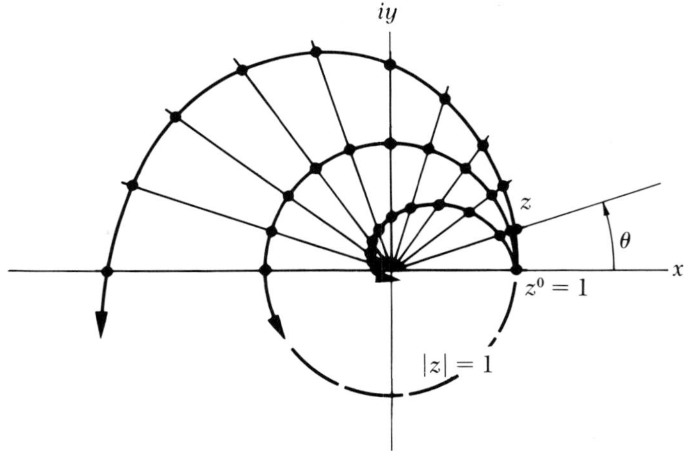

Figure 2.1 The behavior of z^n as $n \to \infty$ for $z = re^{i\theta}$, taking $r < 1$, $r > 1$, and $r = 1$.

(ii) *If α is a fixed complex number, then the sequence $\{\alpha z_n\}$ converges, and*

$$\lim_{n\to\infty} \{\alpha z_n\} = \alpha \cdot \left(\lim_{n\to\infty} z_n\right).$$

(iii) *The sequence $\{z_n \cdot w_n\}$ converges, and*

$$\lim_{n\to\infty} \{z_n \cdot w_n\} = \left(\lim_{n\to\infty} z_n\right) \cdot \left(\lim_{n\to\infty} w_n\right).$$

(iv) *If $\lim_{n\to\infty} w_n$ is non-zero, then the sequence of quotients $\left\{\dfrac{z_n}{w_n}\right\}$ converges, and*

$$\lim_{n\to\infty} \left\{\frac{z_n}{w_n}\right\} = \frac{\lim_{n\to\infty} z_n}{\lim_{n\to\infty} w_n}.$$

Improper Limits of Real Sequences

Here are a few results, summarized from calculus, that will be particularly useful to us.

Certain real sequences that diverge are considered to have the "improper limits" $+\infty$ or $-\infty$. We say that $\lim_{n\to\infty} x_n = +\infty$ if, for any positive number K, the terms x_n eventually exceed this value, so that $x_n \geq K$ for all sufficiently large n. The limit $\lim_{n\to\infty} x_n = -\infty$ is defined similarly. One must be aware that a sequence of real numbers may diverge in ways that do not allow us to assign even these improper values as a limit; we cannot assign values $+\infty$ or $-\infty$ to the divergent sequences whose terms are $x_n = (-1)^{n+1}$ or $x_n = (-1)^n n^2$. However, sequences of real numbers that are **increasing**, in the sense that $x_{n+1} \geq x_n$ for all n, are more restricted in their behavior.

Theorem 2.3 *If $\{x_n\}$ is an increasing sequence of real numbers, then* $\lim_{n\to\infty} x_n$ *exists if the terms x_n are bounded from above, in the sense that there is some constant K such that $x_n \leq K$ for $n = 1, 2, \ldots$. If the terms are not bounded, they must increase without bound, and* $\lim_{n\to\infty} x_n = +\infty$.

There is a similar result for decreasing sequences of real numbers; they are either bounded from below, and converge, or else $\lim_{n\to\infty} x_n = -\infty$.

Improper limits for sequences of complex numbers will be discussed in Chapter 4.

EXERCISES

1. To prove that $\lim_{n\to\infty} z^n = 0$, provided $|z| < 1$, we must show that $|z^n - 0| = |z|^n < \varepsilon$ eventually, for every $\varepsilon > 0$. Consider $z = (i + 1)/4$; if $\varepsilon = 1/100$, how large must one take $N(\varepsilon)$ to insure that $|z|^n < \varepsilon$ for all $n \geq N(\varepsilon)$? What if $\varepsilon = 100$ or $\varepsilon = 10^{-6}$?

2. Prove the following convergence statements:

(i) $\lim_{n\to\infty} \left(\dfrac{i+1}{\sqrt{\pi}}\right)^n = 0$

(ii) $\lim_{n\to\infty} \dfrac{in}{n^2 + i} = 0$

(iii) $\lim_{n\to\infty} \dfrac{in + n^2}{n^2 + i} = 1$

(iv) $\lim_{n\to\infty} \exp\left(\dfrac{-1}{n^2 + i}\right) = 1$

3. If the sequence $\{z_n + w_n\}$ converges, is it necessarily true that one of the sequences $\{z_n\}$ or $\{w_n\}$ must converge?

Answer: No.

4. Prove the convergence statements in Theorem 2.2.

5. Prove that if $z = \lim_{n\to\infty} z_n$, then $|z| = \lim_{n\to\infty} |z_n|$. Give an example in which the sequence $\{z_n\}$ diverges, while the sequence of absolute values converges.

Hint: Use the triangle inequality $0 \leq d(|z_n|, |z|) = \big||z_n| - |z|\big| \leq |z_n - z|$.

6. Prove that if $\lim_{n\to\infty} z_n = z_0$, then:

(i) The sequence is *bounded;* i.e., there is some constant M such that $d(z_n, 0) = |z_n| \leq M$ for all n.

(ii) The sequence of conjugates converges: $\lim_{n\to\infty} \bar{z}_n = \bar{z}_0$.

(iii) If $f(z) = z^2 + z + 1$, the sequence of values $w_n = f(z_n)$ converges to $f(z_0) = (z_0)^2 + z_0 + 1$.

(iv) $\lim_{n\to\infty} \dfrac{1}{z_n - z_0}$ does not exist.

7. If $\{s_n\}$ is a convergent sequence of complex numbers, say $s_n \to s$, then define $t_n = s_{n+1}$ for all n. Prove that

$$\lim_{n\to\infty} s_n = \lim_{n\to\infty} t_n = s.$$

Then use Theorem 2.2 to prove that the difference between successive terms converges to zero

$$\lim_{n\to\infty} s_{n+1} - s_n = 0$$

for any convergent sequence $\{s_n\}$ of complex numbers.

8. If $|z| = 1$ and $z \neq 1 + i0$, use Exercise 7 to prove that the sequence of powers $\{z^n\}$ diverges.

Hint: Can the differences $|z^{n+1} - z^n|$ go to zero?

2.2 INFINITE SERIES OF COMPLEX NUMBERS

Now we turn to the problem of summing an infinite series $\sum_{n=1}^{\infty} a_n$ of complex (or real) numbers. So far, only *finite* sums have any meaning, and an infinite sum $\sum_{n=1}^{\infty} a_n = a_1 + a_2 + \cdots$ must be defined as some kind of limit. Given a sequence $\{a_n\}$, we can add up the first k terms that appear, and this can be done for each integer $k = 1, 2, \ldots$; only finitely many operations are required to form each of these sums. Thus, from the original sequence of terms in the series, we obtain a new sequence, the sequence $\{s_k : k = 1, 2, \ldots\}$ of **partial sums** of the infinite series $\sum_{n=1}^{\infty} a_n$; these are the numbers

$$\begin{aligned} s_1 &= a_1 \\ s_2 &= a_1 + a_2 \\ &\vdots \\ s_k &= a_1 + a_2 + \cdots + a_k = \sum_{n=1}^{k} a_n \\ &\vdots \end{aligned}$$

If the sequence $\{a_n\}$ is indexed starting with $n = 0$ (or some other integer) we still define s_n as the sum of the *first k terms*, so that

$$s_1 = a_0; \quad s_2 = a_0 + a_1; \quad \cdots; \quad s_k = \sum_{n=0}^{k-1} a_n; \quad \cdots$$

in this situation.

Definition 2.2 *We say that an infinite series* $\sum_{n=1}^{\infty} a_n$ **converges** *if and only if the sequence* $\{s_n\}$ *of partial sums converges to some limit, and we say that* $\sum_{n=1}^{\infty} a_n$ **diverges** *if the sequence* $\{s_n\}$ *diverges. If the sequence* $\{s_n\}$ *converges, we assign the value*

$$\sum_{n=1}^{\infty} a_n = \lim_{n\to\infty} s_n = \lim_{n\to\infty} (a_1 + \cdots + a_n) = \lim_{n\to\infty} \left\{\sum_{k=1}^{n} a_k\right\}$$

as the "sum" of the series.

If $\{s_n\}$ diverges, there is simply no reasonable value we can assign for $\sum_{n=1}^{\infty} a_n$. Thus, the series $\sum_{n=1}^{\infty} (-1)^{n+1}$, whose terms are alternately $+1$ and -1, has a sequence of partial sums $\{s_n\}$ alternating between the values $+1$ and 0, and the series diverges.

A series $\sum_{n=1}^{\infty} a_n$ of real or complex numbers is **absolutely convergent** if the associated series of absolute values $\sum_{n=1}^{\infty} |a_n|$ converges. Since the partial sums of $\sum_{n=1}^{\infty} |a_n|$ form an increasing sequence of positive numbers, they must either converge to a finite limit or increase without bound. In the latter case it is customary to write $\sum_{n=1}^{\infty} |a_n| = +\infty$ (an improper limit); convergence is indicated by writing $\sum_{n=1}^{\infty} |a_n| < +\infty$.

Theorem 2.4 *If* $\sum_{n=1}^{\infty} a_n$ *is a real or complex series that is absolutely convergent, then the series* $\sum_{n=1}^{\infty} a_n$ *converges.*

It is not obvious that convergence of the series of absolute values, $\sum_{n=1}^{\infty} |a_n| < +\infty$, implies convergence of the original series $\sum_{n=1}^{\infty} a_n$. The proof requires consideration of the "least upper bound" property of real numbers, or the "Cauchy criterion" for convergence of complex sequences. These matters are somewhat technical and will not be used later on; we will rely on Theorem 2.4 as our basic result, and refer the interested reader to Kaplan [12], 6.1 to 6.5 and 6.19, for the details of its proof.

Certain series $\sum_{n=1}^{\infty} a_n$ are convergent, but are not absolutely convergent; absolute convergence is a sufficient condition for convergence, but is not a *necessary* condition. Such a series is said to be **conditionally convergent**. Various examples will be given.

Example 2.6 (The harmonic series) It is well known that the real series

$$\sum_{n=1}^{\infty} \frac{1}{n} = 1 + \frac{1}{2} + \frac{1}{3} + \cdots$$

diverges (the integral test settles this question); thus, the series

$$\sum_{n=1}^{\infty} \frac{(-1)^{n+1}}{n} = 1 - \frac{1}{2} + \frac{1}{3} - \cdots$$

is *not* absolutely convergent. It *does* converge, since it is an alternating series of real numbers decreasing to zero.

Example 2.7 (The geometric series) Let z be any complex number such that $|z| < 1$, and consider the series

$$\sum_{n=0}^{\infty} z^n = 1 + z + z^2 + \cdots.$$

The nth partial sum is

$$s_n = 1 + z + \cdots + z^{n-1} = \frac{1 - z^n}{1 - z} \quad \text{for} \quad n = 1, 2, \ldots.$$

If we set $\alpha = 1/(1 - z)$, which is well defined since $z \neq 1$, we get

$$0 \leq |s_n - \alpha| = \left| \frac{1 - z^n}{1 - z} - \frac{1}{1 - z} \right|$$

$$= \left| \frac{1}{1 - z} \right| \cdot |z^n| = |\alpha| \cdot |z|^n.$$

Now $|z|^n \to 0$ as $n \to \infty$, since $|z| < 1$; thus, the terms on the right converge to zero as $n \to \infty$, and $|s_n - \alpha| \to 0$ as $n \to \infty$. This implies that $\lim_{n\to\infty} s_n = \alpha$, so the series converges:

$$\sum_{n=0}^{\infty} z^n = \alpha = \frac{1}{1 - z}$$

for all z such that $|z| < 1$. Actually, the series is *absolutely convergent* if $|z| < 1$, because

$$\sum_{n=0}^{\infty} |z^n| = \sum_{n=0}^{\infty} |z|^n = 1/(1 - |z|) < +\infty.$$

Below we assemble basic facts about convergent series of complex numbers.

Theorem 2.5 *Let* $\sum_{n=1}^{\infty} a_n$ *be a series of complex numbers. If* $\sum_{n=1}^{\infty} a_n$ *converges, the size of the* n^{th} *term* a_n *approaches zero as* $n \to \infty$*; that is,* $\lim_{n\to\infty} |a_n| = 0$.

PROOF: For $n = 1, 2, \ldots$ we have $a_n = \sum_{k=1}^{n} a_k - \sum_{k=1}^{n-1} a_k = s_n - s_{n-1}$. But if $\alpha = \sum_{n=1}^{\infty} a_n$, then

$$\alpha = \lim_{n\to\infty} \{s_n\} \quad \text{and} \quad \alpha = \lim_{n\to\infty} \{s_{n-1}\},$$

and by applying Theorem 2.2 we conclude that

$$a_n = s_n - s_{n-1} \to \alpha - \alpha = 0 \quad \text{as} \quad n \to \infty.$$

Now recall that $a_n \to 0$ as $n \to \infty$ if and only if $|a_n| \to 0$ as $n \to \infty$, because $|a_n| = |a_n - 0|$. ▌

The series $\sum_{n=1}^{\infty} a_n = \sum_{n=1}^{\infty} (1/n)$ has $|a_n| = 1/n \to 0$ as $n \to \infty$, but the series diverges. Thus, to insure convergence of a series $\sum_{n=1}^{\infty} a_n$, it is not enough to know that $\lim_{n\to\infty} |a_n| = 0$; we must also know that $|a_n| \to 0$ sufficiently rapidly that the cumulative sum of all the a_n is finite.

Theorem 2.6 *If $\sum_{n=1}^{\infty} a_n$ and $\sum_{n=1}^{\infty} b_n$ are convergent series, then so are the series*

$$\sum_{n=1}^{\infty} (a_n + b_n)$$

and

$$\sum_{n=1}^{\infty} (\alpha a_n) \qquad (\alpha \textit{ a fixed complex number});$$

furthermore,

$$\sum_{n=1}^{\infty} (a_n + b_n) = \sum_{n=1}^{\infty} a_n + \sum_{n=1}^{\infty} b_n$$

$$\sum_{n=1}^{\infty} (\alpha a_n) = \alpha \cdot \left(\sum_{n=1}^{\infty} a_n\right).$$

PROOF: These formulas follow by applying Theorem 2.2 to the sequences of partial sums. ▌

If the series

$$\text{(2)} \qquad \begin{aligned} \sum_{n=0}^{\infty} a_n &= a_0 + a_1 + a_2 + \cdots \\ \sum_{n=0}^{\infty} b_n &= b_0 + b_1 + b_2 + \cdots \end{aligned}$$

both converge, the product of their values $\left(\sum_{n=0}^{\infty} a_n\right) \cdot \left(\sum_{n=0}^{\infty} b_n\right)$ is *not* given by $\sum_{n=0}^{\infty} a_n b_n$. Actually, the collection of all products of pairs of terms of the form

$a_l b_m$ can be arranged into an infinite array

(3)
$$\begin{array}{ccccc} a_0b_0 & a_0b_1 & a_0b_2 & a_0b_3 & \cdots \\ a_1b_0 & a_1b_1 & a_1b_2 & a_1b_3 & \cdots \\ a_2b_0 & a_2b_1 & a_2b_2 & a_2b_3 & \cdots \\ a_3b_0 & a_3b_1 & a_3b_2 & a_3b_3 & \cdots \\ \cdot & \cdot & \cdot & \cdot & \\ \cdot & \cdot & \cdot & \cdot & \\ \cdot & \cdot & \cdot & \cdot & \\ \cdot & \cdot & \cdot & & \cdot \;\cdot \end{array}$$

If we group together terms $a_l b_m$ such that $l + m = p$, for $p = 0, 1, 2, \ldots$ (the terms lying along diagonals, as shown above), we might expect the product to be given by the formula

(4)
$$\left(\sum_{n=0}^{\infty} a_n\right) \cdot \left(\sum_{n=0}^{\infty} b_n\right) = \sum_{p=0}^{\infty} \left(\sum_{k=0}^{p} a_{p-k} b_k\right)$$

by analogy with the formula for the product of two *finite* sums.

Theorem 2.7 *If the series* $\sum_{n=1}^{\infty} a_n$ *and* $\sum_{n=1}^{\infty} b_n$ *are absolutely convergent, then the series* $\sum_{n=0}^{\infty} c_n$ *whose terms are*

$$c_n = a_n b_0 + a_{n-1} b_1 + \cdots + a_0 b_n = \sum_{l+m=n} a_l b_m,$$

is absolutely convergent, and equation (4) *is valid. If the series* $\sum_{n=0}^{\infty} a_n$ *and* $\sum_{n=0}^{\infty} b_p$ *are conditionally convergent, the product series* (4) *may diverge.*

The series $\sum_{n=0}^{\infty} c_n$ is called the **Cauchy product** of the two series we started with; it is the Cauchy product of two series which converges to the product of their values. We will not prove Theorem 2.7 here; it leans heavily on the theory of rearrangements of series, and these details would take up too much space to justify their inclusion. However, there is a clear, self-contained account of the Cauchy product formula, together with a discussion of rearrangements, in Hille [9], Sections 5.2–5.3.

We might try to prove formula (4) as follows. The sum $\sum_{n=0}^{\infty} c_n$ is obtained by summing the numbers in the array (3) along diagonals. It seems plausible that we should obtain the same result by summing over rows and then adding up the row-sums:

(5)
$$\sum_{n=0}^{\infty} c_n = \sum_{l=0}^{\infty} \left(\sum_{m=0}^{\infty} a_l b_m\right).$$

The expression on the right is just what we want; by Theorem 2.6 it follows that

$$\sum_{l=0}^{\infty}\left(\sum_{m=0}^{\infty} a_l b_m\right) = \sum_{l=0}^{\infty}\left(a_l \cdot \left(\sum_{m=0}^{\infty} b_m\right)\right) = \left(\sum_{l=0}^{\infty} a_l\right) \cdot \left(\sum_{m=0}^{\infty} b_m\right).$$

The whole point of a legitimate proof lies in justifying the identity (5); this is a problem in rearranging series, and the absolute convergence assumed for the series in Theorem 2.7 is needed to justify this step.

To decide convergence questions we often compare the terms of a given series with those of a well known series, such as the geometric series. This "comparison principle," used in conjunction with Theorem 2.4, will be one of our most effective tools in studying series.

Theorem 2.8 (The comparison test for convergence) *Let* $\sum_{n=1}^{\infty} a_n$ *and* $\sum_{n=1}^{\infty} b_n$ *be series with non-negative real terms. Suppose that*

(i) $\sum_{n=1}^{\infty} b_n$ *converges.*

(ii) There is a constant K *such that* $a_n \leq K \cdot b_n$ *for all sufficiently large indices* n*, say for* $n \geq N$.

(In other words, the terms of $\sum_{n=1}^{\infty} a_n$ *are "dominated" by the terms of a known convergent series* $\sum_{n=1}^{\infty} b_n$*). Then* $\sum_{n=1}^{\infty} a_n$ *converges.*

PROOF: Remember that we are dealing with series of *non-negative* numbers. Obviously $\sum_{n=N}^{k} b_n \leq \sum_{n=1}^{\infty} b_n < +\infty$ for all indices $k \geq N$, and $a_n \leq K \cdot b_n$ for $n \geq N$. Consequently,

$$\sum_{n=N}^{k} a_n \leq K \cdot \sum_{n=N}^{k} b_n \leq K \cdot \sum_{n=1}^{\infty} b_n \quad \text{for} \quad k = N, N+1, \ldots.$$

If we add on the finite number of remaining terms, we get

$$s_k = \sum_{n=1}^{k} a_n \leq K' \quad \text{for} \quad k = N, N+1, \ldots$$

where $K' = |a_1| + \cdots + |a_{N-1}| + K \cdot \left(\sum_{n=1}^{\infty} b_n\right)$. Since $\{s_k\}$ is an increasing sequence of real numbers which is bounded from above, it converges; thus $\sum_{n=1}^{\infty} a_n$ converges. ∎

For another version of the comparison test, see Exercise 13.

We may test for absolute convergence of a complex series $\sum_{n=1}^{\infty} z_n$ by comparing the series of absolute values $\sum_{n=1}^{\infty} |z_n|$ with known convergent (or divergent) series of non-negative real numbers. The Ratio Test, below, is proved using the geometric series $\sum_{n=0}^{\infty} r^n$ as the known series.

Theorem 2.9 (The ratio test) *Suppose $\sum_{n=1}^{\infty} a_n$ is a series of complex numbers whose terms are nonvanishing, and suppose that*

$$L = \lim_{n\to\infty} \frac{|a_{n+1}|}{|a_n|} \tag{6}$$

exists (the improper limit $+\infty$ is allowed). Then the series is absolutely convergent if $L < 1$, divergent if $L > 1$, and we cannot make a definite statement either way if $L = 1$.

Proof: If $L < 1$, the average $\alpha = (L + 1)/2$ lies between L and 1, so $L < \alpha < 1$. In view of (6) we get

$$\frac{|a_{n+1}|}{|a_n|} < \frac{L+1}{2} = \alpha$$

for all large n (say for $n \geq N$). Therefore,

$$|a_{N+k}| < \alpha\, |a_{N+k-1}| < \alpha^2\, |a_{N+k-2}| < \cdots < \alpha^k |a_N| = \frac{|a_N|}{\alpha^N} \cdot \alpha^{N+k},$$

which is the same as saying that

$$|a_j| < \left(\frac{|a_N|}{\alpha^N}\right) \cdot \alpha^j$$

for all large indices j ($j \geq N$). Thus, the series $\sum_{n=1}^{\infty} |a_n|$ is dominated by the geometric series $\sum_{n=1}^{\infty} \alpha^n$ (taking $K = |a_N|/\alpha^N$ in Theorem 2.8), which converges since $\alpha < 1$.

On the other hand, if $L > 1$, then

$$\frac{|a_{n+1}|}{|a_n|} \geq 1$$

for all large n (say $n \geq N$), so that

$$0 < |a_N| \leq |a_{N+1}| \leq |a_{N+2}| \leq \cdots \leq |a_{N+k}| \leq \cdots.$$

This prevents us from having $|a_n| \to 0$ as $n \to \infty$, as would be the case if $\sum_{n=1}^{\infty} a_n$ converged (see Theorem 2.5). ▮

In examining a series $\sum_{n=1}^{\infty} a_n$ we might have various objectives in mind. Often we only need to know whether or not the series converges (or converges absolutely); at other times we must actually determine its *sum*. The convergence question can usually be resolved without actually calculating the sum of the series; the comparison and ratio tests above refer specifically to the convergence of the series, and do not yield the value of its sum. If we are obliged to calculate the sum, other methods must be used, and we may face an extremely difficult (if not impossible) task. For example, it is not very hard to prove that the series $\sum_{n=0}^{\infty} 1/n!$, $\sum_{n=1}^{\infty} (-1)^{n+1}/n$, and $\sum_{n=1}^{\infty} 1/n^2$ converge—the ratio test works on the first, and the usual tests of calculus (for real series) are adequate for the others. Calculating their sums,

$$\sum_{n=0}^{\infty} \frac{1}{n!} = e; \qquad \sum_{n=1}^{\infty} \frac{(-1)^{n+1}}{n} = \log(2); \qquad \sum_{n=1}^{\infty} \frac{1}{n^2} = \frac{\pi^2}{6}$$

is a much more difficult task.

Often we cannot hope for closed formulas for the sum, like the ones above. Then we must resort to numerical approximations. In almost every case the convergence question must be resolved before attempting a numerical calculation of the sum. The comparison test is useful in these approximate calculations because it provides us with error estimates.

Example 2.8 Suppose $\{a_n\}$ is any sequence such that $a_n = \pm 1$, but with the sign chosen "at random" for each n. We may form the "geometric series" with these signs, $\sum_{n=0}^{\infty} a_n z^n$. Now $|a_n z^n| = |z^n| = |z|^n$ for all n, and this series is absolutely convergent for every z such that $|z| < 1$, in spite of the signs we have introduced. However, we cannot expect the sum of the new series to be $1/(1 - z)$; in fact, its sum can be extremely difficult to calculate in closed form.

Example 2.9 The ratio test, and comparison tests in general, are limited to tests of divergence or *absolute* convergence. Thus, the series $\sum_{n=1}^{\infty} (-1)^{n+1}/n$ converges, as the alternating series test (from calculus) shows, but is only conditionally convergent, and comparison tests are not sensitive enough to prove that it converges. The series $\sum_{n=1}^{\infty} 1/n^2$ is not directly comparable to the geometric series, and the ratio test yields no information ($L = 1$). However, it *is* absolutely convergent; we may use the integral test from calculus, and the fact that $\int_1^{\infty} (1/x^2)\, dx = \lim_{N\to\infty} \int_1^N (1/x^2)\, dx = 1 < +\infty$. Now that the convergence of $\sum_{n=1}^{\infty} 1/n^2$ is established, this series may be used as the "known series" in comparison tests. (See also Exercise 12.)

EXERCISES

1. Look up a proof, in any competent calculus text, that

(i) $\sum_{n=1}^{\infty} \frac{1}{n} = 1 + \frac{1}{2} + \frac{1}{3} + \cdots$ diverges (divergent harmonic series)

(ii) $\sum_{n=1}^{\infty} \frac{(-1)^{n+1}}{n} = 1 - \frac{1}{2} + \frac{1}{3} \cdots$ converges (alternating series test)

(iii) $\sum_{n=1}^{\infty} \frac{1}{n^2} = 1 + \frac{1}{4} + \frac{1}{9} + \cdots$ converges (integral test).

Do these proofs give the *values of the series* as well as their convergence?

2. Prove that the series

$$\sum_{n=0}^{\infty} z^n = 1 + z + z^2 + \cdots \quad \text{and} \quad \sum_{n=1}^{\infty} \frac{z^n}{n^2}$$

both *diverge* if $|z| > 1$.

3. Test the following series for convergence and absolute convergence.

(i) $\sum_{n=1}^{\infty} \frac{i^n}{n^2}$

(ii) $\sum_{n=1}^{\infty} \frac{i^n \log n}{n^2 + 1}$

(iii) $\sum_{n=1}^{\infty} (-1)^{n+1} \left(\frac{\log n}{n}\right)$

(iv) $\sum_{n=1}^{\infty} \frac{(1+i)^n}{n}$

(v) $\sum_{n=1}^{\infty} \frac{i^n + 1}{n^{3/2}}$

(vi) $\sum_{n=1}^{\infty} \frac{z^n}{1 - z^n}$ if $|z| < 1$.

Answer: (i) absolutely convergent; (ii) absolutely convergent; (iii) conditionally convergent; (iv) divergent; (v) absolutely convergent; (vi) absolutely convergent.

4. Determine *all* z for which the series below are absolutely convergent.

(i) $\sum_{n=1}^{\infty} \frac{z^n}{n^2}$

(ii) $\sum_{n=1}^{\infty} z/n!$

(iii) $\sum_{n=1}^{\infty} \frac{1}{n!}\left(\frac{1}{z}\right)^n$

(iv) $\sum_{n=1}^{\infty} \frac{1}{2^n}\left(\frac{1}{z+1}\right)^n$

(v) $\sum_{n=1}^{\infty} \left(\frac{z-1}{z+1}\right)^n$

Answer: (i) $|z| \leq 1$; (ii) all z; (iii) all $z \neq 0$; (iv) $|z + 1| > \frac{1}{2}$; (v) $\text{Re}(z) > 0$.

5. Prove that the following series converge absolutely to the limits shown.

(i) $\sum_{n=1}^{\infty} z^n = \dfrac{z}{1 - z}$ if $|z| < 1$

(ii) $\sum_{n=0}^{\infty} (-1)^{n+1} z^{4n} = -1 + z^4 - z^8 + \cdots = \dfrac{1}{z^4 - 1}$

if $|z| < 1$

(iii) $\sum_{n=0}^{\infty} \left(\dfrac{z - 1}{z + 1}\right)^n = \dfrac{z + 1}{2}$ if $\text{Re}(z) > 0$

(iv) $\sum_{n=0}^{\infty} \left(\dfrac{1}{z}\right)^n = \dfrac{z}{z - 1}$ if $|z| > 1$

(v) $\sum_{n=1}^{\infty} \left(\dfrac{1}{z}\right)^n = \dfrac{1}{z - 1}$ if $|z| > 1$

Hint: Remember that $\sum_{n=0}^{\infty} w^n = \dfrac{1}{1 - w}$ if $|w| < 1$.

6. Using the geometric series, define series of the form $\sum_{n=0}^{\infty} a_n z^n$ whose sums have the values

(i) $\dfrac{1}{1 + z^2}$ (ii) $\dfrac{1}{1 + z}$

for $|z| < 1$.

Answer: (i) $(1 + z^2)^{-1} = 1 - z^2 + z^4 - \cdots = \sum_{n=0}^{\infty} (-1)^n z^{2n}$;

(ii) $(1 + z)^{-1} = 1 - z + z^2 - \cdots = \sum_{n=0}^{\infty} (-1)^n z^n$.

7. Write the terms of $\sum_{n=1}^{\infty} z_n$ as $z_n = x_n + y_n$; then consider the (real) series $\sum_{n=1}^{\infty} x_n$ and $\sum_{n=1}^{\infty} y_n$, and prove that:

(i) The complex series converges if and only if both real series converge; moreover,

$$\sum_{n=1}^{\infty} z_n = \left(\sum_{n=1}^{\infty} x_n\right) + i\left(\sum_{n=1}^{\infty} y_n\right).$$

(ii) If the real series are absolutely convergent, so is the complex series.

(iii) If the complex series is absolutely convergent, so are both of the real series.

8. Prove that if the complex series $\sum_{n=1}^{\infty} z_n$ converges, then $\left|\sum_{n=1}^{\infty} z_n\right| \leq \sum_{n=1}^{\infty} |z_n|$.

Note: The series on the right is allowed to have the improper value $+\infty$, in which case the inequality is trivial. What conditions on the terms z_n would make these series equal?

Hint: Use the triangle inequality to compare the partial sums.

9. If $\sum_{n=1}^{\infty} a_n$ converges, with sum α, show that:

(i) Each "remainder" $R_k = \sum_{n=k+1}^{\infty} a_n$ is a convergent series.

(ii) If s_k is the kth partial sum, then $R_k = \alpha - s_k$ for all k.

(iii) $\lim_{k\to\infty} R_k = 0$.

Hint: $R_k = \lim_{m\to\infty} (a_{k+1} + \cdots + a_m) = \lim_{m\to\infty} (s_m - s_k)$, if s_m is the mth partial sum of the given series. The rest follows from Theorem 2.2.

10. Use the Integral Test to decide which real exponents $\alpha \neq 0$ make the sums $\sum_{n=1}^{\infty} n^{\alpha}$ converge or diverge.

Answer: Converges if $-\infty < \alpha < -1$; diverges if $\alpha \geq -1$.

11. Show that the alternating real series $\sum_{n=0}^{\infty} a_n = \sum_{n=1}^{\infty} \frac{(-1)^{n+1}}{\sqrt{n}}$ ($a_0 = 0$) is convergent. Then calculate the terms c_n in the Cauchy product of this series with itself and show that $\sum_{n=0}^{\infty} c_n$ diverges.

12. Use Exercises 8 and 9 to show that the "error" $e_k = \left|s_k - \sum_{n=1}^{\infty} a_n\right|$ may be estimated directly in certain examples:

(i) For the geometric series $\sum_{n=0}^{\infty} z^n$, taking $|z| < 1$, we get

$$e_k = \frac{|z|^k}{1 - |z|} \quad \text{for} \quad k = 1, 2 \cdots.$$

(ii) If $|z| \leq 1$ in the series $\sum_{n=0}^{\infty} z^n/n^2$, we get

$$e_k \leq \int_k^{\infty} t^{-2}\, dt = \frac{1}{k}.$$

By comparing series, or by other means, decide how many terms should be taken in the following series to calculate their values, so that $e_k \leq 0.01$.

(iii) $\sum_{n=0}^{\infty} e^{ni} z^n \quad \text{if } |z| = \dfrac{1}{2}$

(iv) $\sum_{n=1}^{\infty} \dfrac{1}{n^2}$

(v) $\sum_{n=1}^{\infty} \dfrac{(-1)^{n+1}}{n} \left(\dfrac{i}{2}\right)^n.$

Hint: Do not calculate the value $\sum_{n=1}^{\infty} a_n$ unless this is easy; instead, observe that $e_k \leq \sum_{n=k+1}^{\infty} |a_n|$ (Exercise 8) and estimate the size of this remainder by integral tests, comparison with known series, or other means.

Answer: (iii) $k \geq 8$; (iv) $k \geq 100$; (v) $k \geq 8$ $\left(\text{compare with } \sum_{n=0}^{\infty} (\tfrac{1}{2})^n\right)$.

13. Let $\sum_{n=1}^{\infty} a_n$ and $\sum_{n=1}^{\infty} b_n$ be series with non-negative real terms. Suppose that $\sum_{n=1}^{\infty} a_n$ is known to diverge and that $\sum_{n=1}^{\infty} b_n$ "dominates" $\sum_{n=1}^{\infty} a_n$ in the sense that there is a constant K such that $a_n \leq K b_n$ (all large n). Prove that the series $\sum_{n=1}^{\infty} b_n$ diverges. Compare this result with Theorem 2.8.

2.3 GEOMETRIC PROPERTIES OF SETS IN THE PLANE

There are a few geometric concepts, in addition to those mentioned in Section 1.4, which should be examined. A point z is said to be **adherent** to a set A if it is the limit of some sequence of points lying within A. A point z that belongs to A is adherent to A, since the constant sequence $a_n = z$ (for all n) consists of points in A and converges to z; however, there may be additional points that are adherent to A but do not belong to A. If A is the disc $A = \{z : |z| < 1\}$, every point on the boundary circle $|z| = 1$ is the limit of a suitably chosen sequence of points in A, and so is adherent to A; on the other hand, if $|z| > 1$, we can find a small disc of positive radius about z that is disjoint from A and isolates z from A. Thus, no point z such that $|z| > 1$ can be the limit of a sequence of points in A. Clearly, the closed disc $\{z : |z| \leq 1\}$ is exactly the set of adherent points of A.

The set of points adherent to A is referred to as the **closure** of A and will be denoted by A^-; if a set already includes all of its adherent points, we refer to it as a **closed set**. It is obvious that A^- contains A and, with a little effort, it can be seen that we get no new points by taking the closure twice in succession; thus $(A^-)^- = A^-$, and A^- is always a closed set in the plane.

The **boundary** of a set A, denoted by $bdry(A)$, consists of all points in the plane that can be approached by sequences from within A *and* from within the complementary set $A^\sim = \mathbf{C} \sim A$; in some sense, then, the boundary points "stand between" A and its complement $A^\sim$—hence the name "boundary." One can show (see Exercise 5) that z is a boundary point if and only if every disc of positive radius about z meets both A and its complement $A^\sim$. For example, if A is the half plane $\{z:\text{Im}(z) > 0\}$, its boundary is just the real axis: $bdry(A) = \{z:\text{Im}(z) = 0\}$; or, if A is the closed (or open) disc of radius one about the origin, the boundary is (in either case) the unit circle: $bary(A) = \{z:|z| = 1\}$. Some boundary points may belong to the original set A, while others may belong to the complementary set $A^\sim$; consider what happens for the discs $|z| < 1$ and $|z| \le 1$. Since the sets A and $A^\sim$ play symmetrical roles in the definition of $bdry(A)$, it is clear that

(7) $$bdry(A) = bdry(A^\sim) \quad \text{for any set } A.$$

One can also show, by comparing definitions, that the closure A^- of any set A is obtained by adjoining its boundary points:

(8) $$A^- = A \cup bdry(A) \quad \text{for any set } A.$$

The notion of *connectedness* turns out to be particularly important in our work, and we will need a careful discussion of this relatively unfamiliar concept. Let A be a set in the plane. We say that A is **disconnected** if it is possible to find two *open* sets E_1 and E_2 in the plane, such that

(i) The set $A_1 = E_1 \cap A$ is not empty
(ii) The set $A_2 = E_2 \cap A$ is not empty
(iii) The sets E_1 and E_2 are disjoint; that is, they have no points in common.
(iv) The sets E_1 and E_2 together cover the set A; that is, A is the union $A = A_1 \cup A_2$ of the pieces A_1 and A_2.

Loosely speaking, a set A is disconnected if it can be divided into two non-trivial (i.e., non-empty) pieces, each isolated from the other. Indeed, the sets E_1 and E_2 divide A into non-trivial pieces A_1 and A_2, as shown in Figure 2.2. Because E_1 and E_2 are assumed to be *open sets*, the pieces A_1 and A_2 are *isolated* from each other in the following way: for each point a in A_1, there is some disc of positive radius about a that does not meet the other piece A_2 (thus, the disc "isolates" a from the points in A_2). Likewise, each point in A_2 is isolated from the points in A_1.

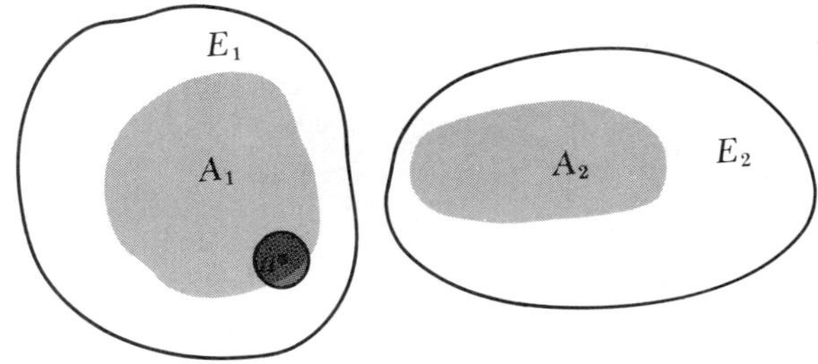

Figure 2.2 A disconnected set separated into disjoint, isolated pieces by open sets E_1 and E_2. Each point in A_1 is isolated by a disc of positive radius from the points in A_2, and vice versa.

Definition 2.3 *The set A is* **connected** *if it is not a disconnected set.*

In these definitions, it is crucial that the sets E_1 and E_2 that separate A be open sets, so that the pieces A_1 and A_2 will be isolated. It is usually possible to divide any set into disjoint pieces, if we do not require the pieces to be isolated as above. For example, we will soon see that the set $A = \mathbf{C}$ (the whole complex plane) is connected; nevertheless, it may be divided into the non-trivial, disjoint pieces

$$A_1 = \{z: \mathrm{Im}(z) \geq 0\} \text{—the upper half plane plus real axis}$$

$$A_2 = \{z: \mathrm{Im}(z) < 0\} \text{—the lower half plane.}$$

The set A_2 is open, but A_1 is not; furthermore, there are points in A_1, namely the points on the real axis, that are not isolated from the set A_2 by any disc of positive radius. Thus, the partitioning $A = A_1 \cup A_2$ does not conflict with the connectedness of A.

We need efficient ways of recognizing when a set is connected, and the definition above is not always convenient for this purpose. For example, if we want to prove that the complex plane $A = \mathbf{C}$ is a connected set, we must show that no choice of disjoint open sets E_1 and E_2, *no matter how cleverly conceived*, can divide the plane into non-trivial, disjoint, isolated pieces. This can be a formidable task, if approached the wrong way. Fortunately, there is a simple, practical test for connectedness; it is based on the notion of "connecting" two points in the plane via parametrized curves.

We define a continuous parametrized curve in the plane by giving a pair of continuous functions $x(t)$ and $y(t)$, defined on some real interval $a \leq t \leq b$; these specify the Cartesian coordinates of a point $z(t) = x(t) + iy(t)$, whose position varies continuously with t. Thus a parametrized curve is a *function*, whose variable is real and whose values are complex. We say that a parametrized curve $z(t)$ **connects the points** p and q if

$$z(a) = p \quad \text{and} \quad z(b) = q,$$

as shown in Figure 2.3. With this definition in hand, one can prove the following result concerning open sets. (We will not go into the details of the proof in this book; see Buck [2], Section 1.5, and Ahlfors [1], Section 1.3 for this result and other elementary comments about connected sets.)

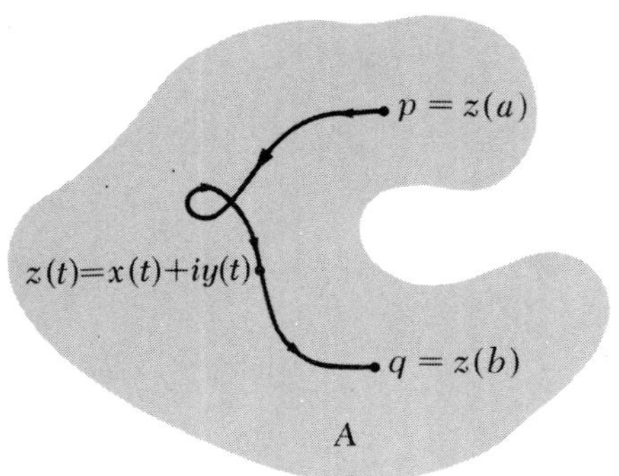

Figure 2.3 A connected open set with a parametrized curve $z(t) = x(t) + iy(t)$ connecting the points p and q.

Theorem 2.10 *Let A be an open set in the plane. Then A is connected if and only if for every pair of points $\{p, q\}$ in A, there is a corresponding continuous parametrized curve $z(t) = x(t) + iy(t)$ that*

(i) *connects p and q*
(ii) *lies entirely within the set A, in the sense that $z(t)$ belongs to A for each t such that $a \leq t \leq b$.*

Figure 2.3 shows a parametrized curve connecting p and q, which lies entirely within the set A. Theorem 2.10 applies only to open sets, but it will suffice for our needs.

Usually it is intuitively obvious whether every pair of points $\{p, q\}$ in A can be connected by a parametrized curve within A, and it is usually not hard to supplement this intuition by providing an explicit construction.

Example 2.10 The unit disc about the origin $A = \{z : |z| < 1\}$ is a connected open set. If $p = x_0 + iy_0$ and $q = u_0 + iv_0$ are typical points in A, let us consider the parametrized curve defined by the linear formula

$$\begin{aligned} z(t) &= (1 - t)p + tq = p + t(q - p) \\ &= [x_0 + t(u_0 - x_0)] + i[y_0 + t(v_0 - y_0)] = x(t) + iy(t). \end{aligned}$$

The coordinate functions $x(t)$ and $y(t)$ are defined for $0 \leq t \leq 1$, and are continuous. It is easy to see that the moving point $z(t)$ traces out the straight line segment from p to q, so $z(t)$ remains within the disc at all times. Since p and q are typical points in A, Theorem 2.10 applies and A is connected. Similar reasoning shows that arbitrary open discs $A = \{z : |z - p| < r\}$, or the whole plane $A = \mathbf{C}$, or open half planes such as $A = \{z : \text{Im}(z) > 0\}$ are all connected sets; if $\{p, q\}$ are typical points, the line segment connecting these points lies within A. More sophisticated arguments, using parametrized circular arcs instead of line segments, show that the punctured plane $A = \{z : z \neq 0\}$ and the ring-shaped set $A = \{z : 1 < |z| < 2\}$ are connected.

Definition 2.4 *A connected open set in the plane will be called a* **domain** *throughout this book.*

In Table 2.1 we present a list of sets in the plane, and indicate how the geometric ideas just presented apply to them. The symbol $\varnothing$ stands for the empty

TABLE 2.1 PROPERTIES OF SELECTED SETS IN $\mathbf{C}$

A	*A*⁻	*bdry (A)*	*A*~	*Closed*	*Open*	*Bounded*	*Connected*	*A Domain*
$\lvert z\rvert \le R$	$\lvert z\rvert \le R$	$\lvert z\rvert = R$	$\lvert z\rvert > R$	YES	NO	YES	YES	NO
$\lvert z\rvert < R$	$\lvert z\rvert \le R$	$\lvert z\rvert = R$	$\lvert z\rvert \ge R$	NO	YES	YES	YES	YES
$\mathrm{Re}(z) > 0$	$\mathrm{Re}(z) \ge 0$	$\mathrm{Re}(z) = 0$	$\mathrm{Re}(z) \le 0$	NO	YES	NO	YES	YES
$\mathrm{Re}(z) \ne 0$	$\mathbf{C}$	$\mathrm{Re}(z) = 0$	$\mathrm{Re}(z) = 0$	NO	YES	NO	NO	NO
$\lvert z\rvert > R$	$\lvert z\rvert \ge R$	$\lvert z\rvert = R$	$\lvert z\rvert \le R$	NO	YES	NO	YES	YES
$\{0\}$	$\{0\}$	$\{0\}$	$z \ne 0$	YES	NO	YES	YES	NO
$\mathbf{C}$	$\mathbf{C}$	$\varnothing$	$\varnothing$	YES	YES	NO	YES	YES
$\varnothing$	$\varnothing$	$\varnothing$	$\mathbf{C}$	YES	YES	YES	YES	YES

set, which has no points in it at all, and $\{p\}$ stands for the one-point set whose only element is the point $z = p$.

EXERCISES

1. Show that the sets E are disconnected; exhibit open sets that separate them into isolated pieces.

(i) $E =$ all "integral points" $z = m + in$ in the plane.
(ii) $E = \{z:\text{Re}(z) < 0\} \cup \{z:\text{Re}(z) > 0 \text{ and } \text{Im}(z) = 0\}$
(iii) $E = \{z:\text{Im}(z) \neq 1\}$
(iv) $E =$ the real axis with the origin removed.

2. Verify the statements displayed in Table 2.1.

3. If $E = \{z:1 < |z| \leq 2\}$, show that $bdry(E)$ consists of the circles $|z| = 1$ and $|z| = 2$, and that the closure $E^- = \{z:1 \leq |z| \leq 2\}$. Demonstrate that E is neither open nor closed.

Note: There are many sets in the plane that are neither open nor closed.

4. Consider the segment of the real axis $E = (-1, +1) = \{z:\text{Im}(z) = 0 \text{ and } -1 < \text{Re}(z) < 1\}$. Sketch the sets

(i) $bdry(E)$ (iii) $E^{\sim} = \mathbf{C} \sim E$
(ii) E^- (iv) $E^- \sim E$

and answer the questions listed in Table 2.1.

Answer: (i) $bdry(E) = [-1, +1]$; (ii) $E^- = [-1, 1]$; (iii) $E^{\sim} = \{z:\text{either } \text{Im}(z) \neq 0 \text{ or } |\text{Re}(z)| \geq 1\}$; (iv) the two point set $\{-1, +1\}$. E is bounded, connected, not open, not closed, and not a domain.

5. Prove that z is adherent to A if and only if every disc of positive radius about z meets A. Prove that z is in $bdry(A)$ if and only if every such disc meets both A and its complement $A^{\sim}$.

2.4 FUNCTIONS OF A COMPLEX VARIABLE

A **function**, or **mapping**, between two sets M and N is a correspondence that assigns to certain points x in M corresponding points $f(x)$ in N, as indicated in Figure 2.4. To specify a function completely, two things are necessary; we must first give its **domain of definition**, the set Dom(f) consisting of all points

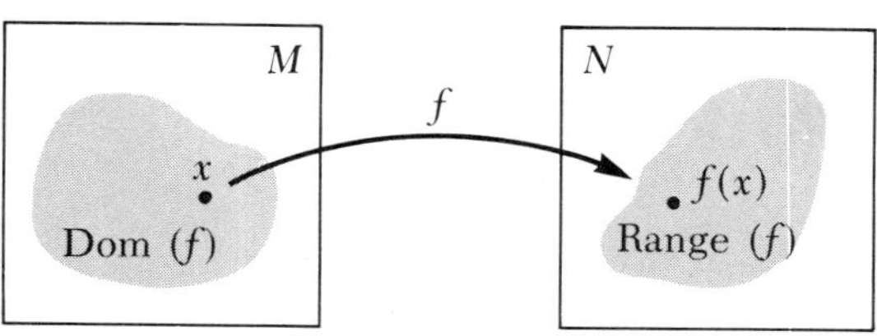

Figure 2.4 A mapping $f:M \to N$ from set M to set N.

in M for which the function is defined.† Then we must give the rule that tells us how to get $f(x)$ from x for each x in $\mathrm{Dom}(f)$. Two functions f and g are regarded as being *identical* if they have the same domain of definition, and assign the same value to each point x in this set:

$$D = \mathrm{Dom}(f) = \mathrm{Dom}(g) \quad \text{and} \quad f(x) = g(x) \quad \text{for all } x \text{ in } D.$$

Complex analysis is mostly concerned with mappings from $\mathbf{C}$ into $\mathbf{C}$, while calculus deals with mappings from $\mathbf{R}$ (or sometimes $\mathbf{R}^n$) into $\mathbf{R}$, functions with real variables and real values. When a function or mapping f is defined throughout a set M, with values in another set N, we usually write $f: M \to N$ to indicate this situation.

Often the domain of definition is not written down explicitly; the domain *must* be specified to fully define f, but it is often clear from the context what the domain $\mathrm{Dom}(f)$ should be. For example, if we speak of the function f given by the rule $f(z) = 1/z$, we take $\mathrm{Dom}(f)$ to be all points z in $\mathbf{C}$ such that the rule makes sense; thus, $\mathrm{Dom}(f) = \{z: z \neq 0\}$. If a function $f: M \to N$ is given, we will sometimes be interested in its **range**, the set $\mathrm{Range}(f) \subseteq N$ consisting of all points in N that are images of points in M.

Example 2.11 Consider $w = f(z) = z^2$, taking $\mathrm{Dom}(f) = \mathbf{C}$. Since every complex number w has at least one square root, every w is the image of one or more points in $\mathrm{Dom}(f)$; thus, $\mathrm{Range}(f) = \mathbf{C}$. Similarly, consider $w = g(z) = 1 + z^2$ defined on $\mathbf{C}$. If w is arbitrary and if $z = \pm\sqrt{w - 1}$, it is clear that $g(z) = w$; thus, $\mathrm{Range}(g) = \mathbf{C}$.

While it is possible to draw sketches and graphs to visualize functions of a real variable, there is a serious difficulty in doing this for functions of a complex variable. Complex numbers are represented as points (or vectors) in the Cartesian plane, having both length and direction, and a function f attaches a complex number $w = f(z)$—a vector—to each point z in $\mathrm{Dom}(f)$. Drawing a graph of this situation would require four dimensions, and is not practical. One way of visualizing f is given below; other methods will be introduced in Chapter 4. Given $f(z)$, we may draw the "level curves" of the function $|f(z)|$. These are the loci $\{z: |f(z)| = c\}$ for various $c \geq 0$. This pattern of curves tells us how large the values of $f(z)$ are, but completely ignores the direction angle associated with the polar form of $f(z)$.

Example 2.12 Take $w = f(z) = z^2$ and $\mathrm{Dom}(f) = \mathbf{C}$. Figure 2.5A shows f as a vector field. If z has length $|z| = r$ and makes an angle θ with the positive real axis, then $f(z) = z^2$ has length $|z^2| = |z|^2 = r^2$ and makes an angle 2θ with the positive real axis (recall the geometric interpretation of multiplication). Thus, all vectors in Figure 2.5A whose base points lie on the circle $|z| = r$ have

† The "domain of definition" of a function may or may not be a "domain" (connected open set). The word "domain" is being used in different ways in these phrases. In practice, it will always be clear from the context which meaning is intended.

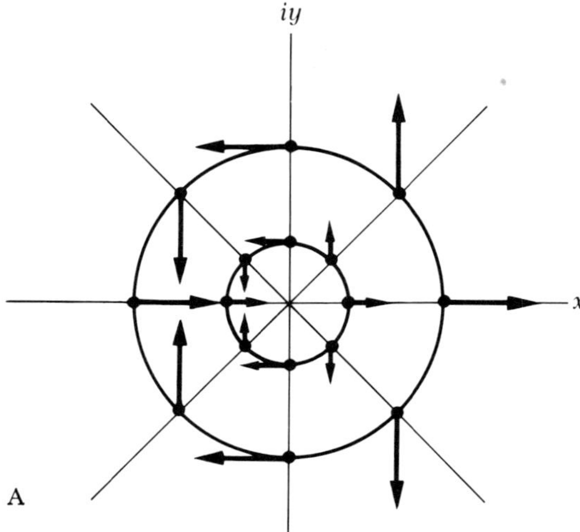

A

Figure 2.5A The complex number $w = f(z)$ shown as a vector attached to the point z.

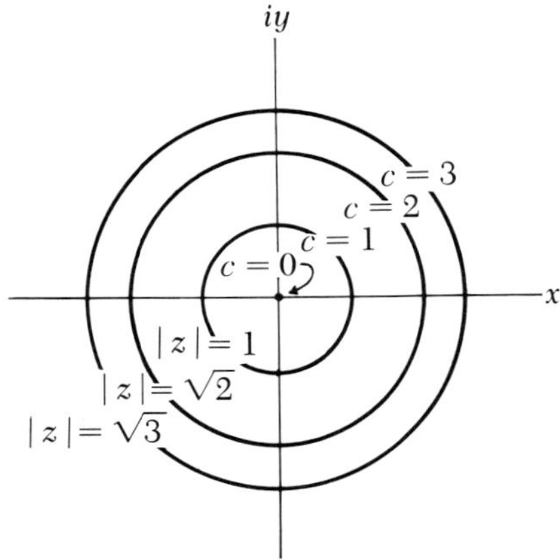

B

Figure 2.5B The level curves $|w| = |f(z)| = c$ for various choices of $c \geq 0$.

length r^2, but different directions, as shown. This kind of diagram is cumbersome, and is not very useful for complicated functions. In Figure 2.5B we show the level curves $|f(z)| = c$, which are circles centered at the origin with radius $c^{1/2}$.

Example 2.13 (The level curves of $f(z) = z^2 + 1$) The polynomial $f(z) = z^2 + 1$ is zero at the two points $z = +i$ and $z = -i$, and $|f(z)| > 0$ at all other points in the plane. For the practical details of plotting the locus $|f(z)| = c$, it is better not to have to deal with square roots, so we should consider the equivalent equation $|f(z)|^2 = c^2$; this is a *quartic* (4th degree) equation in the variables x and y if we write $z = x + iy$:

$$\begin{aligned} |z^2 + 1|^2 &= |(x^2 - y^2 + 1) + i(2xy)|^2 \\ &= (x^2 - y^2 + 1)^2 + 4x^2y^2 = c^2. \end{aligned}$$

It is not practical to solve explicitly for y as a function of x (or vice versa) in such a high degree equation; nevertheless, we may estimate the positions of these loci by numerical calculations and by qualitative arguments. We are aided by

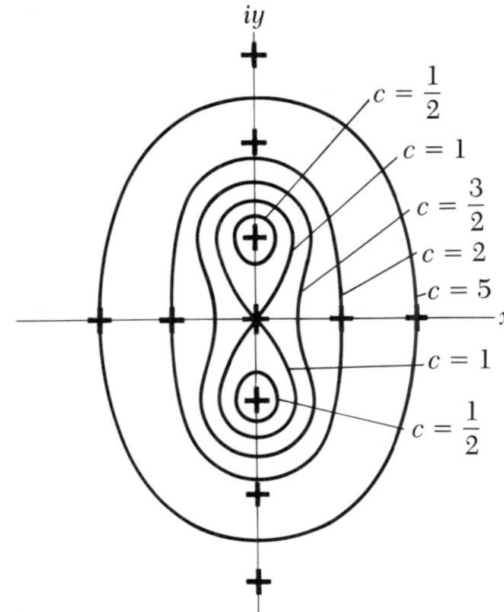

Figure 2.6 The level curves $c = |f(z)|$ for the function $f(z) = z^2 + 1$, which has zeroes at $z = +i$ and $z = -i$.

the knowledge that a quartic locus cannot meet any vertical or horizontal line in more than *four* places because quartic polynomials in x, or in y, cannot have more than four roots.

The locus $|f(z)| = 0$ consists of the two points $+i$ and $-i$. As c increases from $c = 0$ to $c = 1$, the curves $|f(z)| = c$ move away from $+i$ and $-i$, as indicated by the locus $c = \frac{1}{2}$ in Figure 2.6. When $c = 1$, we get the double curve passing through the origin (if $z = 0$, clearly $|z^2 + 1| = |1| = 1$). For $c > 1$ the locus is given by a single smooth curve, and for very large values of $c > 1$ there is very little difference between the level curves for the functions $f(z) = z^2 + 1$ and $g(z) = z^2$; that is, the locus $|f(z)| = c$ is almost indistinguishable from the locus $|g(z)| = c$, a circle of radius $\sqrt{c}$, for large values of c.

The simplest of all the functions we shall encounter are the **polynomials:**

(9) $$f(z) = a_n z^n + a_{n-1} z^{n-1} + \cdots + a_1 z + a_0 \quad \text{(the } a_k \text{ are complex)}.$$

Here n is called the **degree** of the polynomial (if $a_n \neq 0$). Polynomials include the **constant functions,** $f(z) = a_0$ for all z; and also include such functions as

$$f(z) = z, \qquad f(z) = z^2, \ldots.$$

If $P(z)$ and $Q(z)$ are both polynomials, and Q is not the constant function which is equal to zero everywhere, then $Q(z) = 0$ at no more than a finite number of points in $\mathbf{C}$, and except for these points the function $P(z)/Q(z)$ is well defined. Naturally, we take $\text{Dom}(P/Q) = \{z : Q(z) \neq 0\}$. This construction gives us functions like

$$f(z) = \frac{1}{1-z}; \qquad f(z) = \frac{1}{z^2+1}; \qquad f(z) = 1 + \frac{2}{z} + z^2 = \frac{z^3 + z + 2}{z}.$$

Functions obtainable in this way are called **rational functions.**

EXERCISES

1. Prove that $w = f(z) = 1/z$ maps the domain $E = \{z : z \neq 0\}$ onto itself, so that Range$(f) = E$. Prove the same thing for $w = g(z) = 1/z^2$.

2. Prove that $w = f(z) = 1 + z^2$ maps the unit disc $D = \{z : |z| < 1\}$ onto the disc $E = \{w : |w - 1| < 1\}$ in the w plane.

Hint: $w = f(z)$ is obtained by performing the transformation $s = z^2$, followed by $w = s + 1$. It is easy to see how these operations transform sets in the plane.

3. Consider the following mappings $w = f(z)$ defined on $E = \mathbf{C}$. Regarding f as a mapping from the z-plane to the w-plane, determine the image under f of a typical vertical line $x = c$ (c real). Determine the image of a typical horizontal line $y = d$ (d real).

(i) $w = z + 1$ (iv) $w = e^z$

(ii) $w = (1 - i)z$ (v) $w = \left(\dfrac{z - 1}{z + 1}\right)$ $(z \neq -1)$

(iii) $w = z^2$ (vi) $w = 1/z$ $(z \neq 0)$

Are any of these lines transformed in an exceptional way?

4. Use Exercise 3 (iv) to verify that Range(exp z) $= \{w : w \neq 0\}$. Why is $w = 0$ not the exponential of any complex number? Use algebraic methods to prove that the range of $w = f(z) = (z - 1)/(z + 1)$ is the punctured plane $\{w : w \neq 1\}$.

5. Sketch the level curves $|g(z)| = c$, where $g(z) = 1 - z^2$. Use Figure 2.6 and substitute $z = iw$ to relate $|g(z)|$ and $|f(w)| = |1 + w^2|$.

Answer: Rotate Figure 2.6 by $+\pi/2$ radians; zeros of g are at $+1$ and -1.

6. Consider circles $|z| = R$ and the function $f(z) = z^3 + z^2 + z + 2$. Show that the exterior domain $E_R = \{z : |z| > R\}$ is mapped into itself for all sufficiently large values of R.

Hint: Use the triangle inequality; compare $|z^3|$ and $|z^2 + z + 2|$.

Note: Later, in Section 4.8, we will interpret this as demonstrating that f maps points "near infinity" to points "near infinity." There is a similar result for any non-constant polynomial.

2.5 CONTINUITY AND LIMITS OF FUNCTIONS

We shall take time only to generalize the ideas of continuity and limits to deal with functions of a complex variable. Consider a function $w = f(z)$ defined on a set $M = \text{Dom}(f)$ in the complex plane, and let p be a point in the plane.

Definition 2.5 *We say that f has a* **limit at p** *if*

(i) *The point p can be approached by sequences of points $\{z_n\}$, all distinct from p and lying in the domain of definition of f.*
(ii) *The limit*

$$\lim_{n\to\infty} f(z_n) = q$$

exists, and has the same value q, for every sequence of points $\{z_n\}$, distinct from p, that approaches p from within Dom(f).

If f has a limit at p, we indicate the common value of the limits in (ii) by writing

$$q = \lim_{z\to p} f \quad \text{or} \quad q = \lim_{z\to p} f(z),$$

or

$$f(z) \to q \quad \text{as} \quad z \to p.$$

This definition makes precise the idea that the values $f(z)$ approach a value q as z approaches p from within Dom(f), keeping $z \neq p$. We can define $\lim_{z\to p} f$ only if p can be approached by sequences of points lying in the domain of definition of f; otherwise, it makes no sense at all to speak of how the values $f(z)$ behave when z is "near" p. In defining $\lim_{z\to p} f$, it is *not* necessary that f be defined at the point p, and even if f is defined there we ignore the value $f(p)$, which may have no connection with the limit $q = \lim_{z\to p} f$. The limit behavior of f at p refers only to what $f(z)$ is doing as z approaches p, keeping $z \neq p$. The reader should notice how this idea was built into the definition just given. It is important to notice that there are an enormous variety of ways in which a sequence $\{z_n\}$ can approach a point p; existence of $\lim_{z\to p} f$ means that $f(z_n)$ converges to q *no matter how* z_n makes its approach to p. A function may or may not have a limit at a particular point p. Some examples are given in the exercises.

Some authors give an equivalent definition of $\lim_{z\to p} f$. It turns out that $\lim_{z\to p} f = q$ if and only if the following statements hold:

(i) Every disc about p includes points from Dom(f) that are distinct from p.
(ii) Given any number $\varepsilon > 0$, there is a corresponding $\delta > 0$ such that

$$|f(z) - q| < \varepsilon \text{ whenever } z \text{ is in Dom}(f),$$

$$z \neq p, \text{ and } |z - p| < \delta.$$

That is, we get $|f(z) - q| < \varepsilon$ for all z, distinct from p and in Dom(f), which lie sufficiently close to p.

Statements (i) and (ii) also embody the idea that $f(z)$ should get close to q as z gets close to p from within Dom(f), keeping $z \neq p$. We will not take the time to prove that these "ε-δ" statements are equivalent to the definition given above, involving convergent sequences. A discussion of these alternative definitions is given in Buck [2], particularly Section 2.1.

Next consider a function $w = f(z)$, and let p be a point in the plane at which f is defined. We say that f is continuous at p if

(i) The function f is defined at p (so that $f(p)$ is defined).
(ii) The limit $\lim_{z \to p} f$ exists.
(iii) This limit agrees with the value of f at p, so that $f(p) = \lim_{z \to p} f$.

It is not difficult to recast these statements in the following way.

Definition 2.6 *Let f be a function, and let p be a point in* Dom(f). *Then f is* **continuous at p** *if, for every sequence of points $\{z_n\}$ in* Dom(f) *that converges to p, the sequence of values $\{f(z_n)\}$ converges to $f(p)$. If f is continuous at every point in a subset $A \subseteq$* Dom(f), *we say that f is* **continuous on A**.

This behavior is illustrated in Figure 2.7.

Example 2.14 Prove that $f(z) = 1/z$ does not have a well defined limit at the point $p = 0$. Here Dom$(f) = \{z : z \neq 0\}$ and f is not defined at p, but it might have a well defined limit there. The points $z_n = (1/n) + i0$ converge to $z = 0$ as $n \to \infty$, and $z_n \neq 0$; but if q is any complex number we get

$$|f(z_n) - q| \geq \left| f\left(\frac{1}{n}\right) \right| - |q| = n - |q|$$

for all n. Clearly, no choice of q can give $f(z_n) \to q$ as $n \to \infty$; the sequence of distances $|f(z_n) - q| \geq n - |q|$ simply cannot converge to zero. According to Definition 2.5, $\lim_{z \to 0} f(z)$ cannot exist if there is *any* sequence of points $\{z_n\}$ approaching the origin such that $\{f(z_n)\}$ diverges.

Theorem 2.1 is sometimes useful in determining the behavior, or continuity, of functions of a complex variable; a sequence of complex numbers $z_n = x_n + iy_n$ converges to a limit $z = x + iy$ if and only if $\lim_{n \to \infty} x_n = x$ and $\lim_{n \to \infty} y_n = y$ in **R**.

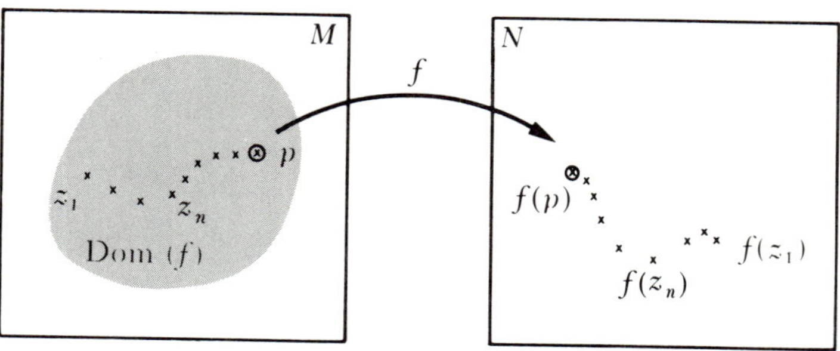

Figure 2.7 Continuity of the function f at $z = p$.

Example 2.15 The function $f(z) = z^2 + 1$ has a limit, and is continuous, at every point in **C**. Consider a point $p = x + iy$ and any sequence $z_n = x_n + iy_n$ that converges to p. Now $f(x + iy) = (x^2 - y^2 + 1) + i(2xy)$, and since $x_n \to x$ and $y_n \to y$ as $n \to \infty$, we get

$$\left.\begin{array}{l}(x_n^2 - y_n^2 + 1) \to (x^2 - y^2 + 1)\\ (2x_n y_n) \to (2xy)\end{array}\right\} \text{ in } \mathbf{R}$$

as $n \to \infty$. Therefore,

$$f(z_n) = (x_n^2 - y_n^2 + 1) + i(2x_n y_n) \to f(p) \quad \text{in } \mathbf{C}$$

as $n \to \infty$. Since this reasoning works for any sequence such that $z_n \to p$, we conclude that $\lim_{z\to p} f(z) = f(p)$, and that f is continuous at p.

Similar arguments show that a polynomial in z is continuous at every point in **C**. However, this follows from a number of general results which say, in effect, that reasonable combinations of continuous functions of a complex variable are still continuous (and likewise for limit behavior). For example, if $f(z)$ and $g(z)$ are both continuous on a set A, so are the combinations

(10)
$$\begin{aligned}(\alpha \cdot f)(z) &= \alpha \cdot (f(z)) \quad (\alpha \text{ a fixed complex number})\\ (f \pm g)(z) &= f(z) \pm g(z)\\ (f \cdot g)(z) &= f(z) \cdot g(z)\\ (f/g)(z) &= f(z)/g(z) \quad (\text{provided } g(z) \neq 0).\end{aligned}$$

Also, if the functions f and g have well defined limits at p, so do these combinations of f and g; furthermore,

(11)
$$\begin{aligned}\lim_{z\to p} \alpha \cdot f &= \alpha \cdot \left(\lim_{z\to p} f\right)\\ \lim_{z\to p} f \pm g &= \left(\lim_{z\to p} f\right) \pm \left(\lim_{z\to p} g\right)\\ \lim_{z\to p} f \cdot g &= \left(\lim_{z\to p} f\right) \cdot \left(\lim_{z\to p} g\right)\\ \lim_{z\to p} f/g &= \left(\lim_{z\to p} f\right) \Big/ \left(\lim_{z\to p} g\right), \qquad \text{provided } \lim_{z\to p} g \neq 0.\end{aligned}$$

These formulas follow directly from the corresponding results for sequences of complex numbers, given in Theorem 2.2.

There is another way to combine functions; we may form a **composite function** $f \circ g$ from f and g by defining

(12)
$$(f \circ g)(z) = f(g(z)).$$

If we write $w = g(z)$ and think of f as a function of w, we get the composite $f \circ g$ by substituting $w = g(z)$ in $f(w)$; thinking of it another way, we get the composite $f \circ g$ by applying the mappings g and then f in succession to points z. The composite function $f \circ g$ is continuous if the functions f and g are continuous (see Exercise 4 for details).

The *composite function* $(f \circ g)(z) = f(g(z))$ should not be confused with the *product* $(f \cdot g)(z) = f(z) \cdot g(z)$ of the functions f and g; the functions are put together in very different ways to form these combinations. For example, if $f(z) = 1 + z^2$ and $g(z) = e^z$, then $(f \cdot g)(z) = (1 + z^2) \cdot e^z$ and $(f \circ g)(z) = 1 + (e^z)^2 = 1 + e^{2z}$, for all complex numbers z.

EXERCISES

1. Prove that $w = f(z)$ is continuous.

(i) $w = \bar{z}$ (conjugation) (iv) $w = \text{Im}(z)$

(ii) $w = |z|$ (v) $w = x^2 + i(y^2 - 1)$

(iii) $w = |z^2 + 1|$ (vi) $w = \dfrac{1}{z} + z$ (except at $z = 0$)

2. Prove the limit relations listed in (10) and (11).

3. Define $f(z) = \exp(-1/z^2)$ on the punctured plane $E = \{z : z \neq 0\}$. Show that $\lim_{z \to 0} f(z)$ does not exist. If we consider the corresponding function of a real variable, $f(x) = e^{-1/x^2}$, defined for $x \neq 0$, show that the limit does exist: $\lim_{x \to 0} e^{-1/x^2} = 0$.

Hint: There are sequences $\{z_n\}$, converging to zero from within E, such that $\{f(z_n)\}$ diverges. For the real variable t, it is well known that $\lim_{t \to +\infty} e^{-t} = 0$.

4. Suppose that f is defined on an open set D, and that g is defined on an open set E which includes the range of f, so that the composite function $(g \circ f)(z) = g(f(z))$ is well defined throughout D. If f is continuous at p (in D) and if g is continuous at $q = f(p)$ (in E), prove that $(g \circ f)(z)$ is continuous at p.

Note: If f is continuous on D and g is continuous on E, then $g \circ f$ is continuous on D (apply the exercise to a typical point p in D).

Hint: If $z_n \to p$ in D, what can be said about $w_n = f(z_n)$ and $q = f(p)$?

5. Describe the limit behavior of the rational function $f(z) = (z - 1)/(z^2 + z - 2)$, defined on $\mathbf{C} \sim \{+1, -2\}$, at the points $p = +1$ and $p = -2$.

Answer: At $+1$, limit exists and is $1/3$; limit at -2 does not exist.

6. If $f(z)$ is continuous on a set A, show that $g(z) = |f(z)|$ is also continuous on A.

7. If $\phi(z)$ is defined near p, and if $\lim_{z \to p} \phi = k$ exists and is *non-zero*, prove that

$$\lim_{z \to p} \frac{\phi(z)}{z - p}$$

does *not* exist.

Note: As a particular example, one might consider e^z/z at $p = 0$. This general result will be useful in many places later on.

8. Consider the functions $f(z) = (z - p)^n$ defined on the punctured plane $E = \{z : z \neq p\}$, where $n = 0, \pm 1, \pm 2, \ldots$ is a fixed integer. For which values of n does $\lim_{z \to p} f$ exist?

9. Consider the function $f(z) = z/|z|$, defined for $z \neq 0$. Construct two sequences $\{z_n'\}$ and $\{z_n''\}$ such that

(i) Both sequences converge to zero.

(ii) The limits $\lim_{n \to \infty} f(z_n')$ and $\lim_{n \to \infty} f(z_n'')$ both exist.

(iii) The limits in (ii) *disagree*.

What does this tell you about the limit $\lim_{z \to p} f$?

10. Prove the following statements.

(i) $\lim_{z \to 0} \dfrac{z^2 e^z}{z} = 0$

(ii) $\lim_{z \to 1} \dfrac{1}{1 - z}$ does not exist

(iii) $\lim_{z \to i} |z| = 1$

(iv) $\lim_{z \to 0} \dfrac{\sin(x + y)}{x + y} + i(y^2 + 1) = 1 + \imath$

(v) $\lim_{z \to 0} \dfrac{e^x \sin y + ie^y \cos x}{y}$ does not exist.

Hint: Sometimes the "ε-δ" definition of limit is easiest to use; show that $|f(z) - \alpha| < \varepsilon$ for all z near p by using the triangle inequality.

11. Prove the following limit relations; the functions $U(x,y)$ or $U(x,y) + iV(x,y)$ are to be regarded as functions of $z = x + iy$.

(i) $\lim_{z\to 0} \dfrac{xy}{x^2 + y^2}$ does not exist.

(ii) $\lim_{z\to 0} \dfrac{x^2y^2}{x^2 + y^2} = 0$

(iii) $\lim_{z\to 0} \dfrac{xy}{\sqrt{x^2 + y^2}} = 0$

(iv) $\lim_{z\to 0} \dfrac{x(\cos y - 1) + iy \sin x}{\sqrt{x^2 + y^2}} = 0$

Hint: In (iv), $|\sin x| \leq |x|$ and $|\cos y - 1| \leq y$ for x and y close to zero. Make liberal use of the triangle inequality to estimate the size of $|f(z)|$ for z near the origin.

2.6 SOME ELEMENTARY FUNCTIONS OF A COMPLEX VARIABLE

In this section we introduce the complex variable analogs of familiar functions of a real variable, such as e^x, $\sin x$, $\cos x$, and $x^{1/2}$. We shall define them carefully, and examine a few of their algebraic properties. At appropriate places we will introduce new techniques for handling all sorts of functions of a complex variable, using these elementary functions to illustrate the ideas.

1. The Exponential Function

This function is defined for all $z = x + iy$ in $\mathbf{C}$ by the formula

$$\exp(z) = e^z = e^x(\cos y + i \sin y) = (e^x \cos y) + i(e^x \sin y), \tag{13}$$

where e^x is the usual exponential function, and $\sin x$ and $\cos x$ are the usual trigonometric functions of a real variable x. We shall use the symbols e^z and $\exp(z)$ interchangeably; here "exp" is the symbol for the exponential function, while $\exp(z) = e^z$ stands for its value at a point z.

Motivating the definition of e^z. If we start with the exponential function for a real variable, e^x, it is not at all obvious what the complex variable analog of this function should be, and the origin of formula (13) may seem obscure. Historically this formula was arrived at in the following natural way. Starting with the real variable exponential function, $f(x) = e^x$, it was first discovered that this function is given by a series (its "Taylor series"):

$$f(x) = \sum_{n=0}^{\infty} \frac{x^n}{n!} = 1 + x + \frac{x^2}{2!} + \cdots \quad \text{all real } x,$$

the series being absolutely convergent for every x in $\mathbf{R}$ (see Exercise 3). This strongly suggests that we examine the complex series

$$E(z) = \sum_{n=0}^{\infty} \frac{z^n}{n!}. \tag{14}$$

For any z this series is absolutely convergent, and for real $z = x + i0$ we get the usual exponential, $E(x + i0) = e^x$. Using the Cauchy product formula, one can show that the function $E(z)$ has the following algebraic property:

$$E(z + w) = E(z) \cdot E(w) \tag{15}$$

for all z and w. In particular, $E(x + iy) = E(x) \cdot E(iy)$. But $E(x) = e^x$, and the series $E(iy)$ may be rewritten as a combination of the Taylor series for $\sin y$ and $\cos y$ (in real variable y). If we make free use of the identity $i^2 = -1$, and Theorem 2.6, we get

$$\begin{aligned} E(iy) &= \left[1 + \frac{(iy)^2}{2!} + \frac{(iy)^4}{4!} + \cdots\right] + \left[\frac{(iy)}{1!} + \frac{(iy)^3}{3!} + \cdots\right] \\ &= \left[1 - \frac{y^2}{2!} + \frac{y^4}{4!} - \cdots\right] + i\left[y - \frac{y^3}{3!} + \frac{y^5}{5!} - \cdots\right] \\ &= \cos y + i \sin y. \end{aligned}$$

Thus the function $E(z)$, defined by the series (14), is the same as the exponential function defined by (15); $E(x + iy) = e^x(\cos y + i \sin y)$. As a special case, we could take $x = 0$ and $y = \pi$, to get the formula $e^{i\pi} = -1$. This was quite startling when it was first discovered as a result of series arguments like the one just presented, since it relates four of the "magic numbers" which keep turning up in mathematics and physics.

The exponential function is continuous at every point $z = x + iy$, since (13) expresses $e^{x+iy} = U(x, y) + iV(x, y)$ as an algebraic combination of continuous functions of the real variables x and y.

As we have indicated in Section 1.5, the complex exponential function converts sums into products:

$$\exp(z + w) = \exp(z) \cdot \exp(w) \quad \text{for all } z \text{ and } w. \tag{16}$$

We have also noted the following formulae, deduced from this basic property.

$$\left.\begin{aligned} e^{-z} &= 1/e^z \\ e^{z-w} &= e^z/e^w \end{aligned}\right\} \quad \text{for all } z \text{ and } w. \tag{17}$$

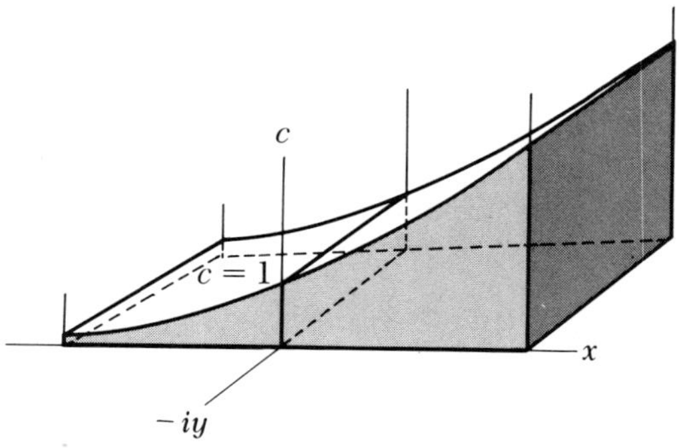

Figure 2.8 Part of the level surface $c = |f(x + iy)|$ for the exponential function.

The absolute value $|e^z|$ is easily calculated: $|e^{x+iy}| = |e^x|\,|e^{iy}|$, and since $|e^{iy}| = \sqrt{\cos^2 y + \sin^2 y} = 1$ for all real y, we get $|e^{x+iy}| = e^x$. This can also be expressed by writing

(18) $$|e^z| = e^{Re(z)} \qquad \text{for all } z.$$

Obviously, the level curves $|e^z| = c$ are precisely the vertical lines in the complex plane. On the line $z = d + iy$ $(-\infty < y < +\infty)$, we have $|e^z| = e^d = c$. Notice that $|e^z|$ increases rapidly as z moves to the right in the plane, and decreases rapidly as z moves to the left. Part of the surface $c = |f(x + iy)|$ is shown in Figure 2.8. From (18) we also see that

(19) $$e^z \text{ is never zero,}$$

since $e^x > 0$ for every real number x.

We have already indicated, in Section 1.5, how complex exponentials are used to express a non-zero complex number z in polar form $z = re^{i\theta}$; the positive part $r = |z|$ is unique, but θ is determined only up to an added constant of the form $2\pi k$ $(k = 0, \pm 1, \pm 2, \ldots)$. From (13) it is obvious that e^z is *periodic* when the variable is translated by $2\pi i$; that is,

(20) $$\exp(z + 2\pi i) = \exp(z) \quad \text{for all } z.$$

A complex number w_0 is called a **period** for a function $f(z)$ if we have $f(z + w_0) = f(z)$ for all z; evidently the numbers $\cdots -2w_0, -w_0, 0, w_0, +2w_0, \ldots$ are also periods. The numbers $w_k = 2\pi ki$ are periods for the exponential function, and we leave it as Exercise 9 for the reader to show that the exponential function has no other periods. This periodicity of $\exp(z)$ means that we know how $\exp(z)$ behaves everywhere in the plane once we know how it acts in any horizontal strip of width 2π, such as the one shown in Figure 2.9.

2. Complex Trigonometric and Hyperbolic Functions

Once we have the complex exponential function in hand, we can define the complex variable analogs of $\sin x$ and $\cos x$ as follows.

(21) $$\cos z = \frac{e^{iz} + e^{-iz}}{2} \qquad \sin z = \frac{e^{iz} - e^{-iz}}{2i}$$

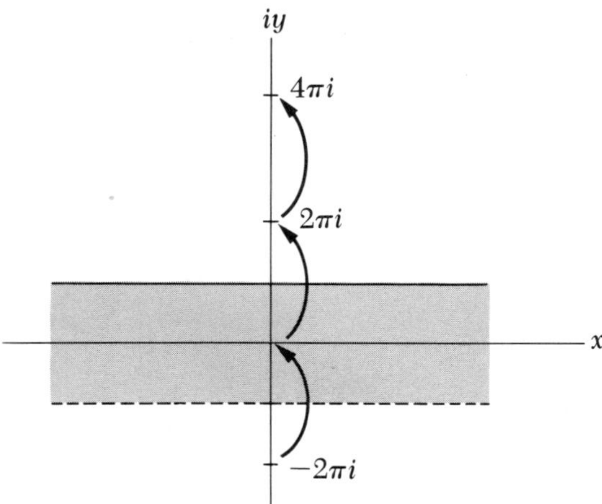

Figure 2.9 The exponential function is periodic under repeated shifts in the vertical direction. Its behavior in the plane is determined by its behavior in any horizontal strip of the kind shown (of width 2π).

These formulas arise naturally if we keep in mind that we are looking for functions on the complex plane that reduce to the usual functions *sin* and *cos* on the real axis. If $z = x + i0$, we get

$$\tfrac{1}{2}\,(e^{ix} + e^{-ix}) = \tfrac{1}{2}\,((\cos x + i \sin x) + (\cos x - i \sin x)) = \cos x$$

$$\frac{1}{2i}\,(e^{ix} - e^{-ix}) = \frac{1}{2i}\,((\cos x + i \sin x) - (\cos x - i \sin x)) = \sin x,$$

so that the combinations (21) achieve this goal.

The **hyperbolic functions** are another important class of elementary functions. We usually define

$$\cosh x = \frac{e^x + e^{-x}}{2} \qquad \sinh x = \frac{e^x - e^{-x}}{2}$$

for a real variable x. If we want functions of a complex variable z that agree with these functions when $z = x + i0$, it is obvious that we should define

(22) $$\cosh z = \frac{e^z + e^{-z}}{2} \qquad \sinh z = \frac{e^z - e^{-z}}{2}$$

For a complex variable, there is a close similarity between trigonometric and hyperbolic functions, which is revealed by comparing formulas (21) and (22). For a real variable, the connection between these two types of functions, and the connection between the functions $\sin x$, $\cos x$, and the exponential function e^x, are not very apparent.

From (21) and (22) we proceed to define the full complement of **trigonometric functions** and **hyperbolic functions:**

(23)
$$\tan z = \frac{\sin z}{\cos z} \qquad \tanh z = \frac{\sinh z}{\cosh z}$$
$$\text{ctn}\, z = \frac{\cos z}{\sin z} \qquad \text{ctnh}\, z = \frac{\cosh z}{\sinh z}$$
$$\sec z = \frac{1}{\cos z} \qquad \text{sech}\, z = \frac{1}{\cosh z}$$
$$\csc z = \frac{1}{\sin z} \qquad \text{csch}\, z = \frac{1}{\sinh z}$$

Naturally we must exclude from the domains of these functions all points at which the denominators equal zero. To see where these functions are undefined, we have to identify the zeros of the functions $\sin z, \ldots, \cosh z$, a task we leave to the reader as Exercise 12. Since e^z, e^{iz}, e^{-z}, and e^{-iz} are all continuous functions of z, the trigonometric and hyperbolic functions are also continuous, where they are defined.

If we restrict z to the real axis, so that $z = x + i0$, various well known identities are valid, such as

$$\sin^2 x + \cos^2 x = 1 \quad \text{or} \quad \cosh^2 x - \sinh^2 x = 1.$$

We shall see in Section 3.8 that validity of a relationship like this for points on the real axis forces the same identity to be true throughout the complex plane. Therefore, *all the usual identities for hyperbolic and trigonometric functions remain valid* if we define these functions for a complex variable as above. These identities could also be proved by direct calculations based on the definitions of $\sin z$, $\cos z, \ldots$ and e^z (see Exercise 11).

In preparation for later applications, we will systematically investigate the following questions.

(i) What are the *periods* of f (if any)?

(ii) What are the *zeros* of f (if any)—the points z such that $f(z) = 0$?

(iii) Plot the level curves $|f(z)| = c$ well enough to determine the radial directions in which $|f(z)|$ increases to $+\infty$ or decreases to zero.

We have done this for $f(z) = e^z$ in equations (18), (19), and (20). Here we work out the answers for $f(z) = \cos z$; the reader should be able to answer these questions himself for other functions.

Example 2.16 Because $\cos z$ is a combination of e^{iz} and e^{-iz}, both of which are periodic under translation by the real numbers $w_k = 2\pi k + i0$, we see that $\cos z$ has periods†

(24)
$$w_k = 2\pi k + i0 \quad \text{for } k = 0, \pm 1, \pm 2, \ldots.$$

We leave it to the reader to show that these are the *only* periods of $\cos z$.

† Since e^z is periodic under translations $w_k' = 0 + 2\pi k i$, then e^{iz} and e^{-iz} must be periodic under translation by $w_k'' = 2\pi k + i0$ for $k = 0, \pm 1, \pm 2 \ldots$

To determine the level curves it is useful to write out the real and imaginary parts of $\cos(x + iy)$ as functions of x and y:

$$\cos(x + iy) = U(x, y) + iV(x, y). \tag{25}$$

By substituting formula (13) into the definition of $\cos z$, we get

$$\begin{aligned}\cos(x + iy) &= \tfrac{1}{2}(e^{ix-y} + e^{-ix+y}) \\ &= \cos(x)\left[\frac{e^{y} + e^{-y}}{2}\right] - i\sin(x)\left[\frac{e^{y} - e^{-y}}{2}\right] \\ &= \cos(x)\cosh(y) - i\sin(x)\sinh(y).\end{aligned}$$

Thus

$$\begin{aligned}U(x, y) &= \operatorname{Re}(\cos(x + iy)) = \cos(x)\cosh(y) \\ V(x, y) &= \operatorname{Im}(\cos(x + iy)) = -\sin(x)\sinh(y)\end{aligned} \tag{26}$$

for all real x and y. This gives us

$$\begin{aligned}|\cos z|^2 &= \cos^2 x \cosh^2 y + \sin^2 x \sinh^2 y \\ &= \cos^2 x \cosh^2 y - \cos^2 x \sinh^2 y + \cos^2 x \sinh^2 y + \sin^2 x \sinh^2 y \\ &= \cos^2 x + \sinh^2 y.\end{aligned} \tag{27}$$

The level curves $|f(z)| = c$ may be interpolated from the level curves of $G(x, y) = \cos x$ (vertical lines) and $H(x, y) = \sinh y$ (horizontal lines), since $|f|^2 = G^2 + H^2$. Since $\cos(z + \pi) = -\cos z$ and $\cos(z + 2\pi n) = \cos z$, we need only plot the level curves in a vertical strip of width π, as shown in Figure 2.10; the pattern is the same in each translate of this strip by $n\pi + i0$. From

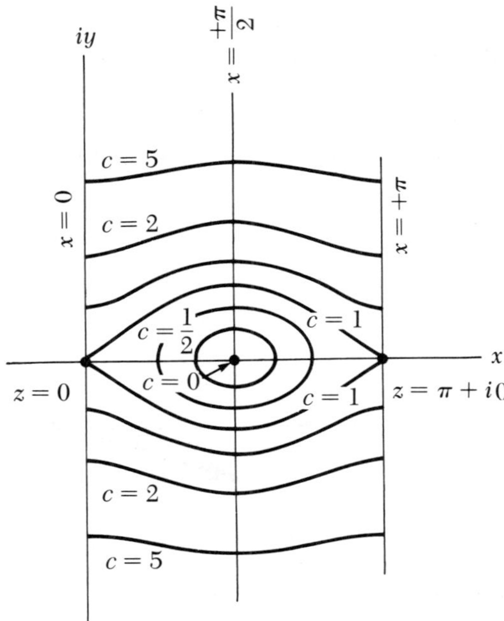

Figure 2.10 The level curves $|\cos(x + iy)| = c$ (periodic outside the strip shown). Remember that $\cos(z + \pi) = -\cos(z)$, so we only need a strip of width π.

this diagram it is evident that $|\cos z|$ grows rapidly as z moves up or down along vertical lines; it is bounded (and periodic) as z moves along any horizontal line.

The zeros of $\cos z$ are easily located using formula (27); clearly, $\cos(x + iy) = 0$ if and only if $\cos x = 0$ *and* $\sinh y = 0$. Since $\sinh y = 0$ only when $y = 0$, it follows that the zeros are the points $z = x + iy$ such that $y = 0$ and $\cos x = 0$:

(28) $$z_k = \left(n\pi + \frac{\pi}{2}\right) + i0 \qquad n = 0, \pm 1, \pm 2, \ldots .$$

3. The Complex Square Root Function $z^{1/2}$

A **determination of the square root** is any function $w = f(z)$ that satisfies the square root equation

(29) $$w^2 = z,$$

so that $(f(z))^2 = z$. We have described the square roots of a typical complex number $z = re^{i\theta}$ in Section 1.6, where we showed that equation (29) has exactly two solutions:

$$w_1 = r^{1/2}e^{i(\theta/2)}$$
$$w_2 = -r^{1/2}e^{i(\theta/2)} = r^{1/2}e^{i(\theta/2)}e^{i\pi} = r^{1/2}e^{i(\pi+\theta/2)}$$

(except when $z = 0$; then $w = 0$ is the only solution). These roots are shown in Figure 2.11. In defining a square root function we are forced to make a selection between these solutions; making the choice at random for each z will produce a highly discontinuous square root function. We will eventually see that there is a fundamental obstacle that defeats any attempt to define $z^{1/2}$ as a *continuous* function *throughout* the plane; however, we can define $z^{1/2}$ on a domain consisting of **C** with a radial line from the origin out to infinity deleted. We shall consider the particular "cut plane" $P = \mathbf{C} \sim J$ obtained by deleting

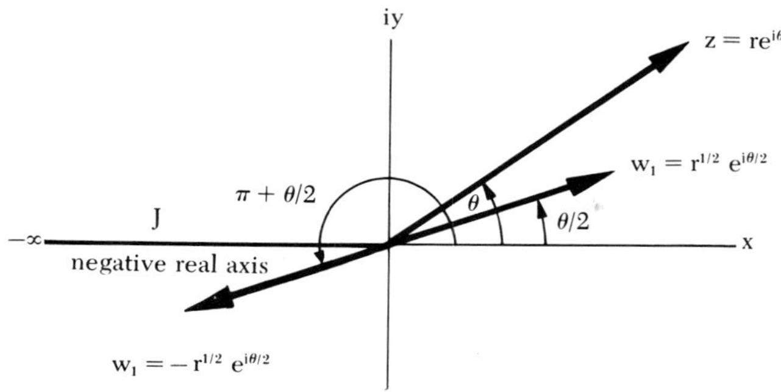

Figure 2.11 Location of the square roots of $z = re^{i\theta}$; the principal square root $z^{1/2}$ is the number w_1 shown.

the negative real axis $J = (-\infty, 0]$, as indicated in Figure 2.11; however, we could carry out our discussion for any other radial line J^*.

A point $z \neq 0$ has a *unique* polar form

$$z = re^{i\theta} = |z|\, e^{i\theta},$$

if we require that $-\pi < \theta \leq +\pi$. We define the **principal determination of the square root function** using this normalization of θ:

(30) $$f(z) = z^{1/2} = r^{1/2}e^{i\theta/2} \quad \text{if} \quad z = re^{i\theta} \neq 0.$$

Since θ ranges from $-\pi$ to $+\pi$, $\theta/2$ ranges from $-\pi/2$ to $+\pi/2$; it is obvious that f maps a radial line extending from zero to infinity onto a radial line making half the angle with the positive real axis. Thus, f maps the cut plane P onto the right half plane $H = \{z\colon \mathrm{Re}(z) > 0\}$. However, f does not "squash" the set P too badly; the mapping $f\colon P \to H$ is actually one-to-one, in the sense that distinct points in P have distinct images in H. Second, it is quite clear that $z^{1/2}$ agrees with the usual square root function for positive real numbers: $(x + i0)^{1/2} = x^{1/2}$. Other determinations of the square root function are obtained by taking other domains of definition; even on the cut plane P, the function

$$g(z) = -z^{1/2} = -f(z)$$

is a square root function, since $(g(z))^2 = z$. However, g does not agree with the usual square root function $x^{1/2}$ on the positive real axis; instead we get $-x^{1/2}$ if $z = x + i0$ is on this half line.

We now examine the continuity of $z^{1/2}$ and its limit behavior at points on the negative real axis J, which is the boundary of the domain P. Consider a point $p = x_0 + i0$ such that $x_0 < 0$, and consider two sequences of points that approach p from within P:

$$p_n = |x_0|\, e^{i\theta_n} \quad \text{and} \quad p_n{}' = |x_0|\, e^{i\phi_n},$$

where $-\pi < \theta_n, \phi_n < +\pi$, and $\theta_n \to +\pi$, $\phi_n \to -\pi$ as $n \to \infty$. Since $\sin\theta$ and $\cos\theta$ are continuous functions of a real variable, we get

$$\begin{aligned} f(p_n) &= |x_0|^{1/2}\, e^{i(\theta_n/2)} = |x_0|^{1/2}\,(\cos(\theta_n/2) + i\sin(\theta_n/2)) \\ &\to |x_0|^{1/2}\,(\cos(+\pi/2) + i\sin(+\pi/2)) = +i\,|x_0|^{1/2} \\ f(p_n{}') &= |x_0|^{1/2}\, e^{i(\phi_n/2)} = |x_0|^{1/2}\,(\cos(\phi_n/2) + i\sin(\phi_n/2)) \\ &\to |x_0|^{1/2}\,(\cos(-\pi/2) + i\sin(-\pi/2)) = -i\,|x_0|^{1/2}. \end{aligned}$$

The limit $\lim_{z\to p} f(z)$ cannot exist, since these values disagree. In fact, $f(z) = z^{1/2}$ suffers a change of sign as z crosses the negative real axis. The reader may verify for himself that $w = z^{1/2}$ is continuous at any point p that lies off of the negative real axis J.

The square root function has no periodicity properties, and is zero only if $z = 0$. Its level curves are extremely simple: concentric circles around the origin. This should be clear from the fact that

(31) $$|z^{1/2}| = |z|^{1/2} \quad \text{for all } z$$

(the right side is the familiar square root of a positive real number).

The square root function is also the first of several "multiple valued functions" we will encounter; that is, it is a particular solution of an equation that has multiple solutions for each z. There are multiple valued functions of a real variable; for example, the inverse trigonometric function $y = \arctan x$ is obtained by solving the equation $x = \tan y$, subject to the normalizing condition $-\pi/2 < y < +\pi/2$. If no normalizing condition is imposed, there are infinitely many solutions y for each x. We may expect that the complex variable analog of any function like $x^{1/2}$ or $\arctan x$, which is already multiple valued as a function of a real variable, will be multiple valued. However, certain functions, like the natural logarithm $\log x$ (defined for $0 < x < +\infty$), which appear to be single valued when we consider only a real variable, turn out to be multiple valued when regarded as functions of a complex variable. This phenomenon will be illustrated in the next two sections.

4. The Argument Function arg(z)

If z is a non-zero complex number, written in polar form as $z = re^{i\theta} = |z|\, e^{i\theta}$, the angle variable θ is often referred to as the **argument of z**. It is natural to try to set up a function $\arg(z) = \theta$, but we will be troubled by the indeterminacy of θ up to an added constant $2\pi k$ unless we normalize θ in some way. To remove this ambiguity, we define the **principal determination of the argument** on the punctured plane $E = \{z : z \neq 0\}$ to be

(32) $$\operatorname{Arg}(z) = \theta$$

where θ is the unique real number such that $-\pi < \theta \leq +\pi$ and $z = |z|\, e^{i\theta}$. We will use the symbol $\arg(z)$ to indicate a determination of the argument that is not necessarily the principal determination. If $z = 0$, the number θ in the polar form $z = re^{i\theta} = 0 \cdot e^{i\theta}$ is completely ambiguous; we shall not try to define $\arg(z)$ if $z = 0$.

If z_1 is a point lying slightly above the negative real axis, so that $z_1 = -R + i\delta$ (with $\delta > 0$), the values of $\operatorname{Arg}(-R + i\delta)$ increase to the value $+\pi = \operatorname{Arg}(-R + i0)$ as $\delta \to 0$, keeping $\delta > 0$ (see Figure 2.12). On the other hand, if we examine points $z_2 = -R - i\delta$ (with $\delta > 0$) lying below the negative real axis, then $\operatorname{Arg}(-R - i\delta)$ approaches the value $-\pi = \operatorname{Arg}(-R + i0) - 2\pi$ as $\delta \to 0$, keeping $\delta > 0$. Thus, the value of $\operatorname{Arg}(z)$ increases by $+2\pi$ as z crosses the negative real axis, moving upwards. Later in this section we will see that $\operatorname{Arg}(z)$ is continuous at any point p lying off of the negative real axis.

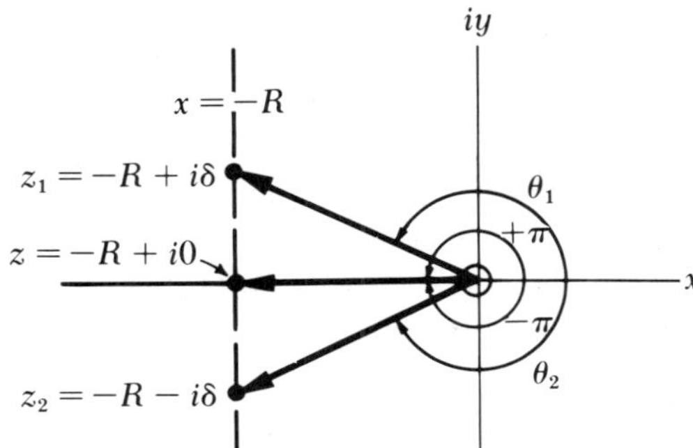

Figure 2.12 The discontinuity of Arg(z) across the negative real axis.

Any function $\phi(z)$ that solves the equation

(33) $$z = |z|\, e^{i\phi(z)} \qquad \text{for all } z \text{ such that } z \neq 0$$

is called a **determination of the argument function**. On the cut plane P, $\phi(z) = \text{Arg}(z)$ is just one of the possible determinations of the argument; others are given by $\arg(z) = \text{Arg}(z) + 2\pi k$ (k a fixed integer). It turns out that there is no way to define a determination of $\arg(z)$ on the whole plane so that it does not have a set of jump discontinuities somewhere. Of course, we could have normalized θ differently, say by requiring that $\alpha - \pi < \theta \leq \alpha + \pi$ for some fixed angle α; this would have shifted the discontinuities so they lie along a different radial line extending from the origin out to infinity. We could also define $\arg(z)$ so that the discontinuities lie along a more general curve extending from the origin to infinity.

Let z and w be two complex numbers. We say that z and w are **congruent modulo 2π,** written

(34) $$z \equiv w \pmod{2\pi},$$

if z and w differ by some integral multiple of 2π; that is, the symbol "$\cdots \equiv \cdots \pmod{2\pi}$" is used to mean that equality holds if we are permitted to adjust the right (or left) side by adding on some multiple $2\pi k$ ($k = 0, \pm 1, \pm 2, \ldots$). The idea of congruence (34) will be very useful in our encounters with exponential functions, and in handling problems with radial symmetry, where troublesome multiples of 2π always appear. Of course, z and w are congruent if they are equal, $z = w$, but the congruence relation $z \equiv w \pmod{2\pi}$ is rather different from ordinary equality $z = w$ of these numbers. For example, we have seen that

(35) $$e^{iz} = 1 \quad \text{if and only if} \quad z \equiv 0 \pmod{2\pi}.$$

Using this fact, and invoking formula (17), we immediately deduce that

(36) $$e^{iz} = e^{iw} \quad \text{if and only if} \quad z \equiv w \pmod{2\pi}.$$

Congruence relations are very convenient in describing properties of the argument function. The function Arg(z), and in fact any determination of

$\arg(z)$, tries to have the algebraic property

(37) $$\operatorname{Arg}(z \cdot w) = \operatorname{Arg}(z) + \operatorname{Arg}(w) \quad \text{for all non-zero } z \text{ and } w.$$

This is actually true for some of the points z and w without qualification (for example, it is true if z and w lie in the right half plane $\operatorname{Re}(z) > 0$), but for certain choices of z and w the equation is off by an added integral multiple of 2π. The true relation is a congruence, rather than an equality:

(38) $$\arg(z \cdot w) \equiv \arg(z) + \arg(w) \pmod{2\pi}$$

for all non-zero z and w, and for any determination of $\arg z$. The adjustment needed to give equality will, of course, depend on which pair of complex numbers we examine. To prove (38), simply multiply by i and take the exponential on both sides of the prospective congruence (38), and use formula (36) above.

It is also useful to notice that

(39) $$\arg(z) \equiv \operatorname{Arg}(z) \pmod{2\pi} \quad \text{for all } z \neq 0$$

for *any* determination $\arg(z)$ of the argument function. Indeed, $\arg(z)$ must satisfy the characteristic equation (33):

$$z = |z|\, e^{i \arg(z)} \quad \text{for all } z \neq 0,$$

so that $z = |z|\, e^{i \arg(z)} = |z|\, e^{i \operatorname{Arg}(z)}$, which means that $e^{i \arg(z)} = e^{i \operatorname{Arg}(z)}$. By (36) this is equivalent to the desired congruence.

Figure 2.13 shows how we may express the function $A(x, y) = \operatorname{Arg}(x + iy)$ in terms of the elementary functions of real variable $\arcsin(t)$ and $\arccos(t)$:

(40) $$\operatorname{Arg}(x + iy) = \begin{cases} \arccos\left(\dfrac{x}{\sqrt{x^2 + y^2}}\right) & \text{in the upper half plane} \\ \arcsin\left(\dfrac{y}{\sqrt{x^2 + y^2}}\right) & \text{in the right half plane} \\ -\arccos\left(\dfrac{x}{\sqrt{x^2 + y^2}}\right) & \text{in the lower half plane} \end{cases}$$

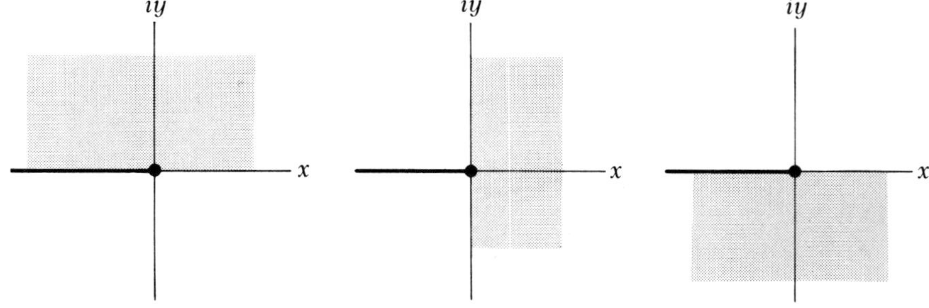

Figure 2.13 The regions in the plane in which $\operatorname{Arg}(z)$ is described by the combinations of arcsin and arccos displayed in equation (40).

The continuity of Arg(z) at points off of the negative real axis follows easily from these formulas, and from the continuity of the functions arcsin(t) and arccos(t).

5. The Logarithm Function log(z)

A **determination of the logarithm function** is any function $f(z)$ such that

(41) $$\exp(f(z)) = e^{f(z)} = z \quad \text{for all } z \text{ in Dom}(f).$$

The value $w = f(z)$ must be selected from among the possible solutions of the equation

(42) $$e^w = z.$$

Here, if $z = 0$ there is no solution at all, because e^w is never zero; therefore, the domain of definition can never include $z = 0$. If $z \neq 0$ and we write z in polar form $z = |z|\, e^{i\theta}$, then $w = u + iv$ is a solution of (42) if and only if

$$e^{u+iv} = e^u e^{iv} = |z|\, e^{i\theta},$$

which means that we must have $|z| = e^u$ and $e^{i\theta} = e^{iv}$. In view of formula (36), the only solutions of the equation $e^{i\theta} = e^{iv}$ are $v = \theta + 2\pi n$ (n any integer). Thus, the only solutions of (42) are of the form

$$u = \log|z|$$
$$v = \theta + 2\pi n \quad (n = 0, \pm 1, \pm 2, \ldots)$$

and we get all possible solutions from

(43) $$w = \log|z| + i\,\text{Arg}(z) + 2\pi i n \quad (n = 0, \pm 1, \pm 2, \ldots).$$

Here $\log x$ is the usual natural logarithm function for real variable $0 < x < +\infty$.

The **principal determination of the logarithm** is the function Log(z) defined on the domain $E = \{z : z \neq 0\}$ by writing

(44) $$\text{Log}(z) = \log|z| + i\theta = \log|z| + i\,\text{Arg}(z).$$

The symbol log(z) will be used to indicate a determination of the logarithm that is not necessarily the principal determination.

Let P be the cut plane obtained by deleting the negative real axis $J = (-\infty, 0]$ from the plane. The function Log(z) is continuous on P, since Arg(z) is continuous on P and $\log|z|$ is readily seen to be continuous on the even larger domain $\{z : z \neq 0\}$. However, $i\,$Arg(z) has a jump of magnitude $2\pi i$ as z crosses the negative real axis, and since Log(z) is the sum of this function and

$\log|z|$, which is continuous at points in J, we conclude that $\text{Log}(z)$ must also exhibit a jump of magnitude $2\pi i$ as z crosses the negative real axis.

From the defining equation for logarithms (41) we see that

$$e^{\log(z)} = z \quad \text{for all } z. \tag{45}$$

On the other hand, logarithms almost give us the converse identity

$$\log(e^z) = z \quad \text{for all } z,$$

but this equation may be off by an added term of the form $2\pi ni$ (n some integer), depending on which z we look at and the particular determination of logarithm being considered. Using the notion of "congruence of two complex numbers (mod $2\pi i$)," we note that

$$\log(e^z) \equiv z \pmod{2\pi i} \quad \text{for all } z. \tag{46}$$

holds for any logarithm function, and in particular

$$\text{Log}(e^z) \equiv z \pmod{2\pi i} \quad \text{for all } z.$$

Here, **congruence (mod 2πi)** means that the numbers involved differ by an integral multiple of $2\pi i$; evidently the statement proved earlier:

$$e^{iz} = e^{iw} \quad \text{if and only if} \quad z \equiv w \pmod{2\pi}$$

is equivalent to

$$e^z = e^w \quad \text{if and only if} \quad z \equiv w \pmod{2\pi i}. \tag{47}$$

To prove (46), take the exponential of $\log(e^z)$; the defining equation for logarithms gives us

$$\exp(\log(e^z)) = \exp(z) \quad \text{for all } z,$$

and from this we get the desired congruence by applying (47).

The exponential function converts sums into products; in so far as possible, the logarithm function reverses this process. Actually,

$$\begin{aligned} \log(z \cdot w) &= \log(e^{\log(z)} \cdot e^{\log(w)}) = \log(e^{\log(z)+\log(w)}) \\ &\equiv \log(z) + \log(w) \pmod{2\pi i} \end{aligned} \tag{48}$$

is a direct consequence of (46), for any determination of $\log(z)$. Equation (48) is the fundamental algebraic property of logarithms, just as equation (16) is the fundamental property of the exponential function. The level curves and other properties of $\text{Log}(z)$ are studied in Exercise 15.

6. The Function $f(z) = z^\alpha$

To define the nth roots $z^{1/n}$ for integers $n = 2, 3, \ldots$ we want a function $w = f(z)$ that is a solution of the nth root equation

$$w^n = z \quad \text{for all } z. \tag{49}$$

In Section 1.6 we saw that this equation has n different solutions for each z, except in the trivial case $z = 0$. Thus the nth root function is multiple valued; to avoid this ambiguity, we define the **principal determination of the $\mathbf{n}^{\text{th}}$ root function** $f(z) = w^{1/n}$ by writing z in polar form $z = |z|\, e^{i\theta}$, normalizing θ so that $-\pi < \theta \leq +\pi$, and taking

$$z^n = r^{1/n} e^{i\theta/n} = |z|^{1/n}\, e^{i\theta/n} = |z|^{1/n}\, e^{i\,\mathrm{Arg}(z)/n} \tag{50}$$

for all $z \neq 0$. We take $f(0) = 0$. This function is discontinuous along the negative real axis (values on opposite sides of the cut differ by a factor $e^{2\pi i/n}$), but it is continuous at all other points. Of course, there are n other continuous functions on the cut plane P that are solutions of (49), but $z^{1/n}$ is the only one of them that agrees with the usual n^{th} root function along the positive real axis: $(x + i0)^{1/n} = x^{1/n}$ for all positive real x.

There is a nice way to represent these fractional powers using exponentials and logarithms. If θ is normalized so that $-\pi < \theta \leq +\pi$, then

$$\frac{1}{n}\,\mathrm{Log}(re^{i\theta}) = \frac{1}{n}\log r + \frac{i}{n}\,\mathrm{Arg}(re^{i\theta}) = \frac{1}{n}\log r + i\,\frac{\theta}{n}\,.$$

For positive real numbers $r > 0$ we know that $e^{(1/n)\log r} = (e^{\log r})^{1/n} = r^{1/n}$, so that

$$\exp\left(\frac{1}{n}\,\mathrm{Log}(z)\right) = \exp\left(\frac{1}{n}\,(\log|z| + i\theta)\right) = \exp\left(\frac{1}{n}\log|z|\right)\cdot e^{i\theta/n}$$
$$= |z|^{1/n}\cdot e^{i\theta/n} = z^{1/n}.$$

We have proved:

$$z^{1/n} = \exp\left(\frac{1}{n}\,\mathrm{Log}(z)\right) \quad \text{for all } z \neq 0. \tag{51}$$

Once the nth root function has been written in this form, we may use the algebraic properties of the exponential and logarithm functions to determine its properties.

However, n^{th} roots are not the only powers that turn up in applications; we often encounter fractional powers like $x^{4/5}$, and even powers like $x^{\sqrt{2}}$. Formula (51) suggests a means of defining arbitrary powers z^α, *even if the exponent is complex.* For a fixed exponent α (any complex number) we define z^α on the set $\{z : z \neq 0\}$ by the formula

$$z^\alpha = e^{\alpha\cdot\mathrm{Log}(z)} \quad \text{for all } z \neq 0. \tag{52}$$

The function we get is referred to as the **principal determination of the αth power** of z. Except when α has the special form $\alpha = n + i0$ for some integer $n = 0, \pm 1, \pm 2, \ldots$, these functions have discontinuities along the negative real axis and are continuous elsewhere in the plane. This reflects the fact that $w = z^\alpha$ is a multiple valued function; its other determinations are obtained by taking different determinations of the logarithm in (52), such as $\log(z) = \text{Log}(z) + 2\pi i n$. In Exercise 25 we indicate why integer exponents do not lead to multiple determinations of z^α.

The family of functions we get by taking various exponents α in (52) includes the usual powers z^n ($n = 0, \pm 1, \pm 2, \ldots$) and n^{th} roots $z^{1/n}$ (for $n \neq 0$), and also more exotic functions like $z^{\sqrt{2}}$ and z^i (see Exercises 20 to 24 for some comments on z^i). The functions in this family obey the usual **laws of exponents**:

(53)

$$\text{(i) } z^0 = 1 \quad \text{for all} \quad z \neq 0; \qquad z^1 = z \quad \text{for all} \quad z \neq 0.$$

$$\text{(ii) } z^{(\alpha+\beta)} = z^\alpha \cdot z^\beta \quad \text{for all} \quad z \neq 0.$$

$$\text{(iii) } z^{-\alpha} = \frac{1}{z^\alpha} \quad \text{for all} \quad z \neq 0.$$

These formulas follow easily from definition (52) and the algebraic properties of $\exp(z)$ and $\text{Log}(z)$. There is some trouble with the formulas

(54) $$(z^\alpha)^\beta = z^{\alpha \cdot \beta} \quad \text{and} \quad (z \cdot w)^\alpha = z^\alpha \cdot w^\alpha.$$

Although they may be true in special cases, they are not true in general.

Exercise 22 is devoted to verifying that

(55) $$(z^2)^{1/2} \neq (z^{1/2})^2$$

for certain z lying off of the negative real axis; thus, the "laws" (54) break down if we take $\alpha = \frac{1}{2}$ and $\beta = 2$. Similarly, if α is rational, so that $\alpha = (m/n) + i0$ with $m, n \neq 0$ integers, we always get

(56) $$z^{(m/n)} = (z^{1/n})^m \quad \text{(take roots first, then powers)}$$

but the alternative formula

$$z^{(m/n)} = (z^m)^{1/n} \quad \text{(take powers first, then roots)}$$

may fail to be true. In particular, the formula $(z^m)^{1/n} = (z^{1/n})^m$ fails to be true, although we might at first expect it to be valid; what really happens is that $(z^m)^{1/n}$ and $(z^{1/n})^m$ may give different determinations of the power $z^{(m/n)}$, and only formula (56) is sure to give the principal determination defined in (52).

7. Exponential Functions $f(z) = \alpha^z$ with Base $\alpha \neq 0$

There is a completely different family of functions $f(z) = \alpha^z$, complementary to the family of functions z^α. If α is a fixed non-zero complex number, we may define the **exponential with base α:**

(57) $$\alpha^z = e^{z \cdot \mathrm{Log}(\alpha)} \qquad \text{for all complex } z.$$

This function of z is well defined and continuous throughout the plane. It is very similar to the ordinary exponential function e^z, and reduces to this function if we take $\alpha = e + i0$. We leave the reader to work out the basic algebraic properties of the functions α^z from formula (57) and the algebraic properties of the usual exponential function. See Exercises 26 and 27.

EXERCISES

1. Verify the following formulas.

(i) $\sin(x + iy) = (\sin x \cosh y) + i(\cos x \sinh y)$

(ii) $\cos(x + iy) = (\cos x \cosh y) - i(\sin x \sinh y)$

(iii) $\sinh(x + iy) = (\sinh x \cos y) + i(\cosh x \sin y)$

(iv) $\cosh(x + iy) = (\cosh x \cos y) + i(\sinh x \sin y)$

(v) $\tan(x + iy) = \left(\dfrac{\sin x \cos x + i \sinh y \cosh y}{\cos^2 x + \sinh^2 y}\right)$

(vi) $\mathrm{sech}(x + iy) = \dfrac{\cos x \cosh y + i \sin x \sinh y}{\cos^2 x + \sinh^2 y}$

What are the natural domains of definition? Are the functions continuous throughout these domains?

2. Show that $|\cos z|^2 = \cosh^2 y - \sin^2 x$ and $|\sin z|^2 = \cosh^2 y - \cos^2 x$.

3. Show that the series $E(z) = \sum_{n=0}^{\infty} z^n/n!$ is absolutely convergent for all z; then use the Cauchy product formula to prove that $E(z + w) = E(z) \cdot E(w)$ for all z and w.

Hint: Recall the binomial expansion $(z + w)^p = \sum_{j=0}^{p} \binom{p}{j} z^{p-j} w^j$, where $\binom{p}{j} = \dfrac{p!}{(p-j)!\,j!}$ are the binomial coefficients.

4. Given a point p, verify that $\arg(z - p)$ is the angle shown in Figure 2.14, for each $z \neq p$. (Use the parallelogram law to relate the directed segment from p to z, and the segment from the origin to $z - p$.)

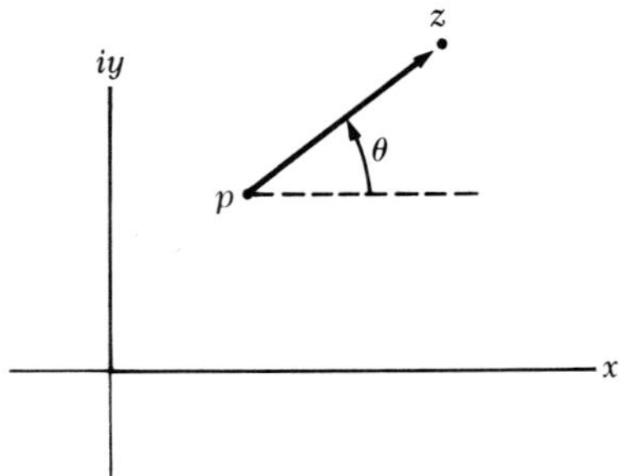

Figure 2.14 Interpretation of $\arg(z - p)$.

5. Given points $p \neq q$ show that

$$\arg\left(\frac{z-p}{z-q}\right) \equiv \arg(z-p) - \arg(z-q) \qquad (\text{mod } 2\pi)$$

corresponds to the angle $\Delta\theta$, measured from the segment $[q, z]$ to the segment $[p, z]$, as indicated in Figure 2.15. What is the value of $\arg[(z-p)/(z-q)]$ (mod 2π) when z lies on various parts of the circle shown in Figure 2.15?

6. Prove that $\text{Arg}(x + iy)$ is given by formula (40) in the parts of the plane indicated in Figure 2.13. In which parts of the cut plane does $\text{Arg}(x + iy) = \arctan(y/x)$?

7. Give explicit formulas for the real and imaginary parts of $z^{1/2}$ as functions of x and y, for $z = x + iy$ in the parts of the plane shown in Figure 2.13.

8. Evaluate the complex numbers

(i) $e^{1+i\pi}$

(ii) $\exp(e^{1+i})$

(iii) $e^{2+7\pi i}$

(iv) $\sqrt{1+i}$

(v) $\sin(1+i)$

(vi) $\cosh(1+i)$

(vii) $\cos(-i)$

9. Determine all complex numbers w (the periods of *exp*) such that: $\exp(z + w) = \exp(z)$ for all z.

Hint: There are the obvious periods $w_k = 2\pi ki$ (k any integer). To see that there are no others, use formula (16) and formula (34) of Section 1.5.

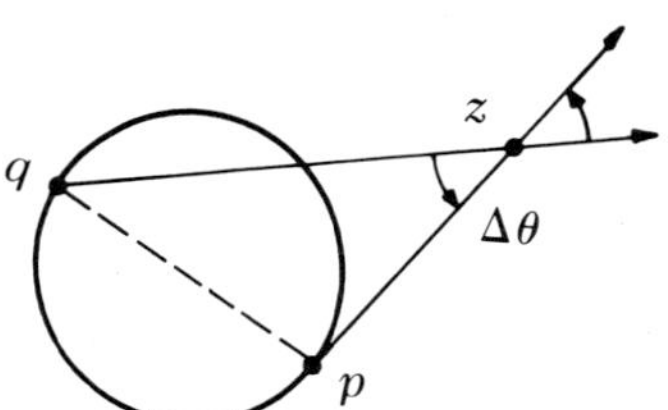

Figure 2.15 Interpretation of $\arg[(z-p)/(z-q)]$.

10. Show that the functions below have precisely the periods indicated (k any integer).

(i) $\cos(z)$ $\quad z_k = 2\pi k + i0$

(ii) $\sinh(z)$ $\quad z_k = 0 + 2\pi k i$

(iii) $\tan(z)$ $\quad z_k = \pi k + i0$

(iv) $\operatorname{sech}(z)$ $\quad z_k = 0 + 2\pi k i$

11. Prove the following identities from the definitions and standard trigonometric identities for functions of a real variable.

(i) $\sin^2 z + \cos^2 z = 1$

(ii) $\cosh^2 z - \sinh^2 z = 1$

(iii) $\cos^2 z - \sin^2 z = \cos 2z$

(iv) $\cos(z \pm w) = \cos z \cos w \mp \sin z \sin w$

(v) $\sinh(iz) = i \sin z$

(vi) $\sin z = \cos\left(z - \frac{\pi}{2}\right)$

12. Verify that

(i) $\sin z = 0$ if and only if $z = \pi k + i0$

(ii) $\cosh z = 0$ if and only if $z = 0 + i\left(\pi k + \frac{\pi}{2}\right)$

(iii) $\tan z = 0$ if and only if $z = \pi k + i0$

for $k = 0, \pm 1, \pm 2, \ldots.$

Hint: Sometimes it is easiest to locate zeros by examining $|f(z)|^2$.

13. Use the formula $e^z = \sum_{n=0}^{\infty} \frac{z^n}{n!}$, basic definitions, and the elementary results on series in Theorem 2.6 to prove that

(i) $\sin z = z - \frac{z^3}{3!} + \frac{z^5}{5!} - \cdots = \sum_{n=0}^{\infty} (-1)^n \frac{z^{2n+1}}{(2n+1)!}$ (all z)

(ii) $\cos z = 1 - \frac{z^2}{2!} + \frac{z^4}{4!} - \cdots = \sum_{n=0}^{\infty} (-1)^n \frac{z^{2n}}{(2n)!}$ (all z)

(iii) $\sinh(z) = z + \frac{z^3}{3!} + \frac{z^5}{5!} + \cdots = \sum_{n=0}^{\infty} \frac{z^{2n+1}}{(2n+1)!}$ (all z)

(iv) $\cosh(z) = 1 + \frac{z^2}{2!} + \frac{z^4}{4!} + \cdots = \sum_{n=0}^{\infty} \frac{z^{2n}}{(2n)!}$ (all z)

(v) $\frac{e^2}{z^2} = \frac{1}{z^2} + \frac{1}{z} + \left(\frac{1}{2!}\right) + \left(\frac{1}{3!}\right)z + \cdots$

$$= \frac{1}{z^2} + \frac{1}{z} + \sum_{n=0}^{\infty} \frac{z^n}{(n+2)!} \qquad \text{(all } z \neq 0\text{)}$$

14. Consider the cut plane obtained by deleting the *negative imaginary axis* $J^* = \{z:\text{Re}(z) = 0 \text{ and } \text{Im}(z) \leq 0\}$ from the plane. How would you define a continuous determination of $\log z$ on this domain such that $\log(x + i0) = \log x$ for positive real x? In which parts of the plane does $\log z$ agree with $\text{Log } z$?

Hint: Use the polar form $z = re^{i\theta}$ and a suitable normalization of θ.

15. Sketch the level curves $|\text{Log } z| = c$ in the cut plane P. Explain why these curves are not closed; relate this to the behavior of $\text{Log } z$ along the cut. Find all the zeros of $\text{Log } z$ in the domain P, and show that $\text{Log } z$ is not periodic in any direction. Show that

(i) $\lim_{z\to 0} \text{Log}(z)$ does not exist

(ii) $\lim_{z\to 0} z \cdot \text{Log}(z) = 0.$

Hint: If x is positive real we know that $\lim_{x\to 0} x \log x = 0$.

Answer: $z = 1$ is the only zero; therefore no periodicity.

16. Prove the following important identities.

(i) $\sinh(z) = -i \sin(iz)$

(ii) $\cosh(z) = \cos(iz)$

(iii) $\tanh(z) = -i \tan(iz)$.

17. Sketch the curves $|\cos z| = c$ in the rest of the plane (refer to Figure 2.10). Then determine the forms of the level curves

(i) $|\sin z| = c$

(ii) $|\cosh z| = c$

(iii) $|\sec z| = c$;

compare these curve patterns with those in Figure 2.10.

Hint: Use the identities of Exercise 16. All results follow easily from the data in Figure 2.10.

18. Prove that the function $f(z) = z^{1/2}$ (principal determination) and $g(z) = \exp[(\frac{1}{2})\text{Log } z]$ agree on the cut plane P.

19. Prove the exponent laws for powers z^α listed in (53).

20. Show that

$$z^i = e^{-\text{Arg}(z)}[\cos(\log |z|) + i \sin(\log |z|)]$$

on the cut plane P. Evaluate i^i, $(-1)^i$, 2^i, 1^i. Show that $|z^i| = e^{-\text{Arg}(z)}$, and sketch the level curves $|z^i| = \text{const.}$

Answer: $i^i = e^{-\pi/2}$; $(-1)^i = e^{-\pi}$; $2^i = \cos(\log 2) + i \sin(\log 2)$; $1^i = 1$.

21. Find all possible determinations of i^i and sketch their locations.
Answer: $e^{-\pi/2} \cdot e^{-2\pi n}$ $(n = 0, \pm 1, \pm 2, \ldots)$, all positive real.

22. Show that $(z^{1/n})^m$ always gives the principal determination of $z^{m/n}$, defined for $\alpha = m/n$ as in (52). Demonstrate that $(z^m)^{1/n}$ does not always give the correct values by calculating $(z^2)^{1/2}$, $(z^{1/2})^2$, and z^1 for $z = i - 1$. Sketch the geometric situation to see why this mismatch occurs.

23. One might expect the function z^α to satisfy the law $(zw)^\alpha = z^\alpha w^\alpha$. Show that this is *not* true throughout P for certain exponents α.
Hint: Take $w = z$ in the last exercise (keep $\alpha = \frac{1}{2}$).

24. Calculate the limit values of z^i along the upper and lower edges of the cut; if $x < 0$ is fixed, show that

$$\lim_{\delta \to 0} f(x + i\delta) = e^{-\pi} e^{i \log|x|}$$

$$\lim_{\delta \to 0} f(x - i\delta) = e^{\pi} e^{i \log|x|}$$

(keeping $\delta > 0$ in both limits). What is the difference $\Delta(x)$ in values of $f(x + i\delta)$ as δ increases from negative to positive values?
Answer: $\Delta(x) = -2 \sinh(\pi)[\cos(\log|x|) + i \sin(\log|x|)]$.

25. For which exponents α do the values of z^α along upper and lower edges of the cut agree?
Answer: $\alpha = k + i0$ for $k = 0, \pm 1, \pm 2, \ldots$.

26. Consider the exponential function α^z having base $\alpha \neq 0$, and prove that:

(i) $\alpha^{z+w} = \alpha^z \cdot \alpha^w$ for all z and w.

(ii) α^z is non-zero for all z.

(iii) $\alpha^{-z} = 1/\alpha^z$ for all z.

(iv) $|\alpha^{x+iy}| = e^{x \log|\alpha|} e^{-y \operatorname{Arg}(\alpha)}$

27. Is it possible for two bases $\alpha \neq \beta$ to give the same exponential functions α^z and β^z on $\mathbf{C}$?
Answer: No.

28. If $\operatorname{Arg} z_n \to \theta$ and $|z_n| \to \rho$ as $n \to \infty$, prove that $z^* = \lim_{n\to\infty} z_n$ exists. How is z^* related to ρ and θ?
Answer: $z^* = \rho e^{i\theta}$, even if $\rho = 0$ and $\theta = \pm\pi$.

29. If $z^* = \lim_{n\to\infty} z_n$, is it necessarily true that $\operatorname{Arg} z^* = \lim_{n\to\infty} \operatorname{Arg} z_n$? What if z^* is in the cut plane P?
Answer: No, if $z^* = x + i0$ such that $x < 0$; otherwise yes—this is the continuity of $\operatorname{Arg}(z)$ on P.

30. Find all solutions of the equations

(i) $e^z = 2$

(ii) $\text{Log } z = +i$

(iii) $\cos z = 7.$

Hint: In (iii) set $Z = e^{iz}$ and show that $Z^2 - 14Z + 1 = 0$.

Answer: (*i*) $z = \log 2 + 2\pi ni$; (ii) $z = \cos 1 + i \sin 1$; (iii) $z = -i \log(7 \pm 4\sqrt{3}) + 2\pi n$. Note that $(7 - 4\sqrt{3}) > 0$.

2.7 DERIVATIVE OF A FUNCTION OF A COMPLEX VARIABLE

Consider a function f defined near a point p (i.e., defined in some disc of positive radius about p). We may imitate the definition of the derivative of a function of a real variable and inquire whether the difference quotients

$$\frac{f(z) - f(p)}{z - p} = \frac{\Delta f}{\Delta z},$$

which are defined for all z near p such that $z \neq p$, have a limit as z approaches p. If so, we call this limit the **derivative of f at p** and denote it by $f'(p)$ or $df/dz(p)$. The function f is then said to be **differentiable** at p; if it is differentiable at various points in its domain of definition, we obtain a new function $f'(z)$, the **derivative of f,** whose domain of definition is the set of points z where $f'(z)$ exists. We shall use the notations f' and df/dz interchangeably.

Since f is defined near p, $f(p + h)$ is defined for all complex h sufficiently close to zero, and if $h \neq 0$ the quotient

$$\frac{f(p+h) - f(p)}{h} = \frac{f(p+h) - f(p)}{(p+h) - p}$$

is defined. It is clear that (writing $z = p + h$)

$$\lim_{z \to p} \frac{f(z) - f(p)}{z - p} = \lim_{h \to 0} \frac{f(p+h) - f(p)}{h},$$

and this gives us a slightly different way of looking at $f'(p)$.

It is possible for a function $w = f(z)$ to be differentiable at a single point, and nowhere else (Exercise 4). Such pathological situations will not be of interest to us, and certainly will not occur if f is defined and differentiable throughout an *open set;* then it must be differentiable on some disc about each point in the set, by the definition of open sets.

Definition 2.7 *Functions that are defined and differentiable on an open set E in the complex plane will be referred to as* **holomorphic functions** *on E.*

One can prove, exactly as one does for functions of a real variable, that all polynomials are holomorphic on **C**, with derivatives:

$$f'(z) = 0 \quad \text{if} \quad f(z) = a_0 \qquad \text{(a constant)}$$

(58)
$$f'(z) = 1 \quad \text{if} \quad f(z) = z$$

$$\frac{d}{dz}(z^n) = nz^{n-1} \quad \text{for} \quad n = 1, 2, 3, \ldots .$$

The αth power function (principal determination) $f(z) = z^\alpha$, defined on the cut plane P, is also holomorphic, and its derivative is given by the familiar equation

(59)
$$\frac{d}{dz}(z^\alpha) = \alpha \cdot z^{\alpha-1} \qquad \text{for all } z \text{ in the cut plane } P,$$

provided $\alpha \neq 0$. (When $\alpha = 0$, $z^\alpha = 1$ for all z and the derivative is *zero*, not z^{-1}.) This formula will be proved in Section 2.9 for arbitrary complex exponents α; here we will only do the simple case $\alpha = (\frac{1}{2}) + i0$ to illustrate direct (i.e., brute force) methods.

Example 2.17 Let $f(z) = z^{1/2}$ on the cut plane P. Let z be a point in P. If $|h|$ is small and $h \neq 0$, then $z + h$ is close to z and is also in P. The algebraic formula $a^2 - b^2 = (a + b)(a - b)$ is valid for complex a and b, and it gives us

$$h = (z + h) - z = ((z + h)^{1/2} + z^{1/2})((z + h)^{1/2} - z^{1/2});$$

thus,

$$\frac{f(z + h) - f(z)}{h} = \frac{(z + h)^{1/2} - z^{1/2}}{((z + h)^{1/2} - z^{1/2})((z + h)^{1/2} + z^{1/2})}$$

$$= \frac{1}{(z + h)^{1/2} + z^{1/2}}$$

The function $f(z) = z^{1/2}$ is continuous at z, so that as $h \to 0$ $(h \neq 0)$, we get $z + h \to z$, and $(z + h)^{1/2} \to z^{1/2}$. Thus,

$$f'(z) = \lim_{h\to 0} \frac{f(z + h) - f(z)}{h} = \lim_{h\to 0} \frac{1}{(z + h)^{1/2} + z^{1/2}}$$

$$= \frac{1}{\lim\limits_{h\to 0} \{(z + h)^{1/2} + z^{1/2}\}} = \frac{1}{2z^{1/2}} = \tfrac{1}{2} z^{-1/2},$$

since the limit in the denominator is not zero.

It is much more difficult for a function of a complex variable to be differentiable than it is for a function of a real variable (how much more so is one of the real surprises of complex analysis). In **C** there are infinitely many ways to approach a point p, and the difference quotients $\Delta f/\Delta z$ must converge to a single limit no matter how we make the approach. In **R** there are only two different directions to contend with, and a function has less to do if it is to be differentiable.

Just as with functions of a real variable, there is a collection of results which tell us that certain combinations of differentiable functions are again differentiable, and at the same time show us how to evaluate the new derivatives. These results are proved in the same way as their real variable counterparts, so we will not prove them here.

Theorem 2.11 *Let f and g be functions defined near a point z in* **C**, *and differentiable at z. Then the following combinations of f and g are also differentiable at z, with derivatives computed as shown below.*

(60)

$$\text{(i)}\quad (f \pm g)'(z) = f'(z) \pm g'(z)$$

$$\text{(ii)}\quad (\alpha f)'(z) = \alpha(f'(z)) \qquad (\alpha \text{ a fixed complex number})$$

$$\text{(iii)}\quad (f \cdot g)'(z) = f'(z) \cdot g(z) + f(z) \cdot g'(z)$$

$$\text{(iv)}\quad \left(\frac{f}{g}\right)'(z) = \frac{f'(z) \cdot g(z) - f(z) \cdot g'(z)}{(g(z))^2} \qquad (\text{provided } g(z) \neq 0)$$

$$\text{(v)}\quad \left(\frac{1}{g}\right)'(z) = \frac{-1}{(g(z))^2} g'(z) \qquad (\text{provided } g(z) \neq 0).$$

There is also a formula for differentiating the composite $(f \circ g)(z) = f(g(z))$ of two differentiable functions; this result, known as the "chain rule," will be proved in the next section.

EXERCISES

1. Prove that $\frac{d}{dz}(z^n) = nz^{n-1}$ for $n = 1, 2, \ldots.$

Hint: Remember that $(a^{n+1} - b^{n+1}) = (a - b)(a^n + a^{n-1}b + \cdots + b^n)$.

2. Prove the following differentiation formulas by directly calculating limits of difference quotients.

$$\text{(i)}\quad \frac{d}{dz}(z^2 + z + 1) = 2z + 1$$

$$\text{(ii)}\quad \frac{d}{dz}\left(\frac{1}{z - p}\right) = \frac{-1}{(z - p)^2} \qquad (p \text{ fixed}; z \neq p)$$

$$\text{(iii)}\quad \frac{d}{dz}(z^{1/4}) = \tfrac{1}{4}z^{-3/4} = \frac{1}{4z^{3/4}} \qquad (\text{principal determinations})$$

$$\text{(iv)}\quad \frac{d}{dz}(e^z) = e^z$$

Note: (iv) is not easy. Later we will get this formula indirectly.

3. Prove the formulas in Theorem 2.11 by calculating limits of difference quotients.

4. Prove that the function $f(z) = |z|^2 = x^2 + y^2$ is differentiable at $z = 0$, and nowhere else in the plane. Do $U = \text{Re}(f)$ and $V = \text{Im}(f)$ have continuous partial derivatives?

2.8 DERIVATIVES AS SOLUTIONS OF A "LINEAR APPROXIMATION" PROBLEM

Let us consider a function $w = f(z)$ defined near a point p. The existence of the derivative $f'(p)$ can be interpreted as giving us a "good linear approximation" to the behavior of $f(z)$ for z near p.

We say that a function $w = F(z)$ is **linear** if it has the very special form

$$F(z) = a \cdot z + b \quad (a \text{ and } b \text{ complex constants}).$$

In order for such a function to agree with $f(z)$ at $z = p$, so that $F(p) = f(p)$, it must have the form

$$F(z) = a \cdot (z - p) + f(p) \quad (a \text{ may be any complex constant}),$$

which has only one undetermined constant left in it. If we also want the derivatives of $F(z)$ and $f(z)$ to agree at p, the only possible choice for $F(z)$ is:

$$F(z) = f(p) + f'(p)(z - p). \tag{61}$$

The degree to which the linear function $F(z)$ approximates the behavior of $f(z)$ for z near p is measured by the size of the *error function*

$$E(z) = f(z) - F(z),$$

which measures the discrepancy between $f(z)$ and $F(z)$. Thus,

$$f(z) = F(z) + E(z) = [f(p) + f'(p)(z - p)] + E(z). \tag{62A}$$

We will prove that (61) gives a "good" linear approximation to $f(z)$ near p in the sense that the error in (62A) has a magnitude $|E(z)|$ which vanishes much faster than $|z - p|$ (the distance from z to p) as z approaches p; thus

$$\frac{|E(z)|}{|z - p|} \to 0 \quad \text{as} \quad z \to p \qquad (\text{with } z \neq p) \tag{62B}$$

in equation (62A).

For functions of a real variable x, the derivative $f'(p) = df/dx(p)$ gives us the approximating linear function $F(x) = f(p) + f'(p)(x - p)$, which is usually interpreted as the equation of the "tangent line" to the graph of f above the point $x = p$. Formula (62) can also be regarded as the first step toward a Taylor formula with remainder; $E(z)$ plays the role of the remainder.

Notice that the coefficients in the linear approximating function (61) will be different if we try to approximate the behavior of $f(z)$ near some other point $z = q$ where f is differentiable; we would have to replace p with q everywhere it appears in (61) and (62).

To prove that the linear function in (61) gives us the rapidly vanishing error exhibited in (62B), we merely look at the definition of the derivative $f'(p)$. If $z \neq p$, then

$$\left|\frac{E(z)}{z-p}\right| = \left|\frac{f(z) - f(p) - f'(p)(z-p)}{z-p}\right| = \left|\frac{f(z) - f(p)}{z-p} - f'(p)\right|$$

tends to zero as $z \to p$, since $f'(p)$ is defined as

$$f'(p) = \lim_{z \to p} \frac{f(z) - f(p)}{z - p}.$$

If $f(z)$ is differentiable at p, the linear function $F(z)$ in (62) that closely approximates $f(z)$ for z near p is called the **best linear approximation** to f near p.

If we write Δz for $z - p$ and Δf for the difference $f(z) - f(p)$, we may recast equations (62) in the form:

(63A) $$\Delta f = f'(p) \cdot \Delta z + E(z) = \frac{df}{dz}(p) \cdot \Delta z + E(z)$$

and

(63B) $$\left|\frac{E(z)}{z-p}\right| \to 0 \quad \text{as} \quad z \to p \qquad (\text{that is, as } |\Delta z| \to 0).$$

If the derivative $f'(p)$ is non-zero, then the non-constant linear term $f'(p) \cdot \Delta z$ is *proportional* to Δz, while the size of the error term becomes negligible in comparison with $f'(p) \cdot \Delta z$, as $\Delta z \to 0$. If $f'(p) = 0$, the formulas (62) and (63) are still valid, but in this event (63A) reduces to the equation $\Delta f = E(z)$ and loses much of its usefulness.

As a simple application of these remarks on linear approximations and derivatives, we prove the following result.

Theorem 2.12 *Let f be a function defined near the point p. If f is differentiable at p, then f must be continuous there.*

PROOF: We must show that $f(z) \to f(p)$ as $z \to p$. If we write

$$f(z) = f(p) + f'(p)(z - p) + E(z),$$

as in (62), we get $|E(z)| \to 0$ as $z \to p$; in fact, $|E(z)|/|z - p| \to 0$. But, it is obvious that $f'(p)(z - p) \to 0$, so the right side has the limit $f(p)$ as z approaches p. ▮

The error term $E(z)$ in (62A) can be written in another way,

(64) $$E(z) = e(z) \cdot (z - p) \quad \text{for all } z \text{ near } p \text{ (including } z = p)$$

where $e(z)$ is a function that is continuous at $z = p$ and has the value $e(p) = 0$ there. To prove this, we simply take

$$e(z) = \begin{cases} 0 & \text{if } \ z = p \\ \dfrac{E(z)}{(z - p)} & \text{if } \ z \neq p. \end{cases}$$

The limit formula (62B) shows that $e(p) = 0 = \lim_{z \to p} \{e(z)\}$, which is the same as saying that $e(z)$ is continuous at $z = p$. The chain rule for complex derivatives may be proved from (62A) by writing the error term in the form (64).

Theorem 2.13 (Chain Rule) *Let $w = g(z)$ be differentiable at p, and let $s = f(w)$ be differentiable at the image point $q = g(p)$. Then the composite function $s = (f \circ g)(z) = f(g(z))$ is differentiable at p, and*

(65A) $$(f \circ g)'(p) = f'(g(p)) \cdot g'(p) = f'(q) \cdot g'(p).$$

This rule can also be written in the form

(65B) $$\frac{d}{dz}(f \circ g)(p) = \frac{df}{dw}(q) \cdot \frac{dg}{dz}(p),$$

which suggests more clearly the variable involved in each differentiation.

PROOF: Let $\{z_n\}$ be any sequence such that $z_n \to p$ (and $z_n \neq p$); we must show that $(f(w_n) - f(q))/(z_n - p)$ approaches $f'(q) \cdot g'(p)$, where $w_n = g(z_n)$. Since $g(z)$ is differentiable at $z = p$, it is also continuous and we must have $w_n = g(z_n) \to g(p) = q$ as $n \to \infty$. Applying formula (62A) to f (the base point is q), we get

$$\begin{aligned} \frac{f(g(z_n)) - f(g(p))}{z_n - p} = \frac{f(w_n) - f(q)}{z_n - p} &= f'(q)\left[\frac{w_n - q}{z_n - p}\right] + \left[\frac{E_f(w_n)}{z_n - p}\right] \\ &= f'(q) \cdot \left[\frac{g(z_n) - g(p)}{z_n - p}\right] + e_f(w_n) \cdot \left[\frac{w_n - q}{z_n - p}\right] \\ &= f'(q) \cdot \left[\frac{g(z_n) - g(p)}{z_n - p}\right] + e_f(w_n) \cdot \left[\frac{g(z_n) - g(p)}{z_n - p}\right]. \end{aligned}$$

From our hypotheses, it is clear that

$$\lim_{n\to\infty} \frac{g(z_n) - g(p)}{z_n - p} = g'(p) \qquad \text{(by differentiability of } g \text{ at } z = p)$$

$$\lim_{n\to\infty} e_f(w_n) = e_f(q) = 0 \qquad \text{(by continuity of } e_f(w) \text{ at } w = q).$$

Now we can invoke standard theorems on limits to see that

$$\lim_{n\to\infty} \frac{f(g(z_n)) - f(g(p))}{z_n - p} = f'(q) \cdot g'(p) + 0$$

for any sequence $z_n \to p$ (such that $z_n \neq p$). This is exactly what we mean by existence of the limit

$$\frac{d}{dz}(f \circ g)(p) = \lim_{z\to p} \frac{(f \circ g)(z) - (f \circ g)(p)}{z - p} = f'(q) \cdot g'(p). \quad \blacksquare$$

Using the chain rule and formulas (60), we may differentiate a large variety of functions, but there are still many functions whose derivatives cannot be evaluated this way; for example, $z^{1/2}$ (whose derivative was calculated by direct examination of difference quotients above), e^z, $\mathrm{Log}(z)$, and $\sin(z)$. Two additional methods are needed: the Cauchy-Riemann equations, and the method of inverse functions.

EXERCISES

1. Write $f(z) = z^2 - 2z + 1$ in the form

$$f(z) = [f(p) + f'(p)(z - p)] + E(z),$$

taking $p = 1 + i0$ and $p = i$. Show that:

(i) $|E(p + \Delta z)| \leq |\Delta z|^2$ for small displacements from $p = 1$.

(ii) $|E(p + \Delta z)| \leq |\Delta z|^2$ for small displacements from $p = i$.

Note: In either case, $|E(z)|/|z - p| \to 0$ as $z \to p$.

2. Show that $f(z) = (1 - z^2)^{1/2}$ is a well defined composite of $w = 1 - z^2$ and $s = w^{1/2}$ (principal determination), if we take as $\mathrm{Dom}(f)$ the doubly cut plane obtained by deleting the segments $(-\infty, -1]$ and $[+1, +\infty)$ from the real axis. Calculate df/dz explicitly.

Answer: $df/dz = -1/(1 - z^2)^{1/2} = -(1 - z^2)^{-1/2}$ (principal determination of $w^{-1/2}$ defined as in Section 2.6).

3. Assuming that we know that $\dfrac{d}{dz}(\sin z) = \cos z$, prove that $\lim\limits_{z\to 0} \dfrac{\sin z}{z} = 1$; then try to calculate this limit directly without using the differentiation formula.

Note: Direct calculation of $\lim\limits_{z\to 0} (\sin z)/z$ is difficult; the differentition formula will be proved, in the next section, by indirect methods.

2.9 THE CAUCHY-RIEMANN EQUATIONS

Any function $w = f(z)$ of a complex variable can be regarded as a mapping from $\mathbf{R}^2$ into $\mathbf{R}^2$. If we identify $\mathbf{C}$ with the Cartesian plane $\mathbf{R}^2$ and write $f(z)$ in the form $f(x + iy) = U(x, y) + iV(x, y)$, then f is identified with the mapping of $\mathbf{R}^2$ given by

$$\Phi(x, y) = (u, v) \quad \text{where} \quad u = U(x, y) \quad \text{and} \quad v = V(x, y).$$

Definition 2.8 *A mapping $w = f(z)$ is said to be* **smooth** *if the component functions $U(x, y)$ and $V(x, y)$ are continuous and have continuous first partial derivatives $\partial U/\partial x, \ldots, \partial V/\partial y$ when we write $f(x + iy) = U(x, y) + iV(x, y)$.*

Smoothness is not enough to give complex differentiability of $w = f(z)$; rather, differentiability of f is equivalent to smoothness plus an important set of relations between the partial derivatives $\partial U/\partial x, \ldots, \partial V/\partial y$, the famed **Cauchy-Riemann equations**:

$$(66) \qquad \frac{\partial U}{\partial x} = \frac{\partial V}{\partial y} \qquad \frac{\partial U}{\partial y} = -\frac{\partial V}{\partial x}$$

This system of partial differential equations plays an important role in the application of complex analysis to potential theory, fluid flow, and other problems in physics. Each of these physical problems has a partial differential equation associated with it that is related to the Cauchy-Riemann equations.

Theorem 2.14 *If $f(z) = U(x, y) + iV(x, y)$ is defined near p, and if df/dz exists at p, then the partial derivatives $\partial U/\partial x, \ldots, \partial V/\partial y$ all exist at p and satisfy the Cauchy-Riemann equations.*

Proof: Write $p = x_0 + iy_0$ and consider points $z = (x_0 + \Delta x) + iy_0$ such that $\Delta x \neq 0$. Then $z - p = \Delta x$ and

$$\frac{f(z) - f(p)}{z - p} = \left[\frac{U(x_0 + \Delta x, y_0) - U(x_0, y_0)}{\Delta x}\right] + i\left[\frac{V(x_0 + \Delta x, y_0) - V(x_0, y_0)}{\Delta x}\right].$$

As $\Delta x \to 0$ we have $z = (x_0 + \Delta x) + iy_0 \to p$, and the left side (hence also the right side) converges to $f'(p)$. This gives us

$$f'(p) = \frac{\partial U}{\partial x}(p) + i\frac{\partial V}{\partial x}(p), \tag{67}$$

together with the existence of both partial derivatives. Notice that the real and imaginary parts of the right side must have limits separately if their sum has a limit as $z \to p$. If we work with z near p of the form $z = x_0 + i(y_0 + \Delta y)$ $(\Delta y \neq 0)$, then $z - p = 0 + i\Delta y$, and by the same reasoning we get

$$f'(p) = \left(\frac{1}{i}\right)\left[\frac{\partial U}{\partial y}(p) + i\frac{\partial V}{\partial y}(p)\right] = \frac{\partial V}{\partial y}(p) - i\frac{\partial U}{\partial y}(p). \tag{68}$$

Comparing the real and imaginary parts of (67) and (68), we obtain the Cauchy-Riemann equations. ▮

We have also shown how to evaluate df/dz in terms of the partial derivatives $\partial U/\partial x, \ldots, \partial V/\partial y$ in the above proof. The basic equation is

$$\frac{df}{dz} = \frac{\partial U}{\partial x} + i\frac{\partial V}{\partial x} = \frac{\partial V}{\partial y} - i\frac{\partial U}{\partial y}, \tag{69}$$

and various others are derived from it by substituting the Cauchy-Riemann equations into this identity.

Theorem 2.15 *Suppose that $f(z)$ is defined on an open set E in the complex plane, and that the component functions U and V in*

$$f(x + iy) = U(x, y) + iV(x, y) \qquad (z = x + iy \text{ in } E)$$

are continuous and have continuous first partial derivatives throughout E. Then f is holomorphic in E if and only if the Cauchy-Riemann equations are satisfied.

Note: In the proof we will have to recall a few facts about partial derivatives. For these details we suggest that the reader consult Kiokemeister and Johnson [11] or Fulks [8].

PROOF: If f is holomorphic, we may apply Theorem 2.14 to each point in E to see that the Cauchy-Riemann equations are satisfied.

For the converse argument we suppose that the Cauchy-Riemann equations are satisfied, and must prove that df/dz exists. The existence and continuity of the partial derivatives $\partial U/\partial x, \ldots, \partial V/\partial y$ near a typical point $p = x_0 + iy_0$ in E means that we can write $U(x, y)$ and $V(x, y)$ in the form

$$\begin{aligned} U(x, y) &= U(p) + \frac{\partial U}{\partial x}(p)(x - x_0) + \frac{\partial U}{\partial y}(p)(y - y_0) + E_1(x, y) \\ V(x, y) &= V(p) + \frac{\partial V}{\partial x}(p)(x - x_0) + \frac{\partial V}{\partial y}(p)(y - y_0) + E_2(x, y), \end{aligned} \tag{70A}$$

where the error terms $E_1(z)$ and $E_2(z)$ vanish rapidly as $z = x + iy$ approaches p:

$$\text{(70B)} \qquad \frac{|E_k(z)|}{|z-p|} = \frac{|E_k(x+iy)|}{|(x+iy)-(x_0+iy_0)|} \to 0 \quad \text{as} \quad z \to p$$

for $k = 1, 2$.† Now consider any point $z = x + iy = (x_0 + \Delta x) + i(y_0 + \Delta y) \neq p$. We may write out the difference quotient as

$$\begin{aligned}\frac{f(z)-f(p)}{z-p} &= \frac{f(z)-f(p)}{\Delta x + i\,\Delta y} \\ &= \frac{1}{\Delta x + i\,\Delta y}[U(z) + iV(z) - U(p) - iV(p)] \\ &= \frac{1}{\Delta x + i\,\Delta y}\left[\frac{\partial U}{\partial x}(p)\,\Delta x + \frac{\partial U}{\partial y}(p)\,\Delta y + i\frac{\partial V}{\partial x}(p)\,\Delta x \right. \\ &\qquad \left. + i\frac{\partial V}{\partial y}(p)\,\Delta y + E_1(z) + iE_2(z)\right].\end{aligned}$$

If we use the Cauchy-Riemann equations, we get

$$\begin{aligned}\frac{f(z)-f(p)}{z-p} &= \frac{1}{\Delta x + i\,\Delta y}\left[\frac{\partial U}{\partial x}(p)(\Delta x + i\,\Delta y)\right. \\ &\qquad \left. + (-i)\frac{\partial U}{\partial y}(p)(\Delta x + i\,\Delta y) + E_1(z) + iE_2(z)\right] \\ &= \frac{\partial U}{\partial x}(p) - i\frac{\partial U}{\partial y}(p) + \frac{E_1(z)}{z-p} + i\frac{E_2(z)}{z-p}.\end{aligned}$$

But as $z \to p$ the last two terms approached zero, by equation (70B), so that this difference quotient has a well defined limit. Thus, f is differentiable at p and

$$f'(p) = \frac{\partial U}{\partial x}(p) - i\frac{\partial U}{\partial y}(p) = \frac{\partial U}{\partial x}(p) + i\frac{\partial V}{\partial x}(p) \quad \blacksquare$$

Example 2.18 The function $\exp(z) = e^x(\cos y + i \sin y)$ is holomorphic on $\mathbf{C}$, and

$$\frac{d}{dz}(e^z) = e^z.$$

Here

$$e^{x+iy} = U(x,y) + iV(x,y) = e^x \cos y + ie^x \sin y,$$

---

† Recall that the distance between two points $z = x + iy$ and $p = x_0 + iy_0$ in the plane is precisely the absolute value $|z - p|$. Thus in formula (70B), the denominator is just the Euclidean distance from p to the nearby point $z = x + iy$.

so that

$$\frac{\partial U}{\partial x} = e^x \cos y \qquad \frac{\partial U}{\partial y} = -e^x \sin y$$

$$\frac{\partial V}{\partial x} = e^x \sin y \qquad \frac{\partial V}{\partial y} = e^x \cos y.$$

Obviously the Cauchy-Riemann equations are satisfied, and

$$\frac{df}{dz} = \frac{\partial U}{\partial x} - i\frac{\partial U}{\partial y} = e^x \cos y + ie^x \sin y = e^z.$$

The Cauchy-Riemann equations are often used to determine whether f is differentiable, since the partials $\partial U/\partial x, \ldots, \partial V/\partial y$ are usually easy to compute. However, many of the functions we are concerned with are presented with the variable z described by its polar coordinates (r, θ) rather than its Cartesian coordinates (x, y); this amounts to writing $z = re^{i\theta}$ and then writing $f(z) = f(r, \theta)$. It is possible, and often desirable, to check the validity of the Cauchy-Riemann equations directly when the variable z has been given in polar coordinates:

$$f(z) = f(r, \theta) = U(r, \theta) + iV(r, \theta)$$

(the *values* of f are still expressed in Cartesian coordinates!). The Cauchy-Riemann equations are equivalent to the set of equations

$$r\frac{\partial U}{\partial r} = \frac{\partial V}{\partial \theta} \quad \text{and} \quad r\frac{\partial V}{\partial r} = -\frac{\partial U}{\partial \theta}, \tag{71}$$

and in this situation the derivative is given by

$$f'(z) = (\cos\theta - i\sin\theta)\left(\frac{\partial U}{\partial r} + i\frac{\partial V}{\partial r}\right) = e^{-i\theta}\left(\frac{\partial U}{\partial r} + i\frac{\partial V}{\partial r}\right). \tag{72}$$

We prove equations (71) and (72) by noticing that if we start with *any* smooth function $g(x, y)$ described in Cartesian coordinates x and y, and convert this into a function $g(r, \theta)$ expressed in polar coordinates via the transformations

$$x = r\cos\theta \quad \text{and} \quad y = r\sin\theta,$$

then we may apply the chain for partial derivatives to get

$$\frac{\partial g}{\partial r} = \frac{\partial g}{\partial x}\frac{\partial x}{\partial r} + \frac{\partial g}{\partial y}\frac{\partial y}{\partial r} = \cos\theta\frac{\partial g}{\partial x} + \sin\theta\frac{\partial g}{\partial y}$$

$$\frac{\partial g}{\partial \theta} = -r\sin\theta\frac{\partial g}{\partial x} + r\cos\theta\frac{\partial g}{\partial y}.$$

On solving these equations for $\frac{\partial g}{\partial x}$ and $\frac{\partial g}{\partial y}$ we see that

$$\frac{\partial g}{\partial x} = \cos\theta \frac{\partial g}{\partial r} - \frac{\sin\theta}{r}\frac{\partial g}{\partial \theta}$$

$$\frac{\partial g}{\partial y} = \sin\theta \frac{\partial g}{\partial r} + \frac{\cos\theta}{r}\frac{\partial g}{\partial \theta}.$$

Since these calculations apply to *any* function $g(x, y)$, the operations $\frac{\partial}{\partial x}$ and $\frac{\partial}{\partial y}$, performed for Cartesian coordinates, correspond to the operations

(73)
$$\frac{\partial}{\partial x} = \cos\theta \frac{\partial}{\partial r} - \frac{\sin\theta}{r}\frac{\partial}{\partial \theta}$$
$$\frac{\partial}{\partial y} = \sin\theta \frac{\partial}{\partial r} + \frac{\cos\theta}{r}\frac{\partial}{\partial \theta}$$

performed for polar coordinates. These considerations, applied to the function $g(x, y) = f(x + iy)$, quickly lead to equations (71) and (72) presented above.

Example 2.19 We shall prove the differentiation formula

(74) $$\frac{d}{dz}(\text{Log } z) = \frac{1}{z}, \qquad \text{for all } z \text{ in the cut plane } P.$$

Start by writing the logarithm in polar coordinates: $\text{Log}(z) = U(r, \theta) + iV(r, \theta) = \log r + i\theta$, if $z = re^{i\theta}$ (θ normalized so $-\pi < \theta < +\pi$). Evidently,

$$\frac{\partial U}{\partial r} = \frac{1}{r} \qquad \frac{\partial U}{\partial \theta} = 0$$
$$\frac{\partial V}{\partial r} = 0 \qquad \frac{\partial V}{\partial \theta} = 1$$
at all points in the cut plane;

the Cauchy-Riemann equations (71) are satisfied and the derivative is given by

$$\frac{d}{dz}(\text{Log } z) = e^{-i\theta}\left(\frac{\partial U}{\partial r} + i\frac{\partial V}{\partial r}\right) = \frac{1}{r}e^{-i\theta} = (re^{i\theta})^{-1} = \frac{1}{z}.$$

We know that other determinations of the logarithm (perhaps defined on other domains in the plane) differ from $\text{Log}(z)$ by an added constant $2\pi ki$, which disappears upon differentiation; thus we conclude that

$$\frac{d}{dz}(\log z) = \frac{1}{z} \quad \text{for} \quad z \neq 0,$$

for any continuous determination of the logarithm. Notice that the derivative $1/z$ is *single valued* (there is no disagreement between the values of $1/z$ on opposite sides of the negative real axis), even though there are many different determinations of $\log z$. We will see that it is quite common for a single valued function to have a multiple valued antiderivative.

EXERCISES

1. The most general *linear* mapping of $\mathbf{R}^2$ into $\mathbf{R}^2$ is given by $w = u + iv = f(x + iy)$, where

$$u = ax + by + s \quad \text{and} \quad v = cx + dy + t \quad \text{(real coefficients)}.$$

Which choices of a, b, c, d and s, t make f holomorphic? If f is holomorphic, show that $w = f(z) = Az + B$, where A and B are complex scalars.

2. Determine whether the following functions are holomorphic.

(i) $(x^2 - y^2) - 2ixy$
(ii) $(x^2 - y^2) + 2ixy$
(iii) $\log\sqrt{x^2 + y^2} + i \arccos(x/\sqrt{x^2 + y^2})$ if $\text{Im}(z) = y > 0$
(iv) $x^2 - i(y^2 + x)$

3. Prove that conjugation $f(z) = \bar{z}$ is not differentiable anywhere.

4. Consider the mapping of the punctured plane $E = \{z : z \neq 0\}$ given by $f(re^{iz}) = (1/r)e^{i\theta}$. Prove that this map is not differentiable. Interpret this map geometrically as a reflection of points through the unit circle $|z| = 1$.

5. Use the Cauchy-Riemann equations, and other techniques such as the chain rule, to prove the following identities.

(i) $\dfrac{d}{dz}(\sin z) = \cos z$

(ii) $\dfrac{d}{dz}(\cos z) = -\sin z$

(iii) $\dfrac{d}{dz}(\tan z) = \sec^2 z \qquad \left(z \neq \dfrac{\pi}{2} + n\pi\right)$

(iv) $\dfrac{d}{dz}(\alpha^z) = (\text{Log}\,\alpha) \cdot \alpha^z \qquad (\alpha \neq 0 \text{ fixed})$

(v) $\dfrac{d}{dz}(z^\alpha) = \alpha z^{\alpha-1} \qquad (\alpha \text{ fixed; } z \text{ in cut plane } P).$

(vi) $\dfrac{d}{dz}(\sinh z) = \cosh z$

6. Show that $A(x, y) = \text{Arg}(x + iy)$ is a smooth function of two real variables on the cut plane P, and that

$$\frac{\partial A}{\partial x} = \frac{-y}{x^2 + y^2} \quad \text{and} \quad \frac{\partial A}{\partial y} = \frac{x}{x^2 + y^2}.$$

Hint: Use the formulas in (40), in different parts of P.

7. Prove that $\text{Log}(z)$ is a smooth function on the cut plane P.

8. Suppose that a smooth function $w = f(z) = U(x, y) + iV(x, y)$ is expressed in the form $f(x, y) = (R(x, y), \Theta(x, y)) = R(x, y)e^{i\Theta(x,y)}$. Show that the Cauchy-Riemann equations are equivalent to

$$\frac{1}{R}\frac{\partial R}{\partial x} = \frac{\partial \Theta}{\partial y} \quad \text{and} \quad \frac{1}{R}\frac{\partial R}{\partial y} = -\frac{\partial \Theta}{\partial x}.$$

9. Show that the Cauchy-Riemann equations are equivalent to

$$\frac{r}{R}\frac{\partial R}{\partial r} = \frac{\partial \Theta}{\partial \theta} \quad \text{and} \quad \frac{1}{R}\frac{\partial R}{\partial \theta} = -r\frac{\partial \Theta}{\partial r}$$

if we write both the values and the variable of $w = f(z)$ in polar form, so that $f(r, \theta) = R(r, \theta)e^{i\Theta(r,\theta)}$.

10. Prove that $z^i = e^{i\text{Log}(z)}$ is a smooth function on the cut plane P. Use the Cauchy-Riemann equations to show that

$$\frac{d}{dz}(z^i) = iz^{(i-1)} \qquad \text{for } z \text{ in } P.$$

11. If $f(x + iy) = U(x, y) + iV(x, y)$ is defined on an open set E, let E^* be the image of E under conjugation (reflection through the real axis). On E^* define:

$$g(z) = \overline{f(\bar{z})} \text{ for all } z \text{ in } E^*.$$

Prove that f is holomorphic on E if and only if the "reflected function" g is holomorphic on the "reflected" set E^*. If $f(z) = e^z$ on the upper half plane $E = \{z : \text{Im}(z) > 0\}$, calculate $g(z)$ on $E^* = \{z : \text{Im}(z) < 0\}$.

Answer: $g(z) = e^z$ for z in E^*.

12. Using differentiation formulas, and the fact that $f'(0) = \lim_{z\to 0} (f(z) - f(0))/z$, show that

(i) $\displaystyle\lim_{z\to 0} \frac{\sin z}{z} = 1$

(ii) $\displaystyle\lim_{z\to 0} \frac{e^{iz} - 1}{z} = i$

(iii) $\displaystyle\lim_{z\to 0} \frac{\sinh z}{iz} = -i$

(iv) $\displaystyle\lim_{z\to 0} z \cdot \text{ctn } z = 1.$

*2.10 ANTIDERIVATIVES AND FUNCTIONS WHOSE DERIVATIVE IS ZERO

Let $w = f(z)$ be differentiable throughout a connected open set (domain) E, and assume that $df/dz = 0$ at all points in E. Then f must be constant: $f(z) = a_0$ for all z in E. This is proved in two steps. First, we show that f is **"locally constant"** in the sense that it is constant throughout some disc about each point in E; this part of the proof uses only the Cauchy-Riemann equations. Then the connectedness of E must be used to show that f takes a single value throughout E. Unless E is connected, it is quite possible for f to take on more than one value as z varies. For example, if E is the (disconnected) open set obtained by deleting the real axis from the plane, E splits into two connected half planes $E_1 = \{z:\text{Im}(z) > 0\}$ and $E_2 = \{z:\text{Im}(z) < 0\}$. We could take

$$f(z) = 1 \quad \text{for } z \text{ in } E_1$$

$$f(z) = 0 \quad \text{for } z \text{ in } E_2.$$

This function is clearly holomorphic on E; $df/dz = 0$ everywhere, and f is locally constant, but f does not take on a unique value throughout E.

Theorem 2.16 *Let E be a domain in the complex plane. Let f be differentiable, with $df/dz = 0$, throughout E. Then there is a constant c such that $f(z) = c$ for all z in E.*

PROOF: First let us show that $f(z)$ is "locally constant." If $f(x + iy) = U(x,y) + iV(x,y)$, this means that

$$0 + i0 = \frac{df}{dz} = \frac{\partial U}{\partial x} + i\frac{\partial V}{\partial x} = \frac{\partial V}{\partial y} - i\frac{\partial U}{\partial y},$$

so that these partial derivatives exist and are zero on E. Now consider what happens near a typical point $p = x_0 + iy_0$ in E. Since E is an open set, there is a disc D about p that lies entirely within E, as shown in Figure 2.16. If $z = x + iy$ is any other point in D, consider the horizontal line segment from (x_0, y_0) to (x, y_0) shown in the figure. By the Fundamental Theorem of Calculus,

$$U(x', y_0) - U(x_0, y_0) = \int_{x_0}^{x'} \frac{\partial U}{\partial x}(t, y_0)\, dt = \int_{x_0}^{x'} 0\, dt = 0$$

for every x' such that $x_0 \leq x' \leq x$, so that U is constant along this line segment. Likewise, if we examine the vertical segment from (x, y_0) to (x, y), we see that

$$U(x, y') - U(x, y_0) = \int_{y_0}^{y'} \frac{\partial U}{\partial y}(x, t)\, dt = \int_{y_0}^{y'} 0\, dt = 0$$

for y' such that $y_0 \leq y' \leq y$, so that U is constant along this segment too; thus, $U(z) = U(x, y) = U(x_0, y_0) = U(p)$. This works for every z in D, so U must be

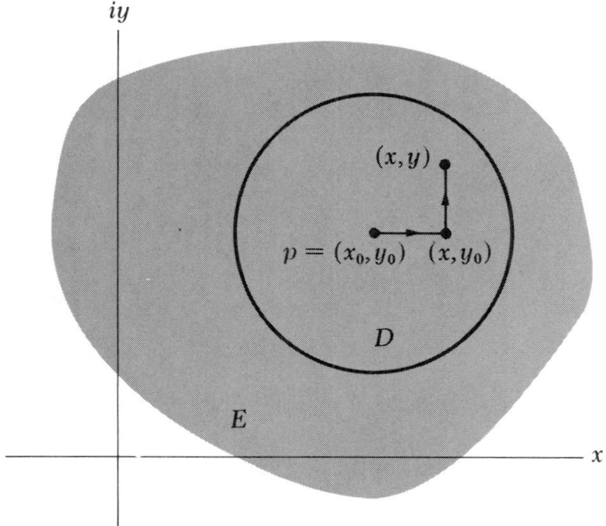

Figure 2.16 The integrations used to prove $f(z) = \text{const}$ on any disc D about $p = x_0 + iy_0$ that lies entirely within the domain of definition E.

constant on this disc. The same reasoning shows that V is constant on D, so that $f(z) = f(p)$ throughout D. Since p was a typical point in E, we have shown that f is locally constant on E.†

To complete the proof we will verify the following useful statement about connected sets.

Theorem 2.17 *Any function $w = f(z)$ that is locally constant on a domain (connected open set) is constant throughout the set.*

PROOF: Pick any point p in E and let $c = f(p)$. Then define these subsets of E:

$$E_1 = \{z : z \text{ is in } E \text{ and } f(z) = c\}$$

$$E_2 = \{z : z \text{ is in } E \text{ and } f(z) \neq c\}.$$

Since f has constant value near each point in E, these must be *open* sets; in fact, if z^* is in E_1, then $f(z) = c$ at z^* and on a disc of positive radius about z^*, and this disc obviously lies within E_1 (by definition). Likewise, if z^* is in E_2, then $f(z^*) = d \neq c$, and $f(z)$ has the same value d for all points near z^*; therefore, all points near z^* are within E_2. The sets E_1 and E_2 together fill up E and they are obviously disjoint. The definition of a connected set is violated if both of these sets are non-empty, so only one can be non-empty. The non-empty set must be E_1 because we already know that it includes at least the point $z = p$. Thus, $E_1 = E$ and $E_2 = \varnothing$ (the empty set, with no points in it), and $f(z) = c$ for all z in E. ▮

† In the example above, where E is the disconnected union of two half planes, why doesn't the reasoning just presented lead to the conclusion that f is constant throughout E?

All of this leads us to the essential uniqueness of antiderivatives for functions of a complex variable (if antiderivatives exist at all). If f is defined on an open set E, we say that a function $g(z)$ is an **antiderivative** of $f(z)$ whenever

(i) $g(z)$ is differentiable throughout E

(ii) $\frac{dg}{dz} = f$ on E.

Theorem 2.18 (Uniqueness of antiderivatives) *If $g_1(z)$ and $g_2(z)$ are antiderivatives of a function $f(z)$ on an open set E, then $g_1(z) - g_2(z)$ is locally constant throughout E. If E is connected, then $g_2(z) = g_1(z) + c$ throughout E, where c is some complex constant.*

PROOF: Obviously, $g_1 - g_2$ is differentiable on E, and

$$\frac{d}{dz}(g_1 - g_2) = \frac{dg_1}{dz} - \frac{dg_2}{dz} = f(z) - f(z) = 0 \qquad \text{for } z \text{ in } E.$$

Now apply Theorem 2.16. ▌

Applying the last result several times in succession, we obtain the following.

Corollary 2.19 *If $f(z)$ has derivatives of orders $k \leq n$ throughout a domain E, and if $f^{(n)}(z) = 0$ on E, then $f(z)$ must be a polynomial in z with degree at most $n - 1$.*

EXERCISES

1. Find antiderivatives for

(i) $\sec^2 z \qquad \left(z \neq \frac{\pi}{2} + n\pi \text{ for } n = 0, \pm 1, +2 \cdots\right).$

(ii) $\sin z + z$

(iii) $z^{1/2}$ (on the cut plane P)

(iv) $z^i + 3z$ (on the cut plane P)

2. Prove that all holomorphic functions on **C** such that $d^2f/dz^2 = e^z$ have the form $e^z + Az + B$ (A, B complex constants).

3. If f is holomorphic on a domain E, and takes on only *real* values, prove that f is constant. Prove the same result if f has purely imaginary values on E.

4. Prove the following result on determinations of $\log(z - p)$.
Theorem: If D is an open disc that excludes the point p, and if $f(z)$ is any *continuous* determination of $\log(z - p)$ on D, then f is necessarily *holomorphic* on D and $df/dz = 1/(z - p)$ throughout D.

The proof is based on the following easy steps.

(i) There exists a holomorphic determination $g(z)$ of $\log(z - p)$ on D whose derivative is $dg/dz = 1/(z - p)$. [*Hint:* Try $g(p + re^{i\theta}) = \log r + i\theta$, normalizing θ suitably.]

(ii) $f(z) - g(z)$ is locally constant on D.

(iii) $f(z) - g(z)$ is constant on D, so that $f(z) = c + g(z)$ is holomorphic. [*Hint:* D is a *domain.*]

(iv) $df/dz = dg/dz = 1/(z - p)$.

5. Use Exercise 4 to prove that any *continuous* determination of $\log(z - p)$, defined on any open set E that excludes p, is necessarily holomorphic on E and has derivative $df/dz = 1/(z - p)$ throughout E.
Hint: Examine a small disc about a typical point in E.

6. Suppose that E is a connected open set that excludes the point p. Suppose that $f(z)$ and $g(z)$ are continuous determinations of $\log(z - p)$ on E. Show that there is a (single) integer N such that $g(z) = f(z) + 2\pi iN$ throughout E. State and prove a similar result for determinations of $\arg(z - p)$ on such a domain.
Hint: $f - g$ is locally constant (why?); E is connected.

7. Let f be a smooth holomorphic function f defined on a domain; prove that $|f(z)|^2 = c$ implies that $f(z) = d$ (d complex constant).
Hint: $|f(z)|^2 = U^2 + V^2$ is constant; use Theorem 2.10 and the Cauchy-Riemann equations.

8. Suppose that E is a connected open set (domain) and that f is a smooth holomorphic function that satisfies the *congruence* $\arg f(z) \equiv c \pmod{2\pi}$ on E, where c is a fixed real constant. Prove that there is a complex constant α such that $f(z) = \alpha$ on E (ordinary equality).

2.11 LAPLACE'S EQUATION AND HARMONIC FUNCTIONS

In this section we consider functions $f(x + iy) = U(x, y) + iV(x, y)$ defined on some open set E in the complex plane, and we shall assume that

(75) The component functions U and V are continuous and have continuous *first and second* partial derivatives on E.†

As is well known, these conditions on U and V are sufficient to guarantee that the mixed second partial derivatives agree:

$$\frac{\partial^2 U}{\partial x\,\partial y} = \frac{\partial^2 U}{\partial y\,\partial x} \quad \text{and} \quad \frac{\partial^2 V}{\partial x\,\partial y} = \frac{\partial^2 V}{\partial y\,\partial x} \tag{76}$$

† This is sometimes expressed by saying that f is **smooth to second order.** Our original definition of smoothness, Definition 2.8, should be regarded as describing functions that are **smooth to order one.**

Theorem 2.20 *Let $f(z)$ be holomorphic on an open set E, and assume that its real and imaginary parts satisfy the smoothness conditions (75). Then both of these components satisfy Laplace's equation on E:*

(77) $$\nabla^2 U = \frac{\partial^2 U}{\partial x^2} + \frac{\partial^2 U}{\partial y^2} = 0 \quad \textit{and} \quad \nabla^2 V = \frac{\partial^2 V}{\partial x^2} + \frac{\partial^2 V}{\partial y^2} = 0.$$

PROOF: Simply notice that U and V satisfy the Cauchy-Riemann equations, so that

$$\frac{\partial^2 U}{\partial x^2} = \frac{\partial}{\partial x}\left(\frac{\partial U}{\partial x}\right) = \frac{\partial}{\partial x}\left(\frac{\partial V}{\partial y}\right) = \frac{\partial^2 V}{\partial x\, \partial y}$$

$$\frac{\partial^2 U}{\partial y^2} = \frac{\partial}{\partial y}\left(\frac{\partial U}{\partial y}\right) = \frac{\partial}{\partial y}\left(-\frac{\partial V}{\partial x}\right) = -\frac{\partial^2 V}{\partial y\, \partial x}$$

and we can apply equations (76). The same reasoning works for V. ∎

In two dimensions **Laplace's equation** is the partial differential equation

(78) $$\nabla^2 H = \frac{\partial^2 H}{\partial x^2} + \frac{\partial^2 H}{\partial y^2} = 0$$

(the higher dimensional analogs will be set down later on), and any function which satisfies this equation is called a **harmonic function.** Harmonic functions arise in physics since equation (78) must be satisfied by electrostatic and gravitational potentials in two dimensional problems; the validity of this equation follows from the basic conservation laws of the physical problems, so the connection between physics and harmonic functions is unavoidable. It would be difficult to underestimate the importance of this partial differential equation, and its higher dimensional analogs, in both pure and applied mathematics.

We have just seen that every differentiable function of complex variable $f(z) = U(x, y) + iV(x, y)$ with sufficiently smooth component functions U and V gives us paired solutions of Laplace's equation, $U = \text{Re}(f)$ and $V = \text{Im}(f)$. But the connection between holomorphic functions and solutions of Laplace's equation is much stronger than this one-way correspondence might suggest. In fact, with one qualification, every harmonic function $H(x, y)$ turns out to be the real part of some holomorphic function of a complex variable. It is this fact which allows us to bring the machinery of complex analysis to bear on problems associated with Laplace's equation in the plane. A detailed account will be given in Chapters 7 and 8.

EXERCISES

1. Show that any linear function $U(x, y) = ax + by + c$ (real coefficients) is harmonic. Find a corresponding function $V(x, y)$ such that $f(x + iy) = U(x, y) + iV(x, y)$ is holomorphic.

2. What choices of real coefficients a, b, c make the quadratic function $U(x, y) = ax^2 + bxy + cy^2$ harmonic?

3. Prove that $A(x, y) = \text{Arg}(x + iy)$ has continuous partial derivatives of first and second orders, and that $\nabla^2 A = 0$ on the cut plane P.
Hint: Recall Exercise 6, Section 2.9.

4. Given $p = x_0 + iy_0$, prove by direct calculations that

$$\log|z - p| = (\tfrac{1}{2}) \log[(x - x_0)^2 + (y - y_0)^2]$$

has continuous first and second partial derivatives, and is harmonic, on the punctured plane $E = \{z : z \neq p\}$.

5. Show that the Laplacian operator has the form

$$\frac{\partial^2}{\partial r^2} + \frac{1}{r}\frac{\partial}{\partial r} + \frac{1}{r^2}\frac{\partial^2}{\partial \theta^2}$$

when expressed in polar coordinates r and θ.

6. Decide which functions below (expressed in polar coordinates) are harmonic.

(i) $U(r, \theta) = r^n \cos n\theta \qquad (n = \pm 1, \pm 2, \ldots)$

(ii) $U(r, \theta) = 1/r$

(iii) $U(r, \theta) = r^n \sin n\theta \qquad (n = \pm 1, \pm 2, \ldots)$

(iv) $U(r, \theta) = \log r$.

7. Assume that $f = U + iV$ has continuous first and second partials, and that f is holomorphic. Show that

(i) $\nabla^2(|f|^2) = 4\,|f'|^2$

(ii) $\log|f(x + iy)|$ is harmonic, except at points z such that $f(z) = 0$ ($\log|f(z)|$ is undefined at these points).

Hint: In (ii) one can use brute force calculations, but there is another proof based on Theorem 2.20.

2.12 INVERSE FUNCTIONS

Let M and N be sets in the complex plane. We say that f is an **invertible mapping between M and N** if

(i) The map f is defined throughout M; that is, $\text{Dom}(f) = M$.
(ii) The map f is one-to-one (**univalent**); distinct points $z_1 \neq z_2$ in M have distinct images $w_1 = f(z_1) \neq w_2 = f(z_2)$ in N.

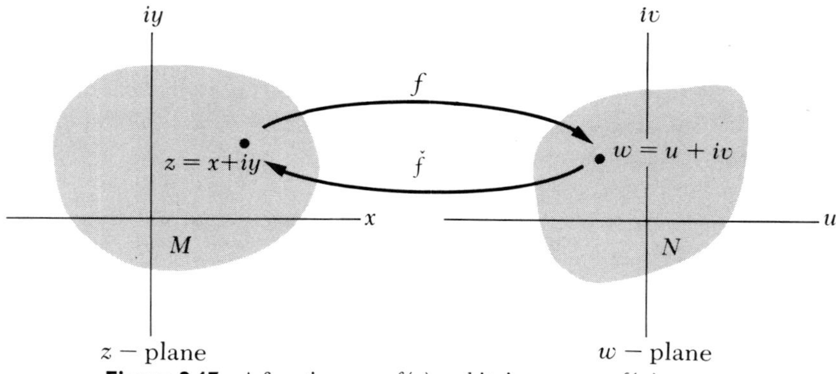

Figure 2.17 A function $w = f(z)$ and its inverse $z = \check{f}(w)$.

(iii) The map f is **onto**; every point w in N is the image $w = f(z)$ of at least one point z in M.

If we systematically write $z = x + iy$ for points in M, and $w = u + iv$ for points in N, to clearly distinguish between the roles of these sets, we see that (i), (ii), and (iii) together mean that

(79) For each w in N there is exactly one point z in M such that $w = f(z)$, and vice versa.

Hence, we can take the equation $w = f(z)$ and uniquely solve it to express z as a (different) function of w. If we write $\check{f}$ for the function that gives $z = \check{f}(w)$, then $\check{f}$ is a mapping $\check{f}: N \to M$ (as shown in Figure 2.17) that *reverses* the original mapping $f: M \to N$. For this reason $\check{f}$ is called the **inverse mapping**, or the **inverse** of f. Obviously f and $\check{f}$ are related by the identities

(80)
$$\begin{aligned} \check{f}(f(z)) &= \check{f}(w) = z \quad \text{for all } z \text{ in } M \\ f(\check{f}(w)) &= f(z) = w \quad \text{for all } w \text{ in } N. \end{aligned}$$

Notice that these mappings play completely symmetrical roles, so that $\check{f}$ is the inverse of f, and at the same time f is the inverse of $\check{f}$. In particular, if we are given one of these mappings, the other is completely determined by (80).

If f is univalent on a set M, it necessarily maps M invertibly onto the set $N = \text{Range}(f)$. We will often say that a function f is **invertible on a set M** if it is univalent on M; if no second set N is specified, it will be understood that f is being regarded as a mapping from M to $N = \text{Range}(f)$.

Example 2.20 The mapping $f: \mathbf{C} \to \mathbf{C}$ given by $w = f(z) = az + b$ ($a \neq 0$) is invertible, and the inverse is given by $z = \check{f}(w) = (1/a)(w - b)$.

Example 2.21 Let $M = \mathbf{C}$ and $N = \mathbf{C}$ and let $w = f(z) = z^2$. This mapping $f: M \to N$ is onto, but it is not univalent on M since $f(z) = f(-z)$, even though $z \neq -z$ if $z \neq 0$. Thus f is not an invertible mapping $f: M \to N$ and it is not possible to define an inverse mapping $\check{f}: N \to M$; in fact, if w is in N, solving the equation $w = f(z) = z^2$ for z in terms of w does not lead to a unique solution z in M.

Let us change the situation a little; take $M = \{z: \text{Re}(z) > 0\}$, and N equal to the cut plane obtained by deleting the negative real axis from **C**. It is not hard to see that $w = f(z) = z^2$ maps M *onto* N, and because of the way we chose M it is no longer possible to find different points in M with the same image under f. Thus $f: M \to N$ is invertible; to get the inverse mapping $\check{f}: N \to M$ we must solve $z^2 = w$ so that z is in M. Obviously, z will be one of the square roots of w, so that

$$z = \check{f}(w) = \pm w^{1/2} \quad (w^{1/2} \text{ is the principal determination});$$

but $z = \check{f}(w)$ has to be in the *right* half plane M for w in N. The principal determination $w^{1/2}$ ends up in the right half plane, while $-w^{1/2}$ is in the left half plane, and we are obliged to take

$$\check{f}(w) = +w^{1/2} \quad (w^{1/2} \text{ the principal determination of square root})$$

for all w in N. We would be forced to exactly the same conclusion about the form of $\check{f}$ by inspecting equations (80), which characterize the inverse function.

If we choose M and N suitably (M a wedge of angle $2\pi/n$, symmetric about the positive real axis, and N the cut plane), the function $f: M \to N$ given by $w = f(z) = z^n$ is invertible and the inverse is the principal determination of the nth root function, $z = \check{f}(w) = w^{1/n}$ for all w in N.

Example 2.22 The exponential function $w = e^z$ is not univalent on $E = \mathbf{C}$, due to the periodicity $\exp(z) = \exp(z + 2\pi ki)$. We leave it to the reader (Exercise 3) to verify that

(i) The exponential mapping $w = e^z$ becomes univalent if we restrict its domain of definition to the horizontal strip $M = \{z: -\pi < \text{Im}(z) < +\pi\}$ (or any other horizontal strip of width 2π).
(ii) The exponential maps the strip M one-to-one onto the cut plane P obtained by deleting the negative real axis from the w-plane.
(iii) The inverse is given by $z = \check{f}(w) = \text{Log } w$, the principal determination of $\log w$.

EXERCISES

1. Prove that $w = f(z) = 4/(z - i)$ maps the domain $E = \{z: z \neq i\}$ one-to-one onto the domain $F = \{w: w \neq 0\}$, and that $z = \check{f}(w) = (4/w) + i = (4 + iw)/w$ is the inverse mapping.

2. Consider $w = f(z) = e^z$ as a mapping of the strip $M = \{z: -\pi < \text{Im}(z) < \pi\}$ into the w-plane. Prove that M is mapped invertibly onto the cut plane P, and that the inverse is $z = \check{f}(w) = \text{Log } w$.

What happens if, instead, we take $w = e^z$ on the strip $M^* = \{z : \pi < \text{Im}(z) < 3\pi\}$?

Hint: Use formula (47) to see that f is invertible on M or M^*.

Answer: The inverse $\check{f}: P \to M^*$ is $z = \check{f}(w) = \text{Log } w + 2\pi i$, another determination of log w.

3. How is the half plane $E = \{z : \text{Re}(z) < -1\}$ mapped by $w = f(z) = 1 + z^2$? Determine the image set $F = f(E)$ explicitly, and show that $f: E \to F$ is univalent. Prove that $z = \check{f}(w) = (-1)(w - 1)^{1/2}$ (principal determination).

Hint: Examine the mapping on the boundary line $\text{Re}(z) = -1$.

4. Consider $f(z) = \dfrac{1}{2}\left(z + \dfrac{1}{z}\right) = (z^2 + 1)/2z$ on the domain $D = \{z : |z| > 1\}$. Prove that f is univalent on D.

Hint: If $f(z) = f(w)$, show that $-zw(w - z) = (w - z)$. This happens only if $w = z$ or $wz = -1$; the latter is impossible if z and w are in D (why?).

5. Show that $f(z) = \sin z$ is univalent on the disc $D = \{z : |z| < 1\}$.

Hint: Let $Z = e^{iz}$ for z in D. If $f(z) = f(w)$, then $-ZW(W - Z) = (W - Z)$, which can only happen if $W = Z$, or $WZ = -1$. For z and w in D, $WZ = -1$ is impossible, and $W = Z$ holds only if $z = w$.

Note: The same methods actually show that sin z is univalent on the the disc $|z| < \pi/2$.

2.13 THREE THEOREMS ON INVERSE FUNCTIONS

The mapping properties of a holomorphic function $w = f(z)$ become degenerate at points where $df/dz = 0$, the **critical points** of f, as we will explain in Chapter 4; other points will be referred to as **regular** points.

Definition 2.9 *A function $w = f(z)$ is* **regular** *on an open set E provided that it is holomorphic and $f'(z)$ is non-zero for all z in E.*

The exponential mapping $w = e^z$ has $dw/dz = e^z \neq 0$ throughout $E = \mathbf{C}$, and so is a regular mapping of the plane. The trigonometric function $w = \sin z$ is holomorphic on the plane, but is only regular on sets that exclude the critical points $z_n = (\pi/2) + n\pi$.

The results of this section apply to regular functions $w = f(z)$ that are also *smooth* (in the sense of Definition 2.8), so that f is continuously differentiable. The theorems below all emerge as corollaries of a single line of reasoning, which analyzes the behavior of a smooth holomorphic function $w = f(z)$ near a particular point $z = p$ where $f'(p) \neq 0$. However, this discussion involves detailed use of the method of successive approximations, and would be rather technical for many students. We will not give the proof here; moreover, its inclusion in an introductory text is not necessary. We will never have to refer to details of the proof, and in specific examples the use of these theorems could be avoided by direct (but perhaps complicated) calculations. The theoretically oriented reader will find a good, self-contained account of these results, and of the successive approximations procedure required for their proof, in Hille [9], Sections 4.5 and 4.6. For an alternative treatment of these theorems as part of advanced calculus, without reference to complex variables, see Buck [2], Sections 5.6 and 5.7.

Theorem 2.21 (Open mapping theorem) *Let f be a smooth, holomorphic function defined on an open set M in the z-plane. Assume that df/dz never vanishes on M. Then the image set $N = f(M) = \{w : w = f(z)$ for some z in $M\}$ is an open set in the w-plane.*

Thus, a smooth regular mapping can not "squash" an open set into a more degenerate kind of set, such as an arc or a set of isolated points. If we are obliged to determine the images of sets when they are transformed from the z-plane to the w-plane (a problem we will encounter frequently), it can be very helpful to know that the image of an open set is open.

Theorem 2.22 (Inverse mapping theorem) *Let f be a smooth, holomorphic function defined on an open set M in the z-plane. Assume that df/dz never vanishes on M. Then f is "locally invertible"; that is, if p is any point in M, there is a small disc D about p on which f is invertible.*

Since f is smooth and regular, any small disc D about p is mapped to an open set in the w-plane, by the open mapping theorem; the point of Theorem 2.22 is that f is univalent if the disc D is made small enough. Once we know that f is univalent on a set D, it must be univalent on any smaller subset of D. The map f is invertible on a small disc about each point in M, but the size of the disc may depend on the location of the point p within M.

Example 2.23 If $M = \mathbf{C}$ and $w = f(z) = e^z$, then $df/dz = e^z$ is never zero and f is a smooth, regular mapping. By referring to formula (47), we see that f is univalent on the disc $D_p = \{z : |z - p| < \pi\}$ about a typical point p; in fact, f is univalent on the horizontal strip $\text{Im}(p) - \pi < \text{Im}(z) < \text{Im}(p) + \pi$, which contains D_p. It is not univalent on any larger disc $|z - p| < r$ centered at p, since $e^{2\pi i + z} = e^z$. Thus, f is not an invertible mapping on M, although it is "locally invertible." The "locally defined" inverse mapping $z = \check{f}(w)$ maps the image set $E_p = f(D_p)$ back to the disc D_p; it is a determination of $z = \log w$ (defined on the set E_p in the w-plane), because $w = f(z) = f(\check{f}(w)) = \exp(\check{f}(w))$.

The open mapping theorem insures that the image set $E_p = f(D_p)$ is an open set in the w-plane. Using direct calculations based on the definition of

$w = e^z$, it would be difficult to calculate the precise shape of E_p, and to verify that E_p is an open set.

Our final result deals with the differentiability of the inverse of an invertible regular mapping.

Theorem 2.23 (Differentiability of inverse functions) *Let f be a smooth, holomorphic mapping defined on an open set M in the z-plane. Assume that $df/dz \neq 0$ on M, and that f is an invertible mapping of M onto the image set $N =$ Range $(f) = f(M)$ in the w-plane. Then the inverse mapping $\check{f}: N \to M$ is holomorphic on N. Furthermore, its derivative is given by*

(81A) $$(\check{f})'(f(z)) = \frac{1}{f'(z)} \qquad \left(\text{i.e., } \frac{d\check{f}}{dw}(f(z)) = \frac{1}{\dfrac{df}{dz}(z)}\right)$$

or all z in M, which is the same as saying that

(81B) $$(\check{f})'(w) = \frac{1}{f'(\check{f}(w))} \qquad \left(\text{i.e., } \frac{d\check{f}}{dw}(w) = \frac{1}{\dfrac{df}{dz}(\check{f}(w))}\right)$$

for all w in N.

The set N is open, by the open mapping theorem; thus, $z = \check{f}(w)$ is defined near each point q in N, and the difference quotients in the formula $(\check{f})'(q) = \lim\limits_{w \to q} (\check{f}(w) - \check{f}(q))/(w - q)$ are well defined for all w near q. Since the reciprocal of df/dz appears on the right side of formula (81), this formula would not make sense if f were not regular. If we start with formula (81A), then (81B) follows by writing $z = \check{f}(w)$ everywhere z appears in (81A); similarly, we can obtain (81A) from (81B). The continuity of $d\check{f}/dw$ is clear from formula (81B), which expresses this function as the reciprocal of a continuous function df/dz that is never zero. It is also clear that $d\check{f}/dw$ is non-vanishing on N; so $\check{f}: N \to M$ is a smooth, regular mapping.

Corollary 2.24 *If $f: M \to N$ is a smooth, regular mapping between open sets that is invertible, then the inverse $\check{f}: N \to M$ is also a smooth, regular mapping.*

Formulas like (81A) and (81B) are derived in calculus for functions of a real variable, and are used to investigate the derivatives of pairs of inverse functions such as e^x and $\log x$, or $\tan x$ and $\arctan x$. However, the proofs given in calculus lean heavily on the order properties of the real number system (by way of the mean value theorem for derivatives), and cannot be adapted to work for functions of a complex variable.

The hard part in the proof of Theorem 2.23 is to establish that the inverse function $\check{f}$ is actually differentiable. Once we are assured of this, we may apply the chain rule to the equation (a composite of differentiable functions)

$$z = \check{f}(f(z)) \quad \text{for all } z \text{ in } M;$$

this "implicit differentiation" gives

$$1 = (\check{f})'(f(z)) \cdot f'(z) = \frac{d\check{f}}{dw}(f(z)) \cdot \frac{df}{dz}(z).$$

Since we assume that $f'(z) \neq 0$, we get (81A), and (81B) follows from this.

Example 2.24 The transformation $w = \tan z$ is not univalent on the full domain $\{z : z \neq ((\pi/2) + n\pi) + i0 \text{ for } n = 0, \pm 1, \pm 2, \ldots\}$ on which it is defined, because it is periodic with $\tan(z) = \tan(z + n\pi)$. There is some hope of obtaining an univalent function if we restrict the domain to a vertical strip of width π, such as $M = \{z : -\pi/2 < \text{Re}(z) < \pi/2\}$. Now $w = \tan z$ is one-to-one on the domain M (see Exercise 3); later on (in Chapter 4), we will show that $w = \tan z$ maps M *onto* the (open set) domain N in the w-plane obtained by removing the two parts of the imaginary axis that lie outside of the unit disc $|w| < 1$; N is a plane with *two* cuts in it, running in opposite directions along the imaginary axis from the points $+i$ and $-i$ out to infinity. Naturally the inverse function $\check{f}: N \to M$ is to be regarded as a determination of $\arctan(w)$. For real numbers $-\infty < u < +\infty$, the familiar function of a real variable $x = \arctan u$ satisfies the conditions (i) $u = \tan x$, and (ii) $-\pi/2 < x < +\pi/2$, so that $z = x + i0$ is in the strip M; we conclude that $\arctan w$ agrees with the familiar arctangent function when w is real:

(82) $$\arctan(u + i0) = \arctan u \quad \text{for all real } u \text{ such that } -\infty < u < +\infty.$$

The particular determination just defined on N will be regarded as the **principal determination of arctangent** and will be indicated by Arctan w hereafter.

Since $d/dz(\tan z) = \sec^2 z = 1/\cos^2 z$ is never zero in M, Theorem 2.23 assures us that $z = \text{Arctan}\, w$ is continuously differentiable throughout N, and gives us the familiar differentiation formula

(83) $$\begin{aligned}\frac{d}{dw}(\text{Arctan}\, w) &= \left[\left.\frac{1}{\dfrac{d}{dz}(\tan z)}\right|_{z=\text{Arctan}\, w}\right] = \left[\left.\frac{1}{\sec^2 z}\right|_{z=\text{Arctan}\, w}\right] \\ &= \left[\left.\frac{1}{1 + \tan^2 z}\right|_{z=\text{Arctan}\, w}\right] = \frac{1}{1 + w^2},\end{aligned}$$

now valid for complex variable, in the doubly cut plane N.

Note: The *inverse* $\check{f}$ of a function f should not be confused with its *reciprocal* $1/f$. For example, if $f(z) = z + 1$ is taken as a mapping $f: \mathbf{C} \to \mathbf{C}$, then $\check{f}(z) = z - 1$, while the reciprocal is $(1/f)(z) = 1/(z + 1)$. Some authors use the symbol f^{-1} to indicate the inverse of f, but this can be confused with the notation for reciprocals (as when we write $z^{-1} = 1/z$), so we shall favor the notation $\check{f}$.

EXERCISES

1. With Figure 2.17 in mind, explain what is meant by expressing $(\check{f})' = d\check{f}/dw$, whose variable is w, as a function of variable z via the formula

$$\frac{d\check{f}}{dw}(z) = \left[\frac{d\check{f}}{dw}\bigg|_{w=f(z)}\right].$$

2. Calculate $d\check{f}/dw$ as a function of z in the following situations. (You need not verify that f is invertible on the domain M indicated).

(i) $f(z) = \text{ctn } z$ on $M = \{z: 0 < \text{Re}(z) < \pi\}$

(ii) $f(z) = \frac{1}{2}\left(z + \frac{1}{z}\right)$ on $M = \{z: |z| > 1\}$.

Note: Sometimes it is difficult to express $d\check{f}/dw$ as a function of w, especially when we cannot calculate $z = \check{f}(w)$ explicitly. Notice how this calculation was avoided in Example 2.24.

3. Show that $w = \tan z$ is a univalent mapping on the strip $M = \{z: -\pi/2 < \text{Re}(z) < +\pi/2\}$. Show that f is regular and holomorphic on M, so that the image set $f(M)$ is an open set in the w-plane.

Hint: Write $\tan z = -i(e^{2iz} - 1)/(e^{2iz} + 1)$. If $f(z') = f(z'')$, set $Z' = e^{2iz'}$ and $Z'' = e^{2iz''}$, and show that $Z' = Z''$. Then verify that $z' = z''$ if both points are in M (use formula (36)).

4. If w is fixed ($w \neq \pm i$), show that all solutions of $w = \tan z$ are given by

$$z = \frac{1}{2i}\text{Log}\left(\frac{1 + iw}{1 - iw}\right) + n\pi \qquad (n = 0, \pm 1, \pm 2, \ldots).$$

These are the values of the various determinations of arctan w. Which one is the principal determination Arctan w? Prove that tan z never takes the values $+i$ or $-i$, for any z.

Hint: Write $Z = e^{iz}$ and $w = -i(Z^2 - 1)/(Z^2 + 1)$; then solve for Z by converting this into a quadratic equation in Z. Use formula (36).

Answer: $n = 0$ gives Arctan w.

5. Prove that $w = f(z) = \sinh(z)$ is univalent on the strip $M = \{z: -\pi/2 < \text{Im}(z) < +\pi/2\}$. Without trying to determine the precise shape of the image domain $N = f(M)$, show that $\check{f}: N \to M$ is differentiable on N and calculate $d\check{f}/dw$ as a function of w.

Answer: $d\check{f}/dw = 1/\sqrt{1 + w^2} = (1 + w^2)^{-1/2}$ (principal determination).

6. Calculate all possible values of the following inverse functions $z = \check{f}(w)$ at the points indicated; that is, find all solutions z of the equation $w = f(z)$.

(i) $\arcsin(1)$ (ii) $\arctan(1)$ (iii) $\arctan(2i)$ (iv) $\operatorname{arctanh}(0)$ (v) $\operatorname{arcsinh}(+1)$ (vi) $\arctan(1 + i)$.

2.14 SMOOTHNESS OF HOLOMORPHIC FUNCTIONS

In certain sections of this chapter we have imposed smoothness conditions on the holomorphic functions $f(z) = U(x,y) + iV(x,y)$ being considered, such as:

(i) The first partials $\partial U/\partial x, \ldots, \partial V/\partial y$ exist and are continuous.

(ii) The second partials $\partial^2 U/\partial x^2, \ldots, \partial^2 V/\partial y^2$ exist and are continuous.

For functions of a real variable, existence of the first derivative df/dx does not insure its continuity (Exercise 1), and even if the derivative is continuous, there is no assurance that higher derivatives will exist; similar remarks apply to the partial derivatives of a function of several real variables.

In Chapter 5 we will be able to show that any function $f(x + iy) = U(x,y) + iV(x,y)$ that is complex differentiable (holomorphic) on an open set E must actually have component functions $U(x,y)$ and $V(x,y)$ whose partial derivatives of *all* orders exist and are continuous. Once this has been established, there is no need to make any special hypotheses about the smoothness of U and V. All reference to conditions (*i*) and (*ii*) may be dropped from the theorems presented above—complex differentiability of f, without further assumptions, is enough.

EXERCISE

1. Define $f(x)$ for real x so that

$$f(x) = x^2 \sin(1/x) \quad \text{if } x \neq 0; \ f(0) = 0.$$

Show that f is differentiable at and near $x = 0$, but that the derivative $f'(x)$ is discontinuous at $x = 0$. Does f'' exist at $x = 0$?

3 POWER SERIES AND ANALYTIC FUNCTIONS

The Taylor series associated with a function of a real variable $f(x)$ and a base point x_0 on the real line is given by

$$\text{(1)}\quad \sum_{n=0}^{\infty} \frac{f^{(n)}(x_0)}{n!}(x-x_0)^n = f(x_0) + f'(x_0)(x-x_0) + \frac{f''(x_0)}{2!}(x-x_0)^2 + \cdots .$$

For example e^x, $\sin x$, and $\cos x$ have the familiar series expansions

$$\begin{aligned} e^x &= \sum_{n=0}^{\infty} \frac{x^n}{n!} = 1 + \frac{x}{1!} + \frac{x^2}{2!} + \frac{x^3}{3!} + \cdots \\ \sin x &= \sum_{n=0}^{\infty} (-1)^n \frac{x^{2n+1}}{(2n+1)!} = x - \frac{x^3}{3!} + \frac{x^5}{5!} - \cdots \\ \cos x &= \sum_{n=0}^{\infty} (-1)^n \frac{x^{2n}}{(2n)!} = 1 - \frac{x^2}{2!} + \frac{x^4}{4!} - \cdots , \end{aligned} \tag{2}$$

valid for all real x. There is also the geometric series

$$\sum_{n=0}^{\infty} x^n = 1 + x + x^2 + \cdots = \frac{1}{1-x} \tag{3}$$

which converges for all x such that $-1 < x < +1$, but diverges otherwise. Our purpose in this chapter is to develop the theory of power series using complex, rather than real, variables. We will examine the unique properties of power series and analytic functions, such as the analytic continuation principle, which appear only when complex variables have been introduced.

There are many reasons for using complex variables in power series. For one thing, as we have already noted, the complex variable analogs of functions like *exp*, *sin*, and *cos* are represented by allowing the complex variable z in the Taylor series, keeping the same (real) coefficients. Moreover, fundamental relationships between these functions are revealed, such as

$$\cos(z) = \frac{e^{iz} + e^{-iz}}{2}; \qquad e^{z} = \cos(iz) - i\sin(iz); \qquad \sinh(iz) = i\sin(z),$$

whose existence would never be suspected if we restricted our view to functions of a real variable. Another benefit is that the use of complex variables allows us to understand the convergence properties of powers series, which might seem strange if we considered only real variables; in particular, the fact that every Taylor series has a "radius of convergence" about its base point x_0 is best understood by introducing complex variables. Finally, since power series are very useful in solving differential equations (for a real variable), complex variables can be introduced into both the differential equation and the series used to solve it. This often reveals surprising relationships between the various families of solutions.

We begin with a few remarks on the notion of convergence of a sequence (or series) of functions to a limit function. Our discussion is arranged so that it applies equally well to both real and complex variables.† Before very long we will restrict our attention to "power series," but for a while we shall take a broader view.

3.1 CONVERGENCE OF SEQUENCES OF FUNCTIONS

Let us consider a sequence $f_1, f_2, f_3, \ldots$ of functions, all defined on a subset E of the complex plane (everything we say is equally valid for functions defined on a subset of $\mathbf{R}$). There are various ways in which this sequence $\{f_n\}$ can "converge" to a limit function f defined on E; we shall mention only two types of convergence: *pointwise convergence* and *uniform convergence.*

Definition 3.1 **Pointwise convergence** *means that*

$$f(z) = \lim_{n\to\infty} f_n(z) \quad (\textit{convergence of a sequence of complex numbers})$$

for each point z in E. We write

$$f_n \to f \text{ (ptwse)} \quad \text{or} \quad f = \lim_{n\to\infty} f_n \text{(ptwse)}$$

to indicate this type of convergence.

† Discussions of convergence of sequences and series of functions in Calculus often lean heavily on the order properties of $\mathbf{R}$. For a complex variable there is no order relation to work with, and more care is required in this context.

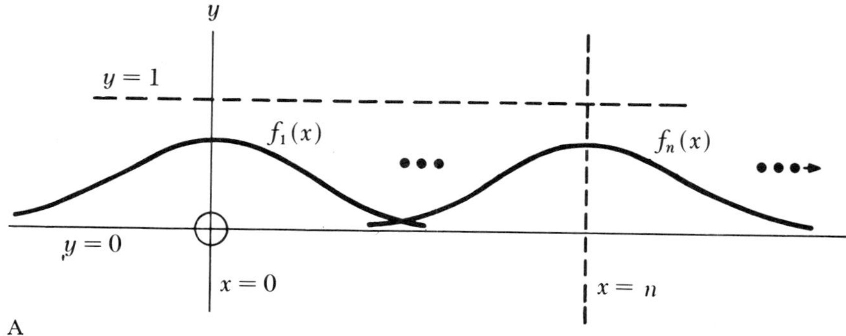

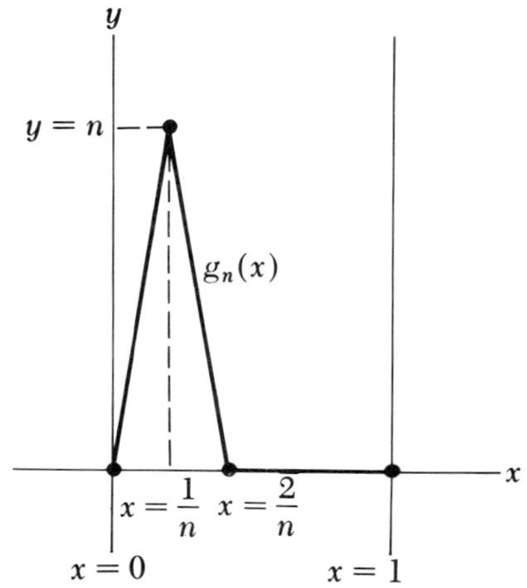

Figure 3.1 Pointwise convergent sequences of functions.

Example 3.1 Here are two examples of pointwise convergence, illustrated with functions of a real variable so that we can draw graphs to help us understand what is going on. The sequence of functions $\{f_n\}$ is defined by taking

$$f_n(x) = \frac{1}{\sqrt{\pi}} e^{-(x-n)^2} \quad \text{for} \quad -\infty < x < +\infty,$$

for $n = 0, 1, 2, \ldots$, as in Figure 3.1A. The area under their graphs moves off toward infinity as n increases; here we are taking $E = \mathbf{R}$. Another sequence of functions $g_n(x)$, defined on the interval $E = [0, 1]$, is shown in Figure 3.1B. Both sequences are pointwise convergent on their respective domains of definition:

$$f = \lim_{n\to\infty} f_n \text{ (ptwse)} \qquad g = \lim_{n\to\infty} g_n \text{ (ptwse)} ,$$

where f and g are the constant functions: $f(x) = 0$ and $g(x) = 0$ for all x.

From these examples one can see that the rate at which the numbers $f_n(x)$ approach their limit $f(x)$ may vary considerably from point to point in E; in other words, the speed of convergence to the limit function f is not "uniform" throughout the set E.

If the functions $\{f_n\}$ converge pointwise (or in some other sense) to the limit function $f = \lim_{n\to\infty} f_n$, we often wish to apply limit operations such as

$$(4) \qquad \frac{d}{dz}(\cdots); \qquad \int_a^b (\cdots)\,dx; \qquad \lim_{z\to p}(\cdots)$$

to f; that is, we wish to calculate "double limits" such as

$$(5) \qquad \begin{aligned} \frac{d}{dz}(f) &= \frac{d}{dz}\left(\lim_{n\to\infty} f_n\right) \\ \int_a^b f(x)\,dx &= \int_a^b \left(\lim_{n\to\infty} f_n\right) dx \\ \lim_{z\to p} f &= \lim_{z\to p}\left(\lim_{n\to\infty} f_n\right) \end{aligned}$$

in which the operation $\lim_{n\to\infty}(\ldots)$ is followed by one of the operations in (4). In practical situations it is usually much easier to evaluate the result if these operations are carried out in the reverse order. Applying the operations in (4) to each term f_n and then calculating $\lim_{n\to\infty}(\ldots)$, we get

$$(6) \qquad \begin{aligned} &\lim_{n\to\infty}\left(\frac{df_n}{dz}\right) \\ &\lim_{n\to\infty}\left(\int_a^b f_n(x)\,dx\right) \\ &\lim_{n\to\infty}\left(\lim_{z\to p} f_n\right). \end{aligned}$$

Is this "interchange of operations" allowable? That is, if $\{f_n\}$ converges, do the quantities calculated in (5) and (6) agree? Unfortunately, if $\{f_n\}$ only converges *pointwise* to its limit, these operations generally cannot be interchanged without seriously affecting the result of the calculation. For example, consider the sequence $\{g_n\}$ of functions whose graphs are shown in Figure 3.1B. Then $\int_0^1 g_n(x)\,dx = 1$ for $n = 1, 2, \ldots$, so that $\lim_{n\to\infty}\left(\int_0^1 g_n(x)\,dx\right) = 1$. On the other hand, the limit function $g(x) = \lim_{n\to\infty} g_n(x) = 0$ on $[0, 1]$, so that

$$\int_0^1 \left(\lim_{n\to\infty} g_n\right) dx = \int_0^1 0\,dx = 0 \neq 1 = \lim_{n\to\infty}\left(\int_0^1 g_n(x)\,dx\right).$$

Thus, we do not get the same result on performing the operations $\lim_{n\to\infty}(\cdots)$ and $\int_0^1(\cdots)\,dx$ in reverse order. Similarly, we cannot interchange the order of the operations $\lim_{n\to\infty}(\cdots)$ and $\int_{-\infty}^{+\infty}(\cdots)\,dx$ for the sequence $\{f_n\}$ shown in Figure 3.1A, because $\int_{-\infty}^{+\infty} f_n(x)\,dx = 1$ for all n, and

$$\lim_{n\to\infty}\left(\int_{-\infty}^{+\infty} f_n(x)\,dx\right) = 1 \neq 0 = \int_{-\infty}^{+\infty}\left(\lim_{n\to\infty} f_n\right)dx = \int_{-\infty}^{+\infty} 0\,dx.$$

These difficulties can often be traced to the nonuniform rate of convergence of the terms f_n to the limit function.

Definition 3.2 *A sequence of functions* $\{f_n\}$ **converges uniformly on E** *to a limit function* f *defined on* E *if: for any number* $\varepsilon > 0$, *we get*

$$|f_n(z) - f(z)| < \varepsilon \quad \textit{for all } z \textit{ in } E \tag{7}$$

once the index n *is beyond some integer* $N(\varepsilon)$ *(so that the formula* (7) *is valid for all large* n*). If this happens, we write*

$$f_n \to f \text{ (unif)} \quad \textit{or} \quad f = \lim_{n\to\infty} f_n \text{ (unif)}.$$

This definition requires that the values $f_n(z)$ approach the limit value $f(z)$ at a uniform rate, for all points z in E, as $n \to \infty$. How far out in the set of indices $n = 1, 2, \ldots,$ we must go in order to get the distance between the values of f_n and the values of f uniformly less than a given number $\varepsilon > 0$ will naturally depend on the particular number $\varepsilon > 0$ being considered in (7); we write $N = N(\varepsilon)$ to indicate this dependence of N on ε.

It is obvious that $f = \lim_{n\to\infty} f_n$(ptwse) if we have $f = \lim_{n\to\infty} f_n$ (unif) on a set E. Furthermore, if $f_n \to f$ (unif) on E, then certainly $f_n \to f$ (unif) on any smaller set $D \subseteq E$. Another basic fact is:

Theorem 3.1 *Let* $z = \phi(w)$ *be a mapping from a set* F *into a set* E *in the plane, and assume that* f *and* $f_1, f_2, \ldots,$ *are functions defined on* E *such that* $f_n \to f$ (unif) *on* E. *Then the composite functions* $g_n(w) = f_n(\phi(w)) = (f_n \circ \phi)(w)$ *converge uniformly to* $g = f \circ \phi$ *on the set* F.

PROOF: If $\varepsilon > 0$ is given, then $|f_n(z) - f(z)| < \varepsilon$ throughout E, for all sufficiently large n, say $n \geq N(\varepsilon)$. For each w in F, the point $z = \phi(w)$ is in E so that

$$|g_n(w) - g(w)| = |f_n(\phi(w)) - f(\phi(w))| < \varepsilon \quad \text{for all } w \text{ in } F,$$

and for each $n \geq N(\varepsilon)$. Now compare this last statement with the definition of uniform convergence on F. ▌

As a general principle, the operation $\lim_{n\to\infty}(\cdots)$ for *uniform limits* of functions may usually be interchanged with other limit operations; this is our reason for introducing the notion of uniform convergence. Here is one result which will be useful in later discussions.

Theorem 3.2 *Assume that the functions $\{f_n\}$ are defined on a subset E in the plane and converge uniformly on E to a limit function f.*

(*i*) *If the functions f_n are each continuous at a point p in E, then f is also continuous at p.*

(*ii*) *If the f_n are each continuous throughout E, so is their uniform limit f.*

Note: Statement (ii) follows directly from statement (i). The latter is really a statement about interchanging the operations $\lim_{n\to\infty}(\cdots)$ and $\lim_{z\to p}(\cdots)$. Continuity at p means that

$$f(p) = \lim_{z\to p} f = \lim_{z\to p}\left(\lim_{n\to\infty} f_n\right). \tag{8}$$

On the other hand, each f_n is continuous at p, so that $\lim_{z\to p} f_n = f_n(p)$, and because $f_n \to f$ we know that $\lim_{n\to\infty} f_n(p) = f(p)$; thus it is clear that

$$\lim_{n\to\infty}\left(\lim_{z\to p} f_n\right) = \lim_{n\to\infty} f_n(p) = f(p).$$

The result (8) we really want would follow immediately if we could interchange the limit operations involved.

Proof of Theorem 3.2: If a number $\varepsilon > 0$ is given, we get

$$|f_n(z) - f(z)| < \varepsilon/3 \quad \text{for all } z \text{ in } E, \tag{9}$$

for all sufficiently large n (say $n \geq N$), due to the uniform convergence $f_n \to f$ on E. We must show that $f(p) = \lim_{z\to p} f$ to establish continuity of f at p. Consider z near p; then by adding and subtracting $f_N(z)$ and $f_N(p)$ we get

$$|f(z) - f(p)| \leq |f(z) - f_N(z)| + |f_N(z) - f_N(p)| + |f_N(p) - f(p)|$$

$$< \frac{\varepsilon}{3} + |f_N(z) - f_N(p)| + \frac{\varepsilon}{3}$$

for all z in E (take N as in (9)). But f_N is continuous at p, so that $|f_N(z) - f_N(p)| < \varepsilon/3$ for all z sufficiently close to p; thus

$$|f(z) - f(p)| < \frac{2}{3}\varepsilon + |f_N(z) - f_N(p)| < \varepsilon$$

for all z sufficiently close to p, say $|z - p| < \delta$. This proves that $f(p) = \lim_{z\to p} f$, as required. ∎

If the functions $\{f_n\}$ only converge *pointwise* on E to the limit function f, continuity of the "approximating functions" $f_1, f_2, \ldots,$ is not always inherited by the limit function f; *uniform* convergence insures that continuity is passed on to the limit function. For other discussion of "interchange of limits," see Rudin [20], Chapter 7.

Note: Every general principle has its exception, and we hasten to point out that (at least for functions of a *real* variable) the operation $\dfrac{d}{dx}(\cdots)$ presents special difficulties when we try to interchange it with a limit of functions $\lim_{n\to\infty}(\cdots)$, even if $f_n \to f$ uniformly. See Exercise 8 of Section 3.2.

Example 3.2 The functions

$$f_n(z) = \frac{1 - z^n}{1 - z} = 1 + z + z^2 + \cdots + z^{n-1}$$

$$f(z) = \frac{1}{1 - z}$$

are well defined on the open disc $E = \{z:|z| < 1\}$ and also on the smaller closed discs $D_r = \{z:|z| \leq r\}$ with radius r *less than one*.† In our earlier remarks on the geometric series we saw that $f = \lim_{n\to\infty} f_n$ (ptwse) on E. Actually,

(10) $\qquad f_n \to f$ (unif) on any closed disc $|z| \leq r$ such that $r < 1$;

however, the sequence $\{f_n\}$ does not converge uniformly to f on the whole disc E (see Exercise 5). To illustrate the handling of a uniform convergence problem, we prove (10). Fixing our attention on the behavior of the functions f_n on a given disc D_r, we notice that

(11) $$|f_n(z) - f(z)| = \frac{|z^n|}{|1 - z|} = \frac{|z|^n}{|1 - z|} \leq \frac{|z|^n}{1 - |z|} \quad \text{for all } z \text{ in } D_r,$$

because the triangle inequality insures that $|1 - z| \geq |1 - |z|\,| = 1 - |z|$ if $|z| < 1$. The expression $|z|^n/(1 - |z|)$ achieves its maximum value on D_r at points where $|z| = r$, so that

(12) $$|f_n(z) - f(z)| \leq \frac{r^n}{1 - r} \quad \text{for all } z \text{ in } D_r.$$

Because $r < 1$, r^n approaches zero as $n \to \infty$, so that $r^n/(1 - r) \to 0$. Thus, if some number $\varepsilon > 0$ is given, there is a corresponding integer $N(\varepsilon)$ such that $0 \leq r^n/(1 - r) < \varepsilon$ for all $n \geq N(\varepsilon)$. For these indices we get

$$|f_n(z) - f(z)| \leq \frac{r^n}{1 - r} < \varepsilon \quad \text{for all } z \text{ in } D_r,$$

and since this reasoning works for any $\varepsilon > 0$, we have proved (10).

---

† Notice that $f_n(z) = 1 + z + \cdots + z^{n-1}$ is just the nth partial sum of the geometric series $1 + z + z^2 + \cdots$.

3.2 SERIES OF FUNCTIONS

Let $f_1, f_2, f_3, \ldots$, be functions defined on a subset E in **C**. To form, or even define, the "sum" of the infinite series $\sum_{n=1}^{\infty} f_n = f_1 + f_2 + \cdots$, we must face the same sort of questions we faced in defining the "sum" of an infinite series of complex numbers. Again, we form the partial sums, which are now functions defined on E,

$$\begin{aligned} s_1(z) &= f_1(z) \\ s_2(z) &= f_1(z) + f_2(z) \\ &\vdots \\ s_n(z) &= f_1(z) + f_2(z) + \cdots + f_n(z) = \sum_{k=1}^{n} f_k(z) \\ &\vdots \end{aligned}$$

and examine the limit behavior of the sequence of functions $s_1, s_2, \ldots$. A new feature now enters the situation—there are various ways in which the sequence of functions $\{s_n\}$ may converge to a limit. If $\{s_n\}$ converges pointwise (or uniformly) to a limit function f, we say that the series $\sum_{n=1}^{\infty} f_n$ **converges pointwise** (or **uniformly**) to f on E, and write

$$f = \sum_{n=1}^{\infty} f_n \text{ (ptwse)} \quad \text{or} \quad f = \sum_{n=1}^{\infty} f_n \text{ (unif)}$$

If the type of convergence is clear from context, we write $f = \sum_{n=1}^{\infty} f_n$.

For *series* of functions there is a very simple, and useful, test for uniform convergence.

Theorem 3.3 (Weierstrass Test) *Let $\sum_{n=1}^{\infty} f_n$ be a series of functions defined on a set E in* **C**, *and suppose we can find non-negative numbers $0 \le M_n < +\infty$ for $n = 1, 2, \ldots$ such that*

(i) $|f_n(z)| \le M_n$ *for all z in E, and all $n = 1, 2, \ldots$*

(ii) $\sum_{n=1}^{\infty} M_n < +\infty$.

Then the series $\sum_{n=1}^{\infty} f_n$ converges uniformly on E to some limit function f.

Theorem 3.3 will be our primary tool for verifying uniform convergence from now on (however, see Exercise 11). Notice that it allows us to decide whether the series is uniformly convergent, without actually evaluating the sum $f = \sum_{n=1}^{\infty} f_n$ of the series.

PROOF OF THEOREM 3.3: If z is a typical point in E, then $\sum_{n=1}^{\infty} |f_n(z)| \leq \sum_{n=1}^{\infty} M_n < +\infty$, so the series of complex numbers $\sum_{n=1}^{\infty} f_n(z)$ is absolutely convergent for each z. In particular, the series *converges* (Theorem 2.4); let us write $f(z)$ for its sum. Then the partial sums of the series of functions $\sum_{n=1}^{\infty} f_n$ obviously converge pointwise to f. We must show that this convergence is actually *uniform* on E.

For any convergent series of complex numbers $\sum_{n=1}^{\infty} a_n$, we know that $\left|\sum_{n=1}^{\infty} a_n\right| \leq \sum_{n=1}^{\infty} |a_n|$ (Exercise 8 of Section 2.2). For the nth partial sum $s_n(z) = f_1(z) + \cdots + f_n(z)$ we conclude that

$$|f(z) - s_n(z)| = \left|\sum_{k=1}^{\infty} f_k(z) - \sum_{k=1}^{n} f_k(z)\right| = \left|\sum_{k=n+1}^{\infty} f_k(z)\right| \leq \sum_{k=n+1}^{\infty} M_k, \tag{13}$$

for all z in E, and all $n = 1, 2, \ldots$. Since the series $\sum_{k=1}^{\infty} M_k$ converges, its "tail end" $R_n = \sum_{k=n+1}^{\infty} M_k$ goes to zero as $n \to \infty$ (Exercise 9 of Section 2.2); $\lim_{n\to\infty} R_n = 0$.

If a number $\varepsilon > 0$ is given, then $0 \leq R_n < \varepsilon$ for all large n, say for $n \geq N(\varepsilon)$. But then formula (13) insures that $0 \leq |f(z) - s_n(z)| \leq R_n < \varepsilon$ for z in E, so that

$$0 \leq |f(z) - s_n(z)| < \varepsilon \qquad \text{for all } z \text{ in } E,$$

for each integer n such that $n \geq N(\varepsilon)$. This is precisely what we mean by uniform convergence $f = \sum_{n=1}^{\infty} f_n$ (unif) on E. ▮

Example 3.3 Consider the series

$$\sum_{n=0}^{\infty} (-1)^n z^{2n} = 1 - z^2 + z^4 - \cdots \tag{14}$$

where $f_0(z) = 1$, $f_1(z) = -z^2$, $f_2(z) = z^4, \ldots$. A simple application of the ratio test (Theorem 2.9) shows that (14) is absolutely convergent provided that $|z| < 1$, and diverges if $|z| > 1$; the series also diverges if $|z| = 1$ because the absolute values $|(-1)^n z^{2n}| = |z|^{2n} = 1$ do not tend to zero as $n \to \infty$. The series (14) is obtained from the geometric series $\sum_{n=0}^{\infty} w^n$ by substituting $w = -z^2$;

if $|z| < 1$, then $|w| = |z|^2 < 1$, so the geometric series converges:

$$\sum_{n=0}^{\infty} (-1)^n z^{2n} = \sum_{n=0}^{\infty} w^n = \frac{1}{1-w} = \frac{1}{1+z^2} \quad \text{for } |z| < 1.$$

Evidently (14) converges pointwise on the unit disc to the function $f(z) = (1 + z^2)^{-1}$. On any smaller disc of the form $D_R = \{z : |z| \leq R\}$, with radius $R < 1$, the series converges *uniformly* to this limit function; indeed, $|f_n(z)| = |z|^{2n} \leq R^{2n}$ throughout D_R, so that the constants $M_n = R^{2n}$ give

$$|f_n(z)| \leq M_n \quad \text{on} \quad D_R; \qquad \sum_{n=0}^{\infty} M_n = \sum_{n=0}^{\infty} (R^2)^n = \frac{1}{1-R^2} < +\infty.$$

Now apply the Weierstrass test to the series $\sum_{n=1}^{\infty} f_n$. The limit function $f(z) = (1 + z^2)^{-1}$ makes sense for all z in the complex plane except $z = +i, -i$, while the series (14) only converges in the disc $|z| < 1$ which has these two points on its boundary. As we shall see later in the chapter, it is the presence of these "singular points" for the function $f(z)$ which prevents the series from converging, and representing the function f, in any larger domain.

EXERCISES

1. Let $z_{mn} = m/(m + n)$ for $m \geq 1$ and $n \geq 1$ integers. Compare the limits

$$\lim_{n\to\infty} \left(\lim_{m\to\infty} z_{mn} \right) \qquad \lim_{m\to\infty} \left(\lim_{n\to\infty} z_{mn} \right)$$

2. Find all points z at which the following sequences of functions $\{f_n\}$ converge. Evaluate the limit if you can.

(i) $\dfrac{z}{z^2 + n^2}$

(ii) $\dfrac{1}{n} \sin(nz)$

(iii) $\dfrac{z}{n^2}$

(iv) ze^{-nz}

(v) $n^z = e^{z \cdot \operatorname{Log} n}$

Hint: In some cases, examine absolute values.

Answers: (i) all z; (ii) $\operatorname{Im}(z) = 0$; (iii) all z; (iv) $\{z : \operatorname{Re}(z) > 0\} \cup \{0\}$; (v) $\{z : \operatorname{Re}(z) < 0\} \cup \{0\}$.

3. Find the points where the following series of functions converge absolutely. (Exclude z if one or more terms are undefined!)

(i) $\sum_{n=1}^{\infty} \frac{(-1)^n}{z+n}$

(ii) $\sum_{n=1}^{\infty} n^p z^n$ (p a fixed integer: $p = 0, \pm 1, \pm 2, \ldots$)

(iii) $\sum_{n=1}^{\infty} \frac{z^n}{1-z^n}$

(iv) $\sum_{n=1}^{\infty} \frac{1}{n^2} \sin(nz)$

(v) $\sum_{n=1}^{\infty} z^n e^{-nz}$

(vi) $\sum_{n=0}^{\infty} \frac{1}{z^2 - n^2}$

(vii) $\sum_{n=0}^{\infty} \frac{1}{(z-n)^2}$

Answers: (i) not absolutely convergent at any z; (ii) $|z| < 1$; (iii) $|z| < 1$; (iv) $\mathrm{Im}(z) = 0$; (v) $\log |z| < \mathrm{Re}(z)$, or $|z| < \exp(\mathrm{Re}(z))$; (vi) undefined at $z = 0, \pm 1, \pm 2, \ldots$, and absolutely convergent elsewhere; (vii) undefined at $z = 0, 1, 2, \ldots$, and absolutely convergent elsewhere.

4. Prove that the following series converge uniformly on the sets indicated.

(i) $\sum_{n=1}^{\infty} \frac{z+1}{z^2+n^2}$ on $\{z: |z| \le R\}$ (assume $0 < R < +\infty$)

(ii) $\sum_{n=0}^{\infty} \frac{z^n}{z^{2n}+1}$ on $\{z: |z| \le r\}$ (assume $0 < r < 1$)

(iii) $\sum_{n=0}^{\infty} \frac{1}{(z^2-1)^n}$ on $\{z: |z| \ge 2\}$

(iv) $\sum_{n=1}^{\infty} \frac{e^{nz}}{n!}$ on $\{z: -\infty < \mathrm{Re}\, z \le +1\}$

Hint: In (iv), show that $e^n/n! \le (e^2/2) \cdot (e/3)^{n-2}$; then notice that $(e/3) < 1$.

5. Prove that $\sum_{n=0}^{\infty} z^n$ does *not* converge uniformly to its limit $f(z) = 1/(1-z)$ on the open disc of convergence $|z| < 1$.

Hint: If $\varepsilon = \frac{1}{2}$ (or less), is there *any* $n = 1, 2, \ldots,$ such that $|s_n(z) - f(z)| < \varepsilon$ throughout the disc $|z| < 1$? (For fixed n, estimate the maximum value of $|s_n(z) - f(z)|$ on the disc, by examining points z near $1 + i0$.)

6. Show that $f_n(x) = nx(1 - x^2)^n$, defined on $E = [0, 1]$, converges pointwise to the constant function $f(x) = 0$. Show that this convergence is *not* uniform on E. Calculate and compare

$$\lim_{n\to\infty}\left(\int_0^1 f_n(x)\,dx\right) \quad \text{and} \quad \int_0^1\left(\lim_{n\to\infty} f_n\right)dx$$

Hint: If $\varepsilon = 1$ (or less) can we ever get $|f_n(x) - f(x)| \leq 1$ *throughout* E by making n large? What is the maximum of f_n on E?

Answers; $$\int_0^1 f_n(x)\,dx = \frac{n}{2n+2} \to \frac{1}{2},$$

while

$$\int_0^1\left(\lim_{n\to\infty} f_n\right)dx = \int_0^1 0\,dx = 0.$$

7. Show that

$$\sum_{n=0}^{\infty}\frac{x^2}{(1+x^2)^n} = \begin{cases} 0 \text{ if } x = 0 \\ 1 + x^2 \text{ if } x \neq 0 \end{cases}$$

for real x such that $-\infty < x < +\infty$. This series of continuous functions converges pointwise (but not uniformly) to a *discontinuous* limit function.

Hint: Bring the common factor x^2 outside; a geometric series remains.

8. Show that the sequence of functions $f_n(x) = \dfrac{\sin nx}{\sqrt{n}}$, defined for all real x, converges uniformly on $\mathbf{R}$ to the constant function $f(x) = 0$. Prove that the sequence $\{df_n/dx\}$ diverges at certain points. Thus, the limit operations $\dfrac{d}{dx}(\cdots)$ and $\lim\limits_{n\to\infty}(\cdots)$ do not give the same result when performed in reverse order, in spite of the uniform convergence of f_n to f. In fact, $\lim\limits_{n\to\infty}\left(\dfrac{df_n}{dx}\right)$ does not even exist at certain points, while

$$\frac{d}{dx}\left(\lim_{n\to\infty} f_n\right) = \frac{df}{dx} = 0 \text{ everywhere.}$$

9. Show that

$$\text{(i)}\ \lim_{n\to\infty} \frac{1 - z^n}{1 + z^n} = +1 \quad \text{if} \quad |z| < 1$$

$$\text{(ii)}\ \lim_{n\to\infty} \frac{1 - z^n}{1 + z^n} = -1 \quad \text{if} \quad |z| > 1.$$

Prove that the convergence (ii) is uniform on $E_r = \{z : r \leq |z| < +\infty\}$ if $r > 1$. Taking $r = 2$ and $\varepsilon = \frac{1}{10}$, find an integer $N(\varepsilon)$ such that $|f_n - f| \leq \varepsilon$ throughout E_r, for all $n \geq N(\varepsilon)$.

Hint: Divide out $(1 - z^n)/(1 + z^n)$ and show that $|f_n - f| = |2/(1 + z^n)|$, if $|z| > 1$.

Answer: $2^n > 21$ guarantees $|f_n - f| \leq \frac{1}{10}$ on E_2; take $N(\varepsilon) = 5$.

10. Let $\sum_{n=p}^{\infty} f_n$ and $\sum_{n=q}^{\infty} g_n$ be series of functions defined on E whose terms agree after a certain point: $f_n = g_n$ for all large n, say $n \geq N$. Show that one series converges uniformly to a limit function on E if and only if the other does. Devise an example showing that the series *sums* need not agree, even though both series converge. Formulate similar results for convergence of *sequences* of functions defined on E.

Note: Altering a finite number of terms in a series cannot affect its convergence (or lack of it), although it *may* change the value of the sum.

11. Show that the series

$$\text{(i)}\ \sum_{n=0}^{\infty} (1 - x)(-x)^n = \frac{1 - x}{1 + x}$$

converges uniformly on $[0, 1]$. Show that the series of absolute values

$$\text{(ii)}\ \sum_{n=0}^{\infty} |(1 - x)(-x)^n| = \sum_{n=0}^{\infty} (1 - x)x^n$$

converges pointwise, but *not* uniformly, on $[0, 1]$.

Note: The Weierstrass test applies to the absolute values of terms in a series; it cannot be used to verify the uniform convergence in (i).

Hint: Partial sum $s_n = \sum_{k=0}^{n-1} (1 - x)(-x)^k = \left(\frac{1 - x}{1 + x}\right)[1 - (-x)^n]$. Find the maximum of $\left|s_n(x) - \left(\frac{1 - x}{1 + x}\right)\right|$ on $[0, 1]$ by Calculus methods.

3.3 POWER SERIES AND OTHER SPECIAL TYPES OF SERIES

In applications we often encounter series whose terms are scalar multiples of the members of some fixed family of functions $f_1, f_2, \ldots,$ defined on a set E.

Thus, if we are given complex scalars (coefficients) $a_1, a_2, \ldots$, we get a typical series

(15) $$\sum_{n=1}^{\infty} a_n f_n = a_1 \cdot f_1(z) + a_2 \cdot f_2(z) + \cdots.$$

Two questions arise.

QUESTION 1. Given a function f defined on E, is there a choice of coefficients $a_1, a_2, \ldots$, such that $f = \sum_{n=1}^{\infty} a_n f_n$ in some sense? That is, which functions f can be represented by a series based on the set of functions $f_1, f_2, \ldots$?

QUESTION 2. If a function f has a representation in the form just described, are the coefficients unique? Or, is it possible to pick two different sequences $\{a_n\}$ and $\{b_n\}$ of numbers such that

$$\sum_{n=1}^{\infty} a_n f_n = f \quad \text{and} \quad \sum_{n=1}^{\infty} b_n f_n = f?$$

Most of our attention will be devoted to the "power series"

(16) $$\sum_{n=0}^{\infty} a_n (z-p)^n = a_0 + a_1(z-p) + a_2(z-p)^2 + \cdots$$

which are formed from the functions $f_0(z) = 1, f_1(z) = (z-p), f_2(z) = (z-p)^2, \ldots$. A series of the form (16), whether it converges or not, is called a **power series about the point p,** or simply a **power series** if the "base point" p is understood. As a special case, if $p = 0$ we obtain series of the form

(17) $$\sum_{n=0}^{\infty} a_n z^n = a_0 + a_1 z + a_2 z^2 + \cdots.$$

The first thing to notice about a power series is that introduction of a base point p different from $p = 0$ only makes the series (in variable w)

$$\sum_{n=0}^{\infty} a_n (w-p)^n$$

have the same behavior, and same sum, at w as the series

$$\sum_{n=0}^{\infty} a_n z^n$$

has at $z = w - p$. As a result, facts about the series (16) are readily translated into corresponding facts about the series (17), which has a different base point but the same sequence of coefficients $\{a_n\}$.

The basic convergence properties of a power series (16) follow from the next result.

Theorem 3.4 *If the series* (16) *converges for some point* $z_0 \neq p$, *then the series is absolutely convergent at every point* z *which lies closer to the base point* p *than* z_0 *does* (*so that* $|z - p| < |z_0 - p|$).

PROOF: We are assuming that the series $\sum_{n=0}^{\infty} a_n(z_0 - p)^n$ converges. This implies that $|a_n(z_0 - p)^n| = |a_n|\,|z_0 - p|^n \to 0$ as $n \to \infty$ (by Theorem 2.5); hence, this sequence is bounded by some constant A:

$$|a_n|\,|z_0 - p|^n \leq A \quad \text{for all } n = 0, 1, 2, \ldots .$$

Now if $|z - p| < |z_0 - p|$, then $\sum_{n=0}^{\infty} a_n(z - p)^n$ must be absolutely convergent since

$$\sum_{n=0}^{\infty} |a_n(z - p)^n| = \sum_{n=0}^{\infty} |a_n|\,|z - p|^n = \sum_{n=0}^{\infty} |a_n| \left(\frac{|z - p|}{|z_0 - p|}\right)^n |z_0 - p|^n$$
$$\leq A \sum_{n=0}^{\infty} \left(\frac{|z - p|}{|z_0 - p|}\right)^n .$$

Since $|z - p|/|z_0 - p| < 1$, the geometric series on the right converges to a finite limit, and $\sum_{n=0}^{\infty} |a_n(z - p)^n| < +\infty$, as required. ∎

We are now able to describe the fundamental property of power series, the existence of a "radius of convergence." Every series of the form (16) converges at the base point $z = p$, to the value a_0, and there are series with no other points of convergence. For example, the ratio test shows that $\sum_{n=0}^{\infty} n!\, z^n$ converges only at the base point $p = 0$. On the other hand, certain series, such as the exponential series $\sum_{n=0}^{\infty} z^n/n!$, are absolutely convergent at every point z (use the ratio test again). Except in these extreme situations, there will be points $z \neq p$ where the series converges, and others where it diverges.

Theorem 3.4 tells us that convergence of the series (16) at a point $z_0 \neq p$ forces absolute convergence throughout the entire disc $\{z : |z - p| < |z_0 - p|\}$ which has z_0 on its boundary. Let r be the largest radius such that (16) is absolutely convergent on the disc $|z - p| < r$; r is finite if there are any points at which the series diverges. The series converges absolutely at z if $|z - p| < r$. On the other hand, the series must diverge at z if $|z - p| > r$; otherwise, it would be absolutely convergent throughout the larger disc of radius $r' = |z - p| > r$, which conflicts with our choice of r. Thus, the circle $|z - p| = r$ divides the plane into domains of absolute convergence and divergence, as shown in Figure 3.2. The radius r is called the **radius of convergence** of the series. In the extreme cases mentioned above, we assign radii of convergence $r = 0$ and $r = +\infty$, respectively. These observations may be summarized as follows.

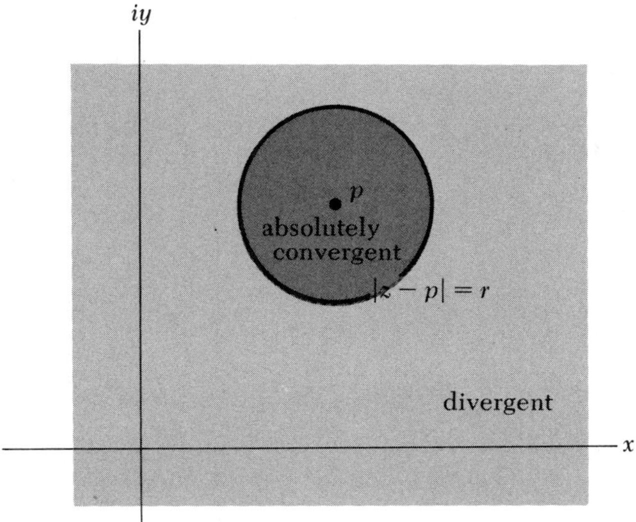

Figure 3.2 Domains of convergence and divergence for a series with radius r about base point $z = p$.

Theorem 3.5 *If a power series* (16) *is given, there is a real number r with $0 \leq r \leq +\infty$, the radius of convergence of the series, such that*

(i) *The series converges absolutely for all z in the open disc $|z - p| < r$.*
(ii) *The series diverges for all z such that $|z - p| > r$.*

Nothing definite can be said about the behavior of the series (16) on the boundary circle $|z - p| = r$. The following examples show that the series may diverge at every point, or converge (absolutely) at every point, or may diverge at some points and converge at others, on the boundary circle. Of course, if $r = 0$ or $r = +\infty$, the boundary circle is either degenerate, or doesn't exist at all, so there is not much to be said unless $0 < r < +\infty$.

Example 3.4 The series

(18)
$$\sum_{n=1}^{\infty} \frac{z^n}{n}$$

is absolutely convergent if $|z| < 1$, since

$$\sum_{n=1}^{\infty} \frac{|z|^n}{n} \leq \sum_{n=1}^{\infty} |z|^n \leq \sum_{n=0}^{\infty} |z|^n = \frac{1}{1 - |z|} < +\infty;$$

it diverges if $|z| > 1$, because the terms $|z|^n/n$ diverge to the improper limit $+\infty$ as $n \to \infty$ (they would converge to zero if the series were convergent). Thus, the radius of convergence is $r = 1$. On the boundary circle the series reduces to the well known divergent series $\sum_{n=1}^{\infty} 1/n$ if $z = 1 + i0$, but if we set $z = (-1) + i0$ we get the convergent alternating (real) series $\sum_{n=1}^{\infty} (-1)^n/n$.

Although (18) converges at certain points on the boundary circle, this convergence can never be *absolute*, because $\sum_{n=1}^{\infty} |z^n|/n = \sum_{n=1}^{\infty} 1/n = +\infty$ if $|z| = 1$.

Example 3.5 The series

$$\sum_{n=1}^{\infty} \frac{z^n}{n^2} \tag{19}$$

is absolutely convergent for all z such that $|z| \leq 1$, because

$$\sum_{n=1}^{\infty} \frac{|z|^n}{n^2} \leq \sum_{n=1}^{\infty} \frac{1}{n^2} < +\infty$$

(convergence follows from the ratio test if $|z| < 1$, and from the integral test of calculus if $|z| = 1$); it diverges if $|z| > 1$ because the terms $|z|^n/n^2$ diverge to the improper limit $+\infty$ as $n \to \infty$. Thus the series (19) has radius of convergence $r = 1$, but in this example the series converges absolutely at every point on the boundary circle $|z| = 1$, as well as within the disc of convergence. By using the Weierstrass test, we can also verify that the series converges *uniformly* to its limit function on the closed disc $\{z:|z| \leq 1\}$. Just take $M_n = 1/n^2$; then $\sum_{n=1}^{\infty} M_n < +\infty$, and we obviously have $|z^n/n^2| \leq M_n$ for all z in this disc. Notice that we have been able to discuss the convergence, and even uniform convergence, of this series without ever evaluating its sum; recall the comments of Section 2.2.

Example 3.6 The geometric series

$$\sum_{n=0}^{\infty} z^n = 1 + z + z^2 + \cdots \tag{20}$$

has radius of convergence $r = 1$, as we have seen. If z lies on the boundary circle, then $|z^n| = |z|^n = 1$ for $n = 0, 1, 2, \ldots$, so the series must diverge; if the series converged, the sequence of terms would converge to zero as $n \to \infty$. The geometric series diverges at every point on the boundary circle.

Before long we will develop systematic methods for computing the radius of convergence, but it is useful to see this done directly in a few cases, as above. Meanwhile, we shall display a number of useful facts about power series.

Theorem 3.6 *The radius of convergence of a power series* (16) *depends only on the absolute values* $|a_0|, |a_1|, |a_2|, \ldots$ *of its coefficients; thus, if* $|a_n| = |b_n|$ *for all n, the series*

$$\sum_{n=0}^{\infty} a_n(z-p)^n \quad \text{and} \quad \sum_{n=0}^{\infty} b_n(z-p)^n$$

have the same radius of convergence (*although their sums will not be related in any reasonable way*). *The radius of convergence is not affected by a change of base point.*

PROOF: We have already pointed out why a change of base point will not affect the radius of convergence. If $|a_n| = |b_n|$ for all n, this implies that:

$$\sum_{n=0}^{\infty} |a_n|\,|z - p|^n < +\infty \quad \text{if and only if} \quad \sum_{n=0}^{\infty} |b_n|\,|z - p|^n < +\infty.$$

Thus, the two series are absolutely convergent at exactly the same points in the plane. From the definition of radius of convergence, it is obvious that these series have the same radius. ▮

Theorem 3.7 *If the power series* (16) *has radius of convergence* r, *then for any smaller radius* r' (*with* $0 < r' < r$) *the series converges uniformly to its limit on the closed disc* $D_{r'} = \{z : |z - p| \le r'\}$.

PROOF: The Weierstrass test for uniform convergence of series is made to order for this task. Take $z = p + (r' + i0)$; then (16) is absolutely convergent at z and $|z - p| = r'$, so that

$$\sum_{n=0}^{\infty} |a_n|\,(r')^n < +\infty.$$

For all z in the closed disc $D_{r'}$ we have $|z - p| \le r'$ and

(21) $$|a_n(z - p)^n| = |a_n|\,|z - p|^n \le |a_n|\,(r')^n,$$

so if $M_n = |a_n|\,(r')^n$ for $n = 0, 1, 2, \ldots$, we get

(22) $$\sum_{n=0}^{\infty} M_n < +\infty.$$

Combining (21) and (22), we have met the requirements of the Weierstrass test, so that series converges uniformly on $D_{r'}$. This reasoning works for any r' less than r (but not for $r' = r$; why?). ▮

Next we apply some properties of uniformly convergent series to prove that the sum of a power series is continuous throughout the (open) disc of convergence.

Theorem 3.8 *A power series* (16) *converges to a function* $f(z)$ *which is continuous at every point in the open disc of convergence* $D = \{z : |z - p| < r\}$.

PROOF: The functions $f_0(z) = a_0$, $f_1(z) = a_1(z - p)$, $f_2(z) = a_2(z - p)^2$, ... which appear in the series are continuous, as are the partial sum $s_n(z) = f_0(z) + \cdots + f_{n-1}(z)$; they are just polynomials in z. These sums converge uniformly to f on any disc of the form $D_{r'} = \{z : |z - p| < r'\}$, if $r' < r$; thus, f is continuous on each of these subdiscs, by Theorem 3.2. If z_0 is a typical point in the disc of convergence D, this point and all nearby points lie within $D_{r'}$ provided r' is greater than the distance $|z_0 - p|$, as shown in Figure 3.3. Thus

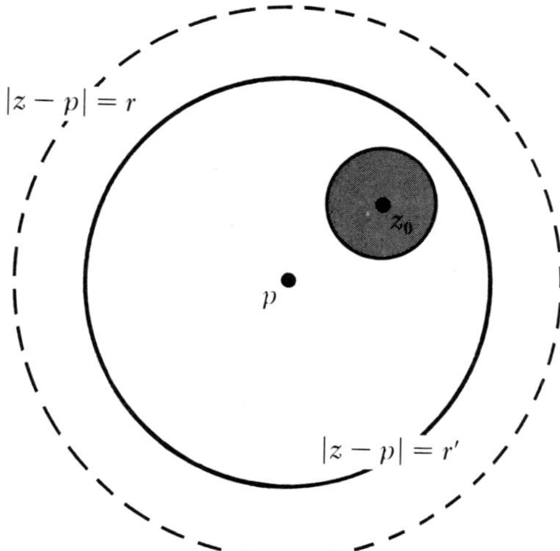

Figure 3.3 In the disc of convergence $D = \{z: |z - p| < r\}$ we consider the smaller closed disc $D_{r'} = \{z: |z - p| \leq r'\}$. If z_0 is given, we want to take r' so that $|z_0 - p| < r' < r$; then $D_{r'}$ includes z_0 and all (shaded) nearby points.

f is continuous at z_0; continuity on D means that f is continuous at each point z_0 in D, so the proof is complete. ▌

There is much more to be said about the "smoothness" of the function

$$f(z) = \sum_{n=0}^{\infty} a_n(z - p) \tag{23}$$

to which the power series (16) converges within its disc of convergence. We now show that f is complex differentiable. Starting with a series (23), we define the **first derived series** to be the power series, with the same base point, which is produced by term-by-term differentiation of (23):

$$\begin{aligned} \sum_{n=1}^{\infty} na_n(z - p)^{n-1} &= \sum_{n=0}^{\infty} (n + 1)a_{n+1}(z - p)^n \\ &= a_1 + 2a_2(z - p) + 3a_3(z - p)^2 + \cdots . \end{aligned} \tag{24}$$

Theorem 3.9 *The derived series* (24) *has the same radius of convergence r as the original series* (23). *If f and g are the sums of these series, then f is differentiable throughout the open disc of convergence $E = \{z: |z - p| < r\}$, and $df/dz = g$ throughout E.*

Proof. The functions f and g are given by the series (23) and (24) respectively; let r and r' be their radii of convergence.

We first show that: if the derived series converges absolutely at z^*, the original series is also absolutely convergent there (this gives $r' \leq r$). To see this, we start with the assumption that

$$\sum_{n=0}^{\infty} (n + 1)\, |a_{n+1}|\, |z^* - p|^n < +\infty; \tag{25}$$

then we rewrite the series associated with f so we can use this information.

$$\begin{aligned}\sum_{n=0}^{\infty}|a_n|\,|z^*-p|^n &= |a_0| + \sum_{n=1}^{\infty}|a_n|\,|z^*-p|^n \\ &= |a_0| + \sum_{n=0}^{\infty}|a_{n+1}|\,|z^*-p|^{n+1} \\ &= |a_0| + \sum_{n=0}^{\infty}(n+1)\,|a_{n+1}|\,|z^*-p|^n\cdot\left(\frac{|z^*-p|}{n+1}\right).\end{aligned}$$

The multiplying factors $|z^*-p|/(n+1)$ are bounded by some constant A for all n, since they converge to zero as $n\to\infty$. Thus the series on the right converges, by comparison with (25).

Next we show that: if the series for f converges absolutely at some point $z^*\neq p$, then the derived series converges absolutely at any z such that $|z-p|<|z^*-p|$. This shows that the (open) disc of convergence for the derived series is at least as large as the (open) disc of convergence for f, so that $r'\geq r$. If $z=p$ the derived series obviously converges, so we need only consider points z such that $0<|z-p|<|z^*-p|$ (this keeps us from dividing by zero in the following calculations). Starting with the information

$$\sum_{n=0}^{\infty}|a_n|\,|z^*-p|^n < +\infty, \tag{26}$$

we can rewrite the series of absolute values associated with g to use this information:

$$\begin{aligned}\sum_{n=0}^{\infty}(n+1)\,|a_{n+1}|\,|z-p|^n &= \sum_{n=0}^{\infty}\left(\frac{n+1}{|z-p|}\right)\cdot|a_{n+1}|\,|z-p|^{n+1} \\ &= \sum_{n=0}^{\infty}\left(\frac{n+1}{|z-p|}\right)\left(\frac{|z-p|}{|z^*-p|}\right)^{n+1}\cdot|a_{n+1}|\,|z^*-p|^{n+1}.\end{aligned}$$

Now

$$\begin{aligned}\sum_{n=0}^{\infty}|a_{n+1}|\,|z^*-p|^{n+1} &= \sum_{n=1}^{\infty}|a_n|\,|z^*-p|^n \\ &\leq \sum_{n=0}^{\infty}|a_n|\,|z^*-p|^n < +\infty.\end{aligned}$$

If we multiply each term in this series by the factor

$$c_n = \left(\frac{n+1}{|z-p|}\right)\left(\frac{|z-p|}{|z^*-p|}\right)^{n+1},$$

the new series converges, by the comparison test, since $c_n\to 0$ as $n\to\infty$ (because $|z-p|/|z^*-p|<1$). Thus the derived series is absolutely convergent at z.

Now that we have established that $r=r'$, so that the derived series converges on E, we must prove that f is differentiable, and that $df/dz=g$. To do this we use the time-honored principle of brute force. If z^* is a point in the disc E, then all points near z^* also lie within E. We want to examine

$$\left|\frac{\Delta f}{\Delta z} - g(z^*)\right| = \left|\frac{f(z) - f(z^*)}{z - z^*} - g(z^*)\right| \tag{27}$$

for $z = z^* + \Delta z \neq z^*$, and prove that this expression converges to the limit *zero* as $z \to z^*$. This is exactly what we mean by the statement that $f'(z^*) = g(z^*)$. For each function appearing in (27) we substitute the corresponding convergent series and collect terms; we get

$$\left|\frac{1}{\Delta z}\sum_{n=0}^{\infty} a_n(z-p)^n - \frac{1}{\Delta z}\sum_{n=0}^{\infty} a_n(z^*-p)^n - \sum_{n=1}^{\infty} na_n(z^*-p)^{n-1}\right|$$
$$= \left|\sum_{n=1}^{\infty} a_n\left[\frac{(z-p)^n - (z^*-p)^n}{\Delta z} - n(z^*-p)^{n-1}\right]\right|.$$

The $n = 0$ terms have cancelled, and for the others we note that

$$\frac{(z-p)^n - (z^*-p)^n}{\Delta z} = \frac{(z-p)^n - (z^*-p)^n}{(z-p) - (z^*-p)}$$
$$= (z-p)^{n-1} + (z-p)^{n-2}(z^*-p) + \cdots + (z^*-p)^{n-1},$$

so the last expression can be written in the form

$$\left|\frac{\Delta f}{\Delta z} - g(z^*)\right|$$
$$= \left|\sum_{n=1}^{\infty} a_n[(z-p)^{n-1} + \cdots + (z^*-p)^{n-1} - n(z^*-p)^{n-1}]\right|$$
$$\le \sum_{n=1}^{\infty} |a_n| \left|\underbrace{(z-p)^{n-1} + \cdots + (z^*-p)^{n-1}}_{n \text{ terms}} - n(z^*-p)^{n-1}\right|. \tag{28}$$

Our limit problem would be resolved if we could interchange the operations $\lim_{z\to z^*}(\cdots)$ and $\sum_{n=1}^{\infty}(\cdots)$; we want $\lim_{z\to z^*}\left(\sum_{n=1}^{\infty}(\cdots)\right) = 0$, while it is obvious that $\sum_{n=1}^{\infty}\left(\lim_{z\to z^*}(\cdots)\right) = \sum_{n=1}^{\infty}|a_n|\,|(z^*-p)^{n-1} + \cdots + (z^*-p)^{n-1} - n(z^*-p)^{n-1}| = \sum_{n=1}^{\infty}|a_n| \cdot 0 = 0$. The use of brute force enters in justifying this manipulation. For our purposes later on, it is enough that the reader be aware of the general idea of the proof up to this point. We include the final details below only for the sake of completeness.

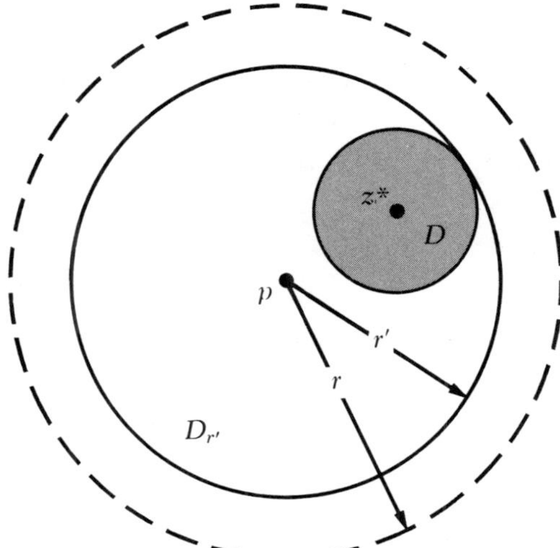

Figure 3.4 The disc D about z^* in which the estimates used to prove Theorem 3.9 are valid.

We are going to need uniform convergence of all series involved in order to make this interchange of limits work, so we take a closed disc about p slightly smaller than the disc of convergence about p, but with radius r' large enough that the disc $D_{r'} = \{z:|z - p| \leq r'\}$ includes z^* in its interior, as indicated in Figure 3.4.

For z close to z^*, say in the small disc D shown in Figure 3.4, we have $|z^* - p| < r'$ and $|z - p| < r'$; therefore, the nth term in the series (28) is bounded by $|a_n| \cdot 2n(r')^{n-1}$ for all z in D. The derived series (24) has radius r; taking $w = p + (r' + i0)$, we get

$$\sum_{n=1}^{\infty} 2\,|a_n|\, n(r')^{n-1} = 2\sum_{n=1}^{\infty} n\,|a_n|\,|w|^{n-1} < +\infty, \tag{29}$$

since $r' < r$.

If any number $\varepsilon > 0$ is given, we will now show that the series in (28) is less than ε for all z near z^*. First, we take an integer $N = N(\varepsilon)$ sufficiently large that the tail end of the convergent series (29) is less than $\varepsilon/2$:

$$\sum_{n=N+1}^{\infty} 2n\,|a_n|\,(r')^{n-1} < \frac{\varepsilon}{2}.$$

Then we break the series in (28) into two pieces,

$$\sum_{n=1}^{N}(\cdots) + \sum_{n=N+1}^{\infty}(\cdots).$$

For any z in D, we know that $|z - p| < r'$ because D is contained in $D_{r'}$; therefore,

$$\sum_{n=N+1}^{\infty}(\cdots) \leq \sum_{n=N+1}^{\infty} 2n\,|a_n|\,(r')^{n-1} < \frac{\varepsilon}{2} \quad \text{for all } z \text{ in } D. \tag{30}$$

On the other hand, the left-hand piece is a *finite* sum,

$$\sum_{n=1}^{N}(\cdots) = \sum_{n=1}^{N}|a_n|\,|(z-p)^{n-1}+\cdots+(z^*-p)^{n-1}-n(z^*-p)^{n-1}|,$$

and as such it goes to zero as z approaches z^*; therefore,

$$\text{(31)}\qquad \sum_{n=1}^{N}(\cdots) < \frac{\varepsilon}{2} \quad \text{for all } z \text{ sufficiently close to } z^*.$$

Combining observations (30) and (31), we conclude that

$$\left|\frac{\Delta f}{\Delta z} - g(z^*)\right| \le \sum_{n=1}^{N}(\cdots) + \sum_{n=N+1}^{\infty}(\cdots)$$

$$< \frac{\varepsilon}{2} + \frac{\varepsilon}{2} = \varepsilon$$

for all z near z^* such that $z \neq z^*$. This reasoning works for every choice of $\varepsilon > 0$, so that

$$\lim_{z\to z^*}\frac{f(z)-f(z^*)}{z-z^*} \text{ exists and is equal to } g(z^*). \quad \blacksquare$$

Now that we know how to differentiate the sum of a power series, we may apply the same technique to the derived series, which converges to df/dz, to see that f is twice differentiable, with

$$\text{(32)}\quad f''(z) = \sum_{n=2}^{\infty} n(n-1)a_n(z-p)^{n-2} = 2\cdot 1\cdot a_2 + 3\cdot 2\cdot a_3(z-p) + \cdots$$

This series, the **second derived series**, has the same radius of convergence as the series for f' and f. By repeated application of Theorem 3.9 we see that, for any $k = 0, 1, 2, \ldots$, the $\boldsymbol{k}^{\text{th}}$ **derived series**

$$\text{(33)}\quad \sum_{n=k}^{\infty} n(n-1)\cdots(n-k+1)a_n(z-p)^{n-k}$$

$$= k!\,a_k + \frac{(k+1)!}{1!}a_{k+1}(z-p) + \frac{(k+2)!}{2!}a_{k+2}(z-p)^2 + \cdots$$

has radius of convergence r and gives us the kth derivative of f throughout the open disc of convergence. (Recall that, by convention, we set $0! = 1$.)

Theorem 3.10 *The sum of a power series* (23) *has derivatives of all orders on the open disc of convergence, and the derivative* $f^{(k)}$ *is the sum of the* k^{th} *derived series* (33), *which has the same radius of convergence as the original series. All of these functions are continuous on the open disc of convergence, being sums of convergent power series.*

On the other hand, we might also look at the **first antiderived series**

$$\sum_{n=0}^{\infty} \frac{1}{n+1} a_n(z-p)^{n+1}. \tag{34}$$

Let r'' be the radius of convergence of the antiderived series. Differentiating (34) term-by-term gives a series whose radius is also r''. But, clearly, we get the original series (23) if we differentiate the terms in (34), so the series (23) and (34) have the same radius of convergence: $r'' = r$. If $h(z)$ is the sum of the antiderived series, then h is differentiable on the disc $D = \{z: |z - p| < r\}$, and

$$\frac{dh}{dz} = \sum_{n=0}^{\infty} \frac{a_n}{n+1} \frac{d}{dz}\{(z-p)^{n+1}\} = \sum_{n=0}^{\infty} a_n(z-p)^n = f(z),$$

so the antiderived series gives us an antiderivative for f on D.

Theorem 3.11 *The kth antiderived series associated with the power series* (23),

$$\sum_{n=0}^{\infty}\left(\frac{1}{n+1}\right)\cdots\left(\frac{1}{n+k}\right) a_n(z-p)^{n+k} \tag{35}$$
$$= \frac{1}{k!} a_0(z-p)^k + \frac{1!}{(k+1)!} a_1(z-p)^{k+1} + \frac{2!}{(k+2)!} a_2(z-p)^{k+2} + \cdots,$$

has the same radius of convergence r as the original series and converges to a function $h(z)$ on the disc $D = \{z: |z-p| < r\}$ such that $h^{(k)} = f$.

We have already noted that, on a *connected* set like the open disc of convergence, the antiderivative of a function (if it exists at all) is uniquely determined up to an added constant. It is also interesting to notice that Theorem 3.9 says, in effect, that the operations $\sum_{n=0}^{\infty}(\cdots)$ and $\frac{d}{dz}(\cdots)$ give the same result when performed in either order on the terms in a power series.

Our final observation is that the coefficients of a power series are closely related to the derivatives of f at the base point $z = p$.

Theorem 3.12 *If f is the sum of a power series* (23) *then*

$$a_n = \frac{f^{(n)}(p)}{n!} \quad \textit{for} \quad n = 0, 1, 2, \ldots,$$

so the series could be rewritten in the form:

$$f(z) = \sum_{n=0}^{\infty} \frac{f^{(n)}(p)}{n!}(z-p)^n.$$

PROOF: Evidently, $f^{(0)}(p) = f(p) = a_0$. From Theorem 3.10 we see that $f'(p) = a_1$; $f''(p) = 2 \cdot 1 \cdot a_2; \ldots ; f^{(n)}(p) = n!\, a_n; \ldots$. ▮

Corollary 3.13 (Uniqueness of coefficients in a power series) *Suppose two power*

$$f(z) = \sum_{n=0}^{\infty} a_n(z-p)^n \quad \text{and} \quad g(z) = \sum_{n=0}^{\infty} b_n(z-p)^n$$

series converge near the base point p and agree on some disc about p. Then the coefficients must be the same: $a_n = b_n$ for $n = 0, 1, 2, \ldots$.

PROOF: By the theorem just proved,

$$a_n = \frac{f^{(n)}(p)}{n!} \quad \text{and} \quad b_n = \frac{g^{(n)}(p)}{n!}$$

for $n = 0, 1, 2, \ldots$. But $f(z) = g(z)$ for all z near p, so the derivatives agree: $f^{(n)}(z) = g^{(n)}(z)$ near p. Set $z = p$ to get $a_n = b_n$ for all n. ▮

Now we turn to the very important problem of effectively calculating the radius of convergence of a power series:

$$\sum_{n=0}^{\infty} a_n(z-p)^n. \tag{36}$$

As we remarked earlier, the information determining this radius is somehow built into the sequence of absolute values of the coefficients, $\{|a_0|, |a_1|, \ldots\}$. One method for reading out this information, which is very useful when it works, is based on the ratio test introduced in Section 2.2.

Theorem 3.14 *If the power series (36) has non-zero coefficients, and if the sequence of ratios $|a_{n+1}|/|a_n|$ has a well defined limit μ as $n \to \infty$ (the improper limit $\mu = +\infty$ is allowed), then the radius of convergence of the series is $r = 1/\mu$. (Take $r = 0$ if $\mu = +\infty$, and $r = +\infty$ if $\mu = 0$.)*

PROOF: The terms $b_n = a_n(z-p)^n$ in the power series have ratios

$$\left|\frac{b_{n+1}}{b_n}\right| = \left|\frac{a_{n+1}(z-p)^{n+1}}{a_n(z-p)^n}\right| = \left|\frac{a_{n+1}}{a_n}\right| |z-p|.$$

These converge to the limit $\mu \cdot |z-p|$ as $n \to \infty$. The ratio test for series convergence shows that the series

$$\sum_{n=0}^{\infty} b_n = \sum_{n=0}^{\infty} a_n(z-p)^n$$

(i) converges absolutely if $\lim_{n\to\infty} |b_{n+1}/b_n| = \mu \cdot |z-p| < 1$

(ii) diverges if $\lim_{n\to\infty} |b_{n+1}/b_n| = \mu \cdot |z - p| > 1.$

Thus, the series converges absolutely if $|z - p| < 1/\mu$ and diverges if $|z - p| > 1/\mu$. Clearly, $r = 1/\mu$. ▮

This test is inadequate in certain situations, especially if the series (36) has plenty of zero coefficients. It is not that these zeros upset the convergence of (36)—rather, they improve its convergence properties—but their presence can make the ratio test inapplicable (the ratios will not be well defined). There are a number of other tests which may be familiar from the study of real power series in Calculus. Many of these can be adapted for complex variables (see Exercise 10).

There is actually an explicit formula for the radius of convergence which always works, but it involves the somewhat subtle notion of the **upper limit** (or **limit superior**)

$$\limsup_{n\to\infty} c_n = \lim_{n\to\infty} (\sup\{c_n, c_{n+1}, c_{n+2}, \ldots\})$$

of a sequence of non-negative real numbers $\{c_n\}$. Unlike the ordinary limit $\lim_{n\to\infty} \{c_n\}$, which may fail to exist, the upper limit $\limsup_{n\to\infty}\{c_n\}$ is always well defined (if we allow the value $+\infty$), and by taking the numbers $c_n = |a_n|^{1/n}$ we get a general formula for the radius of convergence r:

(37) $$\frac{1}{r} = \limsup_{n\to\infty}\{|a_n|^{1/n}\} \qquad \text{(\textbf{Hadamard's formula} for the radius)}$$

which has a meaning, no matter how irregular the sequence $\{|a_1|, |a_2|^{1/2}, |a_3|^{1/3}, \ldots\}$. Since the notion of limsup is fairly technical, we will not invoke formula (37) in this book, but it is worth knowing that such a formula is available. There is a concise account of upper limits, and the Hadamard formula, in Hille [9], Sections 1.2 and 5.4, and in Ahlfors [1], Sections 2.2.1 to 2.2.4.

Some concluding remarks about power series are in order. First, all of our discussion applies equally well to real variables to give the theory of real power series usually discussed in Calculus: for a real power series, there is an *interval of convergence* about the base point, and it is easy to see, by restricting our attention to real variables in the above proofs, that the radius of convergence for real variable is exactly the same as the radius of convergence for complex variable (the coefficients and real base point p being kept the same).

We must also point out that the convergence properties of power series, especially the radius of convergence phenomenon, do not carry over to series of the form $\sum_{n=0}^{\infty} a_n f_n(z)$ associated with some family of functions other than the special family $f_0(z) = 1, f_1(z) = (z - p), f_2(z) = (z - p)^2, \ldots$, we have been working with. For example, we might consider the family of all **complex trigonometric polynomials**

(38)
$$\begin{array}{ll} f_1(t) = 1 & \\ f_2(t) = e^{2\pi it} & f_3(t) = e^{-2\pi it} \\ \vdots & \vdots \\ e_{2n}(t) = e^{2\pi int} & f_{2n+1}(t) = e^{-2\pi int} \\ \vdots & \vdots \end{array}$$

defined either on the interval [0, 1], or on all of **R** (real *variable*; complex *values*). Series of the form

$$\sum_{n=1}^{\infty} a_n f_n(t) \qquad (a_n \text{ complex scalars})$$

are referred to as (complex) **Fourier series,** and the resolution of Questions 1 and 2, stated earlier, is one of the central problems in the theory of Fourier series expansions. It is worth pointing out two features of this family of series. First: the convergence properties of Fourier series are *very* different from those of power series, and are more difficult to handle (see Exercises 12 to 16). The reader should never mistakenly try to carry our ideas on power series over to Fourier series problems. Second, the complex trigonometric polynomials are more or less equivalent to the ordinary trigonometric polynomials, since $e^{2\pi int} = \cos(2\pi nt) + i\sin(2\pi nt)$ and we can get one family of trigonometric functions from the other by taking complex linear combinations. By introducing the functions (38) we make the first steps in connecting the theory of Fourier series with complex analysis. In applications of Fourier series this can be extremely useful (particularly in wave propagation problems), but we will not have space to pursue these paths of inquiry in this text.

EXERCISES

1. If $\sum_{n=0}^{\infty} a_n(z-p)^n$ has radius of convergence $0 < r < \infty$, show that $\sum_{n=0}^{\infty} |a_n|\,(r')^n < +\infty$ for any r' such that $0 < r' < r$.

Hint: Examine $z = p + (r' + i0)$.

2. Determine radii of convergence for the following power series. (What is the base point in each case?)

(i) $\sum_{n=1}^{\infty} \frac{z^n}{n^3}$

(ii) $\sum_{n=1}^{\infty} nz^n$

(iii) $\sum_{n=0}^{\infty} 2^n(z-1)^n$

(iv) $\sum_{n=0}^{\infty} \left(\frac{i^n - 1}{n}\right)(z+i)^n$

(v) $\sum_{n=1}^{\infty} \frac{n^3}{3^n} z^n$

(vi) $\sum_{n=0}^{\infty} \frac{2^n}{n!}(z+1)^n$

(vii) $\sum_{n=1}^{\infty} \frac{z^n}{n^n}$

Answers: (i) $r = 1$; (ii) $r = 1$; (iii) $r = \frac{1}{2}, p = 1$; (iv) $r = 1, p = -i$; (v) $r = 3$; (vi) $r = \infty, p = -1$; (vii) $r = +\infty$.

3. Write out the Taylor series expansions of the polynomial $1 + z^2$ about the base points $p = +i, p = -i, p = +1, p = 0$. For which base points is the constant term a_0 equal to zero?

Answer: Each series is a polynomial; $a_0 = 0$ for $+i$ and $-i$.

4. Calculate the Taylor series (and its radius of convergence) for

(i) Log z about $p = 1$

(ii) $\sin z$ about $p = 0$

(iii) $\text{Log}(1 - z)$ about $p = 0$

(iv) $\sin z$ about $p = \pi/2$

(v) $\frac{1}{1+z^2}$ about $p = 0$

(vi) $\frac{1}{z^2}$ about $p = -1$

(vii) $z^i = e^{i\,\text{Log}(z)}$ about $p = 1$

(viii) $\sinh z$ about $p = +i\pi$.

Note: Later we will determine whether these series agree with the original functions.

5. Prove that

$$-\text{Log}(1-z) = \text{Log}\left(\frac{1}{1-z}\right) = z + \frac{z^2}{2} + \frac{z^3}{3} + \cdots = \sum_{n=1}^{\infty} \frac{z^n}{n}$$

and

$$\text{Log}(1+z) = z - \frac{z^2}{2} + \frac{z^3}{3} - \cdots = \sum_{n=1}^{\infty} (-1)^{n+1} \frac{z^n}{n}$$

for $|z| < 1$, by comparing the derivatives of the functions and the series.

Note: Thus we have evaluated the sum of the *harmonic series* in closed form.

6. Show that the sum $f(z)$ of the series $\sum_{n=1}^{\infty} z^n/n^2$ satisfies the differential equation

$$\frac{d}{dz}\left(z\frac{df}{dz}\right) = \frac{1}{1-z} \quad \text{for } |z| < 1.$$

Show that $f(z)$ is an antiderivative of the holomorphic function

$$g(z) = \frac{-\mathrm{Log}(1-z)}{z} = \frac{1}{z}\,\mathrm{Log}\left(\frac{1}{1-z}\right)$$

on the domain $E = \{z:0 < |z| < 1\}$.

7. In Exercise 5, sketch the cut planes on which these composite functions are well defined, and about $z = 0$ draw in the discs of convergence of the series expansions. Is there a relation between the cuts and the discs of convergence?

Answers: $-\mathrm{Log}(1 - z) = \mathrm{Log}\left(\frac{1}{1-z}\right)$ is defined off of the cut $[+1, +\infty)$ and the disc has $+1$ on its boundary. $\mathrm{Log}(1 + z)$ is defined off of the cut $(-\infty, -1]$ and -1 is on the boundary of the disc.

8. Show that

$$\mathrm{Arctan}\, z = z - \frac{z^3}{3} + \frac{z^5}{5} - \cdots = \sum_{n=0}^{\infty} (-1)^n \frac{z^{2n+1}}{(2n+1)}$$

on the disc $|z| < 1$. (Arctan is defined in Example 2.24).

Hint: Write $f(z) = \mathrm{Arctan}\, z$ and $g(z)$ for the series sum. Compare df/dz and dg/dz. Recall Example 2.24; recall Section 2.10.

9. Suppose that p is real and that the series $\sum_{n=0}^{\infty} a_n(x - p)^n$ has real coefficients and real variable x. Suppose that this series converges on the (real) interval $p - \delta < x < p + \delta$; show that the corresponding complex series $\sum_{n=0}^{\infty} a_n(z - p)^n$ (same real coefficients and base point) converges absolutely on the disc $|z - p| < \delta$.

10. (The root test) Suppose that a power series $\sum_{n=0}^{\infty} a_n(a - p)^n$ has non-zero coefficients (actually, finitely many zero coefficients are allowable), and suppose that $\mu = \lim_{n\to\infty} |a_n|^{1/n}$ exists. Prove that $r = 1/\mu$.

Hint: Consult any Calculus text which discusses the "root test."

11. Show that the following series expansions are valid, and specify the domains of convergence.

(i) $\dfrac{1}{z^2 - 2z + 1} = \dfrac{1}{(z-1)^2} = (-1)(1 + 2z + 3z^2 + \cdots)$

(ii) $\dfrac{1}{1+z^2} = 1 - z^2 + z^4 - \cdots$

(iii) $\dfrac{1}{z} = \dfrac{1}{1-(z+1)} = \sum\limits_{n=0}^{\infty} (z+1)^n$

(iv) $\dfrac{1}{z^2} = 1 + \sum\limits_{n=1}^{\infty} (n+1)(z+1)^n$

(v) $\dfrac{1}{z^2} = \dfrac{1}{4} + \dfrac{1}{4}\sum\limits_{n=1}^{\infty} \dfrac{(-1)^n(n+1)}{2^n}(z-2)^n$

Hint: The function in (i) is the derivative of a function whose series expansion has already been calculated; others can be obtained by working with geometric series.

12. Consider a Fourier series $\sum\limits_{n=0}^{\infty} a_n e^{in\theta}$ (complex coefficients a_n) defined for all real θ. If the series converges at θ_0, show that it must also converge at the points $\theta_0 + 2\pi n$ $(n = \pm 1, \pm 2, \ldots)$.

13. Show that $\sum\limits_{n=0}^{\infty} \sqrt{n} \sin nx$ converges to zero at $x = 0, \pm\pi, \pm 2\pi, \ldots$ and diverges for all other real x. What can be said about the complex Fourier series $\sum\limits_{n=0}^{\infty} \sqrt{n} e^{in\theta}$ defined for real θ?

14. Fix $0 \le r < 1$. Show that the complex Fourier series $\sum\limits_{n=0}^{\infty} r^n e^{in\theta}$ converges absolutely for all real θ in the interval $[0, 2\pi]$. Show that

$$\sum_{n=0}^{\infty} r^n e^{in\theta} = \frac{1 - r\cos\theta}{1 - 2r\cos\theta + r^2} + i\,\frac{-r\sin\theta}{1 - 2r\cos\theta + r^2}.$$

How is this evaluation related to evaluation of the power series $\sum\limits_{n=0}^{\infty} z^n$?

Hint: Set $z = re^{i\theta}$ and use what we know about geometric series. Use standard methods to express $1/(1-z) = u + iv$.

15. Using Exercise 14, evaluate the sums of the *sine* and *cosine* series $\sum\limits_{n=0}^{\infty} r^n \sin n\theta$ and $\sum\limits_{n=0}^{\infty} r^n \cos n\theta$, taking $0 \le r < 1$.

16. One might think of $\sum_{n=0}^{\infty} \frac{1}{n^2} \cos(nz)$ as a "Fourier series" with complex variable z. At which points z does this series converge?

Answer: $\mathrm{Im}(z) = 0$; otherwise $|\cos(nz)|$ behaves like $e^{n \cdot \mathrm{Im}(z)}$.

17. (Dirichlet series). Let $n^z = e^{z \operatorname{Log}(n)}$ for $n = 1, 2, \ldots,$ and consider the **Dirichlet series**

$$\sum_{n=1}^{\infty} \frac{1}{n^z} = \sum_{n=1}^{\infty} n^{-z}$$

Prove that this series

(i) diverges if $\mathrm{Re}(z) \leq 1$

(ii) converges absolutely if $\mathrm{Re}(z) > 1$

(iii) converges uniformly on any rectangular set

$$\left\{z : 1 + \frac{1}{R} \leq \mathrm{Re}(z) < R \quad \text{and} \quad -R \leq \mathrm{Im}(z) \leq +R\right\},$$

if $1 < R < +\infty$.

(iv) does *not* converge uniformly on $E = \{z : \mathrm{Re}(z) > 1\}$.

Note: This series is of great importance in number theory.

3.4 ANALYTIC FUNCTIONS

We say that a function f defined on an open set in the plane has a **local power series expansion** valid near p, if there is some choice of coefficients $a_0, a_1, a_2, \ldots,$ such that the power series

$$\sum_{n=0}^{\infty} a_n (z - p)^n \tag{39}$$

has the properties

(40) (i) The series has positive radius of convergence $r > 0$.

(ii) The sum of the series $g(z) = \sum_{n=0}^{\infty} a_n (z - p)^n$ coincides with the given function f in some disc of positive radius about p (that is, $f = g$ *near* p).

We say that f is **analytic on D** if there is such a local power series expansion about each point p in D. Of course, as p varies, the coefficients in (39) will vary too, so there is a different set of coefficients associated with each point p.

If f has a local power series expansion at a point p, there is only one possible choice for the coefficients of this series: f must be infinitely differentiable at p, by Theorem 3.10, and we must have

(41) $$a_n = \frac{f^{(n)}(p)}{n!} \quad \text{for} \quad n = 0, 1, 2, \ldots .$$

Thus,

(42) $$f(z) = \sum_{n=0}^{\infty} a_n(z-p)^n = \sum_{n=0}^{\infty} \frac{f^{(n)}(p)}{n!}(z-p)^n$$

for all z in some disc about p. Consequently, if f is analytic at p it is represented, at least near p if not at great distances, by the usual Taylor series (42) about p. We sum up these observations as follows.

Theorem 3.15 *If f is defined on an open set E, and if f has a local power series expansion at every point in E, then f is infinitely differentiable at every point in E, the coefficients in the series expansion about p are unique, and they coincide with the usual Taylor series coefficients of f at p.*

The main problem we must look into is the analog of Question 1 in Section 3.3.

(43) When is a function f analytic on an open set E?

If f is analytic, it must have complex derivatives of all orders. One of the great surprises of complex analysis, and the point upon which the theory of function of a complex variable diverges from the theory for a real variable, is that mere differentiability of a function of a complex variable makes it analytic.

Theorem * *A function f of a complex variable is analytic on an open set E if and only if it is holomorphic on E.*

To emphasize that this result is special to functions of a complex variable, we point out that it is possible for a function of a real variable $f(x)$ to be infinitely differentiable without being analytic; or to put it another way, $f(x)$ can be unrelated to its Taylor series, even though it has continuous derivatives of all orders throughout **R**.

Example 3.7 Consider the function $f(x)$ defined for all real x by the formula

(44) $$f(x) = \begin{cases} 0 & \text{if} \quad -\infty < x \le 0 \\ e^{-(1/x^2)} & \text{if} \quad 0 < x < +\infty. \end{cases}$$

This function, sketched in Figure 3.5, is obviously continuous at $x = 0$, since $\lim_{x \to 0} e^{-(1/x^2)} = 0 = f(0)$. It is not too difficult, using L'Hospital's rule, to prove that f is actually infinitely differentiable at $x = 0$, with

$$f^{(n)}(0) = 0 \quad \text{for } n = 0, 1, 2, \ldots ;$$

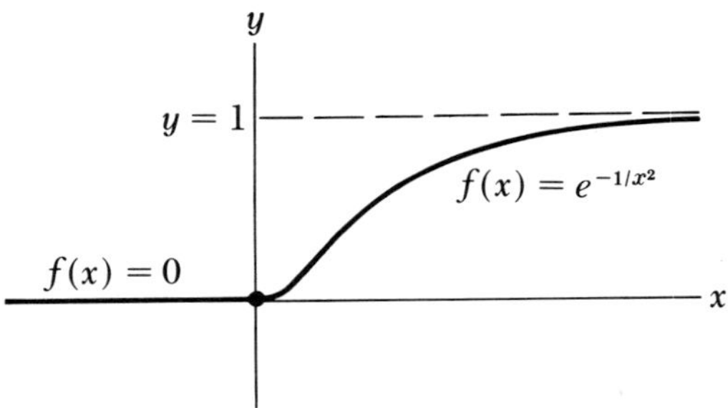

Figure 3.5 The function defined by formula (44).

the function obviously has derivatives of all orders at any point $x \neq 0$ in **R**. Now, if f has a local power series expansion about the point $p = 0$, it must be given by the usual Taylor series for f about p. But this Taylor series is degenerate: it converges to zero

$$f(0) + f'(0)x + \frac{f^{(2)}(0)}{2!}x^2 + \cdots = 0 + 0 \cdot x + 0 \cdot x^2 + \cdots = 0$$

for every x in **R**. Thus, the sum $g(x)$ of the Taylor series of f about p is the function $g(x) = 0$, and the Taylor series certainly does not reproduce the original function for *all* x near $x = 0$, even though f is infinitely differentiable throughout **R**.

If the reader will consult the proofs given in Calculus that functions of a real variable like e^x, *sin* x, and *cos* x coincide with their Taylor series, he will find that this was no simple matter, and special techniques were required in each case; mere differentiability of these functions could not be much help, as the last example shows. Perhaps these remarks will make it clear how remarkable it is that, for functions of a complex variable, mere differentiability is enough to make the function analytic.

We will not be able to prove the missing step in Theorem $*$ (that differentiable implies analytic) until Chapter 5, when we discuss the use of integration methods in complex analysis. But it is certainly not too soon to be aware of this result—this is really why we are devoting a chapter to analytic functions and power series. As we shall see, analytic functions have some very unusual properties; the phenomenon of "analytic continuation" is particularly interesting. One consequence of Theorem $*$ is that holomorphic functions are already analytic and have all these properties. For instance, in the theory of differential equations the solution functions, which are evidently differentiable, must actually be analytic and possess analytic continuation properties, among others, which are useful in understanding the differential equation.

It is clear (see Section 2.2), that if f and g are analytic on an open set, the functions $(f + g)(z) = f(z) + g(z)$ and $(\alpha \cdot f)(z) = \alpha \cdot f(z)$ (α a fixed complex scalar) are also analytic; given series expansions near $z = p$

$$f(z) = \sum_{n=0}^{\infty} a_n(z - p)^n \quad \text{and} \quad g(z) = \sum_{n=0}^{\infty} b_n(z - p)^n,$$

the new functions have series expansions

$$\sum_{n=0}^{\infty}(a_n + b_n)(z - p)^n \quad \text{and} \quad \sum_{n=0}^{\infty}\alpha \cdot a_n(z - p)^n.$$

Furthermore, if f and g are analytic at p, so is the product function $(f \cdot g)(z) = f(z) \cdot g(z)$; the series expansion about p is

$$(f \cdot g)(z) = a_0 b_0 + (a_1 b_0 + a_0 b_1)(z - p) + \cdots + \Big(\sum_{k+l=n} a_k b_l\Big)(z - p)^n + \cdots.$$

This is an easy consequence of the Cauchy formula for products of absolutely convergent series (Theorem 2.7), from which it is clear that the product series has a radius of convergence at least as large as the smaller of the radii for f and g, because the product series will converge absolutely at p if the series for f and g both do so.

The set $\mathscr{H}(E)$ of all functions which are analytic throughout an open set E is an **algebra:** functions may be added, subtracted, scaled by fixed complex numbers, and multiplied together without altering their analyticity. Once we have proved Theorem *, we will see that this class of analytic functions is identical to the class of all holomorphic functions on E, and thus must have several additional properties.

(i) If f is analytic, then so is its reciprocal $g = 1/f$, except at points z such that $f(z) = 0$, where the reciprocal is not well defined.
(ii) If $w = g(z)$ is analytic near $z = p$ and if $s = f(w)$ is analytic near the image point $q = g(p)$, then the composite function $h(z) = (f \circ g)(z) = f(g(z))$ is analytic near $z = p$.
(iii) If $w = f(z)$ is an analytic, invertible mapping between open sets $f:E \to F$, and if the derivative df/dz is non-vanishing on E, then the inverse mapping $\check{f}:F \to E$ is analytic.

If we replace "analytic" by "holomorphic," the validity of these statements is clear from the results on derivatives given in Chapter 2. It is possible to prove each of these results directly from the power series definition of analyticity, without reference to Theorem *, but the proofs are not easy and space limitations preclude our doing this in an introductory text.

These details may be found in the first chapter of the book by H. Cartan [3]. In his account of complex analysis, analytic functions and power series representations play the primary role, so it is natural that he should determine all the properties of analytic functions by working directly with power series. Understanding his book requires a fair amount of experience with mathematical proofs.

In the rest of this chapter Theorem *, and the corollary results (i) $\cdots$ (iii) mentioned above, will be used freely in working out *examples* to illustrate the properties of analytic functions; but it will not be used in proving any of the basic theorems which follow, so its tentative nature will not interfere with our analysis of how analytic functions behave.

Example 3.8 A polynomial $f(z) = a_n z^n + a_{n-1} z^{n-1} + \cdots + a_1 z + a_0$ is analytic at every point p in the plane, because it can always be written as

$$a_n((z-p)+p)^n + \cdots + a_1((z-p)+p) + a_0,$$

a finite sum that is easily rearranged into a (finite) series about the base point p

$$f(z) = c_n(z-p)^n + \cdots + c_1(z-p) + c_0.$$

The reciprocal of this polynomial, $g(z) = 1/f(z)$, is differentiable except at points where $f(z) = 0$, and hence is analytic (by Theorem *). Similarly, any rational function $f/g = f\cdot(1/g)$ (f and g polynomials; g not identically equal to zero) is analytic, except where the denominator is zero and the quotient is undefined.

Another important fact is that the sum of a power series is analytic at every point q in its disc of convergence. As it is, the representation of a function f as the sum of a power series only exhibits a power series expansion of f about the *base point p* of the power series; we want to show that f actually has a local power series expansion about any other point q in the disc of convergence.

Let q be in the disc of convergence $|z-p| < r$, and write $(z-p) = (z-q) + (q-p)$ in the series

$$f(z) = \sum_{n=0}^{\infty} a_n(z-p)^n. \tag{45}$$

Then

$$(z-p)^n = [(z-q)+(q-p)]^n = \sum_{k=0}^{n} \binom{n}{k} (z-q)^k (q-p)^{n-k}$$
$$= \sum_{k=0}^{\infty} \binom{n}{k} (z-q)^k (q-p)^{n-k}$$

for z in the disc $|z-p| < r$. Here $\binom{n}{k}$ are the usual "binomial coefficients"

$$\binom{n}{k} = 0 \qquad \text{if } k > n$$
$$\binom{n}{k} = \frac{n!}{(n-k)!\,k!} \qquad \text{if } 0 \le k \le n,$$

and we are writing $(a+b)^n = \sum_{k=0}^{\infty} \binom{n}{k} a^k b^{n-k}$ (a series with only finitely many non-zero terms). Clearly,

$$f(z) = \sum_{n=0}^{\infty} a_n(z-p)^n = \sum_{n=0}^{\infty} a_n\left[\sum_{k=0}^{\infty} \binom{n}{k} (z-q)^k (q-p)^{n-k}\right]$$
$$= \sum_{n=0}^{\infty}\left[\sum_{k=0}^{\infty} a_n \binom{n}{k} (z-q)^k (q-p)^{n-k}\right].$$

If we could legitimately interchange the order of the summations $\sum_{k=0}^{\infty}(\cdots)$ and $\sum_{n=0}^{\infty}(\cdots)$, we would get the desired result, a representation of $f(z)$ as the sum of a series in powers of $(z-q)$ for z near q; indeed, we would have

(46)
$$f(z) = \sum_{n=0}^{\infty}\left[\sum_{k=0}^{\infty} a_n \binom{n}{k} (z-q)^k (q-p)^{n-k}\right]$$
$$= \sum_{k=0}^{\infty}\left[\sum_{n=0}^{\infty} a_n \binom{n}{k} (z-q)^k (q-p)^{n-k}\right]$$
$$= \sum_{k=0}^{\infty}(z-q)^k \cdot \left[\sum_{n=0}^{\infty} a_n \binom{n}{k} (q-p)^{n-k}\right]$$
$$= \sum_{k=0}^{\infty} b_k (z-q)^k.$$

The legitimacy of such an interchange requires careful examination; we must stress the fact that limit operations are *not* always interchangable, and that naive manipulation of them can lead to the most ridiculous conclusions. We will not go into these details, since they would require a discussion of "double series"; there is a nice, self-contained account of this subject, and its application to prove (46), in Hille [9], Sections 5.3 and 5.6. We will not refer to these details in the rest of this book; after all, once Theorem $*$ has been proved, the analyticity of the sum $f(z)$ in (45) is obvious, because we know that $f(z)$ is complex differentiable. On the other hand, the proof based on manipulating series yields additional information, not obtained by invoking Theorem $*$, and the "double series" techniques introduced in the discussion are very useful in their own right. The facts we shall use are summarized as follows.

Theorem 3.16 *A power series $\sum_{n=0}^{\infty} a_n(z-p)^n$ converges to a function $f(z)$ that is analytic on its open disc of convergence $E = \{z : |z-p| < r\}$. If q is any point in this disc, the Taylor series of f about this new base point*

$$\sum_{k=0}^{\infty} b_k (z-q)^k = \sum_{k=0}^{\infty} \frac{f^{(k)}(q)}{k!}(z-q)^k$$

converges, and agrees with $f(z)$, near q. The radius r' of the new series is bounded between the limits $r_1 \le r' \le r_2$, where

(47)
$$r_1 = r - |q-p| \quad \text{and} \quad r_2 = r + |q-p|.$$

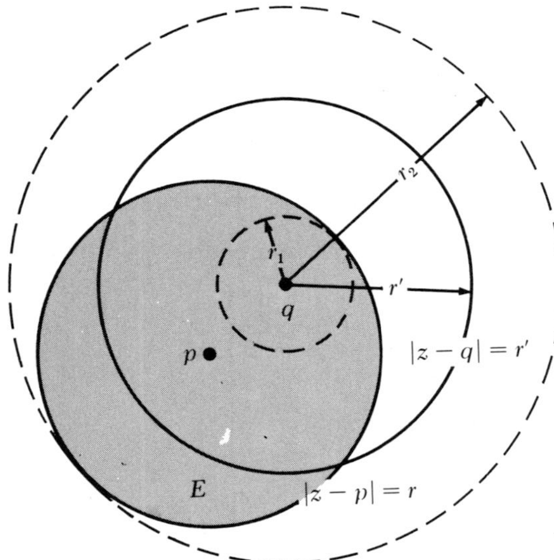

Figure 3.6 The radius of convergence r' of the Taylor series about a new base point $z = q$ lies between the limits $r_1 = r - |p - q|$ and $r_2 = r + |p - q|$. The original disc of convergence $E = \{z\colon |z - p| < r\}$ is shaded.

In Figure 3.6 we show open discs about q which have radii r_1 and r_2. These are, respectively, the largest disc about q lying entirely within E, and the smallest disc about q which entirely contains E. The actual disc of convergence $|z - q| < r'$ must be located more or less as shown in Figure 3.6 (we show the case $r_1 < r' < r_2$). There are series for which $r' = r_1$ and others for which $r' = r_2$ (see Exercise 12). If $r' > r_1$, this means that the new disc of convergence, about q, extends outside of the original disc of convergence $|z - p| < r$. Whenever the Taylor series for f about a new base point q converges at points which lie outside of the disc E on which f was originally defined, the sum of this series provides us with a means of defining f at points where it was not originally defined. This "continuation" or "extension" process will be discussed in Section 3.7.

EXERCISES

1. Use L'Hospital's rule to prove that df/dx exists at $x = 0$ and is equal to zero, in Example 3.7. Calculate df/dx for $x \neq 0$ by direct methods. Verify that df/dx is continuous at $x = 0$.

2. If $p \neq 0$ and if $\log p$ is one of the values of the logarithm of p, show that the series

$$\log(p) + \sum_{n=1}^{\infty} \frac{(-1)^{n+1}}{np^n} (z - p)^n$$

converges on the disc $D = \{z\colon |z - p| < |p|\}$. Show that its sum $f(z)$ is a determination of $\log z$ (recall the definition in Section 2.6) on D such that $f(p) = \log p$. Sketch D, showing its relationship to $z = p$ and $z = 0$.

Hint: Differentiate the series; display a holomorphic determination of $\log z$ on D which has the same derivative; use Theorem 2.18.

3. Use Exercise 2 and the essential uniqueness of $\log z$ (Exercise 6, Section 2.10) to show that any continuous determination of $\log z$, defined on an open set E that excludes $z = 0$, is *analytic* on E.

Note: This proves the analyticity of determinations of $\log z$ directly, without appealing to Theorem $*$.

4. If $a \neq 0$, show that $f(z) = 1/(a + z)$ has a series expansion

$$f(z) = \sum_{n=0}^{\infty} \frac{(-1)^n}{a^{n+1}} z^n$$

valid in the disc $|z| < |a|$ about the origin. What can be said for $g(z) = 1/(a - z)$?

Hint: Use facts about geometric series to sum this series. Could you have determined the form of the series on your own, by noting the resemblance between $f(z)$ and $1/(1 - z)$?

5. Let f be analytic, with $f(z) = \sum_{n=0}^{\infty} a_n(z - p)^n$ near p, and assume that $f(p) = a_0 \neq 0$. Then, by Theorem $*$, the reciprocal $g(z) = 1/f(z)$ is also analytic near p. Show that the coefficients in the series expansion $g(z) = \sum_{n=0}^{\infty} c_n(z - p)^n$ are obtained by recursively solving the simple system of equations

$$\begin{aligned}
&a_0c_0 = 1\\
&a_1c_0 + a_0c_1 = 0\\
&a_2c_0 + a_1c_1 + a_0c_2 = 0\\
&\quad\vdots\\
&\sum_{k+l=n} a_kc_l = 0 \qquad n \geq 1\\
&\quad\vdots
\end{aligned}$$

Hint: Use the Cauchy formula for products of series and the identity $f(z) \cdot g(z) = 1$, valid for all z near p.

6. Use the ideas in Exercise 5 to calculate the first four *non-zero* coefficients in the series expansions.

(i) $\operatorname{sech} z = 1/\cosh z$ about $p = 0$
(ii) $1/(1 + z^2)$ about $p = 0$
(iii) $1/(1 + z + z^2)$ about $p = 1 + i0$.

Note: In Section 3.5 we will show that $\cosh z = 1 + z^2/2! + z^4/4! + \cdots$ for all z.

7. Extend the ideas in Exercise 5 to calculate the first three *non-zero* terms of $\tan z$ about $p = 0$. (Use the identity $\cos z \cdot \tan z = \sin z$, and the Cauchy product formula; set up a system of equations whose solution yields the coefficients of the series for $\tan z$.)

Note: In Section 3.5 we will show that $\sin z = z - z^3/3! + z^5/5! - \cdots$ and $\cos z = 1 - z^2/2! + z^4/4 - \cdots$ for all z.

$$\textit{Answer:}\ \tan z = z + \left(\frac{1}{2!}\right) z^3 + \left(\frac{1}{2!\,2!} - \frac{1}{4!}\right) z^5 + \cdots.$$

8. Calculate the Taylor series about $p = 1$ for $f(z) = z + (1/z)$. Can you show, by direction calculations, that f agrees with its Taylor series near p? What is the radius of convergence? Does f exhibit any exceptional behavior on the boundary circle $|z - p| = r$?

9. Use the series representation $\cos z = 1 - z^2/2! + z^4/4 - \cdots$ for all z (proved in Section 3.5) to calculate the following limits.

(i) $\displaystyle\lim_{z\to 0} \frac{1 - \cos z}{z^2} = \frac{1}{2!}$

(ii) $\displaystyle\lim_{z\to 0} \frac{1 - \cos z}{z^3}$ does not exist

(iii) $\displaystyle\lim_{z\to 0} \frac{1 - \cos z}{z} = 0.$

Note: In Chapter 6 we will work out a detailed L'Hospital rule using similar calculations with series expansions.

10. For the following differential equations, determine series representations for all solutions that are defined and analytic near the origin, and satisfy the boundary conditions shown.

(i) $\displaystyle\frac{d^2f}{dz^2} + f = 0; \qquad f(0) = 1, f'(0) = 0$

(ii) $\displaystyle z\frac{d^2f}{dz^2} + f = 0; \qquad f(0) = 0, f'(0) = 1$

Calculate the radius of convergence of each solution series.

Hint: Write $f(z) = \sum_{n=0}^{\infty} c_n z^n$ and determine the c_n by matching coefficients in the series for f, f'', and zf''.

Answers: (i) $f(z) = 1 - z^2/2! + z^4/4! - \cdots = \cos z$; radius $r = +\infty$; (ii) $c_0 = 0$, $c_1 = 1$, $c_n = (-1)^{n+1}/n!\,(n-1)!$ if $n \geq 1$; radius $r = +\infty$.

11. Show that the function $f(z) = 1/(1 - z)$, defined on $E = \{z : z \neq 1\}$, has the Taylor series about a point p:

$$\sum_{n=0}^{\infty} \frac{f^{(n)}(p)}{n!} (z - p)^n = \left(\frac{1}{1 - p}\right) \sum_{n=0}^{\infty} \left(\frac{z - p}{1 - p}\right)^n.$$

Verify that

(i) Its radius of convergence is $r = |1 - p|$, the distance from the base point p to the "singular point" $z = 1$.

(ii) The sum of the Taylor series is equal to $f(z)$ on the whole disc of convergence $|z - p| < |1 - p|$.

Thus, we have a direct proof that f is analytic on E.

Hint: For (ii), substitute $w = (z - p)/(1 - p)$ in the geometric series $\sum_{n=0}^{\infty} w^n$.

12. Consider the series expansions of $\sum_{n=0}^{\infty} z^n = 1/(1 - z)$ about various new base points q in the disc of convergence $D = \{z : |z| < 1\}$. For which choices of q does the new radius of convergence r' satisfy (i) $r' = r_1$, (ii) $r_1 < r' < r_2$, (iii) $r' = r_2$ (radii r_1, r_2 defined in (47))? Use Exercise 11.

13. Take Log z defined on the cut plane P obtained by deleting the negative real axis $(-\infty, 0]$ from **C**. For p in P, calculate the Taylor series about p, and sketch the disc of convergence D_p, noting particularly the location of the origin with respect to D_p. For which choices of p does D_p extend across the cut? How is the sum of the Taylor series related to Log z? Is Log z analytic on P?

3.5 POWER SERIES EXPANSIONS OF THE EXPONENTIAL AND OTHER ELEMENTARY FUNCTIONS

Starting with the well known Taylor series expansions (in real variable) for the functions

(48) $$e^x = 1 + x + \frac{x^2}{2!} + \cdots \qquad \text{(all real } x)$$

and

(49) $$\sin y = y - \frac{y^3}{3!} + \frac{y^5}{5!} - \cdots$$
$$\cos y = 1 - \frac{y^2}{2!} + \frac{y^4}{4!} - \cdots \qquad \text{(all real } y),$$

we will show that the complex exponential e^z is given by the sum of the absolutely convergent power series

$$e^z = 1 + z + \frac{z^2}{2!} + \cdots = \sum_{n=0}^{\infty} \frac{z^n}{n!} \tag{50}$$

at all points in the complex plane. The fact that this series is everywhere absolutely convergent, so that its sum is analytic, is clear from the ratio test; what is not at all clear is that the sum of this series agrees with the function e^z which, as far as we know, is only holomorphic. The reader might want to recall the difficulties involved in proving formulas (48) and (49) for a real variable.

If we multiply each term of the series for $\sin y$ by i, we get an absolutely convergent series of complex numbers

$$i \sin y = (iy) - i\frac{y^3}{3!} + \cdots = (iy) + \frac{(iy)^3}{3!} + \frac{(iy)^5}{5!} + \cdots;$$

combining this with the series for $\cos y$

$$\cos y = 1 - \frac{y^2}{2!} + \frac{y^4}{4!} - \cdots = 1 + \frac{(iy)^2}{2!} + \frac{(iy)^4}{4!} + \cdots$$

gives us the formula

$$e^{iy} = \cos y + i \sin y = 1 + (iy) + \frac{(iy)^2}{2!} + \frac{(iy)^3}{3!} + \cdots,$$

valid and absolutely convergent for all real y. Now we may apply the Cauchy formula for the product of two series to express $e^{x+iy} = e^x \cdot e^{iy}$ as an absolutely convergent series in z;

$$\begin{aligned} e^z = e^{x+iy} &= \left(1 + x + \frac{x^2}{2!} + \cdots\right) \cdot \left(1 + (iy) + \frac{(iy)^2}{2!} + \cdots\right) \\ &= 1 + (x + iy) + \cdots + \left(\frac{x^n}{n!} + \frac{x^{n-1}}{(n-1)!}\frac{(iy)}{1!} + \cdots + \frac{(iy)^n}{n!}\right) + \cdots. \end{aligned}$$

But the nth term here is the finite sum

$$\sum_{k=0}^{n} \frac{x^{n-k}}{(n-k)!}\frac{(iy)^k}{k!} = \frac{1}{n!}\sum_{k=0}^{n} \frac{n!}{(n-k)!\,k!} x^{n-k}(iy)^k = \frac{1}{n!}\sum_{k=0}^{n} \binom{n}{k} x^{n-k}(iy)^k,$$

where $\binom{n}{k}$ is the usual binomial coefficient; obviously, then, this sum is just the binomial formula for the product $\frac{1}{n!}(x + iy)^n$, and the series takes the desired form

$$e^z = 1 + z + \frac{z^2}{2!} + \cdots = \sum_{n=0}^{\infty} \frac{z^n}{n!}.$$

From the formulas in Theorem 2.6 for combining convergent series, it follows immediately that the functions $\sin z$, $\cos z$, and $\sinh z$, $\cosh z$ agree with the series obtained by substituting the complex variable z in the familiar Taylor series, valid for the real variable x:

$$\sin z = z - \frac{z^3}{3!} + \frac{z^5}{5!} - \cdots$$

$$\cos z = 1 - \frac{z^2}{2!} + \frac{z^4}{4!} - \cdots$$

and

$$\sinh z = z + \frac{z^3}{3!} + \frac{z^5}{5!} = \cdots$$

$$\cosh z = 1 + \frac{z^2}{2!} + \frac{z^4}{4!} + \cdots.$$

Functions like $\tan z = \dfrac{\sin z}{\cos z}$ are analytic throughout their natural domains of definition, but their Taylor series, like the Taylor series of their real variable counterparts, are more difficult to calculate.

EXERCISES

1. Show that the polynomials $s_n(z) = 1 + z + \cdots + z^n/n!$ do not converge uniformly to their limit $f(z) = e^z$ on $E = \mathbf{C}$.

Hint: What is the maximum of $|s_n(z) - e^z|$ on $E = \mathbf{C}$, for fixed n?

2. Exhibit a power series $\sum_{n=0}^{\infty} a_n z^n$ such that

$$\frac{\sin z}{z} = \sum_{n=0}^{\infty} a_n z^n$$

for all $z \neq 0$. Prove that

(i) $\lim_{z\to 0} \dfrac{\sin z}{z} = 1$

(ii) If we define $h(z)$ to be $\dfrac{\sin(z)}{z}$ for $z \neq 0$ and $h(0) = 1$, then h is analytic throughout $\mathbf{C}$.

Note: This is the prototype of a vast family of examples.

3. Show that $\lim_{z\to 0} \operatorname{ctn} z$ does not exist, by writing $\operatorname{ctn} z$ in the form $\varphi(z)/z$ for $z \neq 0$, where φ is continuous at the origin with $\varphi(0) = \lim_{z\to 0} \varphi =$ 1.

Hint: Use series expansion of the denominator $\sin z$. If we remove the common factor z from the series for $\sin z$, what can be said about the convergence of the series which remains?

4. Use series expansions of numerator and denominator to calculate the following limits. (Carry common factors of the form $(z - p)^k$ outside the series.)

(i) $\lim_{z\to 0} z \operatorname{ctn} z = 1$

(ii) $\lim_{z\to 0} \dfrac{1 - \cosh z}{\sinh z} = 1$

(iii) $\lim_{z\to i} \dfrac{(z - i)e^z}{(1 + z^2)} = e^i/2i$

(iv) $\lim_{z\to 0} \dfrac{z}{\sin z} = 1$

(v) $\lim_{z\to 0} \dfrac{1}{\sin z}$ does not exist.

3.6 ANALYTIC CONTINUATION (I)

The "analytic continuation" phenomenon for analytic functions of a complex variable arises from the following seemingly innocent fact.

Theorem 3.17 *Let f be analytic on the open set E and assume that there is a sequence $\{p_n\}$ of points in E such that*

(i) *The sequence converges to some point p in E.*

(ii) *The points p_n are all distinct from p.*

(iii) *Each point is a zero of f, so that $f(p_n) = 0$ for $n = 0, 1, 2, \ldots$.*

Then there is some disc of positive radius about p on which f is identically equal to zero.

This result says that if an analytic function f is zero on an infinite set of points having a limit in the domain of definition of f, then f is identically zero *near* the limit point.

PROOF: Since E is an open set and p is in E, there is some disc of positive radius about p which lies entirely within E. Because f is analytic at p, there is a smaller disc D about p which lies within E, and on which f coincides with the sum of its Taylor series:

$$(51) \quad f(z) = f(p) + f'(p)(z - p) + \frac{f''(p)}{2!}(z - p)^2 + \cdots \quad \text{for all } z \text{ in } D.$$

The geometry of the situation is shown in Figure 3.7. Now f is continuous at p, and $f(p_n) = 0$ for all n, so we get $f(p) = \lim_{n\to\infty} f(p_n) = 0$. This means that the

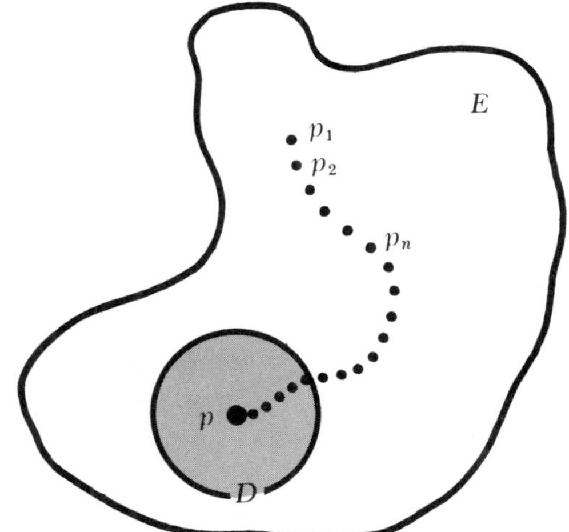

Figure 3.7 The point p is the limit of a sequence of points $\{p_n\}$ where $f(z) = 0$. The disc D lies within the set E where f is defined and analytic.

first coefficient in the series (51) is zero, and we may extract a common factor of $(z - p)$ from the remaining terms. This removal of a common factor cannot affect the convergence properties of the series, so the right-hand series below must converge on D to some function $F_1(z)$:

$$\text{(52)} \quad f(z) = \sum_{n=1}^{\infty} \frac{f^{(n)}(p)}{n!}(z-p)^n = (z-p)\sum_{n=1}^{\infty} \frac{f^{(n)}(p)}{n!}(z-p)^{n-1} = (z-p)F_1(z).$$

But the fact that $f(p_n) = 0$, and $p_n \neq p$, tells us that $F_1(p_n) = f(p_n)/(p_n - p) = 0/(p_n - p) = 0$ for all n. Now F_1, being the sum of a power series

$$\text{(53)} \qquad F_1(z) = f'(p) + \frac{f''(p)}{2!}(z-p) + \cdots$$

is continuous at p, so we see that $F_1(p) = \lim_{n\to\infty} F_1(p_n) = 0$. Thus, the lead coefficient in (53) is zero: $f'(p) = 0$, and the Taylor series (51) must have its first *two* coefficients equal to zero. Thus we may remove the common factor $(z - p)^2$ from the remaining terms, and write

$$\text{(54)} \qquad f(z) = \sum_{n=2}^{\infty} \frac{f^{(n)}(p)}{n!}(z-p)^n = (z-p)^2\sum_{n=2}^{\infty} \frac{f^{(n)}(p)}{n!}(z-p)^{n-2}$$
$$= (z-p)^2F_2(z),$$

where the right hand series converges on D to some function F_2 such that

$$\text{(55)} \qquad F_2(z) = \frac{f''(p)}{2!} + \frac{f'''(p)}{3!}(z-p) + \cdots.$$

Just as with $F_1(z)$, we see that $F_2(p_n) = f(p_n)/(p_n - p)^2 = 0$ for all n, so that $F_2(p) = 0$; this implies that the lead coefficient in (55) is zero, and that

$f''(p)/2! = 0$. Thus, the first *three* terms are zero in the Taylor series (51). Continuing these arguments, we find that

$$\frac{f^{(n)}(p)}{n!} = 0 \quad \text{for} \quad n = 0, 1, 2, \ldots,$$

so that the Taylor series (51) takes the form $f(z) = 0 + 0 \cdot (z - p) + \cdots = 0$ for all z in the disc D. ▌

Corollary 3.18 *If f and g are analytic functions on an open set E, and if $f(p_n) = g(p_n)$ for $n = 1, 2, \ldots$, for some sequence $\{p_n\}$ of distinct points in E which converge to a limit p which is also in E, then $f = g$ throughout some disc of positive radius about p.*

PROOF: The function $h = f - g$ is analytic on E, with $h(p_n) = 0$ for all n. Since the p_n are distinct, we may apply Theorem 3.17. ▌

If we now consider analytic functions defined on *connected* sets (domains), we arrive at a truly remarkable result. If f is zero at an infinite sequence of points which have their limit in the domain, the function is zero *everywhere*. Naturally, the proof makes heavy use of the connectedness of the set on which f is defined.

Theorem 3.19 (Fundamental theorem on analytic continuation) *Let E be a domain (connected open set). Let f and g be analytic functions defined on E that agree at some sequence of distinct points $\{p_n\}$ which converge to a point p that is also in E. Then $f = g$ throughout all of E.*

PROOF: Since $f(p_n) = g(p_n)$ and the points p_n are distinct, we know that $f = g$ on some disc of positive radius about p. In the rest of the proof we will show how the connectedness of E implies that $f = g$ on all of E. The remaining arguments are almost entirely geometric, and are not essential in understanding our applications of this theorem.

Consider the set F_1 of all points z in E such that $f = g$ near z (that is, on some disc about z). Let F_2 be the complementary set $F_2 = E \sim F_1$. Obviously, F_1 and F_2 are disjoint sets whose union is E. The set F_1 is non-empty, since it includes p. It is also an *open* set: if z_0 is in F_1, then $f = g$ on some disc about z_0, which means that $f = g$ near each point in this disc. Therefore, the disc lies within F_1 (since each of its points belongs to F_1), and F_1 is open.

The set F_2 may be empty; if it is non-empty, it is *open*, as we will now prove. Consider a typical point z_0 in F_2. There must be a disc of positive radius about z_0 which does not meet F_1.

If this is not so, points from F_1 will appear in every such disc about z_0. By considering the radii $r_n = 1/n$, we can then select points z_n from F_1 such that $0 < |z_n - z_0| < 1/n$ for $n = 1, 2, \ldots$; these points are distinct from z, since they lie in F_1 while z_0 belongs to F_2. Now $z_n \to z_0$ as $n \to \infty$ and $f(z_n) = g(z_n)$ for all n; in view of Corollary 3.18, we are led to conclude that $f = g$ near z_0, so that z_0 is in F_1. This is impossible; z_0 is in F_2, which is disjoint from F_1. We arrived at this impasse by assuming that no disc about z_0 is disjoint from F_1; thus, there must be some disc of positive radius which does not meet F_1.

By taking a slightly smaller disc, if necessary, we can insure that this disc about z_0 also lies within E, since E is open. This smaller disc must lie within F_2 since it is contained in E and does not meet F_1. Thus, if z_0 is any point in F_2, there is a disc of positive radius about z_0 that lies within F_2, and F_2 is an open set if it is non-empty.

If F_2 is non-empty, then F_1 and F_2 are *open* sets such that

$$F_1 \cap F_2 = \varnothing \text{ (disjoint sets)}$$

$$F_1 \cup F_2 = E.$$

Such a partitioning of E into F_1 and F_2 is inconsistent with its connectedness. Therefore, F_2 must be empty. This implies that $F_1 = E$, so that $f = g$ throughout E, and the theorem is proved. ▌

Example 3.9 On a *disconnected* open set this theorem can fail. For example, $E = \{z : \text{Im}(z) \neq 0\}$ consists of two disjoint open sets, and upper half plane and a lower half plane, and is not connected. In E consider the functions f and g such that $f(z) = 0$ for all z, and

$$g(z) = 0 \text{ on the lower half plane } (\text{Im}(z) < 0)$$

$$g(z) = 1 \text{ on the upper half plane } (\text{Im}(z) > 0).$$

Here, $f \neq g$ on the upper half plane, while $f = g$ on the lower half plane.

A crucial aspect of the hypotheses in these theorems is that the points p_n where $f(p_n) = 0$ (or $f(p_n) = g(p_n)$) must have a limit which lies *within* the domain we are considering. It is not enough to have infinitely many distinct points p_n such that $f(p_n) = 0$ (or $f(p_n) = g(p_n)$); these points may tend toward the boundary of our domain, or toward infinity, and not have any limit within E, as in the next example.

Example 3.10 Let $E = \mathbf{C}$ and let $f(z) = e^z - 1$. Then $f(z) = 0$ at the points $z_n = 2\pi ni$ (n any integer), but f certainly is not zero throughout the plane. Here the sequence $\{z_n\}$ tends toward infinity, along the imaginary axis, and has no limit in E. For a different kind of example, let E be the unit disc $\{z : |z| < 1\}$ and let

$$f(z) = \exp\left(\frac{2\pi i}{1 - z}\right) - 1 \quad \text{for all} \quad z \text{ in } E.$$

This function is certainly not identically zero on E and is analytic, being a composite of analytic functions $w = \dfrac{2\pi i}{1 - z}$ and $s = e^w - 1$. Furthermore, it is zero at an infinite sequence of points $p_n = 1 - (1/n)$ $(n = 1, 2, \ldots)$ which approach the boundary point $p = 1 + i0$ as $n \to \infty$. However, this point p is not in E, so this situation does not conflict with Theorem 3.19.

The last theorem above has the very striking consequence that something happening on a tiny piece of a *connected* open set E completely determines what is happening throughout E, even at great distances. For example, if f and g are analytic, and if $f = g$ on some small disc about a point p, this coincidence "propagates" to be true throughout E. Also, an analytic function f on E cannot be zero on a small piece of E without being identically zero throughout E (compare f with the analytic function $g(z) = 0$ on E). This "propagation" or "continuation" of identities is special to *analytic* functions of a real or a complex variable. Once Theorem $*$ is proved, we will see that all holomorphic functions of a complex variable exhibit this phenomenon. For real variables, even infinitely differentiable (but non-analytic) functions don't do this; we refer to the example given earlier of an infinitely differentiable function of a real variable which was identically zero on the left half of $\mathbf{R}$ and non-zero throughout the right half. The set $E = \mathbf{R}$ is connected, but the function is not (real) analytic, in spite of its differentiability. Here are a few interesting applications of the analytic continuation principle embodied in Theorem 3.19.

Application 1 (The change of base point theorem) If we start with the sum of a power series $f(z) = \sum_{n=0}^{\infty} a_n(z - p)^n$ on its disc of convergence $E = \{z : |z - p| < r\}$ and consider the Taylor series about some new base point q within E, the new series converges to its sum $g(z)$ in a disc D about q whose radius r' has been discussed in Section 3.4. The two discs of convergence E and D meet in a lens-shaped open set, like the one shown in Figure 3.6; we know that f and g agree near the base point $z = q$ by Theorem 3.16, but the discussion in that proof does not tell us whether they will agree throughout the larger set $D \cap E$ where the discs meet. By using the analytic continuation principle it is easy to see that:

(56) The series $f(z)$ about p and $g(z)$ about q must agree throughout the set $D \cap E$ where the two discs of convergence meet.

In fact, the set $D \cap E$ is connected (for two points in the set, the line segment between them lies within the set), and the functions $f(z)$ and $g(z)$, which are both defined on $D \cap E$, are analytic since they are sums of power series. Since they agree near $z = q$, they must agree throughout $D \cap E$.

Application 2 (Permanence of algebraic identities) Suppose that the functions $f, g, \ldots$ are analytic functions defined on the complex plane.

Theorem 3.20 *If $f = g$ on the real axis, $f(x + i0) = g(x + i0)$ for all real x, then $f = g$ throughout $\mathbf{C}$. Also, if $f = 0$ on the real axis, then $f = 0$ throughout $\mathbf{C}$.*

PROOF: Take any distinct sequence of points $z_n = x_n + i0 \to p = x + i0$ in $\mathbf{R}$; then $f = g$ at these points. Since $\mathbf{C}$ is a connected open set, it follows that $f = g$ throughout **C**. For the second statement, take $g(z) = 0$ everywhere. ∎

One very interesting consequence of this idea is that any trigonometric identity, valid on **R** for the trigonometric functions of a real variable, is valid throughout **C** when we use equations (21) and (22) of Chapter 2 to define the complex variable versions of trigonometric functions. For example, the formula $\sin^2 x + \cos^2 x = 1$ is actually valid throughout **C**; $\sin^2 z + \cos^2 z = 1$ for all z (the right side is the constant function $g(z) = 1$, which is certainly analytic). Even formulas such as

$$\cos^2 z - \sin^2 z = \cos(2z)$$

are valid: they are valid for $z = x + i0$ on the real axis, and both sides are analytic functions—the left side, because sums and products of analytic functions are analytic; the right side, because we may exhibit a power series expansion for $\cos(2z)$ by substituting $w = 2z$ in the series for $\cos w$. As another example, suppose that f is analytic on **C** and satisfies a certain differential equation, such as

$$z\frac{d^2f}{dz^2} - f = 0,$$

at all points $z = x + i0$ on the real axis. Since f and its derivatives are analytic, the left hand expression is analytic on **C**; therefore, the identity must be satisfied throughout the plane.

Going one step further, these arguments work equally well for analytic functions f and g defined on any domain which meets the real axis in an interval of positive length, with $f = g$ on this interval. We conclude that $f = g$ throughout E. For example, to prove the identity

$$\tan^2 + 1 = \sec^2 z$$

we must work with the domain $E = \{z : z \neq n\pi + (\pi/2) \text{ for } n = 0, \pm 1, \pm 2, \ldots\}$ on which these functions are well defined. Since $\tan z$ and $\sec z$ are quotients of analytic functions, they are analytic on E, and we know that the identity is true on an interval such as $(-\pi/2, +\pi/2)$ in the real axis. Therefore, the identity must be true on E because E is a domain.

Extending these ideas in another direction, we may take an algebraic combination of analytic functions $f, g, \ldots$, and the variable z (which is given by the analytic function $h(z) = z$) on some domain E. Then this combination is also an analytic function on E. If the combination vanishes on a suitable small subset of E, it must vanish throughout E. For example, suppose the combination

$$f(z)^2 - g(z) + \frac{1}{z}$$

is considered on a domain which excludes $z = 0$ (so that $1/z$ is well defined), and suppose that this combination is zero on an arc or a small disc in E. Then this identity is true throughout the domain E. (The arc need not lie on the real axis.)

***Application 3* (Comparing analytic functions with their Taylor series.)**

Theorem 3.21 *Let f be analytic on an open set E (connected or not), and let p be a point in E. Assume that the domain of convergence of the Taylor series*

(57) $$\sum_{n=0}^{\infty} \frac{f^{(n)}(p)}{n!}(z-p)^n$$

of f about p includes some disc D which lies within E. Then f agrees with its Taylor series not only near p (which is all that is guaranteed by the definition of "analytic at p"), but actually agrees with its Taylor series throughout the disc D.

PROOF: Write $g(z)$ for the sum of the series (57). Then f and g are both defined and analytic on D, and they agree near p since f is assumed to be analytic. Since D is a connected open set, they must actually agree throughout this disc. ▌

In particular, if the whole disc of convergence of (57) lies within E, f agrees everywhere with its Taylor series about p. While this is still fresh in mind, we hasten to present an example in which the sum of the Taylor series *differs* from the original function at points far removed from the base point. It is crucial to notice that, in this example, the disc of convergence does not lie entirely within the domain E on which f is defined (Theorem 3.21 forbids it). Instead, it crosses the boundary of E and then re-enters E. The open set $D \cap E$ where the disc of convergence D meets E is not a connected set; thus, Theorem 3.19 is not violated.

Example 3.11 (*The square root function vs. its Taylor series*)

Let E be the cut plane obtained by deleting the negative real axis $J = (-\infty, 0]$ from the plane, and take $f(z) = z^{1/2}$. We have already seen that f is holomorphic (even infinitely differentiable) on E, and if we could invoke Theorem * we would see that f is analytic on E.

We can show that f is analytic by directly examining the Taylor series about a typical point p. Since

$$\frac{d^n}{dz^n}(z^{1/2}) = (\tfrac{1}{2})(\tfrac{1}{2}-1)\cdot\ldots\cdot(\tfrac{1}{2}-n+1)z^{((1/2)-n)}$$

for $n \geq 1$, this is just the series

(58) $$g(z) = \sum_{n=0}^{\infty} \binom{\frac{1}{2}}{n} p^{((1/2)-n)}(z-p)^n,$$

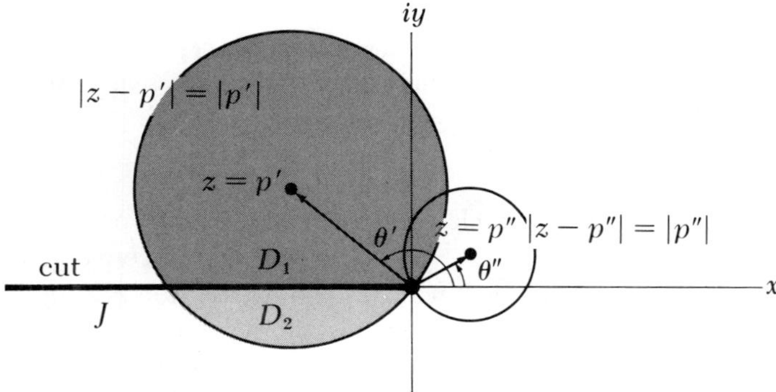

Figure 3.8 Discs of convergence for the Taylor series of $z^{1/2}$ about various points p', p'' in the cut plane E. The intersection $D \cap E$ consists of isolated open sets D_1 and D_2 when $z = p'$.

where $\binom{\frac{1}{2}}{n}$ are the "fractional binomial coefficients:"

$$\binom{\frac{1}{2}}{0} = 1 \qquad \text{if } n = 0$$

$$\binom{\frac{1}{2}}{n} = \frac{(\frac{1}{2})(\frac{1}{2}-1)\cdot\cdots\cdot(\frac{1}{2}-n+1)}{n!} \qquad \text{if } n = 1, 2, \ldots .$$

The ratio test quickly shows that (58) has radius of convergence $r = |p|$, so the point $z = 0$ lies on the boundary circle, as shown in Figure 3.8. For completeness, we indicate how one can prove directly that the series (58) coincides with $f(z) = z^{1/2}$ near p; this amounts to a direct proof that $z^{1/2}$ is analytic, and might give some useful insight into the management of Taylor series. However, we will not refer to these calculations, and they may be omitted if one is willing to invoke Theorem $*$.

Consider the series sum $g(z)$ on a disc about p small enough to be contained entirely within the cut plane. Term-by-term differentiation of the series gives $z \cdot g'(z) = (\frac{1}{2})g(z)$ on D, since

$$\begin{aligned} z \cdot g'(z) &= (z-p)g'(z) + p \cdot g'(z) \\ &= \sum_{n=0}^{\infty}\left[n\binom{\frac{1}{2}}{n} + (n+1)\binom{\frac{1}{2}}{n+1}\right]p^{((1/2)-n)}(z-p)^n \\ &= \frac{1}{2}\sum_{n=0}^{\infty}\binom{\frac{1}{2}}{n}p^{((1/2)-n)}(z-p)^n = \frac{1}{2}g(z). \end{aligned}$$

Hence,

$$\begin{aligned} \frac{d}{dz}\left[z^{-1/2}\cdot g(z)\right] &= (-\tfrac{1}{2})z^{-3/2}g(z) + z^{-1/2}g'(z) \\ &= z^{-3/2}[(-\tfrac{1}{2})g(z) + z\cdot g'(z)] = 0 \end{aligned}$$

vanishes on D. Since D is a domain, our results on the essential uniqueness of antiderivatives insure that $z^{-1/2}g(z) = \alpha$ (α a complex constant) on D; thus, $g(z) = \alpha z^{1/2}$. To determine α, take $z = p$; clearly, $g(p) = p^{1/2}$, so that $\alpha = 1$ and $g(z) = z^{1/2}$ throughout D. This reasoning works for any p in the domain E.

Once analyticity of $z^{1/2}$ is clear, we consider a base point of the form $p'' = re^{i\theta''}$, where $-\pi/2 \leq \theta'' \leq +\pi/2$. Then the disc of convergence of (58) lies within the cut plane, as shown in Figure 3.8, so the sum $g(z)$ of the Taylor series agrees with $f(z) = z^{1/2}$ throughout this disc, by Theorem 3.21. But if we consider $p' = re^{i\theta'}$ with $\pi > \theta' > \pi/2$ (or $-\pi/2 > \theta' > -\pi$), the disc of convergence of (58) crosses over the boundary of the cut plane E (bdry(E) is just the negative real axis J), and E meets the disc of convergence in two disjoint connected pieces D_1 and D_2, as shown in Figure 3.8. The point p' is in D_1, and $g(z) = z^{1/2}$ near p', so that $g(z) = z^{1/2}$ throughout D_1, by Theorem 3.19. But $f(z) = z^{1/2}$ suffers a discontinuity (a change of sign) when z crosses the negative real axis J, while the series (58) sums to a function $g(z)$ which has no discontinuities in the disc D. Thus $g(z)$ cannot coincide with $f(z) = z^{1/2}$ in the domain D_2 where the disc of convergence extends below the axis J. It is not very hard to see that g coincides with $-f(z) = -z^{1/2}$ throughout D_2, since the square root of z is unique up to a factor of ± 1.

What has happened in the last example is a basic phenomenon. If an analytic function f is given on an open set E and if we expand f about a point p in E, the disc of convergence of the Taylor series about p may cross over the boundary of E and its sum

$$g(z) = \sum_{n=0}^{\infty} \frac{f^{(n)}(p)}{n!}(z - p)^n$$

may differ from the original function f if the disc re-enters E after having crossed over the boundary. In the last example, the boundary was the negative real axis. In the next section we collect a few more observations on this phenomenon by which analytic functions may be extended (or "continued") beyond their original domains of definition.

EXERCISES

1. Suppose $f(z)$ is analytic on a domain E that is *symmetric* (if z is in E, so is $-z$) and includes a small interval $(-\delta, +\delta)$ in the real axis. If $f(x) = f(x + i0)$ is an *even function*, so that $f(-x) = f(x)$ for $-\delta < x < \delta$, show that $f(z)$ is an even function on E; that is, $f(-z) = f(z)$ on E. Prove a similar result for *odd functions*, those for which $f(-z) = -f(z)$.

2. Suppose that $f(z)$ is analytic on a domain E that includes the interval $(-1, +1)$ on the real axis. Assume that $f(x + i0) = \sqrt{1 - x^2}$ for $-1 < x < +1$. Prove that $f(z)$ is an analytic determination of the multiple valued function $\sqrt{1 - z^2}$ on E.

Hint: Show f satisfies a certain algebraic identity on $(-1, +1)$.

Note: In Chapter 4 we will take up the problem of explicitly constructing such determinations of $\sqrt{1 - z^2}$.

3. If E is a domain that includes either of the points $+1$ or -1, show that it is impossible to define an analytic determination $f(z)$ of $\sqrt{1 - z^2}$ on E.

Hint: If such an $f(z)$ existed, then $f(z)^2 = 1 - z^2$ on E. Differentiate both sides to calculate df/dz. What goes wrong at $z = +1$?

4. The functions $\sqrt{z}$ and $f(z) = (\sin \sqrt{z})/\sqrt{z}$ are well defined and holomorphic on the cut plane $P = \mathbf{C} \sim (-\infty, 0]$. Show that

$$f(z) = 1 - \frac{1}{3!} z + \frac{1}{5!} z^2 - \cdots = \sum_{n=0}^{\infty} \frac{(-1)^{n+1}}{(2n+1)!} z^n$$

for all z in P. Explain why it is then possible to define $f(z)$ on the cut so f is analytic on $\mathbf{C}$. What values $f(x + i0)$ should be assigned on the cut? Can this analysis be repeated for $g(z) = \sin \sqrt{z}$?

Hint: Set $w = \sqrt{z}$ in the series for $\sin w$; then divide by $\sqrt{z}$.

3.7 ANALYTIC CONTINUATIONS (II)

If $f(z)$ is analytic on some domain E, there may be *different* ways to continue f so that it is analytic on larger domains. We can systematically handle the questions which surround this phenomenon by introducing the following definition.

Definition 3.3 *Let f be analytic on a domain E, and f^* analytic on another domain E^*. We say that f^* is an* **analytic continuation** *of f from E to E^* if*

(i) *E^* is larger than E, so $E^* \supseteq E$*
(ii) *$f^* = f$ throughout the smaller domain E.*

By Theorem 3.19 it follows that the analytic function that extends $f(z)$ is uniquely determined once the larger domain E^* is given (if it is possible to define any continuation at all on E^*); in fact, E^* is connected and every continuation to E^* must agree with f on the subdomain D.

For example, we might think of the analytic function $f(z) = \text{Log}(z)$, the principal determination of the logarithm, defined on a small disc E about the point $p = 1 + i0$. It is possible to piece together various determinations of $\log z$ in different parts of the domains shown in Figure 3.9 to get well defined complex differentiable (hence analytic) continuations of f to analytic functions f^* defined on the larger domains E^*. When these continuations are compared at a point q which is some distance from p, such as $q = 0 + i1$ which appears in each of the domains in Figure 3.9, we may find that they do not agree. Every continuation (f^*, E^*) of $f(z) = \text{Log } z$ must give a determination of the logarithm because the identity

$$\exp(f(z)) = z \quad \text{(valid near } p)$$

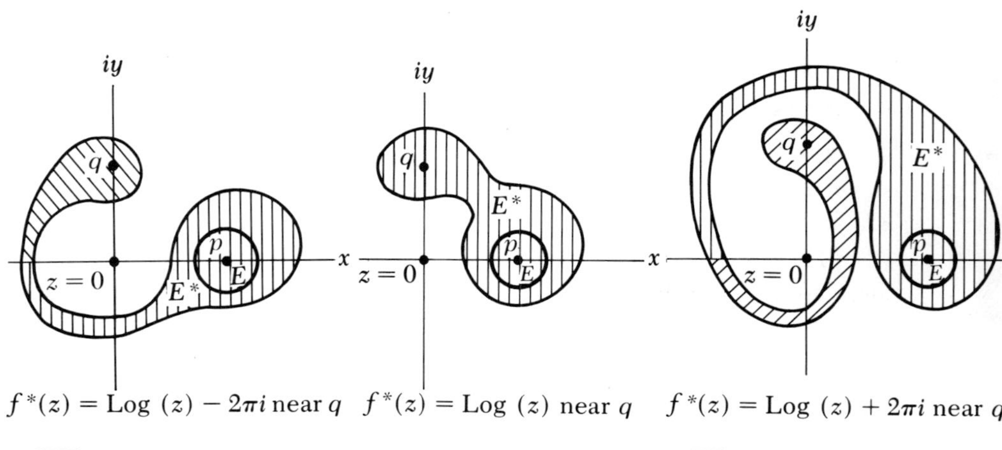

Figure 3.9 Three domains E^* in which we can define analytic continuations of Log z defined in the disc E about $p = 1 + i0$.

"propagates" to be valid throughout the connected domain E^*.† But we produce different determinations of log z near q by continuing Log z through different domains which include q; the particular determination we get for each domain is indicated in Figure 3.9. Of course, we must have $f^*(z) =$ Log $z + 2\pi ki$ for z near q, but the particular integer $k = 0, \pm 1, \pm 2, \ldots,$ which appears depends on the shape of the domain E^* (essentially, k is the number of times E^* makes a complete counterclockwise loop around the origin).

We might expect analytic continuations of $f(z) =$ Log z to give different determinations of the multiple valued function log z, but each continuation is required to be a well defined (single valued) function, and different continuations are needed to produce all possible determinations of log z. Similar comments apply to other functions which are multiple valued—in fact, the appearance of different analytic continuations might be taken as an *intrinsic definition* of what it means for a function to be "multiple valued." Thus, these remarks apply to functions like $f(z) = z^\alpha$ (α not an integer; otherwise the function is single valued), or Log z, or $\sqrt{z^2 - 1}$. See Example 3.13 below for additional comment.

There are certain functions which, when continued from one place to another, always produce the same function regardless of the shape of the domains used to carry out the continuation. Such functions can be thought of as "intrinsically single valued." For example, if we consider $f(z) = \tan z$, or rational functions such as $f(z) = 1/(z^2 + 1)$, defined near some point p, and make analytic continuations through domains which include another point q, these continuations all produce the same function near q (see Example 3.12 below).

The process of sorting out the various functions produced by analytic

† If f^* is holomorphic on E^*, so is $w = \exp(f^*(z))$, by the chain rule for derivatives. Comment (ii) associated with Theorem $*$ insures that this function is analytic, so Theorem 3.19 may be applied.

continuations of a given function $f(z)$ requires a deeper discussion of continuations, and the introduction of "Riemann surfaces" as a device for cataloging the different continuations which can turn up. A detailed account is too involved to include in this introduction to complex analysis. An extensive discussion of analytic continuations (and Riemann surfaces) can be found in more advanced texts, such as De Pree and Oehring [5], Nevanlinna and Paatero [18], or Levinson and Redheffer [15].

Example 3.12 Let points p and q be given (not equal to $+i$ or $-i$), and consider $f(z) = 1/(1 + z^2)$ on some small disc about p. We will show that every analytic continuation of f to a domain E^* which includes this disc and the point q merely reproduces the function $1/(1 + z^2)$ near q (and in fact throughout E^*).

No continuation (f^*, E^*) can have either of the points $+i$ or $-i$ within its domain E^*. In fact, the product $(1 + z^2)f^*(z)$ is analytic on E^*, and $(1 + z^2)f^*(z) = (1 + z^2)f(z) = 1$ near p. This identity must be valid throughout E^* by the analytic continuation theorem. This would be impossible if $+i$ or $-i$ appeared in E^*; the product would be *zero* at these points, no matter what value is assumed by f^*. By the same reasoning we also see that $(1 + z^2)f^*(z) = 1$ on E^*, so that

$$f^*(z) = \frac{1}{1 + z^2} \quad \text{for all} \quad z \text{ in } E^*.$$

A similar analysis can be made for other single valued analytic functions.

Example 3.13 Consider $f(z) = z^\alpha = e^{\alpha \cdot \text{Log}\, z}$, taking the exponent $\alpha = i$; f is defined and analytic on the small disc $D = \{z : |z - 1| < \frac{1}{4}\}$ about $p = 1 + i0$. The function has other determinations on the cut plane P obtained by deleting the negative real axis from the plane; these are obtained by taking other determinations of log z in the defining formula, to get

$$f_k(z) = e^{i(\text{Log}\, z + 2\pi k i)} = e^{i\, \text{Log}\,(z)} e^{-2\pi k} = \frac{z^i}{e^{2\pi k}}$$

for $k = 0, \pm 1, \pm 2, \ldots$. We can easily calculate the limit values of the principal determination $f_0(z) = z^i$ as z approaches a typical point $z_0 = -R + i0$ on the (deleted) negative real axis, from the upper or lower half plane.

$$\lim_{\delta \to 0+} (-R + i\delta)^i = e^{i(\log R + i\pi)} = \left(\frac{1}{e^\pi}\right) e^{i \log R}$$

$$\lim_{\delta \to 0+} (-R - i\delta)^i = e^{i(\log R - i\pi)} = (e^\pi) e^{i \log R}.$$

The values are altered by a multiplied factor $1/e^{2\pi}$ as z moves upwards across the negative real axis; similar calculations show this to be true for each of the other determinations $f_k(z)$, too.

If we line up different determinations f_k on opposite sides of the negative real axis we can make their limit values match up, so they give functions which are defined and differentiable at points on this axis (the pasted-together functions will now disagree along the *positive* real axis). The pairings which match along the cut are indicated in the following diagram.

$$\cdots\quad \frac{f_1}{f_2}\bullet \qquad \frac{f_0}{f_1}\bullet \qquad \frac{f_{-1}}{f_0}\bullet \qquad \frac{f_{-2}}{f_{-1}}\bullet\quad \cdots$$

By piecing together compatible determinations of z^i we can make analytic continuations of z^i from the disc D across the cut to obtain, eventually, the infinite number of distinct determinations of this function.

Notice how the other determinations would be forced to our attention, even if we only started with the principal determination z^i, once we begin to examine its analytic continuations. Once again the intimate connection between analytic continuations and multiple valued functions of complex variable makes its appearance.

EXERCISES

1. If f is a determination of $\log z$ defined on a domain E, show that every analytic continuation (f^*, E^*) to a larger domain is also a determination of $\log z$. Same for determinations of $z^{1/2}$.

Hint: Use the permanence of identities.

2. Let $f(z)$ be a determination of $\log z$ defined on a small disc E about $p \neq 0$. Show that no continuation (f^*, E^*) can be defined in which the enlarged domain $E^* \supseteq E$ includes the *origin*. If $p = +1$ and $q = -1$, display two continuations (f^*, E^*) and (f^{**}, E^{**}) of (f, E) such that (i) E^* and E^{**} both include q; (ii) f^* and f^{**} give *different* determinations of $\log z$ near $z = q$.

3. Show that $f(z) = \tan z$, defined near the origin, cannot be continued analytically so that the new domain E^* includes one of the "singular points" $z_n = (\pi/2) + n\pi$ $(n = 0, \pm 1, \pm 2, \ldots)$. Prove that $f^*(z) = \tan z$ throughout E^* (so $\tan z$ is intrinsically single valued).

Hint: Use the identity $\tan z \cdot \cos z = \sin z$.

3.8 THE MAXIMUM MODULUS THEOREM FOR ANALYTIC FUNCTIONS

As we go along we will see various ways in which the behavior of an analytic function in one place determines behavior elsewhere. We have already discussed the "permanence" of algebraic identities; now we turn to a family of results which show that if an analytic function has small absolute value (modulus) on a certain set, this forces the values to be small on other, related, sets. The basic result deals with the behavior of an analytic function near a single point.

Theorem 3.22 (Local maximum modulus theorem) *Consider a function $f(z)$ which is defined and analytic on some disc D about a point p, and assume that f is not identically constant on D. Then we can find points z' as close to p as we wish, such that*

$$|f(z')| > |f(p)|.$$

If $f(p) \neq 0$, we can also find points z'' as close to p as we wish, such that

$$|f(z'')| < |f(p)|.$$

We will prove this result at the end of the section; present commentary is directed toward its implications. As immediate consequences of Theorem 3.22 we see that an analytic function f defined on a *domain* E must have the following properties if it is not constant on E.

(i) If $|f(p)| \neq 0$, the absolute values $|f(z)|$ cannot assume a relative maximum or relative minimum value at p.
(ii) The only places where $|f(z)|$ is an absolute minimum are those points where $f(z) = 0$, the *zeros of* f.
(iii) The absolute value $|f(z)|$ can never achieve a *relative maximum at any point* within E.

Just consider $|f(z)|$ on a small disc about a typical point p in E. (Since E is a domain, f cannot be constant on such a disc without being constant throughout E.) Observation (iii) tells us that a maximum for $|f(z)|$ can only occur at a boundary point of E, and not within E. A very interesting interpretation of (i) and (ii) can be made by considering the surface $s = |f(x + iy)|$ over the set E in the plane. At each point $z = x + iy$ where $|f(z)| > 0$, there are directions in which $|f(z)|$ increases and others in which $|f(z)|$ decreases. Rain falling onto such a surface would always run off "downhill" in the direction of steepest descent, when it strikes a point on the surface where $|f(z)| \neq 0$. Since the zeros of f are the only relative minima, water would collect at these zeros, or else would run off the edge of the surface (when z reaches the boundary of E). Furthermore, the surface cannot have any "humps," which would correspond to relative maxima and are forbidden by (iii).

In Figure 2.6 we have sketched the level curves $s = |f(z)|$ for the polynomial $f(z) = 1 + z^2$; the reader should try to visualize the surface $s = |f(z)|$ which corresponds to these curves. The only zeros are located at $+i$ and $-i$. Note the presence of a "saddle point" over the origin; the existence of saddle points on the surface $s = |f(z)|$ is not forbidden by (i), (ii), and (iii). At great distances from the origin the functions $|f(z)| = |1 + z^2|$ and $|z^2| = |z|^2$ are practically indistinguishable. In Figure 2.8, we have shown the surface $s = |f(z)|$ for the exponential function $f(z) = e^z$; this surface has *no* zeros. It is of constant height along each line parallel to the imaginary axis, since $|f(x + iy)| = |e^{x+iy}| = e^x \cdot |e^{iy}| = e^x$ for all real x and y. The reader should examine this surface with remarks (i), (ii), and (iii) in mind.

One of the most striking results based on Theorem 3.22 is the following.

Theorem 3.23 (Maximum modulus theorem) *Let E be a domain that is bounded, and let $\bar{E}$ be its closure.*† *Let $f(z)$ be a function that is defined and continuous on $\bar{E}$, and is analytic on E. Then the absolute values $|f(z)|$ have a finite maximum value on the boundary*

$$M = \max\{|f(z)| : z \textit{ in } bdry(E)\},$$

and this maximum value dominates the modulus $|f(z)|$ for all points z within E:

$$|f(z)| \leq M \quad \textit{for all } z \textit{ in } E.$$

If $f(z)$ is not a constant function on E, we actually have $|f(z)| < M$ for all z in E.

Proof: (from Theorem 3.22): If $f(z)$ is not constant on E, it can't be constant on any small disc about a point p in E either, because this constancy would "propagate" throughout the connected set E. The absolute value $|f(z)|$ is a continuous function on $\bar{E}$, since f is continuous, and any continuous real valued function on a *closed bounded set* in the plane, like $\bar{E}$, achieves its maximum value somewhere on $\bar{E}$. (The boundedness of E is needed to draw this conclusion.) If the maximum is achieved at some point z^* which lies within E, f is analytic and non-constant near z^*, and Theorem 3.22 assures us that we can and other points z^{**} near z^* at which $|f(z^{**})| > |f(z^*)|$. This is impossible, since $|f(z)|$ is never larger than the maximum value $|f(z^*)|$; therefore, the points z^* where $|f(z)|$ is maximum must all be located on the boundary of E, and

$$|f(z)| < |f(z^*)| = M \quad \text{for all } z \text{ in } E. \quad \blacksquare$$

Here it is essential that E be *bounded* and *connected*. The behavior of f on the boundary of E imposes restrictions on the behavior of f within E: $|f(z)|$ can never be larger within E than it is on the boundary. If $f(z)$ is actually analytic on a larger open set F which includes E and $bdry(E)$, the size of $|f(z)|$ on the boundary of E only influences the size of $|f(z)|$ *inside* this boundary, and $|f(z)|$ can be quite large outside of $\bar{E}$. For example, Theorem 3.23 restricts the behavior of $f(z) = 1 + z^2$ within the circle $\Gamma = \{z : |z| = 1\}$ (consider $E = \{z : |z| < 1\}$), but $|f(z)|$ behaves much like $|z|^2$ and is unbounded as z is moved far away from the origin.‡

This behavior of the absolute value $|f(x + iy)|$ for analytic functions contrasts strikingly with that of other smooth real valued functions of two real variables. For example, the non-negative function

$$f(x, y) = \frac{1}{1 + x^2 + y^2} \qquad \text{defined for all } (x, y)$$

has continuous partial derivatives of all orders. Its values on the unit circle $\Gamma = \{z : |z| = 1\}$ are all *smaller* than the value achieved at the origin. Obviously,

† In the terminology of Section 2.3, the closure $\overline{E}$ of a set E is obtained by adjoining to E the boundary set bdry(E).

‡ By the triangle inequality, $|z|^2 + 1 \geq |f(z)| \geq |z|^2 - 1$ for all z outside of the unit disc.

this function cannot be the absolute value of any analytic function defined on $E = \{z:|z| < 1\}$, when we regard it as a function of complex variable $f(x + iy) = f(x, y)$.

We conclude this section by taking up the proof of Theorem 3.22. This proof illustrates a very basic idea in the use of power series representations of analytic functions. We will examine the leading non-zero terms in the power series expansion about a typical point p; the sum of the remaining terms becomes negligible compared to the leading terms for z near p, so we may safely consider the *polynomial* made up of the lead terms in place of the actual analytic function.

PROOF OF THEOREM 3.22: Consider a non-constant analytic function defined near p. If $f(p) = 0$, then f cannot be identically zero near p, so we must be able to find points z', as close to p as we wish, such that $|f(z')| > 0 = |f(p)|$. This takes care of the case: $f(p) = 0$.

If $f(p) \neq 0$, there is a power series representation

$$f(z) = a_0 + a_n(z - p)^n + \sum_{k=n+1}^{\infty} a_k(z - p)^k \qquad \text{for } z \text{ near } p; \tag{59}$$

here $a_0 = f(p)$ and a_n is the next non-zero coefficient (corresponding to the first of the derivatives of f which is not zero at p). Write

$$R(z) = \sum_{k=n+1}^{\infty} a_k(z - p)^k \tag{60}$$

for the sum of the remaining terms; this series has the same radius of convergence as the original Taylor series (59), and since each term in (60) has a common factor $(z - p)^{n+1}$, we can bring this factor outside of the series without affecting convergence, to get

$$R(z) = (z - p)^{n+1}g(z),$$

where the series $g(z) = a_{n+1} + a_{n+2}(z - p) + \cdots$ has the same radius of convergence as (59) and gives an analytic function $g(z)$ defined near p. Let us fix our attention on a small disc $D = \{z:|z - p| < r_0\}$ about p on which all of these series converge, and on which $|g(z)|$ is bounded (say $|g(z)| \leq K$ for all z such that $|z - p| \leq r_0$). Equation (59) takes the form

$$f(z) = f(p) + a_n(z - p)^n + (z - p)^{n+1}g(z) \quad \text{all } z \text{ in } D. \tag{61}$$

With these preliminary notation exercises out of the way we can take up the main idea of the proof: since the last term in (61) involves $(z - p)^{n+1}$, which decreases in size much faster than $(z - p)^n$ as z approaches p, the last term is negligible in comparison with the first two, and $f(z)$ is very closely approximated by the polynomial

$$F(z) = f(p) + a_n(z - p)^n, \tag{62}$$

for z near p. It is an easy algebraic exercise to find points $z \neq p$ as close as we wish to p, such that the modulus $|F(z)|$ is greater than, or less than, the value $|f(p)|$. These two ideas can be put together to make the proof. The less theoretically oriented reader can skip over these details, now that we have given the general idea. (Note: A careful examination of these details actually tells us how to locate the directions of steepest ascent (or descent) for $|f(z)|$, starting from $z = p$. Following directions of steepest descent leads us directly to the zeros of f, and this process is sometimes of great interest in practical applications.)

Write a_n in polar form $a_n = |a_n| e^{i\phi}$, and examine points $z = p + re^{i}$ for small $r > 0$ (holding θ fixed). Notice that

$$a_n(z - p)^n = |a_n| e^{i\phi} r^n e^{in\theta} = r^n |a_n| e^{i(\phi+n\theta)}$$

and

$$|a_n(z - p)^n| = r^n |a_n|$$

for all $r > 0$. First take θ so that $\phi + n\theta = \text{Arg}(f(p))$; this insures that the vector in the plane corresponding to $a_n(z - p)^n$ is *collinear* with the vector which describes $f(p)$. Because of this collinearity, we get

(63A)
$$\begin{aligned} |f(p) + a_n(z - p)^n| &= |f(p)| + |a_n(z - p)^n| \\ &= |f(p)| + r^n |a_n| \end{aligned}$$

for all $r > 0$; note that this modulus is greater than $|f(p)|$. If, instead, we fix θ so that $\phi + n\theta = \pi + \text{Arg}(f(p))$, we get

$$e^{i(\phi+n\theta)} = e^{i\pi} e^{i\,\text{Arg}(f(p))} = (-1)e^{i\,\text{Arg}(f(p))}.$$

Thus, $a_n(z - p)^n$ points in a direction *opposite* that of $f(p)$; consequently,

(63B)
$$\begin{aligned} |f(p) + a_n(z - p)^n| &= |f(p)| - |a_n(z - p)^n| \\ &= |f(p)| - r^n |a_n| \end{aligned}$$

for all $r > 0$. This modulus is *less than* $|f(p)|$.

If K is the bound on $|g(z)|$, and if we restrict r to be small enough, we get

$$\frac{r^{n+1}}{r^n} \leq \frac{1}{2} \frac{|a_n|}{K},$$

since $\lim_{r\to 0} r^{n+1}/r^n = \lim_{r\to 0} r = 0$. This gives us the inequality $|R(z)| = |g(z)|\,|z - p|^{n+1} \leq Kr^{n+1} \leq \frac{1}{2}r^n |a_n|$ for all small $r > 0$, regardless of how θ chosen in $z = p + re^{i\theta}$.

For the particular choices of θ discussed in (63A) and (63B), we obtain the estimates

(64A)
$$\begin{aligned} |f(z)| &\geq |f(p) + a_n(z - p)^n| - |R(z)| = |f(p)| + r^n |a_n| - |R(z)| \\ &\geq |f(p)| + r^n |a_n| - \tfrac{1}{2}r^n |a_n| = |f(p)| + (\tfrac{1}{2})r^n |a_n| > |f(p)| \end{aligned}$$

and

$$\begin{aligned} |f(z)| &\le |f(p) + a_n(z-p)^n| + |R(z)| = |f(p)| - r^n\,|a_n| + |R(z)| \\ &\le |f(p)| - r^n\,|a_n| + \tfrac{1}{2}r^n\,|a_n| \\ &= |f(p)| - (\tfrac{1}{2})r^n\,|a_n| < |f(p)|, \end{aligned} \tag{64B}$$

valid for all small $r > 0$. These give the points z' and z'' we seek, and the proof is complete. ▮

We will return to the maximum modulus theorem when we discuss integration methods (Chapter 5), and in applications concerning harmonic functions (Chapter 7).

EXERCISES

1. Show that the maximum modulus theorem is not necessarily true for analytic functions on *unbounded* domains, by examining the behavior of $f(z) = 1 + z^2$ on $E = \{z:|z| > 1\}$. Give an upper bound for the absolute values $|f(z)|$ on bdry(E) $=\{z:|z| = 1\}$ and exhibit at least one point z^* in E where $|f(z^*)|$ exceeds this value.

2. The function $f(z) = 1/z^2$ has modulus $|f(z)| = 1$ on the circle $|z| = 1$. Is $|f(z)|$ bounded by 1 for $|z| \le 1$? Explain why Theorem 3.23 is not violated.

3. Prove that the constant function $f(z) = 0$ is the only analytic function defined throughout the plane which "vanishes at infinity" in the sense that:

Given any number $\varepsilon > 0$, there is a corresponding radius $R = R(\varepsilon)$ such that:

$$|f(z)| < \varepsilon \quad \text{if } |z| \ge R.$$

4. Suppose f and g are continuous on $\bar{E}$ and analytic on E, where E is a bounded domain. If $f(z) = g(z)$ on bdry(E), prove that $f = g$ throughout E.

Hint: Examine $f - g$.

5. Let $p_1, \ldots, p_N$ be fixed points in a bounded domain E. If z is a variable point in E, define

$$d(z) = |z - p_1| \cdot |z - p_2| \cdot \ldots \cdot |z - p_N|$$

(the product of distances). Show that d is maximum at some boundary point, and not at any point in E. Where are the *minimum values achieved*?

Hint: Is $d(z) = \big|f(z)\big|$ for some analytic function $f(z)$?

6. Suppose $f(z)$ is defined and analytic on the unit disc $|z| < 1$. For $0 \leq r < 1$ define $M(r)$ to be the maximum value of $|f(z)|$ on the circle $|z| = r$. Prove that $M(r)$ is a strictly increasing function of r, unless $f(z) = const.$ on the disc.

7. If $f(z)$ is analytic on a domain E (bounded or not) show that $\phi(z) = \text{Re}(f(z))$ attains neither a maximum nor a minimum value at points in E unless ϕ is constant.

Hint: Consider $e^{f(z)}$ in the local maximum modulus theorem.

8. Let E be a bounded domain. Assume that $f(z)$ is defined and continuous on the closure $\bar{E}$, and is analytic at each point in E. *If* $f(z) \neq 0$ *throughout* E, prove that

(i) The *minimum* value of $|f(z)|$ on $\bar{E}$ is attained at some point in bdry(E).

(ii) If the *minimum* value is also attained at a point z^* in E, then $f(z) = const.$

Hint: Examine $1/f$ and use Theorem 3.23.

9. With Exercise 8 in mind, formulate a version of Theorem 3.22 that deals with *minimum* values of $|f(z)|$. If $f(z) = 0$ somewhere in E, show that the minimum value for $|f(z)|$ need not be achieved on bdry(E); give a simple example.

10. Suppose that f and g are analytic on a bounded domain E, and suppose that the *real parts* Re(f) and Re(g) may be defined on bdry(E) to give a continuous function on the closure $\bar{E}$. (We make no assumptions about behavior of Im(f) and Im(g) at the boundary of E.) If $\text{Re}(f(z)) = \text{Re}(g(z))$ on bdry(E), prove that $g(z) = f(z) + ic$ throughout E, where c is some real constant.

Note: If Re(f) is not too irregular at the boundary of E, its behavior (that is, its limit values) on bdry(E) determines f throughout E, up to an added imaginary constant.

11. Suppose $w = f(z)$ is an open mapping defined on an open set E in the z-plane; that is, f maps any open subset of E to an open set in the w-plane. Prove that f satisfies a local maximum modulus property at each point p in E. Formulate and prove an analog of Theorem 3.23 valid if f is continuous on $\bar{E}$, and an open mapping on E.

Note: We are not assuming f is analytic (there are open mappings of the plane that are not even complex differentiable). *Regular* holomorphic mappings (those with f' non-vanishing on E) are known to be open mappings, so the results of this section could be proved for these mappings without involving Theorem $*$.

3.9 SERIES IN NEGATIVE POWERS OF Z

So far we have only considered series of the form $a_0 + a_1(z - p) + a_2(z - p)^2 + \cdots$ which involve positive powers of the variable. Later on we will need to understand what happens if we take instead the negative powers of $(z - p)$ as in

$$\text{(65)} \qquad \sum_{n=0}^{\infty} a_n(z-p)^{-n} = a_0 + \frac{a_1}{(z-p)} + \frac{a_2}{(z-p)^2} + \cdots$$

The functions $f_n(z) = a_n(z-p)^{-n}$ which enter into this series are well defined except at the base point $z = p$. To understand these series we must answer the same sort of questions we examined in studying ordinary power series: (i) What sort of domain of convergence may we expect? (ii) On which sets in the plane is the series uniformly convergent? (iii) Are the sums analytic functions on the domain of convergence? The series (65) is so closely related to the series of positive powers with the same coefficients,

$$\text{(66)} \qquad \sum_{n=0}^{\infty} a_n w^n = a_0 + a_1 w + a_2 w^2 + \cdots,$$

that we will not have to do much work to answer all of these questions. For example, the ordinary series (66) has a definite radius of convergence $0 \leq r \leq +\infty$. It is absolutely convergent at points in the open disc bounded by the circle $|w| = r$, and it diverges at all points outside of this circle. Moreover, it is uniformly convergent on any closed disc whose radius is smaller than r, and the sum $g(w) = \sum_{n=0}^{\infty} a_n w^n$ is analytic on the open disc $|w| < r$.

Now let us substitute $w = \dfrac{1}{z-p}$ in the ordinary series (66). Obviously,

$$|w| < r \quad \text{if and only if} \quad |z-p| > \frac{1}{r}$$

$$|w| = r \quad \text{if and only if} \quad |z-p| = \frac{1}{r}$$

$$|w| > r \quad \text{if and only if} \quad |z-p| < \frac{1}{r}.$$

Therefore, the series (65) diverges if $|z-p| < 1/r$; it converges absolutely if $|z-p| > 1/r$, and its behavior is undecided on the circle $|z-p| = 1/r$. The mapping $w = \phi(z) = 1/(z-p)$ transforms the circle $|z-p| = 1/r$ onto the circle $|w| = r$, and maps the exterior domain $\{z : |z-p| > 1/r\}$ onto the domain $\{w : 0 < |w| < r\}$, as indicated in Figure 3.10. Obviously, the circle of radius $R = 1/r$ divides the z-plane into separate domains on which the series (65) converges or diverges, but now it is the set of points lying *outside of* this circle which is the domain of convergence. If we write $f(z)$ for the sum of the series of negative powers (65), it is well defined on $E = \{z : |z-p| > 1/r = R\}$ and it is clear that

$$f(z) = \left[g(w) \Big|_{w=1/(z-p)} \right] = g\left(\frac{1}{z-p}\right) = g(\phi(z))$$

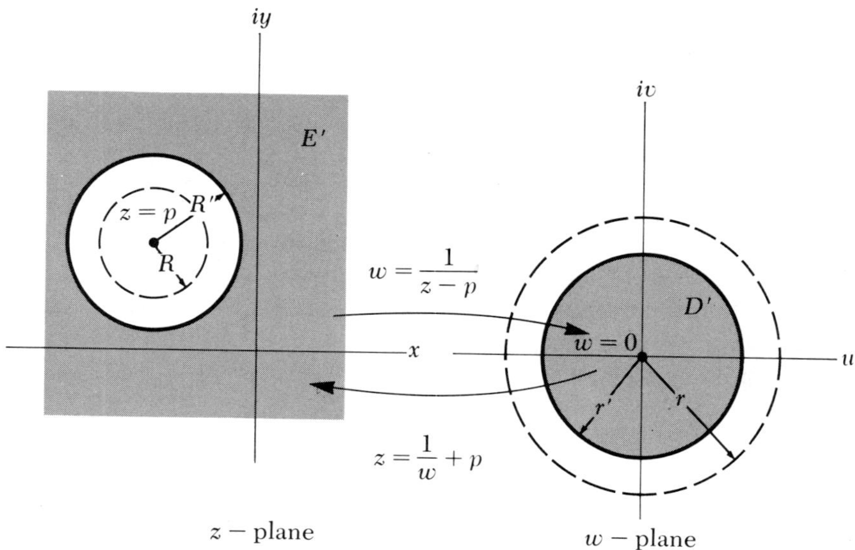

Figure 3.10 The mapping $w = f(z) = \dfrac{1}{z-p}$ maps the shaded domain in the z-plane (inner radius R') to the closed disc $\left(\text{radius } r' = \dfrac{1}{R'}\right)$ in the w-plane. The series are uniformly convergent on the sets E' and D' shown. Dashed circles correspond to the radii of convergence.

for all z in E. But ϕ is holomorphic, and maps E into the domain $D = \{w: |w| < r\}$ on which $g(w)$ is analytic; therefore, the composite function $f(z) = (g \circ \phi)(z)$, the sum of the series of negative powers, is holomorphic throughout the domain of convergence E. Once Theorem $*$ is proved, we will see that this sum is actually analytic on E.

The series of negative powers (65) converges uniformly to its limit on any set of the form $E' = \{z: |z - p| \geq R'\}$, as long as R' is greater than the radius of convergence $R = 1/r$ associated with (65). It is interesting to notice that sets like E' are unbounded; nevertheless, the series is uniformly convergent to its limit on any set of this form. To demonstrate this uniform convergence we start with the known fact that ϕ maps E' into the closed disc $D' = \{w: |w| \leq r' = 1/R'\}$; since $R' > R = 1/r$, we get $r' < r$, so the series (66) converges uniformly on D'. Now use Theorem 3.1 (applied to the sequences of partial sums of the two series here) to see that (65) converges uniformly on E'.

Here is an example which illustrates the contrasting convergence properties of a series of negative powers like (65), and its associated series (66).

Example 3.14 Consider the series

$$1 + \frac{1}{z} + \left(\frac{1}{2!}\right)\frac{1}{z^2} + \left(\frac{1}{3!}\right)\frac{1}{z^3} + \cdots \tag{67}$$

(the base point is $p = 0$). The ordinary series, in variable w, with the same set of coefficients is just the exponential series

$$\exp(w) = 1 + w + \frac{w^2}{2!} + \cdots,$$

which has radius of convergence $r = +\infty$. The series of negative powers (67) then converges absolutely for all z such that $|z| > 0$, since $R = 1/r = 0$; i.e., it converges to an analytic function throughout the punctured plane $E = \{z : z \neq 0\}$. It diverges, because the functions appearing in the series are undefined, at the point $z = 0$. We can evaluate the sum in (67), since we get

$$1 + \frac{1}{z} + \left(\frac{1}{2!}\right)\left(\frac{1}{z}\right)^2 + \cdots = \exp\left(\frac{1}{z}\right) \qquad \text{for all } z \neq 0,$$

when we substitute $w = 1/z$ into the exponential series.

Example 3.15 Consider the series of negative powers analogous to the geometric series:

$$\sum_{n=0}^{\infty} z^{-n} = 1 + \frac{1}{z} + \frac{1}{z^2} + \cdots.$$

The associated series of positive powers is the usual geometric series $\sum_{n=0}^{\infty} w^n$, which converges to $\frac{1}{1-w}$ with radius of convergence $r = 1$. Since $R = 1/r = 1$, the series we are interested in converges absolutely for all points lying *outside of* the circle $|z| = 1$; by substituting $w = 1/z$ in the geometric series, we can evaluate the sum

$$\sum_{n=0}^{\infty} z^{-n} = \frac{1}{1 - \frac{1}{z}} = \frac{z}{z-1} \qquad \text{for all } z \text{ such that } |z| > 1.$$

If we drop the constant term in the original series, it is obvious that

$$\sum_{n=1}^{\infty} z^{-n} = \frac{1}{z} + \frac{1}{z^2} + \frac{1}{z^3} + \cdots = \frac{1}{z-1} = (-1)\frac{1}{1-z}$$

for all z such that $|z| > 1$.

EXERCISES

1. Express the following functions as series involving negative (and possibly positive) powers of $(z - p)$.

(i) $\dfrac{e^z}{z}$ at $p = 0$

(ii) $\dfrac{1 - \cos z}{z}$ at $p = 0$

(iii) $\exp\left(-\dfrac{1}{z^2}\right)$ at $p = 0$

(iv) $\sin\left(\dfrac{1}{z}\right)$ at $p = 0$

(v) $\dfrac{1}{z(z-1)^2}$ at $p = 1$

(vi) $\dfrac{z^2 - z + 1}{z^3}$ at $p = 0$.

2. Prove directly, using the Weierstrass test, that the series

$$\exp\left(-\frac{1}{z^2}\right) = \sum_{n=0}^{\infty} \frac{1}{n!}\left(-\frac{1}{z^2}\right)^n = \sum_{n=0}^{\infty} \frac{(-1)^n}{n!} z^{-2n}$$

converges uniformly on the unbounded set $E_r = \{z : r \leq |z| < +\infty\}$, if $r > 0$.

3. If series of negative powers $f(z) = \sum_{n=0}^{\infty} a_n(z-p)^{-n}$ and $g(z) = \sum_{n=0}^{\infty} b_n(z-p)^{-n}$ have radii $R_1 > 0$ and $R_2 > 0$ respectively, at which points z does the Cauchy product of these series converge to $f(z) \cdot g(z)$?

4. If $f(z) = \sum_{n=0}^{\infty} a_n(z-p)^{-n}$ has radius of convergence $R < +\infty$, give a direct proof that

(i) f is holomorphic on $E = \{z : |z| > R\}$

(ii) the derivative is given by term-by-term differentiation of the series:

$$\frac{df}{dz} = \sum_{n=0}^{\infty} a_n \frac{d}{dz}\{(z-p)^{-n}\} = \sum_{n=0}^{\infty} -na_n(z-p)^{-(n+1)}$$

for z in E.

What goes wrong with the procedure for calculating *antiderivatives* if the coefficient on $(z-p)^{-1}$ is non-zero?

Hint: Use the Chain Rule on $(g \circ \phi)(z)$, where $g(w) = \sum_{n=0}^{\infty} a_n w^n$ and $w = \phi(z) = 1/(z-p)$.

4 MAPPING PROPERTIES OF HOLOMORPHIC FUNCTIONS

In this chapter we will view a function of a complex variable $w = f(z)$ as a mapping from one subset of the complex plane to another. In order to clearly distinguish between the variable z and the values $w = f(z)$ of the function, we shall think of two copies of the complex plane whose points are labeled $z = x + iy$ and $w = u + iv$, respectively, and we will regard f as a mapping from the z-plane into the w-plane.

We will explore the remarkable geometric properties of holomorphic mappings in two ways. First we will study the behavior "in the small" of a mapping $w = f(z)$, by examining the displacements $\Delta w = f(p + \Delta z) - f(p)$ corresponding to small displacements Δz from some base point p in the domain of definition of f. Behavior "in the large" will be studied by examining how f transforms various families of lines, circles, and other curves from the z-plane into the w-plane. These extreme points of view, taken together, provide a reliable intuitive understanding of holomorphic mappings.

There are some passing comments about non-holomorphic mappings of the plane. However, we must emphatically warn the reader that most of the ideas we are going to present fail drastically once we leave the realm of *holomorphic* mappings. Mere smoothness of the mapping $f = U + iV$, in the sense that the first partial derivatives $\partial U/\partial x, \ldots, \partial V/\partial y$ all exist and are continuous, is not enough. The behavior of non-holomorphic mappings, even smooth ones, in the small is much more complicated than the behavior of holomorphic mappings. Restructuring our discussion to include non-holomorphic mappings of the plane would require a lengthy excursion into advanced calculus that would be out of place here. A curious situation arises: in a certain sense *most* smooth mappings of the plane are non-holomorphic, but the mappings that turn up most often in mathematical and physical problems *are* holomorphic. It is this tendency which makes our study of holomorphic mappings so useful in practice.

4.1 BEHAVIOR OF HOLOMORPHIC FUNCTIONS IN THE SMALL

Let $w = f(z)$ be holomorphic on some open set E and consider the behavior of f near a point p in E. Our investigation rests on the simple results proved in Section 2.8. Since f is complex differentiable at p, it is closely approximated near p by the *linear* function of z:

$$\tilde{f}(z) = f(p) + f'(p)(z - p).$$

That is, if we write f in the form

$$f(z) = \tilde{f}(z) + E(z) = [f(p) + f'(p)(z - p)] + E(z), \tag{1}$$

then the error term $E(z) = f(z) - \tilde{f}(z)$ vanishes rapidly as z approaches p:

$$\frac{|E(z)|}{|z - p|} \to 0 \quad \text{as} \quad z \to p. \tag{2}$$

Since we are interested in the limit behavior of f near the base point p, it is convenient to examine the displacements

$$\Delta f = f(z) - f(p)$$
$$\Delta\tilde{f} = \tilde{f}(z) - \tilde{f}(p) = f'(p)(z - p) = f'(p)\,\Delta z$$

from the image point $q = f(p)$ in the w-plane, corresponding to a displacement $\Delta z = z - p$ away from the base point p in the z-plane. Equations (1) and (2) then take the form

$$\Delta f = \Delta\tilde{f} + E(z) = f'(p)\cdot\Delta z + E(z)$$
$$\left|\frac{E(z)}{z - p}\right| \to 0 \quad \text{as} \quad z \to p. \tag{3}$$

If $f'(p) \neq 0$, the linear term $\Delta\tilde{f} = f'(p)\cdot\Delta z$ in equation (3) is directly proportional to the displacement Δz, while the magnitude $|E(z)|$ of the error term approaches zero much faster than $|\Delta z|$ as z approaches p, and becomes negligible in comparison with the linear term $\Delta\tilde{f} = f'(p)\cdot\Delta z$.

It will be easier to read geometric information out of equation (3) if we use the polar form

$$f'(p) = |f'(p)|\, e^{i\alpha}$$

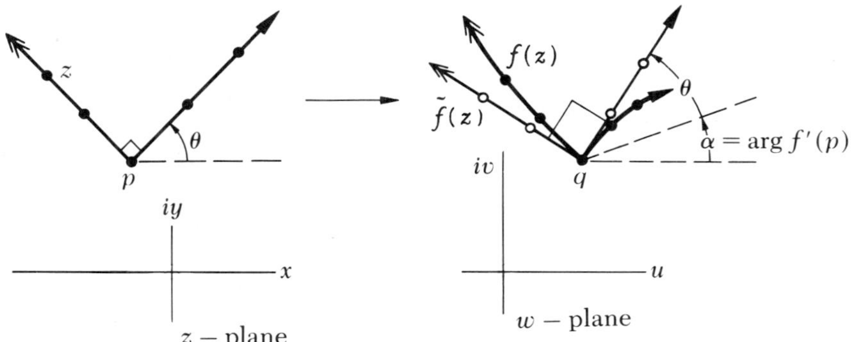

Figure 4.1 The images of points z near the base point p, under f and $\tilde{f}$. Image points $w = f(z)$ are shown as solid dots, and images under the linear map $w = \tilde{f}(z)$ as open dots, in the w-plane.

(α real, any determination of $\arg f'(p)$). Then

$$\Delta f = |f'(p)|\, e^{i\alpha} \Delta z + E(z). \tag{4}$$

The geometric effect of multiplying Δz by the complex number $f'(p)$ to get $\Delta\tilde{f} = f'(p) \cdot \Delta z$ is to scale Δz by the positive factor $|f'(p)|$ and then rotate it counterclockwise from its original orientation by an angle of α radians. For any small displacement Δz from p, Δf and $\Delta\tilde{f} = f'(p)\,\Delta z$ are attached to the point $q = f(p)$ in the w-plane, while Δz is attached to the base point p in the z-plane, as shown in Figure 4.1. It is very important to notice that each displacement Δz from p is acted upon in the same way to get $\Delta\tilde{f}$; we scale it by the factor $|f'(p)|$ and then rotate it by an angle $\alpha = \arg f'(p)$, regardless of the orientation of Δz. For non-holomorphic mappings, the length and direction of Δf can depend in a very complicated way on the length *and* orientation of Δz; it is not even approximately described by simple scaling and rotation operations on Δz. Here is an example that illustrates the geometric interpretation of equation (4).

Example 4.1 The exponential function is a smooth holomorphic mapping of the plane: $w = f(z) = e^z = (e^x \cos y) = i(e^x \sin y) = U(x, y) + iV(x, y)$. Its derivative $f'(z) = e^z = f(z)$ is never zero. If $p = x_0 + iy_0$, the polar form of $f'(p)$ is $|f'(p)|\, e^{i\alpha}$, where

$$|f'(x_0 + iy_0)| = e^{x_0} > 0$$
$$\alpha \equiv \arg f'(x_0 + iy_0) \equiv y_0 \pmod{2\pi}.$$

The linear mapping $\tilde{f}(z)$ that best approximates $f(z)$ for z near p is

$$\tilde{f}(z) = e^p + e^p(z - p) = e^p + e^p \Delta z$$
$$\Delta\tilde{f} = e^p \Delta z = e^{x_0} e^{iy_0} \Delta z.$$

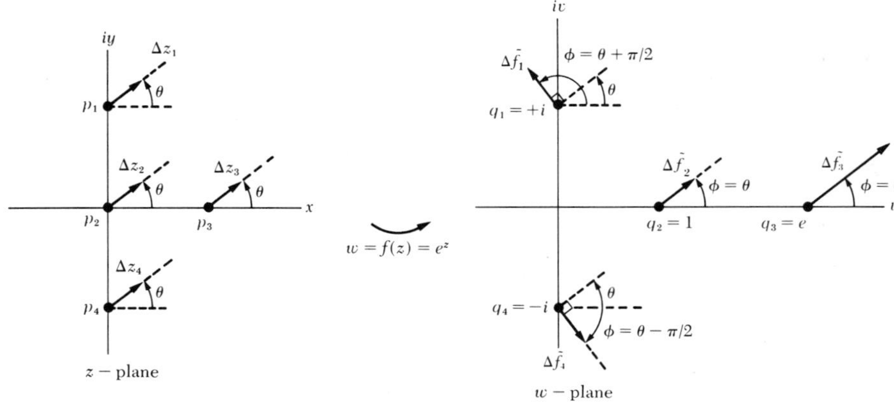

Figure 4.2 The approximate displacements $\Delta \tilde{f} = f'(p) \cdot \Delta z$ for displacements Δz away from various base points in the z-plane. The mapping is $w = e^z$; the data are listed in Table 4.1.

Figure 4.2 shows how Δz and $\Delta \tilde{f}$ are related for various base points p_k, taking Δz so that $|\Delta z| = 1$ and $\theta = \arg \Delta z = \pi/4$. The reader might find it interesting to calculate the positions of the *actual* displacements $\Delta f = f(p + \Delta z) - f(p)$ and compare them with the approximate displacements $\Delta \tilde{f}$ listed in Table 4.1. (Also, repeat these calculations taking $|\Delta z| = 1/10$.)

Table 4.1 Calculation of $\Delta \tilde{f}$ for Base Points Shown in Figure 4.2.

	$p_1 = i\pi/2$	$p_2 = 0$	$p_3 = 1$	$p_4 = -i\pi/2$
$q_j = f(p_j)$	$e^{i\pi/2} = +i$	1	$e^1 = e$	$e^{-i\pi/2} = -i$
$\lvert f'(p_j)\rvert$	1	1	e	1
$\alpha = \arg f'(p_j)$	$+\pi/2$	0	0	$-\pi/2$
$\lvert\Delta \tilde{f}\rvert$	$\lvert\Delta z\rvert$	$\lvert\Delta z\rvert$	$\lvert\Delta z\rvert \cdot e$	$\lvert\Delta z\rvert$

Example 4.2 Consider $w = f(z) = az + b$ (a and b fixed complex constants; assume $a \neq 0$). Then consider a base point p and write out the linear mapping $\tilde{f}$ that best approximates f for z near p. Clearly, $f(p) = ap + b$ and $f'(p) = a$, so formula (1) becomes

$$\tilde{f}(z) = (ap + b) + a \cdot (z - p) = az + b = f(z)$$

for all z. Thus, the approximating linear map $\tilde{f}(z)$ *coincides with* $f(z)$. Since f was a linear function of z to begin with, it should be no great surprise that $\tilde{f} = f$.

EXERCISES

1. Explain how a holomorphic function $w = f(z)$ behaves (approximately) in the small at a base point p in each situation listed below.

(i) $\alpha = \arg f'(p) = 0$ and $|f'(p)| = 1$
(ii) $|f'(p)|$ is arbitrary, but $\alpha = \arg f'(p) = 0$
(iii) $\alpha = \arg f'(p)$ is arbitrary, but $|f'(p)| = 1$.

2. Consider $w = f(z) = \exp z$ near the base points shown in Figure 4.2. Taking Δz so that $|\Delta z| = 1$ and $\theta = \arg \Delta z = \pi/4$, calculate the *actual* displacements $\Delta f = f(p + \Delta z) - f(p)$ for each base point. Compare these with the approximate displacements $\Delta \tilde{f}$ shown in the figure.

3. Prove that the transformation $w = (z + 1)/(z - 1)$ has $dw/dz \neq 0$ on $E = \{z\colon z \neq 1\}$. For the base points

$$p_1 = 0, \qquad p_2 = -1, \qquad p_3 = +2, \qquad p_4 = -100i,$$

calculate: (i) the scale factor $|f'(p)|$ and the rotation angle $\alpha(p) = \arg f'(p)$, and (ii) the new base point $q = f(p)$ in the w-plane. Display the behavior of f in the small near these base points (use Figure 4.2 as a model).

4. For

$$w = e^z, \quad w = \frac{z+1}{z-1}, \quad \text{and} \quad w = z^2 + 1,$$

determine the points p in the z-plane at which the approximate displacement $\Delta \tilde{f}$ is obtained from Δz by the following special kinds of operations.

(i) A pure rotation
(ii) A pure dilation by a factor $0 < a < 1$
(iii) The identity mapping (so that $\Delta \tilde{f} = \Delta z$).

What conditions on the derivative $f'(p)$ correspond to these different types of behavior in the small?

5. Set up a diagram like Figure 4.2 for $w = \sin z$. Show how $\Delta \tilde{f}$ is related to the displacement $\Delta z = (0.1)e^{i\pi/4}$ for the base points $p = 0$, $\pi/2$, $+i$, $-i$, $10i$, and $+\pi$.

4.2 SMOOTH CURVES AND THEIR TRANSFORMATION BY MAPPINGS $w = f(z)$

A **curve** in the complex plane will always be taken to mean a **parametrized curve.** Thus it is determined by specifying a point

$$\gamma(t) = x(t) + iy(t) \tag{5}$$

in the plane for each t within some interval $[a, b]$ of real numbers. The real valued functions $x(t)$ and $y(t)$ specify the coordinates of $\gamma(t)$ for each t. A parametrized curve γ is therefore a *mapping* defined on an interval $I = [a, b]$; each number t corresponds to a well defined point $\gamma(t)$ in the complex plane.

The point set $\Gamma = \Gamma(\gamma)$ traced out by $\gamma(t)$ as t varies will always be referred to as the **trajectory** of the parametrized curve γ. It might seem natural to regard this point set Γ as "the curve," but for many purposes the trajectory, *together with its parametrization*, is what we must study. For example, if we regard t as a time variable and let $\gamma(t)$ stand for the location of some moving particle (or planet) in a two dimensional astronomical problem, it is not enough to know the orbit (trajectory) of the particle; we really want to know how the particle moved along its orbit (the parametrization) in most problems. If we only know the set Γ, we have lost all information about where the moving point $\gamma(t)$ started or finished its motion, the direction (or velocity) with which it moved, and so on; the mapping γ incorporates all of this extra information.

We will always assume that the coordinate functions $x(t)$ and $y(t)$ in (5) are continuous on $[a, b]$. This is the same as saying that the mapping $\gamma : I \to \mathbf{C}$ is continuous, and has the effect of preventing $\gamma(t)$ from making abrupt jumps in position as t increases. We also adopt the convention that $a \leq b$ whenever we write the symbol $[a, b]$ for the set of all real numbers between $t = a$ and $t = b$. The following terminology for parametrized curves is standard:

(1) The point $\gamma(a)$ is called the **initial point** of γ.
(2) The point $\gamma(b)$ is called the **final point** of γ.
(3) If $\gamma(a) = \gamma(b)$, so that $\gamma(t)$ returns to its starting point, we say that γ is a **closed curve.**
(4) If $\gamma(t'') \neq \gamma(t')$ when $t'' \neq t'$, so γ never crosses over itself, we say that γ is a **simple curve.**

Evidently a closed curve cannot be simple, according to these definitions; nevertheless, by time-honored abuse of terminology, we say that γ is a **simple closed curve** if it is closed and satisfies condition (4), except when $t'' = a$ and $t' = b$ (or $t'' = b$ and $t' = a$). Some examples are indicated in Figure 4.3 and in the exercises; others are discussed at the beginning of Chapter 5.

Example 4.3 Consider the curve $\gamma(t) = x(t) + iy(t)$, whose coordinate functions are

$$x(t) = 2 \cos t \qquad \text{and} \qquad y(t) = 2 \sin t,$$

defined for $0 \leq t \leq \pi$. The x and y coordinates of $z = \gamma(t)$ satisfy the equation $x^2 + y^2 = 4(\cos^2 t + \sin^2 t) = 4$ for $0 \leq t \leq \pi$ (eliminate t between the above

Figure 4.3 Various types of curves which may occur.

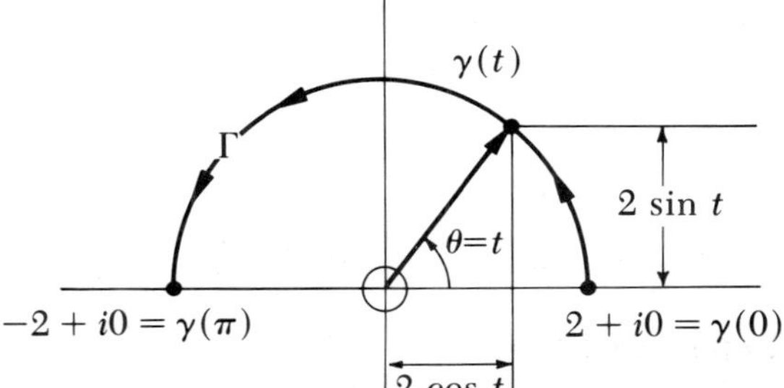

Figure 4.4 The curve $\gamma(t) = (2 \cos t) + i(2 \sin t)$ in Example 4.3. The curve is simple and is not closed.

equalities), so that $\gamma(t)$ lies on the circle of radius $r = 2$ about the origin. Furthermore, the y coordinate is always positive, since $y(t) = 2 \sin t \geq 0$ if $0 \leq t \leq \pi$, so that $\gamma(t)$ lies on the circular arc in the upper half plane shown in Figure 4.4. By tracing the motion of $\gamma(t)$ as t increases, one can see that $\gamma(t)$ starts at the point $2 + i0 = \gamma(0)$ and ends at $\gamma(\pi) = -2 + i0$, and moves smoothly counterclockwise along the arc Γ. Thus Γ is the trajectory of our parametrized curve.

A parametrized curve $\gamma(t) = x(t) + iy(t)$ is a function, with complex values and a real variable t; thus, we can define the **derivative** $\gamma'(t_0)$ at a point t_0 in the interval $[a, b]$ by forming difference quotients (each one a complex number)

$$\frac{\Delta\gamma}{\Delta t} = \frac{\gamma(t) - \gamma(t_0)}{t - t_0} \quad \text{for} \quad t \neq t_0 \text{ in } [a, b],$$

and inquiring whether these approach a limit as t approaches t_0. These quotients may be written in the form

$$\frac{\Delta\gamma}{\Delta t} = \left[\frac{x(t) - x(t_0)}{t - t_0}\right] + i\left[\frac{y(t) - y(t_0)}{t - t_0}\right];$$

clearly, differentiability of the component functions $x(t)$ and $y(t)$ at t_0 insures that the derivative exists, and that

$$\text{(6)} \qquad \gamma'(t_0) = \lim_{t \to t_0} \frac{\Delta\gamma}{\Delta t} = \lim_{t \to t_0} \frac{\Delta x}{\Delta t} + i \lim_{t \to t_0} \frac{\Delta y}{\Delta t} = \frac{dx}{dt}(t_0) + i\frac{dy}{dt}(t_0).$$

We will use the symbols

$$\gamma'(t_0) \quad \text{and} \quad \frac{d\gamma}{dt}(t_0)$$

interchangeably to indicate this derivative.

The derivative of γ at t_0 is a complex number. Most of the curves we shall discuss will be differentiable (or continuously differentiable, twice differentiable, etc.) throughout the interval $[a, b]$, so that the derivatives

$$\gamma'(t) = \frac{d\gamma}{dt}, \frac{d^2\gamma}{dt^2}, \ldots$$

are functions of the same kind as γ.

Definition 4.1 *A parametrized curve $\gamma(t) = x(t) + iy(t)$ defined on an interval $[a, b]$ is said to be* **smooth** *if it is continuously differentiable. This is the same as insisting that the functions $x(t)$ and $y(t)$ be continuously differentiable on $[a, b]$, with well defined one-sided derivatives at the end points a and b.*†

The curve in Example 4.3 has continuous derivatives of all orders:

$$\gamma(t) = (2\cos t) + i(2\sin t)$$

$$\frac{d\gamma}{dt}(t) = (-2\sin t) + i(2\cos t) = i\cdot\gamma(t)$$

$$\frac{d^2\gamma}{dt^2}(t) = (-2\cos t) - i(2\sin t) = (-1)\cdot\gamma(t)$$

$$\vdots$$

If we think of t as a time parameter, then the derivative $\gamma'(t)$ has an important interpretation: the vector corresponding to the complex number $\gamma'(t_0)$ is associated with the point $p = \gamma(t_0)$ on the trajectory, and tells us the **velocity** (speed and direction) of the moving point $\gamma(t)$ when $t = t_0$. The Euclidean length of this "velocity vector" or "tangent vector" is just the absolute value of the corresponding complex number:

$$\left|\frac{d\gamma}{dt}(t_0)\right| = \sqrt{\left(\frac{dx}{dt}\right)^2 + \left(\frac{dy}{dt}\right)^2}$$

and is interpreted as the **speed** (a scalar quantity) of the moving point when $t = t_0$. The speed gives us no information about the direction in which $\gamma(t)$ is moving.

The derivative $\gamma'(t_0)$ may be zero, as in Example 4.4 below. Physically, this means that the moving point has momentarily come to rest when $t = t_0$, an occurrence that is not at all unusual in physical problems. If γ is a smooth curve, then the velocity $\gamma'(t)$ is a continuously varying complex number, and the speed $|\gamma'(t)|$ is also a continuous function of t. However, the *direction* of motion, which we might identify as $\arg\gamma'(t)$, is undefined if $\gamma'(t_0) = 0$;

† In defining the limit of difference quotients, t must remain within the interval $[a, b]$ on which $\gamma(t)$ is defined. Thus, at $t = a$ or $t = b$ we obtain the "right hand" and "left hand" one-sided derivatives.

in this case, its values for $t < t_0$ can be quite unrelated to its values for $t > t_0$ (assuming that $\gamma'(t) \neq 0$ for $t \neq t_0$). Some of the possibilities are illustrated below.

Example 4.4 The parametrized curve $\gamma(t) = t^3 + i0$, defined for $-1 \leq t \leq +1$, traces out the segment $\Gamma = [-1, +1]$ on the real axis as t increases. The component functions $x(t) = t^3$ and $y(t) = 0$ are continuously differentiable and the derivative $\gamma'(t) = 3t^2 + i0$ is zero at $t_0 = 0$. For values of $t \neq 0$, the velocity $\gamma'(t)$ is non-zero, and is directed parallel to the segment Γ. In this situation there is no discrepancy between the direction of motion before and after the momentary halt at the origin.

Next consider $\eta(t)$ defined for $-1 \leq t \leq +1$:

$$\eta(t) = \begin{cases} 0 + it^2 & \text{if} \quad -1 \leq t \leq 0 \\ t^2 + i0 & \text{if} \quad\ \ \ 0 \leq t \leq +1. \end{cases}$$

It is easily verified that η is differentiable, even at $t = 0$, and that

$$\eta'(t) = \begin{cases} 0 + 2it & \text{if} \quad -1 \leq t \leq 0 \\ 2t + i0 & \text{if} \quad\ \ \ 0 \leq t \leq +1. \end{cases}$$

Clearly $\eta'(t)$ is a continuous function of t, and so is the speed $|\eta'(t)| = 2\,|t|$ of the moving point. But $\eta'(0) = 0$, and the direction of motion changes abruptly as t passes through zero. The point $\eta(t)$ moves downwards from $+i$ to 0, slows to a momentary halt at the origin, and then moves to the right from 0 to $+1$, as indicated in Figure 4.5. There is no reasonable tangent line to the curve at the critical point $p = \eta(0) = 0$, where the derivative vanishes. Yet, in spite of the kink in its trajectory, η is a smooth curve and $\eta'(t)$ varies continuously.

Why do we refer to η as a "smooth curve" in the last example, when its trajectory is so obviously angular and "non-smooth"? In defense of this terminology, we point out that the smoothness in the phrase "smooth curve" refers to the nature of the parametrization, and not necessarily the shape of the trajectory. That is, a smooth curve is one that is smoothly parametrized, so that $\eta(t)$ is continuously differentiable. Since we shall deal almost exclusively with parametrized curves in calculating contour integrals and other quantities, it is reasonable to emphasize the nature of the parametrization.

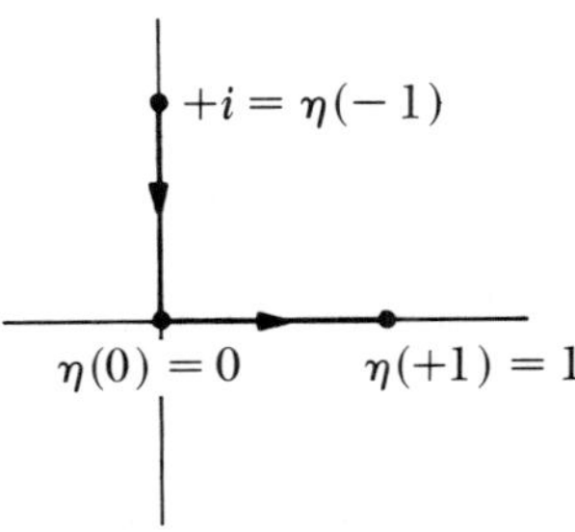

Figure 4.5 The "smooth curve" of Example 4.4. The derivative $\eta'(t)$ vanishes at the origin.

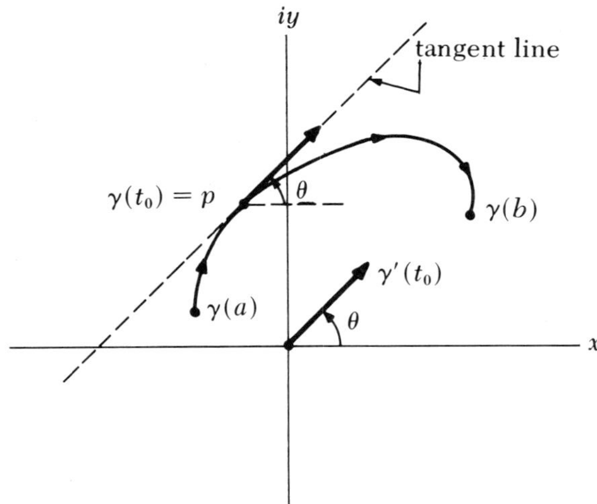

Figure 4.6 A smooth curve $\gamma(t)$ in the complex plane with the velocity vector $\gamma'(t_0)$ shown attached to the corresponding point $p = \gamma(t_0)$ on the trajectory. Since $\gamma'(t_0) \neq 0$, there is a well defined tangent line.

If $\gamma'(t_0) \neq 0$, this derivative determines a well defined direction at $p = \gamma(t_0)$. There is a definite tangent line to γ at p, the line through p that is parallel to $\gamma'(t_0)$, as shown in Figure 4.6. If $\gamma'(t)$ is continuous and *non-vanishing* throughout $[a, b]$, the orientation of the tangent line at $\gamma(t)$ varies continuously as t increases. Furthermore, one can demonstrate that the trajectory cannot have kinks of the sort exhibited in Figure 4.5.

In a few special circumstances we will be interested in the tangent directions at points on the parametrized curves we are examining. Then the curves must not only be smooth, but must also have non-zero derivatives.

This requirement will be mentioned explicitly whenever it arises. In most of our applications, smooth curves will do nicely, whether their derivatives vanish occasionally or not; the main requirement is that γ' should exist and be continuous.

A smooth curve $\gamma(t)$ in the z-plane is transformed to a smooth curve $\eta(t)$ in the w-plane by any mapping $w = f(z)$ whose domain of definition includes the trajectory of γ. The new curve, given by the formula

$$\eta(t) = \left[f(z)\,\big|_{z=\gamma(t)}\right] = f(\gamma(t))$$

is defined on the same interval $[a, b]$ as γ. Since curves are regarded as mappings, the transformed curve η is just the composite map $\eta = f \circ \gamma$. Figure 4.7 shows how the mapping $w = z^2$ transforms a typical curve γ in the z-plane.

Now suppose that $f(z) = U(x, y) + iV(x, y)$ is a smooth mapping of the plane. The components of $\eta(t) = u(t) + iv(t)$ are given by

$$u(t) = U(x(t), y(t)) \quad \text{and} \quad v(t) = V(x(t), y(t));$$

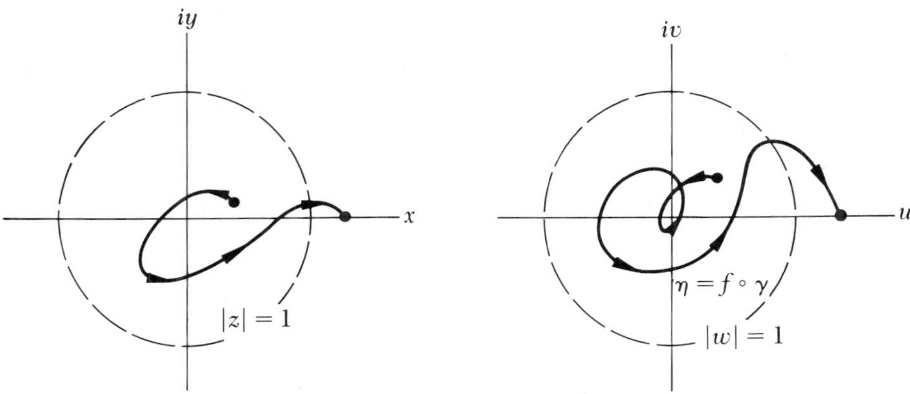

Figure 4.7 Positions of the curves are shown relative to the unit circles $|z| = 1$ and $|w| = 1$ (dashed circles). Transformation of a curve γ by the mapping $w = f(z) = z^2$.

a simple application of the chain rule for partial derivatives shows that these functions of t are continuously differentiable on $[a, b]$, and that

(7)
$$u'(t) = \frac{du}{dt}(t) = \frac{\partial U}{\partial x}(\gamma(t))\frac{dx}{dt} + \frac{\partial U}{\partial y}(\gamma(t))\frac{dy}{dt}$$
$$v'(t) = \frac{dv}{dt}(t) = \frac{\partial V}{\partial x}(\gamma(t))\frac{dx}{dt} + \frac{\partial V}{\partial y}(\gamma(t))\frac{dy}{dt},$$

so we have calculated $d\eta/dt = u'(t) + iv'(t)$.

If f is actually a smooth *holomorphic* function, we may use the Cauchy-Riemann equations to reduce (7) to a simpler form,

(8)
$$\frac{d\eta}{dt} = \frac{df}{dz}(\gamma(t)) \cdot \frac{d\gamma}{dt};$$

in fact,

$$\begin{aligned}\frac{d\eta}{dt} = u'(t) + iv'(t) &= \left(\frac{\partial U}{\partial x} + i\frac{\partial V}{\partial x}\right)\frac{dx}{dt} + \left(\frac{\partial U}{\partial y} + i\frac{\partial V}{\partial y}\right)\frac{dy}{dt} \\ &= \left(\frac{\partial U}{\partial x} + i\frac{\partial V}{\partial x}\right)\frac{dx}{dt} + i\left(\frac{\partial U}{\partial x} + i\frac{\partial V}{\partial x}\right)\frac{dy}{dt} \\ &= \frac{df}{dz}(p)\left(\frac{dx}{dt} + i\frac{dy}{dt}\right) = \frac{df}{dz}(p)\frac{d\gamma}{dt}.\end{aligned}$$

EXERCISES

1. If p and q are complex numbers, show that the contour $\gamma(t) = (1 - t)p + tq = p + t(q - p)$, defined for $0 \le t \le 1$, traces out the line segment that connects p to q. Show that $\gamma(t)$ moves from p to q along this segment. Sketch the orientation of the tangent vector $\gamma'(t)$ at the point $\gamma(t)$ on the segment.

2. Plot the trajectory and indicate the direction of motion for the following parametrized curves. Indicate the initial and final points. Which are simple, closed, or simple closed curves? Write out the coordinate functions $x(t)$ and $y(t)$, if these are not given explicitly.

(i) $\gamma(t) = p + Re^{2\pi it}$ for $0 \leq t \leq 1$ ($R > 0$ real)

(ii) $\gamma(t) = e^{i(t+\alpha)}$ for $0 \leq t \leq 2\pi$ (α fixed, real)

(iii) $\gamma(t) = (2t^2 + 1) + it$ for $-1 \leq t \leq 1$

(iv) $\gamma(t) = e^{it}$ for $0 \leq t \leq 4\pi$

(v) $\gamma(t) = \frac{1}{2} + it$ for $-1 \leq t \leq 1$.

Answer: (i) Circle $|z - p| = R$, simple closed; (ii) circle $|z| = 1$, initial = final = $e^{i\alpha}$; (iii) arc of the parabola $x = 2y^2 + 1$; (iv) circle $|z| = 1$, *traversed twice;* (v) the line segment from $(\frac{1}{2}) - i$ to $(\frac{1}{2}) + i$.

3. Calculate derivatives $d\gamma/dt$ for the curves in Exercise 2. (Are they differentiable at the end points?) Locate all t such that $\gamma'(t) = 0$. Calculate the speed $|\gamma'(t)|$ at typical points on the trajectory.

4. In Example 4.4, prove that $d\eta/dt$ exists at $t = 0$, and that its value is zero. Determine the direction $\theta(t) = \arg \eta'(t)$ of the tangent vector $\eta'(t)$ for $-1 \leq t < 0$ and for $0 < t \leq +1$; how do the values compare when t is near zero?

5. Plot the trajectory of the parametrized curve

$$\gamma(t) = t^2 + i\,|t|^3 \quad \text{for} \quad -1 \leq t \leq +1.$$

Verify that γ is a smooth curve. What happens (geometrically) at $t = 0$?

6. Use $w = z^2$ to transform the following curves $\gamma(t)$ to curves $\eta(t)$ in the w-plane. Calculate the derivatives $d\eta/dt$. Sketch the trajectories of γ and η, in the respective planes, indicating the direction of motion.

(i) $\gamma(t) = e^{it}$ for $0 \leq t \leq \pi$

(ii) $\gamma(t) = \frac{1}{2} + it$ for $-1 \leq t \leq 1$

(iii) $\gamma(t) = (\frac{1}{2} + t) + i0$ for $-1 \leq t \leq +1$

(iv) $\gamma(t) = t \cdot e^{i\alpha}$ for $0 \leq t \leq 10$ (α fixed, real).

7. Carry out the program of Exercise 6 for the following transformations $w = f(z)$.

(i) $w = z + 1$

(ii) $w = z^2 + 1$

(iii) $w = (z + 1)^2 = z^2 + 2z + 1$.

8. Apply the non-holomorphic transformation $w = f(x + iy) = y^2 - ix$ to each of the radial line segments

$$\gamma_\alpha(t) = t \cdot e^{i\alpha} = (\cos \alpha)t + i(\sin \alpha)t \quad \text{for} \quad 0 \leq t \leq 10$$

(each real value $0 \leq \alpha < 2\pi$ gives a different curve γ_α). Write the transformed curves $\eta_\alpha(t) = f(\gamma_\alpha(t))$ in the form $u(t) + iv(t)$ and sketch the trajectories. Calculate the derivatives $d\eta_\alpha/dt$, using formula (7).

Answer: If $\cos \alpha = 0$, η_α traces out the segment $[0, 100]$ on the real axis; otherwise η_α traces out an arc on the parabola $u = \lambda_\alpha v^2$, where $\lambda_\alpha = \tan^2 \alpha$.

4.3 CONTOURS

We will frequently work with curves that are only **piecewise smooth.** This means that the interval $I = [a, b]$ on which $\gamma(t)$ is defined can be broken up into a finite number of consecutive subintervals $I_1, \ldots, I_N$ such that $\gamma(t)$ is smooth in each subinterval. We still require $\gamma(t)$ to be continuous throughout the whole interval $[a, b]$, so the curve will not have any breaks in it, but the derivative $\gamma'(t)$ is allowed to have simple jump discontinuities at the boundary points between intervals. These discontinuities correspond to the presence of "kinks" in the curve, at which the velocity (and direction of the tangent line) may change abruptly. This sort of behavior is illustrated in Figure 4.8. Questions involving piecewise smooth curves can always be reduced to questions about the individual smooth curves we get by restricting t to the various subintervals $I_1, \ldots, I_N$. Throughout this book we will use the word **contour** to refer to a smooth or piecewise smooth parametrized curve in the complex plane.

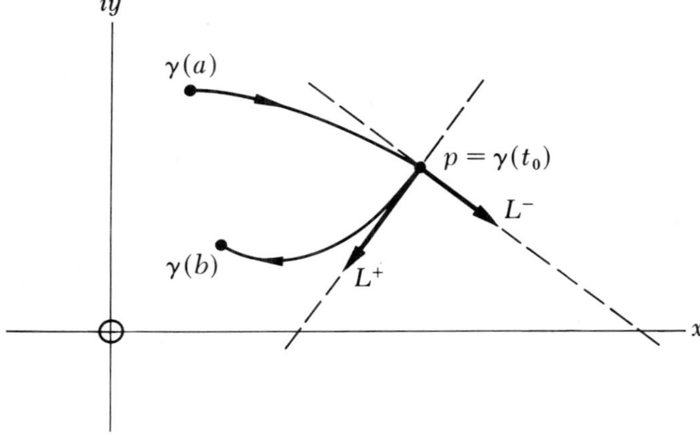

Figure 4.8 A piecewise smooth curve. The tangent changes direction abruptly at $p = \gamma(t_0)$ due to a jump discontinuity of $\frac{d\gamma}{dt}$ when $t = t_0$. Here L^+ and L^- indicate the right- and left-handed derivatives of $\gamma(t)$ at $t = t_0$.

EXERCISES

1. Suppose $I = [a, b]$ and $J = [b, c]$ are adjacent intervals in the real line, and that γ and η are continuous curves parametrized on I and J respectively, such that $\gamma(b) = \eta(b)$. Let us "piece together" γ and η to get a single curve $\phi(t)$ parametrized on $[a, c]$, by taking

$$\phi(t) = \begin{cases} \gamma(t) & \text{for} \quad a \leq t \leq b \\ \eta(t) & \text{for} \quad b \leq t \leq c. \end{cases}$$

Prove that $\phi(t)$ is continuous on $[a, c]$, particularly at $t = b$. If γ and η are smooth curves, is ϕ a contour?

2. Take γ and η as follows in Exercise 1:

$$\gamma(t) = e^{i\pi t} \quad \text{for} \quad -1 \leq t \leq +1$$
$$\eta(t) = t \qquad \text{for} \quad 1 \leq t \leq 3.$$

Sketch the trajectory of the contour ϕ, defined for $-1 \leq t \leq 3$, with particular attention to the behavior when $t = 1$.

4.4 CONFORMALITY OF HOLOMORPHIC FUNCTIONS

Let $w = f(z)$ be a smooth holomorphic mapping of the plane and assume that $df/dz \neq 0$ at p. If γ is a smooth curve that passes through p (say when $t = t_0$) and has non-zero derivative $\gamma'(t_0)$, then $\gamma'(t_0)$ determines a well defined tangent vector at p. The derivative of the transformed curve $\eta = f \circ \gamma$ is also non-zero when $t = t_0$:

$$\frac{d\eta}{dt}(t_0) = \frac{df}{dz}(\gamma(t_0))\frac{d\gamma}{dt}(t_0) = \frac{df}{dz}(p)\frac{d\gamma}{dt}(t_0), \tag{9}$$

since $f'(p) \neq 0$; it determines a new tangent vector $\eta'(t_0)$ at the point $q = f(p) = \eta(t_0)$ on the transformed curve, as indicated in Figure 4.9. This

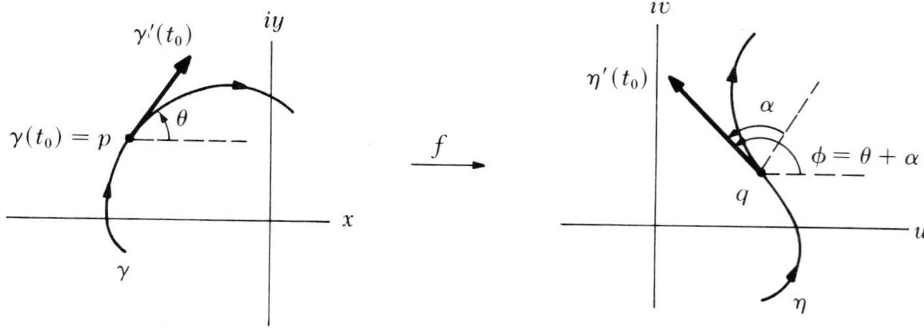

Figure 4.9 The transformation of tangent vectors by $w = f(z)$. Here $\eta(t) = f(\gamma(t))$ and $p = \gamma(t_0)$, so $q = f(p) = \eta(t_0)$. Angles are measured from the positive real axis to the derivatives, as indicated by θ and ϕ, and $\alpha = \arg f'(p)$.

transformation of tangent vectors is effected in the same way for all curves passing through p, regardless of their orientation, by multiplying the original derivative $\gamma'(t_0)$ by the fixed complex number $df/dz(p)$ and shifting the base point from p in the z-plane to $q = f(p)$ in the w-plane.

If θ is the angle measured from the positive real axis to the tangent vector $\gamma'(t_0)$, and if ϕ is the corresponding angle for the transformed tangent vector $\eta'(t_0)$, then

$$\text{(10)} \qquad \phi = \theta + \alpha \quad \text{where} \quad \alpha \equiv \arg \frac{df}{dz}(p) \pmod{2\pi},$$

as indicated in Figure 4.9. The lengths of tangent vectors are proportional,†

$$\text{(11)} \qquad |\eta'(t_0)| = \left| \frac{df}{dz}(p) \right| \cdot |\gamma'(t_0)| \,.$$

Now consider what happens if we take two smooth curves that pass through p when $t = t_0$ and have non-zero tangent vectors there. Let $\Delta\theta$ be the angle between tangent vectors at p, measured from γ_1 to γ_2, as shown in Figure 4.10. The transformed curves $\eta_1 = f \circ \gamma_1$ and $\eta_2 = f \circ \gamma_2$ pass through $q = f(p)$ in the w-plane when $t = t_0$, and have non-zero tangent vectors there since $f'(p) \neq 0$; let $\Delta\phi$ be the angle measured from η_1 to η_2. Using (10), we may calculate $\Delta\phi$ in terms of $\Delta\theta$:

$$\Delta\phi = (\theta_2 + \alpha) - (\theta_1 + \alpha) = (\theta_2 - \theta_1) = \Delta\theta.$$

Here $\alpha = \arg f'(p)$; this proves the following basic result.

Theorem 4.1 *Let $w = f(z)$ be a smooth holomorphic function whose derivative df/dz is non-zero at p. If two smooth curves pass through p and have non-zero tangent vectors which make an angle $\Delta\theta$ (measured from γ_1 to γ_2), then the transformed curves $\eta_1 = f \circ \gamma_1$ and $\eta_2 = f \circ \gamma_2$ have non-zero tangent vectors at $q = f(p)$. The angle $\Delta\phi$ between these tangent vectors has the same magnitude and sense as the original angle; that is, $\Delta\phi = \Delta\theta$.*

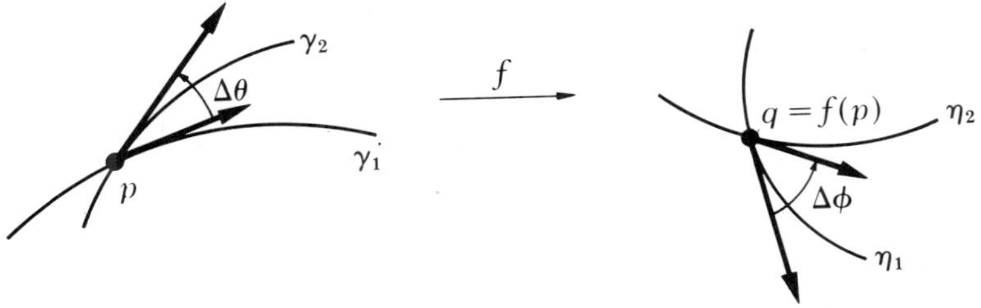

Figure 4.10 A holomorphic mapping preserves the angle between two curves which meet at p, if $f'(p) \neq 0$, so that $\Delta\phi = \Delta\theta$. Here $\Delta\theta$ is measured from γ_1 to γ_2, and $\Delta\phi$ from $\eta_1 = f \circ \gamma_1$ to $\eta_2 = f \circ \gamma_2$, as shown.

† Formulas (9) and (11) would be valid even if $df/dz = 0$ at p, but formula (10) would be meaningless. (Why?) This is why we assumed that $f'(p) \neq 0$ at the beginning of this discussion.

In other words, a smooth holomorphic mapping preserves the size and sense of angles between curves, at any point p where $f'(p) \neq 0$. In general, if we consider a smooth but non-holomorphic mapping $w = f(z)$, the angle $\Delta\phi$ will depend in a very complicated way on both $\Delta\theta$ and the orientation of the original tangent vectors at p (see Example 4.5 below). For holomorphic mappings, the angle-preserving property fails to hold at points where $f'(p) = 0$ (see Exercise 4). This remarkable angle preserving property distinguishes regular holomorphic mappings from other smooth mappings of the plane, and leads us to make an important definition.

Definition 4.2 *Let $w = f(z) = U(z) + iV(z)$ be a smooth mapping of the plane. We say that f is* **conformal at p** *if:*

(i) *The Jacobian determinant $J_f(p)$ is non-zero (this determinant is defined below).*

(ii) *The angle between tangents to any pair of smooth curves passing through p is preserved, so that $\Delta\phi = \Delta\theta$ regardless of how the tangent vectors are oriented.*

If f does this at each point in some open set E, we say that f is **conformal** *on E. Sometimes we shall enlarge this category of mappings of the plane to include maps that preserve the size of the angle but reverse its sense, so that $\Delta\phi = -\Delta\theta$. Such mappings are* **anti-conformal,** *and these two kinds of mappings, taken together, are referred to as* **isogonal** *mappings.*

Whether f is holomorphic or not, formula (7) tells us how tangent vectors are transformed. In this definition we must impose the condition that the Jacobian determinant

$$J_f(p) = \det\begin{bmatrix} \dfrac{\partial U}{\partial x} & \dfrac{\partial U}{\partial y} \\[2ex] \dfrac{\partial V}{\partial x} & \dfrac{\partial V}{\partial y} \end{bmatrix} = \frac{\partial U}{\partial x}\frac{\partial V}{\partial y} - \frac{\partial U}{\partial y}\frac{\partial V}{\partial x} \tag{12}$$

is non-zero; otherwise, non-zero tangent vectors oriented in certain directions would be transformed to tangent vectors of length zero attached to the image point $q = f(p)$, and measurement of the angle $\Delta\phi$ would be meaningless. If f is a smooth holomorphic function of a complex variable, its components $U(x, y)$ and $V(x, y)$ satisfy the Cauchy-Riemann equations and the Jacobian determinant is just

$$J_f(p) = \left(\frac{\partial U}{\partial x}\right)^2 + \left(\frac{\partial V}{\partial x}\right)^2 = \left|\frac{\partial U}{\partial x} + i\frac{\partial V}{\partial x}\right|^2 = \left|\frac{df}{\partial z}(p)\right|^2. \tag{13}$$

For this special class of functions, the Jacobian determinant vanishes if and only if $df/dz = 0$. We leave it as Exercise 2 for the reader to deduce the following result from the Cauchy-Riemann equations and the definitions given above.

Theorem 4.2 *Let $w = f(z)$ be any smooth mapping defined on an open set E in the plane. Then f is a conformal mapping on E if and only if it is holomorphic with non-vanishing derivative on E (that is, f is a regular holomorphic mapping on E).*

Notice that conformality is defined without reference to complex derivatives—it is a purely geometric property of smooth mappings. Theorem 4.2 establishes the fact that conformality is equivalent to complex differentiability (with non-vanishing derivative).

If a mapping $w = f(z)$ is conformal (or isogonal) on a domain E, this means that curves which are orthogonal where they intersect remain orthogonal after the transformation into the w-plane has been performed. This observation will be very useful in studying the global mapping properties of holomorphic functions. Here are a few examples of conformal and non-conformal mappings of the plane.

Example 4.5 The mapping $w = f(x + iy) = (-y^2) + i(xy)$ is smooth since

$$U(x, y) = -y^2 \quad \text{and} \quad V(x, y) = xy$$

have continuous partial derivatives, but the Cauchy-Riemann equations are not satisfied and f is not holomorphic anywhere in the plane. The Jacobian determinant is

$$J_f(p) = \det\begin{bmatrix} 0 & -2y \\ y & x \end{bmatrix} = 2y^2,$$

and is non-zero at points off the locus $y = 0$ (the x-axis); let us see if f is conformal at any of the points where $J_f(p) \neq 0$. If $p = x_0 + iy_0$, equation (7) shows that a non-zero tangent vector to a curve γ through p,

$$\gamma'(t_0) = \frac{dx}{dt}(t_0) + i\frac{dy}{dt}(t_0),$$

is transformed to the tangent vector

$$\eta'(t_0) = \frac{du}{dt}(t_0) + i\frac{dv}{dt}(t_0) = \left(-2y_0\frac{dy}{dt}\right) + i\left(y_0\frac{dx}{dt} + x_0\frac{dy}{dt}\right)$$

at $q = f(p)$. The slope of the tangent to the transformed curve η is

$$\tan\phi = \left(\frac{dv}{dt}\right) \Big/ \left(\frac{du}{dt}\right) = \frac{y_0 + x_0\tan\theta}{-2y_0\tan\theta} = -\frac{1}{2}\left(\frac{1}{\tan\theta}\right) - \frac{x_0}{2y_0}.$$

From this formula it is easy to see that we do not have $|\Delta\phi| = |\Delta\theta|$; in fact, $\Delta\phi$ depends in a very complicated way on $\tan\theta_1$ and $\tan\theta_2$ as well as $\Delta\theta$, when we transform two curves that pass through p and make an angle $\Delta\theta$. The mapping f is certainly not conformal or isogonal. Nor is there any simple analog of formula (3) which tells us how holomorphic functions of z behave for z near p.

Example 4.6 The conjugation mapping $w = f(z) = \bar{z}$ has the form

$$f(x + iy) = \overline{x + iy} = x - iy,$$

with component functions $U(x, y) = x$ and $V(x, y) = -y$. It is certainly a smooth mapping of the plane, but it is not holomorphic since the Cauchy-Riemann equations are not satisfied:

$$\frac{\partial U}{\partial x} = 1 \neq -1 = \frac{\partial V}{\partial y}, \qquad \frac{\partial U}{\partial y} = 0 = -\frac{\partial V}{\partial x}.$$

This non-holomorphic function is much better behaved than the one in the last example. The Jacobian determinant is -1 everywhere and the transformation law (7) for tangent vectors to curves through a point $p = \gamma(t_0) = x_0 + iy_0$ has a particularly simple form:

$$\eta'(t_0) = \frac{du}{dt}(t_0) + i\frac{dv}{dt}(t_0) \quad \text{where} \quad \frac{du}{dt} = \frac{dx}{dt} \quad \text{and} \quad \frac{dv}{dt} = -\frac{dy}{dt}.$$

Thus, a tangent vector $\gamma'(t_0)$ attached to p which makes an angle θ with the positive real axis (measured from the axis to the vector, as usual) is transformed to a tangent vector at $q = f(p)$ which makes an angle $\phi = -\theta$ with the positive real axis in the w-plane. In particular, if we transform two curves, γ_1 and γ_2, whose tangents make an angle $\Delta\theta$ (measured from γ_1 to γ_2), the transformed curves η_1, η_2 make an angle $\Delta\phi = -\Delta\theta$ (measured from η_1 to η_2). Thus f is an isogonal mapping, although it is not conformal or holomorphic.

These properties of f can also be seen from its geometric interpretation: $\bar{z} = x - iy$ is the image of $z = x + iy$ when we reflect points in the complex plane through the real axis.

Example 4.7 (Linear Mappings) Consider the mapping $w = f(z) = az + b$, where a and b are complex constants and $a \neq 0$. To understand the action of such mappings it is sufficient to examine three special cases.

1. *Translations* If we take $a = 1$ and b arbitrary, the position of every point z is shifted by adding on b. Obviously we get a conformal mapping of the plane which is invertible, the inverse map being given by $z = \check{f}(w) = w - b$.

2. *Rotations* Taking $b = 0$ and a such that $|a| = 1$, we may write a in polar form $a = e^{i\alpha}$ (α real). Thus, $w = e^{i\alpha} \cdot z$, and every point z is rotated counterclockwise about the origin by an angle of α radians. This mapping is conformal and invertible, and its inverse is $z = \check{f}(w) = e^{-i\alpha}w$.

3. *Dilations* Taking $b = 0$ and a any positive real number, $a = |a| > 0$, the mapping $w = f(z) = a \cdot z$ acts by scaling each complex number by the factor a, leaving directions unaltered. If $a = 1$ we get the **identity mapping** $f(z) = z$ for all points in the plane; we often refer to f as an *expanding map* if $1 < a < +\infty$ and a *contracting map* if $0 < a < 1$, but the term "dilation" is meant to allow for either possibility. These mappings are all conformal and invertible, with inverse $z = \check{f}(w) = (1/a) \cdot w$.

A general linear map f is obtained by composing these elementary mappings: if $a = |a|\, e^{i\alpha}$, we obtain $w = az + b$ by performing the operations

$$w = e^{\alpha} \cdot z, \qquad w = |a| \cdot z, \qquad w = z + b$$

in succession. Since the elementary mappings are conformal and invertible on the plane, so is their composite $w = f(z)$.

Linear mappings preserve the *shape* of sets. Dilations scale the size of a figure up or down without changing its orientation. Translations and rotations (and composite mappings obtained from them) actually preserve the *size* of a figure as well as its general shape, although they can change its position and orientation; these special mappings are distance preserving in the sense that $|f(z') - f(z'')| = |z' - z''|$ for any pair of points z' and z'' in the plane. The effect of a typical linear transformation

$$w = f(z) = (i - 1)z + (i + 1)$$

on a rectangular set is shown in Figure 4.11; here $a = (i - 1) = \sqrt{2}\, e^{i3\pi/4}$ and $b = (i + 1) = \sqrt{2}\, e^{i\pi/4}$.

Example 4.8 The reciprocal mapping $w = f(z) = 1/z$ is defined and holomorphic on the punctured plane $E = \{z\colon z \neq 0\}$. Since the derivative $f'(z) = -1/z^2$ is never zero, the mapping is conformal everywhere on E. If a base point $p \neq 0$ is fixed, $f(z)$ is closely approximated for z near p by the linear mapping

$$\tilde{f}(z) = f(p) + f'(p)(z - p) = \left(\frac{1}{p}\right) - \left(\frac{1}{p^2}\right)(z - p) = \frac{1}{p} - \frac{\Delta z}{p^2},$$

so that $\Delta\tilde{f} = -\Delta z/p^2$ if $\Delta z = z - p$. The displacement $\Delta f = f(z) - f(p)$ is closely approximated by $\Delta\tilde{f}$ if Δz is small; the size $|\Delta f|$ of this displacement

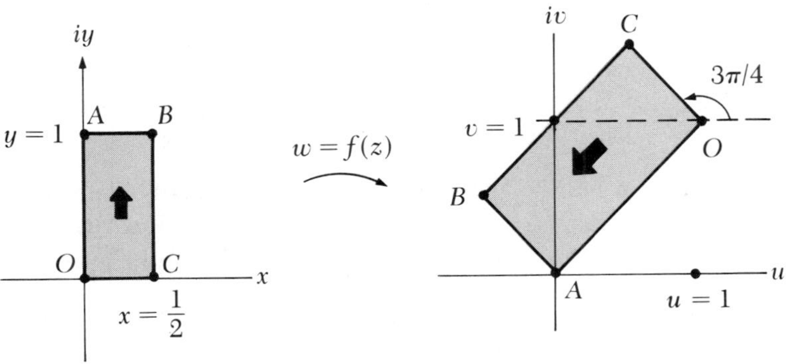

Figure 4.11 Action of $w = (i - 1)z + (i + 1)$ on a rectangular block. Here $\alpha = \arg f'(0) = +3\pi/4$.

in the w-plane is therefore approximately given by

$$|\Delta f| \approx |\Delta \tilde{f}| = \frac{|\Delta z|}{|p^2|} = \frac{1}{|p|^2} \cdot |\Delta z| \, .$$

In particular, if the base point p lies near the origin, lengths are dilated by a large factor and, conversely, the scale factor for lengths is quite small if the base point is far from the origin. The direction $\arg \Delta \tilde{f}$ can be written as $\theta + \alpha$ where $\theta = \arg \Delta z$ and α is some fixed angle associated with the transformation f and the base point p. In fact, $\alpha = \arg f'(p)$; this angle may be calculated explicitly by noting that

$$f'(p) = -\frac{1}{p^2} = -\frac{1}{|p|^2} \cdot (e^{-i \arg p})^2 = +\frac{1}{|p|^2} \, e^{i\pi} e^{-2i \arg p},$$

so that

$$\alpha \equiv \pi - 2 \arg(p) \quad (\text{mod } 2\pi).$$

For example, if $p = 1$, then Δz is scaled by the factor $+1$ and rotated by $+\pi$ radians to obtain $\Delta \tilde{f}$, so displacements from p are (approximately) transformed by a pure rotation counterclockwise by π radians. On the other hand, if $p = +2i$, then Δz is scaled by $1/|p|^2 = 1/4$ and rotated by $\alpha = 0$ radians; thus, displacements from $p = +2i$ are (approximately) transformed by a pure dilation, without rotation. The reader can examine other base points and set up a diagram like the one presented in Figure 4.2 (for the *exp* function); it is also interesting to determine the base points p where f acts on small displacements from p like a pure rotation or like a pure dilation.

If we measure the angle between two curves that have non-zero tangent vectors at p before and after applying the transformation f, the angle α drops out and we get the relation $\Delta\phi = \Delta\theta$ expected of a conformal mapping.

EXERCISES

1. Consider an arbitrary linear transformation $(u, v) = \Phi(x, y)$ of two real variables x and y:

$$\begin{aligned} u &= u_0 + Ax + By \\ v &= v_0 + Cx + Dy \end{aligned} \qquad (A, B, C, D \text{ and } u_0, v_0 \text{ real})$$

Show that the mapping $w = u + iv = f(x + iy)$ is holomorphic if and only if $D = A$ and $C = -B$; show that f is a complex linear mapping ($w = az + b$, where a and b are complex constants) if f is holomorphic.

2. The fact that "holomorphic" implies "conformal," for smooth mappings of the plane, has been established in this section. Prove that "conformal" implies "holomorphic" by demonstrating that the partial derivatives $\partial u/\partial x, \ldots, \partial V/\partial y$ must satisfy the Cauchy-Riemann equations if $f = U + iV$ is conformal. This proves Theorem 4.2.

3. If f is a smooth holomorphic function on an open set E, show that

(i) $g(z) = \overline{f(z)}$ on E
(ii) $h(z) = f(\bar{z})$ on $E^* = \{z\colon \bar{z} \text{ is in } E\}$

are anti-conformal mappings.

4. Consider the action of $w = f(z) = z^2$ on parametrized lines through the origin $\gamma(t) = t \cdot e^{i\alpha}$ (α real; $0 \leq \alpha < 2\pi$). The transformed curves $\eta_\alpha(t) = f(\gamma_\alpha(t))$ have zero derivative at $w = 0$ (why?), so there are no well defined tangent directions to be discussed. However, the "limit tangent directions"

$$\theta_\alpha = \lim_{t \to 0} \arg \eta'_\alpha(t)$$

are well defined (prove this) and may be compared for various choices of α. Show that the angle $\Delta\theta$ between two of these curves through $z = 0$ is *doubled* by the transformation $w = z^2$. Thus $w = z^2$ is not "conformal," even if we interpret $\Delta\phi$ using the limit tangent directions above. Sketch the transformed curves η_α for a few values of α.

5. Repeat Exercise 4 for $w = z \cdot \sin z$.

6. Show that $w = z^2$ maps radial lines that make an angle $\Delta\theta$ at $z = 0$ to radial lines that make an angle $\Delta\phi = 2 \cdot \Delta\theta$ at $w = 0$. Then, sketch the image in the w-plane of the annular sector E whose polar coordinates satisfy $r_1 \leq r \leq r_2$ and $\theta_1 \leq \theta \leq \theta_2$ (assume $\theta_1 - \theta_2 < \pi$, to keep the image from overlapping itself). Do the same, taking $r_1 = 0$ and $r_2 = +\infty$.

7. Sketch the transformed image of the rectangular block $E = \{z\colon 1 \leq \text{Re}(z) \leq 2 \text{ and } 0 \leq \text{Im}(z) \leq 2\}$ under the following linear transformations. (Using Figure 4.11 as a model.)

(i) $w = -z$ (iii) $w = \left(\frac{i}{2} z\right) - i$

(ii) $w = iz - 3$ (iv) $w = \frac{i}{2}(4 - z)$

8. At which points do the following mappings fail to be conformal?

(i) $w = z + z^3$ (iv) $w = \dfrac{z - 1}{z + 1}$ (for $z \neq -1$)

(ii) $w = \sinh z$ (v) $w = e^z - z$

(iii) $w = \cos z$ (vi) $w = \exp\left(\dfrac{1}{z^2}\right)$ (for $z \neq 0$)

9. For $w = \exp z$ determine all points p at which the approximate displacement $\Delta\tilde{f}$ is obtained from Δz by

(i) a pure rotation
(ii) a pure dilation by a factor $a > 1$
(iii) a pure dilation by a factor $a < 1$
(iv) the identity transformation $\Delta\tilde{f} = \Delta z$.

Sketch these loci in the z-plane. How are tangent vectors transformed for base points p on these loci?

10. Repeat Exercise 9 for $w = z^2$.

4.5 MAPPING PROPERTIES OF HOLOMORPHIC FUNCTIONS IN THE LARGE

There are a few really general principles we can use to examine the *global* mapping properties (the properties "in the large") of holomorphic functions. For example, if we want to study the behavior of a mapping by seeing how it transforms a family of curves (lines, circles, etc.) from the z-plane into the w-plane, it is obvious that the curve family should be chosen so it is related to the intrinsic properties of the mapping we are considering. There is no single family of curves that is appropriate in every situation. We also have at our disposal the following conformality principle, which follows directly from the conformality of holomorphic functions, and needs no further comment.

Theorem 4.3 *Let $w = f(z)$ be a holomorphic function defined on an open set E in the complex plane. If p is a point in E where $f'(p) \neq 0$, then any pair of smooth curves that meet at p, and have non-zero orthogonal tangents at p, are mapped by f to curves in the w-plane whose tangents are orthogonal at the image point $q = f(p)$ where the image curves meet.*

The condition $f'(p) \neq 0$ is vital to the result.

We illustrate these principles by examining the global mapping properties of a few very familiar holomorphic functions. The first example is selected to avoid the difficulties that would arise if there were points at which $f'(p) = 0$.

Example 4.9 The map $w = f(z) = e^z$ is holomorphic on the whole z-plane, and its derivative $f'(z) = e^z$ is never zero. Consider the curve families in the z-plane: $\mathscr{A}$ consisting of all vertical lines ($x =$ constant) and $\mathscr{B}$ consisting of all horizontal lines ($y =$ constant). These are **orthogonal families;** curves from $\mathscr{A}$ and $\mathscr{B}$ are orthogonal where they intersect. The conformality principle assures us that the transformed families $f(\mathscr{A})$ and $f(\mathscr{B})$ are orthogonal families in the w-plane. In particular, a vertical line $z = c + iy$ $(-\infty < y < +\infty)$ gets mapped to the set of points

$$w = e^{c+iy} = e^c e^{iy} = e^c(\cos y + i \sin y);$$

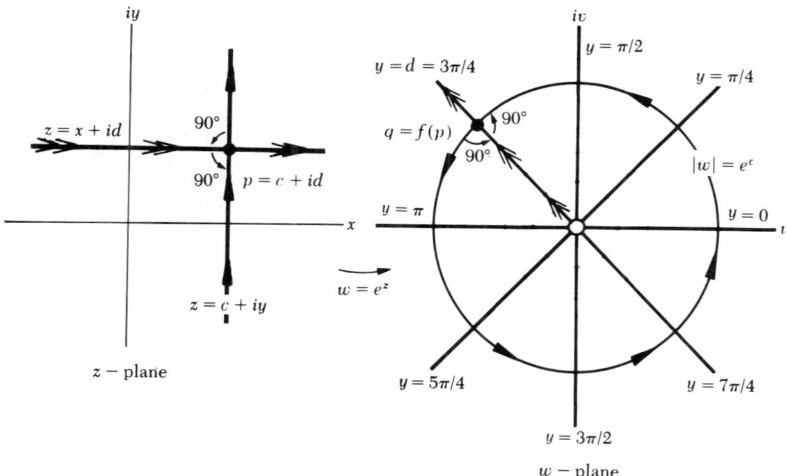

Figure 4.12 Transformations of horizontal and vertical lines $x = constant$ and $y = constant$ to corresponding curves in the w-plane under $w = e^z$. The case $y = d = 3\pi/4$ is shown at left; images of horizontal lines $y = n\pi/4$ are the radial lines shown in w-plane.

as y increases, these points trace out the circle $|w| = e^c$ (note that $e^c > 0$ for any choice of the real number c), covering the circle once each time y moves through an interval of length 2π. A horizontal line $z = x + id$ $(-\infty < x < +\infty)$ gets mapped to the set of points

$$w = e^{x+id} = e^x e^{id}.$$

Thus $\arg(w) = d$, a constant value, while the absolute value is $|w| = e^x$. Clearly, $w = e^{x+id}$ traces out the ray from 0 to infinity (with $w = 0$ excluded) that makes an angle of d radians with the positive real axis. A few of these transformed curves are shown in Figure 4.12. Obviously, the original families $\mathscr{A}$ and $\mathscr{B}$ are mapped to the families: $f(\mathscr{A})$ = all circles about the origin, and $f(\mathscr{B})$ = all radial lines originating at $w = 0$ (with the end point $w = 0$ excluded). The curves in the transformed families are obviously mutually orthogonal.

It is interesting to combine these observations to see how the exponential mapping transforms certain sets, notably rectangular blocks of the form

$$c_1 \leq x \leq c_2 \quad \text{and} \quad d_1 \leq y \leq d_2.$$

As in Figure 4.13, this set is transformed to an annular sector with interior angle $\phi = d_2 - d_1$ and radii $r_1 = e^{c_1} < r_2 = e^{c_2}$. The sector overlaps itself, possibly several times, if the length of the original block in the y direction exceeds 2π: $d_2 - d_1 \geq 2\pi$. From the way the exponential map transforms lines and rectangular domains, it is clear that the range is Range$(exp) = \{w \colon w \neq 0\}$, the punctured plane.

Example 4.10 (The reciprocal map $w = 1/z$) This conformal transformation, defined on the punctured plane $E = \{z : z \neq 0\}$, is in the unusual

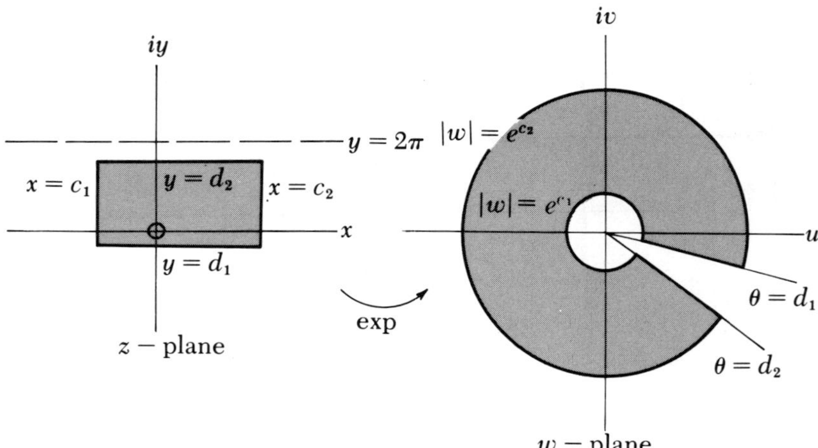

Figure 4.13 Transformation of rectangular blocks to circular sectors by the mapping $w = e^z$.

position of being its own inverse:

$$z = \check{f}(w) = 1/w \qquad \text{for all } w \neq 0.$$

That is, applying the map f twice in succession to a point z brings us back to where we started:

$$f(f(z)) = f(1/z) = \frac{1}{(1/z)} = z \quad \text{for all} \quad z \neq 0.$$

One way of examining the global properties of f is to write z in polar form $z = re^{i\theta}$; the transformed point

$$\frac{1}{z} = \frac{1}{re^{i\theta}} = \frac{1}{r}e^{-i\theta}$$

can be obtained by performing two simpler operations in succession. Thus, $f = C \circ J$, where J is the mapping that reflects points through the unit circle in the following manner (it is defined for $z \neq 0$):

(14) $$J(re^{i\theta}) = \frac{1}{r}e^{i\theta} \quad \text{for all} \quad r > 0 \quad \text{and all real } \theta,$$

and C is the familiar complex conjugation map $C(z) = \bar{z}$, which has the polar form

(15) $$C(re^{i\theta}) = re^{-i\theta} \quad \text{for all} \quad r \geq 0 \quad \text{and all real } \theta.$$

Now f is the composite mapping $f(z) = (C \circ J)(z) = C(J(z))$ for all $z \neq 0$; the action of the individual mappings J and C is very simple, and is illustrated in Figure 4.14. Notice that z and $J(z)$ have the same argument θ, but their

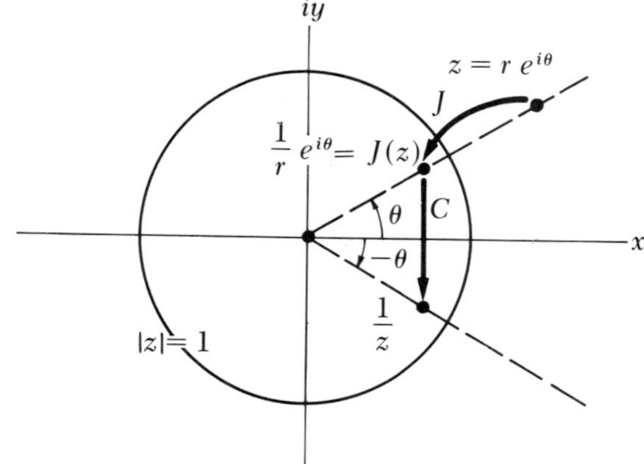

Figure 4.14 The inversion $w = 1/z$ realized as a composition of J = reflection through the unit circle and C = complex conjugation.

absolute values are reciprocal to one another; points on the unit circle $|z| = 1$ are left where they stand, and this is why we refer to J as a "reflection through the unit circle."

> *Note:* We have already pointed out that the function $C(z) = \bar{z}$ is not complex differentiable, and the reader can easily check that $J(z)$ is not differentiable either, by referring to the Cauchy-Riemann equations. It is interesting that these mappings, *neither of which is differentiable*, combine to give us a holomorphic function $J(C(z)) = f(z) = 1/z$. It may be helpful to point out that the mappings J and C are anticonformal; at any point in their respective domains of definition they preserve the size of the angle $\Delta\theta$ between two curves but reverse the sense of this angle so that the image curves make an angle $\Delta\phi = -\Delta\theta$. Obviously, two sense reversing maps performed in succession give a sense preserving (conformal) mapping.

Other global mapping properties are revealed by examining the way $w = 1/z$ transforms horizontal and vertical lines in the z-plane. Points on the horizontal line

$$z = x + id \quad (d \neq 0 \text{ a constant}; \quad -\infty < x < +\infty)$$

are transformed to points

$$w = u + iv = \frac{1}{x + id} = \frac{x}{x^2 + d^2} + i\,\frac{-d}{x^2 + d^2} \quad (-\infty < x < +\infty)$$

and we can eliminate x between the expressions for u and v to see that u and v satisfy the equation:

$$u^2 + \left(v + \frac{1}{2d}\right)^2 = \left(\frac{1}{2d}\right)^2.$$

This equation describes a circle centered at $u = 0$, $v = -1/2d$ on the v-axis; this circle is tangent to the u-axis and passes through the origin (see Figure 4.15).

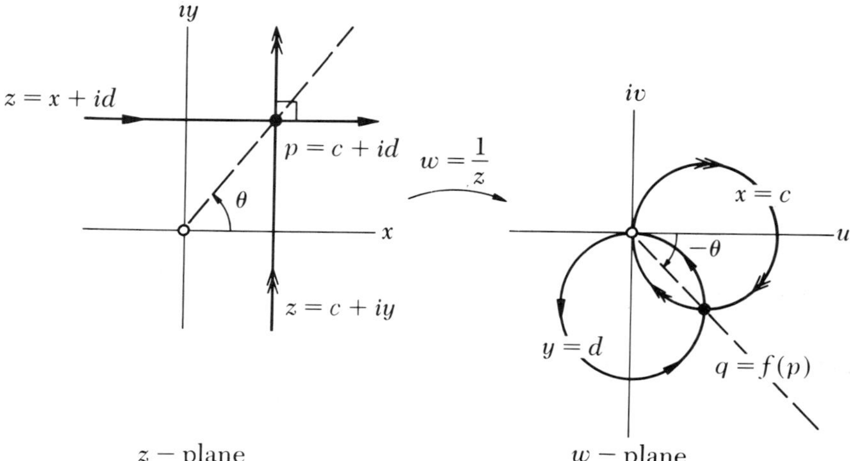

Figure 4.15 Transformation of the horizontal and vertical lines through a point $p = c + id$ by the mapping $w = 1/z$. The image points trace out the circles shown, avoiding the exceptional point $w = 0$.

The points $w = f(x + id)$ trace out all points on this circle except $w = 0$; they approach the origin as $|x| \to +\infty$, as indicated by the arrows in Figure 4.15. Similarly, the vertical line $z = c + iy$ ($c \neq 0$ a constant; $-\infty < y < +\infty$) is mapped one-to-one into the circle centered at the point $u = 1/2c$, $v = 0$ on the u-axis, which passes through the origin:

$$\left(u - \frac{1}{2c}\right)^2 + v^2 = \left(\frac{1}{2c}\right)^2$$

We leave it to the reader to account for the transformation of the exceptional lines

$$z = x + i0 \quad \text{and} \quad z = 0 + iy$$

(with the origin $z = 0$ deleted, since it is not in the domain of definition E); these sets are mapped into the real and imaginary axes in the w-plane, avoiding the point $w = 0$.

The two families of circles we get by transforming the families of horizontal lines and vertical lines in the z-plane to circles in the w-plane are orthogonal where they intersect, as we expect from the conformality of $w = 1/z$. This is perhaps more clearly illustrated in Figure 4.16, where we display the action of this mapping on a checkerboard pattern of rectangles bounded by horizontal and vertical lines in the z-plane. Remember that $w = 1/z$ maps the circle $|z| = 1$ one-to-one onto the circle $|w| = 1$, and interchanges the set of points inside the unit circle with the set of points outside of this circle, as the numbering of squares in our diagram indicates. Consequently, the square labelled with numeral 1 in Figure 4.16 is mapped to a domain which extends out to infinity. The nature of the pattern in the other three quadrants is easily determined

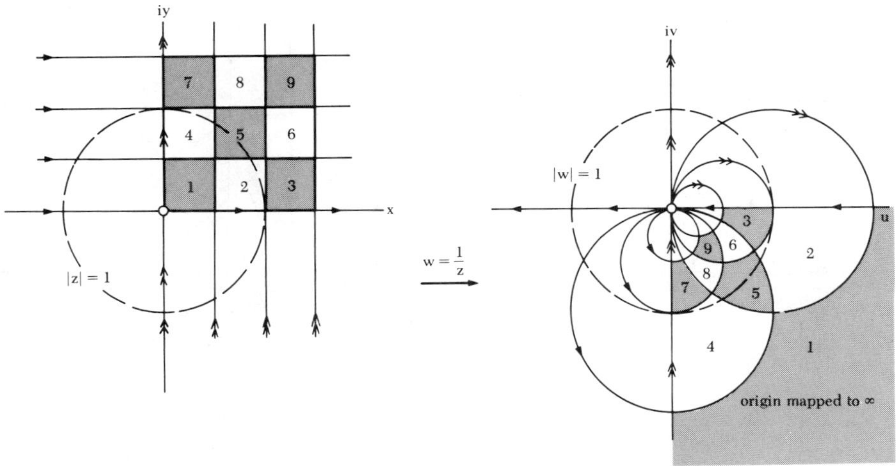

Figure 4.16 Mapping of a pattern of horizontal and vertical lines under the reciprocal mapping $w = 1/z$.

using the symmetry properties

$$f(-z) = -f(z) \qquad f(\bar{z}) = \overline{f(z)} \qquad \text{(all } z \neq 0\text{)}$$

of the reciprocal mapping.

The mappings studied above are all relatively simple; here is a familiar holomorphic transformation of the plane whose global mapping properties present a more substantial challenge.

**Example 4.11 (The Complex Trigonometric Function $w = \sin z$)* This function is analytic throughout the plane, and has the following symmetry properties:

$$\begin{aligned} \sin(z + 2\pi n) &= \sin z && \text{(sin } z \text{ is periodic)} \\ \sin(-z) &= -\sin z && \text{(sin } z \text{ is an } \textit{odd} \text{ function of } z\text{)} \\ \sin(z + \pi) &= -\sin z \end{aligned}$$

Therefore, to fully comprehend its mapping properties, it is enough that we understand how it behaves in a vertical strip of width π, such as the one shown in Figure 4.17.† We will show that $w = \sin z$ maps the open strip $D = \{z: -\pi/2 < \text{Re}(z) < \pi/2\}$ to the doubly cut plane E obtained by deleting the rays $(-\infty, -1]$ and $[+1, +\infty)$ from the real axis in the w-plane. Since $d/dz(\sin z) = \cos z \neq 0$ for z in D, f is a conformal mapping $f: D \to E$. We will see that it is invertible (maps D one-to-one onto the domain E), so there is a well defined inverse $z = \check{f}(w)$ defined on the doubly cut plane E, which will be

† The periods are of length 2π, but the extra symmetry relations tell us how sin z behaves in a strip of width 2π if we know it for strips of width π.

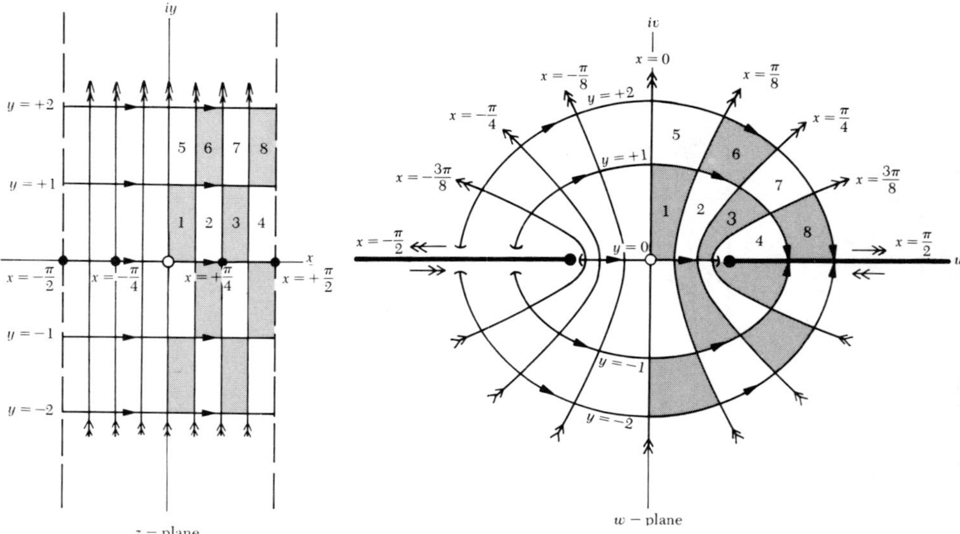

Figure 4.17 The mapping $w = \sin z$ from the strip $-\pi/2 < \text{Re}(z) < +\pi/2$ one-to-one onto the doubly cut w-plane with rays $(-\infty, -1]$ and $[+1, +\infty)$ deleted.

regarded as the principal determination of $z = \arcsin w$ for the complex variable w. We will denote it by $\check{f}(w) = \text{Arcsin } w$ hereafter.

The global nature of $w = \sin z$ is conveniently studied by examining the transformation of horizontal and vertical line segments in the strip D. First write out the real and imaginary parts of the transformation:

$$\sin(x + iy) = U(x, y) + iV(x, y) = (\sin x \cosh y) + i(\cos x \sinh y).$$

The vertical lines

$$z = c + iy \quad \left(c \text{ fixed with } -\frac{\pi}{2} < c < +\frac{\pi}{2}\,;\, -\infty < y < +\infty\right)$$

get mapped to branches of hyperbolas, except in the special case when $c = 0$; then $w = \sin(0 + iy) = i \sinh y$ sweeps out the imaginary axis, in the direction shown in Figure 4.17, as y increases. For $c \neq 0$ we may eliminate y between the formulas for u and v to get

$$\text{(16)} \qquad \frac{u^2}{\sin^2 c} - \frac{v^2}{\cos^2 c} = \cosh^2 y - \sinh^2 y = 1 \quad \text{for all } y,$$

for the points $u + iv$ swept out by $w = \sin(c + iy)$. Equation (16) describes a hyperbola centered at the origin with focal points $\pm 1 + i0$ and asymptotic lines whose slope is $\pm 1/\tan c$. It is not hard to see that $w = \sin(c + iy)$ traces out the right-hand branch of the hyperbola if $c > 0$ and left-hand branch if $c < 0$, as indicated in the figure.

The horizontal lines

$$z = x + id \quad \left(d \text{ fixed}; \quad -\frac{\pi}{2} < x < +\frac{\pi}{2}\right)$$

in our strip transform to complementary halves of an ellipse, except in the special case when $d = 0$; then $w = \sin(x + i0) = \sin x$ sweeps out the segment $(-1, +1)$ on the real axis, moving from left to right as x increases. Otherwise, $d \neq 0$ and

$$\text{(17)} \quad \frac{u^2}{\cosh^2 d} + \frac{v^2}{\sinh^2 d} = \sin^2 x + \cos^2 x = 1 \quad \text{for all} \quad -\frac{\pi}{2} < x < +\frac{\pi}{2}.$$

This is the equation of an ellipse with focal points $\pm 1 + i0$ and axis intercepts $u = \pm\cosh d$, $v = \pm\sinh d$. A little thought reveals that $w = \sin(x + id)$ traces out only the upper half of this ellipse, in the direction shown in Figure 4.17, as x increases when $d > 0$; it traces out the lower half, as shown, when $d < 0$. Notice that no point in the open strip D gets mapped into the deleted segments $(-\infty, -1]$ or $[+1, +\infty)$, and every other point w in the cut plane is the image of exactly one point z in the strip, so $f: D \to E$ is an *invertible* mapping.

The boundary lines for the strip D, consisting of points

$$z = -\frac{\pi}{2} + iy \quad \text{or} \quad z = +\frac{\pi}{2} + iy \quad (-\infty < y < \infty),$$

are mapped onto the rays $(-\infty, -1]$ and $[+1, +\infty)$ in the w-plane in a two-to-one manner. But we are not considering these boundary lines as part of our domain D.

The nature of the mapping is better understood if one considers how the checkerboard pattern shown in the right half of the strip in Figure 4.17 is transformed by our mapping. It should be clear that the transformation $w = \sin z$ becomes somewhat badly behaved as z approaches the boundary points $z = \pm(\pi/2) + i0$; vertical lines in the strip D come closer and closer to being doubled over upon themselves when they pass near these points on the boundary. This is related to the fact that $d/dz(\sin z) = 0$ at the points $z_n = (\pi/2) + n\pi$, so the mapping $w = \sin z$ is not conformal at these points.

From the way $w = \sin z$ transforms vertical and horizontal lines in D, we have seen that there is an inverse mapping $z = \check{f}(w) = \text{Arcsin } w$ defined for w in E. By Theorem 2.23, the inverse map is differentiable throughout E.

The values $x = \text{Re}(\text{Arcsin } w)$ for the inverse map are constant on the hyperbolic arcs, which are images of the lines $x = \textit{constant}$ under the map in the opposite direction, $w = \sin z$; and $y = \text{Im}(\text{Arcsin } w)$ is constant on the elliptical arcs corresponding to $y = \textit{constant}$ shown in Figure 4.17. This observation can be useful in understanding the properties of the inverse mapping $z = \text{Arcsin } w$. For example, it is clear that $x = \text{Re}(\text{Arcsin } w)$ approaches the limit value $+\pi/2$ as w approaches any point $w_0 = u_0 + i0$ on the cut $[+1, +\infty)$, while $y = \text{Im}(\text{Arcsin } w)$ approaches values with opposite sign as w approaches this point from the upper or lower half plane. Similar reasoning applies to the left-hand cut. We leave it as Exercise 6 for the reader to determine the precise limit values when a typical point w_0 is approached from the upper side of the cut:

$$(18)\qquad \lim_{\delta \to 0+} \{\text{Im}(\text{Arcsin}(u_0 + i\delta))\} = \text{arccosh}(u_0) = \log(u_0 + \sqrt{u_0 - 1})$$

if $u_0 \geq +1$. We will have more to say about arcsin z and its various determinations in Section 4.10.

In Chapter 7 we will show that a holomorphic mapping between domains in the complex plane carries solutions of Laplace's equation $\nabla^2 H = 0$ on one domain to solutions of this equation on the other domain, if we make the obvious change of variable. The mapping properties we have just exhibited for the exponential function show how we may reduce the problem of solving Laplace's equation in an annular sector to the much simpler problem of solving this equation on a rectangular domain. The mapping properties of the other functions studied in this chapter can be employed in a similar way for other types of curvilinear domains. We will examine this application in some detail later on.

EXERCISES

1. Divide the strip $E = \{z : -\pi < \text{Im}(z) < +\pi\}$ into a grid of rectangular blocks, numbering a few of the blocks near the origin in the upper half plane. Determine how $w = e^z$ transforms this checkerboard pattern; indicate which curvilinear blocks in the w-plane correspond to the numbered blocks in the z-plane. (Use Figures 4.16 and 4.17 as general models.)

2. Carry out a similar analysis for $w = \text{Log}(z)$, defined on the cut plane P obtained by deleting the segment $(-\infty, 0]$ from the real axis.

3. Determine the images of the circle $|z - i| = 1$ and the line $y = 1$ under the following transformations.

(i) $w = 2iz$ (iv) $w = z^2 + 1$

(ii) $w = 1/z$ (v) $w = z^3$

(iii) $w = z - i + 4$ (vi) $w = \dfrac{z+1}{z-1}$

4. Verify that the transformations below are one-to-one and conformal throughout the domains indicated. Identify the image domains in the w-plane.

(i)	$w = z^2 + 1$	on $\mathrm{Re}(z) > 1$
(ii)	$w = (z + 1)^2$	on $\mathrm{Re}(z) > 1$
(iii)	$w = \mathrm{Log}\, z$	on $\mathrm{Im}(z) > 0$
(iv)	$w = \mathrm{Log}\, z$	on $\mathrm{Im}(z) > 1$
(v)	$w = \sin z$	on the half strip $-\pi/2 < \mathrm{Re}(z) < \pi/2;\ \mathrm{Im}(z) > 0$
(vi)	$w = \sin z$	on the square $-1 < x < +1;\ -1 < y < +1$
(vii)	$w = \sin z$	on the rectangle $-\pi/2 < x < \pi/2;\ 1 < y < 2$

Hint: How is the boundary of the domain transformed?

Answers: (iii) strip $0 < \mathrm{Im}(w) < \pi$; (iv) region bounded by curve $u = \log(\csc v)$ for $0 < v < \pi$; (v) upper half plane; (vi) and (vii) see Figure 4.17.

5. How does $w = \sin z$ transform the boundary lines $x = -\pi/2$ and $x = +\pi/2$ of the strip E in Figure 4.17? Explain how the lack of conformality at $z = +\pi/2$ and $z = -\pi/2$ makes itself felt in the transformation properties of these lines. How do these observations fit into the facts displayed in Figure 4.17?

6. Verify formula (18) of Example 4.11.

Hint: The function of real variable $\cosh u = v$ maps $[0, +\infty)$ one-to-one onto $[1, +\infty)$; demonstrate that the inverse mapping is $\operatorname{arccosh} v = \log(v + \sqrt{v - 1})$.

7. Consider a point $p = x_0 + iy_0$ that does not lie on the real or imaginary axis. Prove that $w = f(z) = z^2$ transforms the horizontal and vertical lines $y = y_0$ and $x = x_0$ to parabolas that meet and are perpendicular at $q = f(p)$, as shown in Figure 4.18. How is this perpendicularity related to conformality of f? Determine how f transforms the exceptional lines $x = 0$ and $y = 0$ that pass through $z = 0$.

8. For $\delta \neq 0$ consider the vertical lines $\gamma_\delta(t) = \delta + it$, defined for $-\infty < t < +\infty$. Show that $w = f(z) = z^2$ transforms these to parabolic curves $\eta_\delta(t) = f(\gamma_\delta(t))$ and determine the equation (in terms of u, v, and δ) of the trajectory. Explain why $\eta_{+\delta}$ and $\eta_{-\delta}$ trace out the same parabola, moving in opposite directions. Sketch the trajectory of η_δ as δ approaches zero through positive and negative values. How is the exceptional line $\gamma_0(t) = it$ (imaginary axis) transformed, and how does η_0 fit into the family $\eta_\delta(\delta \neq 0)$?

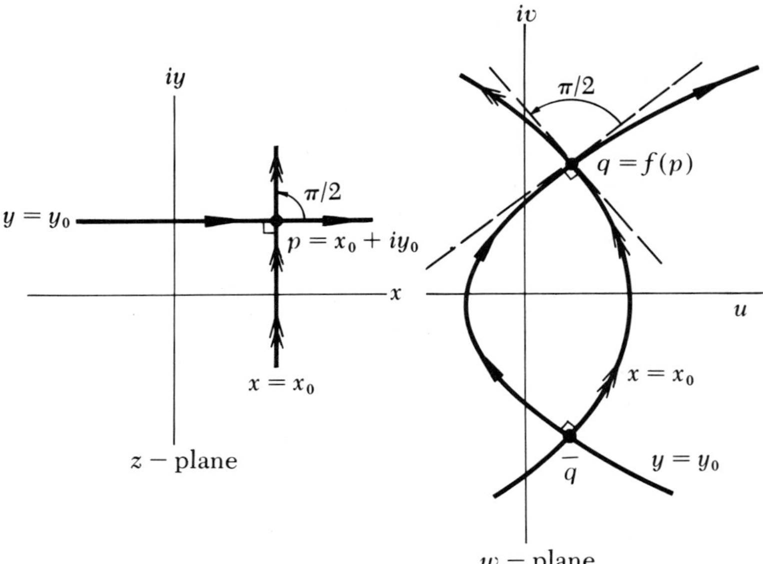

Figure 4.18 Transformation of horizontal lines y = constant and vertical lines x = constant to parabolic curves under $w = z^2$ (provided $p \neq 0$).

9. Repeat Exercise 8, considering the transformation of *horizontal* lines in the z-plane under $w = z^2$. Again, refer to Figure 4.18.

10. Divide the right half plane $E = \{z:\text{Re}(z) > 0\}$ into a checkerboard pattern using *vertical lines* $x = c$ $(c > 0)$ and *horizontal lines* $y = d$ $(-\infty < d < +\infty)$. Sketch the transformed pattern of curvilinear blocks we obtain by applying the mapping $w = f(z) = z^2$. Pay special attention to the blocks adjacent to $z = 0$; are they transformed in an exceptional way?

11. Divide the right half plane $E = \{z:\text{Re}(z) > 0\}$ into a pattern of curvilinear blocks determined by *semi-circles* $|z| = r > 0$ and *radial lines* $\arg z = \alpha$ $(-\pi/2 < \alpha < \pi/2)$. How does $w = f(z) = z^2$ transform this pattern into the w-plane? Identify the image domain $F = f(E)$.

4.6 THE COMPLEX SPHERE AND STEREOGRAPHIC PROJECTION

We will examine other special functions as we proceed, but first we want to discuss the behavior of a function "at infinity." There is no complex number that plays the role of this hypothetical point "infinity;" nevertheless, in various situations it can be very helpful, and enlightening, to think of this number as if it really existed. We shall do this by taking the complex plane **C** and adjoining to it a new point, labeled ∞, to form the **extended complex plane** $\mathbf{C}^* = \mathbf{C} \cup \{\infty\}$. The new point ∞ will be referred to as the **point at infinity.**

We might entertain the idea of defining the usual algebraic operations, sum, product, and quotient, in this enlarged system $\mathbf{C}^*$, and this can be done to a certain extent. Naturally, we take the usual operations for pairs of numbers in the subset $\mathbf{C} \subseteq \mathbf{C}^*$, the **ordinary complex numbers** in $\mathbf{C}^*$. Our intuition also tells us how we should define *some* of the operations between ordinary complex numbers and the exceptional point ∞; the following rules are almost self-evident:

$$\begin{array}{ll}
\infty \cdot z = \infty = z \cdot \infty & \text{for } z \neq 0 \text{ in } \mathbf{C} \\
\infty \pm z = \infty = z \pm \infty & \text{for all } z \text{ in } \mathbf{C} \\
z/0 = \infty & \text{for } z \neq 0 \text{ in } \mathbf{C} \\
z/\infty = 0 & \text{for all } z \text{ in } \mathbf{C} \\
\infty \cdot \infty = \infty. &
\end{array}$$

Unfortunately, there is no reasonable way to define all the familiar operations in the extended number system $\mathbf{C}^*$, due to the fact that ∞ does not really behave like an ordinary complex number. In particular, there is no reasonable value we can assign to the combinations

$$\infty \pm \infty \qquad \infty/\infty \qquad 0/0 \qquad \infty \cdot 0 \qquad 0 \cdot \infty.$$

Even in Calculus these are indeterminate forms, and must be left undefined.†

In this book our real desire is to use the extended number system to understand *geometric* problems. For this purpose we will now set up a simple geometric model of the extended complex number system $\mathbf{C}^*$ in which the exceptional point ∞ appears as an actual point. A natural model is provided by the **stereographic projection** of the complex plane onto a sphere. Let us start with three dimensional Euclidean space $\mathbf{R}^3$ with coordinates (ξ, η, ζ). The plane determined by setting $\zeta = 0$ is identified with the complex plane by letting $z = x + iy$ correspond to the point $(x, y, 0)$, so that $\xi = x$, $\eta = y$, $\zeta = 0$. Let N stand for the point $(0, 0, 1)$, which can be regarded as the north pole of the sphere $\mathbf{S}$ given by the equation

$$\xi^2 + \eta^2 + (\zeta - \tfrac{1}{2})^2 = (\tfrac{1}{2})^2.$$

This sphere S has radius $r = 1/2$, and its south pole $O = (0, 0, 0)$ rests on the origin of the complex plane. The stereographic projection maps a point $Z = (x, y, 0)$ in the complex plane to a point on $\mathbf{S}$ along the straight line segment that connects Z to the polar point N, as shown in Figure 4.19. We write $\Phi(Z)$ for the projected point, and when we identify the point $Z = (x, y, 0)$ with the corresponding complex number $z = x + iy$ in $\mathbf{C}$, we may regard Φ as a mapping from $\mathbf{C}$ into the sphere $\mathbf{S}$, $\Phi : \mathbf{C} \to \mathbf{S}$.

---

† Notice that $0/\infty = 0$ and $\infty/0 = \infty$ are not indeterminate forms according to the rules we have set up. The combination $\infty + \infty$ is indeterminate (and *not* $= \infty$) since our notion of infinity cannot distinguish between $+\infty$ and $-\infty$; thus $\infty + \infty$ is no better than the obviously indeterminate expression $\infty - \infty$.

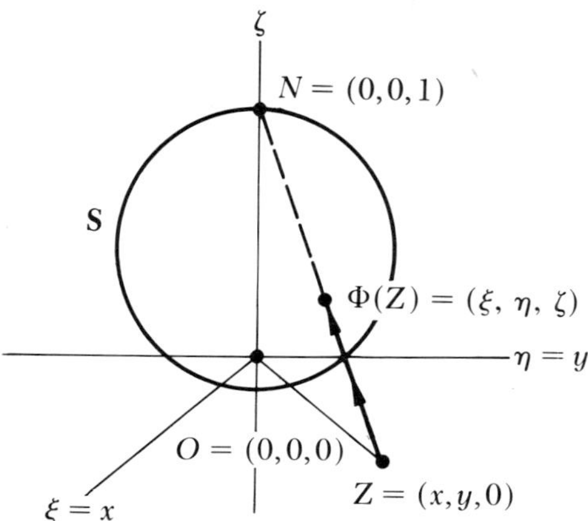

Figure 4.19 Stereographic projection of the plane $\zeta = 0$ onto the sphere **S**.

Each point $z = x + iy$ in the complex plane maps to a unique point on **S**, and by direct calculations with similar triangles (which we leave to the reader) we see that the projected point has coordinates $\Phi(z) = (\xi, \eta, \zeta)$ given by

(19)
$$\xi = \frac{x}{1 + r^2} = \frac{\operatorname{Re}(z)}{1 + |z|^2} \qquad \eta = \frac{y}{1 + r^2} = \frac{\operatorname{Im}(z)}{1 + |z|^2}$$
$$\zeta = \frac{r^2}{1 + r^2} = \frac{|z|^2}{1 + |z|^2}$$

where $r^2 = x^2 + y^2$. Conversely, a point (ξ, η, ζ) on the sphere, *other than the exceptional point* $N = (0, 0, 1)$, corresponds to the point $z = x + iy$ in the complex plane such that

(20)
$$x = \frac{\xi}{1 - \zeta} \qquad y = \frac{\eta}{1 - \zeta} \qquad |z|^2 = \frac{\zeta}{1 - \zeta}.$$

It is important to notice that no point in the complex plane projects to the polar point N, while every point in the "punctured sphere" $\mathbf{S} \sim \{N\}$ corresponds to a unique complex number. The exceptional role of the polar point N allows us to use the sphere **S** as a model for the extended complex plane $\mathbf{C}^* = \mathbf{C} \cup \{\infty\}$, in which the exceptional point ∞ in $\mathbf{C}^*$ is identified with N and an ordinary complex number z with its projection $\Phi(z)$. Because we have chosen to represent $\mathbf{C}^*$ as a sphere, the extended complex number system $\mathbf{C}^*$ is often referred to as the **complex sphere** (or **Riemann sphere**).

Obviously, any line L in the complex plane determines a unique plane in $\mathbf{R}^3$ passing through L and the polar point N. This plane intersects **S** in a circle which passes through N, and stereographic projection maps L to the circular arc in **S** obtained by deleting N from this circle. As particular examples of the way in which lines are transformed, we note that radial lines through the origin in the complex plane are mapped to *meridians* on the sphere, with N deleted (a meridian is any great circle that passes through the polar points O and N). As we will show below, circles on the complex plane get mapped to circles on the sphere that avoid the exceptional point N. Elementary manipulations with similar triangles show that the unit circle $K = \{z: |z| = 1\}$ is mapped to the equatorial circle on the sphere; the closed disc $D = \{z: |z| \leq 1\}$ is mapped one-to-one onto the lower hemisphere of $\mathbf{S}$. Other circles centered at the origin get mapped to latitude circles on the spere, and circles of large radius are mapped to very small circles on $\mathbf{S}$ that are centered at the exceptional point N. These observations are summed up as follows.

Theorem 4.4 *Stereographic projection $\Phi: \mathbf{C} \to \mathbf{S}$ maps the complex plane one-to-one onto the punctured sphere $\mathbf{S} \sim \{N\}$ obtained by deleting the point N from $\mathbf{S}$. Furthermore,*

(i) *Every straight line in the plane is mapped to a circular arc in $\mathbf{S}$ which consists of some circle in $\mathbf{S}$ passing through N, with N deleted.*
(ii) *Every circle in the plane gets mapped to a circle in $\mathbf{S}$ which avoids the exceptional point N.*

Conversely, every circle in the complex sphere corresponds to a unique line or circle in the complex plane under stereographic projection (circles in $\mathbf{S}$ that pass through N correspond to lines in the plane).

PROOF: Statement (i) requires no further comment. A circle in the plane is determined by an equation of the form

$$x^2 + y^2 + Ax + By + C = \left(x + \frac{A}{2}\right)^2 + \left(y + \frac{B}{2}\right)^2 + \left(\frac{4C - A^2 - B^2}{4}\right) = 0; \tag{21}$$

we must take $A^2 + B^2 - 4C > 0$ to get a non-degenerate locus (radius $r > 0$). If we substitute the expressions in (20) for x and y, we see that the projected points on the sphere satisfy

$$\frac{\xi^2}{(1-\zeta)^2} + \frac{\eta^2}{(1-\zeta)^2} + \frac{A\xi}{1-\zeta} + \frac{B\eta}{1-\zeta} + C = 0. \tag{22}$$

But these points must also satisfy the equation that defines $\mathbf{S}$:

$$\xi^2 + \eta^2 + \zeta^2 - \zeta = 0;$$

thus, $\xi^2 + \eta^2 = \zeta(1 - \zeta)$. If we substitute this into (22), we see that the projected point $(\xi, \eta, \zeta) = \Phi(z)$ satisfies the equation

$$\zeta + A\xi + B\eta + C(1 - \zeta) = 0, \tag{23}$$

which is the equation of a *plane* in $\mathbf{R}^3$. This plane meets the sphere non-trivially since it includes all point $\Phi(z)$ with z in the circle defined by equation (21). Furthermore, this plane must meet the sphere in a circle, so this circle is the projected image of the circle (21) in the complex plane. The point $N = (0, 0, 1)$ does not lie in the plane determined by (23); this equation can't be satisfied by N, so the point N cannot lie on the circle we get by intersecting this plane with the sphere. It is not very difficult to reverse this argument to see that every circle on the sphere that misses the point N is actually the projected image of some circle in the complex plane. ▌

Further efforts with analytic geometry reveal that the stereographic projection preserves the angle between circular arcs (or lines) where they meet. For arcs in the sphere this angle is measured as the angle between tangent lines; these tangent lines lie in the plane (in space) that just touches the sphere at the point where the arcs meet, and the angle between tangent lines is measured within this plane. We leave the proof as Exercise 5.

Theorem 4.5 *Let C_1 and C_2 be circular arcs (or line segments) in the complex plane that meet in an angle $\Delta\theta$ at some point p, measured from C_1 to C_2. Stereographic projection maps these to circular arcs K_1 and K_2 in the complex sphere that meet at $P = \Phi(p)$ in an angle $\Delta\phi = \Delta\theta$, measured from K_1 to K_2.*

In order to determine the sense (sign) of angles in the tangent plane to a point P on the complex sphere, we must specify whether we are to view the tangent plane from inside the sphere or outside the sphere; once a definite convention is set, counterclockwise angles are counted positively, as usual. We shall measure angles looking at the tangent plane from within the sphere. This convention insures that stereographic projection preserves the sense as well as the magnitude $|\Delta\theta|$ of angles between two lines. If angles were measured from outside the sphere, we would get $\Delta\phi = -\Delta\theta$ (sense of angles *reversed*).

EXERCISES

1. Prove that stereographic projection maps the unit circle $\{z\colon |z| = 1\}$ to the equatorial circle on the complex sphere.

2. Describe the relative positions of z, $-z$, $\bar{z}$, $-\bar{z}$, $1/z$, $1/\bar{z}$ in the plane and on the Riemann sphere.

3. Verify formulas (19) and (20).

4. A sector $0 \leq |z| \leq R$, $\alpha \leq \arg z \leq \beta$ is projected onto $\mathbf{C}^*$. What is the area of the projected set?

5. Prove Theorem 4.5 using the following geometric ideas (refer to Figure 4.20). If A and B are rays that make an angle $\Delta\theta$ at p in the plane, these project to circular arcs $\Phi(A)$ and $\Phi(B)$ that meet at $\Phi(p)$ in an angle $\Delta\phi$. To show $\Delta\phi = \Delta\theta$, notice that the angle $\Delta\theta'$ (measured in the plane tangent to ∞) is clearly equal to $-\Delta\theta$, since the complex plane is parallel to the tangent plane at ∞. Next, prove that $\Delta\theta' = -\Delta\phi$, by noting how $\Phi(A)$ and $\Phi(B)$ are the intersection with $\mathbf{S}$ of planes through A, B, and ∞.

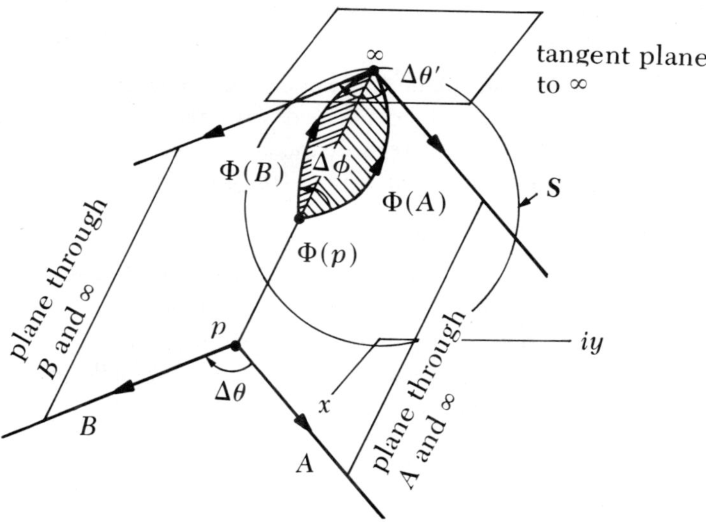

Figure 4.20 The rays A and B project to arcs on **S** determined by the planes through A and ∞, and through B and ∞. Remember: the sign of an angle is determined by viewing it from within **S**.

6. Prove that z and w correspond to diametrically opposite points on the Riemann sphere if and only if $z\bar{w} = -1$.

7. If a circle $\Gamma = \{z : |z - p| = R\}$ in the complex plane is projected onto a circle K in the complex sphere, calculated the radius of K. Does the center of Γ project to the center of K?

4.7 BEHAVIOR OF FUNCTIONS AT INFINITY: IMPROPER LIMITS

It is frequently useful to say that a sequence $\{z_n\}$ of complex numbers "approaches infinity" or "converges to infinity" if its terms are eventually outside of any closed disc $D_r = \{z\colon |z| \le r\}$ of finite radius. If this happens we say that $\{z_n\}$ has the **improper limit** ∞:

$$\lim_{n\to\infty} z_n = \infty.$$

A sequence of complex numbers can diverge (fail to have a limit in the ordinary sense) without converging to the improper limit ∞; the sequence

$$z_n = \begin{cases} 0 & \text{if } n \text{ is even} \\ 2^n & \text{if } n \text{ is odd} \end{cases}$$

diverges, but the absolute value $|z_n|$ frequently have the value zero, so we do not get $\lim_{n\to\infty} z_n = \infty$. Convergence to ∞ refers to a very special kind of divergence.

Notice that we do not attempt to distinguish between $+\infty$ and $-\infty$ in defining $\lim_{n\to\infty} z_n = \infty$, as we would if we were dealing with sequences of real numbers. Since the complex plane does not possess convenient order properties, any attempt to do this would be plagued with technical difficulties. If a sequence of real numbers $\{x_n\}$ is regarded as a sequence of complex numbers, the limits $\lim_{n\to\infty} x_n = +\infty$ and $\lim_{n\to\infty} x_n = -\infty$, defined in Calculus, reduce to the single statement $\lim_{n\to\infty} x_n = \infty$.

Once we have defined what it means for a sequence of complex numbers to converge to ∞, we may then consider limits of a function $f(z)$ of a complex variable at any point p on $\mathbf{C}^*$, including $p = \infty$, and we may also allow the improper value ∞ as the value of such a limit. Thus, we will be able to make sense of improper limit formulas such as

$$\lim_{z\to 0}\left\{\frac{1}{z}\right\} = \infty \qquad \lim_{z\to\infty}\left\{\frac{z}{1+z}\right\} = 1 \qquad \lim_{z\to\infty}\{z^2 + z\} = \infty.$$

Consider a function $w = f(z)$ defined on an open set E in the plane. If p is a point in $\mathbf{C}^*$ ($p = \infty$ is allowed) that can be approached by sequences of points in E, we say that

$$\lim_{z\to p} f = \alpha \quad (\alpha \text{ in } \mathbf{C}^*)$$

if $\lim_{n\to\infty} f(z_n) = \alpha$ for every sequence of points in E, distinct from p, such that $z_n \to p$ as $n \to \infty$.

Example 4.12 Let $w = f(z) = 1/z$ for z in the domain $E = \{z : z \neq 0\}$. To prove that f has a well defined improper limit at the origin, let us consider any sequence of points $z_n \to 0$ as $n \to \infty$. If $R > 0$ is given, we know that

$$|z_n - 0| = |z_n| < 1/R \quad \text{for all large } n$$

(take $\varepsilon = 1/R$ in the definition of the limit $\lim_{n\to\infty} z_n = 0$). This means that $|f(z_n)| = 1/|z_n| > 1/(1/R) = R$ for all large n; thus

$$\lim_{n\to\infty} f(z_n) = \infty$$

for every sequence $\{z_n\}$ that converges to zero ($z_n \neq 0$, all n). This proves

$$\lim_{z\to 0} f = \lim_{z\to 0}\{1/z\} = \infty.$$

It is also easy to show that f has a limit at infinity: $\lim_{z\to\infty}\{1/z\} = 0$. Consequently, f may be regarded as a *continuous* mapping of $\mathbf{C}^*$ into $\mathbf{C}^*$ if we assign the values $f(0) = \infty$ and $f(\infty) = 0$; for any p in $\mathbf{C}^*$ we get

$$\lim_{z\to p} f \text{ exists and is equal to the value } f(p).$$

Example 4.13 Linear mappings such as $w = f(z) = az + b$ $(a \neq 0)$ become continuous mapping of $\mathbf{C}^*$ into $\mathbf{C}^*$ if we assign the value $f(\infty) = \infty$; that is, this choice of value at $p = \infty$ insures that $\lim_{z \to \infty} f = \infty = f(\infty)$. We already know that $\lim_{z \to p} f = f(p)$ for an ordinary point p.

Special types of linear transformations have interesting interpretations as mappings $f : \mathbf{C}^* \to \mathbf{C}^*$. Rotations $w = e^{i\alpha} \cdot z$ rotate the complex sphere by an angle α radians about the axis passing through 0 and ∞; dilations $w = a \cdot z$ act by shifting points along *meridians* on the sphere, moving them toward ∞ if $a > 1$ and toward 0 if $0 < a < 1$. Translations $w = z + b$ have a slightly more complicated interpretation: in the plane points are moved along lines parallel to the line through the points $\{0, b\}$. In the complex sphere these lines correspond to the family of circles through ∞ shown in Figure 4.21, and translation shifts points along these circles in the same direction as b, as shown.

Example 4.14 To show that $\lim_{z \to \infty} \{z^2 + z\} = \infty$, consider a typical sequence that converges to infinity in the plane. Then

$$|z_n^2 + z_n| \geq |z_n|^2 - |z_n| = |z_n| \cdot (|z_n| - 1),$$

by the triangle inequality, and for all large n we get $|z_n| - 1 \geq 2$. Thus, $|f(z_n)| \geq 2 \cdot |z_n|$ for all large n. Now it is clear that the absolute value $|f(z_n)|$ becomes arbitrarily large as n increases, so that $\lim_{n \to \infty} f(z_n) = \infty$.

In closing this discussion we must warn the reader that not every holomorphic function is sufficiently well behaved at infinity, or near its singular points (such as $z = 0$ in Example 4.12), that we can assign a definite value at these points.

Example 4.15 Clearly $w = e^z$ is defined near infinity. If $\{z_n\}$ is a sequence that approaches ∞ along the *positive* real axis, say $z_n = x_n + i0$, then we obviously get $\lim_{n \to \infty} \exp(z_n) = \lim_{n \to \infty} e^{x_n} = \infty$; on the other hand, if z_n

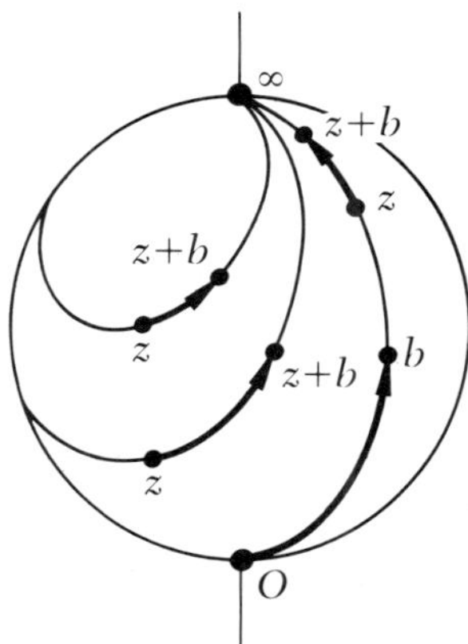

Figure 4.21 The translation $w = z + b$ moves points along the family of circles passing through ∞, in the same direction as b.

approaches ∞ along the *negative* real axis, we get $\lim_{n\to\infty} \exp(z_n) = 0$. Furthermore, if $z_n = 0 + iy_n$ approaches ∞ moving in either direction along the imaginary axis, we cannot expect the sequence of values

$$f(z_n) = \exp(iy_n) = \cos y_n + i \sin y_n$$

to have any limit at all. This should convince the reader that $f(z) = e^z$ is badly behaved as $z \to \infty$, and no definite limit value can be assigned; not even the value $w = \infty$ is acceptable!

EXERCISES

1. Determine the limits $\lim_{z\to p} f$ for

$$f(z) = \frac{1}{2}\left(z + \frac{1}{z}\right)$$

at $p = 0$ and $p = \infty$. Can values $f(0)$ and $f(\infty)$ be assigned so that f gives a continuous mapping $f: \mathbf{C}^* \to \mathbf{C}^*$?

2. Prove the following statements about improper limits.

(i) $\lim_{z\to\infty} \sin z$ does not exist (not even the value ∞ is acceptable)

(ii) $\lim_{z\to 0} \dfrac{\sin z}{z^2} = \infty$

(iii) $\lim_{z\to 0} \dfrac{1}{\sin z} = \infty$

(iv) $\lim_{z\to \pi/2} \tan z = \infty$

(v) $\lim_{z\to \pi/2} \left(z - \dfrac{\pi}{2}\right) \tan z = -1$

(vi) $\lim_{z\to 0} \sin\left(\dfrac{1}{z}\right)$ does not exist.

3. Determine whether $\lim_{z\to\infty} f$ exists, for

(i) $f(z) = \exp(e^z)$
(ii) $f(z) = \tan z$
(iii) $f(z) = \sin(1/z)$

Calculate $|f(x + iy)|$ explicitly for these functions.
Answer: (i) no limit; (ii) no limit; (iii) 0.

4. The limit $\lim_{z\to\infty} e^z$ does not exist if we consider e^z as a function defined on the whole complex plane. However, if we define $f(z) = e^z$ on a smaller domain $E \subseteq \mathbf{C}$, the limit $\lim_{z\to\infty} f(z)$ might exist (then one considers only sequences in E such that $z_n \to \infty$). Prove the following statements, taking E as indicated. Sketch E in each case.

(i) $\lim_{z\to\infty} f = 0$; $f(z) = e^z$ for $3\pi/4 < \arg z < 5\pi/4$.

(ii) $\lim_{z\to\infty} f = \infty$; $f(z) = e^z$ for $-\pi/4 < \arg z < \pi/4$.

(iii) $\lim_{z\to\infty} f$ does not exist; $f(z) = e^z$ for $\operatorname{Re}(z) < 0$.

(iv) $\lim_{z\to\infty} f$ does not exist; $f(z) = e^z$ for $-\pi < \operatorname{Im}(z) < +\pi$.

5. If $\lim_{n\to\infty} z_n = \infty$ and $\lim_{n\to\infty} w_n = \infty$, show by way of an example that the sequence $\zeta_n = z_n + w_n$ need not have any limit at all (not even ∞). Prove that $\lim_{n\to\infty} z_n \cdot w_n = \infty$.

6. Consider a sequence of points $z_n = r_n e^{i\alpha}$, situated on the ray $\arg z = \alpha$ (real α) such that $r_n \to +\infty$ as $n \to \infty$. For which directions α does

(i) $\lim_{n\to\infty} |\sin(z_n)| = \infty$?

(ii) $\lim_{n\to\infty} |\sinh(z_n)| = \infty$?

(iii) $\lim_{n\to\infty} |\exp(z_n)| = \infty$?

4.8 FRACTIONAL LINEAR TRANSFORMATIONS

There is a simple family of holomorphic mappings, the **fractional linear transformations** of the complex plane, which enter into a truly amazing variety of mathematical and physical problems. They play an important role in mathematical fields as diverse as number theory and non-Euclidean geometry, and we find them turning up in a natural way at the heart of things in relativity and in quantum field theory. Indeed, the theory of the famed Lorentz transformations of special relativity coincides almost exactly with the theory surrounding the fractional linear transformations, if we take the right point of view. We will, unfortunately, not be able to go into these diverse applications, since any discussion of this sort would presume detailed knowledge of the physical or mathematical problems involved, as well as the theory we will develop in this section.

Here we will study the mapping properties and some of the theory of this family of mappings. The discussion should help the reader develop his own techniques for using the tools we have assembled for studying the mapping properties of holomorphic functions. The mapping properties of fractional linear transformations will be very useful in solving conformal mapping problems, especially for domains bounded by lines and circular arcs. In some books these transformations are referred to as **bilinear transformations, Möbius transformations,** or **linear transformations,** but we will not use this terminology.†

Definition 4.4 *A* **fractional linear transformation** *is any mapping T of the complex plane that has the form*

$$T(z) = \frac{az + b}{cz + d}, \tag{24}$$

where the complex coefficients a, b, c, and d satisfy the condition $ad - bc \neq 0$.

Of course, T is determined once we specify the matrix of coefficients

$$\begin{bmatrix} a & b \\ c & d \end{bmatrix}, \tag{25}$$

and $ad - bc$ is just the determinant of this matrix. If this determinant were equal to zero, the mapping T either would be constant (all values the same), or would be undefined everywhere due to the presence of a zero in the denominator; in either case we would want to exclude this degenerate situation.

The correspondence between T and the matrix of coefficients is not one-to-one. If every term in the matrix (25) is multiplied by the same non-zero complex scalar $\mu \neq 0$, this constant cancels out of (24) and we get the same transformation. This is actually the *only* way two different matrices can give the same mapping of the plane; i.e., two matrices (with non-zero determinant)

$$\begin{bmatrix} a & b \\ c & d \end{bmatrix} \quad \text{and} \quad \begin{bmatrix} a' & b' \\ c' & d' \end{bmatrix}$$

give the same transformation, in the sense that

$$\frac{az + b}{cz + d} = \frac{a'z + b'}{c'z + d'} \quad \text{for all } z,$$

if and only if there is a non-zero scalar μ such that $a' = \mu a, \ldots, d' = \mu d$. We leave the verification as Exercise 2.

If $c = 0$ in the set of coefficients, we must then have $a \neq 0$ and $d \neq 0$, and the transformation simplifies to a first degree polynomial in z, a *linear* function of z, $T(z) = (a/d)z + (b/d) = a'z + b'$ for all z. These transformations have

† Although some authors use the phrase "linear transformation" to refer to the fractional linear transformations defined here, we will always reserve this term to indicate functions of the form $w = az + b$.

been described in Example 4.7. If such a mapping is regarded as a transformation of the complex sphere $T:\mathbf{C}^* \to \mathbf{C}^*$, it is well behaved at infinity and we get a continuous mapping of $\mathbf{C}^*$ onto itself by assigning the value $T(\infty) = \infty$ at infinity. The linear mappings together with the reciprocal mapping $w = 1/z$, which corresponds to the matrix of coefficients

$$\begin{bmatrix} 0 & 1 \\ 1 & 0 \end{bmatrix},$$

form an easily studied class of "elementary" fractional linear transformations. It turns out that these transformations determine all other fractional linear transformations; however, before we demonstrate this, let us collect a few observations about the reciprocal mapping.

The map $w = T(z) = 1/z$ is a regular invertible mapping on $E = \{z: z \neq 0\}$; its derivative $dT/dz = -1/z^2$ is everywhere non-vanishing. If we assign the values $T(0) = \infty$ and $T(\infty) = 0$, we get a continuous invertible mapping of $\mathbf{C}^*$ (recall Section 4.7). We have already indicated that T can be written as the composite $T = C \circ J$ of two extremely simple anti-conformal mappings:

$$J(re^{i\theta}) = \frac{1}{r}\, e^{i\theta} \qquad \text{(reflection through the unit circle)}$$

$$C(re^{i\theta}) = re^{-i\theta} \qquad \text{(ordinary complex conjugation).}$$

It is enlightening to notice how J and C appear when regarded as transformations of the complex sphere. They are identified with simple reflections of $\mathbf{C}^*$ through certain planes passing through the center of the sphere, as shown in Figure 4.22:

(i) J reflects points in $\mathbf{C}^*$ through the equatorial plane (the equator corresponds to the unit circle $|z| = 1$ under stereographic projection). This is why we refer to J as a "reflection through the unit circle."
(ii) C reflects points through the plane determined by the real axis, which appears as a meridian on $\mathbf{C}^*$.

Notice that we get the same result on performing the operations J and C in either order (the operations *commute*); $T = C \circ J = J \circ C$.

This representation of T as successive reflections in $\mathbf{C}^*$ reveals certain geometric properties which are not so obvious if we regard $w = 1/z$ as a transformation of the plane. Clearly, the reflections J and C act on the sphere to map circles to circles; so does the composite $T = J \circ C$. Interpreting this as a statement about the transformation of lines and circles in the plane under $w = 1/z$, it is not difficult to see that $w = 1/z$ transforms families of lines and circles in the following ways:

(i) Radial lines through the origin map to radial lines through the origin.
(ii) Lines that avoid the origin map to circles that pass through the origin.
(iii) Circles through the origin map to lines that avoid the origin.
(iv) Circles that avoid the origin map to circles that avoid the origin.

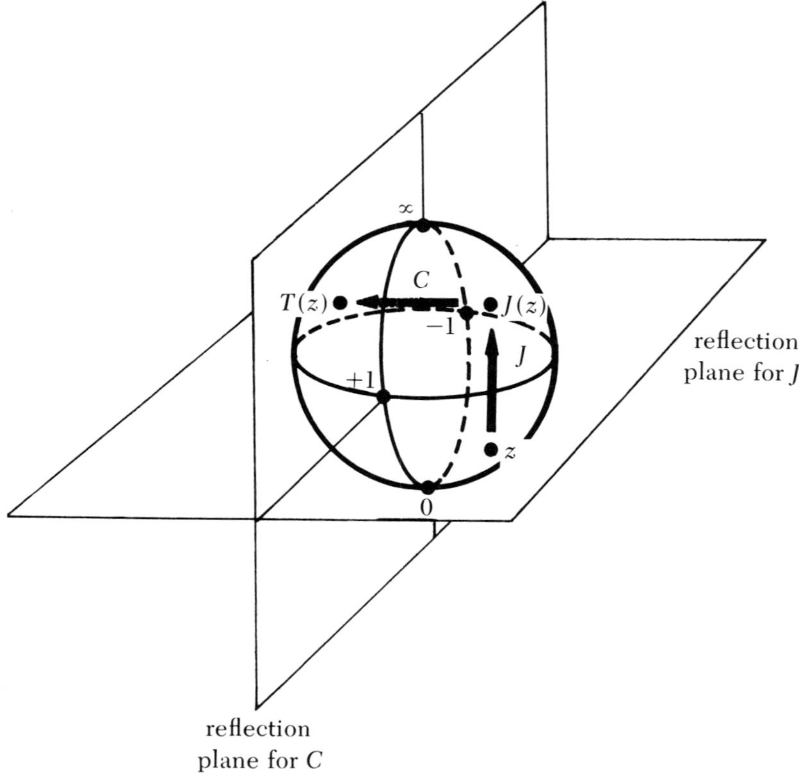

Figure 4.22 The actions of the anti-conformal mappings $w = J(z)$ and $w = C(z)$ in the complex sphere combine to give the inversion map $T = J \circ C = C \circ J$.

Of course, these mapping properties could also be determined by direct algebraic calculations. We have already done such calculations, in Example 4.10, to decide how horizontal and vertical lines are transformed. Sometimes we are obliged to perform detailed algebraic calculations. For example, we might ask which circle or line is the image of (or maps to) a specified circle in the plane under T; or, we might want to know the fine details of how individual points on one circle K are mapped to points on the image curve $T(K)$.

We now turn to general fractional linear transformations. Whatever discussion is necessary will be made for the case when $c \neq 0$ in the coefficient matrix (25); what we have to say is usually obvious for the *linear* transformations that arise if $c = 0$, and we will not bother with the separate treatment they require. If $c \neq 0$, the typical transformation

$$T(z) = \frac{az + b}{cz + d} \tag{26}$$

is undefined at $z = -(d/c)$ due to the appearance of a zero in the denominator, so $T(z)$ has the natural domain of definition $E = \{z : cz + d \neq 0\} = \{z : z \neq$

$-d/c\}$. It is obviously holomorphic throughout E and

$$\frac{dT}{dz} = \frac{ad - bc}{(cz + d)^2}$$

is never zero on E, since $ad - bc \neq 0$. Thus T is conformal at every point in its natural domain of definition. The behavior of $T(z)$ as $z \to \infty$ or $z \to -(d/c)$ is quite orderly; there are well defined improper limits at these points, and T becomes a continuous mapping of $\mathbf{C}^*$ if we assign the values

$$T(-(d/c)) = \infty \quad \text{and} \quad T(\infty) = (a/c)$$

at the exceptional points.

These assignments agree with formula (26) if we apply the algebraic rules for handling ∞ that were introduced in Section 4.6. In the degenerate case when $c = 0$, formula (26) reduces to $T(\infty) = \infty$ which, as we have seen in Section 4.7, is the correct assignment of values for a linear transformation.

Every fractional linear transformation has an inverse which is a transformation of the same kind.

Theorem 4.6 *Each fractional linear transformation $w = T(z)$ maps the complex sphere $\mathbf{C}^*$ one-to-one onto itself, so that there is a well defined inverse mapping $z = \check{T}(w)$ which reverses T. The inverse mapping $\check{T}$ is again a fractional linear transformation, given by*

$$z = T(w) = \frac{-dw + b}{cw - a}. \tag{27}$$

PROOF: If $c = 0$ these statements are clear from previous examples, so we pass to the case when $c \neq 0$. Then T maps the singular points $z = -d/c$ and $z = \infty$ to $w = \infty$ and $w = a/c$, respectively. If w is not one of the latter, it is obvious that $w = T(z) = (az + b)/(cz + d)$ has the *unique* solution

$$z = \frac{-dw + b}{cw - a}.$$

Therefore, T maps $\mathbf{C}^* \sim \{-d/c, \infty\}$ one-to-one onto $\mathbf{C}^* \sim \{\infty, a/c\}$. According to our conventions for defining T at the exceptional points, it follows that T actually maps $\mathbf{C}^*$ one-to-one onto $\mathbf{C}^*$, and that $\check{T}$ is given by

$$z = \check{T}(w) = \frac{-dw + b}{cw - a} \quad \text{for all } w \text{ in } \mathbf{C}^*,$$

even if $w = \infty$ or $w = a/c$. This inverse mapping is obviously fractional linear. ∎

Besides the **inverse** $\check{T}$ of a fractional linear transformation T, we can also define the **product** of two such transformations S and T, which is the composite

(28) $$(S \circ T)(z) = S(T(z)) \quad \text{for all } z \text{ in } \mathbf{C}^*.$$

Note carefully that $S \circ T$ need not be the same mapping as the product $T \circ S$, taken in the opposite order. (See Exercise 1.) The identity transformation $I(z) = z$ (a linear transformation) acts as a unit for this multiplication operation:

(29) $$I \circ T = T = T \circ I$$

for any fractional linear transformation T, and the characteristic equations satisfied by the inverse $\check{T}$ of a given transformation T may be summarized in the form:

(30) $$T \circ \check{T} = I = \check{T} \circ T.$$

We must emphasize that the product $S \circ T$ (a composition of mappings) is again a fractional linear transformation. In fact, if the coefficient matrices of S and T are

$$S = \begin{bmatrix} a & b \\ c & d \end{bmatrix} \quad \text{and} \quad T = \begin{bmatrix} a' & b' \\ c' & d' \end{bmatrix},$$

then direct computations show that $S \circ T$ is the fractional linear transformation whose coefficient matrix is the "matrix product" of the above matrices:

$$S \circ T = \begin{bmatrix} aa' + bc' & ab' + bd' \\ ca' + dc' & cb' + dd' \end{bmatrix} = \begin{bmatrix} a & b \\ c & d \end{bmatrix} \times \begin{bmatrix} a' & b' \\ c' & d' \end{bmatrix}.$$

We will not use this formula, so further comments on matrix multiplication are unnecessary.

We have displayed the fractional linear transformations as a family of mappings on $\mathbf{C}^*$ in which we can form (composition) products and inverses freely, without stepping outside of this family. We have remarked that multiplication is *not commutative;* i.e., we can find S and T such that $S \circ T \neq T \circ S$. It is not hard to see that composition of mappings is **associative**: $R \circ (S \circ T) = (R \circ S) \circ T$. In algebraic terminology this is summed up by saying that the fractional linear transformations form a **group** of transformations of the complex sphere. Many computational arguments can be reduced to trivialities if we make careful use of the group properties of this family of transformations. Extensive discussion of the group theory viewpoint is beyond the scope of this book; yet we must note that the group properties of the fractional linear transformations are the basis for some of their most important applications in physics and mathematics.

Obviously, any product of successive elementary operations, including dilations, rotations, translations, and reciprocals, is a fractional linear transformation; conversely, the elementary transformations "generate" the whole set of fractional linear transformations.

Theorem 4.7 *Every fractional linear transformation T can be written as a product of four elementary fractional linear transformations.*

PROOF: If $c = 0$, the result is obvious. If $c \neq 0$, we can write

$$T(z) = \frac{az+b}{cz+d} = \left(\frac{bc-ad}{c^2}\right) \cdot \left(\frac{1}{z+(d/c)}\right) + (a/c),$$

which expresses T in the form $T = T_4 \circ T_3 \circ T_2 \circ T_1$, where

$$T_1(z) = z + \frac{d}{c}\,; \; T_2(z) = \frac{1}{z}\,; \; T_3(z) = \left(\frac{bc-ad}{c^2}\right)z; \; T_4(z) = z + \frac{a}{c}\,. \quad \blacksquare$$

Important global mapping properties of fractional linear transformations are revealed when we notice how they transform circles and lines. By taking **extended lines** (adjoining the point at infinity) we may regard lines as a special class of circles in the complex sphere, the ones which pass through infinity.

Theorem 4.8 *Let $\mathscr{K}$ be the family of all circles in the complex sphere and let T be any fractional linear transformation. Then*

(i) *The image curve $T(\Gamma)$ is in $\mathscr{K}$ for every circle Γ in $\mathscr{K}$.*

(ii) *Distinct circles $\Gamma_1 \neq \Gamma_2$ in the family $\mathscr{K}$ are mapped to distinct image circles: $T(\Gamma_1) \neq T(\Gamma_2)$.*†

(iii) *Every circle Γ in $\mathscr{K}$ is the image of a uniquely determined circle $\tilde{\Gamma}$ in $\mathscr{K}$, so that $T(\tilde{\Gamma}) = \Gamma$.*

PROOF: We have already indicated that the elementary transformations have property (i). Any composite of elementary mappings must also have this property; now invoke Theorem 4.7. Property (ii) holds because T is a one-to-one onto (invertible) mapping of $\mathbf{C}^*$ onto itself; thus, two sets coincide (or differ) after being mapped by T if and only if they coincide (or differ) to begin with. For (iii), let $\check{T}$ be the inverse of T and take $\tilde{\Gamma} = \check{T}(\Gamma)$. Obviously, by (i), $\tilde{\Gamma}$ is a circle and by definition of inverse mappings we get $T(\tilde{\Gamma}) = T \circ \check{T}(\Gamma) = \Gamma$. By (ii), no other circle in $\mathscr{K}$ can possibly be mapped to Γ under the action of T. $\blacksquare$

† Two point sets in $\mathbf{C}^*$ are said to be *equal*, $A = B$, if and only if they contain exactly the same points. The sets A and B can intersect non-trivially without being equal as sets.

These results are usually summarized by saying that fractional linear transformations map circles to circles in $\mathbf{C}^*$. Some applications will be presented later in this section; then, in Chapter 7, we will apply these ideas to an extensive collection of mapping problems.

Also important in understanding the mapping properties of general fractional linear transformations are the **fixed points** of T, those points z in $\mathbf{C}^*$ that satisfy the equation $T(z) = z$, so that z is left unmoved by T. The identity transformation $I(z) = z$ is exceptional as far as fixed points are concerned, because *every* z in $\mathbf{C}^*$ is a fixed point. Other fractional linear transformations are severely restricted in the number of fixed points they can have.

Theorem 4.9 *A fractional linear transformation with three or more fixed points must be the identity transformation.*

PROOF: If $c = 0$ the transformation is linear, $w = T(z) = Az + B$, and the fixed points in $\mathbf{C}^*$ are

(i) $z = \infty$ and $z = \dfrac{B}{1 - A}$ (if $A \neq 1$)

(ii) $z = \infty$ (if $A = 1$ and $B \neq 0$)

(iii) all z (if $A = 1$ and $B = 0$; then T is the identity).

Clearly, there are at most two fixed points if T is not the identity.

If $c \neq 0$, the point ∞ is definitely *not* a fixed point. An ordinary point z in the plane is left fixed if and only if $z = T(z) = (az + b)/(cz + d)$, which is the same as saying that

$$cz^2 + (d - a)z - b = 0.$$

This quadratic equation has at most two complex roots. Evidently, the only situation that leads to three or more fixed points is the one in which T is the identity. ∎

From this simple result many remarkable consequences follow.

Theorem 4.10 *If S and T are fractional linear transformations that agree at three or more distinct points in the complex sphere, say*

$$S(z_k) = T(z_k) \quad \textit{for} \quad k = 1, 2, 3$$

for distinct points $\{z_1, z_2, z_3\}$, then S and T are identical throughout $\mathbf{C}^*$.

PROOF: We are assuming that

$$S(z_k) = w_k = T(z_k) \quad \text{for} \quad k = 1, 2, 3.$$

Let $\check{S}$ be the inverse of S, so that $\check{S}(w_k) = z_k$ $(k = 1, 2, 3)$. The product $\check{S} \circ T$ is a fractional linear transformation such that

$$\check{S} \circ T(z_k) = \check{S}(T(z_k)) = \check{S}(w_k) = z_k \qquad (k = 1, 2, 3).$$

Thus, $\check{S} \circ T$ has three distinct fixed points, and must be the identity map: $\check{S} \circ T = I$. Multiplying both sides of this equality by the transformation S, we get

$$S = S \circ I = S \circ \check{S} \circ T = I \circ T = T,$$

since $S \circ \check{S} = I$, and this proves the theorem. ▌

Theorem 4.11 *If we are given two sets $\{z_1, z_2, z_3\}$ and $\{w_1, w_2, w_3\}$ each consisting of three distinct points in $\mathbf{C}^*$, there is a uniquely determined fractional linear transformation T that maps the first set of points to the second:*

$$T(z_1) = w_1 \qquad T(z_2) = w_2 \qquad T(z_3) = w_3.$$

Note: Although the points in each triple must be distinct, we do not exclude the possibility that there are some points in common between $\{z_1, z_2, z_3\}$ and $\{w_1, w_2, w_3\}$. For example, the sets $\{-1, 0, +1\}$ and $\{0, \infty, +1\}$ are quite acceptable. This result says that a fractional linear transformation has so few degrees of freedom that its behavior throughout $\mathbf{C}^*$ is completely determined once we know what it does to three distinct points. If we only specify the values $w = T(z)$ at *two* points, there are infinitely many fractional linear transformations that satisfy these requirements (Exercise 14).

PROOF: It is an easy matter to devise a fractional linear transformation R that maps the triple $\{z_1, z_2, z_3\}$ to the "standard triple" $\{0, \infty, 1\}$. In fact, we may take

(31) $$R(z) = \frac{z_3 - z_2}{z_3 - z_1}\left(\frac{z - z_1}{z - z_2}\right).$$

Any transformation of the form $w = \alpha(z - z_1)/(z - z_2)$ maps z_1 to 0 and z_2 to ∞, and we have chosen α to give $R(z_3) = +1$.

If one of the points $\{z_1, z_2, z_3\}$ is ∞, one must use slightly different formulas, but the alterations needed are almost self-evident. In each case we get the correct formula by "canceling out" terms in the numerator and denominator of (31) which involve ∞. For example, if $z_1 = \infty$ the correct formula for $R(z)$ is

$$R(z) = \frac{(z_3 - z_2)}{(\cancel{z_3 - \infty})} \cdot \frac{(\cancel{z - \infty})}{(z - z_2)} = \frac{(z_3 - z_2)}{(z - z_2)}.$$

Likewise, there is another fractional linear transformation S that maps $\{w_1, w_2, w_3\}$ to $\{0, \infty, +1\}$. We get the transformation we really want by taking the product $T = \check{S} \circ R$, where $\check{S}$ is the inverse of S; obviously,

$$T(z_1) = \check{S} \circ R(z_1) = \check{S}(0) = w_1$$
$$T(z_2) = \check{S} \circ R(z_2) = \check{S}(\infty) = w_2$$
$$T(z_3) = \check{S} \circ R(z_3) = \check{S}(1) = w_3.$$

The uniqueness of the transformation T that interpolates these points has already been proved. ▮

Using formula (31) we can explicitly calculate the transformation $w = T(z)$ that carries $\{z_1, z_2, z_3\}$ to $\{w_1, w_2, w_3\}$. For the moment, let us assume that ∞ does not appear in either triple. Define S and R as above; then $T = \check{S} \circ R$, so that $S \circ T = S \circ \check{S} \circ R = I \circ R = R$. Therefore, by formula (31), the points z and $w = T(z)$ satisfy the equation

$$\begin{aligned} \frac{(w - w_1)}{(w - w_2)} \cdot \frac{(w_3 - w_2)}{(w_3 - w_1)} &= S(w) = S(T(z)) = S \circ T(z) \\ &= R(z) = \frac{(z_3 - z_2)}{(z_3 - z_1)} \cdot \frac{(z - z_1)}{(z - z_2)}. \end{aligned} \tag{32}$$

We obtain the formula for T by solving for w as a function of z in this identity. The reader can check that we still get a valid formula by solving this identity, even if ∞ appears in the triples, provided we cancel terms involving ∞ on each side.

If α, β, γ, δ are distinct points in $\mathbf{C}^*$ (∞ is allowed as one of the points), we define their **cross ratio** to be the number in $\mathbf{C}^*$

$$(\alpha, \beta, \gamma, \delta) = \frac{(\alpha - \gamma)(\beta - \delta)}{(\alpha - \delta)(\beta - \gamma)}, \tag{33}$$

calculated according to the rules in Section 4.6 if ∞ appears.† Formula (32) is just a statement about cross ratios; if T is any fractional linear transformation and if distinct points $\{z, z_1, z_2, z_3\}$ are given, then

$$(w, w_3, w_1, w_2) = (z, z_3, z_1, z_2)$$

where $w = T(z)$ and $w_k = T(z_k)$ $(k = 1, 2, 3)$. By labeling points suitably, we get:

Theorem 4.12 *The cross ratio of four distinct points in* $\mathbf{C}^*$ *is invariant under any fractional linear transformation.*

This invariance includes formula (32), and is very useful in calculating the transformations that solve certain mapping problems. It will be used extensively in Chapter 9; meanwhile, we will only give a few examples of its use in mapping problems (see also Exercise 16).

Example 4.16 Determine a fractional linear transformation that maps $\{-1, 0, +1\}$ on the real axis to the points $\{-1, -i, +1\}$ on the unit circle. By

† Since the points are distinct, none of the differences can be zero. If ∞ is one of the points, it appears once in the numerator and once in the denominator of (33); these terms should be canceled to obtain the correct value for the cross ratio in this case. Except for this, an indeterminate form cannot arise.

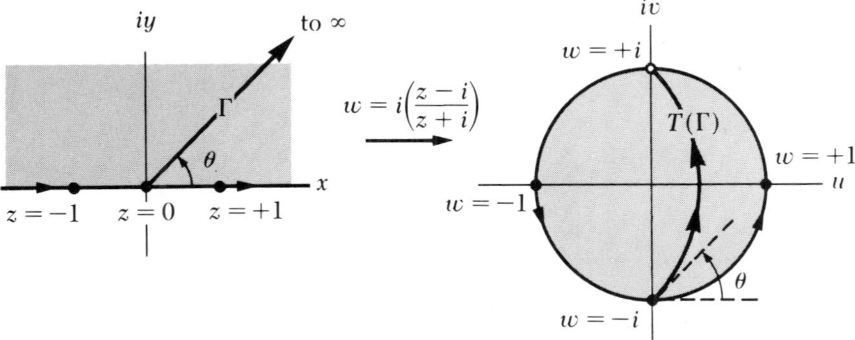

Figure 4.23 The fractional linear transformation $w = i\left(\dfrac{z-i}{z+i}\right)$ in Example 4.16.

substituting these values directly into formula (32) we get

$$\left(\frac{1+i}{2}\right)\left(\frac{w+1}{w+i}\right) = \frac{1}{2}\left(\frac{z+1}{z}\right),$$

and solving for $w = T(z)$ we get

$$w = T(z) = i\left(\frac{z-i}{z+i}\right). \tag{34}$$

Notice that $T(\infty) = +i$, so the point ∞, which serves as a terminal point for the real axis in $\mathbf{C}^*$, is also mapped into the unit circle $|w| = 1$. Now T must map circles to circles, and since three distinct points in $\mathbf{C}^*$ fully determine a circle in the sphere, we are assured that T maps the extended real line onto the unit circle. If we substitute $z = x + i0$ into (34) we know that $T(x + i0) \to +i$ as $x \to +\infty$ or $x \to -\infty$, and a careful examination of (34) shows that $T(x + i0)$ moves *counterclockwise* around the circle $|w| = 1$ as x increases. In effect the real axis is being "bent around" the unit circle, covering all points on the circle except $w = +i$, which corresponds to $z = \infty$.

Because T wraps the real axis around the circle in a counterclockwise direction, it is intuitively obvious that T should map the upper half plane to the *interior* disc $|w| < 1$ and the lower half plane to the *exterior* domain $|w| > 1$. This can be verified by examining how T transforms radial lines extending from the origin to infinity; by Theorem 4.8 these must be mapped to circular arcs extending from $T(0) = -i$ to $T(\infty) = +i$ in the w-plane. If we consider a ray that makes an angle θ with the positive x-axis ($0 < \theta < \pi$), we know that the transformed curves must also meet at an angle θ at $T(0) = -i$ because T is conformal at $z = 0$. The situation is shown in Figure 4.23. These observations completely determine the circular arc in the disc $|w| < 1$ corresponding to any ray from the origin in the upper half plane. Similar considerations may be applied to rays in the lower half plane, to see that T does map upper and lower half planes onto the domains $|w| < 1$ and $|w| > 1$, respectively.

Further use of the principles embodied in Theorem 4.8 allows us to determine how a checkerboard pattern determined by the lines $x =$ constant

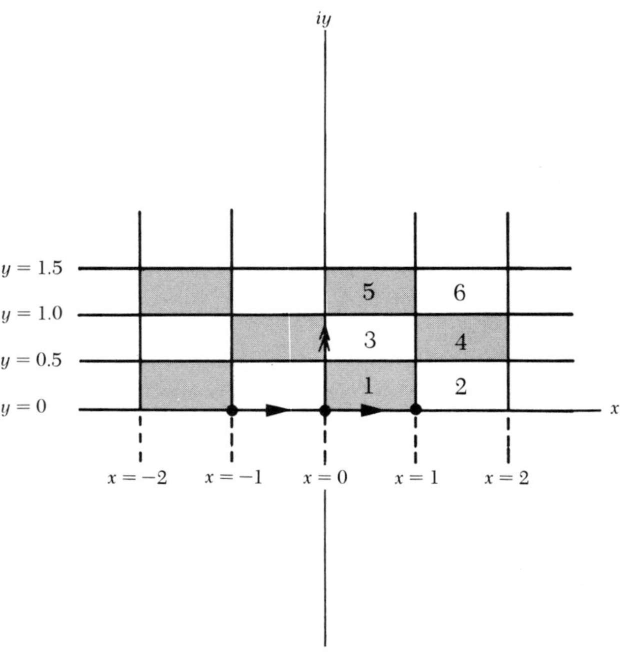

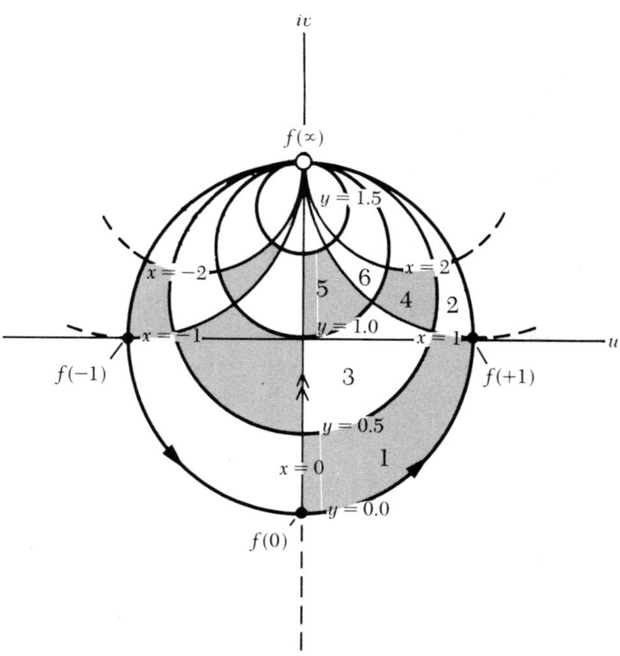

Figure 4.24 The transformation $w = i\left(\dfrac{z - i}{z + i}\right)$ maps the upper half plane Im $(z) > 0$ conformally onto the open unit disc $|w| < 1$. The curves in the w-plane corresponding to $x =$ *constant* and $y =$ *constant* are shown; these describe the behavior of the inverse function $z = \check{f}(w)$ on the disc.

and $y =$ constant is mapped into the w-plane; the effect of T is illustrated in Figure 4.24.

Using similar calculations, we can produce a mapping S that carries $\{-1, 0, +1\}$ to the points $\{-1, +i, +1\}$, namely

$$S(z) = \frac{1}{i}\left(\frac{z+i}{z-i}\right).$$

This map is obtained by performing the operation $w = T(z)$ followed by the inversion operation $w = 1/z$. Clearly, S maps the lower half plane $\text{Im}(z) < 0$ onto the disc $|w| < 1$ and maps the upper half plane $\text{Im}(z) > 0$ to the exterior domain $|w| > 1$. The inverses of these mappings are also interesting in applications if we want to map a disc, or the complementary exterior domain, conformally onto a half plane; we leave the reader to calculate the inverse mappings in Exercise 8.

EXERCISES

1. Prove that the mappings $S(z) = 1/z$ and $T(z) = z + 2$ *do not* commute: $S(T(z)) \neq T(S(z))$ for at least one z on the complex sphere, so that $S \circ T$ and $T \circ S$ are not the same.

2. If two matrices with non-zero determinant,

$$\begin{bmatrix} a & b \\ c & d \end{bmatrix} \quad \text{and} \quad \begin{bmatrix} a' & b' \\ c' & d' \end{bmatrix},$$

determine the same fractional linear transformation, prove that there is a complex scalar $\mu \neq 0$ such that $a' = \mu a, \ldots, d' = \mu d$.

3. If $T(z) = (az + b)/(cz + d)$, with $c \neq 0$, and if we assign values $T(-d/c) = \infty$, $T(\infty) = a/c$, prove that T is a continuous mapping on the complex sphere.

4. Decide how the following curves are transformed under the fractional linear transformations indicated.

(i) $\text{Re}(z) = 1$ under $w = 1/z$
(ii) $|z - 2| = 1$ under $w = 1/z$
(iii) $|z - 3i| = 1$ under $w = i/z$
(iv) $\text{Re}(z) = 1$ under $w = 2/(z - i)$

Hint: Use the ideas in Theorems 4.3 and 4.8, rather than elaborate calculations.

Answers: (i) $|w - (\frac{1}{2})| = \frac{1}{2}$; (ii) circle through $\{\frac{1}{3}, 1\}$, center on real axis; (iii) $|w - (\frac{3}{8})| = \frac{1}{8}$; (iv) $|w - (2i + 1)| = 1$.

5. Decide how the following domains are transformed.

(i) $\mathrm{Re}(z) > 1$ under $w = 1/z$.

(ii) $0 < \mathrm{Re}(z) < \delta$ under $w = 1/z$.

(iii) The part of the disc $|z| < 1$ in the right half plane, under $w = 2/z$.

(iv) The first quadrant under $w = 1/(z + 1)$.

Answers: (i) *Interior* of the circle $|w - \frac{1}{2}| = \frac{1}{2}$; (*ii*) *complement* of disc $\left|w - \left(\frac{1}{2\delta}\right)\right| < \frac{1}{2\delta}$ in the half plane $\mathrm{Re}(w) > 0$; (iii) $\mathrm{Re}(w) > 0$ and $|w| > 2$; (iv) half disc $|w + \frac{1}{2}| < \frac{1}{2}$, $\mathrm{Im}(w) < 0$.

6. If T is a fractional linear transformation and Γ and Γ' are circles in $\mathbf{C}^*$, prove that the following relationships between Γ and Γ' remain valid for the images $T(\Gamma)$ and $T(\Gamma')$.

(i) $\Gamma \cap \Gamma' = \varnothing$ (disjoint)

(ii) Γ and Γ' meet in just one point

(iii) Γ perpendicular to Γ' where they meet

(iv) Γ tangent to Γ'.

7. Show that fractional linear transformations do not always transform the center of a circle Γ to the center of the image circle $\Gamma' = T(\Gamma)$.

Hint: Try $w = 1/z$.

8. Show that the inverse mapping (carrying the unit disc to the half plane) in Example 4.16 is given by

$$z = \check{T}(w) = \frac{1}{i}\left(\frac{w + i}{w - i}\right), \text{ all } w \text{ in } \mathbf{C}^*.$$

Compose T (or $\check{T}$) with familiar *linear* transformations, to produce mappings S with the following properties:

(i) S maps the right half plane $\mathrm{Re}(z) > 0$ onto the unit disc $|w| < 1$ (onto the disc $|w| < r$).

(ii) S maps the unit disc $|z| < 1$ onto the half plane $\mathrm{Im}(w) > 0$ so that $z = 0$ maps to $w = 2 + 2i$.

(iii) S maps the half plane $\mathrm{Re}(z) > -1$ onto the domain $|w - i| > 1$.

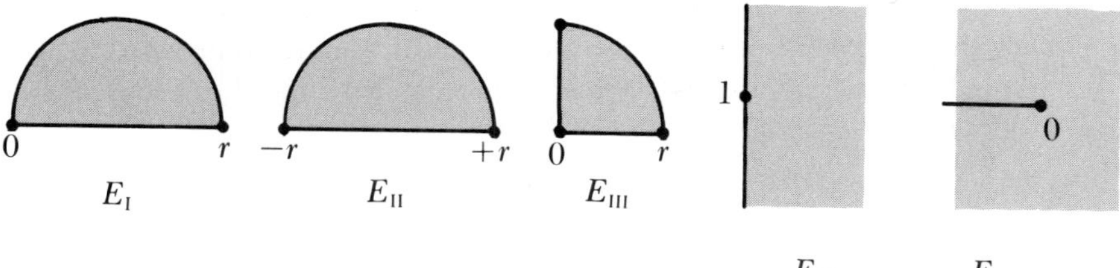

Figure 4.25 The initial domains in Exercises 9, 10, and 11.

Answers: (i) $S(z) = T(+iz) = i\left(\dfrac{z-1}{z+1}\right)$; (ii) $S(z) = 2(\check{T}(z)) + 2 = 2 \cdot \dfrac{(1+i)z + (1+i)}{iz+1}$; (iii) Set $\zeta = i(z+1)$ and take $w = S(\zeta) = \dfrac{1}{i}\left(\dfrac{\zeta + i}{\zeta - i}\right) + i$; direct calculations yield $w = \dfrac{2i}{z+2}$.

9. Prove that no fractional linear transformation $T: \mathbf{C}^* \to \mathbf{C}^*$ can possibly map the domain E_{I} shown in Figure 4.25 onto the unit disc $D = \{w: |w| < 1\}$.

Hint: Fractional linear transformations are conformal. How would T map the "corners" at $z = 0$ and $z = r$?

10. Use the mapping $w = T(z) = i(z - i)/(z + i)$ of the upper half plane onto the disc (Example 4.16) to devise an invertible conformal mapping f of the half disc E_{I} shown in Figure 4.25 onto the disc $D = \{w: |w| < 1\}$. Write an explicit formula for the mapping $f: E_{\mathrm{I}} \to D$.

Hint: Map the corner $z = r$ to ∞ via $w = 1/(r - z)$; this maps E_{I} to a quadrant. Then transform to the first quadrant using $w = z - (1/r)$. Next apply $w = z^2$ to get the upper half plane; then apply T. Take f to be the composite.

Answer: $f(z) = i\dfrac{z^2 - ir(r - z)^2}{z^2 + ir(r - z)^2}$ (a quadratic rational function, *not* a fractional linear transformation).

11. Use the ideas in Exercise 10 to devise conformal invertible mappings that transform the domains E_{II}, E_{III}, E_{IV}, and E_{V} (shown in Figure 4.25) to the disc $|w| < 1$.

Hint: Use $w = z^{1/2}$ with E_{III} and E_{V}, and $w = i(z - 1)$ with E_{IV}.

12. Find a conformal mapping (not necessarily fractional linear) that maps a crescent-shaped domain, bounded by a pair of circular arcs, onto the disc $|w| < 1$.

Hint: Use fractional linear transformations to map one boundary arc to a line segment. Recall Exercise 10.

13. A domain bounded by circular arcs and line segments is mapped to the same kind of domain by fractional linear transformations. Determine the images of the square $E = \{z{:}0 \leq \operatorname{Im}(z) \leq 1 \text{ and } 0 \leq \operatorname{Re}(z) \leq 1\}$ under

(i) $w = 1/z$

(ii) $w = \dfrac{1}{z+i}$

(iii) $w = \dfrac{i}{z+i}$

(iv) $w = \dfrac{z+i}{z} = \dfrac{i}{z} + 1.$

Hint: By conformality, angles at corners are preserved unless the vertex is mapped to ∞.

14. Suppose pairs of points $\{z_1, z_2\}$ and $\{w_1, w_2\}$ are given in $\mathbf{C}^*$. Demonstrate that there are infinitely many different fractional linear transformations $w = T(z)$ such that $w_1 = T(z_1)$ and $w_2 = T(z_2)$. (Compare this with Theorem 4.10).

Hint: Introduce third points z and w.

15. If Γ and Γ' are two distinct circles in the complex sphere, there may be more than one fractional linear transformation that maps Γ onto Γ'. (It is important to realize that Theorem 4.10 does not forbid this.) Calculate two *different* fractional linear transformations that map

(i) $\Gamma = \{z{:}\ |z| = 1\}$ onto $\Gamma' = \{w{:}\ |w| = 1\}$

(ii) $\Gamma = \mathbf{R}^*$ (extended real axis) onto $\Gamma' = \{w{:}\ |w| = 1\}$

(iii) $\Gamma = \{z{:}\ |z| = 2\}$ onto $\Gamma' = \{w{:}\ \operatorname{Re}(w) = 0\} \cup \{\infty\}$.

Hint: Use (i) in (ii). The calculations in Theorem 4.11 and Example 4.16 should suggest ways to solve (iii).

16. Find the fractional linear transformations that carry

(i) $z = \{0, 1, i\}$ to $w = \{-1, -i, 0\}$

(ii) $z = \{0, 3, \infty\}$ to $w = \{-1, -i, 0\}$.

17. Decide how the following domains are transformed by the fractional linear transformations indicated.

(i) $\operatorname{Re}(z) > 1$ under $w = \dfrac{1+z}{1-z}$

(ii) The triangle with vertices $\{0, +i, +1\}$, under

$$w = \frac{1}{i}\left(\frac{z+i}{z-i}\right)$$

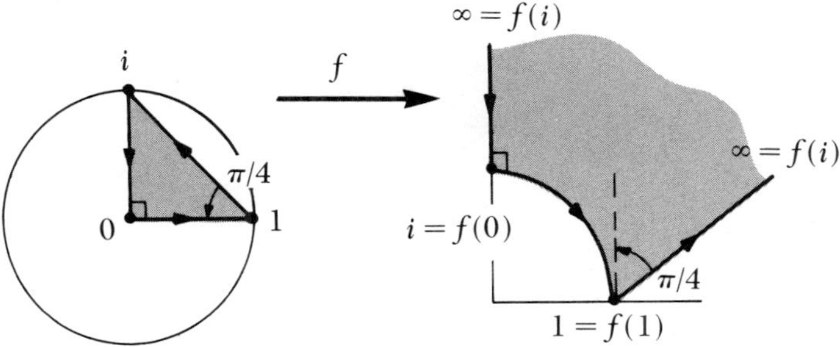

Figure 4.26 Exercise 17(ii).

(iii) The complement of the triangle in (ii), under

$$w = \frac{1}{i}\left(\frac{z+i}{z-i}\right)$$

(iv) The half strip $0 < \text{Re}(z) < 1$ and $\text{Im}(z) > 0$, under

$$w = i\left(\frac{z-i}{z+i}\right)$$

(v) The square $0 < \text{Re}(z) < 1$, $0 < \text{Im}(z) < 1$ under

$$w = i\left(\frac{z-i}{z+i}\right)$$

Answers: (i) $\text{Re}(w) > 1$; (ii) Shaded region in Figure 4.26; (iii) complement of shaded region in Figure 4.26; (iv) shaded region in Figure 4.27; (v) part of shaded region in Figure 4.27 (dashed line).

18. If T is a fractional linear transformation that has fixed points $z_1 = 0$ and $z_2 = \infty$ in $\mathbf{C}^*$, prove that T has the form $T(z) = \mu z$ (μ a non-zero scalar).

19. If T is fractional linear and has $z = \infty$ as its one and only fixed point, prove that T is a translation: $w = z + b$ ($b \neq 0$ a scalar).

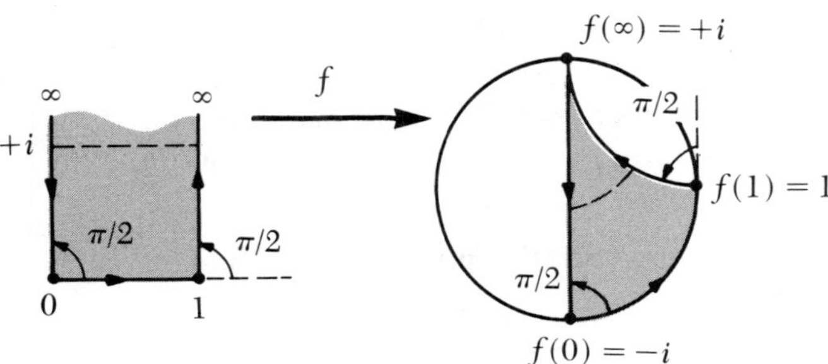

Figure 4.27 Exercise 17(iv).

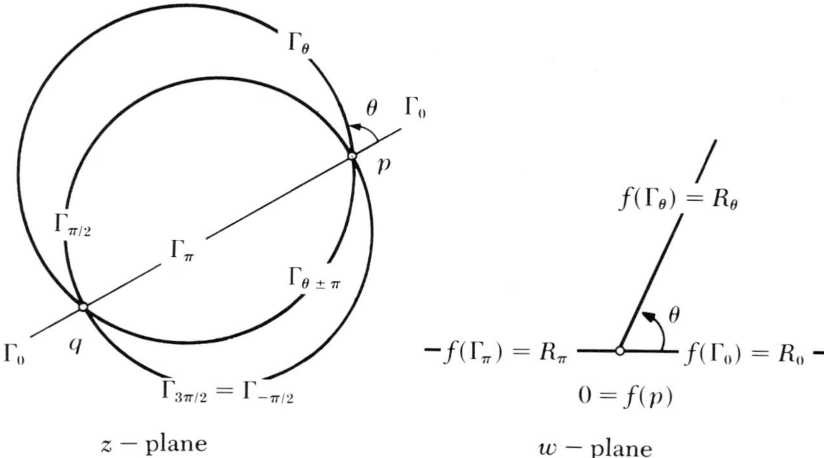

Figure 4.28 Exercise 21.

20. Find a fractional linear transformation T that has $z = +i$ as its one and only fixed point in $\mathbf{C}^*$ (∞ is *not* fixed).

Hint: Find an S that interchanges $+i$ and ∞; then use Exercise 19.

21. Let $p \neq q$ be ordinary points in the plane and let $w = f(z) = \left(\frac{z-p}{z-q}\right)$. Prove that

(i) The circular arcs Γ_θ (θ measured as shown) in Figure 4.28 are mapped to the rays $R_\theta = \{w : \arg w = \theta\}$ in the w-plane.

(ii) $\operatorname{Arg} f(z)$ is constant on these circular arcs.

(iii) $|f(z)|$ is constant on the family of circles

$$K_\lambda = \{z : |z - p|/|z - q| = \lambda\}, \text{ where } 0 < \lambda < \infty.$$

Hint: Γ_0 (an extended line segment) and Γ_π (a segment) are degenerate circular arcs unless we view things in $\mathbf{C}^*$. Notice $f(p) = 0$ and f is conformal; show $\Gamma_0 \cup \{\infty\}$ maps to the ray R_0. Use conformality of T at p to identify the image of Γ_θ as the ray R_θ.

22. If we take the principal determination $\phi(z) = \operatorname{Arg}\left(\frac{z-p}{z-q}\right)$ in Exercise 21, identify the locus of discontinuities of ϕ and determine the values of ϕ along either side of this arc. Prove that ϕ is a smooth real valued function off of this locus.

23. If $p = +1$ and $q = -1$ in Exercises 21 and 22, sketch the loci

(i) $\operatorname{Arg}\left(\frac{z-p}{z-q}\right) = \alpha$ for $\alpha = -\pi, -\pi/2, -\pi/4, 0, \pi/4, \pi/2, \pi$

(ii) $\left|\frac{z-p}{z-q}\right| = \lambda$ for $\lambda = \frac{1}{2}, 1, 2, 100.$

Sketch the locus of discontinuities of $\operatorname{Arg}\left(\frac{z-p}{z-q}\right)$.

24. Show that the cross ratio of four points is real if and only if the points lie on a circle Γ in $\mathbf{C}^*$.

Hint: Transform the circle to the real axis.

*4.9 REFLECTION THROUGH A CIRCLE IN THE COMPLEX SPHERE

If Γ is a circle in the complex sphere, we shall define an anti-conformal mapping J_Γ that reflects points across Γ, interchanging points on either side of Γ and leaving points on Γ fixed. If Γ is an extended line (that is, Γ passes through ∞ on $\mathbf{C}^*$) we take J_Γ to be the usual reflection across Γ. In particular, if $\Gamma = \mathbf{R}^*$, then $J_\Gamma(z) = \bar{z}$, the usual complex conjugate, and $J_\Gamma(\infty) = \infty$. If $\Gamma = \{z: |z - p| = R\}$ is a circle in the plane, our definition is motivated by examining the reflection through the unit circle

$$J(re^{i\theta}) = \frac{1}{r}e^{i\theta} \quad (\text{if } r \neq 0);\ J(0) = \infty;\ J(\infty) = 0$$

defined earlier (recall Figures 4.14 and 4.22). If $z \neq p$, then $z^* = J_\Gamma(z)$ should be defined so that

$$\arg(z^* - p) = \arg(z - p) = \theta;$$

furthermore, the ratio $|z^* - p|/R$ should be the *reciprocal* of $|z - p|/R$, so that

$$|z^* - p| = \frac{R^2}{|z - p|}.$$

Thus, z^* should be given by $z^* - p = \frac{R^2}{|z-p|}e^{i\theta}$ if $\theta = \arg(z - p)$, and

(35) $$J_\Gamma(z) = p + \frac{R^2}{|z-p|}e^{i\arg(z-p)} = p + \frac{R^2}{|z-p|^2}(z - p)$$

$$J_\Gamma(p) = \infty; \qquad J_\Gamma(\infty) = p$$

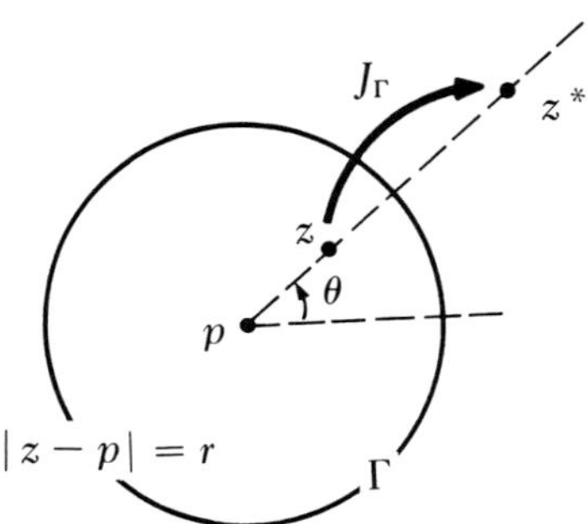

Figure 4.29 The reflection J_Γ of points through the circle $\Gamma = \{z: |z - p| = r\}$.

as indicated in Figure 4.29. Points that are images of each other under J_Γ are said to be **symmetric with respect to Γ.** These reflections are useful in solving mapping problems involving fractional linear transformations because of the following remarkable property.

Theorem 4.13 *Let T be a fractional linear transformation and let Γ be any circle in the complex sphere. Write Γ' for the transformed circle $\Gamma' = T(\Gamma)$, and consider the reflection mappings J_Γ and $J_{\Gamma'}$. Then*

$$T(J_\Gamma(z)) = J_{\Gamma'}(T(z)) \quad \textit{for all } z \textit{ in } \mathbf{C}^*. \tag{36}$$

That is, if $\{z, z^*\}$ are symmetric with respect to Γ, then their images $\{T(z), T(z^*)\}$ must be symmetric with respect to the circle $\Gamma' = T(\Gamma)$.

PROOF: This could be proved by complicated direct calculations, but we shall give a geometric proof. First notice that:

(i) If S and T are transformations such that formula (36) is valid, this formula is also valid for the composite $S \circ T$.

In fact, if $\Gamma' = T(\Gamma)$ and $\Gamma'' = S(\Gamma') = (S \circ T)(\Gamma)$, we get

$$(S \circ T)(J_\Gamma z) = S(T(J_\Gamma z)) = S(J_{\Gamma'}(Tz)) = J_{S(\Gamma')}(S(Tz)) = J_{\Gamma''}((S \circ T)z)$$

for all z in $\mathbf{C}^*$. This proves (i).

If Γ is the circle $|z - p| = R$, it is fairly obvious that the following transformations have property (36):

$$w = \frac{1}{z - p}\text{; rotations, dilations, translations.}$$

The reciprocal mapping $w = T(z) = 1/z$ is the product (composite) $T = R \circ S$ of the mappings $S(z) = z + p$ and $R(z) = 1/(z - p)$:

$$(R \circ S)(z) = R(S(z)) = R(z + p) = \frac{1}{(z + p) - p} = \frac{1}{z},$$

so that the elementary transformations

$$w = 1/z; \text{ rotations, dilations, translations}$$

all satisfy condition (36). Now apply (i) and Theorem 4.7 to see that every fractional linear transformation has this property.

If Γ had been a line instead of a circle, slightly different reasoning would lead to the same conclusion. ▮

These observations can be applied to mapping problems that would be very difficult to solve by other means.

Example 4.17 Suppose we wish to map the half plane $\mathrm{Re}(z) > 0$ onto the unit disc $|w| < 1$ so that $T(+1) = 0$ and $T(\infty) = +1$. The desired mapping $w = T(z)$ should map the boundary Γ of the half plane (the extended imaginary axis) to the boundary circle $\Gamma' = \{w : |w| = 1\}$, subject to the side conditions

$$T(+1) = 0 \quad \text{and} \quad T(\infty) = +1.$$

Notice that $J_\Gamma(+1) = -1$ and $J_{\Gamma'}(0) = \infty$; by Theorem 4.13, T must also map -1 to ∞. Thus, a typical point z and the special points 1, ∞, and -1 should have the images indicated in the following table.

z	z	1	∞	$-1 = J_\Gamma(+1)$
$w = T(z)$	w	0	$+1$	$\infty = J_{\Gamma'}(0)$

Now T leaves cross ratios invariant; thus, $(z, 1, \infty, -1) = (w, 0, +1, \infty)$ for all z, which means that

$$\frac{\cancel{(z-\infty)}}{z+1}\,\frac{1+1}{\cancel{(1-\infty)}} = \frac{w-1}{\cancel{(w-\infty)}}\,\frac{\cancel{(0-\infty)}}{0-1}.$$

On solving this equation, we get

$$w = T(z) = \frac{z-1}{z+1}.$$

It should be clear that similar methods can be used for any choice of $\{\Gamma;\ z_1, z_2\}$ and $\{\Gamma';\ w_1, w_2\}$.

The reader should re-examine Example 4.16 and see how that mapping problem could have been solved using these principles; it is also interesting to see how much effort would be required to solve the above problem using the methods of Section 4.9.

EXERCISES

1. Verify that the reflection through the unit circle $|z| = 1$ is given by $w = 1/\bar{z}$.

2. Prove that the reflections J_Γ are anticonformal mappings. (Handle the cases Γ a circle, Γ a line separately.)

3. If Γ and Γ' are circles in $\mathbf{C}^*$, show that the composite $w = f(z) = (J_\Gamma \circ J_{\Gamma'})(z)$ is holomorphic (it may be undefined at two points if these are not extended lines). Show that f is actually fractional linear.

4. If $\Gamma = \mathbf{R}^*$ and Γ' is the unit circle, express the composites $J_\Gamma \circ J_{\Gamma'}$ and $J_{\Gamma'} \circ J_\Gamma$ as functions of z. Where are they undefined as mappings of $\mathbf{C}$ to $\mathbf{C}$?

5. Transform the domain $-\pi/6 < \arg z < \pi/6$ conformally onto the unit disc $|w| < 1$, arranging that $z = 1, 0$ map to the points $w = 0, -1$.

Hint: First map onto a half plane via $\zeta = z^3$.

Answer: $w = (z^3 - 1)/(z^3 + 1)$.

6. If $|a| < 1$ is given, construct a fractional linear transformation that maps the unit disc $|z| < 1$ onto itself and carries $z = 0$ to $z = a$.

Hint: The circle $|z| = 1$ should be mapped to itself. The map is not uniquely determined, and we may impose a side condition such as $T(1) = 1$ to get a definite mapping, as in Example 4.17.

7. Using Exercise 6, produce a fractional linear transformation that maps $H = \{z: \mathrm{Re}(z) > 0\}$ onto the disc $D = \{w: |w| < 2\}$, carrying $z = +1$ to $w = i/2$. Can you produce two different maps that do this?

8. Let $\{z_1, z_2, z_3\}$ be points in the complex plane. Prove that z and z^* are symmetric with respect to the circle (or extended line) Γ determined by z_1, z_2, and z_3 if and only if $(z^*, z_1, z_2, z_3) = (z, z_1, z_2, z_3)$.

9. If Γ is a circle in the plane and if z is any point lying off Γ, prove that $z^* = J_\Gamma(z)$ is obtained by the geometric construction shown in Figure 4.30.

10. Find all fractional linear transformations that correspond to a rotation of $\mathbf{C}^*$ about some axis.

Hint: How are the points $\{z, z^*\}$, where the axis meets $\mathbf{C}^*$, related to the equatorial circle corresponding to this axis?

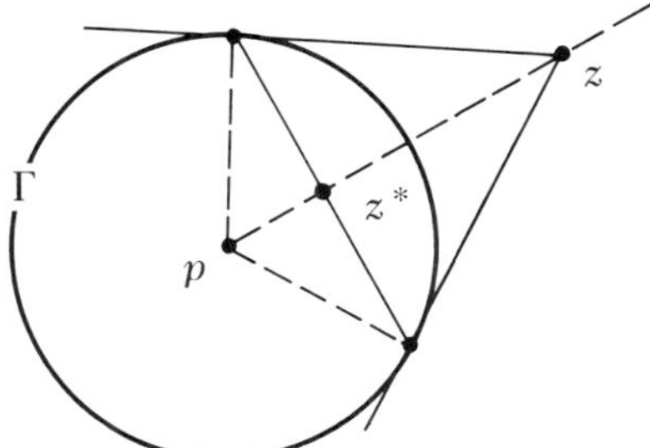

Figure 4.30 z and z^* are symmetric with respect to the circle Γ.

4.10 INVERSE MAPPINGS

We have already encountered inverse mappings in several places. In this section we shall try to fit these ideas into a more coherent whole, emphasizing the techniques that will be most useful in working out inverse mapping problems later on. The reader may rest assured that such problems will confront us in a number of important situations.

Generally, our problem is to decide whether a holomorphic function $w = f(z)$ has an inverse, and to work out explicit formulas for the inverse function $z = \check{f}(w)$, if possible. This task is complicated by the fact that many functions become invertible (univalent mappings) when restricted to subsets, although they might not be univalent mappings on their natural domains of definition. Thus, $w = e^z$ is not univalent on $E = \mathbf{C}$, but is univalent (with inverse functions of great practical significance) if z is restricted to any horizontal strip of width 2π. We obtain the principal determination of the logarithm as the inverse of $w = e^z$ by restricting z to the strip $E_0 = \{z: -\pi < \text{Im}(z) \leq +\pi\}$, and we get other determinations $\text{Log}_n\, z = \text{Log}\, z + 2\pi ni$ by taking parallel strips $E_n = \{z: -\pi + 2\pi n < \text{Im}(z) \leq \pi + 2\pi n\}$ for $n = \pm 1, \pm 2, \ldots$. If we partition the z-plane into disjoint strips E_n, the mapping $w = e^z$ transforms each of these one-to-one onto the domain $F = \{w: w \neq 0\}$; the inverse function $\text{Log}_n = \text{Log} + 2\pi ni$ maps F back onto the proper strip, $\text{Log}_n: F \to E_n$ as indicated in Figure 4.31. This figure provides some insight into the cause of the discontinuous behavior of the inverse functions $\text{Log}\, w + 2\pi ni$ along the negative real axis (which is normally excluded from the domain of definition of these logarithm functions, to make them holomorphic functions of w). The inverse function $\text{Log}_n\, w$ maps points near the lower side of the cut to points $z = \text{Log}_n\, w$ lying close to the *lower* edge of the strip E_n, and it maps points lying slightly above the cut to points z which lie close to the upper edge of the strip E_n. To put it briefly, the inverse mapping is intrinsically multiple valued, and we get the various inverse functions $\text{Log}_n\, w$ just described by sorting out the possible (multiple) values of $\log w$ to give coherently defined holomorphic functions of w; the scheme shown in Figure 4.31 shows how these mappings arise naturally, in a purely geometric way, when we try to find inverses to the exponential function by cutting down the size of its domain of definition in various ways.

The general problem of producing an inverse for any holomorphic function

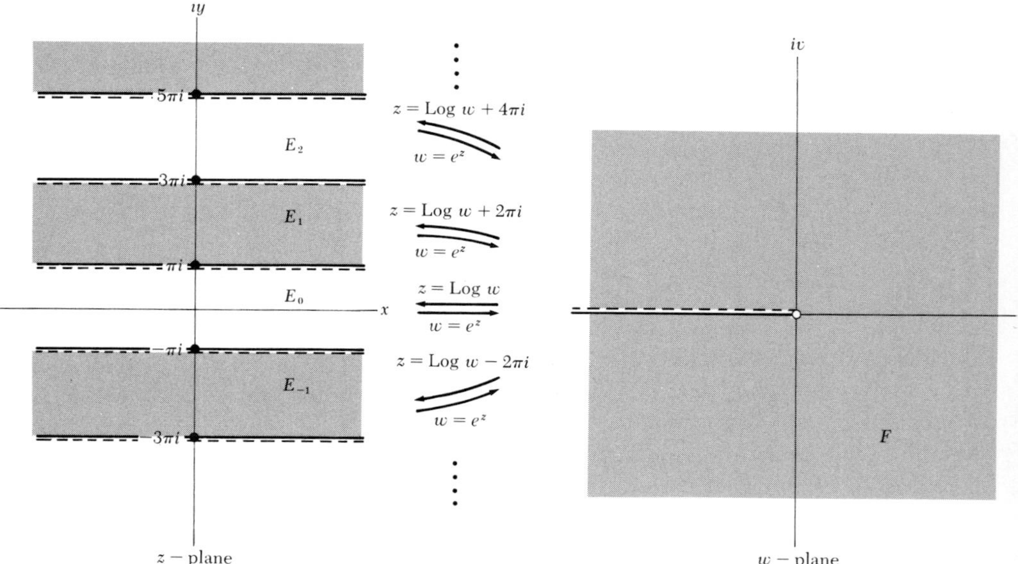

Figure 4.31 The mapping $w = e^z$ and its inverses when restricted to the strips E_n. The dashed and solid lines show how the boundary lines of each strip are transformed.

on its total domain of definition (not just a piece of it) was solved by Riemann with the introduction of Riemann surfaces. This construction allows us to "paste together" locally defined inverses, such as the functions $\text{Log}_n(w)$ associated with $w = e^z$, to define an "inverse", even for functions that are not invertible in the usual sense. Riemann's method also sheds light on the subject of multiple valued functions. However, these considerations must be left to more advanced courses.

Evidently, the question, "Does $w = f(z)$ have an inverse?" should be interpreted in the broad sense, "Are there substantial subsets on which $w = f(z)$ is an invertible mapping?" Our program in dealing with inverse mappings has two steps.

Step 1. Determine which domains in the z-plane are mapped invertibly into the w-plane; that is, determine domains (the larger the better) on which $w = f(z)$ is a univalent mapping.

Step 2. Determine explicit formulas for the inverse mapping(s) associated with $w = f(z)$ as in Step 1. (This will usually require a determination of all possible solutions of the equation $w = f(z)$ for fixed w.)

We have outlined answers to these questions for the exponential function, and the reader can review for himself the information we have acquired about the functions $w = z^n$ to see that these are invertible when z is restricted to wedges with opening $\Delta\theta = 2\pi/n$, bounded by rays emanating from the origin. It is then easy to set up a diagram similar to Figure 4.31, in which the z-plane is partitioned into wedges $E_0, \ldots, E_{n-1}$ (instead of strips); the function $w = z^n$

maps the open wedge E_k invertibly onto the cut w-plane, and the inverse functions $z = f_k(w)$ are the various determinations of the nth root function $w^{1/n}$. We shall work out an inverse mapping problem that is less familiar, to illustrate our program.

Example 4.18 (Inverting $w = \tan z$) The periodicity and symmetry properties

$$\tan(z + \pi) = \tan z \qquad \tan(-z) = -\tan z \qquad \tan \bar{z} = \overline{\tan z}$$

strongly suggest that we can only expect $w = \tan z$ to be univalent on vertical strips of width π (or smaller sets contained in these). We demonstrated in Exercise 3 of Section 2.13 that $w = \tan z$ is indeed univalent on each of the strips $E_n = \{z : -(\pi/2) + n\pi < \text{Re}(z) < (\pi/2) + n\pi\}$ for $n = 0, \pm 1, \pm 2, \ldots$.† Each of these strips is mapped to a domain F in the w-plane, by the open mapping theorem, and the image domain F is the same for each of the strips due to the periodicity of $\tan z$. Determining the precise shape of F is a non-trivial global mapping problem. Here is a new approach to problems of this kind. Notice that

$$w = \tan z = \frac{\sin z}{\cos z} = \frac{1}{i}\left(\frac{e^{iz} - e^{-iz}}{e^{iz} + e^{-iz}}\right) = -i \cdot \left(\frac{e^{2iz} - 1}{e^{2iz} + 1}\right)$$

can be obtained by performing a series of quite elementary operations in succession:

$$w = 2iz \qquad w = e^z \qquad w = \frac{z - 1}{z + 1} \qquad w = \frac{1}{i}\, z = -iz;$$

the only non-trivial map here can be broken down into even simpler operations by writing $w = \dfrac{z - 1}{z + 1} = 1 - \dfrac{2}{z + 1}$, so $w = \dfrac{z - 1}{z + 1}$ is obtained by performing the successive operations

$$w = \frac{2}{z + 1} \quad \text{and} \quad w = 1 - z.$$

It is easy to determine the shape of the successive image domains, and each map gives an invertible transformation from one domain to the next. The shape of the transformed strip in successive steps is indicated in Figure 4.32; we leave the reader to check that we have drawn each of the intermediate domains correctly. After all operations have been performed, the image of E_0 under the complete mapping $w = \tan z$ is the doubly cut w-plane obtained by deleting the parts of the imaginary axis that lie outside of the unit disc $|w| < 1$.

† We may adjoin either the right-hand or left-hand boundary line without losing the one-to-one property; however, the points $z_n = (\pi/2) + n\pi + i0$ must be excluded, because $\tan z$ is not defined there.

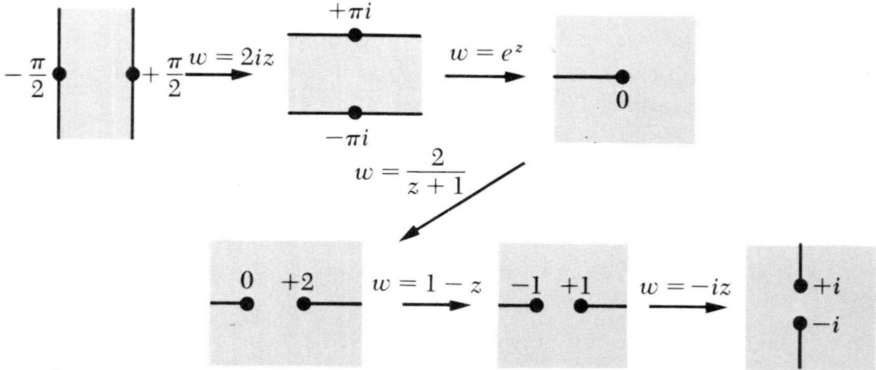

Figure 4.32 The successive elementary transformations of the strip $-\pi/2 < \text{Re}(z) < +\pi/2$ which combine to give $w = \tan z$.

We could also determine the shape of the image domain by the familiar technique of examining the transforms of horizontal and vertical lines in the basic strip E_0, but the calculations are fairly difficult (more difficult than the corresponding problem we faced in answering this question for $w = \sin z$ in Example 4.11). However, if we carried out these calculations, we would find that $w = \tan z$ has the following global mapping properties.

(i) Vertical lines $z = c + iy$ are transformed to circular arcs passing through the points $\{+i, -i\}$, with these end points excluded ($+i$ and $-i$ are not the images of any point in the z-plane). The point $w = \tan(c + iy)$ approaches $+i$ as $y \to +\infty$ and approaches $-i$ as $y \to -\infty$, as indicated in Figure 4.33.

(ii) Horizontal lines $z = x + id$ $\left(\text{with } -\frac{\pi}{2} < x < +\frac{\pi}{2}\right)$ are mapped to a family of circles satisfying the equations

$$\left|\frac{w - i}{w + i}\right| = \alpha \qquad (0 < \alpha < +\infty).$$

These are shown in Figure 4.33 (we get a line, the real axis, when $\alpha = 1$; this should be regarded as a degenerate circle). However, we must exclude the point where each circle meets the cut in the w-plane, because these correspond to $x = \pm\pi/2$ (the boundary of the strip E_0).

Similar calculations apply to each of the strips E_n as a result of the periodicity of tan z. The more detailed analysis summarized in (i) and (ii) of course demonstrates that $w = \tan z$ maps each strip one-to-one onto the doubly cut w-plane, but it provides a great deal of additional information about this transformation, as one can see by studying Figure 4.33.

Since $w = \tan z$ is univalent on each strip E_n, there is a well defined inverse mapping $z = \check{f}_n(w)$ defined on F which maps $\check{f}_n : F \to E_n$; these give different determinations of $z = \arctan w$. The inverse mapping associated with

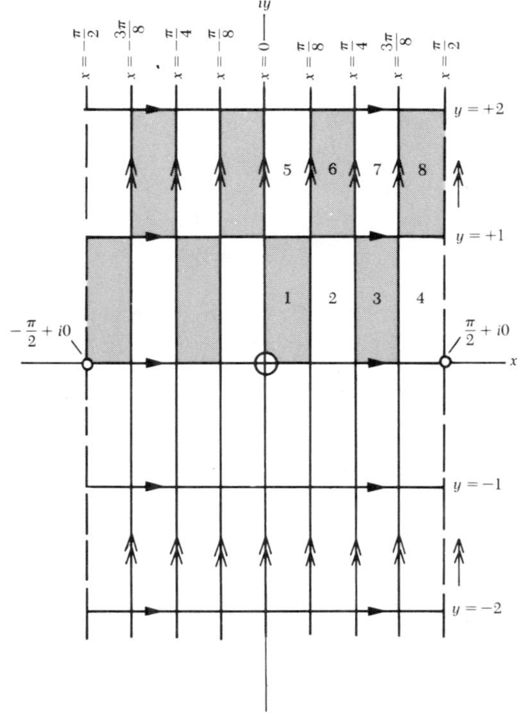

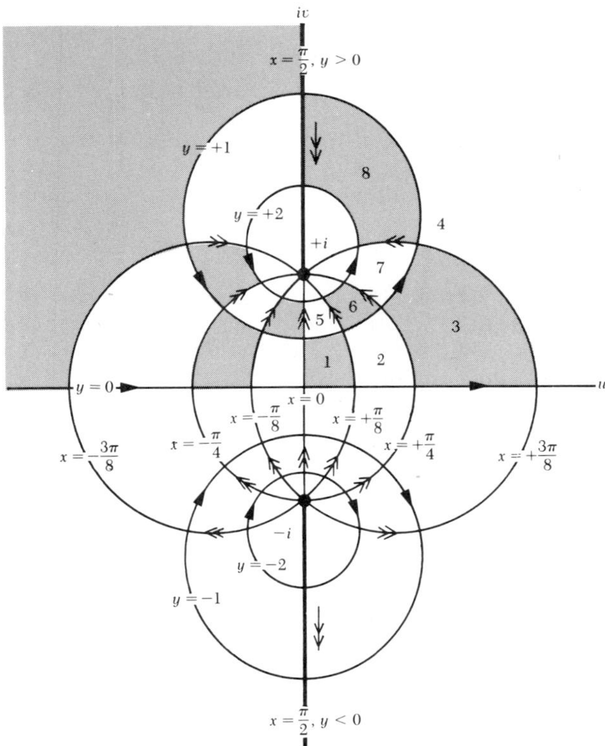

Figure 4.33 The transformation $w = \tan z$ maps the strip $-\pi/2 < \text{Re}(z) < +\pi/2$ to the doubly cut w-plane with rays from $+i$ and $-i$ to infinity deleted.

the strip E_0 is the only one assigning values x with $-\pi/2 < x < \pi/2$ to points $w = u + i0$ on the real axis in F, so this is the determination of arctan w which agrees with the usual function of a real variable: $\operatorname{Arctan}(u + i0) = \arctan u$ for all real u (see Exercise 27).

Finally, let us solve the equation $w = \tan z$ explicitly. If w is not equal to $+i$ or $-i$, we may solve

$$w = \tan z = -i\left(\frac{e^{2iz} - 1}{e^{2iz} + 1}\right) \quad (w \text{ fixed})$$

by writing $Z = e^{iz}$ and examining the equation we get upon solving for Z:

$$e^{2iz} = Z^2 = -\left(\frac{w - i}{w + i}\right).$$

Taking logarithms of both sides we obtain certain congruences (mod $2\pi i$) by employing the functional equation of the logarithm:

$$2iz \equiv \log\left((-1)\,\frac{w - i}{w + i}\right) \quad (\text{mod } 2\pi i).$$

Therefore, the complete set of solutions is given by

$$\text{(37)} \qquad \arctan w = z = \frac{1}{2i}\operatorname{Log}\left(-\frac{w - i}{w + i}\right) + n\pi \qquad n = 0, \pm 1, \pm 2, \ldots$$

if w is not equal to $+i$ or $-i$. Simple calculations (Exercise 4) show that we never have $\tan z = +i$ or $\tan z = -i$, no matter how z is taken in the natural domain of definition of $\tan z$, $E = \left\{z : z \neq \left(\frac{\pi}{2} + n\pi\right) + i0\right\}$. We can actually sort out the holomorphic determinations of arctan w on the doubly cut plane F using formula (37); in fact, the fractional linear transformation $\zeta = -\dfrac{w - i}{w + i} = \dfrac{2i}{w + i} - 1$ maps F one-to-one onto the cut z-plane obtained by deleting the ray $(-\infty, 0]$ from the real axis. Log ζ is well defined on this domain, and we get well defined holomorphic functions

$$\text{(38)} \qquad \check{f}_n(w) = \frac{1}{2i}\operatorname{Log}\left(-\frac{w - i}{w + i}\right) + n\pi \quad \text{defined for } w \text{ in } F$$

for $n = 0, \pm 1, \pm 2, \ldots$. In particular, $\operatorname{Arctan} w = \dfrac{1}{2i}\operatorname{Log}\left(-\dfrac{w - i}{w + i}\right)$ on F.

There are equally remarkable formulas for the various determinations of arcsin w and arccos w:

$$\text{(39)} \qquad \begin{aligned} \arcsin w &= \frac{1}{i}\operatorname{Log}(iw \pm \sqrt{1 - w^2}) + 2\pi n \\ \arccos w &= \frac{1}{i}\operatorname{Log}(w \pm \sqrt{w^2 - 1}) + 2\pi n. \end{aligned}$$

To sort these out into holomorphic functions defined on suitable domains in the w-plane, we would of course have to carry out a detailed analysis of the global mapping properties, and invertibility, of the "forward mappings" $w = \sin z$ and $w = \cos z$ from which these are derived. Much of this has already been done for $w = \sin z$ in Example 4.11, and further details are provided in Exercise 9.

We now introduce two diagrams that are useful in studying the global mapping properties of a function $w = f(z)$. In the w-plane we display the image curves that correspond to the horizontal lines y = constant and the vertical lines x = constant in the z-plane; conversely, in the z-plane we display the curves in this plane that correspond to the lines u = constant and v = constant in the w-plane. The latter have a very natural interpretation if $w = f(z)$ is an *invertible* mapping $f: E \to F$ between domains in the two planes; they are just the images of the horizontal lines v = constant and the vertical lines u = constant in the w-plane under the *inverse mapping* $z = \check{f}(w)$. For reference purposes we display in Figure 4.34 the diagrams corresponding to the function $w = \sin z$, taken as an invertible mapping of the strip $E = \{z: -\pi/2 < \mathrm{Re}(z) < +\pi/2\}$ onto the doubly cut w-plane F obtained by removing the segments $(-\infty, -1]$ and $[+1, +\infty)$.

The pattern of curves u = *constant* and v = *constant* in the rest of the z-plane is obtained by repeating and reflecting the pattern shown in Figure 4.34, using the symmetry properties $\sin(z + \pi) = -\sin z$, $\sin(z + 2\pi n) = \sin z$, and $\sin(-z) = -\sin z$.

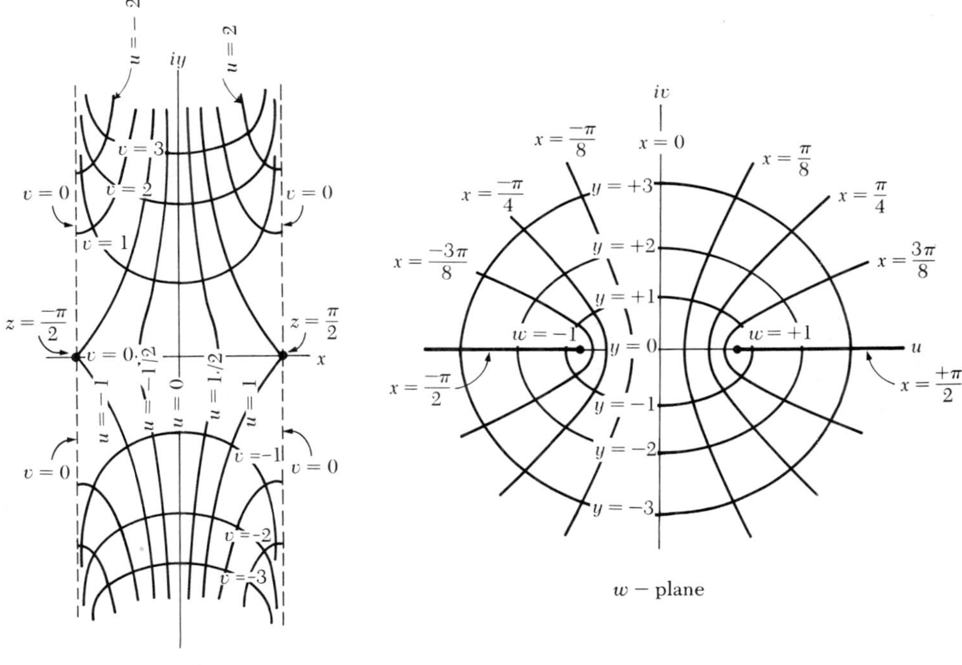

Figure 4.34 The curves u = *constant* and v = *constant* (in the z-plane) and x = *constant* and y = *constant* (in the w-plane) for $w = \sin z$ and $z = \mathrm{Arcsin}\, w$.

Mapping properties of the inverse function $z = \check{f}(w)$, in this case $z = \text{Arcsin } w$, can be read directly out of the family of curves in the w-plane. On the curve corresponding to $x = c$ the real part of $\check{f}(w)$ is constant,

$$c = x = \text{Re}(z) = \text{Re}(\check{f}(w)),$$

and similarly

$$d = y = \text{Im}(z) = \text{Im}(\check{f}(w))$$

on the curve corresponding to $y = d$. Thus the value $\check{f}(w)$ at any point in F is determined by noting which curves of the form $x = \textit{constant} = c$ and $y = \textit{constant} = d$ pass through w, the value of the function being $\check{f}(w) = x + iy = c + id$. By inspecting the family of curves in the w-plane we can determine the values of $\check{f}(w)$ at typical points in F, and we can read out all sorts of limit behavior at the boundary. For the function $z = \text{Arcsin}(w)$, we see that

$$\lim_{w \to 1} \{\text{Arcsin}(w)\} = \frac{\pi}{2} + i0 \quad \text{and} \quad \lim_{w \to \infty} \{\text{Arcsin}(w)\} = \infty\dagger$$

and there is a discontinuity involving a change of sign in $y = \text{Im}(\text{Arcsin } w)$ when we cross either of the cuts.

Since the functions f and $\check{f}$ play symmetrical roles in this situation, the behavior of $w = f(z)$ can be read out of the curve families $u = \textit{constant}$ and $v = \textit{constant}$ in the z-plane; keep in mind that

$$\text{Re}(f(z)) = u = \textit{constant}, \text{ on curves corresponding to } u = \textit{constant}$$

$$\text{Im}(f(z)) = v = \textit{constant}, \text{ on curves corresponding to } v = \textit{constant}.$$

There are a few other comments to be made concerning these diagrams. In Figure 4.34 the curve families in the z-plane, as well as those in the w-plane, are mutually perpendicular. This is so because the mappings $w = \sin z$ and $z = \text{Arcsin } w$, inverses of one another, are holomorphic with non-vanishing derivatives on their respective domains. Although the derivatives do vanish at certain boundary points, these lie outside the domains in question; for example, the points $z = \pm\pi/2 + i0$ on the boundary of the strip E. Therefore, each mapping is conformal and the orthogonal families of horizontal and vertical lines, in either plane, are transformed into orthogonal curve families in the other plane. Another use of these diagrams is to decide how various domains bounded by horizontal and vertical lines are transformed in either direction; this sort of domain mapping problem will be encountered in Chapters 7 and 8. Finally, the functions

$$u = U(x, y) = \text{Re}(\sin z) \quad \text{and} \quad v = V(x, y) = \text{Im}(\sin z) \quad \text{on } E$$

and

$$x = X(u, v) = \text{Re}(\text{Arcsin } w) \quad \text{and} \quad y = Y(u, v) = \text{Im}(\text{Arcsin } w) \quad \text{on } F$$

† An improper limit, defined as in Section 4.7; here we examine $\text{Arcsin}(w)$ for w in F as $|w| \to +\infty$. Notice that $|\text{Arcsin } w|$ is uniformly large outside of the elliptical contours $|y| = \textit{constant}$, because $|\text{Arcsin } w| \geq |\text{Im}(\text{Arcsin } w)| = |y|$.

are harmonic functions on these domains, and thus are relevant to applied problems. The curves in the diagrams give us the level curves of these real valued functions and (most important!) make the boundary behavior of these functions quite clear. This will be very useful in handling boundary value problems associated with Laplace's equation in the plane, as explained in Chapter 7.

EXERCISES

1. Verify the formulas (39). Show that $\sin z$ and $\cos z$ assume every complex value as z varies in $\mathbf{C}$. (Contrast this with the behavior of $\tan z$, which never takes the values $+i$ or $-i$.)

2. Explain why $w = \cos z$ is *not* invertible on the strip $\{z: -\pi/2 < \operatorname{Re}(z) < +\pi/2\}$, used in Example 4.11 for $\sin z$. The basic strip must be correctly located with respect to the symmetries of the function being considered. Show that $w = \cos z$ maps $E = \{z: 0 < \operatorname{Re}(z) < \pi\}$ invertibly onto the doubly cut w-plane F shown in Figure 4.17. Sketch the curve families u = constant and v = constant, and x = constant and y = constant in the respective domains E and F.

Hint: Use $\cos z = \sin[z + (\pi/2)]$ and Example 4.11; no elaborate calculations are needed.

3. Prove directly that $w = \tan z$ is an invertible (univalent) mapping on the strip $E = \{z: -\pi/2 < \operatorname{Re}(z) < +\pi/2\}$. If w' and w'' are on opposite sides of the upper cut in the image domain, where are these points mapped by $z = \operatorname{Arctan} w$? (by $\arctan w = \operatorname{Arctan} w + n\pi$?) What happens near the lower cut? Set up a diagram similar to Figure 4.31.

4. Prove that $w = \tan z$ never assumes the values $+i$ or $-i$ on its domain of definition.

5. Calculate $\operatorname{Arctan}(iy)$ for $-1 < y < +1$. Do the same for $\operatorname{Arcsin}(iy)$, for $-\infty < y < +\infty$.

6. For $r > 0$, draw a circle of this radius about each of the singular points $z_n = (\pi/2 + n\pi)$ of $w = \tan z$. If we remove the discs $|z - z_n| \leq r$ from the plane, show $|\tan z|$ is bounded on the remaining domain.

7. Consider $w = f(z) = z^3$ and $p = e^{i\pi/3}$. How large may we take r so that f is invertible on $D = \{z: |z - p| < r\}$? Let E be the image domain in the w-plane; calculate $z = \check{f}(w)$ explicitly.

Hint: The principal determination $z = w^{1/3}$ does not work. You need not make a complete determination of the shape of the image domain E to answer the question.

8. Determine the largest radius such that f is invertible on $|z - p| < r$, taking

(i) $w = e^z$; p any point;

(ii) $w = \tan z$; $p = 0$

(iii) $w = \sin z$; $p = 0, +i, +\pi/2$.

9. Prove that Arcsin(w), as defined in Example 4.11, is given by

$$\text{Arcsin}(w) = \frac{1}{i}\,\text{Log}\,[iw + \sqrt{1 - w^2}] \text{ for all } w \text{ in } F$$

where $\sqrt{1 - w^2}$ is the principal determination of square root. Show that Arcsin$(u + i0)$ = arcsin u for $-1 < u < +1$, by direct calculations based on this formula. Verify that $\sqrt{1 - w^2}$ is a well defined holomorphic function on the doubly cut plane F.

Hint: If $-1 < u < +1$, then $iu + \sqrt{1 - u^2}$ is in the right half plane; it would be in the left half plane if we took $(-)$ instead of $(+)$. Recall the formulas for Arg$(u + iv)$ associated with Figure 2.13, Section 2.6. The analytic continuation principle might be used to circumvent intricate geometric reasoning.

10. Consider Arcsin w defined on the upper half plane $\text{Im}(w) > 0$, and show that its real part may be expressed explicitly as a function of u and v in the following way:

$$(40)\quad H(u, v) = \text{Re}[\text{Arcsin } w] = \arcsin \tfrac{1}{2}[\sqrt{(u + 1)^2 + v^2} - \sqrt{(u - 1)^2 + v^2}\,],$$

where we take the usual determination of $s = \arcsin(t)$, defined for $-1 \leq t \leq +1$. Use the following steps to arrive at this formula.

(i) If $z = \text{Arcsin } w$, then $w = \sin z$, which means that $u = \sin x \cdot \cosh y$ and $v = \cos x \cdot \sinh y$. Thus,

$$\frac{u^2}{\sin^2 x} - \frac{v^2}{\cos^2 x} = 1.$$

(ii) For fixed x, equation (i) describes a branch of a hyperbola that has focal points $+1$ and -1. A point $w = u + iv$ is on this branch whenever the difference of the distances from w to the focal points is equal to $2 \sin x$; that is,

$$|w + 1| - |w - 1| = 2 \sin x.$$

The difference must be taken in the order shown to get w in the right half plane whenever $x > 0$ (so that $\sin x > 0$).

(iii) Since $x = \text{Re}[\text{Arcsin } w]$ lies in the interval $(-\pi/2, \pi/2)$ for all w in the doubly cut w-plane on which Arcsin w is defined, we get formula (40) if we take the usual determination of $\arcsin(t)$ for $-1 \leq t \leq +1$.

Note: It would be difficult to obtain this from the formula

Arcsin $w = -i$ Log$[iw + \sqrt{1 - w^2}]$ (for $\text{Im}(w) > 0$) in Exercise 9; we have examined the equation $w = \sin z$ from another point of view to get (40).

11. Calculate Arccos(w) explicitly, referring to Exercise 2 for the definition and mapping properties of this principal determination.

12. Use Exercise 9 to show that Arcsin(w) has the following limit values along the cuts if w approaches the cut from within the upper half plane $\text{Im}(w) > 0$:

$$\lim_{w \to -R+i0} \text{Arcsin}(w) = \frac{1}{i} \log |R - \sqrt{R^2 - 1}| - \frac{\pi}{2}$$

$$\lim_{w \to R+i0} \text{Arcsin}(w) = \frac{1}{i} \log |R - \sqrt{R^2 - 1}| + \frac{\pi}{2},$$

for R real and $R > 1$. We get the limit values on the lower side of the cuts if we replace i with $-i$ in the right-hand formulas.

Note: $R - \sqrt{R^2 - 1} \geq 0$ if $R \geq 1$; also, $(R - \sqrt{R^2 - 1})^{-1} = R + \sqrt{R^2 - 1}$.

13. Consider $z = \check{f}(w) = \text{Arcsin}(w)$ defined on the doubly cut plane shown in Figure 4.34. By examining this figure, show that:

(i) $\lim_{w \to p} \text{Re}(f) = -\pi/2$ if p is on the left-hand cut;

(ii) $\lim_{w \to 1} f = +\pi/2$

(iii) $\lim_{w \to p} \text{Im}(f)$ does not exist if p is real and $|p| > 1$.

(iv) $\lim_{w \to \infty} \text{Im}(f) = \infty$

(v) $\lim_{w \to \infty} \text{Re}(f)$ and $\lim_{w \to \infty} f$ do not exist.

If we considered f to be defined only on the upper half plane $E = \{w : \text{Im}(w) > 0\}$, which of your answers would change?

14. Show that $w = z^2$ maps the right half plane $E = \{z : \text{Re}(z) > 0\}$ invertibly onto the cut w-plane $F = \mathbf{C} \sim (-\infty, 0]$. Sketch the pattern of curves $u =$ constant and $v =$ constant in E and $x =$ constant and $y =$ constant in F.

15. Repeat Exercise 14, taking $w = z^2$ as a mapping of the upper half plane $E = \{z : \text{Im}(z) > 0\}$ onto the cut w-plane $F = \mathbf{C} \sim [0, +\infty)$. Compare the sketches in these two situations.

16. For $w = e^z$, determine the pattern of curves $u =$ constant and $v =$ constant in the strip $E = \{z : -\pi < \text{Im}(z) < +\pi\}$, and $x =$ constant and $y =$ constant in the cut w-plane $F = \mathbf{C} \sim (-\infty, 0]$.

17. Consider $w = f(z) = \sinh(z)$ on the strip $E = \{z: -\pi/2 < \mathrm{Im}(z) < \pi/2\}$. Prove that f is invertible, determine its range, and work out a diagram like Figure 4.34 using this function in place of $w = \sin z$. Explicitly calculate the inverse $z = \check{f}(w) = \mathrm{Arcsinh}(w)$.

Hint: Use $\sinh z = -i \sin(iz)$ to deduce your results from the facts concerning $\sin z$; extensive direct calculations are unnecessary.

18. Repeat the program of Exercise 17 for $w = f(z) = \tanh(z)$ defined on $E = \{z: -\pi/2 < \mathrm{Im}(z) < \pi/2\}$. Refer to Example 4.18 and the accompanying figure.

19. Show that $w = f(z) = \dfrac{1}{2}\left(z + \dfrac{1}{z}\right)$ is conformal and invertible on $E = \{z: |z| > 1\}$. Prove that circles $K_r = \{z: |z| = r\}$ $(r > 1)$ are mapped onto ellipses

$$\left(\frac{2u}{r + \frac{1}{r}}\right)^2 + \left(\frac{2v}{r - \frac{1}{r}}\right)^2 = 1$$

in the w-plane. Verify that these ellipses are disjoint, and are focused at $w = +1$ and $w = -1$. Show that the image domain $F = f(E)$ is the cut plane obtained by deleting $[-1, +1]$ from the w-plane. How is the boundary circle $|z| = 1$ transformed?

Hint: Write $z = re^{i\theta}$; hold r fixed (θ real).

20. Show that $w = f(z) = \dfrac{1}{2}\left(z + \dfrac{1}{z}\right)$ maps the rays $L_\theta = \{z: |z| > 1 \text{ and } \arg(z) = \theta\}$ onto segments of hyperbolas

$$\left(\frac{u}{\cos\theta}\right)^2 - \left(\frac{v}{\sin\theta}\right)^2 = 1$$

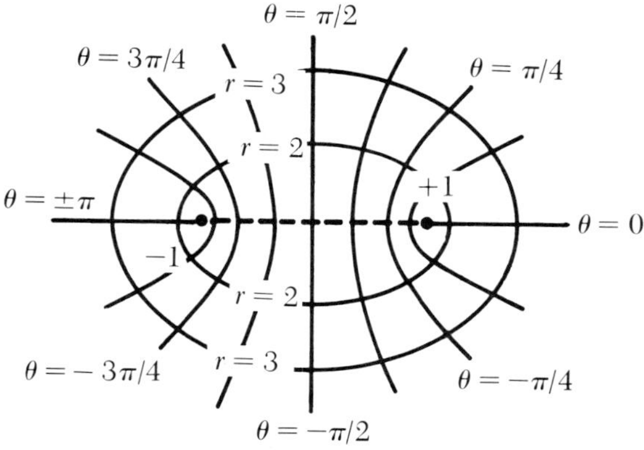

Figure 4.35 The transformation $w = \frac{1}{2}\left(z + \frac{1}{z}\right)$.

as indicated in Figure 4.35. (The rays $\theta = 0, \pm\pi/2, \pm\pi$ require separate treatment; they are mapped to the line segments shown.)

Hint: Write $z = re^{i\theta}$; hold θ fixed $(1 < r < +\infty)$.

21. Prove that the principal determination $\sqrt{\zeta}$ gives a well defined holomorphic function $w = f(z) = \sqrt{1 - z^2}$ on the doubly cut plane F obtained by deleting $(-\infty, -1]$ and $[+1, +\infty)$ from the plane. Show that this particular choice of square root *does not* give a holomorphic function on the cut plane $E = \mathbf{C} \sim [-1, +1]$ (where are the discontinuities located?).

Hint: How are the cut segments mapped by $\zeta = 1 - z^2$?

22. Repeat Exercise 21 using

$$f(z) = \sqrt{\frac{1 - z}{1 + z}} \quad \text{(principal determination of } \sqrt{w}\text{)}$$

in place of $\sqrt{1 - z^2}$.

Hint: For which z does $w = (1 - z)/(1 + z)$ lie on the ray $(-\infty, 0]$?

23. Use the following ideas to show that the functions $\sqrt{1 - z^2}$ and $\sqrt{z^2 - 1}$ both admit holomorphic determinations on the cut plane $E = \mathbf{C} \sim [-1, +1]$. (One cannot use the principal determination of square root for all z in the domain E; cf. Exercise 21.)

(i) Taking principal determinations of square root in $f_1(z) = \sqrt{z - 1}$ and $f_2(z) = \sqrt{z + 1}$, both are holomorphic off the cut $(-\infty, +1]$. Their product $g(z) = \sqrt{z - 1}\sqrt{z + 1}$ gives a holomorphic determination of $\sqrt{z^2 - 1}$ off this cut.

(ii) Calculate the limit values of f_1 and f_2 along the upper and lower edges of the segment $(-\infty, -1)$, showing that

$$\lim_{\delta \to 0} f_1(x + i\delta) = +i\sqrt{|x| + 1}$$

$$\lim_{\delta \to 0} f_1(x - i\delta) = -i\sqrt{|x| + 1}$$

$$\lim_{\delta \to 0} f_2(x + i\delta) = +i\sqrt{|x| - 1}$$

$$\lim_{\delta \to 0} f_2(x - i\delta) = -i\sqrt{|x| - 1}$$

where $-\infty < x < -1$ and $\delta > 0$.

(iii) Use (ii) to verify that $g(z)$ is *continuous* on the segment $(-\infty, -1)$ if we define $g(x + i0) = -\sqrt{x^2 - 1}$ for $-\infty < x < -1$. (However, g is discontinuous on the segment $(-1, +1)$.)

(iv) Show that g must be *holomorphic* at points on $(-\infty, -1)$ as a consequence of its continuity there; thus g is holomorphic on E. (Recall Example 2.17; if $p = x + i0$ for $-\infty < x < -1$, multiply and divide the difference quotient $(g(z) - g(p))/(z - p)$ by $g(z) + g(p)$ to conclude that these quotients approach the limit $dg/dz = -x/\sqrt{x^2 - 1}$).

(v) Since $g(z)$ is a holomorphic determination of $\sqrt{z^2 - 1}$ on E, both $ig(z)$ and $-ig(z)$ give holomorphic determinations of $\sqrt{1 - z^2}$ on E.

24. Show that the determination of $f(z) = \sqrt{1 - z^2}$ in Exercise 23 can be chosen to agree with $\sqrt{1 - x^2} = \sqrt{(-1)(x^2 - 1)} = +i\sqrt{x^2 - 1}$ for $1 < x < +\infty$. Now $f(z)$ is unique, by the analytic continuation principle. Evaluate the limit values along the cut

$$L^+(x) = \lim_{\delta \to 0} f(x + i\delta)$$

(take limit keeping $\delta > 0$)

$$L^-(x) = \lim_{\delta \to 0} f(x - i\delta)$$

for $-1 < x < +1$. Does f have well defined limits at $+1$ and -1 (for z in the domain E)? What are the values $f(x + i0)$ for $-\infty < x < -1$?

Answer: $L^+(x) = -\sqrt{1 - x^2}$; $L^-(x) = \sqrt{1 - x^2}$; $f(x + i0) = -i\sqrt{x^2 - 1}$ if $x < -1$. The limits are zero at $+1$ and -1.

25. Use the methods outlined in Exercises 21 through 24 to produce holomorphic determinations of the following functions.

(i) $\sqrt{z^2 - 1}$ on the plane with $(-\infty, +1]$ and $[+1, +\infty)$ removed, so that $f(x + i0) = +i\sqrt{1 - x^2}$ for $-1 < x < +1$.

(ii) $\sqrt{z^2 - 1}$ on the plane with $[-1, +1]$ removed, so that $f(x + i0) = \sqrt{x^2 - 1}$ for $1 < x < +\infty$.

(iii) $\sqrt{1 + z^2}$ on the plane with $[-i, +i]$ removed, so that

$$\lim_{\delta \to 0_+} f(\delta + iy) = \sqrt{1 - y^2} \text{ for } -1 < y < +1.$$

(iv) $\sqrt{\dfrac{1 - z}{1 + z}}$ on the plane with $[-1, +1]$ removed, so that

$$\lim_{\delta \to 0+} f(x + i\delta) = +\sqrt{\frac{1 - x}{1 + x}} \quad \text{for} \quad -1 < x < +1.$$

For each function, determine the values on opposite sides of the cut(s) and explain how $\arg f(z)$ varies as z moves around a circle $|z| = r$.

26. Use the results of Exercises 21 through 23 of Section 4.8, and the formula

$$w = \tan z = \frac{1}{i}\left(\frac{Z - 1}{Z + 1}\right) \quad \text{where} \quad Z = e^{2iz},$$

to verify the detailed statements concerning the transformations of vertical and horizontal lines in the strip $-\pi/2 < \operatorname{Re}(z) < \pi/2$, given in Example 4.18.

27. Reconcile the formula

$$\operatorname{Arctan} w = \frac{1}{2i} \operatorname{Log}\left(-\frac{w - i}{w + i}\right)$$

with the fact that $\operatorname{Arctan}(u + i0)$ gives the usual arctangent function of a real variable for real values of u. Is $\operatorname{Log}\left(-\dfrac{w - i}{w + i}\right)$ holomorphic on the doubly cut plane on which $\operatorname{Arctan} w$ is defined?

5 INTEGRATION IN THE COMPLEX PLANE

We are going to explain how a continuous function of a complex variable $f(z)$ can be integrated along a curve γ in the complex plane to get a complex number

$$\int_\gamma f(z)\,dz \qquad \left(\text{often written as } \int_\gamma f\,dz\right),$$

the line integral of f along γ. These integrals are indispensable in understanding analytic functions of a complex variable, and are the basis for the most important applications of complex analysis. It would be impossible to exaggerate the importance of understanding how these integrals work.

These integrals are similar to the "line integrals" discussed in advanced calculus for functions of several real variables. However, many special properties arise because both the variable z and values $f(z)$ of the integrand are complex numbers, and it is best to develop our integration theory from the beginning. Our discussion will be self-contained, in that we assume no previous knowledge of line integrals; familiarity with the definite integral $\int_a^b f(t)\,dt$ (Riemann integral) of a real valued function of a real variable is enough. The reader will soon discover that integration theory for functions of a complex variable has a very different feel to it when compared with the integration theory he learned in calculus.

5.1 REPARAMETRIZING A CONTOUR

The reader should recall that a *contour* is a piecewise smooth parametrized curve.

$$\gamma(t) = x(t) + iy(t)$$

defined for real t in some interval $a \leq t \leq b$ (review Section 4.3 for definitions). We must examine the underlying similarities which can exist between two curves presented in parametrized form. Having the same trajectory is the most elementary of these similarities, but we will need a deeper understanding of the possibilities later on, so we shall discuss the idea of reparametrizing a curve. The following examples will help orient our discussion.

Example 5.1 Consider the smooth curves

$$\gamma_1(t) = \cos(2\pi t) + i \sin(2\pi t) = e^{2\pi i t}$$
$$\gamma_2(t) = \cos(2\pi t^2) + i \sin(2\pi t^2) = e^{2\pi i t^2}$$
$$\gamma_3(t) = \cos(4\pi t) + i \sin(4\pi t) = e^{4\pi i t}$$
$$\gamma_4(t) = \cos 2\pi(1 - t) + i \sin 2\pi(1 - t) = e^{2\pi i(1-t)},$$

all defined for $0 \leq t \leq 1$. The trajectory is just the unit circle $\Gamma = \{z: |z| = 1\}$, but the way $\gamma(t)$ transverses Γ is different in each case. All are closed curves with base point (initial/final point) $\gamma(0) = \gamma(1) = 1 + i0$. The curves γ_1, γ_2, γ_4 are simple, but $\gamma_3(t)$ moves around Γ *twice* in a counterclockwise direction with constant speed. The point $\gamma_4(t)$ moves around Γ in a clockwise direction, and thus differs from the other curves, which move counterclockwise.

Now let us consider a few subtle points. It is helpful to think of t as a time parameter. Consider a fixed point p on the common trajectory Γ of these curves; the curves will pass through p at different times: $\gamma_k(t_k) = p$ (and in fact, $\gamma_3(t)$ passes through p at *two* different times). It is interesting to compare the tangent vectors $\gamma_k'(t_k)$, and the tangent lines L_k, associated with the various curves as they pass through p. We leave it as Exercise 8 for the reader to compute the various derivatives; although the tangent vectors $\gamma_k'(t_k)$ attached to p have different lengths (which means that $\gamma_k(t)$ moves past p with different *speeds* for $k = 1, 2, 3, 4$) their orientation, and the tangent lines L_k, all coincide. It should not be surprising that the tangent lines all agree with the line L which passes through p and is tangent to the circle Γ at this point. The calculations of Exercise 6 show that the curves γ_1, γ_3, γ_4 traverse Γ with constant speed, while $\gamma_2(t)$ moves around Γ with steadily increasing speed $|\gamma'(t)| = 2t$.

Example 5.2 (Parametrized line segment from z_1 to z_2) Let z_1 and z_2 be distinct points in the plane, and for $0 \leq t \leq 1$ define

(1) $$\gamma(t) = (1 - t)z_1 + tz_2 = z_1 + t(z_2 - z_1).$$

Obviously,

$$\gamma'(t) = z_2 - z_1 \quad \text{(a constant!)}$$

for all t, and $\gamma(t)$ moves with constant speed from z_1 to z_2, along the line segment Γ connecting these points (see Figure 5.1). If we reverse the roles of z_1 and z_2 in (1), we get a new curve

(2) $$\eta(t) = tz_1 + (1 - t)z_2 = z_2 + t(z_1 - z_2) \quad \text{for} \quad 0 \leq t \leq 1,$$

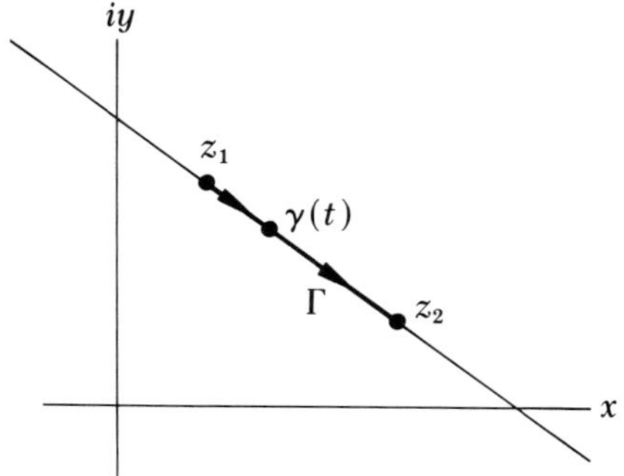

Figure 5.1 The curve in Example 5.2.

and now $\eta(t)$ moves in the opposite direction. The new curve η is related to γ by the change of variable $t \to 1 - t$, so that $\eta(t) = \gamma(1 - t)$ for $0 \le t \le 1$. The reader should consider what happens if we drop the restraint that t remain in the interval $[0, 1]$ in equations (1) and (2); see Exercise 3.

If $\gamma(t) = x(t) + iy(t)$, defined for $a \le t \le b$, is a smooth curve in the plane, its **length** is given by the familiar arc length formula

$$\text{Length}(\gamma) = \int_a^b \sqrt{\left(\frac{dx}{dt}\right)^2 + \left(\frac{dy}{dt}\right)^2}\, dt = \int_a^b \sqrt{x'(t)^2 + y'(t)^2}\, dt.$$

The integrand is just the absolute value of the derivative $\gamma'(t) = x'(t) + iy'(t)$, so that the arc length formula may be written

(3) $$\text{Length}(\gamma) = \int_a^b |\gamma'(t)|\, dt.$$

If t is interpreted as a time parameter, $|\gamma'(t)|$ is the speed of the moving point $\gamma(t)$ and formula (3) agrees nicely with our geometric intuition. These formulas work equally well for piecewise smooth curves (contours); simply break the interval of integration I into consecutive subintervals $I_1, \ldots, I_m$ on which $\gamma(t)$ is smooth.

In many situations we want to change the variable in a curve $\gamma(t)$ by writing the original variable t as a function $t = \phi(s)$ of some new variable s. Let $\gamma(t) = x(t) + iy(t)$ be defined on $I = [a, b]$, and let the new variable s run through some other interval $J = [c, d]$; thus, ϕ is really a mapping $\phi: J \to I$. Substituting $t = \phi(s)$ gives us a new curve

(4) $$\eta(s) = \gamma(\phi(s)) = u(s) + iv(s) \quad \text{for } s \text{ in } J = [c, d],$$

with coordinate functions $u(s) = x(\phi(s))$ and $v(s) = y(\phi(s))$; the new curve η is called a **reparametrization** of γ. It is actually a composite function, obtained by performing the mappings $\phi: J \to I$ and $\gamma: I \to \mathbf{C}$ in succession, as indicated in Figure 5.2; in the usual notation for composite mappings, we have $\eta = \gamma \circ \phi$. We will only be interested in smooth or piecewise smooth

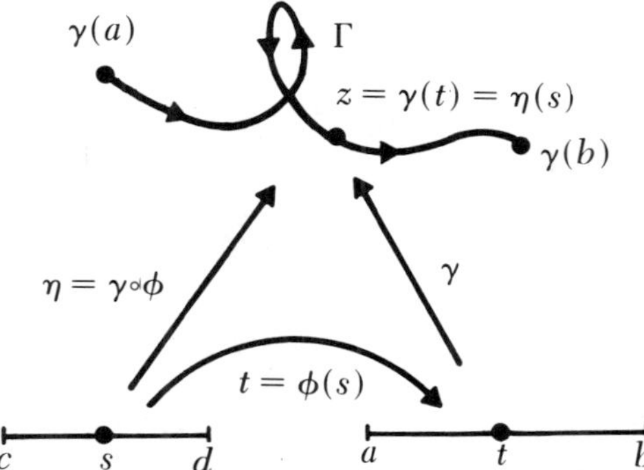

Figure 5.2 Relation between a curve $\gamma(t)$ and its reparametrization $\eta(s)$ by a change of variable $t = \phi(s)$.

curves, so we will always place the following natural restrictions on any reparametrization of a curve.

1. The function $t = \phi(s)$ must be continuous on $[c, d]$ so that $\eta(s)$ will be a continuous function of s and will give us a curve without breaks in it.
2. The function $t = \phi(s)$ must be smooth ($d\phi/ds$ exists and is continuous on the interval $J = [c, d]$), so that the new curve $\eta(s)$ will be piecewise smooth, as γ is.
3. The function $\phi: J \to I$ must map the interval $J = [c, d]$ one-to-one onto the interval $I = [a, b]$, so that it is an *invertible* mapping from J to I.

If we did not introduce this last requirement, $t = \phi(s)$ might not traverse all of I as s runs through J, or else it might run through I several times, causing $\eta(s)$ to double back upon itself. This sort of behavior would occur if we took $J = [0, 1]$ and tried to use one of the following changes of variable:

$$t = \phi(s) = s/4 \qquad 0 \leq s \leq 1$$

or

$$t = \phi(s) = 4s(1 - s) \qquad 0 \leq s \leq 1$$

in the curve $\gamma_1(t)$ of Example 5.1 (see Exercise 12 for a detailed analysis of these inadmissible changes of variable). In either case, the "reparametrized" curve does not even resemble the original curve γ_1 to the extent of having the same trajectory, or remaining a simple closed curve. We have excluded changes of variable this drastic by imposing condition (3).

Definition 5.1 *Hereafter, when we refer to a* **reparametrization** *of a contour γ, we will mean that the map $t = \phi(s)$ effecting the change of variable satisfies conditions* (1), (2), *and* (3) *listed above.*

If we apply the chain rule to differentiate the new coordinate functions $u(s) = x(\phi(s))$ and $v(s) = y(\phi(s))$, we get

$$\frac{du}{ds} = \frac{dx}{dt}(\phi(s)) \cdot \frac{d\phi}{ds} \qquad \frac{dv}{ds} = \frac{dy}{dt}(\phi(s)) \cdot \frac{d\phi}{ds},$$

so that

$$\frac{d\eta}{ds} = \frac{d\gamma}{dt}(\phi(s)) \cdot \frac{d\phi}{ds}(s) \qquad \text{for all } s \text{ in } J = [c, d]. \tag{5}$$

Example 5.3 In Example 5.1, consider

$$\gamma_1(t) = \cos 2\pi t + i \sin 2\pi t = e^{2\pi i t},$$

defined for $0 \leq t \leq 1$. If we make the change of variable $t = \phi(s) = s^2$, with s in $J = [0, 1]$, this mapping $\phi\colon J \to I$ evidently satisfies conditions (1), (2), and (3), and the reparametrized curve

$$\eta(s) = \gamma_1(\phi(s)) = \cos(2\pi s^2) + i \sin(2\pi s^2) = e^{2\pi i s^2} \tag{6}$$

defined on J coincides with the curve $\gamma_2(t)$ in Example 5.1, except that the variable in γ_2 has been labelled s instead of t; however, changing the *name* of the variable makes no difference at all in the nature of the parametrized curve, so η and γ_2 are the same.

On the other hand, if we start with $\gamma_1(t)$ defined on $I = [0, 1]$, and set

$$t = \phi(s) = 1 - s \quad \text{for} \quad 0 \leq s \leq 1,$$

we get a reparametrization defined on $J = [0, 1]$

$$\eta(s) = \gamma_1(\phi(s)) = \gamma_1(1 - s) = \cos 2\pi(1 - s) + i \sin 2\pi(1 - s) = \gamma_4(s) \tag{7}$$

for all s such that $0 \leq s \leq 1$. Therefore, the reparametrized curve $\eta(s)$ is just $\gamma_4(t)$ with the name of the variable changed. Again, the name or symbol we use for the variable is totally irrelevant; these functions $\eta(s)$ and $\gamma_4(t)$ define exactly the same mapping of the interval $J = [0, 1]$ into the complex plane, and must be regarded as identical parametrized curves.

Notice that the reparametrized curves in Example 5.3 differ significantly from the original curve γ_1; for one thing, $\gamma_1(t)$ moves with constant speed, while the reparametrized curve (6) moves with increasing speed, and the curve (7) moves around the trajectory in a direction opposite that of $\gamma_1(t)$. On the other hand, many important features are unaffected by a reparametrization:

1. The trajectories are the same (Exercise 9; this follows from condition (3)).
2. The arc lengths are the same (this will be proved as Theorem 5.1 below).

We will soon prove that line integrals are essentially unaffected by reparametrizing the contour involved in the integration. This very important result has the effect of allowing us to select the most advantageous parametrization of a curve when we set out to calculate arc lengths, line integrals, etc. The outcome of our calculations will not be affected by the choice of parametrization, but the complexity of the calculations can often be reduced by choosing the parametrization carefully.

We can distinguish two different kinds of reparametrization. Condition (3) in the definition of a reparametrization implies that $t = \phi(s)$ must either be

(8A) strictly increasing on J (so that $\phi(s'') > \phi(s')$ if $s'' > s'$)

or else is

(8B) strictly decreasing on J (so that $\phi(s'') < \phi(s')$ if $s'' > s'$).

Definition 5.2 *We say that a reparametrization $t = \phi(s)$ is* **order preserving** *or* **order reversing** *according to whether (8A) or (8B) holds.*

Remember that we always mean $a \leq b$ and $c \leq d$ if we write intervals in the form $I = [a, b]$ and $J = [c, d]$. If $\phi: J \to I$ is a strictly increasing function of s, it is evident that an order preserving reparametrization gives us a new curve $\eta \circ \phi$ whose initial and final points agree with those of γ; we get $\phi(c) = a$ and $\phi(d) = b$, so that

$$p = \gamma(a) = \eta(c) \quad \text{and} \quad q = \gamma(b) = \eta(d).$$

Under an order reversing reparametrization, we get $\phi(c) = b$ and $\phi(d) = a$, so that the initial point of γ is the final point of $\eta = \gamma \circ \phi$ and the final point of γ is the initial point of η:

$$p = \gamma(a) = \eta(d) \quad \text{and} \quad q = \gamma(b) = \eta(c).$$

The intuitive idea is that an order preserving reparametrization $\eta = \gamma \circ \phi$ of a contour γ traverses the common trajectory in the same direction as γ, while an order reversing reparametrization moves in the opposite direction. This is clearly illustrated in Example 5.2, where the curve $\eta(s) = \gamma(1 - s)$ is an order reversing reparametrization.

Example 5.4 (Changing the interval of definition) Let γ be a contour defined on some interval $I = [a, b]$. We can make an order preserving reparametrization to get this curve defined on the standard interval $J = [0, 1]$, or on any other interval if we wish. For the order preserving change of variable, take

$$t = \phi(s) = (1 - s)a + sb = a + s(b - a) \quad \text{for} \quad 0 \leq s \leq 1.$$

Obviously $d\phi/ds = (b - a) > 0$, since $b > a$, and ϕ is order preserving. There is a similar order reversing reparametrization, namely

$$t = \phi(s) = sa + (1 - s)b = b + s(a - b) \quad \text{for} \quad 0 \leq s \leq 1,$$

for which $d\phi/ds = -(b-a)$; ϕ is strictly decreasing on $J = [0, 1]$. We leave it to the reader to work out the simple modifications that enable us to reparametrize γ to be defined on an aribitrary interval $J = [c, d]$ instead of the standard interval $J = [0, 1]$.

Example 5.5 A problem that often arises in practice is to recognize whether two given curves are related by a reparametrization. Consider

(9)
$$\begin{aligned} \gamma(t) &= x(t) + iy(t) = (2t^2 - 1) + it && -1 \le t \le +1 \\ \eta(\theta) &= x(\theta) + iy(\theta) = \cos(2\theta) + i\cos\theta && 0 \le \theta \le \pi. \end{aligned}$$

Both trajectories lie within the solution set of the equation $x = 2y^2 - 1$, a parabola in the x,y-plane; in fact, we can eliminate θ between $x(\theta)$ and $y(\theta)$ by noticing that $x(\theta) = \cos 2\theta = 2\cos^2\theta - 1 = 2y(\theta)^2 - 1$. By inspecting equations (9) we see that we should take $t = \phi(\theta) = \cos\theta$ for $0 \le \theta \le \pi$. Then, $d\phi/d\theta = -\sin\theta \le 0$, and ϕ is a strictly decreasing map of $[0, \pi]$ onto $[-1, +1]$, so that $\eta(\theta) = \gamma(\phi(\theta))$ is an order reversing reparametrization of $\gamma(t)$.

Theorem 5.1 *If $\gamma(t) = x(t) + iy(t)$ is a smooth curve defined on $I = [a, b]$, and if $\eta(s) = \gamma(\phi(s)) = u(s) + iv(s)$ is any reparametrization of γ (order preserving or order reversing), then the arc lengths of γ and η are equal:*

$$L(\gamma) = \int_a^b \sqrt{\left(\frac{dx}{dt}\right)^2 + \left(\frac{dy}{dt}\right)^2}\, dt = L(\eta) = \int_c^d \sqrt{\left(\frac{du}{ds}\right)^2 + \left(\frac{dv}{ds}\right)^2}\, ds.$$

PROOF: This follows from the change of variable formula of calculus. In the case where $t = \phi(s)$ reverses order, we have $d\phi/ds \le 0$, so that

$$\frac{d\phi}{ds} = (-1)\left|\frac{d\phi}{ds}\right| \qquad \text{on } J = [c, d].$$

The reparametrized curve η has coordinate functions $u(s) = x(\phi(s))$ and $v(s) = y(\phi(s))$, and derivative $\eta'(s) = \gamma'(\phi(s)) \cdot \dfrac{d\phi}{ds}(s)$ for $c \le s \le d$ (recall formula (5)). The change of variable formula does the rest of the work:

$$\begin{aligned} L(\eta) &= \int_c^d |\eta'(s)|\, ds = \int_c^d \left|\frac{d\phi}{ds}\right| \cdot \sqrt{x'(\phi(s))^2 + y'(\phi(s))^2}\, ds \\ &= (-1)\int_c^d \frac{d\phi}{ds}(s) \cdot \sqrt{x'(\phi(s))^2 + y'(\phi(s))^2}\, ds \\ &= (-1)\int_{b=\phi(c)}^{a=\phi(d)} \sqrt{x'(t)^2 + y'(t)^2}\, dt = (-1)\int_b^a |\gamma'(t)|\, dt \\ &= \int_a^b |\gamma'(t)|\, dt = \text{Length}(\gamma). \end{aligned}$$

There is no minus sign to manipulate in the case where ϕ is order preserving, since $d\phi/ds = |d\phi/ds| \geq 0$; otherwise, the proof is the same. ▮

There is little difficulty extending this result to piecewise smooth curves.

EXERCISES

1. Write out parametrized curves $\gamma(t)$ tracing out the following loci in the manner indicated, taking $0 \leq t \leq 1$.

(i) The line segment from $z = 0$ to $z = 1$.
(ii) The line segment from $z = 1$ to $z = i$.
(iii) The line segment from $z = +i$ to $z = i - 3$.
(iv) The part of the circle $|z| = 2$ in the right half plane, moving from $+2i$ to $-2i$.

2. Write out parametrized curves $\gamma_1(t), \ldots, \gamma_4(t)$ which trace out the four edges of the square $-1 \leq x, y \leq +1$, moving counterclockwise.

3. How does $\gamma(t) = (1 - t)p + tq$ behave as t varies through $-\infty < t < +\infty$? In a sketch, indicate the direction of motion and location of $\gamma(t)$ for $t = -1$, $t = 0$, $t = +1$, $t = +2$.

4. Set up equations for a parametrized curve $\gamma(t) = x(t) + iy(t)$ that moves along the parabola $y = x^2 + 1$, starting at $(0, 1)$ and ending at $(1, 2)$.

5. Parametrize the following loci as counterclockwise oriented simple curves by expressing them in polar coordinates.

(i) The ellipse $x^2 + 4y^2 = 1$.
(ii) The circle $x^2 + (y - 1)^2 = 4$.
(iii) The parabolic arc $y = x^2 - 1$ for $-1 \leq x \leq +1$.

6. Calculate the speed of the moving point $\gamma(t)$ as a function of t for each parametrized curve in Example 5.1. Which parametrizations cause $\gamma(t)$ to move with constant speed?

7. Calculate the arc length traversed by each parametrized curve in Example 5.1. Explain why the length traversed by γ_4 is *twice* the usual length of the circle $|z| = 1$.

8. In Example 5.1 determine the values of $t = t_k$ for which each curve $\gamma_k(t)$ passes through the point $p = e^{i\pi/4}$ on the unit circle. Calculate the velocity of $\gamma_k(t)$ as it passes through p and show that the velocity vectors $\left[\left.\dfrac{d\gamma_k}{dt}\right|_{t=t_k}\right]$ are all *collinear*, even though they are not of equal length.

9. If $\gamma(t)$ is a smooth curve in the plane, and if $t = \phi(s)$ is an admissible change of variable, verify the following relationships between γ and the reparametrized curve $\eta(s) = \gamma(\phi(s))$.

(i) If γ is a closed curve, so is η.

(ii) If γ is a simple curve, so in η.

(iii) The trajectories of γ and η are the same.

10. If $\gamma(\theta) = e^{2\pi i\theta}$ for $0 \leq \theta \leq 1$, the function $\theta = \sqrt{t} = \phi(t)$ defined in interval $0 \leq t \leq 1$ is *not* an allowable reparametrization of $\gamma(t)$. Which of the requirements (1), (2), and (3) is not satisfied? Show that the function

$$\eta(t) = [\gamma(\theta)|_{\theta=\phi(t)}] = e^{2\pi i\sqrt{t}} \qquad 0 \leq t \leq 1$$

gives a continuous parametrized curve. Explain why this curve is not *smooth*.

Hint: Examine differentiability at $t = 0$.

11. If η is a reparametrization of γ, show that the following properties of γ need not be inherited by η; give counterexamples for each case.

(i) The speed $\left|\dfrac{d\gamma}{dt}\right|$ is constant.

(ii) The interval of definition is [0, 1].

(iii) $z = +i$ is the initial point of γ.

12. In the circular contour $\gamma(t) = e^{2\pi it}$ defined for $0 \leq t \leq 1$, substitute the changes of variable

(i) $t = \phi(s) = s/4$ (ii) $t = \phi(s) = 4s(1 - s)$

defined for s in the interval $J = [0, 1]$. Explain why each of these is inadmissible.

13. In Exercise 12, the contour $\eta(s) = [\gamma(t)|_{t=\phi(s)}]$ is a well defined smooth contour defined for $0 \leq s \leq 1$, even though the substitutions $t = \phi(s)$ do not give admissible reparametrizations of $\gamma(t)$. Compare trajectories, end points, and direction of motion for η and γ in each case.

Note: This should demonstrate the necessity of excluding certain changes of variable $t = \phi(s)$ in discussing reparametrizations. In these examples the curves η and γ bear little resemblance to each other.

*5.2 LINE INTEGRALS IN THE PLANE: PRELIMINARIES FROM CALCULUS

In this section we recall, very briefly, some facts about the definite integral (Riemann integral) $\int_a^b f(x)\,dx$ defined in calculus. Our construction of line integrals for functions of a complex variable, in Section 5.3, will be based on this material.

The familiar Riemann integral operates by taking a function $f(t)$ defined on an interval $I = [a, b]$ in $\mathbf{R}$, and assigning to it a number

$$\int_a^b f(t)\, dt \qquad \left(\text{sometimes written } \int_a^b f\, dt\right),$$

which is called the **definite** (or **Riemann**) **integral** of f over the interval $[a, b]$. The functions f have real variable and real values in calculus; however, we can easily allow integrands f with complex *values* (as long as we insist that the *variable* t remain real) by making the simple definition

$$\int_a^b f(t)\, dt = \int_a^b \mathrm{Re}(f(t))\, dt + i \int_a^b \mathrm{Im}(f(t))\, dt. \tag{10}$$

This introduction of complex valued integrands causes very little trouble; only the definition of a useful integral for functions whose *variable* is complex requires real innovations. In particular, when we integrate functions $z = f(t)$ defined on $I = [a, b]$, the integral (10) is well defined if f is continuous (or bounded and piecewise continuous) on $[a, b]$. Most properties of the Riemann integral carry over to the integrals (10). We adhere to the usual convention for reversing the endpoints of an integral:

$$\int_a^b f(t)\, dt = (-1) \int_b^a f(t)\, dt.$$

The usual algebraic rules for handling integrals can then be verified for complex valued integrands by referring to definition (10) and the analogous properties for real valued integrands; thus,

(11)

(i) $\displaystyle\int_a^b f(t)\, dt = \int_a^c f(t)\, dt + \int_c^b f(t)\, dt$ if $a < c < b$

(ii) $\displaystyle\int_a^b (\alpha \cdot f)(t)\, dt = \alpha \cdot \int_a^b f(t)\, dt$ if α is any complex number

(iii) $\displaystyle\int_a^b (f \pm g)(t)\, dt = \int_a^b f(t)\, dt \pm \int_a^b g(t)\, dt.$

Only one family of properties of the Riemann integral must be re-examined when we allow complex valued integrands. These are the *order properties*, which are based on the fact that

> If f and g are real valued functions on $[a, b]$ such that $f(t) \geq g(t)$ everywhere, then
>
> $$\int_a^b f\, dt \geq \int_a^b g\, dt. \tag{12}$$

In other words, the definite integral *preserves inequalities* between functions if we consider $f \geq g$ to mean that $f(t) \geq g(t)$ for all t in $[a, b]$. From (12) we can easily derive another property (see Exercise 7):

If f is a real valued function $[a, b]$, then

(13)
$$\left| \int_a^b f \, dt \right| \leq \int_a^b |f| \, dt.$$

Here $|f|$ stands for the function with value $|f(t)|$ for each t in $[a, b]$.

Equation (12) has no analog which makes sense for complex valued integrands, since there is no meaningful "$\leq$" relation between the complex values of such functions, but formula (13) cannot be rejected on these grounds. In fact, the inequality (13) is valid for real or complex integrands, and provides the backbone for most of our efforts to estimate the size of integrals.

Theorem 5.2 *If f is a complex valued function on $[a, b]$ that is continuous (or piecewise continuous and bounded), then*

$$\left| \int_a^b f \, dt \right| \leq \int_a^b |f| \, dt.$$

PROOF: There is a choice of real number θ such that $e^{i\theta} \cdot (\int_a^b f \, dt) = |\int_a^b f \, dt|$; notice that

$$\left| \int_a^b f \, dt \right| = e^{i\theta} \int_a^b f \, dt = \int_a^b e^{i\theta} \cdot f \, dt.$$

Since the integral of $g(t) = e^{i\theta} \cdot f(t)$ is *real*, and $\int_a^b g \, dt = \int_a^b \mathrm{Re}(g) \, dt + i \int_a^b \mathrm{Im}(g) \, dt$, we conclude that the imaginary term in this sum is zero, and that

$$\left| \int_a^b f \, dt \right| = \int_a^b g \, dt = \int_a^b \mathrm{Re}(g) \, dt + i0 = \int_a^b \mathrm{Re}(g) \, dt.$$

Now apply (12) to the real valued function $\mathrm{Re}(g(t))$, remembering that $|\mathrm{Re}(z)| \leq |z|$ for any complex number z; we get

$$\left| \int_a^b f \, dt \right| = \int_a^b \mathrm{Re}(g) \, dt \leq \int_a^b |\mathrm{Re}(g)| \, dt \leq \int_a^b |g| \, dt.$$

Obviously, $|g(t)| = |e^{i\theta} f(t)| = |f(t)|$ for $a \leq t \leq b$, so that

$$\left| \int_a^b f \, dt \right| \leq \int_a^b |g| \, dt = \int_a^b |f| \, dt. \quad \blacksquare$$

Here are a few facts that will be particularly important to us. Their validity for complex valued integrands is established in a routine way by using definition (10).

Theorem 5.3 (Fundamental Theorem of Calculus) *Every continuous function $z = f(t)$ defined on the interval $I = [a, b]$ has an antiderivative $F(t)$, a function $F: I \to \mathbf{C}$ that is differentiable, with $dF/dt = f(t)$, for all t in I. The integral with variable upper limit of integration*

$$F(x) = \int_a^x f\,dt \qquad \text{defined for } a \leq x \leq b$$

is an antiderivative of f.

Theorem 5.4 *If $f(t)$ is continuous on $I = [a, b]$ and if $F(t)$ is any antiderivative of f on I, then*

$$\int_a^b f\,dt = F(b) - F(a).$$

Thus, if an antiderivative of f is known, we may evaluate the Riemann integral $\int_a^b f\,dt$ without calculating limits.

Corollary 5.5 *If $f(t)$ is continuously differentiable on $I = [a, b]$, then*

$$\int_a^b \frac{df}{dt}\,dt = \int_a^b f'\,dt = f(b) - f(a).$$

PROOF: $G(t) = f(t)$ is an antiderivative of the continuous function $g(t) = df/dt$ on $[a, b]$. ▮

Theorem 5.6 (Change of variable formula) *Let $I = [a, b]$ and $J = [c, d]$ be intervals in $\mathbf{R}$ and let $t = \phi(s)$ be a function with continuous first derivative that maps J one-to-one onto I, $\phi: J \to I$. If $f(t)$ is continuous on I, then $f(\phi(s))$ is a continuous function on J and*

$$\int_{\phi(c)}^{\phi(d)} f(t)\,dt = \int_c^d f(\phi(s))\,\frac{d\phi}{ds}(s)\,ds. \tag{14}$$

Note: $\phi(c)$ and $\phi(d)$ must be the end points of the interval $I = [a, b]$, but ϕ may be an order preserving or order reversing change of variable. Formula (14) is set up to be true in either eventuality, using the convention that

$$\int_b^a f\,dt = (-1) \cdot \int_a^b f\,dt.$$

PROOF: Take $F(t)$ to be any antiderivative of $f(t)$ defined on I. By the chain rule,

$$\frac{d}{ds}(F \circ \phi)(s) = \frac{dF}{dt}(\phi(s)) \cdot \frac{d\phi}{ds} = f(\phi(s))\,\frac{d\phi}{ds}$$

for all $c \leq s \leq d$. Theorem 5.5, applied to $(F \circ \phi)(s)$ for $c \leq s \leq d$, leads to the desired conclusion:

$$\int_c^d \frac{d}{ds}(F \circ \phi)(s)\,ds = F(\phi(d)) - F(\phi(c)) = \int_{\phi(c)}^{\phi(d)} \frac{dF}{dt}(t)\,dt = \int_{\phi(c)}^{\phi(d)} f(t)\,dt. \quad ▮$$

EXERCISES

1. Verify the formulas in (11) when $f(t) = u(t) + iv(t)$ is a complex valued function of a real variable t.

2. Evaluate the integrals $\int_a^b f(t)\, dt$ in the following situations.

(i) $f(t) = e^{int}$; $[a, b] = [0, 2\pi]$ and $n = 0, \pm 1, \pm 2, \ldots$

(ii) $f(t) = 1 + e^{it} + 2e^{2it} + 3e^{3it}$; $[a, b] = [0, 2\pi]$

(iii) $f(t) = e^{it}$; $[a, b] = [0, \pi]$

Hint: In (i) distinguish cases $n \neq 0$ and $n = 0$.

Answer: (i) 2π if $n = 0$; 0 if $n \neq 0$; (ii) 2π; (iii) $2i$.

3. In a Riemann integral $\int_a^b f(t)\, dt$, prove that $\int_a^b f(t)\, dt = \lim_{n\to\infty} \int_a^b f_n(t)\, dt$ provided the functions $f_n(t)$ converge uniformly to $f(t)$ on the interval of integration $[a, b]$.

Hint: Use

$$\left| \int_a^b f_n\, dt - \int_a^b f\, dt \right| \leq \int_a^b |f_n - f|\, dt$$

and the definition of uniform convergence.

Note: This is true even for complex-valued integrands.

4. Prove Theorem 5.3 starting with definition (10) and the corresponding result for real-valued integrands.

5. Prove Theorem 5.4 starting from Theorem 5.3.

6. If $\gamma(t)$ is a smooth curve defined for $a \leq t \leq b$, it gives a complex valued function of real variable $\gamma(t) = x(t) + iy(t)$. Show that

$$\int_a^b \frac{d\gamma}{dt}\, dt = \gamma(b) - \gamma(a).$$

7. For integrals of real valued functions on the interval $[a, b]$ show that formula (13) follows from the basic inequality (12).

Hint: Write the integrand f as the difference of two non-negative functions $f = f_+ - f_-$, taking

$$f_+(x) = \begin{cases} f(x) & \text{if value } f(x) \text{ is positive} \\ 0 & \text{otherwise} \end{cases}$$

$$f_-(x) = \begin{cases} |f(x)| & \text{if value } f(x) \text{ is negative} \\ 0 & \text{otherwise} \end{cases}$$

Then $f = f_+ - f_-$ and $|f| = f_+ + f_-$.

5.3 COMPLEX LINE INTEGRALS

To define a useful integral for functions $f(z)$ of a complex variable, we shall integrate $f(z)$ along various contours $z = \gamma(t)$. Let $f(z)$ be continuous on an open set E in the complex plane. If $\gamma(t)$, defined for $a \le t \le b$, is any contour (piecewise smooth curve) lying in E, we define the **integral of f along γ**

$$\int_\gamma f(z)\,dz \qquad \left(\text{sometimes written as } \int_\gamma f\,dz\right)$$

to be the number given by the Riemann integral

(15) $$\int_\gamma f(z)\,dz = \int_a^b f(\gamma(t)) \cdot \frac{d\gamma}{dt}(t)\,dt.$$

The right-hand integral has a complex valued integrand and real variable $a \le t \le b$; obviously if $f = U + iV$ and if $\gamma(t) = x(t) + iy(t)$, we get

$$f(\gamma(t)) \cdot \frac{d\gamma}{dt} = (U + iV)(x' + iy') = (Ux' - Vy') + i(Uy' + Vx')$$

so (15) can be rewritten as

$$\int_\gamma f(z)\,dz = \int_a^b (Ux' - Vy')\,dt + i\int_a^b (Uy' + Vx')\,dt.$$

The latter formula is often written in the suggestive form

$$\int_a^b U\,dx - V\,dy + i\int_a^b U\,dy + V\,dx$$

or

$$\int_a^b (U + iV)\,dz,$$

which gives the correct Riemann integral if we set $dx = x'(t)\,dt$ and $dy = y'(t)\,dt$, and $dz = \dfrac{d\gamma}{dt}(t)\,dt = dx + i\,dy$.

This definition may seem a bit formal; we have framed it to mimic the change of variable formula (write $dz = d\gamma/dt$ and take an ordinary Riemann integral in place of $\int_\gamma \cdots dz$), but the natural geometric origin of the complex line integral and its true similarity to the Riemann integral are obscured. In more advanced texts the integral $\int_\gamma f(z)\,dz$ is defined by a limit process that is very similar to the construction of Riemann integrals as limits of "Riemann sums"; formula (15) is then deduced from the limit definition as a *theorem*, valid when γ is piecewise smooth and the integrand $f(z)$ is continuous. In calculating the values of line integrals we would face serious difficulties if we had to evaluate these integrals as limits of partial sums; formula (15) determines the line integral in terms of already familiar objects (Riemann integrals), and all results discussed in this book are derived from this formula, so we will not be hampered by taking it as the definition of line integral. A brief outline of the "Riemann sum" definition of a line integral, and its connection with formula (15), may be found in De Pree and Oehring [5], Sections 24 and 25, or Nevanlinna and Paatero [18], Sections 8.1 and 8.7.

Using formula (15) we now calculate the values of an important family of line integrals.

Example 5.6 Let $E = \{z : z \neq 0\}$ and consider the function $f(z) = 1/z$. Let γ_R be the parametrized circle

$$\gamma_R(t) = Re^{it} \quad \text{defined for} \quad 0 \leq t \leq 2\pi.$$

The component functions are $x(t) = R\cos t$ and $y(t) = R\sin t$, so that

$$\frac{d\gamma_R}{dt}(t) = iRe^{it}$$

and

$$f(\gamma_R(t)) = \frac{1}{\gamma_R(t)} = \frac{1}{Re^{it}} = \frac{1}{R}e^{-it}$$

for $0 \leq t \leq 2\pi$. Equation (15) gives

$$\begin{aligned}\int_{\gamma_R} \frac{1}{z}\,dz = \int_{\gamma_R} f(z)\,dz &= \int_0^{2\pi} f(\gamma_R(t)) \cdot \gamma_R'(t)\,dt \\ &= \int_0^{2\pi} \frac{1}{R}e^{-it} \cdot iRe^{it}\,dt \\ &= i\int_0^{2\pi} dt = 2\pi i.\end{aligned}$$

The value of the integral is the same for every choice of $R > 0$.

In an integral $\int_\gamma f(z)\,dz$ we must have a *parametrized* curve $\gamma(t)$ in mind in order to use formula (15). Since it is only the values

$$f(\gamma(t)) \quad \text{for} \quad a \leq t \leq b$$

that enter into formula (15), the values of $f(z)$ for points $z = \gamma(t)$ lying on the trajectory Γ alone determine the value of this integral; thus, the following result is clear.

Theorem 5.7 *The behavior of the integrand $f(z)$ off the trajectory of γ cannot influence the value of the integral $\int_\gamma f(z)\,dz$. In particular, if $f_1(z)$ and $f_2(z)$ are continuous* *functions that agree at every point on the trajectory* Γ, *we get*

$$\int_\gamma f_1\,dz = \int_\gamma f_2\,dz.$$

The following algebraic properties of line integrals arise directly from corresponding properties of the Riemann integral appearing on the right side of (15).

Theorem 5.8 *Let* f *and* g *be continuous functions of a complex variable defined on a set that includes the trajectory of the contour* γ. *Then*

(i) $\displaystyle\int_\gamma (f \pm g)\,dz = \left(\int_\gamma f\,dz\right) \pm \left(\int_\gamma g\,dz\right)$

(ii) *If* α *is any complex number and we define* (αf) *so that* $(\alpha f)(z) = \alpha \cdot f(z)$, *then*

$$\int_\gamma (\alpha f)\,dz = \alpha \cdot \left(\int_\gamma f\,dz\right).$$

(iii) *If we break the interval* $I = [a, b]$ *into two subintervals* $I_1 = [a, c]$ *and* $I_2 = [c, b]$, *where* $a \le c \le b$, *we get two contours* γ_1 *and* γ_2 *from* γ *by restricting the variable* t *to* I_1 *and* I_2, *respectively. Then*

$$\int_\gamma f\,dz = \int_{\gamma_1} f\,dz + \int_{\gamma_2} f\,dz.$$

It is appropriate to think of the line integral as an operation that puts together a contour γ and a function $f(z)$ to give a complex number $\int_\gamma f\,dz$ (subject to the requirement that f be defined on the trajectory of γ, of course). This operation depends *linearly* on the integrand (equations (i) and (ii)) and *additively* on the contour γ, if we regard γ as a "sum" of γ_1 and γ_2 in equation (iii); i.e., $\gamma = \gamma_1 + \gamma_2$.

As we proceed, it will be interesting to see how information about the integrand $f(z)$ or the contour γ corresponds to information concerning the integral $\int_\gamma f\,dz$. For example, if we know the arc length of γ and we know that $|f(z)|$ is bounded as z varies within the trajectory Γ, say $|f(z)| \le M$ for all z on Γ, then there is a corresponding limitation on the size of the integral. The exact estimate, given next, will be an important tool for studying line integrals.

Theorem 5.9 *Suppose that* γ *is a contour in the complex plane and that* $f(z)$ *is a continuous function defined on the trajectory* Γ, *such that* $|f(z)| \le M$ *for all points* z *on this trajectory. Then*

$$\left|\int_\gamma f\,dz\right| \le M \cdot \text{Length}(\gamma). \tag{16}$$

PROOF: Formula (13) for Riemann integrals gives

$$\left|\int_\gamma f\,dz\right| = \left|\int_a^b f(\gamma(t)) \cdot \gamma'(t)\,dt\right| \le \int_a^b |f(\gamma(t))| \cdot |\gamma'(t)|\,dt.$$

But for every t in $[a, b]$, the point $z = \gamma(t)$ is on the trajectory Γ, so that $|f(\gamma(t))| \leq M$. Therefore,

$$|f(\gamma(t))| \cdot |\gamma'(t)| \leq M \cdot |\gamma'(t)| \quad \text{for all } t \text{ in } [a, b].$$

Now apply the basic order inequality (12) to get

$$\left| \int_\gamma f\, dz \right| \leq \int_a^b M \cdot |\gamma'(t)|\, dt = M \cdot \int_a^b |\gamma'(t)|\, dt = M \cdot \text{Length}(\gamma). \quad \blacksquare$$

By using this estimate we can demonstrate a useful fact. If the contour γ is fixed, the integral $\int_\gamma f\, dz$ varies "continuously" as we change the integrand.

Theorem 5.10 *Let $f_1, f_2, \ldots$ be a sequence of continuous functions defined on the trajectory Γ of a contour γ, and assume that they converge uniformly on Γ to some continuous function $f(z)$:*

$$f = \lim_{n\to\infty} f_n \qquad (\textit{uniformly on } \Gamma).$$

Then

$$\int_\gamma f_n\, dz \to \int_\gamma f\, dz \quad \textit{as} \quad n \to \infty. \tag{17}$$

Note: Uniform convergence of a sequence $\{f_n\}$ of functions to a limit function f on a set E was discussed in Sections 3.1 and 3.2. The reader should take a moment to recall what this means. As we mentioned above, the behavior of the integrands $f_n(z)$ and $f(z)$ on Γ alone determines the values of these integrals, and this is why we only need to know that f_n converges uniformly to f *on the set* Γ, to get (17).

Proof: By definition of uniform convergence, if $\varepsilon > 0$ is given we get

$$|f_n(z) - f(z)| \leq \varepsilon \quad \text{for all } z \text{ on } \Gamma \text{ (i.e., uniformly on } \Gamma)$$

for all large n, say $n \geq N(\varepsilon)$. Since line integrals are linear operations, we get

$$\left| \int_\gamma f_n\, dz - \int_\gamma f\, dz \right| = \left| \int_\gamma (f_n - f)\, dz \right| \leq \varepsilon \cdot \text{Length}(\gamma),$$

for all $n \geq N(\varepsilon)$. This works for any choice of $\varepsilon > 0$, which is exactly what we mean when we say that

$$\int_\gamma f_n\, dz \to \int_\gamma f\, dz \quad \text{as} \quad n \to \infty. \quad \blacksquare$$

The above result shows that a *uniform* limit may be interchanged with the operation $\int_\gamma (\cdots)\, dz$; that is,

$$\lim_{n\to\infty} \left(\int_\gamma f_n\, dz \right) = \int_\gamma \left(\lim_{n\to\infty} f_n \right) dz = \int_\gamma f\, dz$$

if f_n converges uniformly to f on the trajectory (recall discussion of Section 3.1).

Another important fact about line integrals is their invariance when we hold the integrand fixed and make an order preserving change of variable in the contour.

Theorem 5.11 *Let $\gamma(t)$ and $\eta(s)$ be contours in the plane that are defined on $I = [a, b]$ and $J = [c, d]$ respectively. Assume that η is a reparametrization of γ:*

$$\eta(s) = (\gamma \circ \phi)(s) = \gamma(\phi(s)) \quad \text{for} \quad c \leq s \leq d$$

via the change of variable $t = \phi(s)$. Then γ and η have the same trajectory Γ, and if $f(z)$ is any continuous function defined on Γ we get

(i) $\displaystyle\int_\eta f\,dz = \int_\gamma f\,dz$ *if ϕ is order preserving*

(ii) $\displaystyle\int_\eta f\,dz = (-1)\cdot\int_\gamma f\,dz$ *if ϕ is order reversing,*

PROOF: First write $\int_\eta f\,dz$ as a Riemann integral:

$$\int_\eta f\,dz = \int_c^d f(\eta(s))\cdot\eta'(s)\,ds = \int_c^d (f\circ\gamma)(\phi(s))\cdot(\gamma\circ\phi)'(s)\,ds$$

$$= \int_c^d (f\circ\gamma)(\phi(s))\cdot\gamma'(\phi(s))\cdot\frac{d\phi}{ds}(s)\,ds;$$

here we use the differentiation rule (5) to calculate $(\gamma\circ\phi)'(s)$. If ϕ is order reversing, then ϕ reverses end points in mapping J to I, so that $b = \phi(c)$ and $a = \phi(d)$, and the integral above has the form

$$\int_\eta f\,dz = \int_{b=\phi(c)}^{a=\phi(d)} (f\circ\gamma)(t)\cdot\gamma'(t)\,dt$$

$$= \int_b^a f(\gamma(t))\cdot\gamma'(t)\,dt$$

$$= (-1)\int_\gamma f(z)\,dz.$$

In the case when ϕ is order preserving, we get $a = \phi(c)$ and $b = \phi(d)$, and there is no need to reverse the order of endpoints in these integrals; thus,

$$\int_\eta f\,dz = \int_\gamma f\,dz. \quad ▮$$

Example 5.7 Consider the contours

$$\gamma(t) = \begin{cases} e^{it} = \cos t + i\sin t & 0 \leq t \leq 2\pi \\ e^{2\pi it} = \cos(2\pi t) + i\sin(2\pi t) & 0 \leq t \leq 1 \\ e^{it^2} = \cos(t^2) + i\sin(t^2) & 0 \leq t \leq \sqrt{2\pi} \end{cases}$$

which traverse the unit circle $|z| = 1$. They are related by order preserving reparametrizations, and we have already calculated $\int_\gamma 1/z\, dz = 2\pi i$ for the first contour in this list. Theorem 5.11 assures us that the other contours must give the same value. If we integrate $f(z) = 1/z$ along

$$\eta(t) = e^{i(2\pi - t)} = \gamma(2\pi - t) \qquad 0 \le t \le 2\pi,$$

which is an order reversing reparametrization of the first contour above, we get

$$\int_\eta \frac{1}{z}\, dz = -2\pi i.$$

The integrals of $f(z) = 1/z$ along each of the contours listed above could be calculated directly, but the advantages of an appeal to Theorem 5.11 should be obvious.

Example 5.8 Let $\gamma(t) = e^{it}$ for $0 \le t \le 2\pi$, and let $n = 0, \pm 1, \pm 2, \ldots$. Integrating $f(z) = z^n$ along γ, we get

$$\begin{aligned} \int_\gamma z^n\, dz &= \int_0^{2\pi} f(\gamma(t)) \cdot \gamma'(t)\, dt = \int_0^{2\pi} \gamma(t)^n \cdot \gamma'(t)\, dt \\ &= \int_0^{2\pi} e^{int} \cdot i \cdot e^{it}\, dt \\ &= i\int_0^{2\pi} e^{i(n+1)t}\, dt. \end{aligned}$$

If $k = \pm 1, \pm 2, \ldots$ we have

$$\int_0^{2\pi} e^{ikt}\, dt = \int_0^{2\pi} \cos(kt)\, dt + i\int_0^{2\pi} \sin(kt)\, dt = 0 + i0;$$

but if $k = 0$, this integral reduces to $\int_0^{2\pi} dt = 2\pi$. Therefore, letting $k = n + 1$, we get

$$\int_\gamma z^n\, dz = \begin{cases} 0 & \text{for } n \ne -1 \\ 2\pi i & \text{for } n = -1 \end{cases} \qquad (n \text{ an integer}).$$

The same conclusion holds for all curves η that are order preserving reparametrizations of γ.

Example 5.9 Consider the contours $\gamma(t)$ and $\eta(t)$ which connect $p = 0$ and $q = i$, as shown in Figure 5.3. These curves can be given definite parametrizations as follows:

$$\gamma(t) = x(t) + iy(t) = \begin{cases} 2t + i0 & \text{for } 0 \le t \le \frac{1}{2} \\ 1 + i(2t - 1) & \text{for } \frac{1}{2} \le t \le 1 \end{cases}$$

$$\eta(t) = u(t) + iv(t) = \begin{cases} 0 + i(2t) & \text{for } 0 \le t \le \frac{1}{2} \\ (2t - 1) + i & \text{for } \frac{1}{2} \le t \le 1. \end{cases}$$

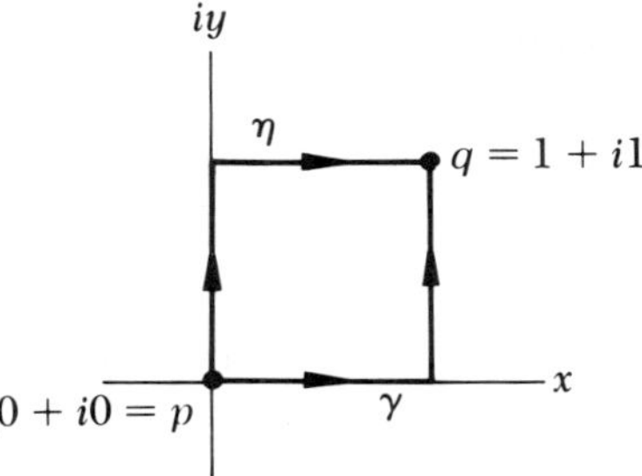

Figure 5.3 The contours of integration in the integral $\int_\gamma z^2\,dz$ of Example 5.9.

Let us integrate $f(z) = z^2$ along each of these contours. At the present stage of our discussion this must be done by direct computations based on formula (15).

$$\begin{aligned}\int_\gamma f(z)\,dz &= \int_0^{1/2} f(\gamma(t))\cdot\gamma'(t)\,dt + \int_{1/2}^1 f(\gamma(t))\cdot\gamma'(t)\,dt\\ &= \int_0^{1/2} 4t^2\cdot 2\,dt + \int_{1/2}^1 (1+i(2t-1))^2\cdot 2i\,dt\end{aligned}$$

The first integral here has value $\frac{1}{3}$; substituting $s = (1 + i(2t - 1))$ and $ds = 2i\,dt$ in the second, we can evaluate it as

$$\left[\int s^2\,ds\Big|_{t=1/2}^{t=1}\right] = \left[\frac{s^3}{3}\Big|_{t=1/2}^{t=1}\right] = \left[\frac{(1+i(2t-1))^3}{3}\Big|_{t=1/2}^{t=1}\right] = \frac{(1+i)^3}{3} - \frac{1}{3}.$$

Thus, $\int_\gamma z^2\,dz = (1+i)^3/3 = (\frac{1}{3})(q^3 - p^3)$. On the other hand,

$$\begin{aligned}\int_\eta f(z)\,dz &= \int_0^{1/2} f(\eta(t))\cdot\eta'(t)\,dt + \int_{1/2}^1 f(\eta(t))\cdot\eta'(t)\,dt\\ &= \int_0^{1/2} -4t^2\cdot(2i)\,dt + \int_{1/2}^1 ((2t-1)+i)^2\cdot 2\,dt.\end{aligned}$$

The first integral is just $-i/3$; in the second integral, we substitute $s = (2t - 1) + i$ and $ds = 2\,dt$, to get the value $(\frac{1}{3})((i+1)^3 - i^3) = (i+1)^3/3 + i/3$. Combining the values of these pieces, we get

$$\int_\eta z^2\,dz = \frac{(i+1)^3}{3} = \tfrac{1}{3}(q^3 - p^3) = \int_\gamma z^2\,dz.$$

The curves γ and η are quite different and are not reparametrizations of each other since they have different trajectories. Nevertheless, it turns out that the integrals have the same value; this is not an accident, as we shall explain in Section 5.6.

EXERCISES

1. Calculate $\int_\gamma z^2\,dz$ along the contour $\gamma(t) = (1 - t)i + t$, defined for $0 \le t \le 1$. What is the trajectory?

Answer: $(i + 1)/3$; trajectory is the segment from $+i$ to $+1$.

2. Calculate $\int_\gamma z^2\,dz$ along the semicircular contour $\gamma(t) = e^{i(\pi-2t)/2}$ from $0 \le t \le \pi/2$. Sketch the trajectory and the direction of motion of $\gamma(t)$.

Answer: $(i + 1)/3$.

3. Evaluate the following integrals.

(i) $\int_\gamma e^z\,dz$ along the line segment from -1 to $+1$

(ii) $\int_\gamma (e^z + z)\,dz$ along the line segment from 1 to $1 + i\pi$

(iii) $\int_\gamma |z|\,dz$ along the counter-clockwise circle $|z| = 2$

(iv) $\int_\gamma \frac{\text{Log } z}{z}\,dz$ along the semi-circular piece of $|z| = 1$ in the right half plane, moving from $-i$ to $+i$

(v) $\int_\gamma \text{Log } z\,dz$ along the contour in (iv).

Answer: (i) $e - e^{-1}$; (ii) $-2e + i\pi - \pi^2/2$; (iii) 0; (iv) $-\pi$; (v) $-2i$.

4. For the polygonal curves joining the points $\{p_1, \ldots, p_n\}$ indicated, calculate the following line integrals. Sketch the trajectory of γ in each case.

(i) $\int_\gamma 3z - 1\,dz$ for $\{-i, +i\}$

(ii) $\int_\gamma 3z - 1\,dz$ for $\{-i, +i, +1, -i\}$

(iii) $\int_\gamma x\,dz = \int_\gamma \text{Re}(z)\,dz$ for $\{-i, -i + 1, i + 1, +i\}$

(iv) $\int_\gamma |z|\,dz$ for $\{-i + 1, i + 1\}$.

Answer: (i) $6i$; (ii) 0; (iii) $+2$; (iv) $i\sqrt{2} + \frac{i}{2}\ln\left|\frac{1 + \sqrt{2}}{1 - \sqrt{2}}\right|$.

5. Verify the properties of line integrals in Theorem 5.8, using the definition (15) and the facts about Riemann integrals.

6. Show that $\int_\gamma z^n\,dz = 0$ for each integer n with $n \neq -1$, if γ is a circular contour about the origin of the form $\gamma(t) = Re^{it}$ for $0 \leq t \leq 2\pi$.

Note: For $n = -1$ see Example 5.6; also note that the answer does not depend on the particular radius $R > 0$ used.

7. If γ is the parametrized circular contour about $z = p$ given by $\gamma(t) = p + re^{it}$ for $0 \leq t \leq 2\pi$, show that

$$\text{(i)} \quad \int_\gamma \frac{1}{z-p}\,dz = 2\pi i$$

$$\text{(ii)} \quad \int_\gamma (z-p)^n\,dz = 0 \qquad \text{if } n \neq -1; \quad n \text{ an integer.}$$

8. If $\gamma(t) = a + t(b - a)$ for $0 \leq t \leq 1$, its trajectory is the segment $I = [a, b]$ in the real axis. Let $f(z)$ be any continuous function of a complex variable defined on this segment, and prove that $\int_\gamma f(z)\,dz$ is precisely the Riemann integral $\int_a^b f(x + i0)\,dx$. What happens if you take an order-reversed reparametrization, such as $\eta(s) = b - t(b - a)$?

9. Calculate the values $\int_\gamma z^n\,dz$ for $n = 0, \pm 1, \pm 2, \ldots$ along the semicircular contour $\gamma(t) = Re^{it}$ defined for $-\pi/2 \leq t \leq +\pi/2$. Do the values depend on the radius R?

Hint: Case $n = -1$ is exceptional.

Answer: $i\pi$ if $n = -1$; $2R^{n+1}i^{n+1}\left[\dfrac{1-(-1)^{n+1}}{n+1}\right]$ if $n \neq -1$.

10. Calculate $\int_\gamma \bar{z}\,dz$ and $\int_\gamma (\bar{z})^n\,dz$ where $\bar{z}$ is the complex conjugate and $\gamma = Re^{it}$, defined for $0 \leq t \leq 2\pi$.

Answer: (i) $2\pi i R^2$; (ii) 0 if $n \neq +1$.

11. Write out in full the Riemann integrals you would have to evaluate to calculate the value $\int_\gamma f(z)\,dz$, taking $\gamma(t) = Re^{it}$ for $0 \leq t \leq 2\pi$ and

(i) $f(z) = \sin z$

(ii) $f(z) = e^z$

(iii) $f(z) = z^2 + 1$

(iv) $f(z) = \dfrac{1}{z - \frac{1}{2}}$ (for $R > \frac{1}{2}$)

Do not evaluate the integrals, unless the definite integrals are particularly simple.

12. If $\gamma_R(t) = Re^{it}$ for $0 \leq t \leq \pi$ (a semi-circle), use Theorem 5.9 to prove:

$$\text{(i)} \quad \left| \int_{\gamma_R} \frac{1}{z^2+1}\, dz \right| \leq \frac{\pi R}{R^2 - 1} \qquad \text{for } R \geq 1$$

$$\text{(ii)} \quad \lim_{R\to\infty} \int_{\gamma_R} \frac{1}{z^2+1}\, dz = 0.$$

13. If γ_R is the parametrized circle $|z| = R$ with $\gamma_R(t) = Re^{it}$ for $0 \leq t \leq 2\pi$, prove that:

$$\text{(i)} \quad \lim_{R\to\infty} \left\{ \int_{\gamma_R} \frac{\text{Log}(z)}{z^2}\, dz \right\} = 0$$

$$\text{(ii)} \quad \lim_{R\to 0} \left\{ \int_{\gamma_R} z^{1/2}\, \text{Log } z\, dz \right\} = 0.$$

Note: The integrands have a simple discontinuity at $z = -R + i0$ but the line integral is still well defined.

Hint: Use estimates as in Theorem 5.9.

14. If a series of continuous functions $\sum_{n=0}^{\infty} f_n(z)$ converges uniformly on the trajectory of contour γ to a continuous function $f(z)$, prove that

$$\int_\gamma f(z)\, dz = \sum_{n=0}^{\infty} \left(\int_\gamma f_n(z)\, dz \right).$$

Hint: A direct consequence of definition and Theorem 5.10.

15. Use the series-expansion principle of Exercise 14 to evaluate $\int_\gamma f(z)\, dz$ for the contour $\gamma(t) = Re^{it}$ with $0 \leq t \leq 2\pi$.

(i) $f(z) = \sin z$

(ii) $f(z) = e^z$

(iii) $f(z)$ any polynomial in z

Answer: (i) 0; (ii) 0; (iii) 0.

16. Use the series-expansion principle of Exercise 14 to evaluate the following integrals.

(i) $\displaystyle\int_\gamma \frac{\sin z}{z}\, dz = 0$, where $\gamma(t) = e^{it}$ for $-\pi/2 \leq t \leq \pi/2$ (a semicircle).

(ii) $\displaystyle\int_\gamma \frac{1}{z - \frac{1}{2}}\, dz = 2\pi i$, where $\gamma(t) = e^{it}$ for $0 \leq t \leq 2\pi$.

17. (Deforming the contour) Let $\gamma(t)$ and $\gamma_1(t)$, $\gamma_2(t), \ldots$ be continuous curves defined on the standard interval $[0, 1]$.

Definition. We say that the curves γ_n **approach** γ **as a limit**, written $\gamma = \lim\limits_{n\to\infty} \gamma_n$, if the complex valued functions of real variable $\gamma_n(t)$ converge uniformly to $\gamma(t)$ for t in $[0, 1]$, and the derivatives $\gamma_n'(t)$ converge uniformly to the derivative $\gamma'(t)$ on $[0, 1]$.

Sketch the curves involved in the following situations and prove that $\gamma = \lim\limits_{n\to\infty} \gamma_n$, as defined above.

(i) $\gamma_n(t) = (1 - t)\left(\dfrac{1}{n}\right) + t\left(1 - \dfrac{1}{n}\right)$ and $\gamma(t) = t.$

(ii) $\gamma_n(t) = r_n e^{2\pi i t}$ and $\gamma(t) = re^{2\pi i t}$, where $\{r_n\}$ is a sequence of positive numbers with $r_n \to r$.

(iii) $\gamma_n(t) = e^{i[(1-t)(1/n)+t(2\pi-(1/n))]}$ and $\gamma(t) = e^{2\pi i t}$

(iv) $\gamma_n(t) = t + i\left(\dfrac{1}{n}\right)$ and $\gamma(t) = t$

(v) $\gamma_n(t) = t\left(1 - \dfrac{1}{n}\right) + it^2\left(1 - \dfrac{1}{n}\right)^2$ and $\gamma(t) = t + it^2$

Hint: For any pair of numbers α, β (real or complex) the combinations $(1 - t)\alpha + t\beta$ with $0 \leq t \leq 1$ lie on the line segment between α and β.

18. Prove the following important theorem, which shows that a contour integral $\int_\gamma f(z)\, dz$ varies continuously as the *contour* is varied (compare with discussion of Theorem 5.10).

Theorem: If contours γ_n approach a limit contour γ, so $\gamma = \lim\limits_{n\to\infty} \gamma_n$, then $\int_\gamma f(z)\, dz = \lim\limits_{n\to\infty} \int_{\gamma_n} f(z)\, dz$, for any continuous function $f(z)$ defined in a region that includes the trajectories of γ_n and γ.

19. Prove that initial and final points converge, $\gamma(0) = \lim\limits_{n\to\infty} \gamma_n(0)$ and $\gamma(1) = \lim\limits_{n\to\infty} \gamma(1)$, if the contours γ_n approach a limit γ.

20. Let $\gamma(t) = e^{it}$ for $0 \leq t \leq 2\pi$ and let $p = re^{i\phi}$ be a point lying inside the trajectory $|z| = 1$. Prove that

$$\int_\gamma \frac{1}{z - p}\, dz = 2\pi i$$

no matter where the singular point p is located, provided $|p| < 1$, starting from the definition (15) of line integrals.

Note: This leads to a difficult Riemann integral, except when $p = 0$. Other methods for calculating this integral appear in Section 5.6.

*5.4 CHANGING A LINE INTEGRAL BY TRANSFORMING THE CONTOUR

Let $w = \psi(z)$ be a holomorphic mapping between sets, $\psi: E \to F$, as indicated in Figure 5.4 (we do not require that ψ map E *onto* F). This mapping transforms each contour γ in E to a contour $\eta = \psi \circ \gamma$ in F, as shown. On the other hand, each continuous function $f(w)$ defined on F is transformed (shifted in the reverse direction) to give a corresponding continuous function $(f \circ \psi)(z) = f(\psi(z))$ defined on E. It should be no surprise that there is a simple formula relating the integral $\int_{\psi\circ\gamma} f(w)\, dw$ of $f(w)$ along the transformed contour $w = \psi(\gamma(t))$ to the integral of the transformed function along the original contour $z = \gamma(t)$. The relation looks very much like the change of variable formula for integrals; notice in particular that the two integrals just described are *not equal*—we must introduce a compensating factor $d\psi/dz$ into the second integral to get the values to agree:

$$\int_\gamma f(\psi(z))\, \frac{d\psi}{dz}\, dz = \int_{\eta=\psi\circ\gamma} f(w)\, dw. \tag{18}$$

The proof is easy; by definition,

$$\int_\gamma f(\psi(z))\, \frac{d\psi}{dz}\, dz = \int_a^b f(\psi(\gamma(t))) \cdot \frac{d\psi}{dz}(\gamma(t)) \cdot \frac{d\gamma}{dt}\, dt$$

$$\int_{\psi\circ\gamma} f(w)\, dw = \int_a^b f(\psi(\gamma(t)))\, \frac{d}{dt}(\psi\circ\gamma)\, dt.$$

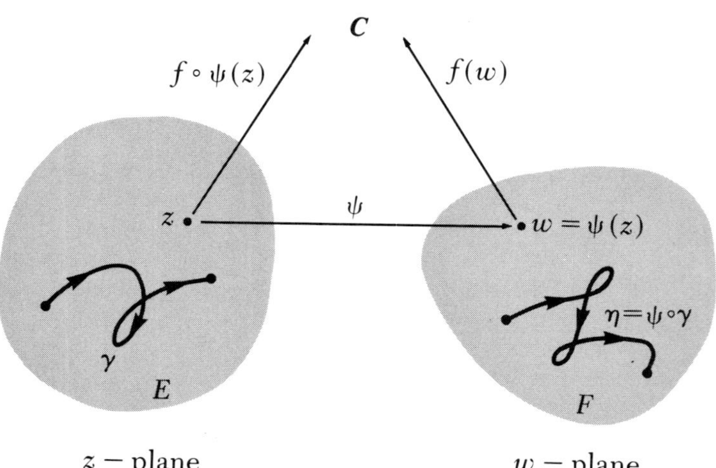

Figure 5.4 Transforming contours and functions via a mapping $w = \psi(z)$.

The derivative of the transformed contour $\eta = \psi \circ \gamma$ is just

$$\frac{d}{dt}(\psi \circ \gamma) = \frac{d\psi}{dz}(\gamma(t)) \cdot \frac{d\gamma}{dt}(t) \qquad \text{for } a \leq t \leq b,$$

and when this chain rule formula is substituted into the second integral we see that the integrands are the same, proving (18).

The reader should consider the distinction between *reparametrizing* a contour and *transforming* it. In the first case we are making a substitution $t = \phi(s)$ for the *variable* in the contour $z = \gamma(t)$, but when we transform a contour by some mapping of the plane $w = \psi(z)$ we leave the variable t alone and act on the *values* to get the new contour $\eta(t) = \psi(\gamma(t)) = (\psi \circ \gamma)(t)$. Line integrals are unaffected by a reparametrization (except that they may be multiplied by -1) but can be changed substantially by a transformation of contour and integrand.

EXERCISES

1. Use $w = \psi(z) = 1/z$ to transform the following contours γ. Sketch the trajectories of $\gamma(t)$ and $\eta(t) = \psi(\gamma(t)) = 1/\gamma(t)$.

(i) $\gamma(t) = Re^{2\pi i t}$ $\quad 0 \leq t \leq 2\pi \quad$ (with $R > 0$)

(ii) $\gamma(t) = \frac{1}{2} + t$ $\quad 0 \leq t \leq 1$

(iii) $\gamma(t) = 1 + i(1 - t)$ $\quad 0 \leq t \leq 1.$

Hint: The fractional linear transformation $w = 1/z$ maps γ in (iii) to a circular arc (part of which circle in the plane?). Recall Section 4.8.

2. Consider the parametrized line segment connecting $p = -i\pi/2$ to $q = i\pi/2$ given by $\gamma(t) = (1 - t)p + tq$ for $0 \leq t \leq 1$. If γ is transformed by the mapping $w = \psi(z) = e^{2z}$, write out the components of the transformed contour $\eta(t) = [\psi(z)\,|_{z=\gamma(t)}] = \psi(\gamma(t))$. Sketch the trajectories of $\gamma(t)$ and $\eta(t)$, indicating the direction of motion. Compare the following aspects of γ and η.

(i) Are the curves closed (simple)?
(ii) Are arc lengths the same?

3. If γ is a contour and $w = \psi(z)$ is a mapping defined and holomorphic on the trajectory, prove that the arc length of the transformed contour $\eta(t) = \psi(\gamma(t))$ is given by

$$L(\eta) = \int_a^b \left| \frac{d\psi}{dt}(\gamma(t)) \right| \left| \frac{d\gamma}{dt} \right| dt.$$

Give an example of a transformation that changes the length of the curve γ.

4. If $w = \psi(z)$ is a mapping defined on a set E in the complex plane it is customary to refer to the set $\psi(E) = \{w : w = \psi(z)$ for at least one z in $E\}$ as the *image* of E under the mapping. If $w = \psi(z)$ is defined on the trajectory Γ of a contour $\gamma(t)$, show that the transformed contour $\eta(t) = \psi(\gamma(t))$ has the trajectory $\Gamma' = \psi(\Gamma)$.

5.5 THE NOTATIONS $\gamma_1 + \cdots + \gamma_n$ AND $-\gamma$ FOR CONTOURS

If we start with a single contour γ defined on an interval I and then break I into successive subintervals $I_1, \ldots, I_m$, we obtain several new contours $\gamma_1, \ldots, \gamma_m$ by restricting the variable t to the different intervals. The new curve γ_k starts from the point where the previous curve γ_{k-1} ended, and we get

$$\int_\gamma f\,dz = \sum_{k=1}^{m} \left(\int_{\gamma_k} f\,dz \right)$$

by applying Theorem 5.8 several times. Often we wish to go the other way: let us start with contours $\gamma_k(t)$ defined on intervals $I_k = [a_k, b_k]$, and let us assume that the successive initial and final points of these contours match up, in the sense that

(*) the initial point of γ_k is the final point of γ_{k-1} for $2 \leq k \leq m$.

We would like to traverse the contours $\gamma_1, \ldots, \gamma_m$ in succession to obtain a single piecewise smooth curve γ, as shown in Figure 5.5. But the intervals $I_1, \ldots, I_m$ might be unrelated, so that they do not fit together end to end (they may even overlap) to give us a single interval on which we can define the new curve γ. However, since we are primarily interested in line integrals, which are not affected by reparametrizations, we can easily reparametrize $\gamma_1, \ldots, \gamma_m$ via order preserving reparametrizations to get

$$I_1 = [0, 1], \qquad I_1 = [1, 2], \ldots, I_m = [m - 1, m];$$

the reparametrized contours fit together to give a single curve defined on the interval $I = [0, m]$. There are other ways of doing this reparametrization, but in every case we will get

$$\int_\gamma f\,dz = \sum_{k=1}^{m} \left(\int_{\gamma_k} f\,dz \right), \tag{19}$$

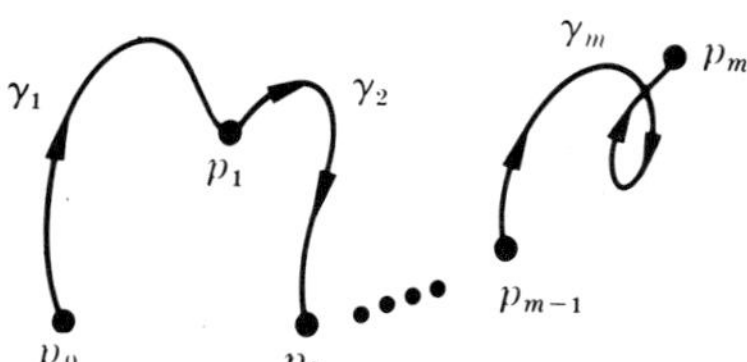

Figure 5.5 Forming the contour $\gamma = \gamma_1 + \cdots + \gamma_m$ from curves $\gamma_1, \ldots, \gamma_m$ that satisfy (*).

so these differences will be irrelevant in computing line integrals. We often write $\gamma = \gamma_1 + \cdots + \gamma_m$ for the contour gotten by traversing the contours $\gamma_1, \ldots, \gamma_m$ in succession; only the requirement $(*)$ must be satisfied for this to make sense.

Similarly, when we are dealing with line integrals we often want to consider arbitrary order reversing reparametrizations η_1, η_2 of a given contour γ. Theorem 5.11 insures that

$$\int_{\eta_1} f\,dz = \int_{\eta_2} f\,dz = (-1)\int_{\gamma} f\,dz,$$

so there is really little reason to write down an explicit reparametrized form of $\gamma(t)$; instead, we use the symbol $-\gamma$ to indicate the contour γ taken with any order reversing reparametrization. This, of course, leaves uncertain which change of variable is meant, but in any case we get

$$\int_{-\gamma} f\,dz = (-1)\cdot\int_{\gamma} f\,dz. \tag{20}$$

Notice that $-\gamma$ starts where γ ended, traverses the trajectory Γ in a direction opposite that of γ, and ends where γ started.

EXERCISES

1. Parametrize the line segments $[0, 1]$ and $[1, 1 + i]$ via

$$\gamma_1(t) = t \qquad 0 \le t \le 1$$

$$\gamma_2(t) = (1 - t) + it \qquad 0 \le t \le 1$$

respectively. How would you interpret the contour $\gamma_1 + \gamma_2$? Does the symbol $\gamma_2 + \gamma_1$ represent a contour? What is the relationship between the contour $\gamma_1 + \gamma_2$ and the contour γ in Example 5.9? What is the value of $\int_{\gamma_1+\gamma_2} z^2\,dz$?

Answer: $\frac{2}{3}(i - 1)$.

2. Consider two parametrizations of the circle $|z - p| = R$,

$$\gamma_1(\theta) = p + Re^{i\theta} \qquad 0 \le \theta \le 2\pi$$

$$\gamma_2(\theta) = p + Re^{i\theta} \qquad \alpha \le \theta \le \alpha + 2\pi \qquad (\alpha \text{ a fixed real number})$$

Show that $\int_{\gamma_1} f(z)\,dz = \int_{\gamma_2} f(z)\,dz$ for any function $f(z)$ defined and continuous on $|z - p| = R$.

Hint: Break both contours into two comparable arcs.

5.6 PATH INDEPENDENCE OF LINE INTEGRALS

In Figure 5.6, E is a fixed domain (open connected set) in the complex plane and z_1 and z_2 are two points in E. Since E is connected, there are various

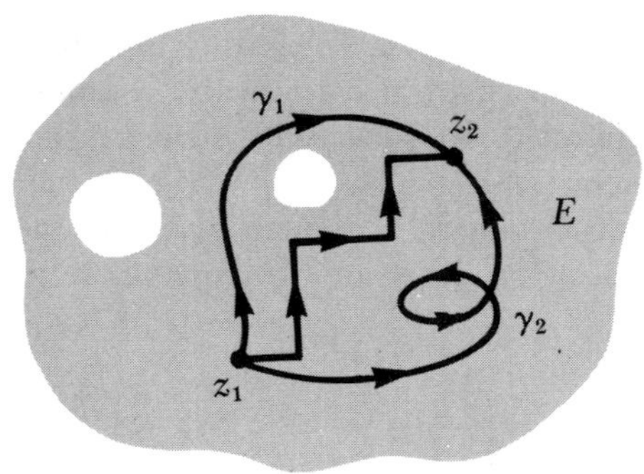

Figure 5.6 Various contours in a domain E which connect z_1 to z_2 within E.

ways we can move from z_1 to z_2 along contours which lie entirely within E; for example, we could always connect z_1 to z_2 via a "polygonal curve" such as the one shown in Figure 5.6.

QUESTION: For a fixed integrand $f(z)$ that is continuous on E, under what circumstances is $\int_\gamma f\,dz$ independent of the path from z_1 to z_2 along which we integrate?

For example, this occurs when the integrand $f(z)$ has an antiderivative $F(z)$ that is defined throughout E; then $dF/dz = f(z)$ for all z in E and if $\gamma(t)$, defined for $a \leq t \leq b$, is any contour lying within E such that $\gamma(a) = z_1$ and $\gamma(b) = z_2$, we get

$$\int_\gamma f\,dz = \int_a^b f(\gamma(t)) \cdot \gamma'(t)\,dt.$$

Since $dF/dz = f(z)$ everywhere, the composite function $(F \circ \gamma)(t)$ is differentiable,

$$\frac{d}{dt}(F \circ \gamma)(t) = \frac{dF}{dz}(\gamma(t)) \cdot \frac{d\gamma}{dt}(t) \qquad \text{for } a \leq t \leq b.$$

Thus, our integral has the form

$$\int_\gamma f\,dz = \int_a^b \frac{d}{dt}(F \circ \gamma)\,dt = F(\gamma(b)) - F(\gamma(a)) = F(z_2) - F(z_1),$$

by the Fundamental Theorem of Calculus. This reasoning works without reference to the form of the contour that connects z_1 to z_2, and also tells us how to evaluate the line integral if an antiderivative is known.

Theorem 5.12 *If $f(z)$ is a continuous function defined on a domain E in the complex plane, and if $f(z)$ has an antiderivative $F(z)$ defined throughout E, so that*

$$\frac{dF}{dz} = f(z) \qquad \textit{for all } z \textit{ in } E,$$

then

$$\int_\gamma f\,dz = F(z_2) - F(z_1) \tag{21}$$

for any contour in E that connects z_1 to z_2. Hence, if γ_1 and γ_2 are two such contours, it follows that $\int_{\gamma_1} f\,dz = \int_{\gamma_2} f\,dz$.

Example 5.10 On any domain, $f(z) = z^2$ has a well defined antiderivative; we could take $F(z) = (\frac{1}{3})z^3$ (other antiderivatives differ by an added constant which cancels in formula (21)). If z_1 and z_2 are two points in E and if γ is a contour from z_1 to z_2, we get

$$\int_\gamma z^2\,dz = \left[\tfrac{1}{3}z^3\Big|_{z=z_1}^{z=z_2}\right] = \tfrac{1}{3}(z_2^{\,3} - z_1^{\,3}).$$

This explains the result obtained by tiresome direct calculations in Example 5.9.

The path independence property will be of very great importance when we have to calculate line integrals in practical applications. The following definition of terms will be convenient.

Definition 5.3 *Let $f(z)$ be continuous on a domain E in the complex plane. Line integrals of f are* **globally path independent** *in E if*

$$\int_{\gamma_1} f\,dz = \int_{\gamma_2} f\,dz$$

for any pair of contours γ_1 and γ_2 in E that have the same initial point and the same final point.

If line integrals of $f(z)$ are globally path independent on E, the following result shows how we may construct an antiderivative $F(z)$ from the original function $f(z)$ almost exactly as we construct the antiderivative

$$F(t) = \int_a^t f\,dx \qquad \text{for } a \le t \le b$$

of a continuous function of a real variable $f(t)$ in calculus.

Theorem 5.13 *Let $f(z)$ be continuous on a domain E in the complex plane and assume that line integrals of $f(z)$ are globally path independent in E. Fix some point p in E. For any other point z in E let us define*

$$F(z) = \int_{\gamma(p,z)} f\,dz \tag{22}$$

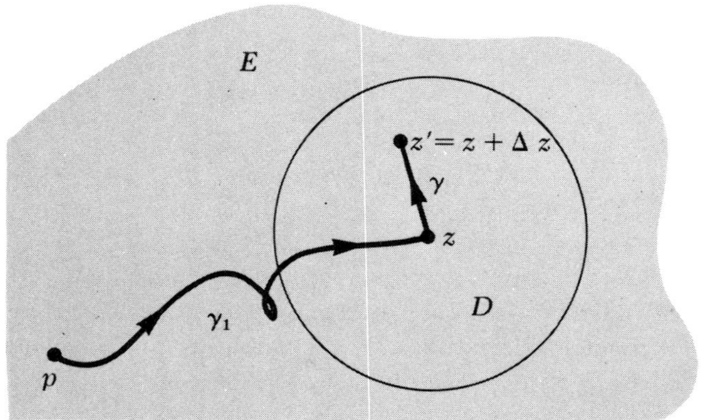

Figure 5.7 The contours used in proving Theorem 5.13.

where $\gamma(p, z)$ is any contour in E that starts at p and ends at z. Then

(i) *$F(z)$ is holomorphic (hence continuous) on E, and*

(ii) *$\dfrac{dF}{dz}(z) = f(z)$ for all z in E.*

Proof: Due to the path independence of line integrals, the right side of (22) gives us the same complex number $F(z)$ no matter how we take the contour $\gamma(p, z)$ connecting p with z. We will show that dF/dz exists and is equal to $f(z)$ by making a frontal assault on the problem, examining difference quotients and their limits. Let D be a small disc about a typical point z in E; we may choose D so that it lies within E, as shown in Figure 5.7. If $z' = z + \Delta z$ is any point in D with $z' \neq z$, let us examine the difference quotient

$$\frac{\Delta F}{\Delta z} = \frac{F(z') - F(z)}{z' - z} = \frac{1}{\Delta z}\left[\int_{\gamma(p,z')} f\,dz - \int_{\gamma(p,z)} f\,dz\right]$$

as $z' \to z$. We may take the contours $\gamma_1 = \gamma(p, z)$ and $\gamma_2 = \gamma(p, z')$ in any way we wish as we examine the right side of this expression. Therefore, let us take γ_1 to be any contour from p to z in E, and let γ be the parametrized line segment from z to z'

$$\gamma(t) = tz' + (1 - t)z = z + t\,\Delta z \quad \text{for} \quad 0 \leq t \leq 1.$$

Obviously, $\gamma'(t) = \Delta z$ (a constant) for $0 \leq t \leq 1$. We may take γ_2 to be the contour obtained by traversing γ_1 and then γ in succession, $\gamma_2 = \gamma_1 + \gamma$, as shown in Figure 5.7. Now

$$\int_{\gamma_2} f\,dz = \int_{\gamma_1+\gamma} f\,dz = \int_{\gamma_1} f\,dz + \int_{\gamma} f\,dz$$

and

$$\frac{\Delta F}{\Delta z} = \frac{1}{\Delta z}\left[\int_{\gamma_2} f\,dz - \int_{\gamma_1} f\,dz\right] = \frac{1}{\Delta z}\int_{\gamma} f\,dz$$

$$= \frac{1}{\Delta z}\int_0^1 f(\gamma(t)) \cdot \Delta z\,dt = \int_0^1 f(\gamma(t))\,dt.$$

We are left with the technical problem of showing that this integral approaches the value $f(z)$ as $z' \to z$. Now f is continuous at z, so $f(z) = \lim_{w \to z} f$ and all values $f(w)$ are close to $f(z)$ if w is sufficiently close to z; i.e., we can make the difference in values smaller than a prescribed number $\varepsilon > 0$ by restricting w to lie within a disc of sufficiently small radius $r = \delta(\varepsilon)$ about z. Thus,

$$|f(w) - f(z)| < \varepsilon \quad \text{provided that} \quad |w - z| < \delta(\varepsilon).$$

If z' is in this disc, so $|\Delta z| = |z' - z| < \delta(\varepsilon)$, the points $w = \gamma(t)$ on the segment connecting z' to z lie even closer to z; therefore, $|\gamma(t) - z| < \delta(e)$ and

$$\text{(23)} \qquad |f(\gamma(t)) - f(z)| < \varepsilon \quad \text{for} \quad 0 \le t \le 1.$$

Next, notice that $f(z) = f(z) \cdot 1 = f(z) \cdot \int_0^1 dt = \int_0^1 f(z)\,dt$ (here $f(z)$ is a complex *constant*; the variable of integration is t). Thus, for any z' such that $|z' - z| = |\Delta z| < \delta(\varepsilon)$, we get

$$\left|\frac{\Delta F}{\Delta z} - f(z)\right| = \left|\int_0^1 f(\gamma(t))\,dt - f(z)\right|$$

$$= \left|\int_0^1 [f(\gamma(t)) - f(z)]\,dt\right|$$

$$\le \int_0^1 |f(\gamma(t)) - f(z)|\,dt < \int_0^1 \varepsilon\,dt = \varepsilon$$

(the last inequality follows from (12)). Since this argument may be carried out for any choice of $\varepsilon > 0$, we have shown that

$$\frac{dF}{dz}(z) = \lim_{z' \to z}\frac{\Delta F}{\Delta z} \quad \text{exists and is equal to } f(z)$$

for any point z in E. ∎

The last two theorems may be combined to give:

Corollary 5.14 *If $f(z)$ is continuous on a domain E, then line integrals of $f(z)$ are globally path independent in E if and only if $f(z)$ has an antiderivative $F(z)$ defined throughout E.*

In a set E where $f(z)$ has path independent line integrals, it is useful to employ the notation

$$\int_p^z f(w)\,dw$$

to indicate integration, along any contour in E, from p to z. If p is held fixed, this function of z is just the antiderivative of f produced in Theorem 5.13. For example, $\int_0^z \cos w\,dw = \sin z$, and there are many other formulas like this, reminiscent of the indefinite integrals one learns in calculus. Here is another useful criterion for global path independence.

Theorem 5.15 *Let $f(z)$ be continuous on a domain E in the complex plane. Then line integrals of $f(z)$ are globally path independent in E if and only if*

$$\int_\gamma f\,dz = 0 \tag{24}$$

for every closed contour γ that lies within E.

PROOF: If line integrals are globally path independent in E, there is an antiderivative $F(z)$ defined on E and we get $\int_\gamma f\,dz = F(\gamma(b)) - F(\gamma(a)) = F(p) - F(p) = 0$ if γ is a closed contour and p is its initial/final point.

Conversely, if (24) holds for *all closed contours* in E, let us consider two (not necessarily closed) contours γ_1 and γ_2 in E which start at a common point p and end at some other common point q; we want to show that

$$\int_{\gamma_1} f\,dz = \int_{\gamma_2} f\,dz.$$

Let us write $-\gamma_2$ for any order reversed reparametrization of γ_2, so $-\gamma_2$ starts at $z = q$ and ends at $z = p$. We may now move along the contours γ_1 and $-\gamma_2$ in succession to get a *closed* contour based at $z = p$, $\gamma = \gamma_1 + (-\gamma_2)$, as shown in Figure 5.8. For this contour we get

$$0 = \int_\gamma f\,dz = \int_{\gamma_1} f\,dz + \int_{-\gamma_2} f\,dz = \int_{\gamma_1} f\,dz - \int_{\gamma_2} f\,dz,$$

which is what we wanted to prove. ∎

Path independence of line integrals depends both on the nature of the integrand $f(z)$, and on the geometric features of the set E. The role played by the "shape" of E will be considered in more detail later on. When line integrals of $f(z)$ fail to be globally path independent, it quite often happens that path independence appears if we restrict our contours to lie within certain subsets of E. This often happens when E has a "hole" in it, and curves that loop around this hole must be excluded if we are to have path independence of line integrals. For this reason we introduce the notion of local (as opposed to global) path independence of line integrals.

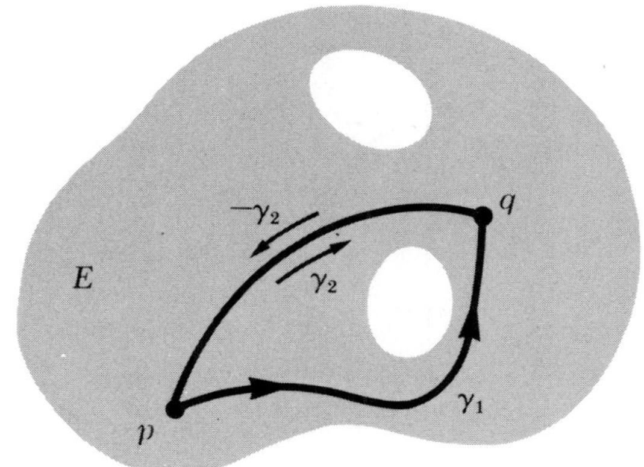

Figure 5.8 Constructing the closed contour $\gamma = \gamma_1 + (-\gamma_2)$ in Theorem 5.15.

Definition 5.4 *Let $f(z)$ be continuous on a domain E. We say that line integrals of $f(z)$ are* **locally path independent** *in E if, for every point p in E, we can find a corresponding disc D_p about p such that*

(i) *The disc D_p lies within E*
(ii) *Line integrals of $f(z)$ are path independent for all contours that lie within D_p.*

Of course, the size of the disc D_p may vary as p is changed. By Theorem 5.14, (ii) is equivalent to the requirement

(ii*) *The function $f(z)$ has an antiderivative defined throughout the disc D_p.*

In other words, local path independence of line integrals for $f(z)$ means that we are able to devise antiderivatives for $f(z)$ near any point p in E, although there may not be an antiderivative defined throughout E. Since the validity of (ii*) is completely decided by examining the behavior of $f(z)$ near individual points p in E, it follows that the property of local path independence depends only on the function $f(z)$, and not at all on the shape of the domain E.

Example 5.11 If $f(z)$ is an *analytic* function defined on a domain E, then line integrals of $f(z)$ are locally path independent. To see this, consider a typical point p in E; then $f(z)$ has a convergent power series expansion in some disc D_p about p. An antiderivative $F(z)$ can be defined on D_p by formally antidifferentiating this series term-by-term, as explained in Section 3.3. Thus, condition (ii*) is verified.

Line integrals of analytic functions need not be globally path independent.

Example 5.12 Let E_1 be the open half plane $E_1 = \{z : \text{Re}(z) > 0\}$ and let $E_2 = \{z : z \neq 0\}$, the punctured plane. These are both connected open sets (domains), and any two points in them may be connected by piecewise smooth curves lying within the domain. Notice that E_2 has a "hole" in it, while E_1 has no such feature. Now $f(z) = 1/z$ is continuous (and even analytic) throughout both domains, although it behaves badly as z approaches the boundary point

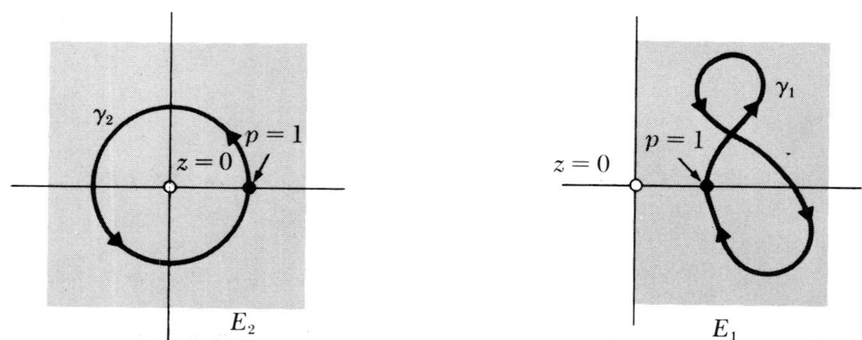

Figure 5.9 The domains in Example 5.12.

$z = 0$. On E_1 take $F(z) = \text{Log } z$ (principal determination); then

$$\frac{dF}{dz}(z) = \frac{1}{z} = f(z) \quad \text{for all } z \text{ in } E_1,$$

so that line integrals of $f(z)$ along contours in E_1 are *globally* path independent. If γ_1 is a closed contour in E_1, such as the one shown in Figure 5.9, we get

$$\int_{\gamma_1} f\,dz = \int_{\gamma_1} \frac{1}{z}\,dz = 0.$$

In E_2 things go rather differently; the prospective antiderivative $\log(z)$ cannot be defined as a continuous (let alone holomorphic) function on E_2—a locus of discontinuities always appears when we attempt to define a determination of $\log z$ on a set like E_2—and we suspect that global path independence may fail. Local path independence is guaranteed by the fact that $1/z$ is analytic on E_2. To confirm our suspicions about global path independence, let us integrate $f(z)$ along the closed contour

$$\gamma_2(t) = e^{2\pi i t} \quad \text{defined for } 0 \leq t \leq 1,$$

shown in Figure 5.9. This integral was evaluated earlier, in Example 5.8:

$$\int_{\gamma_2} \frac{1}{z}\,dz = 2\pi i \neq 0.$$

This result conflicts with formula (24), so the global path independence property cannot hold for $f(z) = 1/z$ on the domain E_2. This failure of global path independence is caused by the relation between f and the special geometric features of E_2 (namely, the hole where $z = 0$ was removed; $z = 0$ must be deleted since $f(z)$ is not analytic at the origin).

Next we apply these ideas to evaluate an important family of contour integrals that are quite difficult to handle by direct methods.

Example 5.13 Consider a circle $\Gamma = \{z\colon |z - p| = r\}$, parametrized counterclockwise in the usual way: $\gamma(t) = p + re^{it}$ for $0 \le t \le 2\pi$. Now let ζ be a point in the open disc $D = \{z\colon |z - p| < r\}$, *not* necessarily the center point as in Example 5.8. The function $f(z) = 1/(z - \zeta)$ is holomorphic, except at $z = \zeta$, which lies off the trajectory Γ. We assert that

$$\int_{\gamma} \frac{1}{z - \zeta}\, dz = 2\pi i \tag{25}$$

no matter where ζ is located within the disc D.

Unfortunately there is no antiderivative $F(z)$ defined throughout the punctured plane $E = \{z : z \neq \zeta\}$ where $f(z)$ is defined (if there were, the integral would have the value $0 = F(\gamma(2\pi)) - F(\gamma(0))$ since γ is a *closed* contour). But there are antiderivatives defined on rather substantial subsets of E, all of the form $\log(z - \zeta)$. Let us draw a vertical line L through ζ. The initial/final point q and the points z_1 and z_2 where L meets Γ divide γ into three successive circular contours $\gamma = \gamma_1' + \gamma_2 + \gamma_1''$ in an obvious way, as shown in Figure 5.10, and

$$\int_{\gamma} \frac{1}{z - \zeta}\, dz = \int_{\gamma_1'} \frac{1}{z - \zeta}\, dz + \int_{\gamma_2} \frac{1}{z - \zeta}\, dz + \int_{\gamma_1''} \frac{1}{z - \zeta}\, dz.$$

We will be able to evaluate the separate integrals on the right using formula (21).

To do this we set up the domains E_1 and E_2 shown in Figure 5.10. Now $F_1(z) = \text{Log}(z - \zeta)$ is differentiable on the domain we get by removing the horizontal ray J' that extends to the left of ζ; furthermore,

$$\frac{dF_1}{dz} = \frac{1}{z - \zeta} \quad \text{all } z \text{ in } E_1.$$

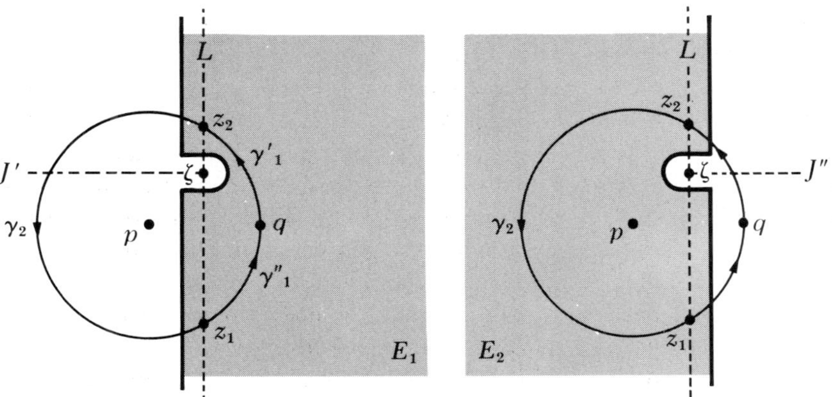

Figure 5.10 The domains E_1 and E_2 in Example 5.13.

Therefore, if we write $q = p + (r + i0)$ for the initial/final point of γ, we get

$$\int_{\gamma_1'} \frac{1}{z-\zeta}\,dz + \int_{\gamma_1''} \frac{1}{z-\zeta}\,dz = F_1(z_2) - F_1(q) + F_1(q) - F_1(z_1)$$
$$= \mathrm{Log}(z_2 - \zeta) - \mathrm{Log}(z_1 - \zeta).$$

On E_2 it is also possible to set up a differentiable determination of $\log(z - \zeta)$; just take a determination $\mathrm{Log}^*(w)$ which has its discontinuities along the *positive* real axis in the w-plane:

$$\mathrm{Log}^*(w) = \log|w| + i\,\mathrm{Arg}^*(w)$$

where $\theta = \mathrm{Arg}^*(w)$ is normalized so that $0 < \theta < 2\pi$. Then define $F_2(z) = \mathrm{Log}^*(z - \zeta)$, substituting $w = z - \zeta$. This gives us a function with $F_2'(z) = 1/(z - \zeta)$ on E_2, so that

$$\int_{\gamma_2} \frac{1}{z-\zeta}\,dz = F_2(z_1) - F_2(z_2) = \mathrm{Log}^*(z_1 - \zeta) - \mathrm{Log}^*(z_2 - \zeta).$$

Therefore,

$$\int_{\gamma} \frac{1}{z-\zeta}\,dz = [\mathrm{Log}(z_2 - \zeta) - \mathrm{Log}^*(z_2 - \zeta)]$$
$$+ [\mathrm{Log}^*(z_1 - \zeta) - \mathrm{Log}(z_1 - \zeta)]$$
$$= i[\mathrm{Arg}(z_2 - \zeta) - \mathrm{Arg}^*(z_2 - \zeta)]$$
$$+ i[\mathrm{Arg}^*(z_1 - \zeta) - \mathrm{Arg}(z_1 - \zeta)]$$
$$= i[+\pi/2 - \pi/2] + i[+3\pi/2 - (-\pi/2)] = 2\pi i$$

as we expected.

Local path independence of line integrals is easy to verify in practice. In the next section we will show that line integrals of $f(z)$ are locally path independent on E if $f(z)$ is holomorphic on E. Complex differentiability of $f(z)$ in turn is easily checked by appeal to the Cauchy-Riemann equations. Unfortunately, in complex analysis what we usually want to know is whether line integrals are *globally* path independent in a set E, so that one integration contour may be replaced by another (between the same two initial and final points) without affecting the value of the integral we are examining. It turns out that, if $f(z)$ has *locally* path independent line integrals in the domain E (i.e., $f(z)$ is holomorphic on E), and if E has the geometric property of being "simply connected" (this will be defined later), then line integrals of $f(z)$ are automatically globally path independent in E.

EXERCISES

1. Calculate the following integrals using Theorem 5.12.

(i) $\int_{-2i}^{2i} \frac{1}{z}\,dz$ for any contour in the right half plane

(ii) $\int_\gamma e^z\,dz$ along the circular arc (part of $|z| = 1$) from $-i$ to $+i$.

(iii) $\int_{-i}^{+i} \sec^2 z\,dz$ for any contour in the plane that avoids the points $z_k = \pi/2 + k\pi$ for $k = 0, \pm 1, \pm 2, \ldots$.

Answer: (i) $i\pi$; (ii) $2i \cdot \sin(1)$; (iii) $2i \cdot \tanh(1)$.

2. Explain why $\int_\gamma z^n\,dz = 0$ for any positive power $n = 0, 1, 2, \ldots$, if γ is a *closed* contour. This is not true for contours that are not closed (Example 5.10).

3. Show that $\int_\gamma \sin z\,dz = 0$ for any *closed* contour γ.

4. Prove that line integrals of $f(z) = 1/(z - p)^n$ are globally path independent on the domain $E = \{z : z \neq p\}$, provided that $n \geq 2$ (n an integer).

5. If $f(z)$ has power series expansion $f(z) = \sum_{n=0}^{\infty} a_n(z - p)^n$ convergent throughout the complex plane, prove that $\int_\gamma f(z)\,dz = 0$ for any closed contour.

6. Prove that line integrals are not locally path independent for the function $f(z) = \bar{z}$ by demonstrating that this continuous function cannot have an antiderivative in the plane (or in any subdomain!).

7. Prove the following antidifferentiation formulas (valid in the domains shown).

(i) $\int_1^z w^{1/2}\,dw = (2/3)z^{3/2} + c$; $E = \mathbf{C} \sim (-\infty, 0]$.

(ii) $\int_0^z e^w\,dw = e^z + c$; $E = \mathbf{C}$

(iii) $\int_i^z \frac{1}{w}\,dw = \operatorname{Log} z + c$; $E = \{z : \operatorname{Im}(z) > 0\}$

(iv) $\int_1^z \frac{1}{1 + w^2}\,dw = \arctan z + c$; $E = \{z : |\operatorname{Im}(z)| < 1\}$.

8. Prove the analog of the **integration-by-parts formula:** If $f(z)$ and $g(z)$ are continuously differentiable on the trajectory of a contour

γ, then

$$\int_\gamma f'(z)g(z)\,dz = f(q)g(q) - f(p)g(p) - \int_\gamma f(z)g'(z)\,dz,$$

where $\gamma(a) = p$ and $\gamma(b) = q$. Notice what happens if γ is a *closed* contour.

Hint: Use Theorem 5.12 on $\dfrac{d}{dz}(f \cdot g)$.

9. Evaluate antiderivatives, in appropriate domains, for the following functions using integration by parts.

(i) ze^z

(ii) $\cos^2 z$

(iii) $\operatorname{Log} z$ in the cut plane $\mathbf{C} \sim (-\infty, 0]$

(iv) $\operatorname{Arctan} z$ in the doubly cut plane obtained by deleting the parts of the imaginary axis such that $|z| \geq 1$.

Answer: (i) $ze^z - e^z$; (ii) $\frac{1}{2}\sin z \cos z + \dfrac{z}{2}$; (iii) $z \cdot (\operatorname{Log}(z) - 1)$; (iv) $z \operatorname{Arctan} z - \frac{1}{2}\operatorname{Log}(z^2 + 1)$.

5.7 CAUCHY'S THEOREM

We have reached a crucial turning point in our discussion as we take up Cauchy's Theorem and the Cauchy Integral Formula. These two results lie at the heart of complex analysis. Many applications, such as the Calculus of Residues, are based directly upon these theorems; furthermore, these results lead directly to a proof of the remarkable fact (discussed in Section 3.4) that holomorphic functions of a complex variable are automatically *analytic*, and are represented by power series expansions near any point. The latter result allows us to bring all of our results on analytic continuation, etc. to bear on problems in which differentiable functions arise. The theory of functions of a complex variable would hardly differ from the theory of functions of several real variables, if this were not true.

We will show that a holomorphic function must have a globally defined antiderivative if its domain of definition is *convex*. Later we will show that this is true of holomorphic functions on a much larger class of domains, those which are simply connected.

Definition 5.5 *A set E in the plane is said to be* **convex** *if and only if, for any pair of points $\{p, q\}$ in the set, the whole line segment between these points*

$$[p, q] = \{z : z = (1 - \alpha)p + \alpha q \quad \textit{for all} \quad 0 \leq \alpha \leq 1\}$$

lies within E.

The complex plane $E = \mathbf{C}$, any half plane (with any orientation), and any disc, are all convex sets in the plane. Notice that the point

$$z_\alpha = (1 - \alpha)p + \alpha q = p + \alpha(q - p)$$

traces out the segment $[p, q]$ starting at p and ending at q, as the parameter α increases from $\alpha = 0$ to $\alpha = 1$. Convex sets are obviously connected.

Theorem 5.16 (Cauchy's Theorem) *Let $f(z)$ be defined on a convex open set E, and assume that*

$$\frac{df}{dz} \quad \text{exists at every point in } E.$$

Then f has an antiderivative on E, and line integrals of f are globally path independent in E.

If f is complex differentiable on an *arbitrary* open set E in the plane, we may apply Theorem 5.16 to discs about typical points in E (discs being convex) to get the following result.

Corollary 5.17 *If df/dz exists at every point in an arbitrary open set E, then f has locally defined antiderivatives, and line integrals of f are locally path independent in E.*

If $f(t)$ is a function of a real variable, the Fundamental Theorem of Calculus says that $f(t)$ has an antiderivative if the function $f(t)$ is merely *continuous*. This is not the case for functions of a complex variable; $f(z) = \bar{z}$ (conjugation function) is continuous, but it does not have antiderivatives near any point in the complex plane (see Exercise 6, Section 5.6). Evidently, we must require more than continuity from $f(z)$ to be sure that this function has locally defined antiderivatives; by Cauchy's Theorem, complex differentiability of $f(z)$ insures the existence of antiderivatives near any point.

In Theorem 5.16 we require nothing more than existence of the derivative df/dz; for example, we do not ask that df/dz be continuous, or that any higher derivatives exist. Since Cauchy's Theorem is the result upon which all later developments rest, it is important to keep the hypotheses on $f(z)$ to a minimum, so we can apply the theorem to the largest possible class of integrands. If we were willing to add the hypothesis that df/dz is *continuous* to our assumption that $f(z)$ is holomorphic on E, and to apply methods from advanced calculus, the proof would be much easier. The component functions U and V in $f(z) = U(x, y) + iV(x, y)$ would then have continuous partial derivatives and would satisfy the Cauchy-Riemann equations; the continuity of these partial derivatives would allow us to apply Green's Theorem to conclude that line integrals are locally path independent. (For this approach to Cauchy's Theorem, cf. Nehari [16], Section 3.3, or Kaplan [13], Sections 9.10 to 9.12.) The proof of Cauchy's Theorem in its full generality, as we have stated it here, requires different methods and somewhat more effort; however, the proof does not use special results from advanced calculus, and is self-contained.

Note: Cauchy's original result was proved assuming continuity of df/dz. The modifications necessary to prove the theorem in its general form are due to Goursat, and Theorem 5.16 is often referred to as the *Cauchy-Goursat Theorem.*

There are two main steps in the proof of Theorem 5.16. First we will show how Cauchy's Theorem can be proved if we can establish the following simplified version of the main result.

Assertion *Suppose $f(z)$ is holomorphic on an open set that includes the boundary and interior of some triangle Δ. If we parametrize the sides of Δ to form smooth curves γ_1, γ_2, and γ_3 joining the vertices p_1, p_2, and p_3, we get*

$$\int_\gamma f(z)\,dz = 0 \tag{26}$$

for the closed contour $\gamma = \gamma_1 + \gamma_2 + \gamma_3$ obtained by traversing γ_1, γ_2, and γ_3 in succession.

The second step in the proof consists of establishing this simple assertion. It is rather technical, and of secondary interest to readers interested in applications, so these details may be omitted without danger of misunderstanding later discussion.

Step 1. Once the *Assertion* has been established, our task is to use it to construct an antiderivative for $f(z)$ on a typical convex open set E. The idea is similar to the one used in Theorem 5.13.

Let us fix a base point p in E and for each z in E let us define

$$F(z) = \int_{\gamma(p,z)} f(w)\,dw, \tag{27}$$

where $\gamma(p, z)$ is the parametrized line segment from p to z. Since E is convex, this segment lies within E. The function $F(z)$ is unambiguously defined, since we are taking a very definite contour for each z in E to get the value (27). We will show directly that $F(z)$ has derivative $f(z)$ on E, and this will complete our work in Step 1.

Consider

$$\frac{\Delta F}{\Delta z} = \frac{F(z') - F(z)}{z' - z} = \frac{1}{\Delta z}\left(\int_{\gamma(p,z')} f\,dz - \int_{\gamma(p,z)} f\,dz\right) \tag{28}$$

where $z' = z + \Delta z \neq z$. If $\gamma(t)$ is the parametrized line segment from z to z', as in Figure 5.11, then

$$\gamma(t) = tz' + (1 - t) = z + t\,\Delta z \quad \text{for} \quad 0 \leq t \leq 1$$

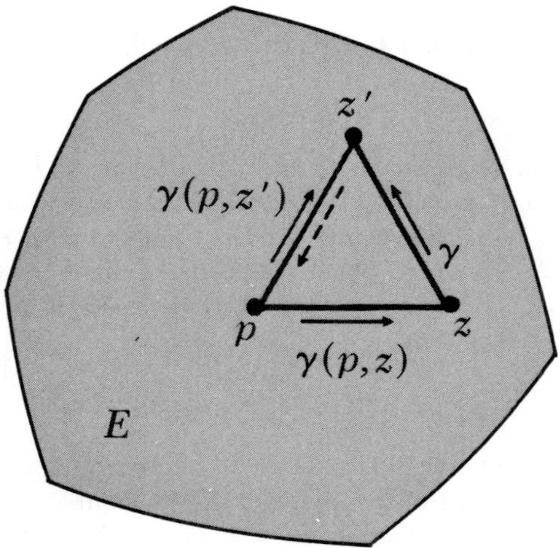

Figure 5.11 Constructing the antiderivative $\int_{\gamma(p,z)} f(w)\, dw$ on a convex set E.

and $\gamma'(t) = \Delta z$ (a constant). The pieced-together contour $\eta = \gamma(p, z) + \gamma + (-\gamma(p, z'))$ is the parametrized boundary of the triangle Δ shown in Figure 5.11, and this triangle (including its interior) lies entirely within the convex set E.† The *Assertion* is applied to give

$$0 = \int_\eta f\, dz = \int_{\gamma(p,z)} f\, dz + \int_\gamma f\, dz - \int_{\gamma(p,z')} f\, dz,$$

so that, by substituting this in (28),

$$\frac{\Delta F}{\Delta z} = \frac{1}{\Delta z}\int_\gamma f\ dz = \frac{1}{\Delta z}\int_0^1 f(\gamma(t)) \cdot \gamma'(t)\ dt = \int_0^1 f(\gamma(t))\ dt.$$

Just as in the proof of Theorem 5.13, the values $f(\gamma(t))$ for $0 \le t \le 1$ are all close to the value $f(\gamma(0)) = f(z)$ provided that $|z' - z| = |\Delta z|$ is sufficiently small; this forces the integral

$$\frac{\Delta F}{\Delta z} = \int_0^1 f(\gamma(t))\ dt$$

to be very close to $\int_0^1 f(z)\ dt = f(z) \cdot \int_0^1 dt = f(z) \cdot 1 = f(z)$. Thus,

$$\frac{dF}{dz}(z) = \lim_{z' \to z} \frac{F(z') - F(z)}{z' - z} \quad \text{exists and is equal to } f(z)$$

at every point z in E, so that $F(z)$ is an antiderivative for $f(z)$ on E. ▮

---

† The vertices p, z, and z' are in E, so the boundary segments of the triangle must also be in E. Any point in the interior of the triangle lies on a segment whose end points are on the boundary of Δ, so by invoking the definition of convexity once more, we see that the interior points of Δ are also in E.

STEP 2. We prove the *Assertion* by examining the consequences which accumulate if we actually had $\int_\gamma f(z)\,dz = \alpha \neq 0$.

It makes no difference in the reasoning whether we label the vertices of Δ so the contour γ traverses the boundary in a clockwise or counterclockwise direction; let us assume that γ moves counterclockwise, to be definite. The triangle Δ is a closed bounded set in the plane, so that any continuous function such as $g(z) = |f(z)|$ must be bounded on Δ; i.e., there is a constant K such that

$$|f(z)| \le K \quad \text{for all } z \text{ in } \Delta. \tag{29}$$

Let us subdivide Δ into four triangles Δ_{11}, Δ_{12}, Δ_{13}, and Δ_{14} by connecting midpoints of the sides of Δ, as shown in Figure 5.12. Let γ_{1k} ($k = 1, 2, 3, 4$) be a contour which traverses the boundary of the triangle Δ_{1k} in a counterclockwise direction. Then

$$\int_\gamma f\,dz = \int_{\gamma_{11}} f\,dz + \int_{\gamma_{12}} f\,dz + \int_{\gamma_{13}} f\,dz + \int_{\gamma_{14}} f\,dz. \tag{30}$$

To see this, we may break each contour γ_{1k} into three parametrized straight line segments; the segments forming the sides of the inner triangle Δ_{14} are traversed in directions opposite those along the sides of the outer triangles Δ_{11}, Δ_{12}, and Δ_{13}. The line integrals in opposite directions along these segments cancel, while the integrals along the remaining segments clearly add up to the integral along the sides of Δ, giving formula (30). Note that we are making free use of the formulas set forth in Section 5.5.

For at least one of the triangles Δ_{11}, Δ_{12}, Δ_{13}, and Δ_{14}, the inequality

$$\left|\int_{\gamma_{1k}} f\,dz\right| \ge \frac{1}{4}\left|\int_\gamma f\,dz\right| = \frac{|\alpha|}{4} \tag{31}$$

must hold, because four complex numbers w_1, w_2, w_3, and w_4 cannot have $w = w_1 + \cdots + w_4$ and also $|w_k| < |w|/4$ for $k = 1, 2, 3, 4$; this, together

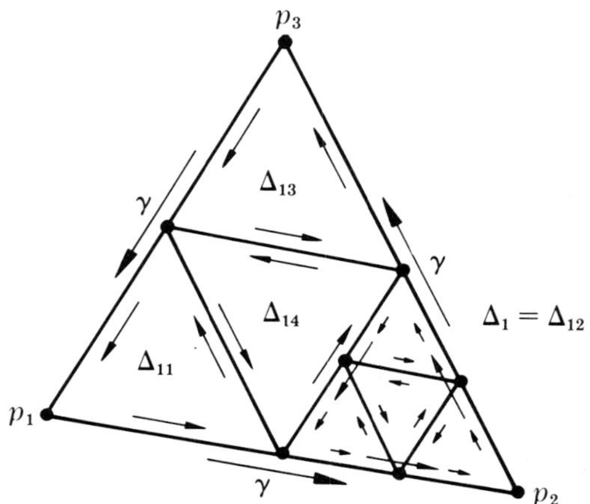

Figure 5.12 The subdivision procedure in Δ.

with the triangle inequality, would give the absurdity $|w| \leq |w_1| + \cdots + |w_4| < |w|$. Pick one of the triangles $\Delta_{11}, \ldots, \Delta_{14}$ for which inequality (31) holds and label it Δ_1; at the same time, label its parametrized boundary as γ_1. Figure 5.12 shows the possibility $\Delta_1 = \Delta_{12}$. Now we repeat the subdivision process, starting with the triangle Δ_1 instead of Δ; thus Δ_1 splits into triangles Δ_{21}, Δ_{22}, Δ_{23}, and Δ_{24} with counterclockwise parametrized boundaries $\gamma_{21}, \gamma_{22}, \gamma_{23}$, and γ_{24}, and we have

$$\int_{\gamma_1} f\,dz = \int_{\gamma_{21}} f\,dz + \cdots + \int_{\gamma_{24}} f\,dz.$$

Again, we must have

$$\textbf{(32)} \qquad \left|\int_{\gamma_{2k}} f\,dz\right| \geq \left(\frac{1}{4}\right)\left|\int_{\gamma_1} f\,dz\right| \geq \left(\frac{1}{4}\right)^2\left|\int_{\gamma} f\,dz\right| = \frac{|\alpha|}{4^2}$$

for at least one of the triangles Δ_{2k}; otherwise we would violate the triangle inequality. Let us pick one of the triangles Δ_{2k} that satisfies (32) and label it Δ_2 (with counterclockwise parametrized boundary γ_2).

Obviously, we may continue this subdivision process to get a nested sequence of progressively smaller triangles $\Delta \supseteq \Delta_1 \supseteq \Delta_2 \supseteq \cdots \supseteq \Delta_n \supseteq \Delta_{n+1} \supseteq \cdots$. The triangle Δ_n and its parametrized boundary curve γ_n have diameter and length

$$\textbf{(33)} \qquad \begin{aligned} \operatorname{diam}(\Delta_n) &= \frac{1}{2}\operatorname{diam}(\Delta_{n-1}) &= \cdots &= \frac{1}{2^n}\operatorname{diam}(\Delta) \\ \operatorname{length}(\gamma_n) &= \frac{1}{2}\operatorname{length}(\gamma_{n-1}) &= \cdots &= \frac{1}{2^n}\operatorname{length}(\gamma) \end{aligned}$$

for $n = 1, 2, 3, \ldots$ due to the nature of the subdivision process. These triangles have been selected so that

$$\textbf{(34)} \qquad \left|\int_{\gamma_n} f\,dz\right| \geq \frac{1}{4}\left|\int_{\gamma_{n-1}} f\,dz\right| \geq \cdots \geq \frac{1}{4^n}\left|\int_{\gamma} f\,dz\right| = \frac{|\alpha|}{4^n}$$

for $n = 1, 2, \ldots$.

Now we shall obtain another estimate on these integrals, which gives us *upper* bounds

$$\textbf{(35)} \qquad \left|\int_{\gamma_n} f\,dz\right| \leq \frac{M_n}{4^n}\cdot\operatorname{diam}(\Delta)\cdot\operatorname{length}(\gamma),$$

where $M_n \to 0$ as $n \to \infty$. Estimates (34) and (35) together will lead us to the desired conclusion that $\alpha = \int_\gamma f\,dz = 0$.

The diameters of the triangles $\Delta_1 \supseteq \Delta_2 \supseteq \Delta_3 \supseteq \cdots$ go to zero as $n \to \infty$ in view of (33), so that these sets close down upon a single point z^*

in the sense that $\{z^*\} = \bigcap_{n=1}^{\infty} \Delta_n$.† Obviously, the point z^* is in Δ since it is in each triangle Δ_n and $\Delta_n \subseteq \Delta$; thus $f(z)$ is differentiable at z^*. This allows us to write $f(z)$ in the form

$$f(z) = f(z^*) + f'(z^*)(z - z^*) + E(z) \quad \text{for all } z \text{ near } z^*$$

where $|E(z)|/|z - z^*| \to 0$ as $z \to z^*$ (recall Section 2.8). Since $f(z^*)$ and $f'(z^*)$ are constants in this formula, we get

$$\int_{\gamma_n} f\,dz = f(z^*)\int_{\gamma_n} 1\,dz + f'(z^*)\int_{\gamma_n} (z - z^*)\,dz + \int_{\gamma_n} E(z)\,dz$$

for $n = 1, 2, \ldots$. The integral of a polynomial along any closed contour is zero, so the first two terms drop out, and we get

$$\int_{\gamma_n} f\,dz = \int_{\gamma_n} E(z)\,dz \quad \text{for } n = 1, 2, \ldots.$$

Since $|E(z)|/|z - z^*| \to 0$ as $z \to z^*$, and the triangles Δ_n shrink down to the point z^*, the maximum value of this quotient on $\Gamma_n = \text{bdry}(\Delta_n)$,

$$M_n = \text{maximum of} \left|\frac{E(z)}{z - z^*}\right| \text{ for } z \text{ on } \Gamma_n,$$

must approach zero as $n \to \infty$; thus, $M_n \to 0$ as $n \to \infty$. For each triangle Δ_n, the boundary Γ_n is just the trajectory of γ_n and

$$\left|\frac{E(z)}{z - z^*}\right| \leq M_n \text{ (or equivalently, } |E(z)| \leq M_n \cdot |z - z^*|)$$

for z on Γ_n, $n = 1, 2, \ldots$. Therefore,

$$|E(z)| \leq M_n \cdot |z - z^*| \leq M_n \cdot \text{diam}(\Delta_n) = \frac{M_n}{2^n}\,\text{diam}(\Delta)$$

for all z on Γ_n, and by Theorem 5.9 we get

$$\left|\int_{\gamma_n} f\,dz\right| = \left|\int_{\gamma_n} E(z)\,dz\right| \leq \frac{1}{2^n} \cdot M_n \cdot \text{diam}(\Delta) \cdot \text{length}(\gamma_n)$$

(36)

$$\leq \frac{1}{4^n} \cdot M_n \cdot \text{diam}(\Delta) \cdot \text{length}(\gamma).$$

† The sets Δ_n are all closed bounded sets in the plane. The intersection $\Delta_\infty = \bigcap_{n=1}^{\infty} \Delta_n$ (the set of points common to all of these triangles) is a subset of Δ_n for $n = 1, 2, \ldots$, so its diameter must be zero; thus Δ_∞ has *at most one* point in it. It might, however, be empty (with no points in it at all). To see that this can't happen, we must invoke a result from advanced calculus, the Heine-Borel Theorem, which states that the intersection of any nested sequence of sets in the plane $A_1 \supset A_2 \supset A_3 \supset \ldots$ is *non-empty* if the A_n are each closed and bounded. We will not have time to go into this theoretical point in this book. (Reference: Buck [2], Section 1.6, or Ahlfors [1], Section 3.1.4.)

Now compare the estimates (34) and (35); clearly

$$0 \le \frac{|\alpha|}{4^n} \le \left|\int_{\gamma_n} f\,dz\right| \le \frac{\text{diam}(\Delta)\cdot\text{length}(\gamma)}{4^n}\cdot M_n$$

for $n = 1, 2, \ldots$. After both sides are multiplied by 4^n, the right side goes to zero as $n \to \infty$, while the left side is always equal to $|\alpha| = |\int_\gamma f\,dz|$. Thus we are forced (indirectly, but inevitably) to conclude that $\int_\gamma f\,dz = 0$. ▮

The proof of the Cauchy Theorem is now complete.

EXERCISES

1. Which of the following sets are convex?

(i) $|z| < 1$ (open disc).

(ii) $|z| > 1$ (exterior of disc).

(iii) The cut plane with $(-\infty, 0]$ removed.

(iv) The wedge $-\phi < \arg z < +\phi$ with $\phi = 3\pi/4$

(v) The semi-infinite strip $Im(z) > 0$ and $-\pi < Re(z) < +\pi$.

2. Explain why the following integrals are zero. Parametrize each locus as a counterclockwise oriented simple closed contour

(i) $\int_\gamma \sin z\,dz$, $\quad |z| = R$ with $R > 0$

(ii) $\int_\gamma ze^z\,dz$, $\quad |z| = R$ with $R > 0$.

(iii) $\int_\gamma \frac{e^z}{(z-2i)}\,dz$, $\quad z$ on the boundary of the square $-1 \le x, y \le 1$.

(iv) $\int_\gamma \frac{1}{1+z^2}\,dz$, $\quad |z| = R$ with $0 < R < 1$.

(v) $\int_\gamma \tan z\,dz$, $\quad z$ on the ellipse $4x^2 + y^2 = 4$.

3. Verify that $f(z) = \dfrac{1}{1+z^2}$ has locally defined antiderivatives on its natural domain $E = \mathbf{C} \sim \{+i, -i\}$ using the Cauchy Theorem. How would you prove this without use of Cauchy's Theorem? How would you prove it without invoking Cauchy's Theorem or the (yet unproved) Theorem $*$ of Chapter 3?

Note: From Theorem $*$ we know that f is analytic and there are locally defined antiderivatives, but invoking Theorem $*$ is not entirely legitimate because Theorem $*$ will not be proved until later in this chapter. Its proof is based on Cauchy's Theorem.

4. A domain E is said to be **star-shaped** if there is a point p in E with the property that, for every other point z in E, the connecting segment $[p, z]$ lies entirely within E. Check through the details of Cauchy's Theorem, making slight modifications in the reasoning where necessary, to prove the more comprehensive result below.

Theorem: If $f(z)$ is complex differentiable throughout a star-shaped domain E, there is a globally defined antiderivative on E, and line integrals along contours in E are globally path independent.

5. Exhibit four domains which are star-shaped but *not* convex. Is every convex set star-shaped?
Hint: Try the cut plane $E = \mathbf{C} \sim (-\infty, 0]$ for one example.

5.8 THE CAUCHY INTEGRAL FORMULA

If $f(z)$ is *analytic* on a domain E, the behavior of $f(z)$ in one part of E determines the behavior of $f(z)$ elsewhere. Cauchy's Integral Formula, proved below, shows that functions of a complex variable that are merely holomorphic on E have a similar property. This foreshadows the even more remarkable fact that holomorphic functions are actually analytic (which will be proved using the Cauchy Integral Formula).

Theorem 5.18 (Cauchy Integral Formula) *Let $f(z)$ be holomorphic on an open set E that includes a disc $D = \{z: |z - p| < r\}$ and also its boundary circle Γ. Let us parametrize Γ as a closed contour $\gamma(t) = p + re^{it}$, defined for $0 \leq t \leq 2\pi$, which traverses Γ once in the counterclockwise direction. If ζ is any point in the disc, the value of f at ζ is given by the integral*

(37) $$f(\zeta) = \frac{1}{2\pi i}\int_\gamma \frac{f(z)}{z - \zeta}\, dz \quad \textit{for all } \zeta \textit{ such that } |\zeta - p| < r.$$

Since line integrals along γ depend only on the behavior of the integrand $h(z) = f(z)/(z - \zeta)$ at points z on the trajectory Γ, the value of the integral for fixed ζ is completely determined once we know the values of $f(z)$ on the boundary circle Γ. Therefore, the behavior of $f(z)$ on Γ completely determines $f(\zeta)$ for each ζ in the disc bounded by Γ—to get $f(\zeta)$ we simply multiply $f(z)$ by the standard function $1/(z - \zeta)$ and integrate along γ. Not only is the behavior of $f(z)$ "propagating" from the boundary circle Γ into the enclosed disc, but formula (37) plainly exhibits the integration process by which this is accomplished.

Before we take up the proof of Theorem 5.18, let us indicate a few variants of the result we have stated. Any contour obtained from γ by an order preserving reparametrization must give the same result, by Theorem 5.11, so the particular parametrization of Γ we use is not very important. Furthermore, we have taken γ to have $p + (r + i0)$ as its base point (initial/final point), but the proof is unaffected if we take another point on Γ as the base point; the value of the integral is the same for the closed contour $\gamma(t) = p + re^{i(t+\theta)}$, defined for $0 \leq t \leq 2\pi$, which has $p + re^{i\theta}$ as its base point, or for any contour obtained from this one by making an order preserving change of variable. Notice that the integrand $f(z)/(z - \zeta)$ in (37) is unbounded as the variable z approaches the point ζ, since the denominator approaches zero. But ζ lies a positive distance from the circle Γ, so that $1/(z - \zeta)$ and $f(z)/(z - \zeta)$ are continuous functions of the variable z as z varies on Γ, and the integral is well defined.

Naturally Cauchy's Theorem, used with a little sophistication, will be the key to the integral formula. We will prove the following intermediate result.

Theorem 5.19 *Let E be any open convex set and $f(z)$ a function which is differentiable throughout E except at one point z^*, where $f(z)$ is only assumed to be continuous. Then $f(z)$ has an antiderivative on E, and the line integrals of $f(z)$ are path independent for contours in E, in spite of the presence of the exceptional point z^*.*

Once this has been established, the integral formula emerges from the following reasoning. Let us assume, for the moment, that Theorem 5.19 is established; then consider the function

$$g(z) = \begin{cases} \dfrac{f(z) - f(\zeta)}{z - \zeta} & \text{for} \quad z \neq \zeta \\ f'(\zeta) & \text{for} \quad z = \zeta \end{cases}$$

and apply Theorem 5.19 to it. This function is obviously differentiable at points other than ζ, and although we cannot say it is differentiable at ζ, it is certainly continuous there by definition of the derivative at ζ; $\lim_{z \to \zeta} \{g(z)\} = f'(\zeta) = g(\zeta)$. We may write

$$\frac{f(z)}{z - \zeta} = \frac{f(z) - f(\zeta)}{z - \zeta} + f(\zeta)\frac{1}{z - \zeta} = g(z) + f(\zeta)\frac{1}{z - \zeta}$$

for all $z \neq \zeta$ in E, and since ζ does not lie on the trajectory of γ we have

$$\frac{1}{2\pi i}\int_\gamma \frac{f(z)}{z - \zeta}\, dz = \frac{1}{2\pi i}\int_\gamma g(z)\, dz + f(\zeta)\frac{1}{2\pi i}\int_\gamma \frac{1}{z - \zeta}\, dz$$

$$= 0 + f(\zeta) \cdot \frac{2\pi i}{2\pi i} = f(\zeta).$$

The first integral on the right vanishes, by Theorem 5.19, and the last integral (of $1/(z - \zeta)$ around a circular contour) has been evaluated in Example 5.13. This proves the Cauchy Integral Formula in full.

Proof of Theorem 5.19: We want to show that $f(z)$ has an antiderivative $F(z)$ on E. Let us define

$$F(z) = \int_{\gamma(z^*, z)} f(w)\, dw \quad \text{for each } z \text{ in } E,$$

where $\gamma(z^*, z)$ is the parametrized line segment from z^* to z (it lies in E since E is convex). Thus, $F(z)$ is well defined for each z. Exactly as in Step 1 of Cauchy's theorem (we will not repeat these details here), we can show that dF/dz exists and is equal to $f(z)$ once the following assertion is proved.

Assertion $*$ *If Δ is any triangle with the exceptional point z^* and two other points z and z' in E for its vertices (so Δ lies within E), then $\int_\gamma f(z)\,dz = 0$, where γ is obtained by traversing the parametrized line segments connecting $\{z^*, z, z'\}$ in succession.*

PROOF OF ASSERTION $*$: Consider such a triangle Δ, as shown in Figure 5.13; on the segments $[z^*, z]$ and $[z^*, z']$ mark off points z_n and z_n' such that

$$\left|\frac{z_n' - z^*}{z' - z^*}\right| = \left|\frac{z_n - z^*}{z - z^*}\right| = \frac{1}{n}.$$

The triangle Δ_n with vertices $\{z^*, z_n, z_n'\}$ is similar to Δ, but with dimensions scaled by the factor $1/n$. If γ_n traverses the sides of Δ_n the same way γ traverses the sides of Δ, we get

$$\int_\gamma f(z)\,dz = \int_{\gamma n} f(z)\,dz \quad \text{for} \quad n = 1, 2, \ldots;$$

for, if we subdivide Δ into Δ_n and a trapezoidal region D_n, as in Figure 5.13, we get

$$\int_\gamma f(z)\,dz = \int_{\gamma n} f(z)\,dz + \int_{\phi n} f(z)\,dz$$

due to cancellation of integrals in opposite directions along the segment $[z_n, z_n']$. We can find a slightly larger open trapezoid that contains D_n and excludes z^*; then $f(z)$ is holomorphic on this larger domain, which is a convex

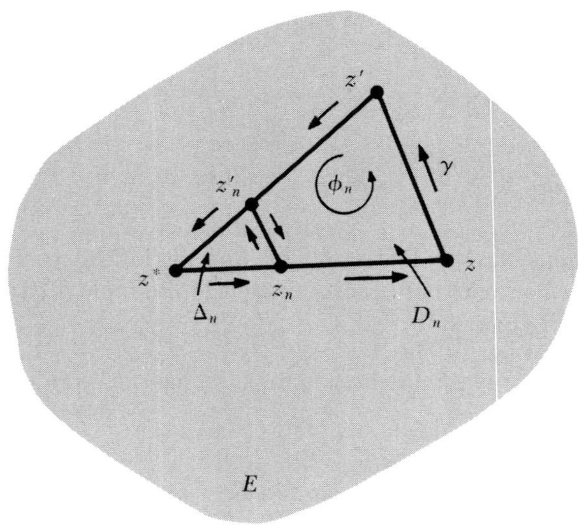

Figure 5.13 The situation in Assertion $*$.

set. Cauchy's theorem guarantees that

$$\int_{\phi_n} f(z)\, dz = 0 \quad \text{for} \quad n = 1, 2, \ldots .$$

Now f, being continuous at z^*, is bounded there, so that $|f(z)| \leq M$ for all z near z^*; consequently,

$$0 \leq \left| \int_\gamma f(z)\, dz \right| = \left| \int_{\gamma_n} f(z)\, dz \right| \leq M \cdot \text{length}(\gamma_n) = \frac{M \cdot \text{length}(\gamma)}{n}$$

for $n = 1, 2, \ldots$. Clearly, the integral along γ must have the value zero. This reasoning works for every triangle Δ, so *Assertion* $*$ is proved. ▮

Suppose that $f(z)$ is holomorphic on an open set that includes a disc $D = \{z\colon |z - p| < r\}$ and the boundary circle $\Gamma = \{z\colon |z - p| = r\}$. What happens to the integral

$$\frac{1}{2\pi i} \int_\gamma \frac{f(z)}{z - \zeta}\, dz \tag{38}$$

as we let ζ vary *throughout* the complex plane? If ζ happens to lie in the open disc D, this integral reproduces the value of f at ζ

$$f(\zeta) = \frac{1}{2\pi i} \int_\gamma \frac{f(z)}{z - \zeta}\, dz \quad \text{for all } \zeta \text{ such that } |\zeta - p| < r.$$

If ζ is on the circle Γ, the function $1/(z - \zeta)$ blows up as z approaches ζ, so the integrand in (38) is not continuous, and the integral is not defined. If ζ lies outside of Γ, so that $|\zeta - p| > r$, the integrand is again a continuous function of z on the trajectory Γ, and (38) is well defined. But for ζ exterior to Γ, formula (38) does not reproduce the value $f(\zeta)$. Our suspicions should be aroused by the fact that the integral makes sense for *all* ζ exterior to Γ, even ones where the original function f is not defined! In fact, if $|\zeta - p| > r$, then ζ is a positive distance from the closed disc $\bar{D} = D \cup \Gamma = \{z\colon |z - p| \leq r\}$ and we can set up a slightly larger open disc D' centered at p, that contains D and Γ but excludes ζ. Then the functions (of variable z) $1/(z - \zeta)$ and $f(z)/(z - \zeta)$ are holomorphic on D', which is a convex domain containing the integration contour γ. Cauchy's Theorem assures us that

$$\frac{1}{2\pi i} \int_\gamma \frac{f(z)}{z - \zeta}\, dz = 0 \quad \text{for all } \zeta \text{ such that } |\zeta - p| > r.$$

Notice that Theorem 5.16 was quite adequate for the last computation; the Cauchy Integral Formula was not required.

Example 5.14 Let Γ be the unit circle and let $\gamma(t)$ be a closed contour that traverses Γ once in a counterclockwise direction. If we take $f(z) = e^z$, the integral (38) starts with the values of $f(z)$ on Γ and gives us a new function $\tilde{f}(\zeta)$ defined at every point ζ in the complement $E = \mathbf{C} \sim \Gamma = \{z: |z| \neq 1\}$,

$$\tilde{f}(\zeta) = \frac{1}{2\pi i} \int_\gamma \frac{f(z)}{z - \zeta}\, dz.$$

Clearly,

$$\begin{aligned} \tilde{f}(\zeta) &= f(\zeta) = e^\zeta && \text{if } |\zeta| < 1 \\ \tilde{f}(\zeta) &= 0 && \text{if } |\zeta| > 1. \end{aligned}$$

The reader will find it instructive to try evaluating these integrals by direct calculations based on the defining formula (15) for line integrals. This exercise should provide a keen appreciation of how much effort is saved by an appeal to Cauchy's Integral Formula (or Cauchy's Theorem) in evaluating line integrals.

The integral (38) can be viewed as an "integral operator" by writing it in the form

$$\tilde{f}(\zeta) = \int_\gamma K(\zeta, z) f(z)\, dz, \tag{39}$$

where

$$K(\zeta, z) = \frac{1}{2\pi i} \frac{1}{z - \zeta}.$$

Here $K(\zeta, z)$ is a function of *two* complex variables z and ζ, defined except when $z = \zeta$. Formula (39) transforms a function $f(z)$ defined on Γ to a function $\tilde{f}(\zeta)$ defined on the complement of Γ. This operation is called an **integral operator** because the transformation from f to $\tilde{f}$ is effected by integrating $f(z)$ against the fixed function $K(\zeta, z)$ of two variables, holding the first variable fixed during the integration. The function $K(\zeta, z)$ is called the **kernel function,** and the particular function $K(\zeta, z) = \dfrac{1}{2\pi i} \dfrac{1}{z - \zeta}$ is called the **Cauchy kernel function**. Transformations that have the general form (39), but perhaps a different kernel function, frequently turn up in applied mathematics. Further comments will be presented in Section 5.10; for the moment the student should appreciate the general way in which such an operator acts. We present it with certain data, namely the values of f on Γ; it takes these data and converts them into a new function $\tilde{f}$ defined on a large portion of the complex plane.

EXERCISES

1. Write out the Riemann integral you would have to evaluate to prove that $\displaystyle\int_\gamma \frac{1}{z - \zeta}\, dz = 2\pi i$ for points ζ such that $|\zeta| < R$, where $\gamma(t) = Re^{it}$. How have we avoided this difficult direct calculation in Theorem 5.18?

2. Calculate the contour integrals shown, using Cauchy's integral formula or Cauchy's Theorem. The contour is to be a counter-clockwise parametrization of the trajectory indicated.

(i) $\displaystyle\int_\gamma \frac{1}{z+\frac{1}{2}}\,dz, \quad |z| = 10$

(ii) $\displaystyle\int_\gamma \frac{\sin z}{z+\frac{1}{2}}\,dz, \quad |z| = 10$

(iii) $\displaystyle\int_\gamma \frac{\text{Arctan } z}{z-1}\,dz,$ along the circle $|z-1| = 1$

(iv) $\displaystyle\int_\gamma \frac{e^z + z}{z+1}\,dz, \quad |z-i| = 3/2$

(v) $\displaystyle\int_\gamma \frac{\cosh z + 1}{z+i}\,dz, \quad |z+2i| = 3/2$

Answer: (i) $2\pi i$; (ii) $\sin(-\frac{1}{2})$; (iii) Arctan $(1) = \pi/4$; (iv) 0; (v) $1 + \cosh(-i) = 1 + \cos(1)$

3. Verify that $f(z) = 1/(z-\zeta)$ is continuous on the circle $|z-p| = R$, provided the point ζ does not lie on the circle. Give an upper bound $M = M(\zeta)$ for the values $|f(z)|$ as z varies through the circle (the bound will depend on the location of ζ).

Answer: $M(\zeta) = 1/(R - |\zeta - p|)$.

4. Calculate $\displaystyle\int_\gamma \frac{\sin z}{z^2+1}\,dz$ along the counterclockwise oriented circles $|z| = 2$ and $|z| = \frac{1}{2}$.

Hint: For $|z| = 2$ write $\dfrac{1}{z^2+1} = \dfrac{A}{z+i} + \dfrac{B}{z-i}$ with suitably chosen constants (this is the *partial fractions decomposition* of the rational function $1/(1+z^2)$).

Answer: $2\pi i \sinh(1)$ for $|z| = 2$; 0 for $|z| = \frac{1}{2}$.

5. Evaluate the following integrals using Cauchy's Theorem or the Cauchy integral formula, as appropriate.

(i) $\displaystyle\int_\gamma \frac{1}{z^2+z+1}\,dz \quad |z| = \frac{1}{2}$

(ii) $\displaystyle\int_\gamma \frac{1}{z^2+1}\,dz \quad |z-i| = 1$

(iii) $\displaystyle\int_\gamma \frac{\sqrt{z}}{z^2-1}\,dz \quad |z-1| = 2/3$

Hint: In (ii) set $(z^2 + 1)^{-1} = h(z)/(z - i)$, taking $h(z) = 1/(z + i)$. In (iii) write integrand as $h(z)/(z - 1)$, where $h(z) = \sqrt{z}/(z + 1)$.

Answer: (i) 0; (ii) $+\pi$; (iii) $\frac{1}{2}$.

6. Calculate the integral $\int_\gamma \tan z\, dz$ counterclockwise along circle $|z - \pi/2| = 1$. Note the presence of a singularity of $\tan z$ at the center point $p = \pi/2 + i0$.

Hint: Write $\cos z = h(z)/(z - \pi/2)$ where h is holomorphic on the plane and nonvanishing at $\pi/2$. (Use series expansion $\sin z = z - z^3/3! + \ldots$ and the relation $\cos(z) = -\sin(z - \pi/2)$ to get a series expansion of $\cos z$ about $\pi/2$.) Then $\tan z = g(z)/(z - \pi/2)$, where $g(z) = \sin(z)/h(z)$ is holomorphic for $|z - \pi/2| \le 1$. Use the Cauchy Integral Formula.

Answer: $-2\pi i$.

7. Using the idea introduced in Exercise 6 above, calculate the integrals

$$\text{(i)}\quad \int_\gamma \sqrt{z+1}\tan z\, dz$$

$$\text{(ii)}\quad \int_\gamma \operatorname{Log}(z - i)\sec z\, dz.$$

counterclockwise along the circle $|z - \pi/2| = \frac{1}{2}$. Sketch the location of discontinuities of $\sqrt{z+1}$ and $\operatorname{Log}(z - i)$ and exhibit a convex set containing the disc $|z - \pi/2| \le 1$ on which these functions are holomorphic.

Hint: Again write $\cos z = h(z)\cdot(z - \pi/2)$ with $h(z)$ holomorphic and nonvanishing on the disc $|z - \pi/2| \le 1$.

$$\textit{Answer:}\ \text{(i)}\ -2\pi i\sqrt{\frac{\pi}{2} + 1};\ \text{(ii)}\ -2\pi i \operatorname{Log}\left(\frac{\pi}{2} - i\right).$$

5.9 APPLICATION: THE PROOF OF THEOREM * (HOLOMORPHIC FUNCTIONS ARE ANALYTIC)

We shall prove Theorem * and several other useful results at the same time.

Theorem 5.20 *Let $f(z)$ be differentiable throughout an open set E and let p be a point in E. Then*

(i) *The function f is analytic at p.*

(ii) *Let d be the distance from p to the boundary set bdry(E), so that $0 < d \le +\infty$. The Taylor series for f about the base point p converges at least on the open disc of radius d about p.*

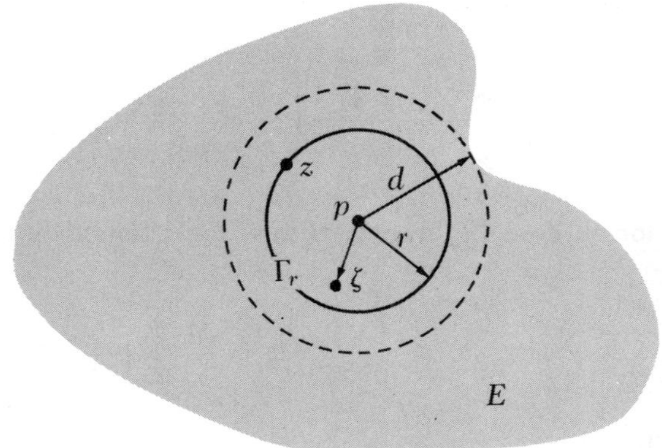

Figure 5.14 Locations of the variables z and ζ in the proof of Theorem 5.20.

Corollary 5.21 *From (i) we get Theorem * of Chapter 3.*

PROOF OF THEOREM 5.20: If d is the distance from p to the boundary of E, then open discs D_r about p of smaller radius $r < d$ all lie within E; in fact, D_d is the largest open disc about p lying entirely within E, by definition of d (Figure 5.14). Also, the boundary circles Γ_r lie within E since $r < d$ (this might not be true if $r = d$), and we may parametrize them to get closed contours γ_r which move once counterclockwise around Γ_r.

Let us fix our attention on one of these discs D_r. If ζ is in this disc, then $|\zeta - p| < r$ and

$$f(\zeta) = \frac{1}{2\pi i}\int_{\gamma_r} \frac{f(z)}{z - \zeta}\, dz.$$

But $1/(z - \zeta)$ can be written as the sum of a geometric series

$$\frac{1}{z-\zeta} = \frac{1}{z-p}\left[\frac{1}{1-\left(\dfrac{\zeta-p}{z-p}\right)}\right] = \frac{1}{z-p}\sum_{n=0}^{\infty}\left(\frac{\zeta-p}{z-p}\right)^n$$

$$= \sum_{n=0}^{\infty}\frac{(\zeta-p)^n}{(z-p)^{n+1}},$$

which converges because $\left|\dfrac{\zeta-p}{z-p}\right| = \dfrac{|\zeta-p|}{r} < 1$ for z on Γ_r. Multiplying each term in this series by the number $f(z)$ gives

(40) $$\frac{f(z)}{z-\zeta} = \sum_{n=0}^{\infty} f(z)\frac{(\zeta-p)^n}{(z-p)^{n+1}} \quad \text{for all } z \text{ on } \Gamma_r.$$

If we keep ζ fixed, the series (40), taken as a function of z, converges *uniformly* on the circle Γ_r. This is easily verified via the Weierstrass test: since f is continuous on Γ_r it is bounded, so that $|f(z)| \leq M$ for all z on Γ_r. The terms in series (40) have corresponding bounds

$$\left| f(z) \frac{(\zeta - p)^n}{(z - p)^{n+1}} \right| \le M \left| \frac{(\zeta - p)^n}{(z - p)^{n+1}} \right| = \frac{M}{r^{n+1}} |\zeta - p|^n = M_n$$

for all z on Γ_r, and the sum of these bounds is finite,

$$\sum_{n=0}^{\infty} M_n = \frac{M}{r} \sum_{n=0}^{\infty} \left(\frac{|\zeta - p|}{r} \right)^n < +\infty$$

because $|\zeta - p| < r$.

Since the series (40) converges uniformly to its limit on Γ_r, we may substitute this series into the integral and interchange the operations $\int_{\gamma_r} (\cdots)\, dz$ and $\sum_{n=0}^{\infty} (\cdots)$; we showed that such an interchange is valid earlier (see Theorem 5.10). Thus we obtain

(41)
$$\begin{aligned} f(\zeta) &= \frac{1}{2\pi i} \int_{\gamma_r} \left[\sum_{n=0}^{\infty} \frac{f(z)}{(z - p)^{n+1}} (\zeta - p)^n \right] dz \\ &= \sum_{n=0}^{\infty} \left(\frac{1}{2\pi i} \int_{\gamma_r} (\zeta - p)^n \frac{f(z)}{(z - p)^{n+1}}\, dz \right) \\ &= \sum_{n=0}^{\infty} \left(\frac{1}{2\pi i} \int_{\gamma_r} \frac{f(z)}{(z - p)^{n+1}}\, dz \right) (\zeta - p)^n \\ &= \sum_{n=0}^{\infty} a_n (\zeta - p)^n. \end{aligned}$$

The variable z is being integrated out of existence, so the coefficients of the terms $(\zeta - p)^n$ are just *constants* that do not depend on ζ. This formula is valid for each ζ in the disc D_r.

Formula (41) exhibits a power series expansion for $f(\zeta)$ about the base point p, convergent at least on D_r. This proves (i). The coefficients of the terms $(\zeta - p)^n$ in (41) must agree with the usual Taylor series coefficients, by uniqueness of power series coefficients, so that

$$a_n = \frac{f^{(n)}(p)}{n!} \quad \text{for} \quad n = 0, 1, 2, \ldots.$$

We get the same conclusion for each of the discs D_r such that $r < d$; the Taylor series for $f(\zeta)$ about the base point p converges to $f(\zeta)$ at every point in each disc D_r. Obviously, this Taylor series must then converge (to f) at each point in the disc D_d which is the union of the smaller discs D_r as r increases with $r < d$. This proves (ii). ▮

EXERCISES

1. Calculate the radius of convergence of the Taylor series for the following holomorphic functions, taking base points p as indicated.

(i) $\dfrac{1}{z^2+1}$ about $p = +1$

(ii) $\sqrt{z}$ (principal determination) about $p = 1 + i$

(iii) $\sqrt{z}$ (principal determination) about $p = -1 - i$

(iv) e^z ctn z about $p = \pi/2 + i0$ and about $p = +i$

(v) $\text{Log}(z + i)$ about $p = -1 + i0$.

In each case explain why the radius of convergence cannot be any larger.
Answer: (i) $\sqrt{2}$; (ii) $\sqrt{2}$; (iii) $\sqrt{2}$; (iv) $\pi/2$ and 1; (v) $\sqrt{2}$.

2. Prove the following result, a converse to Cauchy's Theorem, frequently used to prove that functions resulting from certain constructions are analytic.

Theorem (Morera's theorem): Suppose $f(z)$ is continuous on a domain E and we have $\int_\gamma f(z)\,dz = 0$ on integrating around the parametrized boundary of every triangle whose sides and interior lie within E. Then f is analytic in E.

Hint: First prove f has an antiderivative F near any point p in E; imitate the proof of Cauchy's Theorem, using our hypotheses in place of the *Assertion.* Then use Theorem $*$ on the antiderivative F; obviously $f = dF/dz$ is analytic too.

3. Prove the following useful theorem:

Theorem: If functions $f_n(z)$ are analytic on an open set E and converge uniformly on E to a function $f(z)$, then $f(z)$ is analytic on E.

Hint: We showed that $f(z)$ must be continuous in Chapter 3. Use Morera's theorem to show that f is analytic on each disc contained in E, hence analytic on E. Recall Theorem 5.10.

5.10 USING INTEGRAL FORMULAS TO GENERATE ANALYTIC FUNCTIONS

Let us consider a contour γ that is *not necessarily closed.* Any continuous function $f(z)$ defined on the trajectory Γ gives us a function

$$\tilde{f}(\zeta) = \frac{1}{2\pi i}\int_\gamma \frac{f(z)}{z-\zeta}\,dz \tag{42}$$

that is well defined for every point lying off Γ. Because ζ is a positive distance from the closed bounded set Γ, the integrand is continuous for z on Γ, and $\tilde{f}(\zeta)$ is well defined unless ζ lies on Γ. There is no need for the function $f(z)$ we start with to be differentiable or analytic at points on Γ, as was the case in the Cauchy Integral Formula, which has obviously inspired our choice of integral in (42); the function $\tilde{f}(\zeta)$ is well defined whether $f(z)$ is holomorphic or not. Nor is there any need for $f(z)$ to be defined off of Γ. We will now show that the new function $\tilde{f}(\zeta)$ is analytic on the complement $E = \mathbf{C} \sim \Gamma$. Thus, the integral formula (42) provides us with an important means of systematically generating analytic functions.

Theorem 5.22 *If $f(z)$ is continuous on the trajectory Γ of a contour γ, then the function*

$$\tilde{f}(\zeta) = \frac{1}{2\pi i}\int_\gamma \frac{f(z)}{z-\zeta}\,dz$$

is defined and holomorphic (hence analytic) on the complement $E = \mathbf{C} \sim \Gamma$. Its derivative $d\tilde{f}/d\zeta$ is given by a similar integral formula

$$\frac{d\tilde{f}}{d\zeta}(\zeta) = \frac{1}{2\pi i}\int_\gamma \frac{f(z)}{(z-\zeta)^2}\,dz$$

for all ζ in E, and the higher derivatives are given by

$$\tilde{f}^{(n)}(\zeta) = \frac{n!}{2\pi i}\int_\gamma \frac{f(z)}{(z-\zeta)^{n+1}}\,dz \quad \text{for} \quad n = 0, 1, 2, \ldots \tag{43}$$

for all ζ in E.

PROOF: The case $n = 0$ is just the definition of $\tilde{f}$ (remember that $0! = 1$). To derive the case $n = 1$ from $n = 0$, consider a typical point ζ in E and examine difference quotients for $\zeta' = \zeta + \Delta\zeta$ near ζ,

$$\begin{aligned}\frac{\Delta\tilde{f}}{\Delta\zeta} = \frac{\tilde{f}(\zeta') - \tilde{f}(\zeta)}{\Delta\zeta} &= \frac{1}{\Delta\zeta}\left[\frac{1}{2\pi i}\int_\gamma \frac{f(z)}{z-\zeta'}\,dz - \frac{1}{2\pi i}\int_\gamma \frac{f(z)}{z-\zeta}\,dz\right]\\ &= \frac{1}{2\pi i}\int_\gamma \frac{f(z)}{\Delta\zeta}\left[\frac{1}{z-\zeta'} - \frac{1}{z-\zeta}\right]dz\\ &= \frac{1}{2\pi i}\int_\gamma f(z)\left[\frac{1}{(z-\zeta')(z-\zeta)}\right]dz.\end{aligned} \tag{44}$$

Let us consider any sequence of points ζ_n', distinct from ζ, that approach ζ from within $E = \mathbf{C} \sim \Gamma$. Since ζ is a positive distance δ away from the trajectory Γ, the functions $h_n(z) = 1/(z - \zeta_n')$ converge uniformly on Γ to the limit function $h(z) = 1/(z - \zeta)$ as $n \to \infty$ (details in Exercises 5 to 8).† It follows that the product

$$h_n(z) \cdot h(z) = \frac{1}{(z - \zeta_n')(z - \zeta)} \quad \text{converges to} \quad h(z) \cdot h(z) = \frac{1}{(z - \zeta)^2}$$

uniformly for z on Γ as $n \to \infty$. Therefore, the integrand in (44) converges uniformly on Γ to the limit function $f(z)/(z - \zeta)^2$, so that

$$\frac{\tilde{f}(\zeta_n) - \tilde{f}(\zeta)}{\zeta_n' - \zeta} = \frac{1}{2\pi i} \int_\gamma \frac{f(z)}{(z - \zeta_n')(z - \zeta)} \, dz \to \frac{1}{2\pi i} \int_\gamma \frac{f(z)}{(z - \zeta)^2} \, dz$$

as $n \to \infty$, by Theorem 5.10. We get the same result for every sequence of points $\zeta_n' \to \zeta$; therefore,

$$\tilde{f}'(\zeta) = \lim_{\zeta' \to \zeta} \frac{\Delta \tilde{f}}{\Delta \zeta} \quad \text{exists and is equal to} \quad \frac{1!}{2\pi i} \int_\gamma \frac{f(z)}{(z - \zeta)^2} \, dz$$

for every point ζ in the complement of the arc Γ.

The formulas for the higher derivatives are obtained in much the same way. For example, to obtain the formula when $n = 2$, we start with the formula for $n = 1$, examine difference quotients, and take limits using the same kind of reasoning just presented. ▮

If γ is a closed circular contour and f is holomorphic on the closed disc enclosed by γ, the function $\tilde{f}(\zeta)$ agrees with $f(\zeta)$ within the disc and $\tilde{f}(\zeta) = 0$ outside the disc, by the Cauchy Integral Formula. But we can use any contour γ, even one that is not closed, and any continuous function defined on Γ, in (42). If $f(z)$ is not related to a holomorphic function in the plane, we can expect that the transformed function $\tilde{f}(\zeta)$, defined off Γ, will bear little resemblance to $f(z)$. This illustrated in Exercise 4.

EXERCISES

1. Outline the steps needed to prove the differentiation formula (43) for $n = 2$, once it has been established for $n = 0, 1$.

2. If $f(z)$ is continuous on the circle $|z| = 1$, prove that the contour integral

$$F(\zeta) = \frac{1}{2\pi i} \int_\gamma \frac{f(z)}{z} \left(\frac{z + \zeta}{z - \zeta} \right) dz$$

along $\gamma(t) = e^{it}$ $(0 \le t \le 2\pi)$ is an analytic function of ζ in the domains $|\zeta| < 1$ and $|\zeta| > 1$.

† In fact, the convergence $h_n \to h$ is uniform on the set $S = \{z : |z - \zeta| \ge \delta\}$, which contains Γ. The precise shape of Γ is irrelevant; uniform convergence on S insures uniform convergence on any subset $X \subseteq S$ (Section 3.1).

Note: This is essentially the *Poisson integral,* used to solve boundary value problems for harmonic functions (in Chapter 7).

Hint: Divide out $\dfrac{z+\zeta}{z-\zeta} = -1 + \dfrac{2z}{z-\zeta}$; $\dfrac{f(z)}{z}$ is continuous on the trajectory of γ.

3. Let γ be the semicircular (non-closed!) arc $\gamma(t) = e^{it}$ defined for $-\pi/2 \leq t \leq \pi/2$, and consider the analytic function

$$f(\zeta) = \frac{1}{2\pi i}\int_\gamma \frac{1}{z-\zeta}\,dz$$

defined on the complement $\mathbf{C} \sim \Gamma$ of this arc. Use the general principles for handling line integrals outlined in this chapter to verify the following properties of this function.

(i) $\lim_{r\to\infty} M(r) = 0$, where $M(r) =$ maximum of $|f(\zeta)|$ on the circle $|\zeta| = r$.

(ii) $f(0) = \frac{1}{2}$

(iii) $\dfrac{1}{n!}f^{(n)}(0) = \dfrac{i^n}{2\pi ni}[1-(-1)^n]$ for $n = 1, 2, \ldots$.

(iv) f has a series expansion about the origin given by

$$f(\zeta) = \frac{1}{2} + \frac{1}{\pi}\zeta - \frac{1}{3\pi}\zeta^3 + \frac{1}{5\pi}\zeta^5 - \cdots.$$

What is its radius of convergence?

4. Consider the analytic function $f(\zeta) = \dfrac{1}{2\pi i}\displaystyle\int_\gamma \frac{1}{z-\zeta}\,dz$ defined in Exercise 3 above. Using the analytic continuation principle, show that f is a determination of $\log\left(\dfrac{\zeta-i}{\zeta+i}\right)$ defined throughout the domain $\mathbf{C} \sim \Gamma$. Prove that $\operatorname{Log}\left(\dfrac{\zeta-i}{\zeta+i}\right)$ is well defined on the disc $|\zeta| < 1$, and agrees with $f(\zeta)$ there.

Hint: For $|\zeta| < 1$, $H(z) = \operatorname{Log}(z-\zeta)$ is analytic near Γ, with $dH/dz = 1/(z-\zeta)$. Then $f(\zeta) = \operatorname{Log}(i-\zeta) - \operatorname{Log}(-i-\zeta)$ (why?), which is *congruent* to $\operatorname{Log}\left(\dfrac{\zeta-i}{\zeta+i}\right)$ modulo $2\pi i$. Show the congruence is an equality for $|\zeta| < 1$.

5. Let ζ be a point in the plane and let $\delta > 0$ be given. If a sequence $\{\zeta_n\}$ converges to ζ, show that

$$h_n(z) = \frac{1}{z-\zeta_n} \quad \text{converges uniformly to} \quad h(z) = \frac{1}{z-\zeta}$$

(as functions of z) on the set $E = \{z: |z-\zeta| \geq \delta\}$.

6. Let A be any set in the plane and let $f_n(z)$ be functions that are well defined and converge uniformly on A to $f(z)$. If $g(z)$ is any bounded function on A, show that $f_n(z) \cdot g(z)$ converges uniformly to $f(z) \cdot g(z)$ on A, as $n \to \infty$.

7. If A is a closed bounded set in the plane, and if ζ is a point outside of A, show that

$$\frac{1}{(z-\zeta_n)(z-\zeta)} \quad \text{converges uniformly to} \quad \frac{1}{(z-\zeta)^2} \text{ on } A$$

as $n \to \infty$, if $\zeta_n \to \zeta$.

Hint: ζ lies a positive distance δ from A; now use Exercises 5 and 6.

8. The trajectory Γ of a contour is a closed bounded set in the plane, and any continuous function $f(z)$ defined on Γ must be bounded, so that $|f(z)| \leq M$ for z on Γ. If ζ lies off Γ, and if $\zeta = \lim_{n\to\infty} \zeta_n$, show that

$$\text{(i)} \quad \frac{f(z)}{(z-\zeta_n)} \to \frac{f(z)}{z-\zeta} \quad \text{as} \quad n \to \infty$$

$$\text{(ii)} \quad \frac{f(z)}{(z-\zeta_n)(z-\zeta)} \to \frac{f(z)}{(z-\zeta)^2} \quad \text{as} \quad n \to \infty$$

uniformly for z on Γ.

Hint: Use Exercises 5 to 7.

5.11 INTEGRAL FORMULAS FOR DERIVATIVES

The Cauchy Integral Formula determines values $f(\zeta)$ of a holomorphic function defined on a disc $|z - p| \leq r$ from its values on the boundary circle. Actually, there are similar formulas which determine the values of each derivative $f^{(n)}(\zeta)$ on the disc, starting with the boundary values of $f(z)$.

Theorem 5.23 (Cauchy's formula for derivatives) *Let $f(z)$ be holomorphic on an open set that contains the disc $D = \{z: |z - p| < r\}$ and its boundary circle $\Gamma = \{z: |z - p| = r\}$, and let γ be a parametrization of Γ which traverses this circle once in the counterclockwise direction. Then*

$$f^{(n)}(\zeta) = \frac{n!}{2\pi i} \int_\gamma \frac{f(z)}{(z-\zeta)^{n+1}}\, dz \qquad n = 0, 1, 2, \ldots \tag{45}$$

for all ζ such that $|\zeta - p| < r$.

PROOF: Since $f(z)$ is continuous on Γ, the formula

$$\tilde{f}(\zeta) = \frac{1}{2\pi i} \int_\gamma \frac{f(z)}{z-\zeta}\, dz$$

defines a function that is analytic on the complement $E = \mathbf{C} \sim \Gamma$, as explained in Section 5.10. The derivatives of $\tilde{f}(\zeta)$ have already been calculated by direct methods:

$$\tilde{f}^{(n)}(\zeta) = \frac{n!}{2\pi i}\int_\gamma \frac{f(z)}{(z-\zeta)^{n+1}}\,dz \quad \text{for } \zeta \text{ lying off } \Gamma,$$

for $n = 0, 1, 2, \ldots$. However, γ is a circular contour, and $f(z)$ is actually analytic on the disc $|z - p| \le r$, so the Cauchy Integral Formula insures that

$$f(\zeta) = \tilde{f}(\zeta) \quad \text{for} \quad |\zeta - p| < r.$$

Obviously, then, we also have $f^{(n)}(\zeta) = \tilde{f}^{(n)}(\zeta)$ on the disc $|\zeta - p| < r$, and formula (45) is proved. ▮

Notice that the boundary values $f(z)$ are multiplied by a different kernel function

$$K_n(\zeta, z) = \frac{n!}{2\pi i}\frac{1}{(z-\zeta)^{n+1}} \qquad n = 0, 1, 2, \ldots$$

in the integral (45) to get each derivative. The integrals (45) in Theorem 5.23 also make sense if ζ lies exterior to the circle Γ (but not if ζ lies *on* Γ). The comments presented after the proof of Cauchy's Integral Formula apply just as well to the present situation to show that

$$\text{(46)} \qquad \frac{1}{2\pi i}\int_\gamma \frac{f(z)}{(z-\zeta)^{n+1}}\,dz = 0 \quad \text{for} \quad n = 0, 1, 2, \ldots$$

for all ζ such that $|\zeta - p| > r$. Why is there this difference between points lying interior to and exterior to Γ? Qualitatively, the integrals differ in that γ loops once around the point ζ where the integrand is singular when $|\zeta - p| < r$, while γ does not enclose this singular point otherwise. More will be said about this when we discuss the "winding number" of a closed curve with respect to a point ζ (Section 5.15).

Example 5.15 Integrate $f(z) = (\sin z)/z^2$ along the counterclockwise parametrized circle $|z| = R$. This function is holomorphic (analytic!) on the punctured plane $E = \{z : z \ne 0\}$. The circle can be parametrized in the form $\gamma(t) = Re^{2\pi it}$, for $0 \le t \le 1$. Now

$$\frac{1}{2\pi i}\int_\gamma \frac{\sin z}{z^2}\,dz$$

is a special case of Cauchy's formula for the first derivative $\frac{d}{dz}(\sin z) = \cos z$, taking $\zeta = 0$; thus,

$$\int_\gamma \frac{\sin z}{z^2}\,dz = 2\pi i \cdot \cos(0) = 2\pi i.$$

The reader might try to obtain this result by direct calculations not involving the Cauchy formulas.

These formulas for derivatives give us a set of inequalities which show that the size of the derivatives $f^{(n)}(p)$ of an analytic function at the center of a circle are limited in a systematic way by the size of the values $|f(z)|$ on the boundary circle.

Theorem 5.24 (Cauchy's estimate for derivatives) *If $f(z)$ is analytic on an open set that includes a circle Γ centered at a point p, together with the disc bounded by Γ, then*

$$|f^{(n)}(p)| \le \frac{n!\,M}{r^n} \quad \text{for} \quad n = 0, 1, 2, \ldots \tag{47}$$

where r is the radius of the circle and M is the maximum of $|f(z)|$ on Γ.

PROOF: Since

$$\left|\frac{f(z)}{(z-p)^{n+1}}\right| = \frac{|f(z)|}{r^{n+1}} \le \frac{M}{r^{n+1}} \quad \text{for all } z \text{ on } \Gamma,$$

and length$(\Gamma) = 2\pi r$, we may apply Theorem 5.9 to the Cauchy formula for derivatives to get

$$|f^{(n)}(p)| \le \frac{2\pi r n!\,M}{2\pi r^{n+1}} = \frac{n!\,M}{r^n}, \quad \text{for} \quad n = 0, 1, 2, \ldots. \quad ∎$$

If the circle Γ is fixed, these estimates also show that the derivatives $f^{(n)}(p)$, and the associated Taylor series coefficients $a_n = f^{(n)}(p)/n!$, cannot grow too rapidly as $n \to \infty$.

1. Use Theorem 5.23 to evaluate the following integrals along counterclockwise oriented circular contours.

(i) $\displaystyle\int_\gamma \frac{\sin z}{z^2}\,dz \qquad |z| = 1$

(ii) $\displaystyle\int_\gamma \frac{\sqrt{z}}{(z-1)^3}\,dz \qquad |z-1| = \tfrac{1}{2}$

(iii) $\displaystyle\int_\gamma \frac{\tan z}{z^2+2z+1}\,dz \qquad |z+1| = \tfrac{1}{10}.$

Answer: (i) $i\pi$; (ii) $-i\pi/12$; (iii) $\pi i/\cos^2(1)$.

2. Calculate integrals along the counterclockwise parametrized circles shown. Sketch the location of the contour and the singularities of the integrand.

(i) $\int_\gamma \frac{\sinh z}{z^6}\,dz$ $\qquad |z-1| = 2$

(ii) $\int_\gamma \frac{\sqrt{z+1}}{z^3}\,dz$ $\qquad |z-\frac{1}{2}| = \frac{2}{3}$

(iii) $\int_\gamma \frac{\sin z}{(z-1)(z^2+z+1)}\,dz$ $\qquad |z-1| = 1$

(iv) $\int_\gamma \frac{e^z}{z^4-1}\,dz$ $\qquad |z| = \frac{1}{2}$

(v) $\int_\gamma \frac{e^z}{z^4-1}\,dz$ $\qquad |z| = 10$

Hint: In (v) write out a "partial fractions" expansion $\frac{1}{z^4-1} = \sum_{k=1}^{4} \frac{A_k}{z-p_k}$ where $\{p_k\}$ are the fourth roots of 1, and $\{A_k\}$ are suitable constants.

Answer: (i) $\frac{2\pi i}{5!}$; (ii) $-\frac{i\pi}{4}$; (iii) $\frac{1}{3}\sin(1)$; (iv) 0; (v) $i\pi\sinh(1) - i\pi\sin(1)$.

3. Evaluate the function of ζ defined by

$$F(\zeta) = \int_\gamma \frac{e^z}{(z+2)(z-\zeta)^2}\,dz$$

for $|\zeta| < 1$ and $|\zeta| > 1$. Here γ is the circle $|z| = 1$ parametrized counterclockwise.

Answer:

$$F(\zeta) = 2\pi i e^\zeta \frac{(\zeta+1)}{(\zeta+2)^2} \quad \text{for} \quad |\zeta| < 1; \; F(\zeta) = 0 \quad \text{if} \quad |\zeta| > 1.$$

4. Suppose that the series of holomorphic functions $\sum_{n=1}^{\infty} f_n(z)$ converges *uniformly* on an open set E to a function $f(z)$. Prove that

(i) $f(z)$ is holomorphic on E

(ii) The derivative of f is given by term-by-term differentiation of the series,

$$\frac{df}{dz} = \sum_{n=1}^{\infty} \frac{df_n}{dz} \quad \text{at every point in } E.$$

The series is not assumed to be related to a power series; the f_n can be arbitrary holomorphic functions. Thus, it is not obvious that the sum $f(z)$ is holomorphic.

Hint: Let p be a typical point in E, let the points z such that $|z - p| \leq r$ be a small disc in E about p, and let $\gamma(\theta) = p + re^{i\theta}$ for $0 \leq \theta \leq 2\pi$. The series converges uniformly on the circle $|z - p| = r$; justify the following equalities for $|\zeta - p| < r$:

$$f(\zeta) = \sum_{n=1}^{\infty} f_n(\zeta) = \sum_{n=1}^{\infty} \frac{1}{2\pi i} \int_\gamma \frac{f_n(z)}{z - \zeta}\, dz$$

$$= \frac{1}{2\pi i} \int_\gamma \frac{1}{z - \zeta} \left(\sum_{n=1}^{\infty} f_n(z) \right) dz$$

$$= \frac{1}{2\pi i} \int_\gamma \frac{f(z)}{z - \zeta}\, dz.$$

The right-hand integral always produces a function that is holomorphic on the disc (Theorem 5.22); thus, f is holomorphic at any point p in E. Once this is established, use the Cauchy formula for first derivatives on f to get (ii).

5. In Exercise 4 we do not really need uniform convergence of the series $\sum_{n=i}^{\infty} f_n$ to the function f *uniformly on E.* It would be sufficient that the series converge uniformly to f on every closed subdisc in E. Why?

Note: For example, if $f_n(z) = z^n/n!$, the series $e^z = \sum_{n=0}^{\infty} f_n(z)$ converges uniformly to its limit on every closed subdisc in the set $E = \mathbf{C}$, but it does not converge uniformly on E itself.

6. Show that the series $f = \sum_{n=1}^{\infty} f_n$ are uniformly convergent on every closed subdisc in the domains E indicated. Applying Exercises 4 and 5, explain why the derivative df/dz exists and calculate df/dz as a series. Can you verify that df/dz exists, in any of these examples, without appealing to Exercises 4 and 5?

(i) $\displaystyle\sum_{n=1}^{\infty} \frac{1}{z^2 - n^2}$ $\qquad E = \{z : z \neq n;\ n = \pm 1, \pm 2, \ldots\}$

(ii) $\displaystyle\sum_{n=1}^{\infty} \frac{e^{inz}}{n^2}$ $\qquad E = \{z : \text{Im}(z) > 0\}$

(iii) $\displaystyle\sum_{n=1}^{\infty} n^{-z}$ $\qquad E = \{z : \text{Re}(z) > 1\}$(Riemann zeta function)

Hint: For (iii) remember that $n^{-z} = e^{-z \cdot \text{Log}\, n}$; recall Exercise 17, Section 3.3.

5.12 APPLICATION: THE MEAN VALUE PROPERTY OF ANALYTIC FUNCTIONS

Let $f(z)$ be a continuous function defined at and near p, and consider the small circular contours around p given by

$$\gamma_r(t) = p + re^{it} \quad \text{for} \quad 0 \leq t \leq 2\pi.$$

Evidently, $f(z)$ is defined on the circular trajectory $\Gamma_r = \{z: |z - p| = r\}$ and on the interior disc $D_r = \{z: |z - p| < r\}$ for all small radii $r > 0$. It is natural to regard the Riemann integral

$$\frac{1}{2\pi}\int_0^{2\pi} f(\gamma_r(t))\, dt \tag{48}$$

as the **average value** of $f(z)$ on the circle Γ_r. This interpretation of (48) as an average is clear if we examine the Riemann sums associated with (48). For the partition π_N which divides $[0, 2\pi]$ into N equal length segments $I_1, \ldots, I_N$ we select points t_k^* in I_k so that t_k^* is the midpoint of the subinterval I_k, and then write $z_k^* = \gamma_r(t_k^*)$. The points $\{z_1^*, \ldots, z_N^*\}$ are equally spaced around Γ_r, and the average of the values of $f(z)$ at these points is just

$$\frac{1}{N}\sum_{k=1}^{N} f(z_k^*) = \frac{1}{N}(f(z_1^*) + \cdots + f(z_N^*)). \tag{49}$$

But this is also the Riemann sum associated with the partition π_N (and selected points $t_1^*, \ldots, t_N^*$):

$$\frac{1}{N}\sum_{k=1}^{N} f(z_k^*) = \frac{1}{2\pi}\sum_{k=1}^{N} f(z_k^*) \cdot \frac{2\pi}{N} = \frac{1}{2\pi}\sum_{k=1}^{N} f(\gamma_r(t_k^*)) \cdot \text{length}(I_k),$$

and these Riemann sums converge to the integral (48) as $N \to \infty$. The integral (48) is thus the limit of the averages (49) as the evenly distributed sampling points $\{z_1^*, \ldots, z_N^*\}$ are increased in number without limit. There seems to be no other choice than to assign the value of this limit as the average of the values of $f(z)$ on the circle Γ_r.

Analytic functions have a remarkable average value property; if we average over any small circle about a given point p, these averages all agree and always give the value at the center point $f(p)$.

Theorem 5.25 *Let $f(z)$ be analytic on an open set that includes p. Then for all small $r > 0$ we get*

$$f(p) = \frac{1}{2\pi}\int_0^{2\pi} f(\gamma_r(t))\, dt$$

where γ_r is the circular contour around p given by $\gamma_r(t) = p + re^{it}$, for $0 \leq t \leq 2\pi$.

PROOF: This is a simple variant of the Cauchy Integral Formula. Since $\gamma_r'(t) = ire^{it}$ and $\gamma_r(t) - p = re^{it}$, we get

$$f(p) = \frac{1}{2\pi i}\int_{\gamma_r} \frac{f(z)}{(z-p)}\,dz = \frac{1}{2\pi i}\int_0^{2\pi} f(\gamma_r(t))\,\frac{ire^{it}}{re^{it}}\,dt \cdot$$

$$= \frac{1}{2\pi}\int_0^{2\pi} f(\gamma_r(t))\,dt. \quad ∎$$

5.13 APPLICATION: ENTIRE FUNCTIONS AND LIOUVILLE'S THEOREM

Functions that are analytic throughout the complex plane, such as e^z and $\sin z$, are of special interest in complex analysis and are referred to as **entire functions** or **integral functions.** These functions suffer a curious dichotomy; if we let the variable z approach infinity, then the absolute values $|f(z)|$ either grow very rapidly (like the exponential function $e^{|z|}$) as $z \to \infty$ in certain ways, or else $f(z)$ is a *polynomial* in the variable z. There are no intermediate types of behavior for entire functions. This contrasts strongly with the behavior of analytic functions of a real variable; the function $f(x) = \cos x$ is one of many analytic functions defined on the real line that do not grow rapidly as $x \to \infty$, and yet are not polynomials in the variable x. A detailed study of entire functions is left to more advanced texts. We will only prove one basic result, Liouville's theorem; an extension of Liouville's theorem will be found in Exercises 4 and 5.

If $f(z)$ is analytic throughout the plane, the function $M(r)$ defined by taking

$$M(r) = \text{maximum value of } |f(z)| \text{ for } z \text{ on the circle } |z| = r$$

gives a convenient measure of the growth of $|f(z)|$ as z approaches infinity.

Theorem 5.26 (Liouville's theorem) *If f is entire and bounded, so that*

$$|f(z)| \le M \quad \textit{for all } z,$$

then $f(z)$ is a constant.

PROOF: Here $|f(z)| \le M$ for all z, so that $M(r) \le M$ for all $r > 0$. Take $\gamma_r(t) = re^{it}$ for $0 \le t \le 2\pi$, and apply the obvious estimates to the Cauchy Integral Formula for $f^{(n)}(0)$, using the contours γ_r. We get

$$\left|\frac{f^{(n)}(0)}{n!}\right| = \left|\frac{1}{2\pi i}\int_{\gamma_r}\frac{f(z)}{z^{n+1}}\,dz\right| \le \frac{M(r)}{r^n} \le \frac{M}{r^n} \quad \text{for all} \quad r > 0$$

for each $n = 0, 1, 2, \ldots$. Except when $n = 0$, the right side decreases to zero as $r \to +\infty$, while the left side does not depend on r. Obviously this means that $f^{(n)}(0) = 0$ for $n \ge 1$. Since f is analytic throughout the plane, it is represented everywhere by its Taylor series about $p = 0$,

$$f(z) = \sum_{n=0}^{\infty} \frac{f^{(n)}(0)}{n!} z^n = f(0) \quad \text{for all} \quad z,$$

and f is constant. ▮

EXERCISES

1. The growth indicator, $M(r)$ = maximum value of $|f(z)|$ on the circle $|z| = r$, is extremely useful in discussing the behavior of $|f(z)|$ as $z \to \infty$. Calculate $M(r)$ as a function of r for the functions listed, and indicate where on the circle $|z| = r$ the absolute value $|f(z)|$ is equal to $M(r)$.

(i) $f(z) = e^z$

(ii) $f(z) = z^n \qquad (n = 1, 2, \ldots)$

(iii) $f(z) = z^2 + 1 \qquad (\text{for } r \geq 1)$

(iv) $f(z) = \cos z$

Answer: (i) e^r; (ii) r^n; (iii) $r^2 + 1$ at $z = r + i0$; (iv) $\cosh(r)$ at $z = 0 + ir$.

2. Define the lower bound $m(r)$ = minimum value of $|f(z)|$ on the circle $|z| = r$. Obviously $m(r) \leq |f(z)| \leq M(r)$ on the circle $|z| = r$, so $m(r)$ is also useful in studying the growth of function values as $z \to \infty$. Calculate $m(r)$ for functions (i) to (iii) in the last example, and indicate where on the circle $|z| = r$ the value $m(r)$ is achieved. What can be said for $f(z) = \cos z$?

Answer: (i) e^{-r}; (ii) r^n; (iii) $r^2 - 1$ at $z = ir$; (iv) $m(r) = 0$ frequently; $m(r) = |\cos r|$.

3. Let $P(z) = a_n z^n + a_{n-1} z^{n-1} + \cdots + a_1 z + a_0$ be an nth degree polynomial (so $a_n \neq 0$); write $P(z) = a_n z^n + (\text{lower order terms}) = a_n z^n + Q(z)$. Prove that

(i) $$\tfrac{1}{2}\,|a_n z^n| \geq |Q(z)| \quad \text{for all} \quad |z| \geq R$$

if the radius R is chosen large enough. Use this (and the same radius R) to prove the following estimate on the size of $|P(z)|$ for large $|z|$.

(ii) $$\tfrac{1}{2}\,|a_n| r^n \leq |P(z)| \leq \tfrac{3}{2}\,|a_n| r^n \quad \text{on the circle} \quad |z| = r$$

provided $r \geq R$.

Hint: Use L'Hospital's rule to compare the real polynomials $|a_n| r^n$ and $|a_{n-1}| r^{n-1} + \cdots + |a_1| r + |a_0|$ as $r \to \infty$.

Note: Obviously $\tfrac{1}{2}\,|a_n| r^n \leq m(r) \leq |P(z)| \leq M(r) \leq \tfrac{3}{2}\,|a_n| r^n$ for $r \geq R$.

4. We say that a function $f(z)$ defined for all large $|z|$ has **polynomial growth at infinity** provided its growth indicator $M(r)$ is bounded by the function r^n for some n; that is, we can find $A > 0$ and an integer $n = 1, 2, \ldots$ such that $M(r) \leq Ar^n$ for all large r. Which functions in Exercise 1 have polynomial growth at infinity?

5. If $f(z)$ is entire, and if f has polynomial growth of degree at most n at infinity, prove that $f(z)$ is a polynomial of degree less than or equal to n.

Hint: Use the Cauchy estimates to estimate the size of $|f^{(n)}(0)/n!|$ in terms of $M(r)$; then let $r \to \infty$ to show that all higher order derivatives are zero in the Taylor series.

6. Exhibit functions defined and holomorphic on the half plane $\text{Im}(z) > 0$ which are not polynomials, but have polynomial growth at infinity in the sense that $|f(z)| \leq A\,|z|^n$ for $|z| \geq R$ in the half plane.

Answer: e^{iz}; Log z.

7. If $f(z)$ is defined for all large $|z|$ we could define a growth indicator with respect to a base point p other than the origin, $M'(r) =$ maximum of $|f(z)|$ on the circle $|z - p| = r$. This is particularly useful if $f(z)$ behaves symmetrically with respect to p. Calculate $M(r)$ and $M'(r)$ for $f(z) = 1/(z - p)^2$. Compare the behavior of these indicator functions as $r \to \infty$.

5.14 APPLICATION: THE FUNDAMENTAL THEOREM OF ALGEBRA

A polynomial $p(x)$ with real coefficients and a real variable may fail to have any real roots (real numbers x solving the equation $p(x) = 0$); for example, the polynomial $p(x) = 1 + x^2$ has no roots in the real number system. But it does have two roots, $z = +i$ and $z = -i$, if we regard p as a function of a complex variable. The reader is probably familiar with this idea of looking for complex roots of a polynomial, even if the polynomial has real coefficients, and he may recall that, in some sense, the number of complex roots of a polynomial is always equal to its degree. This can be easily established once we prove a simpler fact: *every polynomial that is not constant (degree zero) has at least one root in the complex number system.* Using this result, one can then extract successive roots to obtain them all.

Theorem 5.27 (Fundamental Theorem of Algebra) *A non-constant polynomial $p(z)$ with real or complex coefficients has at least one root in the complex plane.*

While this is a statement about a purely algebraic situation (existence of solutions of a polynomial equation), no purely algebraic proof exists for this important result. All proofs require a command of the calculus of functions of several real variables, or of complex analysis. Our complex variable proof is based on Liouville's theorem.

PROOF: Consider what happens when we form $g(z) = 1/p(z)$ if $p(z)$ has no roots (is never zero). Evidently, $g(z)$ is differentiable, hence analytic, throughout the plane. We investigate the growth of $|g(z)|$ as $|z| \to +\infty$ by writing $p(z)$ in the form

$$p(z) = a_n z^n + \cdots + a_1 z + a_0 \qquad (n \geq 0,\ a_n \neq 0).$$

The lead term $a_n z^n$ plays the decisive role in determining the growth of $|p(z)|$; it is not hard to prove that, for sufficiently large choice of R_0, we have

$$|p(z)| \geq \tfrac{1}{2}\,|a_n z^n| = \tfrac{1}{2}\,|a_n|\,|z|^n \quad \text{for all } z \text{ such that } |z| \geq R_0,$$

so that

$$|g(z)| = \frac{1}{|p(z)|} \leq \frac{2}{|a_n|\,|z|^n} \leq \frac{2}{|a_n|\,R_0{}^n},$$

provided that $|z| \geq R_0$. Thus $|g(z)|$ is bounded by the constant $2/|a_n|R_0^n$ outside of the closed disc $|z| \leq R_0$. On the other hand, any continuous function is bounded on a closed bounded set in the plane, so that $|g(z)|$ is bounded both on the disc and on its complement; therefore, $|g(z)|$ is bounded throughout the plane. By Liouville's theorem, $g(z) = c$ and $p(z) = 1/c$ (a constant) if $p(z)$ is a polynomial without roots. ▌

Given any root ζ, we can split a factor $(z - \zeta)$ out of the polynomial $p(z)$ and write $p(z) = (z - \zeta)q(z)$ for all z, where $q(z)$ is a polynomial one degree lower than $p(z)$. To see that this splitting is possible (and how to compute $q(z)$) we write the variable z as $z = (z - \zeta) + \zeta$, which gives

$$\begin{aligned} p(z) &= a_n((z - \zeta) + \zeta)^n + \cdots + a_1((z - \zeta) + \zeta) + a_0 \\ &= a_n(z - \zeta)^n + \cdots + (a_n\zeta^n + \cdots + a_1\zeta + a_0). \end{aligned}$$

The constant term left after this rearrangement of terms is zero, $a_n\zeta^n + \cdots + a_1\zeta + a_0 = p(\zeta) = 0$ since ζ is a root, and all other terms have a common factor $(z - \zeta)$, which can be split off. The lead term in the polynomial $q(z)$ which remains is $a_n(z - \zeta)^{n-1}$, and $a_n \neq 0$, so that degree(q) = degree$(p) - 1$.

For complex polynomials, every polynomial of degree greater than zero has a root, so we can split another factor off of $q(z)$ unless degree$(q) = 0$. Repeating this process n times, we get a factorization of $p(z)$ into linear polynomials

$$p(z) = a_n \cdot (z - \zeta_1) \cdot (z - \zeta_2) \cdot \cdots \cdot (z - \zeta_n) \quad \text{for all } z.$$

The numbers $\{\zeta_1, \ldots, \zeta_n\}$ we get are each a root of $p(z)$, since at least one factor in this product is zero if we set $z = \zeta_k$; furthermore, these numbers ζ_k include *all* roots of $p(z)$, because if $z \neq \zeta_k$ for $k = 1, 2, \ldots, n$ this insures that each factor is nonzero, so that $p(z) \neq 0$. The collection of numbers $\{\zeta_1, \ldots, \zeta_n\}$ may include certain roots several times; if we gather together the

ζ_k's which are the same, we can write the product in the form

$$p(z) = a_n \cdot (z - \zeta_1)^{m_1} \cdot \cdots \cdot (z - \zeta_l)^{m_l} \quad \text{for all } z$$

where the numbers $\{\zeta_1, \ldots, \zeta_l\}$ are the *distinct* roots of $p(z)$. The number m_k associated with the root ζ_k is called the **multiplicity** of the root, because ζ_k must appear m_k times in the factorization. Obviously, $m_1 + \cdots + m_l = \text{degree}(p) = n$, and if we agree to the convention of counting each distinct root of $p(z)$ as many times as its multiplicity, we arrive at the result:

(50) Every polynomial of a complex variable has as many roots as its degree.

This would be false if we counted only distinct roots, or if we allowed only a real variable in the polynomial.

EXERCISES

1. Calculate the roots of the following polynomials.

(i) $z^2 + iz - 1$
(ii) $z^3 + i$
(iii) $z^3 + z^2 - z - 1$
(iv) $z^4 + z^2 - 2iz - 1$
(v) $z^4 + 1$

What are the multiplicities of the roots? Write out each polynomial in factored form.

2. If a polynomial $P(z) = a_n z^n + \cdots + a_1 z + a_0$ has real coefficients, show that:

(i) If ζ is a non-real root of $P(z)$, so is its conjugate $\bar{\zeta}$.

(ii) There are an even number of non-real roots (regard $n = 0$ as even, if the roots are all *real*).

Thus the non-real roots occur in **conjugate pairs,** if there are any at all.

Hint: $a_n(\bar{\zeta})^n = \overline{(a_n \zeta^n)}$ if a_n is real; recall definition of a root of P.

3. Prove that every polynomial with *real* variable and *real* coefficients $P(x) = a_n x^n + \cdots + a_1 x + a_0$ can be factored as $P(x) = A \cdot f_1(x) \cdot f_2(x) \cdots f_m(x)$ where each factor is a linear or quadratic polynomial with real coefficients:

$$f(x) = x - r \qquad \text{(linear factor)}$$

$$f(x) = x^2 + rx + s \qquad \text{(quadratic factor)}$$

At the same time, show that $\text{degree}(P) = \sum_{k=1}^{m} \text{degree}(f_n)$.

Hint: Factor the corresponding complex polynomial with real coefficients $P(z)$. For each conjugate pair of roots ζ, $\bar{\zeta}$ we have a factor $(z - \zeta)(z - \bar{\zeta})$; multiply this out and verify that it has real coefficients. Once $P(z)$ has been written as a product of linear and quadratic factors with real coefficients, restrict attention to real z.

4. Show that every cubic polynomial $ax^3 + bx^2 + cx + d$ with real coefficients has at least one real root. Exhibit a real cubic polynomial with just one real root.

Hint: Use Exercise 2 above.

5.15 THE WINDING NUMBER (OR INDEX) OF A CONTOUR

If γ is a *closed* contour it is very helpful to take into account the number of times $\gamma(t)$ winds around various points ζ lying off its trajectory Γ. If $\gamma(t)$ is defined on an interval $I = [a, b]$ and if ζ is fixed, then for each t such that $a \leq t \leq b$ the function $\arg(\gamma(t) - \zeta)$ is well defined up to an added integral multiple of 2π, and measures the angle between the directed ray from ζ parallel to the positive real axis and the directed ray from ζ through the point $z = \gamma(t)$, as shown in Figure 5.15. We would like to define the **winding number,** or **index,** $I(\gamma, \zeta)$ of ζ with respect to γ to be $1/2\pi$ times the cumulative increment in $\arg(z - \zeta)$ as $z = \gamma(t)$ moves through the closed curve. This index is often indicated by the suggestive symbol

$$I(\gamma, \zeta) = \frac{1}{2\pi}\Delta_\gamma[\arg(z - \zeta)].$$

A geometric interpretation of the winding number is obtained if we think of an observer stationed at the point ζ who keeps turning to face the moving point $z = \gamma(t)$ as t increases. The index is the total number of rotations he has made when t reaches $t = b$ and $\gamma(t)$ returns to its starting point $\gamma(a) = \gamma(b)$. We adhere to the convention that counterclockwise rotations are counted positively and clockwise ones negatively. Thus, a negative value for $I(\gamma, \zeta)$ means that

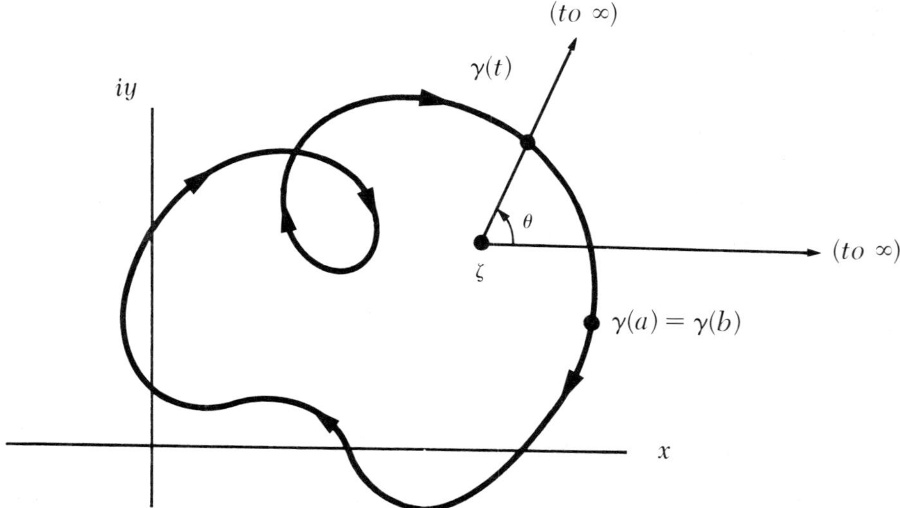

Figure 5.15 The significance of $\arg(\gamma(t) - \zeta) = \theta$. This angle, measured in radians, is ambiguous up to an added term $2\pi n$.

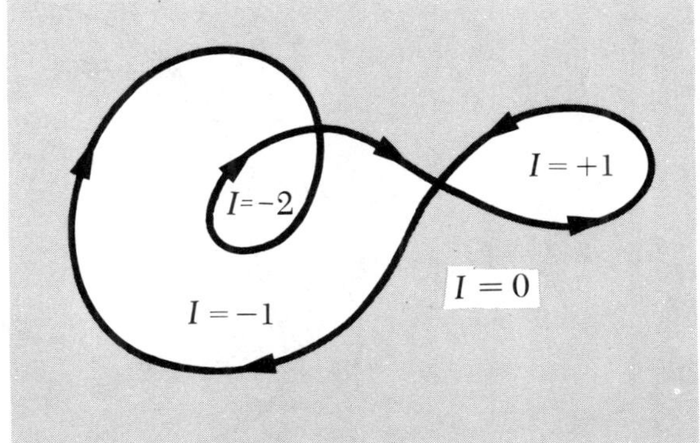

Figure 5.16 A contour and the index $I = I(\gamma, \zeta)$ in various parts of the complementary set $E = \mathbf{C} \sim \Gamma$.

$\gamma(t)$ loops around ζ more times in the clockwise direction than in the counterclockwise direction; zero index will signify that the contour does not loop around ζ at all, or does so an equal number of times in each direction. Figure 5.16 shows how the winding number $I(\gamma, \zeta)$ should be assigned for various points in the complement of Γ, and serves as a warning that the notion of winding number can be somewhat confusing if the contour is at all complicated. Therefore, our first task is to find alternative ways of calculating winding numbers.

The contour γ must be closed, so $\gamma(a) = \gamma(b)$, and we will never attempt to define a winding number for points ζ which lie on the trajectory Γ. We might try to define the index to be $1/2\pi$ times the difference in arguments when $t = a$ and $t = b$:

$$I(\gamma, \zeta) = \frac{1}{2\pi}\,[\arg(\gamma(b) - \zeta) - \arg(\gamma(a) - \zeta)], \tag{51}$$

but we immediately run into the difficulty that arguments are indeterminate up to added terms $2\pi n$ (n an integer). Nevertheless, our intuition has suggested a way to extract a definite number out of this formula; once we decide upon a definite choice for the value of $\arg(\gamma(t) - \zeta)$ at the initial point $t = a$, there is only one way left to choose the values of $\arg(\gamma(t) - \zeta)$ for $t > a$ to get a *continuous* function of t. This is the function we should use in (51). As indicated below, it is fairly easy to construct such a continuous determination of $\arg(\zeta(t) - \zeta)$ for $a \leq t \leq b$; moreover, the construction process yields an important dividend, a surprising and extremely useful integral formula for the index.

Theorem 5.28 *Let γ be a closed contour in the plane and let E be its complement, $E = \mathbf{C} \sim \Gamma$. Then for every point ζ in E we get*

$$I(\gamma, \zeta) = \frac{1}{2\pi i}\int_\gamma \frac{1}{z - \zeta}\,dz. \tag{52}$$

The details of the proof are somewhat technical and can be skipped without hindering the reader's progress through the remaining discussion, and applications, of winding numbers.

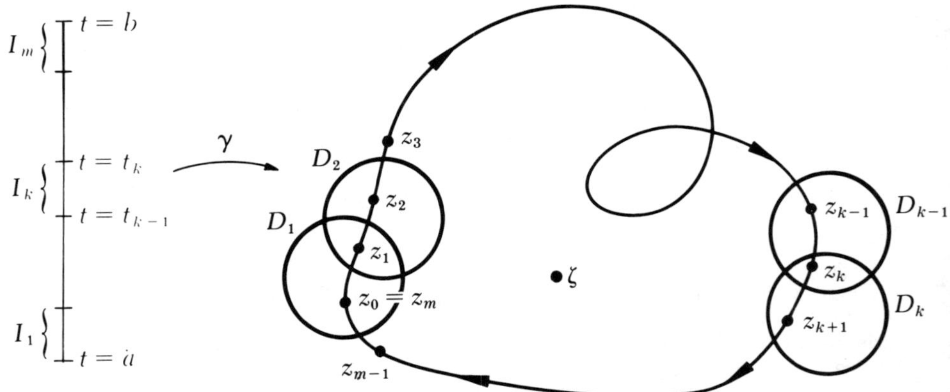

Figure 5.17 The proof of Theorem 5.28. Here $z_k = \gamma(t_k)$ and the base point is $z_0 = z_m$; disc D_k contains the arc from z_k to z_{k+1}.

PROOF: Fix ζ. We will work from elementary facts about $\log(z - \zeta)$ and $\arg(z - \zeta)$ which should be clear from our discussion in Chapter 2.

1. If D is any open disc that excludes the point ζ, there is a continuous determination of $\arg(z - \zeta)$ defined on D.

2. If $\text{Arg}^*(z - \zeta)$ is a continuous determination of $\arg(z - \zeta)$ defined on an open set E that excludes ζ, then $\text{Log}^*(z - \zeta) = \log|z - \zeta| + i\,\text{Arg}^*(z - \zeta)$ is a holomorphic (hence analytic) determination of $\log(z - \zeta)$ on E.

This is all very trivial if $\zeta = 0$, and for $\zeta \neq 0$ we reduce it to this case by examining $\arg(w)$ and $\log(w)$ where $w = z - \zeta$.

Now let us partition the interval $I = [a, b]$ on which γ is defined into successive intervals $I_1 = [t_0, t_1], \ldots, I_m = [t_{m-1}, t_m]$. By making these subintervals sufficiently small we can insure that the corresponding trajectory segments Γ_k, swept out by $\gamma(t)$ for t in I_k, each lie within a corresponding open disc D_k which excludes ζ (this is possible since $\gamma(t)$ is continuous and ζ lies a positive distance from the trajectory Γ). The situation is illustrated in Figure 5.17.

On the first disc D_1 we take any continuous determination of $\arg(z - \zeta)$ and call it

$$\text{Arg}_1(z - \zeta) \quad \text{for } z \text{ in } D_1.$$

On the next disc D_2 there are various continuous determinations of $\arg(z - \zeta)$, any two of which differ by an added constant $2\pi k$, but only one of these will agree with $\text{Arg}_1(z - \zeta)$ at the point $z_1 = \gamma(t_1)$, which is common to the discs D_1 and D_2. Let us call the appropriate determination

$$\text{Arg}_2(z - \zeta) \quad \text{defined for } z \text{ in } D_2.$$

Continuing this process, we get continuous determinations of $\arg(z - \zeta)$

$$\text{Arg}_k(z - \zeta) \quad \text{defined for } z \text{ in } D_k \ (k = 1, 2, \ldots, m)$$

which have the following compatibility property:

$$\text{(53)} \qquad \operatorname{Arg}_{k-1}(z - \zeta) = \operatorname{Arg}_k(z - \zeta) \text{ at the point } z_k = \gamma(t_k).$$

Note that z_k is common to the successive discs D_{k-1} and D_k.

Now consider the corresponding analytic determinations of $\log(z - \zeta)$,

$$\text{(54)} \quad \operatorname{Log}_k(z - \zeta) = \log|z - \zeta| + i \operatorname{Arg}_k(z - \zeta) \text{ defined for } z \text{ in } D_k;$$

these satisfy a similar compatibility condition:

$$\text{(55)} \qquad \operatorname{Log}_{k-1}(z - \zeta) = \operatorname{Log}_k(z - \zeta) \text{ at the point } z_k = \gamma(t_k).$$

With these functions in hand we may construct a continuous determination of $\arg(\gamma(t) - \zeta)$ for $a \leq t \leq b$ by taking

$$A(t) = \operatorname{Arg}_k(\gamma(t) - \zeta) \text{ for } t \text{ in the interval } I_k.$$

We might be troubled by the fact that we have two different prescriptions for $A(t)$ at the points $t = t_k$ where successive intervals overlap, but the compatibility condition (53) insures that they both give the same value. Since each of the functions $\operatorname{Arg}_k(z - \zeta)$ is continuous, $A(t)$ will be a continuous function of t, and it is certainly clear that $A(t)$ is one of the values of $\arg(\gamma(t) - \zeta)$ for each t in $[a, b]$. Thus, $A(t)$ is a continuous determination of $\arg(\gamma(t) - \zeta)$ for $a \leq t \leq b$, and $I(\gamma, \zeta) = (1/2\pi)(A(b) - A(a))$.

To prove (52), write $\gamma = \gamma_1 + \gamma_2 + \cdots + \gamma_m$ corresponding to our partitioning of the interval $I = [a, b]$; it is clear that

$$\frac{1}{2\pi i}\int_\gamma \frac{1}{z-\zeta}\,dz = \frac{1}{2\pi i}\int_{\gamma_1} \frac{1}{z-\zeta}\,dz + \cdots + \frac{1}{2\pi i}\int_{\gamma_m} \frac{1}{z-\zeta}\,dz.$$

But

$$\frac{d}{dz}\{\operatorname{Log}_k(z - \zeta)\} = \operatorname{Log}_k'(z - \zeta) \cdot \frac{d}{dz}(z - \zeta) = \frac{1}{z - \zeta} \text{ on } D_k,$$

and the trajectory of γ_k is just the curve segment Γ_k, which lies within D_k. Thus,

$$\frac{1}{2\pi i}\int_{\gamma_k} \frac{1}{z-\zeta}\,dz = \frac{1}{2\pi i}[\operatorname{Log}_k(z_k - \zeta) - \operatorname{Log}_k(z_{k-1} - \zeta)],$$

and on adding up these terms we get a long sum in which successive terms (except for the first and last) cancel out because of the compatibility condition (55).

$$\begin{aligned}
\frac{1}{2\pi i}\int_\gamma \frac{1}{z-\zeta}\,dz &= \frac{1}{2\pi i}[-\operatorname{Log}_1(z_0 - \zeta) + \operatorname{Log}_1(z_1 - \zeta) - \operatorname{Log}_2(z_1 - \zeta) \\
&\qquad + \operatorname{Log}_2(z_2 - \zeta) - \cdots] \\
&= \frac{1}{2\pi i}[-\operatorname{Log}_1(z_0 - \zeta) + \operatorname{Log}_1(z_1 - \zeta) - \operatorname{Log}_1(z_1 - \zeta) \\
&\qquad + \cdots + \operatorname{Log}_m(z_m - \zeta)] \\
&= \frac{1}{2\pi i}[\operatorname{Log}_m(p - \zeta) - \operatorname{Log}_1(p - \zeta)].
\end{aligned}$$

Here $p = \gamma(a) = \gamma(b)$ is the base point of γ (recall that $t_0 = a$ and $t_n = b$, so that $p = z_0 = z_m$ in this formula). Now the integral takes the form:

$$\frac{1}{2\pi i}[\mathrm{Log}\,|p-\zeta| + i\,\mathrm{Arg}_m(p-\zeta) - \log|p-\zeta| - i\,\mathrm{Arg}_1(p-\zeta)]$$

$$= \frac{1}{2\pi i}\cdot i\cdot[\mathrm{Arg}_m(p-\zeta) - \mathrm{Arg}_1(p-\zeta)]$$

$$= \frac{1}{2\pi}[\mathrm{Arg}_m(\gamma(b)-\zeta) - \mathrm{Arg}_1(\gamma(a)-\zeta)].$$

But, defining $A(t)$ as above, this is just the identity we want:

$$\frac{1}{2\pi i}\int_\gamma \frac{1}{z-\zeta}\,dz = \frac{1}{2\pi}[A(b) - A(a)] = I((\gamma, \zeta). \quad \blacksquare$$

Before we turn to the uses of the winding number we list a few of its elementary properties.

PROPERTY 1. If $-\gamma$ is an order reversed reparametrization of γ, then $I(-\gamma, \zeta) = -I(\gamma, \zeta)$ for all ζ (this is clear from the integral formula of Theorem 5.28).

PROPERTY 2. $I(\gamma, \zeta)$, defined on the complement $E = \mathbf{C} \sim \Gamma$, is holomorphic (hence analytic) on E (see Theorem 5.22).

PROPERTY 3. Since $I(\gamma, \zeta)$ is analytic and can take only *integer* values, it must be locally constant—taking the same value at all points near a given point ζ_0—and $I(\gamma, \zeta)$ must be *constant* on any *domain* that does not meet Γ (see Theorem 2.17).

Example 5.16 Consider the contour $\gamma(t) = p + re^{it}$, defined for $0 \le t \le 2\pi$, which moves once counterclockwise around the circle $\Gamma = \{z: |z-p| = r\}$. We could determine the index for points off Γ by noting how the angle $\theta(t)$, shown in Figure 5.18, varies as $\gamma(t)$ traverses Γ. It should be clear that the

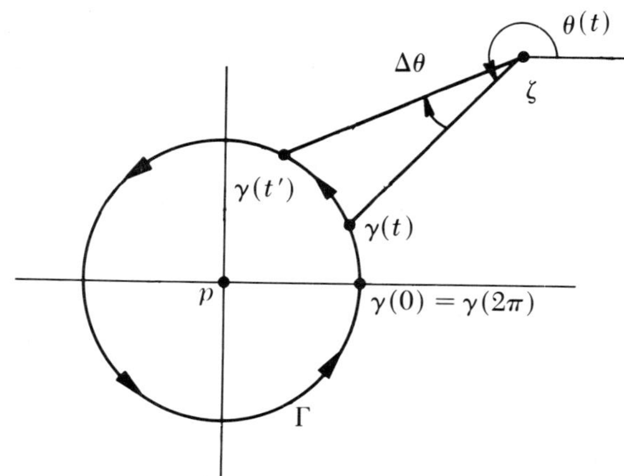

Figure 5.18 The cumulative change in angle $\theta(t)$ as t increases is $I(\gamma, \zeta)$, which is zero if $|\zeta - p| > r$ as shown. Here $t' > t$ and $\Delta\theta$ is negative (clockwise).

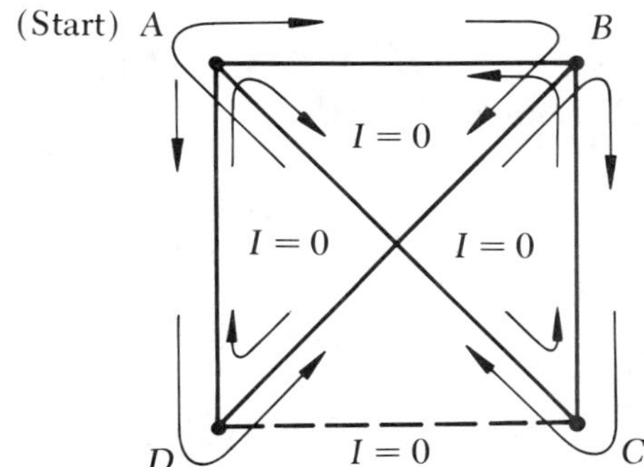

Figure 5.19 The closed contour of Example 5.17. $I(\gamma, \zeta) = 0$ for every point off of Γ.

net change in this angle as t increases is

$$I(\gamma, \zeta) = \frac{1}{2\pi} \Delta_\gamma[\arg(z - \zeta)] = \begin{cases} 0 & \text{if } |\zeta - p| > r \\ +1 & \text{if } |\zeta - p| < r. \end{cases}$$

On the other hand, we would also obtain this result using the integration formula (52); for circular contours, the integral

$$I(\gamma, \zeta) = \frac{1}{2\pi i} \int_\gamma \frac{1}{z - \zeta} \, dz$$

is a special case of the Cauchy Integral Formula.

Example 5.17 Perhaps the following contour will give the reader more of a challenge. Form γ by piecing together the parametrized line segments made up of certain sides and diagonals of the square ABCD shown in Figure 5.19, moving through vertices in the order ADBCABDACBA (the initial/final point is A). This contour has zero index at every ζ in the complement $E = \mathbf{C} \sim \Gamma$. Notice that each segment in γ is covered once in each direction, so it is obvious that integrals cancel, giving

$$\frac{1}{2\pi i} \int_\gamma \frac{1}{z - \zeta} \, dz = 0 \quad \text{all } \zeta \text{ in } E.$$

We leave the reader to arrive at the same conclusion by keeping track of increments of $\arg(\gamma(t) - \zeta)$ as t increases, as in the last example. In this particular example it is clearly advantageous to calculate winding numbers via the integral formula (52), although there are other situations in which the integral is much harder to evaluate than the cumulative increase in $\arg(\gamma(t) - \zeta)$.

A contour has a trajectory that is a bounded set in the plane,† and therefore it can be enclosed within a sufficiently large disc, say $D = \{z : |z| < R\}$. For

† A contour is a continuous mapping $\gamma : I \to \mathbf{C}$ of a closed interval $I = [a, b]$ in $\mathbf{R}$ into the complex plane. The function $f(t) = |\gamma(t)|$ is continuous and real valued on $[a, b]$, and therefore must have a (finite) maximum, achieved at some point in the interval. Take R greater than this maximum.

all ζ outside D we have $I(\gamma, \zeta) = 0$ because the integrand $1/(z - \zeta)$ appearing in (52) is holomorphic throughout D; D is a convex open set and Cauchy's Theorem gives

$$I(\gamma, \zeta) = \frac{1}{2\pi i} \int_\gamma \frac{1}{z - \zeta}\, dz = 0 \quad \text{if} \quad |\zeta| > R.$$

In other words, winding numbers have another basic property:

PROPERTY 4. We have $I(\gamma, \zeta) = 0$ for all points ζ sufficiently far removed from the origin.

This allows us to define the winding number of the point at infinity to be $I(\gamma, \infty) = 0$, and makes the winding number a continuous function of ζ on the complement of Γ in the complex sphere.

EXERCISES

1. Verify the winding numbers assigned to points in different parts of the plane in Figure 5.16, by keeping track of angle increments.

2. Let E be any domain that includes the closed disc $|z - p| \leq r$. Let γ be the boundary circle $|z - p| = r$, parametrized as $\gamma(t) = p + re^{it}$ for $0 \leq t \leq 2\pi$. Prove that γ has *zero index* $I(\gamma, \zeta) = 0$ at any point γ lying outside of E.

Hint: Use formula (52) for winding numbers. Invoke Cauchy's Theorem as in the discussion of Property 4.

3. If E is any domain that includes the closed disc $|z - p| \leq r$, prove that *every* closed contour γ whose trajectory lies inside this disc has the property: $I(\gamma, \zeta) = 0$ for all ζ in the complement $\mathbf{C} \sim E$ of E.

4. Show that a closed contour lying within the square $-1 \leq x$, $y \leq 1$ has zero winding number about any point ζ lying outside of this square.

Hint: If ζ is not in S, we can choose a determination of $\log(z - \zeta)$ that is analytic at least on S.

5.16 THE JORDAN CURVE THEOREM

If γ is a *simple* closed contour, so that γ never crosses over itself except at the initial/final point where $\gamma(a) = \gamma(b)$, our intuition strongly suggests that Γ must divide the plane into two disjoint domains (connected open sets) E_1 and E_2, which consist of the points which lie "inside of" Γ and the points which lie "outside of" Γ, respectively. Of course this is only to be expected for *simple* contours; the contour shown earlier in Figure 5.16 defies such analysis.

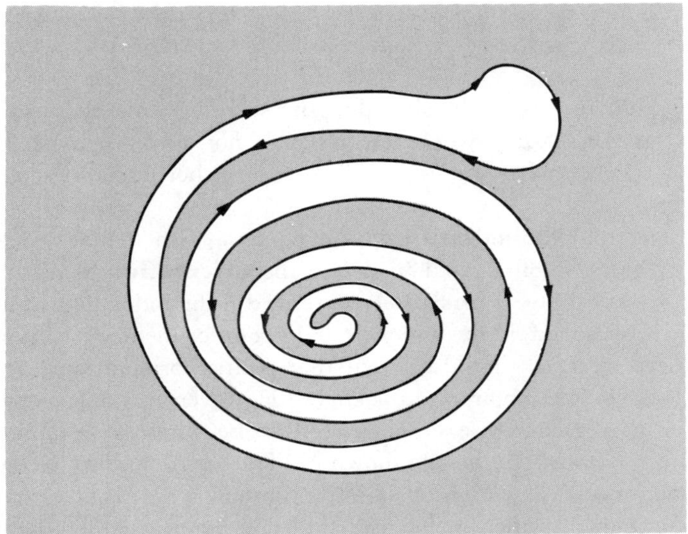

Figure 5.20 The curve in the discussion of "interior" points for a simple closed curve in the plane. The shaded parts are the exterior points.

Although our expectation is borne out by experience, it is surprisingly difficult to prove this result. Part of the problem is to define just what we mean by saying that a point ζ is "inside" or "outside" of the simple closed curve γ; another difficulty is that our considerations must work for *every* conceivable simple closed contour, no matter how intricately it moves around in the plane (the reader might consider the contour shown in Figure 5.20 with this in mind). The notion of winding number provides the key to this classification of points lying off Γ into categories of "inside" and "outside" points. Even with this, there is still too much to be done for us to attempt a proof of this result; we will occasionally refer to the Jordan Curve Theorem later on, but these uses will be rather superficial, and could be avoided entirely by modest rewording of the text here and there. In fact, one can give a full account of complex variable theory without ever invoking the Jordan Curve Theorem, and so there is no compelling reason to go into the difficulties of its proof; the much more elementary notion of winding number (and the idea that a curve "encloses" a point ζ if the index is nonzero) is adequate for all of complex analysis.

Theorem 5.29 (Jordan Curve Theorem) *Let γ be a simple closed contour in the plane and let E be the complement of its trajectory, $E = \mathbf{C} \sim \Gamma$. Let us define the interior and exterior points with respect to γ to be*

$$E_{int}(\gamma) = \{z : I(\gamma, z) \neq 0\}$$

$$E_{ext}(\gamma) = \{z : I(\gamma, z) = 0\}.$$

Then these sets are disjoint and together fill up E. The set of exterior points $E_{ext}(\gamma)$ is an unbounded set which contains all points sufficiently far removed from the origin, while the set of interior points $E_{int}(\gamma)$ is a bounded set. Both sets are domains (connected), so *that* Γ *divides the plane into disjoint connected pieces, one bounded and the other unbounded. On* E_{int} *the index is either* $+1$ *everywhere, or else is* -1 *everywhere.*

The main difficulty is in showing that the sets E_{int} and E_{ext} are connected and that on the set E_{int} where the index is non-zero we must have either $I(\gamma, \zeta) = +1$ everywhere or $I(\gamma, \zeta) = -1$ everywhere, and these are the only possibilities.

The fact that the index is constant on E_{int}, with $+1$ or -1 as its only allowable values, is often used to define the **orientation** of a simple closed contour; we say that γ is **positively oriented** if the index is $+1$ and that it is **negatively oriented** if the index is -1. For elementary contours, like a parametrized circle, it should be clear that positive orientation corresponds to counterclockwise motion around the curve. However, it is not so easy to decide whether γ moves clockwise or counterclockwise when we examine a really complicated contour like the one shown in Figure 5.20, and so we must make a precise (non-intuitive) definition of "orientation" if we want to handle quite general contours. We leave the reader to determine whether the contour in Figure 5.20 is positively or negatively oriented. Determining which points are in E_{int} and which are in E_{ext} also requires some thought in this example; but since E_{ext} is connected and includes all points far from the origin, one sees that the exterior points make up the shaded area in Figure 5.20 (simply propagate the shading as far as it will go without crossing the trajectory Γ).

5.17 WINDING NUMBERS AND THE CAUCHY THEOREMS

We have already explained what we mean by $-\gamma$, this being any order reversed reparametrization of γ, and by the sum $\gamma = \gamma_1 + \gamma_2 + \cdots + \gamma_n$ of several contours whose successive initial and final points match up (recall Section 5.5). These definitions are somewhat ambiguous, but their ambiguities (which involve choosing reparametrizations) are irrelevant when we use these expressions in the line integral $\int_\gamma f(z)\,dz$, which is really the only place we have used them. As long as line integrals are the objects of primary interest to us, we will often find that several isolated contours $\gamma_1, \ldots, \gamma_m$ lying within the domain of definition of a function $f(z)$ should be treated collectively as a single object, a sort of "generalized contour" $\phi = \gamma_1 + \cdots + \gamma_m$; then we may define the integral of $f(z)$ along this combination ϕ by the formula

$$\int_\phi f(z)\,dz = \int_{\gamma_1} f(z)\,dz + \cdots + \int_{\gamma_m} f(z)\,dz.$$

For example, we might consider the set $\mathscr{H}(E)$ of functions that are defined and analytic on the domain $E = \mathbf{C} \sim \{+i, -i\}$. We will soon find that there are natural reasons for regarding the contours γ, η_1, and η_2 shown in Figure 5.21 as a single generalized contour $\phi = \gamma + \eta_1 + \eta_2$; indeed, we will soon demonstrate (in the general Cauchy Theorem below) that

$$(56) \qquad 0 = \int_\phi f(z)\,dz = \int_\gamma f(z)\,dz + \int_{\eta_1} f(z)\,dz + \int_{\eta_2} f(z)\,dz$$

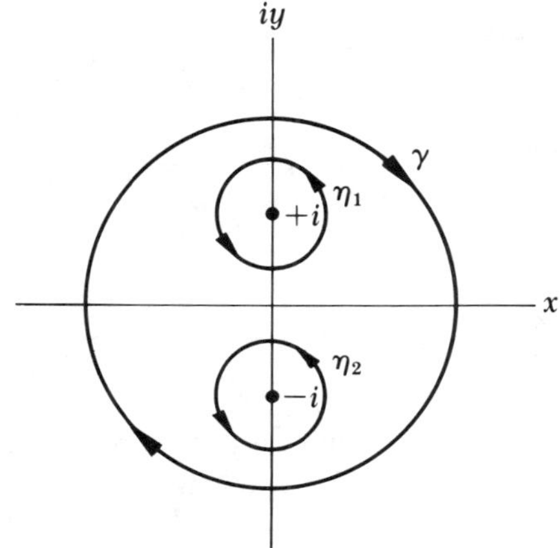

Figure 5.21 Three contours which are being considered as a single "generalized contour" $\phi = \gamma + \eta_1 + \eta_2$ in the domain $E = \mathbf{C} \sim \{+i, -i\}$.

for *every* analytic function on E. We can only get this surprising null result by combining the contours γ, η_1, and η_2 as a whole into the generalized contour ϕ; none of the ordinary contours which make up ϕ can have the property that *every* analytic function on E has integral *zero* (see Exercise 7). We will give examples below, but first we take a moment to define our terms.

A **generalized contour** (sometimes referred to as a **chain** in the literature) in a domain E is any expression of the form

(57) $$\gamma = n_1\gamma_1 + \cdots + n_m\gamma_m$$

where $\gamma_1, \ldots, \gamma_m$ are contours lying within E and $n_1, \ldots, n_m$ are integers. The end points of the contours γ_k are not required to match up; the integer n_k indicates the "multiplicity" of γ_k in the formal combinations (57)—the number of times it is to be counted—and a negative multiplicity is interpreted as counting the order reversed contour $-\gamma_k$ a positive number of times. These generalized contours will only appear in line integrals:

$$\int_{n_1\gamma_1 + \ldots + n_m\gamma_m} f(z)\, dz.$$

We may integrate any function $f(z)$ that is defined and continuous on E along a generalized contour γ, by defining

(58) $$\int_{n_1\gamma_1 + \cdots + n_m\gamma_m} f(z)\, dz = n_1\int_{\gamma_1} f(z)\, dz + \cdots + n_m\int_{\gamma_m} f(z)\, dz.$$

The **trajectory** of a generalized contour $\gamma = n_1\gamma_1 + \cdots + n_m\gamma_m$ is interpreted as the union of the points in the individual trajectories $\Gamma_1, \ldots, \Gamma_m$ of the ordinary contours involved. Furthermore, we call a generalized contour γ a **cycle** (or **generalized closed contour**) if each contour γ_k appearing in the expression (57) is a *closed* contour. In applications, our main interest will be in integrating functions along cycles.

If $\gamma = n_1\gamma_1 + \cdots + n_m\gamma_m$ is a cycle, we can define its **winding number** or **index** at points lying off its trajectory to be the sum of the individual winding numbers, taken with coefficients n_k:

$$I(\gamma, \zeta) = n_1 I(\gamma_1, \zeta) + \cdots + n_m I(\gamma_m, \zeta). \tag{59}$$

Obviously,

$$I(\gamma, \zeta) = \sum_{k=1}^{m} n_k I(\gamma_k, \zeta) = \sum_{k=1}^{m} \frac{n_k}{2\pi i} \int_{\gamma_k} \frac{1}{z - \zeta}\, dz = \frac{1}{2\pi i} \int_{\gamma} \frac{1}{z - \zeta}\, dz \tag{60}$$

(we use our definition of the integral of a function along a cycle on the right side), so the integral formula for winding numbers has the same appearance whether we apply it to cycles or to ordinary closed contours.

After a few computational examples we shall pursue our primary goal, which is to use generalized contours and cycles in order to state Cauchy's theorems in their most powerful and useful form. In its ultimate form Cauchy's Theorem says this: if E is any domain and γ any cycle which does not wind around points outside of E with nonzero index, then $\int_\gamma f(z)\, dz = 0$ for *every* analytic function $f(z)$ defined on E.

Example 5.18 Let γ_1 and γ_2 be the counterclockwise parametrized circular contours

$$\gamma_1(t) = R_1 e^{it} \qquad \gamma_2(t) = R_2 e^{it} \qquad (0 \le t \le 2\pi)$$

centered at the origin, as shown in Figure 5.22. As examples of generalized contours (in fact, *cycles*) that can be formed by combining γ_1 and γ_2, we might consider

$$\phi_1 = \gamma_1 + \gamma_2 \qquad (\text{where } n_1 = +1,\ n_2 = +1)$$
$$\phi_2 = \gamma_1 - \gamma_2 = \gamma_1 + (-\gamma_2) \qquad (\text{where } n_1 = +1,\ n_2 = -1)$$
$$\phi_2 = 2\gamma_1 \qquad (\text{where } n_1 = +2,\ n_2 = 0)$$

The winding numbers are well defined in the open set E we get by removing the circles $|z| = R_1$ and $|z| = R_2$ from the plane; the values of the winding number

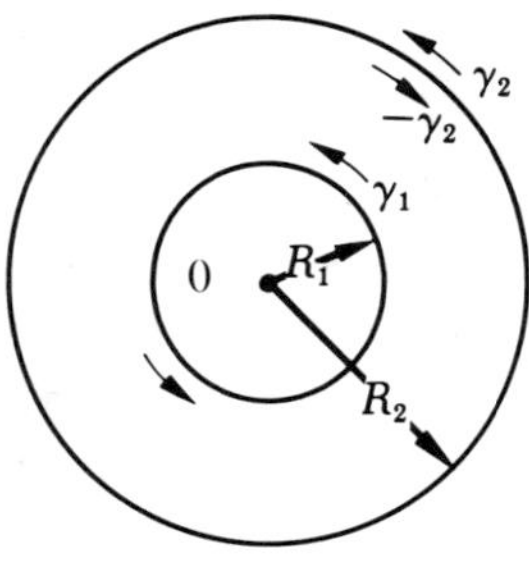

Figure 5.22 The contours in the cycles $\phi = \gamma_1 \pm \gamma_2$ in Example 5.18.

Table 5.1 The Index $I(\gamma, \zeta)$ For Various Contours and Cycles

	$0 \leq \|\zeta\| < R_1$	$R_1 < \|\zeta\| < R_2$	$R_2 < \|\zeta\| < +\infty$
γ_1	$+1$	0	0
γ_2	$+1$	$+1$	0
$\phi_1 = \gamma_1 + \gamma_2$	$+2$	$+1$	0
$\phi_2 = \gamma_1 - \gamma_2$	0	-1	0
$\phi_3 = 2\gamma_1$	$+2$	0	0

for each of these cycles, and for the original closed contours γ_1 and γ_2, are shown in Table 5.1.

To illustrate the calculation of integrals along these generalized contours, we will calculate the integral of $f(z) = 1/z$, which is continuous on the trajectories Γ_1 and Γ_2. From previous calculations we know that

$$\int_{\gamma_1} \frac{1}{z}\, dz = 2\pi i \quad \text{and} \quad \int_{\gamma_2} \frac{1}{z}\, dz = 2\pi i,$$

and by using the defining formula (58) we see that

$$\int_{\phi_1} \frac{1}{z}\, dz = \int_{\gamma_1} \frac{1}{z}\, dz + \int_{\gamma_2} \frac{1}{z}\, dz = 4\pi i$$

$$\int_{\phi_2} \frac{1}{z}\, dz = \int_{\gamma_1} \frac{1}{z}\, dz - \int_{\gamma_2} \frac{1}{z}\, dz = 0$$

$$\int_{\phi_3} \frac{1}{z}\, dz = 2 \cdot \int_{\gamma_1} \frac{1}{z}\, dz = 4\pi i.$$

We could perform similar calculations to evaluate the integral, along any one of these cycles, of any function $f(z)$ that is defined and continuous on the circles Γ_1 and Γ_2. Notice that these cycles cannot be regarded as ordinary contours because they consist of two isolated arcs.

Definition 5.6 *Let E be an open set in the complex plane. We say that a cycle γ is* **homologous to zero within** *E if $I(\gamma, \zeta) = 0$ for all points ζ lying outside of E.*

Intuitively, γ is homologous to zero in E if γ does not wind around any point ζ in the complement of E with nonzero index.

If we are given two cycles in an open set E, say

$$\gamma = p_1\gamma_1 + \cdots + p_m\gamma_m$$

$$\eta = q_1\eta_1 + \cdots + q_l\eta_l,$$

and if $I(\gamma, \zeta) = I(\eta, \zeta)$ for all points outside of E, this means that γ and η have the same winding properties with respect to every point outside of E, and

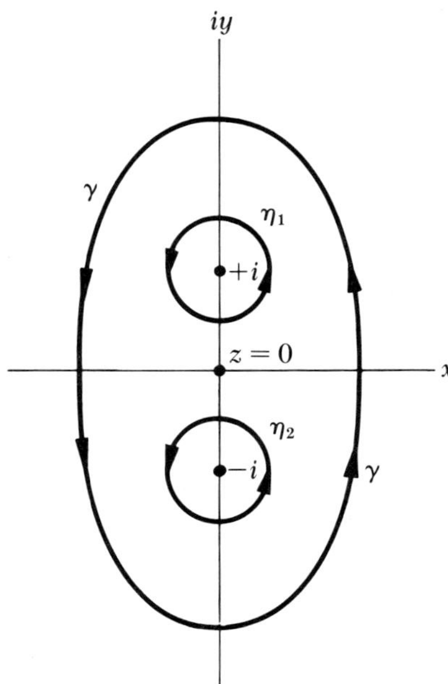

Figure 5.23 The contours in Example 5.19. Here $E = \mathbf{C} \sim \{+i, -i\}$ and $F = \mathbf{C} \sim \{+i, -i, 0\}$.

we say that the cycles γ and η are **homologous** (to each other) **within** E. Since the combined cycle $\phi = \gamma - \eta$ has index $I(\phi, \zeta) = I(\gamma, \zeta) - I(\eta, \zeta) = 0$ for every point ζ it follows that γ and η are homologous within E if and only if the cycle $\phi = \gamma - \eta$ is homologous to zero in E. In the special case when our set E is the whole complex plane, there are no points at all outside of E and we agree, by definition, that every cycle (or closed contour) is homologous to zero in this particular domain.

Example 5.19 Consider the domain E consisting of the complex plane with the points $+i$ and $-i$ deleted, $E = \mathbf{C} \sim \{+i, -i\}$. The parametrized ellipse shown in Figure 5.23 is homologous, within E, to the cycle $\eta = \eta_1 + \eta_2$ consisting of two small counter-clockwise oriented circles η_1 and η_2 centered at the points $+i$ and $-i$, respectively. In fact, the complement $\mathbf{C} \sim E$ is just the pair of points $\{+i, -i\}$, and by keeping track of angle increments as $\gamma(t)$, $\eta_1(t)$, and $\eta_2(t)$ move around their respective trajectories, the reader can easily verify that we have the winding numbers shown in Table 5.2 at $\zeta = +i, -i$. Since the cycles γ and η have the same winding properties with respect to each of the (two) points outside of E, these cycles are homologous within E.

It is not really necessary for the circles η_1 and η_2 to be centered at $+i$ and

Table 5.2 The winding numbers of the cycles in Figure 5.23

	$\zeta = +i$	$\zeta = -i$
η_1	+1	0
η_2	0	+1
$\eta = \eta_1 + \eta_2$	+1	+1
γ	+1	+1

$-i$, as long as they loop around these points with index $+1$; it isn't even necessary to use circles for the contours η_1 and η_2, and we could have used various other simple closed contours that stay inside of γ and loop around the respective points $+i$ and $-i$ with index $+1$. In this example it is best to calculate winding numbers by keeping track of angle increments; evaluating integrals like (52) for an elliptical contour can lead to Riemann integrals which are extremely difficult to evaluate.

The same contours γ, η_1, η_2 when placed in a different domain F might not be homologous in F. Thus, the property of two cycles being homologous depends on the shape of the domain being considered (this should be clear from the definition). For example, if we also delete the origin in Figure 5.23, and consider γ and η as cycles within the smaller domain $F = \mathbf{C} \sim \{+i, -i, 0\}$, then these cycles are *not homologous within* F because the point $\zeta = 0$ is in the complement of F, and the cycles have different winding numbers there;

$$I(\gamma, 0) = +1 \quad \text{while} \quad I(\eta, 0) = 0 + 0 = 0.$$

EXERCISES

1. Consider the circular contours shown in Figure 5.24. Calculate winding numbers at points in the sets E_1, E_2, and E_3 for the following cycles.

(i) $\phi = \gamma_1 + \gamma_2$
(ii) $\phi = \gamma_1 - 2\gamma_2$
(iii) $\phi = 2\gamma_2$

Answer: (i) $+1, +1, 0$; (ii) $+1, -2, 0$; (iii) $0, +2, 0$.

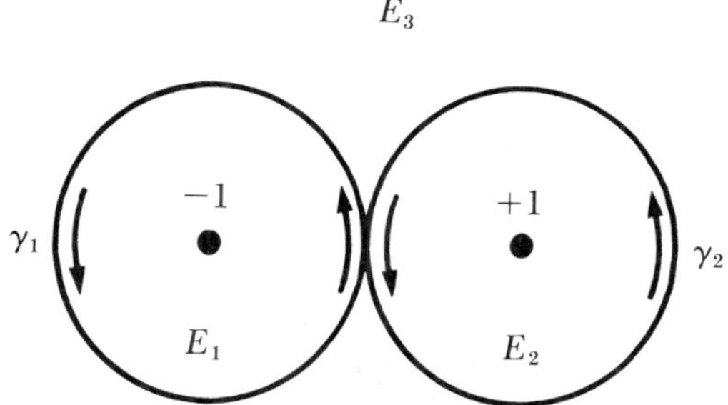

Figure 5.24

2. Let γ_1 be the counterclockwise parametrized circle $|z| = 1$ and γ_2 the ellipse $x^2 + 16y^2 = 4$, parametrized as in Figure 5.25. Tabulate the winding numbers $I(\phi, \zeta)$ for points ζ in the domains $E_1, \ldots, E_6$ for each cycle ϕ listed below.

(i) $\phi = \gamma_1 + \gamma_2$
(ii) $\phi = \gamma_1 - \gamma_2$
(iii) $\phi = \gamma_2 - \gamma_1$
(iv) $\phi = 2\gamma_1 - 2\gamma_2$
(v) $\phi = \gamma_2 + \gamma_1$
(vi) $\phi = \gamma_2 - 3\gamma_1$

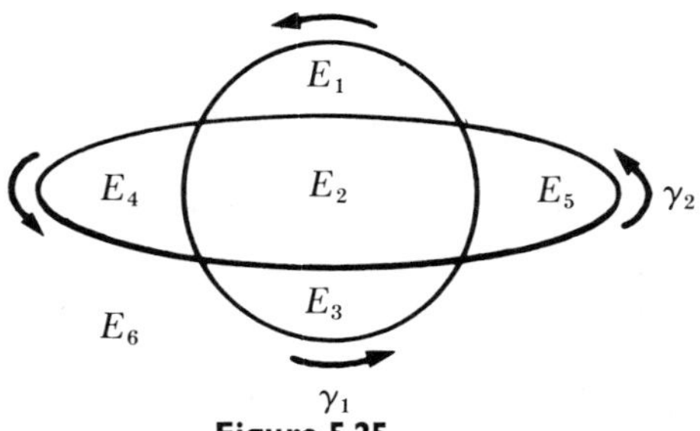

Figure 5.25

Answer: (i) $+1, +2, +1, +1, +1, 0$; (ii) $+1, 0, +1, -1, -1, 0$; (iii) $-1, 0, -1, +1, +1, 0$.

3. Calculate a table of values for the integrals

$$\text{(i)} \int_\phi z^2 \, dz$$

$$\text{(ii)} \int_\phi \frac{1}{z - \frac{3}{2}} \, dz$$

for the cycles ϕ in Exercise 2 above.

Answer: First integrals always zero; (i) $2\pi i$; (ii) $-2\pi i$; (iii) $2\pi i$; (iv) $-4\pi i$; (v) $2\pi i$; (vi) $2\pi i$.

4. Given (non-closed) contours shown in Figure 5.26, calculate $\int_\phi z^2 \, dz$ for the generalized contours ϕ indicated below.

(i) $\phi = \gamma_1$

(ii) $\phi = \gamma_3$

(iii) $\phi = \gamma_1 + \gamma_2 + \gamma_3 + \gamma_4$

(iv) $\phi = \gamma_1 - 5\gamma_3$

(v) $\phi = \gamma_1 - \gamma_2$

(vi) $\phi = \gamma_3 - \gamma_4$

(vii) $\phi = \gamma_4 - \gamma_3$

(viii) $\phi = \gamma_4 + \gamma_3$

(ix) $\phi = \gamma_3 - \gamma_4 + 2\gamma_1$

Answer: (i) $\frac{7}{3}$; (ii) $-\frac{2}{3}$; (iii) $\frac{10}{3}$; (iv) $+\frac{17}{3}$; (v) 0; (vi) 0; (vii) 0; (viii) $-\frac{4}{3}$; (ix) $\frac{14}{3}$.

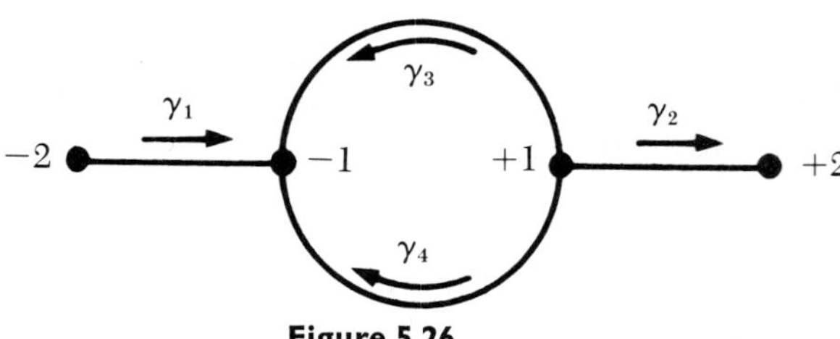

Figure 5.26

5. If $\gamma_1 = e^{i\theta}$; $\gamma_2 = 2e^{3i\theta}$; $\gamma_3 = 3e^{2i\theta}$ for $0 \leq \theta \leq 2\pi$, verify that these contours have the winding numbers indicated in Table 5.3 on the domains $E_1 = \{z: |z| < 1\}$; $E_2 = \{z: 1 < |z| < 2\}$; $E_3 = \{z: 2 < |z| < 3\}$; $E_4 = \{z: |z| > 3\}$.

Table 5.3 Winding Numbers in Exercise 5

	ζ in E_1	ζ in E_2	ζ in E_3	ζ in E_4
γ_1	1	0	0	0
γ_2	3	3	0	0
γ_3	2	2	2	0

Then tabulate the winding numbers for the following cycles ϕ.

(i) $\phi = 2\gamma_1 - \gamma_2$

(ii) $\phi = \gamma_1 + \gamma_2 + \gamma_3$

(iii) $\phi = \gamma_2 - \gamma_1 - \gamma_3$

(iv) $\phi = \gamma_2 - 3\gamma_1$

(v) $\phi = 2\gamma_3$

6. Which choices of integers l, m, and n give a cycle $\phi = l \cdot \gamma + m\eta_1 + n\eta_2$ (formed from the contours shown in Figure 5.21) that is homologous to zero in the domain $E = \mathbf{C} \sim \{+i, -i\}$?

7. For the contours γ, η_1, and η_2 shown in Figure 5.21, verify the values for $\int_\phi f(z)\,dz$ for the functions and contours given in Table 5.4. The integrands $f(z)$ are analytic on $E = \mathbf{C} \sim \{+i, -i\}$; for each individual contour, there is an analytic function on E whose integral is non-zero.

Table 5.4 Values of the integral of f(z) over ϕ

	$\dfrac{1}{z-i}$	$\dfrac{1}{z+i}$	$\dfrac{1}{z^2+1}$
η_1	$2\pi i$	0	$+\pi$
η_2	0	$2\pi i$	$-\pi$
γ	$-2\pi i$	$-2\pi i$	0

Note: We will soon show that the integral of every analytic function on E is zero for the cycle $\phi = \gamma + \eta_1 + \eta_2$.

Hint: These integrals all reduce to versions of the Cauchy integral formula for circular contours.

8. Using the contours γ, η_1, and η_2 from Exercise 7, tabulate the values of these integrals along the following cycles.

(i) $\phi = \gamma + \eta_1 + \eta_2$

(ii) $\phi = \gamma + \eta_1$

(iii) $\phi = \gamma - \eta_2$

(iv) $\phi = \eta_1 + \eta_2$

(v) $\phi = \eta_1 - \eta_2$

(vi) $\phi = \gamma - \eta_1 - \eta_2$

(vii) $\phi = 2\gamma - \eta_1 - \eta_2$

(viii) $\phi = \gamma - 2\eta_1$

Could any of these cycles give the null result $\int_\phi f(z)\,dz = 0$ for *all* analytic functions $f(z)$ on the domain $E = \mathbf{C} \sim \{+i, -i\}$?

9. Prove that any closed contour (simple or not) in the domain $E = \{z : \text{Im}(z) > 0\}$ is necessarily homologous to zero in E. Explain why it follows that

(i) Every cycle ϕ in E is homologous to zero in E;
(ii) If ϕ and ψ are cycles in E they must be homologous in E.

10. Repeat the discussion in Exercise 9 for the more complicated domain $E = \{z : \text{Im}(z) > 0 \text{ and } |z| > 1\}$—a half plane with a semi-disc removed.

Hint: If ζ is in the complement of E, there is a well defined determination of $\text{Log}(z - \zeta)$ in E. Then use Theorems 5.15 and 5.28.

11. Suppose we have two domains E and F with $E \supseteq F$, and let γ be a closed contour lying within F. Prove that if γ is homologous to zero within F, it is homologous to zero in the larger domain E.

12. The function tan z is analytic except at the points $\left\{z_k = \frac{\pi}{2} + k\pi : k = 0, \pm 1, \pm 2, \ldots\right\}$. In the domain E obtained by deleting these singular points from the plane, consider the rectangular contour γ and the small circular contours γ_k about z_k shown in Figure 5.27. Which circles γ_k, and multiplicities n_k, should be used to form a cycle $\phi = \Sigma\, n_k \gamma_k$ which is homologous to γ in the domain E? If η is the circular contour $\eta(s) = 10e^{-4\pi i s}$ defined for $0 \le s \le 1$, how should the γ_k be combined to get a cycle homologous to η in E?

Note: η has winding number -2 in places.

Answer: (i) $\phi = \gamma_{-3} + \gamma_{-2} + \gamma_{-1} + \gamma_0 + \gamma_1 + \gamma_2$;
(ii) $\phi = -2\gamma_{-3} - 2\gamma_{-2} - 2\gamma_{-1} - 2\gamma_0 - 2\gamma_1 - 2\gamma_2$.

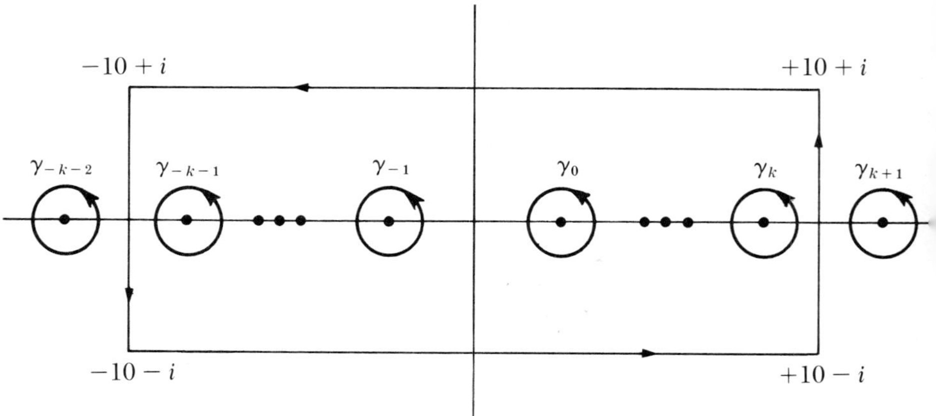

Figure 5.27

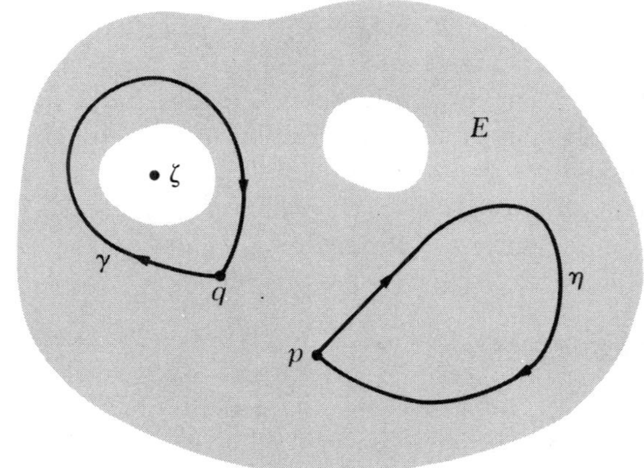

Figure 5.28 A multiply connected domain E. The closed contour γ winds around certain points ζ in the complement of E with non-trivial index; contour η is homologous to zero in E.

5.18 SIMPLY CONNECTED SETS IN THE PLANE

The geometric property of simple connectedness for an open set E makes precise the intuitive notion that E has no "holes" in it. Observe that if E is a domain with a "hole", such as the domain shown in Figure 5.28, it should be possible to devise a closed contour in E which loops around the hole to give nonzero index at points ζ in the hole (these points lie *outside* of E). With this in mind, we shall take the following *definition* of simply connected sets.

Definition 5.7 *An open set E in the plane is* **simply connected** *if*

(i) *The set E is connected (a domain), and*
(ii) *Every closed contour γ in E is homologous to zero; thus, no closed contour in E can wind around points outside of E.*

If a domain is not simply connected, we say it is **multiply connected.** The set shown in Figure 5.28 is multiply connected; notice that some closed contours in E may be homologous to zero within E (such as the contour η), but there must be at least one contour which is not.

Example 5.20 Every *convex* domain is simply connected. In fact, if γ is a closed contour in E we may calculate its index at a typical point ζ outside of E by formula (52):

$$I(\gamma, \zeta) = \frac{1}{2\pi_i} \int_\gamma \frac{1}{z - \zeta}\, dz.$$

But the integrand $f(z) = 1/(z - \zeta)$ is complex differentiable (analytic) on E, and must have a well defined antiderivative since E is convex (by our original version of Cauchy's theorem—Theorem 5.16). Thus line integrals of $f(z)$ are path independent in E and $I(\gamma, \zeta) = 0$ for every point ζ outside of E; i.e., γ is homologous to zero in E.

Example 5.21 The domain we get by deleting a single point p from the complex plane, $E = \mathbf{C} \sim \{p\}$, is connected but is not simply connected because the parametrized circles about p of the form $\gamma_R(t) = p + Re^{it}$ (defined for $0 \leq t \leq 2\pi$) evidently have non-zero index at the (one) point $\zeta = p$ in the complement of E. The deletion of finitely many more points only makes the situation worse, as far as simple connectedness goes. Similar reasoning shows that an **annulus**—a ring-shaped domain like $E = \{z : \frac{1}{2} < |z| < 1\}$—is not simply connected, although it is connected.

EXERCISES

1. Prove that the domain obtained by removing the interval $[\frac{1}{2}, 1]$ from the open disc $|z| < 1$ gives a domain E (a slit disc) that is not convex. Prove that E is, nevertheless, simply connected. Do likewise for the cut plane $E = \mathbf{C} \sim (-\infty, 0]$.

Hint: Find determinations of $\operatorname{Log}(z - \zeta)$ on E when ζ is in $\mathbf{C} \sim E$.

2. Verify that the semi-annulus $E = \{z : \operatorname{Im}(z) > 0 \text{ and } 1 < |z| < 2\}$ is a simply connected domain.

3. Prove that the domain exterior to the ellipse $x^2 + 4y^2 = 1$ is *not* simply connected.

Hint: Do it by integrations along circular, not elliptical, contours. Use the index formula in Theorem 5.28.

4. Prove that every star-shaped domain (definition in Exercise 4 of Section 5.7) is simply connected.

Hint: Use the more powerful version of Cauchy's theorem cited in Exercise 4 of Section 5.7.

5.19 THE GENERALIZED CAUCHY THEOREMS AND THEIR APPLICATIONS

The notion of two cycles (or closed contours) being homologous within some open set E may seem a little subtle at first glance. But once the reader has mastered the notion of winding number he will find it quite easy in practice to verify, by inspection, whether cycles in a given set E are homologous or not. Not only are the concepts of winding number and homologous cycles easy to handle on an intuitive level, but they are exactly the notions we need to coherently state Cauchy's Theorem and the Cauchy Integral Formulas in their most general and useful form. It turns out that these notions allow us to state precisely, in a practically useful way, when two closed contours (or cycles) in a given set E *give the same line integral for all analytic functions defined on* E. Therefore, they tell us precisely when we may replace one contour by another contour (or cycle) in evaluating a line integral, a piece of information of the greatest practical importance.

Theorem 5.30 (Generalized Cauchy theorem) *Let E be any open set in the plane and let γ be a cycle (or closed contour) which is homologous to zero within E. Then*

$$\int_\gamma f(z)\,dz = 0 \tag{61}$$

for every function $f(z)$ that is defined and analytic throughout E. Hence if γ and η are homologous cycles within E, then

$$\int_\gamma f(z)\,dz = \int_\eta f(z)\,dz \tag{62}$$

for every function $f(z)$ that is defined and analytic throughout E.

It is clear from the definitions that (62) follows from (61). The proof of this theorem involves a sequence of manipulations of integrals over cycles that is too complicated to include in this text; to the reader interested in the details of this result we recommend the account given in Pennisi [19], Sections 5.8 and 5.9. Just one point about the proof will be relevant in the following discussion. The proof rests directly on the elementary Cauchy theorems (Theorems 5.16 and 5.18), and so applies *only* to integrals of functions $f(z)$ that are holomorphic (hence analytic) throughout the set E we are considering. It cannot be applied to the wider class of integrands that are merely continuous, but not differentiable.

For simply connected domains we obtain a far-reaching path independence result for line integrals along *arbitrary contours* in E (not necessarily closed contours).

Corollary 5.31 *Let E be a simply connected domain. Then line integrals are globally path independent in E for any function $f(z)$ that is analytic throughout E. Thus if γ and η are contours in E with common initial and final points, we get*

$$\int_\gamma f(z)\,dz = \int_\eta f(z)\,dz \quad \textit{for every analytic function on } E.$$

The proof from Theorem 5.30 is simple; $\phi = \gamma - \eta$ is a *closed* contour in E since the initial/final points of γ and η match up. Also, ϕ is homologous to zero in E because E is assumed simply connected. Therefore,

$$0 = \int_\phi f(z)\,dz \quad \text{(because } E \text{ is simply connected)}$$

and

$$\int_\phi f(z)\,dz = \int_\gamma f(z)\,dz - \int_\eta f(z)\,dz \quad \text{(by formula (58), Section 5.17)}$$

for every analytic function on E. ▮

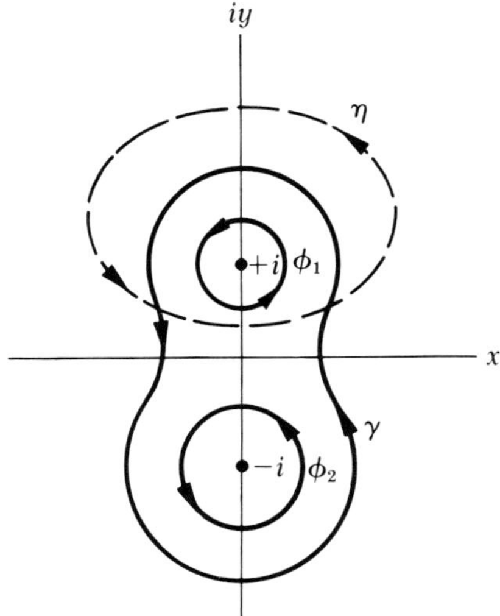

Figure 5.29 The contour γ in

$$\int_\gamma \frac{1}{z^2 + 1}\,dz.$$

Another interesting contour is shown as a dotted line.

The generalized Cauchy theorem can be used to reduce the calculation of integrals along complicated contours to the use of Cauchy's Integral Formula for elementary circular contours. This will be explained in great detail in the next chapter, so we only give a brief example of this reduction here.

Example 5.22 Let $f(z) = 1/(1 + z^2)$, which is analytic except at the singular points $+i$ and $-i$, and let us compute the integral

$$\int_\gamma \frac{1}{1 + z^2}\,dz$$

for various closed contours γ. The integrand may be factored as a product of two simpler functions $g(z) = 1/(z + i) = 1/(z - (-i))$ and $h(z) = 1/(z - i)$, which are analytic except at the points $-i$ and $+i$ respectively. Let γ be the irregular closed contour shown in Figure 5.29; we can devise a cycle ϕ which is homologous to γ in the domain $E = \mathbf{C} \sim \{+i, -i\}$ by taking two small counterclockwise circular contours ϕ_1 and ϕ_2 about the points $+i$ and $-i$, and forming $\phi = \phi_1 + \phi_2$. From Table 5.5 of winding numbers at the (two) points $\zeta = +i, -i$ which lie outside of E, we see that γ and ϕ are homologous within E. Therefore,

$$\int_\gamma f(z)\,dz = \int_\phi f(z)\,dz = \int_{\phi_1} f(z)\,dz + \int_{\phi_2} f(z)\,dz$$

Table 5.5 Winding numbers at $+i$ and $-i$

	$\zeta = +i$	$\zeta = -i$
ϕ_1	$+1$	0
ϕ_2	0	$+1$
$\phi = \phi_1 + \phi_2$	$+1$	$+1$
γ	$+1$	$+1$

for any analytic function $f(z)$ on E, by Theorem 5.30. The integrals along ϕ_1 and ϕ_2 are simple examples of Cauchy's Integral Formula (for circular contours):

$$\int_{\phi_2} \frac{1}{z^2+1}\,dz = \int_{\phi_2} \frac{h(z)}{z-(-i)}\,dz = 2\pi i\left[h(\zeta)\Big|_{\zeta=-i}\right] = -\pi$$

$$\int_{\phi_1} \frac{1}{z^2+1}\,dz = \int_{\phi_1} \frac{g(z)}{z-(+i)}\,dz = 2\pi i\left[g(\zeta)\Big|_{\zeta=+i}\right] = +\pi.$$

Combining these results we see that

$$\int_{\gamma} \frac{1}{z^2+1}\,dz = 2\pi i \cdot h(-i) + 2\pi i \cdot g(+i) = 0.$$

The result would be rather different if we had taken a contour that enclosed only one of the points $+i$ and $-i$ (such as the dashed contour η in Figure 5.29). Since the methods used to reduce the computation to a case of Cauchy's Integral Formula for circular contours would be the same as those above, we leave the reader to verify that

$$\int_{\eta} \frac{1}{1+z^2}\,dz = 2\pi i \cdot g(+i) = +\pi.$$

In the last example, Theorem 5.30 was coupled with the Cauchy Integral Formula (for circular contours) to calculate the values of certain integrals along irregular contours. It can also be coupled with the Cauchy Integral Formula for higher derivatives, in the same way.

Example 5.23 Let us take $E = \mathbf{C}$. Using the rectangular contour γ shown in Figure 5.30, we shall evaluate the integral

(63) $$\int_{\gamma} \frac{e^z}{(z+1)^2}\,dz = \int_{\gamma} \frac{e^z}{z^2+2z+1}\,dz.$$

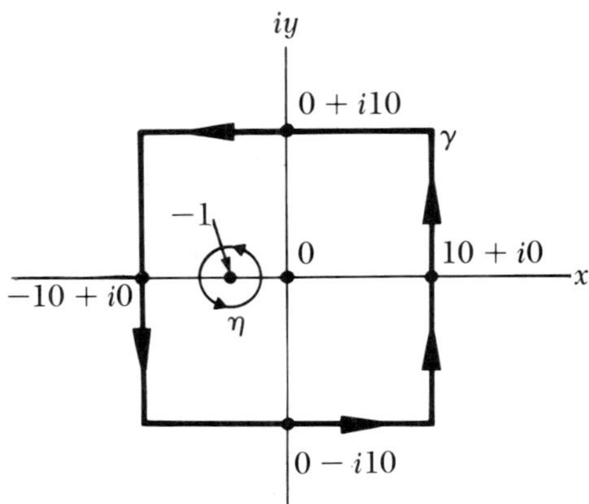

Figure 5.30 The contours γ and η in Example 5.23.

The integrand is not analytic at $\zeta = -1$, since $(z - \zeta)^2$ appears in the denominator. Let us introduce a small counterclockwise oriented circular contour η about ζ, as shown in the figure. Regarding γ and η as contours in the domain $E^* = E \sim \{\zeta\} = \{z : z \neq -1\}$, it is obvious that γ and η are homologous within E^*; both contours have index $+1$ at ζ, the only point in the complement of E^*. Furthermore, the integrand in (63) is analytic on E^*, since we have removed the point ζ where it fails to be analytic. Applying Theorem 5.30, and the Cauchy Integral Formula,

$$\int_\gamma \frac{e^z}{z^2 + 2z + 1}\, dz = \int_\eta \frac{e^z}{(z - \zeta)^2}\, dz$$
$$= 2\pi i \left[\frac{d}{dz}(e^z)\Big|_{z=-1}\right] = 2\pi i e^{-1} = \frac{2\pi i}{e}.$$

This result is not a consequence of the elementary Cauchy Integral Formula, valid only for *circular* contours, and would be exceedingly difficult to obtain by direct calculations. This should illustrate the desirability of having the general Cauchy Theorem in hand as we turn to the applications of integration theory in the next chapter.

Application 1 (A General Cauchy Integral Formula) Let $f(z)$ be analytic on a *simply connected* domain E, and let γ be a closed contour in E. Then the following generalization of Cauchy's Integral Formula for derivatives is valid.

$$\text{(64)} \qquad \frac{1}{2\pi i}\int_\gamma \frac{f(z)}{(z - \zeta)^{n+1}}\, dz = I(\gamma, \zeta) \cdot \frac{f^{(n)}(\zeta)}{n!} \qquad n = 0, 1, 2, \ldots$$

for all ζ in $E \sim \Gamma$. The Cauchy-type integral (64) is analytic for ζ in $\mathbf{C} \sim \Gamma$ (recall Section 5.10); however, for points ζ outside of E the integral cannot be compared with $f^{(n)}$, which is not defined off E.

For a concrete example, the reader might consider the contour in the right half plane E, shown in Figure 5.31, taking $f(z) = \tanh(z)$.

To prove (64) for a typical point ζ in $E \sim \Gamma$, take a small counterclockwise oriented circular contour η centered at ζ, and regard γ and η as closed contours in $E^* = E \sim \{\zeta\}$. Now $f(z)/(z - \zeta)^{n+1}$ is analytic on E^* (since ζ is not in E^*). Our basic task is to find a simple cycle ϕ in E^* that is homologous to γ in E^*,

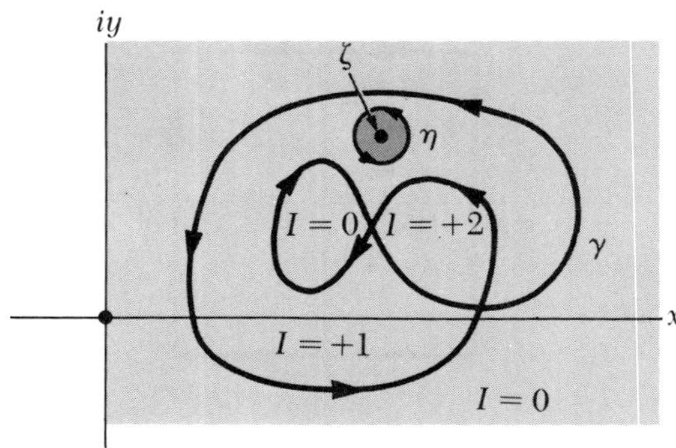

Figure 5.31 An example of the situation described in Application 1. Here $E = \{z : \mathrm{Re}(z) > 0\}$ and $f(z) = \tanh(z)$. Singularities of $f(z)$ lie on the imaginary axis, out of E.

Table 5.6 Index of the Cycles in Figure 5.31 at Points Outside the Domain E*

	$w = \zeta$	$w \neq \zeta$ w outside of E
η	$+1$	0
$\phi = k\eta$	k	0
γ	$I(\gamma, \zeta)$	0

so that the integral (64) along γ can be replaced by the (hopefully easier to calculate) integral along ϕ.

The indices of γ and η are listed in Table 5.6 for points in the complement of E^*. Obviously $I(\eta, \zeta) = +1$ and $I(\eta, w) = 0$ for points w lying outside of E; in fact, $I(\eta, w) = 0$ for all points outside the disc enclosed by η. Since E is simply connected, $I(\gamma, w) = 0$ for w outside of E, but the value $k = I(\gamma, \zeta)$ is not specified in our hypotheses and may be arbitrary, since ζ is in E. Obviously, the cycles γ and $\phi = k\eta = I(\gamma, \zeta) \cdot \eta$ are homologous within E^*. The general Cauchy Theorem tells us that

$$\frac{1}{2\pi i}\int_\gamma \frac{f(z)}{(z-\zeta)^{n+1}}\,dz = \frac{k}{2\pi i}\int_\eta \frac{f(z)}{(z-\zeta)^{n+1}}\,dz = I(\gamma, \zeta)\cdot\frac{f^{(n)}(\zeta)}{n!},$$

which is the formula we want.

Application 2 (Determinations of log z on simply connected domains) Let E be any simply connected domain that excludes $z = 0$. Then $f(z) = 1/z$ is analytic on E, and line integrals of $f(z)$ are globally path independent, by Corollary 5.31. Therefore, just as in Theorem 5.13, we can define an antiderivative throughout E by taking

(65) $$F(\zeta) = \int_p^\zeta f(z)\,dz \quad \text{for all } \zeta \text{ in } E,$$

where p is a fixed base point in E and $\int_p^\zeta (\cdots)\,dz$ stands for the integral along any contour in E that connects p to ζ. The function $F(\zeta)$ is well defined in spite of the ambiguity of the particular contour to be used in (65), and $F(z)$ is differentiable (hence analytic) on E, with

(66) $$\frac{dF}{dz}(z) = \frac{1}{z} \quad \text{for all } z \text{ in } E.$$

Now we want to see how close $F(z)$ comes to being a determination of $\log z$ defined throughout E. This would be accomplished if $e^{F(z)} = z$ on E. Computing the derivative,

$$\frac{d}{dz}\left\{\frac{e^{F(z)}}{z}\right\} = \frac{F'(z)e^{F(z)}}{z} - \frac{e^{F(z)}}{z^2} = \frac{e^{F(z)} - e^{F(z)}}{z^2} = 0 \quad \text{all } z \text{ in } E,$$

we see that $e^{F(z)}/z$ is constant on the domain E, so that

$$e^{F(z)} = c \cdot z \quad \text{for all } z \text{ in } E,$$

for some complex constant c ($c \neq 0$ because the exponential on the left side can never equal zero). Evidently, $F(z)$ comes very close to being a determination of $\log z$; to get an actual determination of $\log z$ that is analytic on E, let us adjust $F(z)$ by adding to it a complex constant α such that $e^{\alpha} = 1/c$; this is possible since e^{z} takes on every complex value except zero. The new function $G(z) = F(z) + \alpha$ is analytic too, and satisfies the characteristic equation

$$e^{G(z)} = e^{F(z)+\alpha} = e^{F(z)} \cdot e^{\alpha} = z \quad \text{for all } z \text{ in } E.$$

This result may be summarized as follows.

Theorem 5.32 *If E is any simply connected domain that excludes $z = 0$, we can then define an analytic determination of $\log z$ on E. Any pair of determinations of $\log z$ will differ by an added constant term of the form $2\pi ni$, where n is some integer.*

By taking the imaginary part of such a determination of $\log z$ we get a continuous determination of $\arg z$ defined on any simply connected domain that excludes the origin.

With this result in hand it is easy to show that there are analytic determinations of various other multiple valued functions on simply connected domains. For example, we can set up analytic determinations of z^{μ} on any simply connected domain that excludes zero by taking a determination of $\log z$ and defining $f(z) = e^{\mu \cdot \log z}$. This function is obviously analytic. There are many useful variations of this construction.

The use of the generalized Cauchy Theorem in practical applications will be explained further in the next chapter. We conclude this chapter by observing that its uses all fit into the following archetypal situation, which may help clarify the reader's thinking on these matters.

Archetye Exapmple Consider a domain E consisting of the complex plane, or any other simply connected set, and let us delete a number of pieces $H_1, \ldots,$ H_m to create "holes" in the original domain E. The contours $\gamma, \gamma_1, \ldots, \gamma_m$ shown in Figure 5.32 can be taken together to form a cycle $\phi = \gamma - \gamma_1 - \gamma_2 - \cdots - \gamma_m$ that has winding number zero with respect to every ζ in the

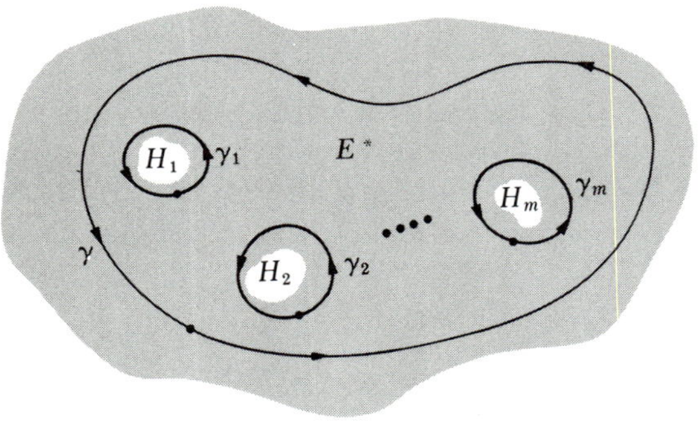

Figure 5.32 The region E^* is shaded; the domain E consists of $E^* \cup H_1 \cup \cdots \cup H_m$, and is simply connected.

complement of E or in one of the holes H_k. If E^* is the domain we get by removing the subsets $H_1, \ldots, H_m$ from E, then ϕ is homologous to zero in E^*. The general Cauchy Theorem assures us that

$$\int_\phi f(z)\,dz = \int_\gamma f(z)\,dz - \sum_{k=1}^{m}\left(\int_{\gamma_k} f(z)\,dz\right) = 0$$

for every function that is defined and analytic on E^*. This is more interesting when expressed in the form

$$\int_\gamma f(z)\,dz = \sum_{k=1}^{m}\left(\int_{\gamma_k} f(z)\,dz\right),$$

which shows clearly that integration along γ can be replaced by integration along the separate contours $\gamma_1, \ldots, \gamma_m$ (or around the cycle $\gamma_1 + \cdots + \gamma_m$) as long as we are considering integrands that are analytic on E^*.

EXERCISES

1. The exercises of Section 5.8 were more or less limited to *circular* contours because they depended on the elementary Cauchy integral formula. Evaluate the following integrals using the general Cauchy theorems.

(i) $\displaystyle\int_\gamma \frac{\sin z}{z^2}\,dz$ along the polygonal arc passing through vertices $\{+1, -i\,\tfrac{1}{2}, -i - \tfrac{1}{2}, +1\}$.

(ii) $\displaystyle\int_\gamma \frac{\text{Arctan } z}{z - 1}\,dz$ counterclockwise along the ellipse $x^2 + 16y^2 = 4$.

(iii) $\displaystyle\int_\gamma \frac{1}{z^2 + z + 1}\,dz$ along the rectangle passing through $\left\{1 + \frac{i}{2}, -\frac{1}{2} + \frac{i}{2}, -\frac{1}{2} - \frac{i}{2}, 1 - \frac{i}{2}, 1 + \frac{i}{2}\right\}$.

(iv) $\displaystyle\int_\gamma \tan z\,dz$ along the rectangle $\{-i, +i, i + 2, -i + 2, -i\}$.

(v) $\displaystyle\int_\gamma \frac{e^z}{z^4 - 1}\,dz$ along the polygonal arc $\{\tfrac{1}{2} - 10i, \tfrac{1}{2} + 10i, -\tfrac{1}{2} + 10i, -\tfrac{1}{2} - 10i, \tfrac{1}{2} - 10i\}$.

Hint: Replace these contours by cycles made up of circular contours which are homologous to γ in the domain where the integrand is holomorphic. Integrals for circular contours may be (or have been) calculated as in Sections 5.8 to 5.10.

Answer: (i) $2\pi i$, (ii) Arctan (1) $= \pi/4$; (iii) 0; (iv) $-2\pi i$; (v) $-i\pi$ sin (1)

2. Evaluate the integral $\dfrac{1}{2\pi i}\displaystyle\int_{\phi}\frac{e^z+1}{z-\zeta}\,dz$ for $\phi = \gamma_2 - \gamma_1$, where $\gamma_2(\theta) = R_2e^{i\theta}$ and $\gamma_1(\theta) = R_1e^{i\theta}$ for $0 \le \theta \le 2\pi$. Assume $R_2 > R_1 > 0$ and consider the cases $|\zeta| < R_1$; $R_1 < |\zeta| < R_2$; $R_2 < |\zeta| < \infty$.

Answer: 0 for $|\zeta| < R_1$; $e^{\zeta} + 1$ for $R_1 < |\zeta| < R_2$; 0 for $|\zeta| > R_2$.

3. The general Cauchy Integral Formula (64) is not valid if we try to apply it to a domain E that is *multiply* connected. In $E = \{z: |z| > 2\}$, take $\gamma(t) = 3e^{it}$ for $0 \le t \le 2\pi$, and $f(z) = 1/(z-1)$ (analytic on E). Show that the integral

$$F(\zeta) = \frac{1}{2\pi i}\int_{\gamma}\frac{f(z)}{z-\zeta}\,dz$$

does *not* agree with $I(\gamma, \zeta)\cdot f(\zeta)$ for ζ in $E \sim \Gamma$. Explicitly calculate $F(\zeta)$ for ζ in the domains $|\zeta| < 3$ and $3 < |\zeta| < +\infty$ which make up the complement of Γ in the plane. (Remember that $F(\zeta)$ is defined and analytic on $\mathbf{C} \sim \Gamma$, by Section 5.10).

4. If E is a multiply connected domain (one that is *not* simply connected) show that there must be at least one *closed* contour γ in E, and one analytic function $f(z)$ on E such that $\int_\gamma f(z)\,dz \neq 0$.

Hint: Consider $\dfrac{1}{z-p}$ where p is a point in the complement $\mathbf{C} \sim E$ such that $I(\gamma, p) \neq 0$.

5. Use the last result to prove that simply connected domains E are the *only* domains with the property $\int_\gamma f(z)\,dz = 0$ for every closed contour γ in E and every function $f(z)$ defined and analytic on E.

6. Show that there are analytic determinations of

(i) $(z-p)^{1/2}$

(ii) $\log(z-p)$

(iii) $(z-p)^\alpha$ (α any complex exponent)

on any simply connected domain E that excludes the point p.

7. Use the general ideas of Application 2 to show that there is an analytic determination of arctan z on any simply connected domain that excludes the points $+i$ and $-i$.

Hint: Start with the (single valued) derivative $1/(1 + z^2)$ in place of the derivative $1/z$ associated with log z.

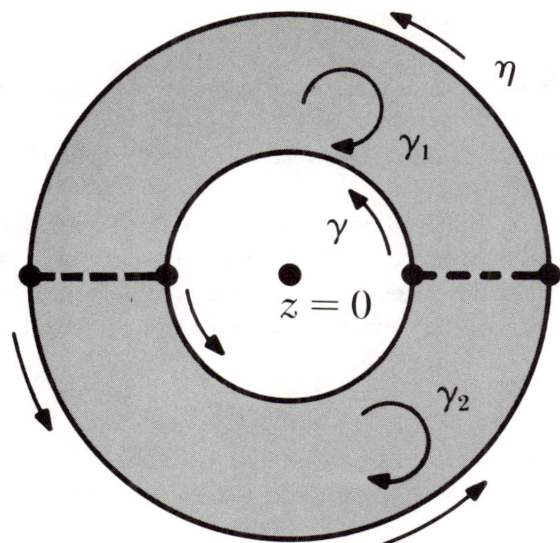

Figure 5.33

8. On the domain $E = \{z : z \neq 0\}$, the contours $\gamma(t) = R_1 e^{it}$ and $\eta(t) = R_2 e^{it}$, defined for $0 \leq t \leq 2\pi$, are already homologous (refer to Figure 5.22). Use the elementary Cauchy Theorem to prove that $\int_\gamma f(z)\,dz = \int_\eta f(z)\,dz$ for every function $f(z)$ that is analytic on E. (Do not invoke the general Cauchy Theorem; see if you can give a direct proof based on Theorem 5.16.)

Hint: Set up related closed contours γ_1 and γ_2 such that $\int_{\gamma_1} f(z)\,dz + \int_{\gamma_2} f(z)\,dz = \int_\gamma f(z)\,dz - \int_\eta f(z)\,dz$ after cancellations; the contours shown in Figure 5.33 do this. Now apply Exercise 4 of Section 5.7 (a useful improvement on Theorem 5.16). Each contour γ_k is contained in a star-shaped domain on which $f(z)$ is holomorphic (the set avoids $z = 0$).

Note: This calculation is a first step toward proving the general Cauchy Theorem. It can be varied to give direct proofs that $\int_\gamma f(z)\,dz = \int_\eta f(z)\,dz$ in many other situations where $f(z)$ is analytic on a domain containing γ and η.

9. Derive the following corollary of the generalized Cauchy Theorem.

Theorem: A domain E is simply connected if and only if

$$\int_\gamma f(z)\,dz = 0$$

for every function $f(z)$ analytic on E, and every closed contour γ in E.

Hint: That (s. conn.) implies (null integrals) is immediate. Conversely, if integrals are null, the integrals that give $I(\gamma, \zeta)$ are zero for ζ not in E (why?).

10. Let $f:E \to F$ be an invertible analytic mapping between two open sets, which has an analytic inverse $\check{f}:F \to E$. Prove that:

Theorem 1: *E is connected if and only if F is connected.*

Theorem 2: *E is simply connected if and only if F is simply connected.*

Hint: E and F play symmetrical roles, so it suffices to prove that F inherits each property if E has the property. In Theorem 1, define connectedness in terms of connecting typical points by continuous parametrized curves. In Theorem 2, use Exercise 9 and the change of variable formula for line integrals.

6 SINGULARITIES AND RESIDUES

The evaluation of integrals along closed contours can be reduced to an almost entirely algebraic procedure by introducing the notion of the *residue* of a function at its isolated singular points. The definition of residue is based on the Cauchy Integral Formula for circular contours. We will use the theory of residues to evaluate various classes of integrals that turn up in Calculus. For example, we will be able to evaluate

$$\int_0^\infty \frac{\sin x}{x}\, dx = \pi/2$$

by relating this integral to certain contour integrals of the analytic function e^{iz}/z†; this integral, which is of importance in diffraction problems in optics, cannot be computed by any of the methods ordinarily used in Calculus. Finally, we will examine the general behavior of analytic functions near isolated singular points, and will develop the *Laurent series* expansion to represent analytic functions near such singular points (these series involve both positive and negative powers of z).

6.1 ISOLATED SINGULARITIES AND THEIR RESIDUES

If $f(z)$ is analytic on a punctured disc $0 < |z - p| < r$, but is undefined at p, we say that p is an **isolated singularity** of f. For example, if we take any polynomial, such as $z^2 + 1$, then the rational function $f(z) = 1/(z^2 + 1)$ is defined and analytic except at the points $z = +i, -i$ where the denominator is zero. Likewise, the function $f(z) = (\sin z)/z$ is analytic everywhere except at $z = 0$, where it is not well defined. These are examples of isolated singular

† The connection between the integrand $(\sin x)/x$ and the function e^{iz}/z is more apparent if we recall that $\sin z = (e^{iz} - e^{-iz})/2i$.

points. Analytic functions may exhibit other kinds of singularities, but we will study only *isolated* singularities in this chapter. For example, $f(z) = \tan\left(\dfrac{\pi}{2} + \dfrac{1}{z}\right)$ is undefined at $z = 0$ and also at the points $z_n = 1/\pi n$, which converge to zero. Each point z_n is an isolated singularity, since f is analytic on a punctured disc about z_n. But $z = 0$ is not isolated from the singular points z_n, and is not an isolated singularity for f. There are also **branch singularities;** the functions $z^{1/2}$ and $\log z$ cannot be defined as single valued analytic functions on any deleted neighborhood $0 < |z| < r$ about $z = 0$. To get well defined analytic functions we must introduce a cut originating at $z = 0$, and the values along opposite sides of the cut will differ.

Hereafter, "singularity" or "singular point" will mean an isolated singularity, unless otherwise specified. In certain cases we may assign a value to f at an isolated singularity p, say $f(p) = \alpha$, so that f becomes analytic at p. Whenever this is possible, f fails to be analytic at p only because it was not defined there. In such cases, p is said to be a **removable singularity** for f.

Example 6.1 Let $f(z) = (\sin z)/z$ for $z \neq 0$. This quotient makes no sense at $z = 0$. Let us try to remedy this by *defining* the value at the origin to be $f(0) = 1$ (this choice is suggested by examining real values of z and using L'Hospital's rule). There is a series expansion $\sin z = z - z^3/3! + z^5/5! - \cdots$ for all z, and if $z \neq 0$ we can multiply each term in this series by $1/z$ without affecting the convergence. Therefore,

$$\frac{\sin z}{z} = 1 - \frac{z^2}{3!} + \frac{z^4}{5!} - \cdots = \sum_{n=0}^{\infty} (-1)^{n+1} \frac{z^{2n}}{(2n+1)!}$$

for $z \neq 0$. The series on the right converges everywhere, even at $z = 0$, by the ratio test; therefore, its sum $g(z)$ is analytic on the plane and agrees with $f(z)$ for $z \neq 0$. Obviously, then, $z = 0$ is a removable singularity for f.

Example 6.2 The singularities of the rational function $f(z) = 1/(z^2 + 1)$ at $z = +i, -i$ are not removable. In fact, if $f(z)$ could be made into an analytic function at $z = +i$ by properly assigning a value $f(+i) = \alpha$, then f would be continuous at $+i$ and we would have $\lim_{z\to+i} f = \alpha$. However, if $z_n \to +i$, then $(z_n + i) \to 2i$ and $(z_n - i) \to 0$; since $(z_n^2 + 1) \to 0$, it follows that

$$|f(z_n)| = \left|\frac{1}{{z_n}^2 + 1}\right| = \left|\frac{1}{z_n + i}\right| \cdot \left|\frac{1}{z_n - i}\right| \to +\infty.$$

Since this happens no matter how the z_n approach $+i$, f can not have a finite limit α at $z = +i$. A similar argument applies to the other singular point $z = -i$.

Let p be an isolated singularity for an analytic function $f(z)$, and let γ_r be the small counterclockwise oriented circular contour centered at p, with radius $r > 0$. For sufficiently small radii r the trajectories of the γ_r all lie within a punctured disc $E = \{z : 0 < |z - p| < R\}$ on which f is analytic. The closed contours γ_r are all homologous within E, so the integrals

$$\frac{1}{2\pi i}\int_{\gamma_r} f(z)\,dz \qquad (0 < r < R)$$

all have the same value. This common value we shall call the **residue of f at the singular point p;** it will be indicated by the symbol $\operatorname{Res}(f, p)$. If p happens to be a removable singularity, the elementary Cauchy Theorem shows that $\operatorname{Res}(f, p) = 0$; however, it is possible to have $\operatorname{Res}(f, p) = 0$ even if p is not a removable singularity.

Now let us show how the evaluation of an integral $\int_\gamma f(z)\,dz$ along a closed contour can be reduced to the evaluation of the residues of the integrand at the various singular points $p_1, \ldots, p_N$ enclosed by γ. The calculation is accomplished by adding up the residues:

$$(1) \qquad \frac{1}{2\pi i}\int_{\gamma} f(z)\,dz = \sum_{k=1}^{N} \operatorname{Res}(f, p_k),$$

if γ is a positively oriented simple closed contour, and by a slightly more complicated formula for more general closed contours. To derive this result we consider the following situation, illustrated in Figure 6.1.

(i) Let $f(z)$ be defined and analytic except at a finite number of points $p_1, \ldots, p_N$ in a domain E. These are clearly isolated singular points; about p_k there is a disc $|z - p_k| < r_k$ that lies entirely within E, and f is analytic on the punctured disc $0 < |z - p_k| < r_k$.

(ii) Let γ be a closed contour in E that is homologous to zero, so that γ does not wind around any points in the complement of E. Let us assume that none of the points p_k lie on the trajectory of γ.

Notice that we regard $p_1, \ldots p_N$ as points in E.

Theorem 6.1 *If conditions* (i) *and* (ii) *are satisfied, then*

$$(2) \qquad \frac{1}{2\pi i}\int_{\gamma} f(z)\,dz = \sum_{k=1}^{N} \operatorname{Res}(f, p_k) \cdot I(\gamma, p_k)$$

where $I(\gamma, p_k)$ is the index of p_k with respect to the contour γ.

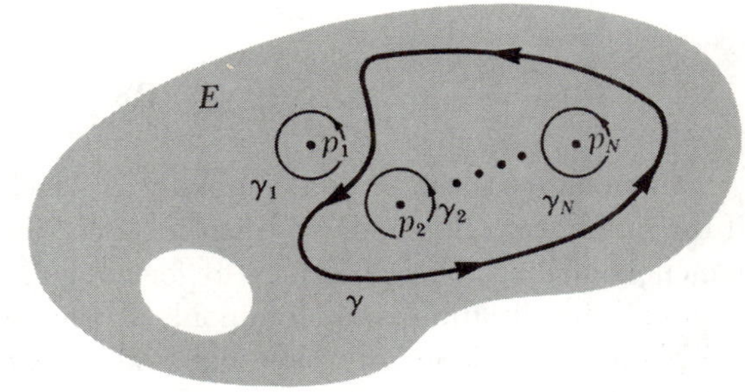

Figure 6.1 The contours in the residue formula. E^* is E with the points $\{p_1, \ldots, p_N\}$ deleted.

Table 6.1 Winding Numbers of the Contours in Figure 6.1 at Points in the Complement of E^*

	$\zeta = p_1$	$\zeta = p_2$	....	$\zeta = p_N$	$\zeta \neq p_k$ (ζ not in E)
γ_1	+1	0	....	0	0
γ_2	0	+1	....	0	0
.	.	.	.	.	.
.	.	.	.	.	.
.	.	.	.	.	.
.	.	.	.	.	.
.	.	.	.	.	.
.	.	.	.	.	.
.	.	.	.	.	.
.	.	.	.	.	.
γ_N	0	0	....	+1	0
η	$I(\gamma, p_1)$	$I(\gamma, p_2)$	....	$I(\gamma, p_N)$	0
γ	$I(\gamma, p_1)$	$I(\gamma, p_2)$	....	$I(\gamma, p_N)$	0

To verify this we form small counterclockwise oriented circular contours $\gamma_1, \ldots, \gamma_N$ centered at the singular points $p_1, \ldots, p_N$. If the radii of the trajectories Γ_k are sufficiently small, these circles will be disjoint from Γ and from each other, and the circles Γ_k together with the small discs they bound will lie entirely within E as indicated in Figure 6.1. Now consider the smaller domain $E^* = E \sim \{p_1, \ldots, p_N\}$ obtained by deleting the singular points from the original domain E. By deleting the singular points we get $f(z)$ to be analytic throughout E^*. Because the disc bounded by Γ_k lies within E^* except for its center point, it is clear that we have the winding numbers indicated in Table 6.1 at the points $\zeta = p_1, \ldots, \zeta = p_N$ and at points ζ outside of the original domain E.

By taking the indices of the original contour γ at the singular points as multiplicities, we may combine $\gamma_1, \ldots, \gamma_N$ into an appropriate cycle

$$\eta = I(\gamma, p_1)\gamma_1 + \cdots + I(\gamma, p_N)\gamma_N$$

which has the same winding numbers as γ at each point outside of E^*. Thus γ and η are homologous in E^*, and f is analytic on E^*, so the generalized Cauchy Theorem insures that line integrals along these cycles in E^* are the same.

$$\frac{1}{2\pi i}\int_\gamma f(z)\,dz = \frac{1}{2\pi i}\int_\eta f(z)\,dz = \sum_{k=1}^{N} I(\gamma_k, p_k)\frac{1}{2\pi i}\int_{\gamma_k} f(z)\,dz$$

$$= \sum_{k=1}^{N} I(\gamma_k, p_k)\operatorname{Res}(f, p_k)$$

If we agree to speak of a contour γ as "enclosing" a singular point p_k only if the index $I(\gamma, p_k)$ is nonzero, then it is only the singular points enclosed by γ which contribute their residues to formula (2). When γ is a *simple* closed contour, the situation most commonly encountered in applications, it divides the plane into disjoint connected domains: the points exterior to γ (those with zero index) and the points interior to γ (those with nonzero index).

The Jordan Curve Theorem says that the index is either $+1$ everywhere, or is -1 everywhere, on the set of interior points. We say that γ is **positively oriented** if it has positive index with respect to its interior points, and that γ is **negatively oriented** if these indices are negative. For simple closed contours, we get a simplified version of formula (2).

Theorem 6.2 (The residue theorem) *Let γ be a positively oriented simple closed contour and assume that $f(z)$ is analytic on the trajectory Γ and on the set $E_{int}(\gamma)$ of interior points enclosed by γ, except for a finite number of isolated singular points $p_1, \ldots, p_N$ which lie interior to γ. Then*

$$\frac{1}{2\pi i}\int_\gamma f(z)\,dz = \sum_{k=1}^{N} \operatorname{Res}(f, p_k). \tag{3}$$

Here there is a domain E that contains Γ and $E_{int}(\gamma)$, and f is analytic on E except at the singular points $p_1, \ldots, p_N$. Now γ has index $+1$ at each p_k, since these points are interior to γ and γ is positively oriented. If γ were negatively oriented, we would have to multiply the right side of (3) by -1.

EXERCISES

1. Show that $f(z) = \sin(z)/z^2$ has a non-removable singularity at $z = 0$.

2. Decide whether the following isolated singularities are removable. (Use series expansions where appropriate.)

(i) $\exp(1/z^2)$ at $z = 0$

(ii) $\dfrac{\sin z}{z - \pi}$ at $z = \pi$

(iii) $z \operatorname{ctn} z$ at $z = 0$

(iv) $z \operatorname{ctn} z$ at $z = \pi$

Answers: (i) No; (ii) yes; (iii) yes; (iv) no.

3. Calculate the residues at the following isolated singularities.

(i) $\dfrac{1}{z} - \dfrac{1}{z^2}$ at $z = 0$

(ii) $\dfrac{\sin z}{z}$ at $z = 0$

(iii) $\dfrac{1 + \sin z}{z}$ at $z = 0$

(iv) $\dfrac{e^z}{z^2 + 2z + 1}$ at $z = -1$

(v) $\dfrac{1}{z+1} + \dfrac{1}{z-1}$ at $z = +1$.

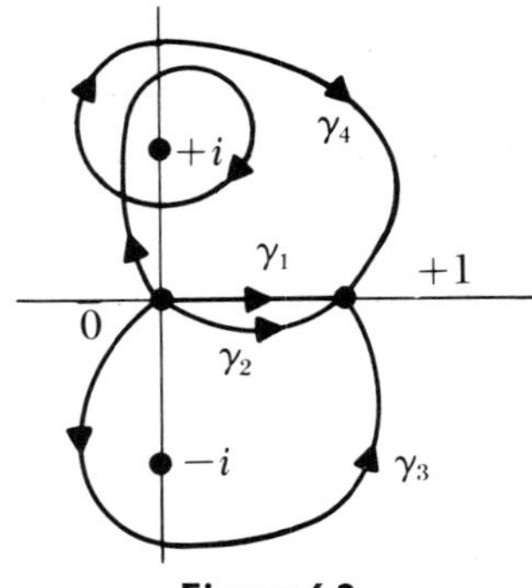

Figure 6.2

Hint: In (iv), recall the Cauchy formula for derivatives.
Answers: (i) $+1$; (ii) 0; (iii) $+1$; (iv) $1/e = e^{-1}$; (v) $+1$.

4. Give an example of a non-removable singularity, at $z = 0$, whose residue is zero.

5. Re-examine the details of Example 5.22, Chapter 5, to deduce that

$$\operatorname{Res}\left(\frac{1}{z^2+1}, +i\right) = \frac{+1}{2i} \quad \text{and} \quad \operatorname{Res}\left(\frac{1}{z^2+1}, -i\right) = \frac{-1}{2i}.$$

6. The function arctan z can be defined by

$$(4) \quad \arctan z = \int_0^z \frac{1}{w^2+1}\, dw \text{ (integral along any contour from 0 to } z).$$

Calculate the values at $z = +1$ obtained by integrating along the contours γ_1, γ_2, γ_3, and γ_4 in Figure 6.2. Explain how the residues of $1/(z^2+1)$ at $+i$ and $-i$ determine the systematic relationship between possible values of arctan z at $z = 1$. Note how they are responsible for the multiple valued nature of arctan z. Devise domains on which formula (4) must give a single valued analytic function; how are $+i$, $-i$ related to such domains?

Answer: Values are $\pi/4 = \text{Arctan}(1) - \text{Arctan}(0)$; $\pi/4$; $(\pi/4) - \pi$; $(\pi/4) - 2\pi$ (residue at $+i$ is $+\pi$; index of $\gamma_4 - \gamma_1$ is -2).

6.2 EVALUATING RESIDUES: CLASSIFYING ISOLATED SINGULARITIES

In order to calculate contour integrals we must develop practical methods for evaluating residues at isolated singularities. These calculations can often be done algebraically, in effect reducing the task of evaluating a contour integral along γ to an algebraic problem of computing its residues at the singular points enclosed by γ. It is this phenomenon that makes the "Calculus of Residues"—that is, the systematic use of the residue formula—so useful in practical problems. We will develop several methods for evaluating residues. All are based on the following analysis of just what can happen to the values $f(z)$ as z approaches an isolated singular point p.

We know what happens if the singularity is removable; there is a well defined limit $\lim_{z\to p} \{f(z)\} = \alpha$ and $\text{Res}(f, p) = 0$. In particular, if we take a small disc about a removable singularity p the values $|f(z)|$ within this disc must be bounded, say $|f(z)| \leq M$, on this disc. We can actually show that the converse is true; if the values of $|f(z)|$ are bounded on some punctured disc about an isolated singular point p, then the singularity at p must be removable! In other words, the characteristic sign that we are dealing with a *non-removable* singularity is that $|f(z)| \to +\infty$ as $z \to p$ in certain ways.

Theorem 6.3 (Riemann's Theorem) *Suppose that $f(z)$ is analytic and bounded on a punctured disc $0 < |z - p| < R$. Then $z = p$ is a removable singularity for f.*

PROOF: We assume that $p = 0$ for simplicity. First multiply by z^2 to get a new function $g(z) = z^2 f(z)$ defined for $0 < |z| < R$, and define the value of g at $z = 0$ by setting $g(0) = 0$. Clearly, g is differentiable at z if $0 < |z - p| < R$. We assert that g is also differentiable at $z = 0$, with $g'(0) = 0$; indeed,

$$\frac{\Delta g}{\Delta z} = \frac{z^2 f(z) - 0}{z} = z \cdot f(z) \to 0 \quad \text{as} \quad z \to 0$$

because f is bounded near the origin. Therefore $g(z)$ is differentiable, and thus analytic, throughout the disc $|z| < R$. The first two terms of its Taylor series about the origin are zero since $g(0) = g'(0) = 0$, so the Taylor series has the form

$$g(z) = a_2 z^2 + a_3 z^3 + \cdots = \sum_{n=2}^{\infty} a_n z^n \quad \text{for all } z \text{ near } z = 0.$$

From the definition of g we see that

$$f(z) = \frac{1}{z^2} \cdot g(z) = a_2 + a_3 z + \cdots$$

for all z near the origin such that $z \neq 0$. In particular, the right-hand series converges at all points where the series for g converges, and also at the origin. Evidently, then, $F(z) = a_2 + a_3 z + \cdots$ is defined and analytic on the disc $|z| < R$ and agrees with $f(z)$ for all $z \neq 0$. The singularity of f at $z = 0$ must be removable. ∎

The non-removable singularities of analytic functions can be classified by examining the growth of $|f(z)|$ as $z \to p$ in various ways. If Γ_r is the circle $|z - p| = r$, let us consider the "growth indicator"

$$M(f, r) = \text{maximum of } |f(z)| \text{ for } z \text{ on } \Gamma_r.$$

We classify the behavior of f at p by comparing $M(f, r)$ with the standard family of functions $1/r$, $1/r^2$, $1/r^3$, . . . which are singular as $r \to 0$. Clearly, the singularity is removable if $M(f, r)$ is bounded as $r \to 0$ (that is, there is a constant K such that $M(f, r) \leq K$ for all small $r > 0$).

Definition 6.1 *An isolated singularity of an analytic function is a* **pole** *if*

(i) *$M(f, r)$ is not bounded as $r \to 0$.*
(ii) *$r^m M(f, r) = M(f, r)/r^{-m}$ is bounded as $r \to 0$ for some exponent $m \geq 1$.*

If $r^m M(f, r)$ is bounded as $r \to 0$ for the exponent $m = m_0$, it is obviously bounded for any exponent $m > m_0$. The smallest exponent $m = 1, 2, \ldots$ that works in (ii) *is called the* **order of the pole.**

Definition 6.2 *An isolated singularity of an analytic function is an* **essential singularity** *if $r^m M(f, r) = M(f, r)/r^{-m}$ is not bounded as $r \to 0$ for any choice of exponent $m = 1, 2, \ldots$. (Roughly speaking, this means that the growth indicator $M(f, r)$ "grows faster" than any of the functions $r^{-m} = 1/r^m$ as $r \to 0$.)*

An isolated singularity p must fit into one of the three mutually exclusive categories we have just defined; either $M(f, r)$ is bounded (p is a removable singularity), or $M(f, r)$ behaves like one of the functions $1/r^m$ (p is a pole of order m), or $M(f, r)$ grows faster than any of the functions $1/r^m$ (p is an essential singularity).

The simplest example of a pole of order m is given by a function such as $f(z) = \alpha/(z - p)^m$, where α is a nonzero constant. Then $M(f, r) = |\alpha|\, r^{-m}$ for $r > 0$, so that $r^m M(f, r) \to |\alpha|$ as $r \to 0$. More generally, if $g(z)$ is analytic at p *and nonzero* there, then $f(z) = g(z)/(z - p)^m = g(z)(z - p)^{-m}$ is defined for $z \neq p$ and has a pole of order m at p (we leave the simple verification as Exercise 1). It is important that $g(p)$ be nonzero; otherwise, the presence of a zero in the numerator might compensate for the singular behavior introduced by dividing by $(z - p)^m$. Actually *every* function with a pole of order m at a point p can be written in this form for a suitably chosen function $g(z)$ which is defined, analytic, and nonzero at p.

Theorem 6.4 *An analytic function with an isolated singularity at $z = p$ has a pole of order m at p if and only if we can write $f(z)$ in the form*

(5)
$$f(z) = \frac{g(z)}{(z - p)^m} \quad \textit{for all} \quad z \neq p$$

where $g(z)$ is analytic and nonzero at $z = p$.

This result, proved below, gives a complete description of how $f(z)$ behaves near a pole; the behavior is largely determined by the behavior of the elementary singular function $(z - p)^{-m}$ which multiplies $g(z)$.

PROOF: The only thing left to verify is that a function with a pole of order m can be written in the form (5). First notice that the function $(z - p)^m f(z)$, defined and analytic at points $z \neq p$, is *bounded* and therefore has a removable singularity at p. On a circle $|z - p| = r$ we have

$$|(z - p)^m f(z)| = |z - p|^m |f(z)| \leq r^m M(f, r),$$

which is bounded as $r \to 0$ by definition of a pole of order m. Now apply Theorem 6.3; by assigning the correct value at $z = p$ we get a function $g(z)$ analytic at p, such that $f(z) = g(z)/(z - p)^m$ for $z \neq p$. This expresses $f(z)$ in the desired form. There is only one point left unresolved; we must also show that the function $g(z)$ is nonzero at $z = p$. But if g has a zero there, its Taylor series about $z = p$ has no constant term and we can write $g(z) = (z - p)h(z)$, where $h(z)$ is analytic at p. Thus,

$$(z - p)^{m-1} f(z) = \frac{(z - p)^m f(z)}{(z - p)} = \frac{g(z)}{(z - p)} = h(z) \quad \text{for} \quad z \neq p$$

is bounded as $z \to p$, which means that $r^{m-1}M(f, r)$ is bounded as $r \to 0$. This is impossible; by definition of a pole of order m, no smaller exponent $k = m - 1$ can make $r^k M(f, r)$ bounded as $r \to 0$. We therefore conclude that $g(z)$ cannot have a zero at $z = p$. ∎

Example 6.3 Let $f(z) = 1/(z^2 + 1)$. Then $z^2 + 1 = (z + i)(z - i)$, and $+i$ and $-i$ are isolated singularities. At $+i$ the function $g(z) = 1/(z + i)$ is analytic, and $g(+i) = 1/2i$ is nonzero. We may write f in the form (5),

$$f(z) = \frac{g(z)}{(z - i)} \text{ for all } z \text{ near } +i \; (z \neq +i),$$

which makes it obvious that f has a pole of order one at $+i$. Likewise there is a pole of order one at $-i$.

More generally, let $P(z)$ be any non-constant polynomial; factor $P(z) = \alpha(z - p_1)^{m_1} \cdot \cdots \cdot (z - p_N)^{m_N}$ as indicated in our discussion of the Fundamental Theorem of Algebra, where $\{p_k\}$ are the distinct roots of $P(z)$ and m_k are their orders, or multiplicities. The same kind of argument we used in the special case $P(z) = z^2 + 1$ shows us that the rational function $f(z) = 1/P(z)$ is analytic except at the isolated singular points $p_1, \ldots, p_N$, and p_k is a pole of order m_k.

Example 6.4 Consider the function $f(z) = (1/z) + (2/z^2)$. Our intuitive feeling is that the function $1/z^2$ has a stronger singularity at $z = 0$ than the function $1/z$, so its behavior should determine the nature of the singularity and make $z = 0$ a pole of order *two*. In fact, for $z \neq 0$ we can write

$$f(z) = \frac{1}{z} + \frac{2}{z^2} = \frac{z + 2}{z^2} = \frac{g(z)}{z^2}$$

where $g(z) = z + 2$ is analytic and nonzero at the origin. Likewise, in any sum

$$\frac{a_1}{z} + \cdots + \frac{a_N}{z^N} \qquad \left(\text{or } \frac{a_1}{(z-p)} + \cdots + \frac{a_N}{(z-p)^N}\right),$$

it is the highest order term, involving z^{-N} or $(z-p)^{-N}$, which determines the nature of the singularity; thus, such functions have poles of order N.

One enlightening consequence of Theorem 6.4 is that every function with a pole at $z = p$ can be represented near p by a series in powers of $(z - p)$ which includes a finite number of *negative* powers to account for the singular behavior of f near p. In fact, if we substitute the Taylor series expansion

$$g(z) = a_0 + a_1(z-p) + \cdots = \sum_{n=0}^{\infty} a_n(z-p)^n \tag{6}$$

into (5) we find that the quotient has the form

$$f(z) = \frac{a_0}{(z-p)^m} + \cdots + \frac{a_{m-1}}{(z-p)} + [a_m + a_{m+1}(z-p) + \cdots] \tag{7}$$

for z near p with $z \neq p$. The series $[\cdot\cdot\cdot]$ converges at $z = p$, and it converges at every other point where the series for g converges; it is obtained by dropping the first m terms from (6) and multiplying each of the remaining terms by the common factor $(z-p)^{-m}$, which cannot alter its convergence. The series $[\cdot\cdot\cdot]$ gives us an analytic function $h(z) = \sum_{n=0}^{\infty} b_n(z-p)^n$ defined near p, and after relabeling the coefficients on negative powers in (7) we see that

$$\begin{aligned} f(z) &= b_{-m}(z-p)^{-m} + \cdots + b_{-1}(z-p)^{-1} + h(z) \\ &= b_{-m}(z-p)^{-m} + \cdots + b_{-1}(z-p)^{-1} + \sum_{n=0}^{\infty} b_n(z-p)^n \\ &= \sum_{n=-m}^{\infty} b_n(z-p)^n \end{aligned} \tag{8}$$

for all z near p with $z \neq p$. It should be clear that the negative power terms are entirely responsible for the singularity at p. Furthermore, if $(z-p)^{-m}$ is the highest order negative power, then $f(z)$ has a pole of order m (see Example 6.4 above). We will have more to say about representing functions in negative powers of $(z-p)$ as we go along.

An isolated singularity that is not removable or a pole is, by default, an essential singularity. The behavior of $f(z)$ near an essential singularity is quite bizarre. We will only indicate a few aspects of this behavior here, leaving the proofs as exercises. If p is an essential singularity, then $|f(z)|$ grows very rapidly (faster than $|z-p|^{-m}$ for any exponent $m = 1, 2, \ldots$) as z approaches p in *certain* ways; however, from Definition 6.2 we are not able to conclude that

Table 6.2 Behavior of $|f(z)|$ near an Isolated Singularity

Removable singularity	There is a bound $M < +\infty$ such that $\|f(z_n)\| \leq M$ for any sequence $z_n \to p$.
Pole	For every sequence $z_n \to p$ we have $\|f(z_n)\| \to +\infty$.
Essential singularity	There are sequences $z'_n \to p$ such that $\|f(z'_n)\| \to +\infty$ and other sequences $z''_n \to p$ for which $\|f(z''_n)\| \to 0$.

$|f(z_n)| \to +\infty$ for *every* sequence of points distinct from p such that $z_n \to p$. In fact, Theorem 6.5 will show that $|f(z)|$ must approach zero as z approaches p in certain ways. This behavior is very different from that of a pole; the kinds of behavior we may expect at isolated singularities are cataloged in Table 6.2.

In particular, we cannot have $|f(z_n)| \to +\infty$ for all sequences $z_n \to p$ at an essential singularity; this can only happen at a pole. These remarks are based on a more general result which shows that the values $f(z)$ near an essential singularity can be made to approach *any* prescribed complex number α if we let z approach p in a suitable way; taking $\alpha = 0$ gives $|f(z_n)| \to |\alpha| = 0$, as required in the last entry of the table.

Theorem 6.5 (Casorati-Weierstrass) *Let f be analytic with an essential singularity at p. Let α be any complex number. Then by choosing a suitable sequence of points $z_n \to p$, distinct from p, we can insure that $f(z_n) \to \alpha$ as $n \to \infty$.*

Of course, the sequence $z_n \to p$ will depend on which complex number α we are trying to approach with the values $f(z_n)$. The proof is outlined in Exercise 14. Here are some illustrative examples.

Example 6.5 The function $f(z) = \exp(1/z)$ is analytic except at the origin. If z approaches the origin along the positive real axis, we have $z = x + i0$ $(x > 0)$ and $\lim_{z\to 0} f = \lim_{x\to 0} e^{1/x} = +\infty$, so that $|f(z)|$ increases rapidly as $z \to 0$ in this direction. On the other hand, if z approaches zero along the negative real axis, then $z = -x + i0$ $(x > 0)$ and $\lim_{z\to 0} f = \lim_{x\to 0} e^{-1/x} = 0$. The function $g(z) = \exp(1/z^2)$ exhibits similar behavior near $z = 0$. Both must have essential singularities at $z = 0$.

Example 6.6 The function $f(z) = \sin(1/z)$ is analytic except at $z = 0$. We leave it to the reader (Exercise 10) to show that there are sequences $z'_n \to 0$ such that $|f(z'_n)| \to 0$, and other sequences $z''_n \to 0$ for which $|f(z''_n)| \to +\infty$. Thus, $z = 0$ is an essential singularity. In contrast, let us consider $f(z) = 1/(\sin z)$, which is singular at the isolated points $z_n = \pi n + i0$, where the denominator vanishes. Each of these points is a pole of order one, as can be seen by examining the Taylor series of $\sin z$ about these points. For example, at $p = 0$,

$$\sin z = z - \frac{z^3}{3!} + \frac{z^5}{5!} - \cdots = z\left(1 - \frac{z^2}{3!} + \cdots\right) = z \cdot h(z),$$

where $h(z)$ is analytic and nonzero at the origin. Therefore, $g(z) = 1/h(z)$ is analytic and nonzero at the origin, so that

$$f(z) = \frac{1}{\sin z} = \frac{1}{z \cdot h(z)} = \frac{g(z)}{z} \quad \text{for} \quad z \neq 0.$$

This shows that the origin is a pole of order one (by Theorem 6.4).

Functions that are analytic throughout the plane, except for isolated singularities *that are all poles* (or removable), are important in complex analysis and are referred to as **meromorphic functions**. Rational functions, of the form P/Q where P and Q are polynomials, and functions such as $\tan z$ and $(\tan z)/(z^2 + 1)$ are meromorphic; but $\exp(1/z)$ is not meromorphic, due to the presence of an essential singularity at $z = 0$.

EXERCISES

1. Suppose $g(z)$ is analytic and *nonzero* at $z = p$. Prove that

$$f(z) = \frac{g(z)}{(z - p)^m} \quad \text{defined for} \quad z \neq p,\ m = 1, 2 \cdots$$

has a pole of order m at p.

Hint: $M(f, r) = r^{-m} \cdot M(g, r)$ for small $r > 0$, and $\lim_{r \to 0} M(g, r) = |g(p)| \neq 0$.

2. Show that the conclusion of Exercise 1 is not valid if g has a zero at p, by calculating the order of the following poles:

$$\text{(i)}\ \frac{1 - \cos z}{z^2} \text{ at } p = 0 \qquad \text{(ii)}\ \frac{\sin z}{z^2} \text{ at } p = 0.$$

Answer: (i) removable singularity, $m = 0$; (ii) $m = 1$.

3. Suppose $g(z)$ is analytic at $z = p$. If $f(z)$ is analytic, with an isolated singularity at p, show that the sum $h(z) = f(z) + g(z)$ has the same kind of singularity at p as f. If p is a pole, show that its *order* is the same for h and f. (The residues need *not* be equal.)

Hint: $g(z)$ is bounded near p, say $|g(z)| \leq K$ for $|z - p| \leq R$. Thus $M(f, r) - K \leq M(h, r) \leq M(f, r) + K$ for small $r > 0$.

4. Use Exercise 3 to determine the type of singularity for

(i) $\dfrac{1}{z} + e^z$ at $p = 0$

(ii) $\dfrac{1}{z+i} + \dfrac{1}{z-i}$ at $p = +i, -i$

(iii) $\dfrac{1}{z} + \dfrac{1}{z^2+1}$ at $p = 0, +i, -i$

(iv) $\dfrac{\cos z}{z - (\pi/2)}$ at $p = \dfrac{\pi}{2}$

(v) $\dfrac{1}{z^4} + \dfrac{1}{z^2} + \sin z$ at $p = 0.$

5. If $f(z)$ is analytic and has a zero at $z = p$, the **order of the zero** is determined by the first nonzero Taylor coefficient; order $= N$ means that

$$f(z) = a_N(z-p)^N + a_{N+1}(z-p)^{N+1} + \cdots \quad (a_N \neq 0).$$

Assuming f has a zero of order $N \geq 1$ at p, show that there is a radius $r > 0$ such that $f(z) \neq 0$ for all z such that $0 < |z - p| < r$. Thus $g(z) = 1/f(z)$ has an isolated singularity at p.

Hint: If no such r exists, there are $z_n \neq p$ converging to p such that $f(z_n) = 0$. By Theorem 3.17, this would imply $f(z) = 0$ identically, which is impossible since $f^{(N)}(p) = N!\, a_N \neq 0$. Now write $f(z) = (z-p)^N F(z)$; then $G = 1/F$ is analytic at p and $g(z) = G(z)/(z-p)^N$.

6. If $f(z)$ is analytic at p and has a zero of order $N \geq 1$ at p, show that $1/f$ has a pole of order N at p. Discuss the limit behavior of $f(z)/(z-p)^k$ as $z \to p$ for exponents $k = 0, 1, 2, \ldots$. Which exponents k give removable singularities?

7. If $f(z)$ is an entire function and is not identically zero on the plane, prove that its reciprocal $1/f$ is meromorphic (all singularities are isolated and are poles). Use Exercise 5.

8. Suppose $g(z)$ is analytic and *nonzero* at $z = p$. Let $f(z)$ be any analytic function that has an isolated singularity at p. Prove that $h(z) = g(z) \cdot f(z)$ has the same type of singularity at p as $f(z)$.

Hint: There are constants $0 < K_1 < K_2$ such that $K_1 \leq |g(z)| \leq K_2$ for all z near p; thus $K_1 \cdot M(f, r) \leq M(h, r) \leq K_2 \cdot M(f, r)$ for small $r > 0$.

9. Show that the following functions are meromorphic (only isolated singularities, all poles). List the singular points and calculate the order m of each pole.

(i) $\operatorname{ctn} z$ (iv) $e^z/\sinh^2 z$

(ii) $\tanh z$ (v) $1/(1 - \cos z)$

(iii) $1 + \tan^2 z$

Answers: (i) $z_k = \pi k$, $m_k = 1$; (ii) $z_k = (i\pi/2) + i\pi k$, $m_k = 1$; (iii) $z_k = (\pi/2) + \pi k$, $m_k = 2$; (iv) $z_k = i\pi k$, $m_k = 2$; (v) $z_k = 2\pi k$, $m_0 = 2$, $m_k = 1$ for $k \neq 0$.

10. Consider $f(z) = \sin(1/z)$ near $z = 0$. Devise sequences $z'_n \to 0$ and $z''_n \to 0$ such that

$$\lim_{n\to\infty} f(z'_n) = \infty \quad \text{and} \quad \lim_{n\to\infty} f(z''_n) = 0.$$

There is an essential singularity at $z = 0$ (why?).

11. Let us introduce the *lower bounds*

$$m(f, r) = \text{minimum of } |f(z)| \text{ on the circle } |z - p| = r.$$

Obviously, $m(f, r) \leq |f(z)| \leq M(f, r)$ on the circle $|z - p| = r$. If p is a pole of order k, prove that these upper and lower bounds have the same limit behavior,

$$\text{(i)} \lim_{r\to 0} \frac{m(f, r)}{r^{-k}} = \lim_{r\to 0} \frac{M(f, r)}{r^{-k}}$$

so that

$$\text{(ii)} \lim_{z\to p} \frac{|f(z)|}{|z - p|^{-k}} \text{ exists and is finite.}$$

Hint: Use formula (5).

12. Calculate $m(f, r)$ and $M(f, r)$ explicitly for $f(z) = \exp(1/z)$, at the essential singularity $p = 0$. Do these upper and lower bounds for $|f(z)|$ on the circle $|z| = r$ have the same limit behavior as $r \to 0$? Compare with the behavior at a pole (Exercise 11).
Answer: $m(f, r) = e^{-1/r}$, $M(f, r) = e^{+1/r}$, all $r > 0$.

13. Locate and classify the singular points for $f(z) = e^{\operatorname{ctn} z}$.
Hint: Work on $z = 0$, then use periodicity of ctn z.

14. Prove the *Casorati-Weierstrass Theorem* by examining the consequences if there is a complex number α that *cannot* be approximated by values $f(z_n)$ where $z_n \neq p$, $z_n \to p$.

(i) There must be radii $r > 0$ and $\delta > 0$ such that $|f(z) - \alpha| \geq \delta$ for all z such that $r > |z - p| > 0$.

[If (i) fails for every choice of r and δ, it fails for the particular choices $r = \delta = 1/n$, so there are points $\{z_n\}$ such that $|f(z_n) - \alpha| < \delta = 1/n$, while $0 < |z_n - p| < r = 1/n$. Obviously $z_n \to p$ and $f(z_n) \to \alpha$. This is impossible by our choice of α.]

(ii) If $E = \{z : 0 < |z - p| < r\}$ as in (i), then $f(z) - \alpha$ is analytic and nonzero on E. Thus, $g(z) = 1/(f(z) - \alpha)$ is analytic on E with $|g(z)| \leq 1/\delta$ in E.

(iii) By Theorem 6.3, g has a removable singularity at p. Thus

$$g(z) = a_N(z - p)^N + a_{N+1}(z - p)^{N+1} + \cdots = (z - p)^N h(z)$$

where $N \geq 0$, $h(z)$ is analytic and nonzero at p.

(iv) $H(z) = 1/h(z)$ is analytic at p and $f(z) - \alpha = H(z)/(z - p)^N$ for z in E; thus f has a removable singularity ($N = 0$) or a pole ($N \geq 1$) at p.

Consequently, f cannot have an essential singularity at p if there is an α that cannot be approximated. This proves Theorem 6.5.

6.3 EVALUATING RESIDUES: PRACTICAL CALCULATIONS

If p is a pole of an analytic function $f(z)$, there are several ways we can evaluate $\mathrm{Res}(f, p)$ without resorting to direct calculations involving line integrals.

Case 1: The function f(z) has a pole of order m = 1

Then $f(z) = g(z)/(z - p)$ for $z \neq p$, where $g(z)$ is analytic and nonzero at p. On one hand, the value of g at $z = p$ is given by the limit

$$\lim_{z \to p} g = \lim_{z \to p} (z - p) f(z) = g(p);$$

the limit exists since $g(z)$ is continuous. On the other hand, the integral along a small circular contour $\gamma_r(t) = p + re^{it}$ is nothing other than the Cauchy Integral Formula,

$$\text{(9)} \qquad \mathrm{Res}(f, p) = \frac{1}{2\pi i} \int_{\gamma_r} f(z)\, dz = \frac{1}{2\pi i} \int_{\gamma_r} \frac{g(z)}{z - p}\, dz = g(p) = \lim_{z \to p} \{(z - p) f(z)\}.$$

The right-hand limit does not involve integrations, or the intermediary function $g(z)$, and can often be evaluated by inspection. Thus we have established our first method of evaluating residues.

Method 1 If $z = p$ is a pole of order one, we may compute its residue by the formula

$$\text{Res}(f, p) = \lim_{z \to p} \{(z - p)f(z)\}. \tag{10}$$

Example 6.7 Let $f(z) = e^{iz}/z$. Since $f(0) = 1$ is nonzero, the origin is a pole of order one. Clearly, for $p = 0$, we have $(z - p)f(z) = z \cdot f(z) = e^{iz}$ and we get $\text{Res}(f, 0) = 1$ by formula (9). In this case the value of the limit is quite easy to compute.

Example 6.8 Let $f(z) = 1/(z^2 + 1)$. Each of the singular points $z = +i, -i$ is a pole of order one. For $p = +i$ we get $(z - p)f(z) = (z - i)/(z^2 + 1) = 1/(z + i)$, which approaches the value $+1/2i$ as $z \to +i$; thus, $\text{Res}(f, +i) = +\frac{1}{2}i$. Similarly, $\text{Res}(f, -i) = -\frac{1}{2}i$.

The success of Method 1 rests on our ability to evaluate the limit in (9). This is not always such a simple matter. Here is another method for dealing with poles of order one, which is often more direct than Method 1.

Method 2 Suppose that $f(z)$ is presented as a quotient $f(z) = G(z)/H(z)$ of functions that are analytic at p, such that (i) G is nonzero at p, and (ii) $H(z)$ has a zero of order one at p. Then f has a pole of order one (Exercise 1, Section 6.2). Its residue is given by the formula

$$\text{Res}(f, p) = \frac{G(p)}{H'(p)} = G(p) \Big/ \frac{dH}{dz}(p). \tag{11}$$

Note: If an analytic function is zero at a point p, the zero is of **order N** if the first N coefficients of the Taylor series about p are zero, $a_0 = \cdots = a_{N-1} = 0$, and $a_N \neq 0$. This means that the derivatives $f^{(0)}(p) = f^{(1)}(p) = \cdots = f^{(N-1)}(p) = 0$ and the next derivative $f^{(N)}(p)$ is nonzero.

To prove (11) we notice that $H'(p)$ must be nonzero, since H has a zero of order one at p. But $H'(p)$ is defined as a limit,

$$H'(p) = \lim_{z \to p} \left\{\frac{H(z) - H(p)}{z - p}\right\} = \lim_{z \to p} \left\{\frac{H(z)}{z - p}\right\},$$

and formula (10) can be interpreted in a new way:

$$\text{Res}(f, p) = \lim_{z \to p} \left\{\frac{G(z)}{\left(\dfrac{H(z)}{z - p}\right)}\right\} = \frac{G(p)}{H'(p)}.$$

The advantages of Method 2 are displayed when we try to evaluate residues for reasonably complicated functions presented in the form $f = G/H$.

Example 6.9 Let $f(z) = 1/(z^6 + 1)$; there are poles of order one at each root of the equation $z^6 + 1 = 0$. These are just the sixth roots of -1, namely

$$p_1 = e^{i\pi/6},\ p_2 = e^{i\pi/2},\ p_3 = e^{i5\pi/6},\ p_4 = e^{-i5\pi/6},\ p_5 = e^{-i\pi/2},\ p_6 = e^{-i\pi/6}.$$

At p_1 we may use Method 2 to calculate

$$\operatorname{Res}(f, p_1) = \left(\frac{1}{6z^5}\bigg|_{z=p_1}\right) = \tfrac{1}{6}e^{-i5\pi/6},$$

and likewise, with equal simplicity, for the other residues. Using Method 1 and the factorization $z^6 + 1 = (z - p_1)\cdot(z - p_2)\cdot\ldots\cdot(z - p_6)$, we would get

$$\begin{aligned}\operatorname{Res}(f, p_1) &= \lim_{z\to p_1}\left\{\frac{z - p_1}{z^6 + 1}\right\} = \lim_{z\to p_1}\left\{\frac{1}{(z - p_2)\cdot\ldots\cdot(z - p_6)}\right\}\\ &= \frac{1}{(p_1 - p_2)\cdot(p_1 - p_3)\cdot\ldots\cdot(p_1 - p_6)}\\ &= \frac{1}{(e^{i\pi/6} - e^{i\pi/2})\cdot\ldots\cdot(e^{i\pi/6} - e^{-i\pi/6})}\,.\end{aligned}$$

Some messy calculations are required to put the answer into the simple polar form $(e^{-i5\pi/6})/6$.

The formulas we have just derived all fail if we try to apply them to poles of higher order or to an essential singularity. Here are some methods we can use to deal with more complicated situations.

Case II: The Function Has a Pole of Order $m \geq 2$

If we can conveniently express $f(z)$ in the form

(12) $$f(z) = \frac{g(z)}{(z - p)^m} \quad \text{for } z \text{ near } p, \quad z \neq p$$

where $g(z)$ is analytic and nonzero at p, it is a simple matter to calculate the residue at p. Indeed, when f is written this way the integrals

$$\operatorname{Res}(f, p) = \frac{1}{2\pi i}\int_{\gamma_r} f(z)\, dz = \frac{1}{2\pi i}\int_{\gamma_r} \frac{g(z)}{(z - p)^m}\, dz$$

are just the integrals that appear in Cauchy's Integral Formula for the $(m - 1)^{st}$ derivative of g, so that

(13) $$\operatorname{Res}(f, p) = \frac{1}{(m - 1)!}\, g^{(m-1)}(p).$$

Theorem 6.4 tells us that it is always *possible* to express f in this form; however, Theorem 6.4 does not guarantee that it will be *easy* to do this, and as a practical matter this might turn out to be very difficult.

For another approach, suppose we can represent f near p using a series involving a finite number of negative powers of $(z - p)$,

$$\begin{aligned} f(z) &= \frac{b_{-m}}{(z-p)^m} + \cdots + \frac{b_{-1}}{(z-p)} \\ &\quad + [b_0 + b_1(z-p) + b_2(z-p)^2 + \cdots] \end{aligned} \tag{14}$$

for all z near p with $z \neq p$. The series $[\cdots]$ on the right gives a function $h(z)$ which is defined and analytic near p, so that by integrating along small circular contours about p we get

$$\begin{aligned} \operatorname{Res}(f, p) &= \frac{b_{-m}}{2\pi i}\int_\gamma \frac{1}{(z-p)^m}\,dz + \cdots + \frac{b_{-1}}{2\pi i}\int_\gamma \frac{1}{z-p}\,dz \\ &\quad + \frac{1}{2\pi i}\int_\gamma h(z)\,dz \\ &= 0 + \cdots + 0 + b_{-1}\cdot 1 + 0 \\ &= b_{-1}. \end{aligned} \tag{15}$$

In other words, if f is represented by a series involving a finite number of negative powers of $(z-p)$, the residue is just the coefficient of $(z-p)^{-1}$. The procedures (13) and (15) may be summed up as follows.

Method 3 If f has a pole of order m at p, and we can express f in the form $f(z) = g(z)/(z-p)^m$ for $z \neq p$, where $g(z)$ is analytic and nonzero at p, then

$$\operatorname{Res}(f, p) = \frac{1}{(m-1)!}\,g^{(m-1)}(p) \quad \text{(the } (m-1)^{st} \text{ Taylor coefficient of } g\text{)}. \tag{16}$$

If, on the other hand, we can express f as a series about p that involves a finite number of negative powers of $(z-p)$, as in (14), then the residue is the coefficient of $(z-p)^{-1}$ in this series,

$$\operatorname{Res}(f, p) = b_{-1}. \tag{17}$$

It can be shown that the coefficients in any representation of f near a pole are uniquely determined (see Exercise 1, Section 6.5). It should also be clear that the degree of the highest order negative power term appearing in a representation of the form (14) is equal to the order m of the pole appearing at $z = p$. This method works best when f is presented in the form (12). For example, if $f(z) = (\cosh z)/z^3$ for $z \neq 0$ then

$$g(z) = 1 + \frac{z^2}{2!} + \frac{z^4}{4!} + \cdots \quad \text{and} \quad f(z) = \frac{1}{z^3} + \frac{1}{2!}\frac{1}{z} + \left(\frac{1}{4!}z + \cdots\right),$$

so f has a pole of order $m = 3$ at the origin and

$$\operatorname{Res}(f, 0) = \frac{1}{2!} = \frac{1}{2!}\frac{d^2g}{dz^2}(0).$$

Method 4 Matters are a little more complicated in another common situation. We are given analytic functions G and H where G is nonzero at p and H has a zero of order $m \geq 1$. We wish to calculate the residue of $f(z) = G(z)/H(z)$ at the mth order pole $z = p$. One of the most direct ways to calculate this residue is to represent $f(z)$ in the form (14) with *undetermined coefficients* $c_{-m}, \ldots, c_{-1}, c_0, \ldots$:

$$f(z) = \frac{c_{-m}}{(z-p)^m} + \cdots + \frac{c_{-1}}{(z-p)} + (c_0 + c_1(z-p) + \cdots);$$

we want to calculate the coefficient c_{-1}. But if we write out the Taylor series for G and H:

$$G(z) = a_0 + a_1(z-p) + \cdots \quad (a_0 \neq 0 \text{ since } G(p) \neq 0)$$
$$H(z) = b_m(z-p)^m + b_{m+1}(z-p)^{m+1} + \cdots \quad (H \text{ has a zero of order } m)$$

we may set up a system of equations that allows us to solve recursively for the unknown coefficients $\{c_{-m}, \ldots, c_{-1}, \ldots\}$ in terms of the known coefficients $\{a_k\}$ and $\{b_k\}$. The idea is to take the identity

$$G(z)/H(z) = f(z) = \frac{c_{-m}}{(z-p)^m} + \cdots + \frac{c_{-1}}{(z-p)} + c_0 + c_1(z-p) + \cdots$$

and multiply both sides by $(z-p)^m H(z)$; this gives us an identity in which every function that appears is analytic at p (or has a removable singularity),

$$\begin{aligned} G(z)(z-p)^m &= H(z) \cdot (z-p)^m f(z) \\ &= H(z) \cdot (c_{-m} + \cdots + c_{-1}(z-p)^{m-1} + c_0(z-p)^m + \cdots). \end{aligned}$$

Now we may use the Cauchy product formula for multiplying series to express both sides of this identity as power series in $(z-p)$:

$$(z-p)^m G(z) = a_0(z-p)^m + a_1(z-p)^{m+1} + \cdots$$

$$\begin{aligned} H(z) \cdot (z-p)^m f(z) &= (b_m(z-p)^m + \cdots)(c_{-m} + c_{-m+1}(z-p) + \cdots) \\ &= c_{-m}b_m(z-p)^m + (b_{m+1}c_{-m} + b_m c_{-m+1})(z-p)^{m+1} \\ &\quad + \cdots \end{aligned}$$

for all z near p. The coefficients on the various powers of $(z - p)$ must agree, so we obtain a system of algebraic equations

(18)
$$\begin{aligned} c_{-m}b_m &= a_0 \\ c_{-m}b_{m+1} + c_{-m+1}b_m &= a_1 \\ &\vdots \\ \sum \{c_k b_l : k + l = n\} &= a_n \\ &\vdots \end{aligned}$$

which can be solved recursively; the mth step yields the coefficient $c_{-1} = \operatorname{Res}(f, p)$ that we want.

Example 6.10 Let $f(z) = e^z/(1 - \cos z)$; f has a pole of order $m = 2$ at the origin, since $1 - \cos z = z^2/2! - z^4/4! + \ldots$. Using the method of undetermined coefficients we write

$$f(z) = \frac{c_{-2}}{z^2} + \frac{c_{-1}}{z} + c_0 + c_1 z + c_2 z^2 + \cdots \qquad \text{for } z \neq 0.$$

We must have $z^2 e^z = (1 - \cos z)(c_{-2} + c_{-1}z + c_0 z^2 + \cdots)$, so that the coefficients c_n can be determined from the relation

$$z^2\left(1 + \frac{z}{1!} + \frac{z^2}{2!} + \cdots\right) = \left(\frac{z^2}{2!} - \frac{z^4}{4!} + \cdots\right)(c_{-2} + c_{-1}z + c_0 z^2 + \cdots)$$

and we get the desired coefficient c_{-1} by solving the system of equations

$$\begin{aligned} 1 &= \frac{c_{-2}}{2!} \\ \frac{1}{1!} &= \frac{c_{-1}}{2!} \\ \frac{1}{2!} &= \frac{c_0}{2!} - \frac{c_{-2}}{4!} \\ &\vdots \end{aligned}$$

Clearly,

$$\operatorname{Res}(f, 0) = c_{-1} = +2$$

These methods, persistently applied, generally allow us to evaluate residues at poles of any order.

Case III: The Function has an Essential Singularity

The preceding methods fail for an essential singularity, and we might have to resort to direct calculation of the line integrals appearing in the definition of residue. However, such evaluations are usually very difficult. Another method is to represent f near the singular point p as a series of *negative* powers in $(z - p)$; these series are related to the Laurent series expansions discussed later in this chapter.

As in Section 3.9, it is obvious that $f(z) = \exp(-1/z^2)$ has a series representation obtained by substituting $w = -1/z^2$ into the usual exponential series; thus,

$$\exp(-1/z^2) = 1 - \frac{1}{z^2} + \frac{1}{2!}\frac{1}{z^4} - \cdots = \sum_{n=0}^{\infty} \frac{(-1)^n}{n!} z^{-2n} \quad \text{for all } z \neq 0,$$

and this series converges *uniformly* on any circle $|z| = R$. Therefore, we may integrate this series term-by-term along the contour $\gamma(t) = Re^{it}$ $(0 \leq t \leq 2\pi)$ (justified by Theorem 5.10), to get

$$\begin{aligned}\operatorname{Res}(f, 0) &= \frac{1}{2\pi i}\int_\gamma \exp(-1/z^2)\, dz \\ &= \sum_{n=0}^{\infty} \frac{(-1)^n}{n!}\left(\frac{1}{2\pi i}\int_\gamma z^{-2n}\, dz\right) = 0\end{aligned}$$

since each of the integrals is zero, by direct calculations. Other examples of such calculations at essential singularities are given in Exercise 2.

EXERCISES

1. Evaluate the residues $\operatorname{Res}(f, p)$, using the most suitable method in each case. Classify the singularity.

(i) $\dfrac{1}{z^2(z-1)}$ at $z = +1$

(ii) $\dfrac{1}{z^2(z^2+i)}$ at $z = 0$

(iii) $\dfrac{1-e^z}{z^3}$ at $z = 0$

(iv) $\dfrac{1-e^z}{z(z-i)^2}$ at $z = +i$

(v) $\tan z$ at $z = \pi/2$

(vi) $\dfrac{e^z}{\tan z}$ at $z = 0$

(vii) $\dfrac{e^z}{\tan z}$ at $z = \pi/2$

(viii) $\dfrac{1}{\operatorname{Log}^2 z}$ at $z = 1$

(ix) $\dfrac{1}{e^z - 1}$ at $z = 0$.

Answers: (i) $+1$; (ii) 0; (iii) -2; (iv) $1 + (i-1)e^i$; (v) -1; (vi) $+1$; (vii) 0 (removable); (viii) $+1$; (ix) $+1$.

2. Calculate the residue of $f(z)$ for the following essential singularities at $z = 0$.

(i) $\sin(1/z)$

(ii) $z^4 \sin(1/z)$

(iii) $(z^2 - z + 1)\sin(1/z)$

(iv) $e^z \cdot \sin(1/z)$

(v) $\exp\left(\dfrac{1}{z^2 + z}\right)$

Hint: Use Cauchy product formula on (iv). In (v), show

$$f(z) = \sum_{n=0}^{\infty} \frac{1}{n!} \frac{1}{z^n(z+1)^n}$$

uniformly on circles about $z = 0$; integrate term-by-term.

Answers: (i) $+1$; (ii) $1/5!$; (iii) $+\frac{5}{6}$; (iv) $+\frac{11}{12}$;

(v) $\displaystyle\sum_{n=0}^{\infty} \frac{(-1)^{n-1}(2n-2)!}{n!\,(n-1)!\,(n-1)!}$ (abs. convergent, by ratio test).

3. Show that the following trigonometric and hyperbolic functions are meromorphic (poles only). List, and calculate the order of, each pole. Calculate all residues.

(i) $\tan z$

(ii) $\text{ctn } z$

(iii) $\sec z$

(iv) $\csc z$

(v) $\tanh z$

(vi) $\text{ctnh } z$

(vii) $\text{sech } z$

(viii) $\text{csch } z$

Hint: Use periodicity properties, and the results in Chapter 2.

4. Calculate the following integrals along the counterclockwise oriented circles indicated.

(i) $\displaystyle\int_\gamma \frac{e^z}{z^2 - 1}\,dz \qquad |z| = 2$

(ii) $\displaystyle\int_\gamma \frac{1}{2z^2 + 3z - 2}\,dz \quad |z| = 1$

(iii) $\displaystyle\int_\gamma \frac{z - 1}{z + 1}\,dz \qquad |z + 1| = 3$

(iv) $\displaystyle\int_\gamma \frac{4z^3 + 2z}{z^4 + z^2 + 1}\,dz \quad |z| = 2$

(v) $\displaystyle\int_\gamma \frac{\tan z}{z}\,dz \qquad |z| = 2.$

Answers: (i) $2\pi i \sinh(1)$; (ii) $2\pi i/5$; (iii) $-4\pi i$; (iv) $8\pi i$; (v) 0.

5. Calculate the integral counterclockwise along the circle $|z| = 2$,

$$\frac{1}{2\pi i}\int_{\gamma} \frac{z^3 + 5}{z(z-1)^3}\, dz.$$

Classify the singularities of the integrand.

Answer: Integral $= +1$; poles at $z = 0$ $(m = 1)$, $z = 1$ $(m = 3)$.

6. Calculate the following integrals along the counterclockwise oriented circles $|z| = 2$ and $|z| = \frac{1}{2}$.

(i) $\displaystyle\int_{\gamma} \frac{e^z}{z(z^2 - 2z + 1)}\, dz$ (ii) $\displaystyle\int_{\gamma} \frac{\sin z}{(z^2 + 2z + 1)(z^2 - 2z + 1)}\, dz$

Hint: Residue integrals resemble Cauchy formulas for derivatives.

Answer: (i) $2\pi i$, $2\pi i$; (ii) $i\pi(\cos(1) - \sin(1))$, 0.

6.4 APPLICATIONS OF THE RESIDUE CALCULUS

We have developed a substantial variety of methods for evaluating residues. Now let us turn to the numerous applications of residue theory. Our first use will be in the calculation of certain types of definite integrals that do not yield to the ordinary methods of calculus. The reader should recall the definition of improper integrals of the form $\int_{-\infty}^{+\infty} f(x)\, dx$. Such an integral exists if and only if the one-sided improper integrals

$$\int_0^{+\infty} f(x)\, dx = \lim_{r\to+\infty}\int_0^r f(x)\, dx \quad\text{and}\quad \int_{-\infty}^0 f(x)\, dx = \lim_{s\to-\infty}\int_s^0 f(x)\, dx$$

exist separately; then we define

$$\int_{-\infty}^{+\infty} f(x)\, dx = \int_0^{+\infty} f(x)\, dx + \int_{-\infty}^0 f(z)\, dx.$$

It is easy to see that the double-ended integral is given, if it exists, by the single limit

(19) $$\int_{-\infty}^{+\infty} f(x)\, dx = \lim_{R\to+\infty}\int_{-R}^{+R} f(x)\, dx.$$

However, the limit on the right-hand side can exist, and have great practical significance, even though the improper integral $\int_{-\infty}^{+\infty} f(x)\, dx$ is not defined. In such cases, let us agree to call the limit in (19) the **Cauchy Principle Value** of the integral; this is indicated by writing

(20) $$PV\int_{-\infty}^{+\infty} f(x)\, dx = \lim_{R\to+\infty}\int_{-R}^{+R} f(x)\, dx.$$

For example, the integrand $f(x) = x$ gives us $\int_0^{+\infty} x\,dx = +\infty$ and $\int_{-\infty}^{0} x\,dx = -\infty$. The improper integral cannot be defined because we are left with an indeterminate form $\infty - \infty$ when we combine these results. Nevertheless, the Cauchy principal value does exist, because

$$PV\int_{-\infty}^{+\infty} x\,dx = \lim_{R\to+\infty}\int_{-R}^{+R} x\,dx = \lim_{R\to+\infty}\left\{\frac{R^2}{2} - \frac{(-R)^2}{2}\right\} = 0.$$

The Cauchy principal value acts like a "generalized integral"; it agrees with $\int_{-\infty}^{+\infty} f(x)\,dx$ when this improper integral makes sense, but it extends the meaning of the operation $\int_{-\infty}^{+\infty}(\cdots)\,dx$ so that it makes sense for a much larger class of integrands. Residue methods not only allow us to calculate improper integrals; they yield the Cauchy principal values when the corresponding improper integral fails to exist.

Application 1 Integrals of the form $\displaystyle\int_{-\infty}^{+\infty} \frac{P(x)}{Q(x)}\,dx.$

If P and Q are polynomials, this improper integral exists whenever

(i) $Q(x)$ has no real roots
(ii) Degree$(Q) \geq$ degree$(P) + 2$.

Then the function of real variable $f(x) = P(x)/Q(x)$ is continuous for all x and vanishes as $1/x^2$ when $|x| \to +\infty$ (which insures that the improper integral has a finite value). The improper integral is evaluated by noticing that the truncated integrals $\int_{-R}^{+R} P(x)/Q(x)\,dx$ in (19) are closely related to contour integrals of the rational function of complex variable $f(z) = P(z)/Q(z)$.

Example 6.11 Show that

$$\int_{-\infty}^{+\infty} \frac{1}{x^2+1}\,dx = \pi.$$

This improper integral could be calculated directly; there is a known indefinite integral $\int (x^2+1)^{-1}\,dx = \arctan x + c$, so that

$$\int_{-R}^{+R} \frac{1}{x^2+1}\,dx = 2\arctan(R) \to +\pi \quad \text{as} \quad R \to +\infty.$$

However, our purpose is to illustrate the use of residue methods in a simple case. In more complicated examples, the indefinite integral cannot be written in elementary form.

In Figure 6.3 the contour γ_R' moves from $-R + i0$ to $+R + i0$ along the real axis, and γ_R'' returns to $-R + i0$ along the semicircle of radius R. We integrate $f(z) = 1/(z^2+1)$ along the closed contour $\gamma_R = \gamma_R' + \gamma_R''$. Once R is greater than one, γ_R encloses the singular point $z = +i$ and, by the

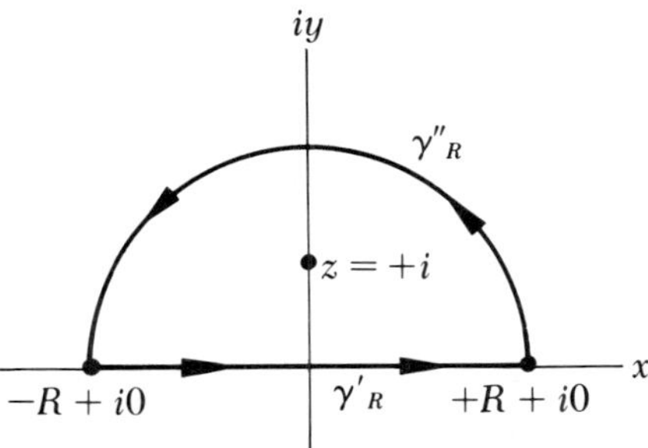

Figure 6.3 The contour $\gamma_R = \gamma'_R + \gamma''_R$ in Example 6.11 and the discussion which follows.

residue theorem,

$$\int_{\gamma_R} \frac{1}{z^2+1}\,dz = 2\pi i \operatorname{Res}(f, +i) = \pi \qquad \text{for all } R > 1.$$

But it is clear that

$$\int_{\gamma'_R} \frac{1}{z^2+1}\,dz = \int_{-R}^{+R} \frac{1}{x^2+1}\,dx,$$

so that

$$\int_{-R}^{+R} \frac{1}{x^2+1}\,dx = \int_{\gamma_R} \frac{1}{z^2+1}\,dz - \int_{\gamma''_R} \frac{1}{z^2+1}\,dz$$

$$= \pi - \int_{\gamma''_R} \frac{1}{z^2+1}\,dz$$

for all large R. We also have an obvious estimate on the size of the integral along γ''_R; if $|z| = R$, then $|z^2+1| \geq |z|^2 - 1 = R^2 - 1$, so that $|f(z)| \leq 1/(|z|^2 - 1) \leq 1/(R^2-1)$ on the trajectory of γ''_R, and

$$\left| \int_{\gamma''_R} \frac{1}{z^2+1}\,dz \right| \leq \frac{\text{length}(\gamma''_R)}{R^2-1} = \frac{\pi R}{R^2-1}.$$

Since this approaches zero as $R \to +\infty$, we see that

$$\int_{-\infty}^{+\infty} \frac{1}{x^2+1}\,dx = \lim_{R\to+\infty} \int_{-R}^{+R} \frac{1}{x^2+1}\,dx = \pi.$$

The size of the integral along γ''_R becomes negligible as $R \to +\infty$ because the maximum value of $|f(z)|$ on the circle $|z| = R$ goes to zero much faster (as $1/R^2$, to be precise) than $\text{length}(\gamma''_R) = \pi R$ increases.

In general, we can evaluate $\int_{-\infty}^{+\infty} P(x)/Q(x)\,dx$ by considering the function of a complex variable $f(z) = P(z)/Q(z)$. This is analytic except where $Q(z) = 0$;

let $p_1, \ldots, p_N$ be the zeros of $Q(z)$ which lie in the *upper half plane*. Then,

$$\int_{-\infty}^{+\infty} \frac{P(x)}{Q(x)}\,dx = 2\pi i \sum_{k=1}^{N} \text{Res}(f, p_k)$$

(21) $$= 2\pi i \sum \{\text{Res}(f, p_k) : \text{all poles } p_k \text{ in the upper half plane}\}$$

Since each p_k is a pole for $f(z)$, we can evaluate the residues at these singular points by the methods described in the last section. The proof of (21) is almost word-for-word the same as the argument we have given in the last example; we only need to observe that the requirement degree$(Q) \geq$ degree$(P) + 2$ means that the maximum value of $|f(z)|$ on the circle $|z| = R$ decreases at least as fast as $1/R^2$, and this insures that the integral along the circular arc γ''_R goes to zero as $R \to +\infty$.

Example 6.12 Show that

$$\int_{0}^{+\infty} \frac{x^2}{x^6 + 1}\,dx = \pi/6.$$

The integrand is an even function of x, so that

$$\int_{0}^{+\infty} \frac{x^2}{x^6 + 1}\,dx = \frac{1}{2}\int_{-\infty}^{+\infty} \frac{x^2}{x^6 + 1}\,dx.$$

The singularities of $f(z) = z^2/(z^6 + 1)$ are the roots of the equation $z^6 + 1 = 0$, and those in the upper half plane are $p_1 = e^{i\pi/6}$, $p_2 = e^{i\pi/2}$, and $p_3 = e^{i5\pi/6}$. These are poles of order one and have residues which are easily evaluated by Method 2 described in the last section;

$$\text{Res}(f, p_k) = \left[\frac{z^2}{6z^5}\bigg|_{z=p_k}\right] = \left[\frac{1}{6z^3}\bigg|_{z=p_k}\right]$$

so that

$$\int_{-\infty}^{+\infty} \frac{x^2}{x^6 + 1}\,dx = \frac{2\pi i}{6}\left(\frac{1}{{p_1}^3} + \frac{1}{{p_2}^3} + \frac{1}{{p_3}^3}\right)$$

$$= \frac{i\pi}{3}\left(e^{-i\pi/2} + e^{-i3\pi/2} + e^{-i5\pi/2}\right)$$

$$= \frac{i\pi}{3}(-i + i - i) = \frac{\pi}{3},$$

as desired.

Application 2 Integrals of the type $\int_{-\infty}^{+\infty} f(x)e^{ix}\,dx$ or $PV\int_{-\infty}^{+\infty} f(x)e^{ix}\,dx$.

The function $f(z)$ may have a finite number of isolated singularities in the upper half plane, none of them lying on the real axis, and should decrease to

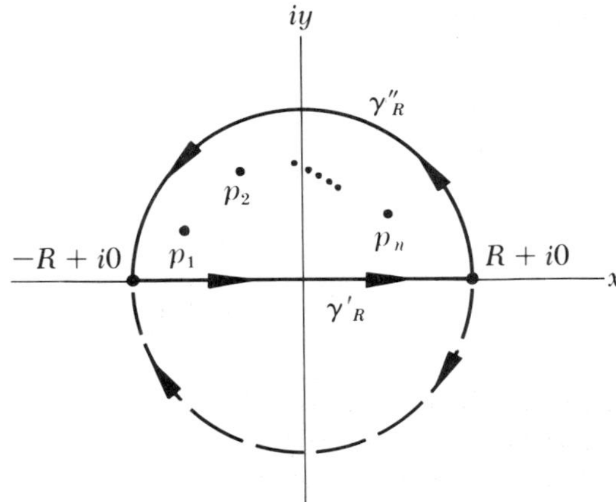

Figure 6.4 The contour in Application 2. The upper contour is used for e^{+ix} and the lower one for e^{-ix}.

zero at infinity within the upper half plane, in the following sense.

(22) If M_R is the maximum value of $|f(z)|$ for z on the semicircular arc $\Gamma_R = \{z : |z| = R \quad \text{and} \quad \operatorname{Im}(z) \geq 0\}$, then $\lim_{R \to +\infty} M_R = 0$.

For example, the functions $1/(z^2 + 1)$ or $e^{-iz}/(z^2 + z + 1)$ do this, but $\sin z$ and e^z do not have this property. We do not restrict the growth of $f(z)$, or the number of singularities, in the *lower* half plane.

The function $f(z)e^{iz}$ has the same singularities as $f(z)$, and once R is large enough, all singularities of $f(z)e^{iz}$ in the upper half plane are enclosed by the contour γ_R shown in Figure 6.4. Thus

$$\int_{\gamma_R} f(z)e^{iz}\, dz = 2\pi i \sum_{k=1}^{N} \operatorname{Res}(f(z)e^{iz}, p_k) \qquad \text{for all large } R > 0,$$

where $p_1, \ldots, p_N$ are the singularities of $f(z)$ in the upper half plane. It is also clear that $\int_{\gamma_R'} f(z)e^{iz}\, dz = \int_{-R}^{+R} f(x)e^{ix}\, dx$ for all R. *If we can show that*

(23)
$$\lim_{R \to +\infty} \left| \int_{\gamma_R''} e^{iz} f(z)\, dz \right| = 0,$$

it follows that the integral we are interested in can be evaluated by residues in the following way:

(24)
$$PV \int_{-\infty}^{+\infty} f(x)e^{ix}\, dx = 2\pi i \sum_{k=1}^{N} \operatorname{Res}(f(z)e^{iz}, p_k)$$

where $\{p_k\}$ are the singularities of $f(z)$ in the upper half plane. This follows from (23) since

$$PV\int_{-\infty}^{+\infty} f(x)e^{ix}\,dx = \lim_{R\to+\infty}\left\{\int_{-R}^{+R} f(x)e^{ix}\,dx\right\}$$

$$= \lim_{R\to+\infty}\left\{\int_{\gamma_R} f(z)e^{iz}\,dz\right\} - \lim_{R\to+\infty}\left\{\int_{\gamma_R''} f(z)e^{iz}\,dz\right\}$$

$$= 2\pi i\sum_{k=1}^{N}\mathrm{Res}(f(z)e^{iz}, p_k) - 0.$$

To verify the missing fact (23) we must use a more ingenious size estimate than the one employed in Application 1; the method used there, based on Theorem 5.9, only gives us the estimate

$$\left|\int_{\gamma_R''} f(z)e^{iz}\,dz\right| \le \mathrm{length}(\gamma_R'')\cdot M_R\cdot \max\{|e^{iz}|: z \quad \text{on} \quad \Gamma_R\}$$

where M_R is the maximum value of $|f(z)|$ on the semicircle Γ_R in the upper half plane (the trajectory of γ_R''). Although $|e^{iz}|$ decreases rapidly as z moves upwards, the function $|e^{iz}|$ achieves its maximum at the end points of Γ_R where we have $|e^{i(\pm R+i0)}| = 1$, so we may only conclude that

$$\left|\int_{\gamma_R''} f(z)e^{iz}\,dz\right| \le \pi R\cdot M_R$$

from the usual size estimates on line integrals. From our hypotheses we only know that $M_R \to 0$ as $R\to+\infty$, but we do not know how rapidly this limit is approached, and so it is not clear whether or not $R\cdot M_R \to 0$.

For a more delicate estimate we shall write $\int_{\gamma_R''} f(z)e^{iz}\,dz$ as a Riemann integral, and then use the formula $|\int_a^b \phi(t)\,dt| \le \int_a^b |\phi(t)|\,dt$. Thus,

$$\left|\int_{\gamma_R''} e^{iz}f(z)\,dz\right| = \left|\int_0^{\pi} f(Re^{i\theta})e^{iR(\cos\theta+i\sin\theta)}iRe^{i\theta}\,d\theta\right|$$

$$\le \int_0^{\pi} |f(Re^{i\theta})|\,e^{-R\sin\theta}R\,d\theta$$

$$\le M_R\cdot\int_0^{\pi} Re^{-R\sin\theta}\,d\theta.$$

We can use elementary methods from Calculus to show that $\int_0^{\pi} Re^{-R\sin\theta}\,d\theta \le \pi$ for all choices of $R > 0$; this is all we need to finish the discussion since it insures that

$$\left|\int_{\gamma_R''} e^{iz}f(z)\,dz\right| \le \pi M_R \to 0 \quad \text{as} \quad R\to+\infty.$$

Now

$$\int_0^{\pi} Re^{-R\sin\theta}\,d\theta = 2\int_0^{\pi/2} Re^{-R\sin\theta}\,d\theta,$$

and on the interval $[0, \pi/2]$ we have $2/\pi \leq (\sin \theta)/\theta \leq 1$ by an elementary use of L'Hospital's rule. This gives us the desired estimate,

$$\int_0^{\pi/2} Re^{-R\sin\theta}\, d\theta \leq \int_0^{\pi/2} Re^{-2R\theta/\pi}\, d\theta \leq \int_0^{+\infty} Re^{-2R\theta/\pi}\, d\theta = \frac{\pi}{2}.$$

Example 6.13 Show that

$$\int_0^{+\infty} \frac{x \sin x}{x^2 + 4}\, dx = \frac{\pi}{2}\, e^{-2}.$$

This integral is related to the double-ended integral

$$\int_{-\infty}^{+\infty} \frac{xe^{ix}}{x^2 + 4}\, dx;$$

since the integrand is an even function of x, it is clear that

$$\int_0^{+\infty} \frac{x \sin x}{x^2 + 4}\, dx = \frac{1}{2} \int_{-\infty}^{+\infty} \frac{x \sin x}{x^2 + 4}\, dx = \frac{1}{2}\, \mathrm{Im}\left[\int_{-\infty}^{+\infty} \frac{xe^{ix}}{x^2 + 4}\, dx\right].$$

Obviously the function $f(z) = z/(z^2 + 4)$ vanishes at infinity, as in (22), so the construction leading to formula (24) is justified. Furthermore, $z = +2i$ is the only singularity in the upper half plane. Method 2 for evaluating residues gives us

$$2\pi i\, \mathrm{Res}(e^{iz}f(z), +2i) = 2\pi i\left[\frac{ze^{iz}}{2z}\bigg|_{z=+2i}\right] = i\pi e^{-2}.$$

Thus,

$$\int_{-\infty}^{+\infty} \frac{x \sin x}{x^2 + 1}\, dx = \mathrm{Im}(i\pi e^{-2}) = \pi e^{-2}.$$

At the same time we have also evaluated the integral

$$\int_{-\infty}^{+\infty} \frac{x \cos x}{x^2 + 1}\, dx = \mathrm{Re}\left[\int_{-\infty}^{+\infty} \frac{xe^{ix}}{x^2 + 1}\, dx\right] = \mathrm{Re}(i\pi e^{-2}) = 0.$$

By using similar methods we may compute integrals of the form $\int_{-\infty}^{+\infty} f(x)e^{-ix}\, dx$, but now we must modify the discussion by making the contour extend into the *lower half plane*, as shown by the dashed curve in Figure 6.4. We allow $f(z)$ to have a finite number of isolated singularities in the lower half plane, none of them on the real axis, and we assume that $|f(z)|$ vanishes at infinity in the lower half plane, in the sense that

(22*) If M_R^* is the maximum value of $|f(z)|$ for z on the semicircular arc $\Gamma_R^* = \{z : |z| = R \text{ and } \mathrm{Im}(z) \leq 0\}$, then $\lim_{R\to+\infty} M_R^* = 0$.

Now the integral is given by

(25) $$PV\int_{-\infty}^{+\infty} f(x)e^{-ix}\,dx = -2\pi i \sum_k \operatorname{Res}(e^{-iz}f(z), q_k)$$

where q_k are the singularities of $f(z)$ in the lower half plane. The reason that we work with the lower half plane when e^{-ix} appears in the integral is that e^{-iz} is bounded throughout the lower half plane, but grows rapidly in the upper half plane; this boundedness insures that the integral along the semicircular arc Γ_R^* goes to zero as $R \to +\infty$. The minus sign in (25) appears because our contour in the lower half plane must be oriented *clockwise* so that the segment $[-R, +R]$ in the real axis is traversed from left to right. When e^{+ix} appears in the integral, instead of e^{-ix}, then e^{iz} is bounded on the upper half plane, and so this is then the appropriate half plane to work with.

This discussion applies equally well to integrals of the form $\int_{-\infty}^{+\infty} f(x)e^{\pm iax}\,dx$ (a real, $a > 0$). We leave it to the reader to determine how the conditions (22) and (22*) should be altered; in formulas (24) and (25) we would take the residues of $f(z)e^{\pm iaz}$ at the poles of $f(z)$. See Exercise 5 for details.

Example 6.14 Show that

$$\int_{-\infty}^{+\infty} \frac{\cos mx}{x^2 + x + 1}\,dx = \frac{2\pi}{\sqrt{3}} \cos\left(\frac{m}{2}\right) e^{-m\sqrt{3}/2} \quad \text{for} \quad m = 1, 2, \ldots.$$

The improper integral is convergent, since the integrand behaves like $|x|^{-2}$ at infinity. Clearly,

$$\int_{-\infty}^{+\infty} \frac{\cos mx}{x^2 + x + 1}\,dx = \operatorname{Re}\left[\int_{-\infty}^{+\infty} \frac{e^{imx}}{x^2 + x + 1}\,dx\right],$$

and the integral involving e^{imx} ($m > 0$ an integer) is given by

$$+2\pi i \sum \{\operatorname{Res}(e^{imz}f(z), p_k) : p_k \text{ the poles of } f(z) \text{ in upper half plane}\}$$

The function $f(z) = 1/(z^2 + z + 1)$ has poles of order one at the points $p_\pm = (-1 \pm i\sqrt{3})/2$. At the singular point p_+ in the upper half plane, we can use Method 2 to get

$$2\pi i \operatorname{Res}(e^{imz}f(z), p_+) = \left[\frac{2\pi i e^{imz}}{2z + 1}\bigg|_{z=p_+}\right] = 2\pi \frac{e^{-im/2}e^{-m\sqrt{3}/2}}{\sqrt{3}}.$$

The value of our integral is the real part of this expression.

Application 3 Integrals of the type $\int_{-\infty}^{+\infty} f(x)e^{\pm ix}\,dx$, where $f(x)$ has a singular point on the real axis.

These integrals are dealt with much as in the last section, except that the contours must also be deformed to avoid the singularities that occur on the real axis. As above, the methods work when $f(z)$ is analytic except at a finite number of isolated singular points, and vanishes at infinity as in (22) or (22*), depending on whether we work with the upper or lower half plane. Formulas (24) and (25) must be altered to include contributions due to the singular points that lie on the real axis. We will work a few typical examples to illustrate the modifications which should be made.

Example 6.15 Show that

$$\int_0^{+\infty} \frac{\sin x}{x}\,dx = \pi/2.$$

This integral is not improper at the end point $x = 0$ since $\lim_{x\to 0} (\sin x)/x = 1$. It is related to integrals involving e^{+iz}/z and e^{-iz}/z because

$$\int_0^{+\infty} \frac{\sin x}{x}\,dx = \frac{1}{2i}\int_0^{+\infty} \frac{e^{ix} - e^{-ix}}{x}\,dx = \lim_{\delta\to 0}\left\{\lim_{R\to+\infty} \frac{1}{2i}\int_\delta^R \frac{e^{ix} - e^{-ix}}{x}\,dx\right\}.$$

By the change of variable $x' = -x$, $dx' = -dx$, we see that

$$\begin{aligned}\int_\delta^R \frac{e^{ix} - e^{-ix}}{x}\,dx &= \int_\delta^R \frac{e^{ix}}{x}\,dx - \int_\delta^R \frac{e^{-ix}}{x}\,dx\\ &= \int_\delta^R \frac{e^{ix}}{x}\,dx - \int_{-\delta}^{-R} \frac{e^{ix}}{x}\,dx\\ &= \int_\delta^R \frac{e^{ix}}{x}\,dx + \int_{-R}^{-\delta} \frac{e^{ix}}{x}\,dx,\end{aligned}$$

so that

$$\int_0^{+\infty} \frac{\sin x}{x}\,dx = \frac{1}{2i}\lim_{\delta\to 0}\left\{\lim_{R\to+\infty}\left\{\int_{-R}^{-\delta} \frac{e^{ix}}{x}\,dx + \int_\delta^R \frac{e^{ix}}{x}\,dx\right\}\right\}. \tag{26}$$

Now e^{iz}/z has a pole at the origin. If we integrate e^{iz}/z along the contour $\gamma_{R,\delta}$ shown in Figure 6.5, which is deformed into the upper half plane to avoid the singular point $z = 0$, the integrals along horizontal segments γ_R' and γ_R'' are equal to the Riemann integrals on the right-hand side of (26). Since e^{iz}/z is

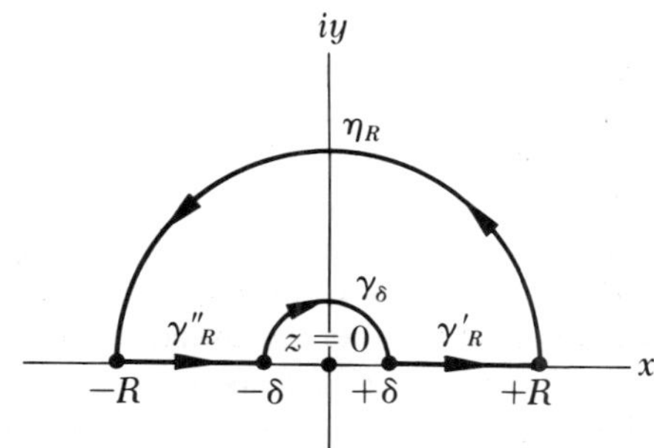

Figure 6.5 The closed contour $\gamma_{R,\delta} = \gamma_R'' + \gamma_\delta + \gamma_R' + \eta_R$.

analytic on $\gamma_{R,\delta}$ and on the set of points enclosed by this contour, the integral along $\gamma_{R,\delta}$ is zero, so that

$$\int_{-R}^{-\delta} \frac{e^{ix}}{x}\,dx + \int_{\delta}^{R} \frac{e^{ix}}{x}\,dx = -\int_{\gamma_\delta} \frac{e^{iz}}{z}\,dz - \int_{\eta_R} \frac{e^{iz}}{z}\,dz$$

for all small $\delta > 0$ and all large R.

Let $R \to +\infty$; the integral along the semicircular contour η_R approaches zero, by the same arguments used to justify equation (23), so we get

$$(27) \qquad \int_{-\infty}^{-\delta} \frac{e^{ix}}{x}\,dx + \int_{\delta}^{+\infty} \frac{e^{ix}}{x}\,dx = -\int_{\gamma_\delta} \frac{e^{iz}}{z}\,dz \quad \text{for all small} \quad \delta > 0.$$

We will now show that

$$(28) \qquad \lim_{\delta\to 0} \int_{\gamma_\delta} \frac{e^{iz}}{z}\,dz = -i\pi;$$

then formulas (26), (27), and (28) give the desired conclusion that

$$\int_0^{+\infty} \frac{\sin x}{x}\,dx = \frac{(-1)}{2i} \lim_{\delta\to 0} \int_{\gamma_\delta} \frac{e^{iz}}{z}\,dz = \frac{i\pi}{2i} = \frac{\pi}{2}.$$

To prove (28), write the numerator $f(z) = e^{iz}$ in the form $f(z) = f(0) + f'(0)\cdot z + e(z)\cdot z$, where $|e(z)| \to 0$ as $z \to 0$ (recall Section 2.8). Then

$$\int_{\gamma_\delta} \frac{e^{iz}}{z}\,dz = f(0)\int_{\gamma_\delta} \frac{1}{z}\,dz + f'(0)\int_{\gamma_\delta} 1\,dz + \int_{\gamma_\delta} e(z)\,dz.$$

By direct calculations, we get

$$f(0)\cdot\int_{\gamma_\delta} \frac{1}{z}\,dz = -i\pi\cdot f(0) = -i\pi;$$

the last two integrals go to zero as $\delta \to 0$, since the integrands are bounded near the origin and $\text{length}(\gamma_\delta) = \pi\delta \to 0$. This proves (28).

***Application 4** Integrals of the form $\int_0^{2\pi} F(\sin x, \cos x)\,dx$, where $F(u, v)$ is a rational function of two real variables u, v that has no singularities on the unit circle.

That is, $F(u, v)$ is a quotient of two polynomials in the real variables u and v; the integrand is then obtained by substituting $u = \sin x$ and $v = \cos x$.

For example, if $F(u, v) = u/v$, then the integrand is $\sin x/\cos x = \tan x$. These integrals are handled by recognizing that the Riemann integral above is just the line integral of a certain function $f(z)$ of a complex variable, along the contour $\gamma(t) = e^{it}$ defined for $0 \leq t \leq 2\pi$ (parametrized unit circle). Recall that if $f(z)$ is defined and continuous on the unit circle $|z| = 1$, we have

$$\int_\gamma f(z)\,dz = \int_0^{2\pi} \left[f(z)\Big|_{z=e^{it}} \right] ie^{it}\,dt.$$

Note that

$$\cos t = \frac{1}{2}(e^{it} + e^{-it}) = \left[\frac{1}{2}\left(z + \frac{1}{z}\right)\Big|_{z=e^{it}} \right]$$

$$\sin t = \frac{1}{2i}(e^{it} - e^{-it}) = \left[\frac{1}{2i}\left(z - \frac{1}{z}\right)\Big|_{z=e^{it}} \right]$$

for $0 \leq t \leq 2\pi$. Thus, we can make $\int_0^{2\pi} F(\sin t, \cos t)\,dt$ into a contour integral by writing

$$\int_0^{2\pi} F(\sin t, \cos t)\,dt = \int_0^{2\pi} \left[F\left(\frac{1}{2i}\left(z - \frac{1}{z}\right), \frac{1}{2}\left(z + \frac{1}{z}\right)\right)\Big|_{z=e^{it}} \right] \frac{ie^{it}}{ie^{it}}\,dt$$

$$= \int_0^{2\pi} [f(z)|_{z=e^{it}}] ie^{it}\,dt = \int_\gamma f(z)\,dz$$

where

(29) $$f(z) = \frac{1}{iz} F\left(\frac{1}{2i}\left(z - \frac{1}{z}\right), \frac{1}{2}\left(z + \frac{1}{z}\right)\right).$$

As an example, we will work out the value of $\displaystyle\int_0^{2\pi} \frac{1}{a + \sin t}\,dt$, where $a > 1$ is a fixed constant. Here we are dealing with the rational function $F(u, v) = 1/(a + u)$ and formula (29) indicates that we should integrate

$$f(z) = \frac{1}{iz}\left[\frac{1}{a + \dfrac{1}{2i}\left(z - \dfrac{1}{z}\right)} \right] = \frac{2}{z^2 + 2iaz - 1}$$

along γ to get the value of the desired integral. The assumption that $F(u, v)$ has no singularities on the unit circle $u^2 + v^2 = 1$ is equivalent to saying that the function $f(z)$, defined by (29), has no singular points on the circle $|z| = 1$. Furthermore, $f(z)$ will always be a rational function of z (a quotient of polynomials in z), so it can only have poles of finite order (and finitely many of them, at that). The residue theorem leads immediately to the conclusion

(30) $$\int_0^{2\pi} F(\sin t, \cos t)\,dt = \int_\gamma f(z)\,dz = 2\pi i \sum_{k=1}^{N} \text{Res}(f(z), p_k)$$

where $\{p_k\}$ are the poles of $f(z)$ that lie *inside* the circle $|z| = 1$ (the trajectory of γ).

Example 6.16 Evaluate

$$\int_0^{2\pi} \frac{1}{a + \sin t}\, dt \quad \text{(assuming } a > 1\text{)}.$$

This is just the contour integral $\int_\gamma \frac{2}{z^2 + 2iaz - 1}\, dz$ along the unit circle. The integrand $f(z)$ has two poles p_+ and p_- given by the quadratic formula

$$p_+ = i(-a + \sqrt{a^2 - 1}) \qquad p_- = i(-a - \sqrt{a^2 - 1}).$$

Clearly $|p_-| > |a| > 1$ so p_- lies outside of the unit circle and does not contribute to the sum of residues (30). On the other hand,

$$|p_+| = |-a + \sqrt{a^2 - 1}\,| \cdot \left| \frac{a + \sqrt{a^2 - 1}}{a + \sqrt{a^2 - 1}} \right|$$

$$= \frac{1}{|a + \sqrt{a^2 - 1}|} < \frac{1}{|a|} < 1.$$

The residue at p_+ is just

$$\operatorname{Res}(f, p_+) = \left[\frac{2}{2z + 2ia} \bigg|_{z=p_+} \right] = \frac{1}{i\sqrt{a^2 - 1}}$$

and we get

$$\int_0^{2\pi} \frac{1}{a + \sin t}\, dt = 2\pi i \operatorname{Res}(f(z), p_+) = \frac{2\pi}{\sqrt{a^2 - 1}}.$$

Notice that we never really have to vary contours in this discussion. We just apply the residue theorem directly to the integral along γ.

Application 5 Branch singularities.

In this situation we will consider a function $f(z)$ that is analytic on a cut plane, such as $f(z) = z^\alpha$ (α real) or $f(z) = \operatorname{Log} z$. The point $z = 0$ is called a **branch point** or **branch singularity** for these functions since we must cut the plane from 0 to ∞ in order to get a single valued analytic function. Along the cut there is usually a systematic relation between the values of $f(z)$ on opposite sides of the cut; for example, the function $f(z) = z^{1/2}$ merely changes sign, while $f(z) = \operatorname{Log} z$ changes by $-2\pi i$ as $\arg z$ increases past $+\pi$. By setting up closed contours with parts running along opposite sides of a cut, we can evaluate new types of integrals. The idea should become clear with the following examples.

Example 6.17 Evaluate $\int_0^{+\infty} \frac{1}{(1 + x)x^{1/2}}\, dx$. (The discussion works if $1/2$ is replaced by any other exponent $0 < \alpha < 1$). The integrand behaves like

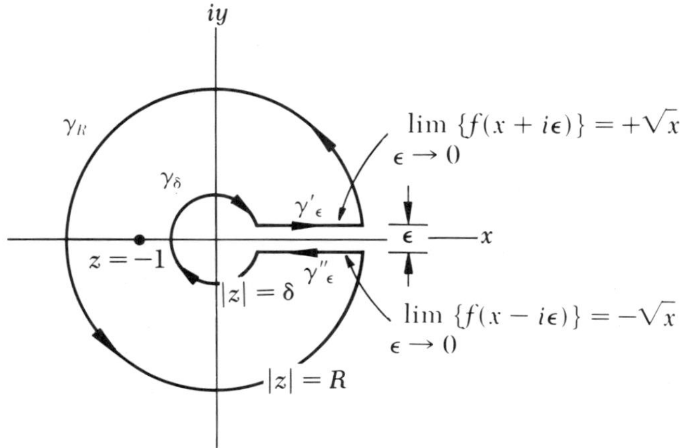

Figure 6.6 Here $f(z)(1 + z) = +x^{1/2}(1 + x)$ along the upper edge of the interval $[0, +\infty)$ and is equal to $-x^{1/2}(1 + x)$ along the lower edge.

$x^{-1/2}$ near $x = 0$, and like $x^{-3/2}$ as $x \to +\infty$, so the improper integral converges. Let us take a determination $f(z)$ of $z^{1/2}$ for which the discontinuity appears along the *positive* real axis; we can do this by defining

$$f(re^{i\theta}) = r^{1/2}e^{i(\theta/2)} \quad \text{(normalize } \theta \text{ so } 0 < \theta < 2\pi).$$

As we approach x on the real axis from the upper half plane, $f(z)$ approaches $+x^{1/2}$, while from the lower half plane $f(z)$ approaches $-x^{1/2}$. Now take the simple closed contour shown in Figure 6.6. The integrand $1/f(z)(1 + z)$ is analytic on the trajectory, and at all points enclosed by this contour, except at the pole of order one located at $z = -1 + i0$. Thus, for all contours of this type we get

$$\int_\gamma \frac{1}{f(z)\cdot(1+z)}\,dz = 2\pi i \operatorname{Res}\left(\frac{1}{f(z)\cdot(1+z)}, -1\right).$$

If we write the integrand in the form $G(z)/H(z)$ near $z = -1$, where $G(z) = 1/f(z)$ and $H(z) = 1 + z$, we can get the residue at $z = -1$ by applying Method 2 of the last section and conclude that:

$$\int_\gamma \frac{1}{f(z)(1+z)}\,dz = 2\pi i\left[\frac{1}{f(z)}\bigg|_{z=-1}\right] = \frac{1}{i}\cdot 2\pi i = 2\pi.$$

Let us see what happens when we vary the parameters $\delta > 0$, $\varepsilon > 0$, $R < +\infty$ that enter into this contour. First let $\varepsilon \to 0$. Clearly the integrals along the segments γ'_ε and γ''_ε converge to the respective values

$$\int_\delta^R \frac{1}{(x+1)x^{1/2}}\,dx \quad \text{and} \quad \int_R^\delta \frac{1}{(x+1)(-x^{1/2})}\,dx = \int_\delta^R \frac{1}{(x+1)x^{1/2}}\,dx,$$

since $f(x + it) \to +\sqrt{x}$ and $f(x - it) \to -\sqrt{x}$ as $t \to 0$ (keeping $t > 0$) for the determination $f(z) = z^{1/2}$ we have chosen. (Also, recall Exercises 17 to 19,

Section 5.3). Meanwhile, the integrals along the inner and outer circular arcs approach the integrals along the full circles. Let $\gamma_R(t) = Re^{it}$ and $\gamma_\delta(t) = e^{it}$ for $0 \le t \le 2\pi$; then, for all $\delta > 0$ and $R < +\infty$, we get

$$2\pi = 2\int_\delta^R \frac{1}{(x+1)x^{1/2}}\,dx + \int_{\gamma_\delta} \frac{1}{f(z)(z+1)}\,dz + \int_{\gamma_R} \frac{1}{f(z)(z+1)}\,dz.$$

For all large $|z|$ the integrand is dominated by $C_1\,|z|^{-3/2}$, where C_1 is a fixed constant, so the size of the integral along γ_R is dominated by $C_1 \cdot 2\pi R \cdot R^{-3/2} = C_1' R^{-1/2}$, and vanishes as $R \to +\infty$. For z near the origin, the term $1/(z+1)$ in the integrand approaches the value 1, so the integrand is dominated by $C_2\,|z|^{-1/2}$, where C_2 is a fixed constant, and this means that the size of the integral along γ_δ is dominated by $C_2 \cdot 2\pi\delta \cdot \delta^{-1/2} = C_2'\delta^{1/2}$, which approaches zero as $\delta \to 0$. Thus,

$$\int_0^{+\infty} \frac{1}{(x+1)x^{1/2}}\,dx = \lim_{\delta\to 0}\left\{\lim_{R\to+\infty}\int_\delta^R \frac{1}{(x+1)x^{1/2}}\,dx\right\} = \pi.$$

By similar arguments, using x^α $(0 < \alpha < 1)$ in place of $x^{1/2}$, we may prove that

$$\int_0^{+\infty} \frac{1}{x^\alpha(x+1)}\,dx = \frac{\pi}{\sin \pi\alpha} \qquad (\text{for} \quad 0 < \alpha < 1).$$

The appropriate analytic determination of z^α is obtained by letting $z^\alpha = e^{\alpha\cdot\log z}$ and taking a determination of $\log z$ defined on the cut plane with the *positive* real axis removed; thus, we can use the definition

$$\log(re^{i\theta}) = \log r + i\theta \text{ (normalize } \theta \text{ so that } 0 < \theta < 2\pi).$$

With this choice, z^α agrees with x^α (respectively, $e^{2\pi i\alpha}x^\alpha$) as we approach the positive real axis from the upper (respectively, lower) half plane. Similarly, we can evaluate integrals of the form $\int_0^{+\infty} x^{-\alpha}R(x)\,dx$ where $R(z)$ is a rational function whose singularities lie off the real axis; we must also assume that

$$\lim_{z\to\infty} |R(z)| = 0,$$

and that $0 < \alpha < 1$, to be sure that our improper integral converges.

Example 6.18 Show that

$$\int_1^{+\infty} \frac{1}{x\sqrt{x^2-1}}\,dx = \pi/2.$$

The integrand behaves like x^{-2} as $x \to +\infty$, and like $(x-1)^{-1/2}$ as $x \to 1$, so that the improper integral is well defined. The function $f(z) = (z^2-1)^{1/2}$ is double valued; near any point p (except the points $+1$ and -1 where its

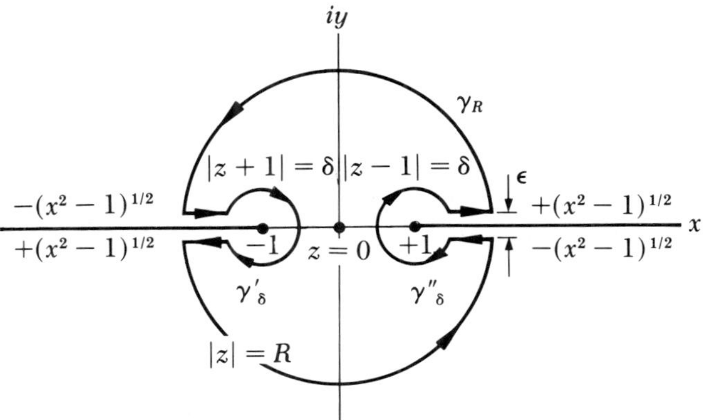

Figure 6.7 The closed contour γ in Example 6.18. The limit values of $(z^2 - 1)^{1/2}$ along the cuts are indicated; note that $(x^2 - 1) > 0$ if $|x| > 1$.

behavior is highly singular) there are two analytic determinations of $(z^2 - 1)^{1/2}$. After some thought (see Exercise 25, Section 4.10) we see that by removing the line segments $(-\infty, -1]$ and $[+1, +\infty)$ to get a doubly cut plane, we can define a single valued analytic determination of $(z^2 - 1)^{1/2}$ on the remaining domain D. Indeed, the mapping $w = z^2 - 1$ maps the doubly cut plane D onto the singly cut plane E obtained by removing the positive real axis from the w-plane. On E there is obviously a determination of $w^{1/2}$ that approaches the value $+u^{1/2}$ at $w = u + i0$ along the upper edge of the positive real axis, and approaches $-u^{1/2}$ at $w = u + i0$ along the lower edge. Thus, the analytic function $(z^2 - 1)^{1/2} = w^{1/2}$, defined on D, approaches the values shown in Figure 6.7 as z approaches a point $x + i0$ on one of the cuts from the upper or lower half plane. Now $f(z) = 1/z \cdot (z^2 - 1)^{1/2}$ is analytic on the doubly cut plane D, except at $z = 0$, where there is a pole of order one with residue

$$\operatorname{Res}(f, 0) = \left[\frac{1}{(z^2 - 1)^{1/2}}\bigg|_{z=0}\right] = \frac{1}{i} = -i.$$

Evidently, $f(z)$ is analytic on the contour γ shown in Figure 6.7 and at all points enclosed by γ, except for the pole $z = 0$; thus, $\int_\gamma f(z)\,dz = 2\pi i \operatorname{Res}(f, 0) = 2\pi$ for every choice of the parameters $\delta > 0$, $\varepsilon > 0$, and $R < +\infty$ which determine γ. If we let $\varepsilon \to 0$ and keep in mind the limit values of $(z^2 - 1)^{1/2}$ along the upper and lower edges of the cuts, we see that the sum of the integrals along the horizontal segments of γ approaches the value

$$-\int_{-R}^{-\delta-1} \frac{1}{x\sqrt{x^2 - 1}}\,dx + \int_{-\delta-1}^{-R} \cdots dx + \int_{1+\delta}^{+R} \cdots dx - \int_{+R}^{1+\delta} \cdots dx$$

$$= 2\int_{1+\delta}^{+R} \frac{1}{x\sqrt{x^2 - 1}}\,dx - 2\int_{-R}^{-\delta-1} \frac{1}{x\sqrt{x^2 - 1}}\,dx.$$

(For limits in which the *contour* varies, recall Exercise 17 of Section 5.3.) Since the integrand is an *odd* function of x, the change of variable $x' = -x$, $dx' = -dx$ in the right-hand integral shows that the sum is equal to $4\int_{1+\delta}^{R} \frac{1}{x\sqrt{x^2-1}}\, dx$. Meanwhile, as $\varepsilon \to 0$ the integral along γ keeps the fixed value $+2\pi$. The integral along the outer circular arcs approaches the value $\int_{\gamma_R} f(z)\, dz$, where γ_R is the counterclockwise oriented circle $|z| = R$; the integrals along the two inner circular arcs approach the value

$$\int_{\gamma_\delta'} f(z)\, dz + \int_{\gamma_\delta''} f(z)\, dz,$$

where γ_δ' and γ_δ'' are parametrized circles of radius δ about $z = -1$ and $z = +1$ (both parametrized *clockwise*, as indicated in Figure 6.7).† Clearly, for all small $\delta > 0$ and all large $R < +\infty$, we get

$$\begin{aligned} 2\pi &= \int_\gamma f(z)\, dz \\ &= 4\int_{1+\delta}^{R} \frac{1}{x\sqrt{x^2-1}}\, dx + \int_{\gamma_R} f(z)\, dz + \int_{\gamma_\delta'} f(z)\, dz + \int_{\gamma_\delta''} f(z)\, dz. \end{aligned} \tag{31}$$

For large values of $|z| = R$, we have $|f(z)| \leq C_1 |z|^{-2}$, for some fixed constant C_1, which insures that the integral of $f(z)$ along γ_R approaches zero as $R \to +\infty$. Because $+1$ and -1 are *branch singularities* for $f(z)$, the integrals along γ_δ' and γ_δ'' cannot be evaluated by residues, and a little finesse is needed in determining their limits as $\delta \to 0$. Notice that

$$|z(z^2 - 1)^{1/2}| = |z|\, |z + 1|^{1/2}\, |z - 1|^{1/2},$$

so that $|f(z)| \leq C_2 |z - 1|^{-1/2}$ for z near $+1$, where C_2 is some fixed constant; the absolute value of the integrand on the circle $|z - 1| = \delta$ is dominated by $C_2 \cdot \delta^{-1/2} \cdot \text{length}(\gamma_\delta'') = C_2' \cdot \delta^{+1/2}$, which approaches zero as $\delta \to 0$. Likewise, we can prove that the integral along the contour γ_δ', around $z = -1$, approaches zero as $\delta \to 0$. By taking limits in (31) as $\delta \to 0$ and $R \to +\infty$, we obtain

$$\int_1^{+\infty} \frac{1}{x\sqrt{x^2-1}}\, dx = \lim_{\delta\to 0}\left\{\lim_{R\to+\infty} \int_{1+\delta}^{R} \frac{1}{x\sqrt{x^2-1}}\, dx\right\} = \pi/2.$$

There are even more elaborate methods for calculating definite integrals by relating them to certain families of contour integrals.

---

† Notice that the integrand $f(z)$ in these integrals is discontinuous where the circles $|z + 1| = \delta$ and $|z - 1| = \delta$ cross the cuts; however, the integrand has only a simple jump discontinuity there, and is continuous and bounded otherwise, so that the contour integrals along γ_δ' and γ_δ'' are well defined.

EXERCISES

1. Assume that $f(x)$ is bounded and continuous on $(0, +\infty)$. If $x^\alpha |f(x)| \to 0$ as $x \to +\infty$ for an exponent $\alpha > 1$, show that the improper integral $\int_0^{+\infty} f(x)\,dx$ exists and is finite. This is not necessarily so if $\alpha = 1$ (give an example).

2. In defining Cauchy Principal Values,

$$PV\int_{-\infty}^{\infty} f(x)\,dx = \lim_{R\to\infty}\left\{\int_{-R}^{R} f(x)\,dx\right\},$$

it is important that we use *symmetrical* end points $-R$ and $+R$ in the limit. Compare

$$PV\int_{-\infty}^{\infty} x\,dx \quad \text{and} \quad \lim_{R\to\infty}\int_{-R}^{+2R} x\,dx.$$

3. Use the methods of Application 1 to evaluate

(i) $\displaystyle\int_{-\infty}^{\infty} \frac{1}{x^6+1}\,dx$

(ii) $\displaystyle\int_{-\infty}^{\infty} \frac{1}{x^2+x+1}\,dx$

(iii) $\displaystyle\int_{0}^{\infty} \frac{x^2}{(x^2+4)(x^2+9)}\,dx$

(iv) $\displaystyle\int_{0}^{\infty} \frac{x^2-1}{(x^2+1)^2}\,dx$

(v) $\displaystyle\int_{-\infty}^{\infty} \frac{x^2+2}{(x^2+2x+2)(x^2-2x+2)}\,dx$

4. The integrals $\displaystyle\int_0^{\infty} \frac{1}{1+x^n}\,dx$ may be evaluated as in Application 1 for even powers $n = 2, 4, \ldots.$ For odd powers, such as $n = 3$, there are several difficulties with this approach (why?). Prove that

$$\int_0^{\infty} \frac{1}{1+x^n}\,dx \quad \text{exists and is equal to} \quad \frac{\pi/n}{\sin(\pi/n)}$$

for $n = 1, 2, \ldots$ by examining integrals along the boundaries of circular sectors bounded by the rays $\arg z \equiv 0$, $\arg z \equiv 2\pi/n$, and the arc $|z| = R$.

5. If $f(z)$ satisfies condition (22) on the upper half plane, as in Application 2, and if $a > 0$, prove that

$$\int_{-\infty}^{+\infty} f(x)e^{iax}\,dx = \sum_k \operatorname{Res}(f(z)e^{iaz}, p_k)$$

where p_k are the poles in the upper half plane. If $a < 0$, and $f(z)$ satisfies condition (22*), work out the corresponding formula involving the lower half plane.

6. Using the methods of Application 2, calculate

(i) $$\int_{-\infty}^{+\infty} \frac{e^{iax}}{1+x^2}\,dx = \pi e^{-a} \quad (a > 0)$$

(ii) $$\int_{-\infty}^{+\infty} \frac{e^{iax}}{1+x^2}\,dx = \pi e^{a} \quad (a < 0)$$

(iii) $$\int_{0}^{\infty} \frac{x^3 \sin mx}{x^4+1}\,dx = (\pi/2)e^{-m/\sqrt{2}} \cos(m/\sqrt{2})$$

$$(m = 1, 2, \ldots)$$

(iv) $$\int_{0}^{\infty} \frac{\cos x}{x^2+a^2}\,dx = \frac{\pi e^{-a}}{2a} \quad (a > 0)$$

(v) $$\int_{-\infty}^{\infty} \frac{x \sin 2x}{x^2-2x+2}\,dx = \frac{\pi}{e^2}(\cos(2) + \sin(2))$$

7. For $0 < a < 1$ show that

$$\int_{-\infty}^{\infty} \frac{e^{ax}}{e^x+1}\,dx = \frac{\pi}{\sin \pi a}$$

by integrating along rectangles with vertices $-R$, $+R$, $R + 2\pi i$, $-R + 2\pi i$.

Hint: Integrals along vertical segments go to zero as $R \to +\infty$.

8. If $f(x)$ has a finite integral, $\int_{\infty}^{-\infty} |f(x)|\,dx < +\infty$, its **Fourier transform** $\mathscr{F}f(s)$ is the complex valued function of a (new) real variable s, given by

$$\mathscr{F}f(s) = \int_{-\infty}^{\infty} e^{isx}f(x)\,dx \quad \text{for} \quad -\infty < s < +\infty.$$

Certain Fourier transforms may be calculated by residue methods. If $f(x) = 1/(1+x^2)$, show that

$$\mathscr{F}f(s) = \pi e^{-|s|} \quad \text{for} \quad -\infty < s < +\infty.$$

Hint: Most of the work appears in previous exercises.

9. Use the methods of Application 3 to prove that

(i) $$\int_0^\infty \frac{1 - \cos x}{x^2}\, dx = \frac{\pi}{2}$$

(ii) $$\int_{-\infty}^\infty \frac{\cos(\pi x/2)}{x^2 - 1}\, dx = \pi$$

(iii) $$\int_{-\infty}^\infty \frac{\sin \pi x}{x^2 - 1}\, dx = 0$$

(iv) $$\int_0^\infty \frac{\sin^2 x}{x^2}\, dx = \frac{\pi}{2}$$

(v) $$\int_0^\infty \frac{\sin x}{x(x^2 + 1)}\, dx = \frac{\pi}{2}\,(1 + \cos(1))$$

Hint: In (i), integrate $f(z) = (1 - e^{iz})/z^2$; in (ii) integrate $f(z) = (e^{i\pi z/2})/(z^2 - 1)$. In (iii), are residue calculations necessary at all? (Integrand is an odd function of x.) In (iv) write $2 \sin^2 x = 1 - \cos 2x$, etc.

10. Use the methods of Application 4 to prove that

(i) $$\int_{-\pi}^{+\pi} \frac{1}{1 + \cos^2 \theta}\, d\theta = \pi\sqrt{2}$$

(ii) $$\int_0^{2\pi} \frac{1}{1 + \alpha \sin \theta}\, d\theta = \frac{2\pi}{\sqrt{1 - \alpha^2}} \qquad (\alpha \text{ real};\ -1 < \alpha < +1)$$

(iii) $$\int_0^{2\pi} \frac{1}{1 + \alpha \cos \theta}\, d\theta = \frac{2\pi}{\sqrt{1 - \alpha^2}} \qquad (\alpha \text{ real};\ -1 < \alpha < +1)$$

(iv) $$\int_0^{\pi} \cos^{2n} \theta\, d\theta = \pi\, \frac{(2n)!}{2^{2n}(n!)^2} \qquad (n = 1, 2, \ldots)$$

11. If $\sqrt{\zeta}$ is the principal determination of square root, show that $f(w) = \sqrt{1 - w^2}$ is holomorphic on $D = \{w : |w| < 1\}$. By calculating images of D under $-\sqrt{1 - w^2}$ and $+\sqrt{1 - w^2}$, show that

$$|1 - \sqrt{1 - w^2}| < 1 \quad \text{and} \quad |1 + \sqrt{1 - w^2}| > 1 \quad \text{for} \quad |w| < 1.$$

Then prove that $2\pi/\sqrt{1 - w^2}$ is produced on D by the integral formula

$$\int_0^{2\pi} \frac{-1}{1 + w \cos \theta}\, d\theta = \frac{2\pi}{\sqrt{1 - w^2}} \qquad (w \text{ a complex parameter; } |w| < 1).$$

Hint: Regard $z = \sqrt{1 - w^2}$ as the composite of $\zeta = 1 - w^2$ and $z = \sqrt{\zeta}$, to determine images of D.

12. Prove that the following integrals agree:

$$\int_0^\infty \sin t^2\,dt = \int_0^\infty \cos t^2\,dt = \frac{1}{\sqrt{2}}\int_0^\infty e^{-t^2}\,dt$$

by integrating $\exp(-z^2)$ along the boundary of the sector $|z| < R$, $0 < \arg z < \pi/4$. The right-hand integral is calculated by various methods of advanced calculus to be $\sqrt{\pi}/2$. This application's objective is to show that certain integrals *agree*, rather than to calculate the values; it has many generalizations.

Hint: Show that the integral along the circular arc goes to zero as $R \to \infty$. (Recall methods of Application 2.)

13. Use methods of Application 5 to show that

$$\text{(i)}\quad \int_0^\infty \frac{x^{1/2}}{x^2+2x+1}\,dx = \frac{\pi}{2} \qquad \text{(ii)}\quad \int_0^\infty \frac{\log^2 x}{1+x^2}\,dx = \frac{\pi^3}{8}.$$

Hint: In (ii) the integrand is an even function of x, so we can work with the integral from $-\infty$ to $+\infty$.

14. Generalize Example 6.17 by proving that

$$\int_0^\infty \frac{1}{x^\alpha(x+1)}\,dx = \frac{\pi}{\sin \pi\alpha} \quad \text{for} \quad 0 < \alpha < 1.$$

Does the integral converge for other real choices of the parameter of α (examine convergence at *both* end points)?

15. Show that

$$\int_0^\infty \frac{1}{x^w(x+1)}\,dx \quad \text{exists and is equal to} \quad \frac{\pi}{\sin \pi w}$$

for *complex* w in the strip $0 < \operatorname{Re}(w) < +1$. (Set $x^w = e^{w\cdot\log x}$ for $0 < x < +\infty$, taking the usual determination of $\log x$.) This integral generates $\pi \csc(\pi w)$ on the strip; many other analytic functions can be generated by similar formulas involving improper integrals with a complex parameter (cf. Exercise 11).

Hint: Define $\log z$ with values $\log x$ and $2\pi i + \log x$ along the upper and lower sides of $[0, +\infty)$; define $z^w = e^{w\cdot\log z}$. What are the values of z^w along the sides of this cut? Now proceed as in Example 6.17.

16. Generalize Example 6.18 to prove that

$$\int_1^\infty \frac{1}{x^n\sqrt{x^2-1}}\,dx = \frac{c_n}{4} = \frac{1}{4}\left[\frac{2\pi}{(n-1)!}(2n-5)(2n-7)\cdot\ldots\cdot3\cdot1\right]$$

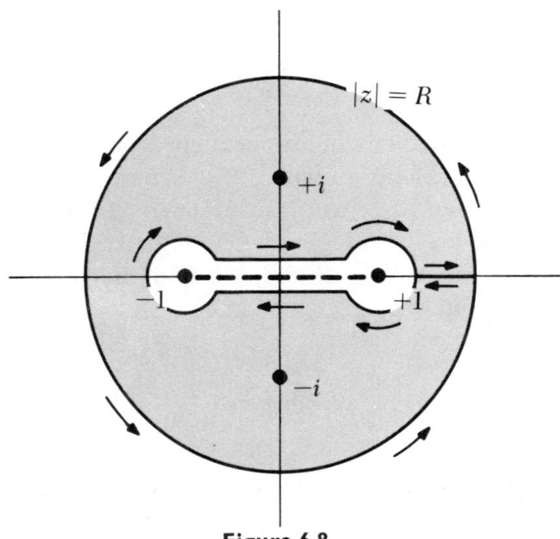

Figure 6.8

for odd powers $n = 3, 5, \ldots$. Where would the discussion break down for an even power, such as $n = 2$?

Hint: Take $f(z) = 1/\sqrt{z^2 - 1}$ and integrate $f(z)/z^n$ around $z = 0$; the value is $c_n = 2\pi i f^{(n-1)}(0)/(n-1)!$ (by the Cauchy formula for derivatives).

17. There is an analytic determination of $f(z) = \sqrt{1 - z^2}$ on the cut plane $E = \mathbf{C} \sim [-1, +1]$ that has limit values $+\sqrt{1 - x^2}$ and $-\sqrt{1 - x^2}$ along the upper and lower edges of the cut (multiply the function in Exercise 23, Section 4.11 by -1). Using the contour in Figure 6.8, calculate

$$\text{(i)} \quad \int_{-1}^{+1} \frac{1}{(1 + x^2)\sqrt{1 - x^2}}\, dx = \frac{\pi}{\sqrt{2}}$$

$$\text{(ii)} \quad \int_{-1}^{+\infty} \frac{1}{(1 + x^2)^2\sqrt{1 - x^2}}\, dx = \frac{3\pi}{4\sqrt{2}}.$$

18. The function $w = \sqrt{\dfrac{1 - z}{1 + z}}$ (principal determination of square root) is analytic off the cut $[+1, +\infty)$. Determine its values along the upper and lower edges of the cut and calculate

$$\int_1^{\infty} \frac{1}{x^2}\sqrt{\frac{x - 1}{x + 1}}\, dx$$

Hint: How is $[+1, +\infty)$ transformed by the *fractional linear* map $w = (z + 1)/(z - 1)$? Where are points above and below the cut mapped? The values are $\pm i\sqrt{\dfrac{x+1}{x-1}}$ along the cut, but the constants $+i$ and $-i$ can be handled outside the integral.

19. The methods of Application 5 cannot be applied directly to calculate $\int_0^\infty R(x)\log x\,dx$, where $R(x)$ is a rational function (the integrals we want, along $[0, +\infty)$, cancel out). Instead, apply the ideas of Application 5 to the integrand $R(x)(\log x)^2$; integrating along contours such as the one in Figure 6.6, we find that a certain combination of residues is equal to

$$4\pi i\int_0^\infty R(x)\log x\,dx - 4\pi^2\int_0^\infty R(x)\,dx.$$

The integrals involving $R(x)(\log x)^2$ cancel out; taking imaginary parts yields the desired integral $\int_0^\infty R(x)\log x\,dx$. Use these ideas to calculate

$$\text{(i)}\quad \int_0^\infty \frac{\log x}{(1+x)^3}\,dx = -\frac{1}{2} \qquad \text{(ii)}\quad \int_0^\infty \frac{\log x}{1+x^2}\,dx = 0.$$

20. Calculate

$$\int_0^\infty \frac{\log x}{x^2-1}\,dx = \pi^2/4.$$

Follow ideas from Exercise 19. However, you must introduce an additional deformation in the contour to avoid the extra singular point $z = +1$. Why is the integral convergent in spite of the singularities at $x = 0$ and $x = 1$?

6.5 LAURENT SERIES EXPANSIONS

Analytic functions can be described near an isolated singular point by means of a power series which includes both positive and negative powers of $(z - p)$. Such an expansion about a point p will have the form

$$\sum_{n=-\infty}^{+\infty} a_n(z-p)^n \tag{32}$$

and is called a **Laurent series about the base point p.** Series of this kind obviously include ordinary power series about p as a special case (in which $a_n = 0$ for $n = -1, -2, \ldots$), and also the series of negative powers discussed in Section 3.9. A word or two is needed to explain what we mean by "convergence" of a series like (32). We have already discussed the convergence properties of series in which the powers are all positive or all negative, such as

(33A) $$\sum_{n=-1}^{-\infty} a_n(z-p)^n = \sum_{n=1}^{\infty} a_{-n}(z-p)^{-n}$$

(33B) $$\sum_{n=0}^{+\infty} a_n(z-p)^n$$

We say that the **series** (32) **converges** at a point z (or **converges absolutely** at z; or **converges uniformly** on some set X; etc.) if the separate series in (33A) and (33B) both converge (or both converge absolutely at z, etc.). If both series converge at z, we assign the obvious value to the series, namely

(34) $$\sum_{n=-\infty}^{+\infty} a_n(z-p)^n = \sum_{n=-1}^{+\infty} a_n(z-p)^n + \sum_{n=0}^{\infty} a_n(z-p)^n.$$

The following properties of series (33A) and (33B) should be recalled. Series (33A) has a radius of convergence r_1, and it is absolutely convergent for points exterior to the circle $|z-p| = r_1$ and divergent for points interior to this circle; series (33B) has a radius of convergence r_2 and this series converges absolutely for points interior to the circle $|z-p| = r_2$ and diverges for points outside of this circle. From these facts we can determine the convergence properties of a typical Laurent series (32):

(i) If $r_1 > r_2$ at least one of the series (33A) or (33B) is divergent for each point z in the plane, and the Laurent series (32) must be regarded as diverging everywhere.

(ii) If $r_1 < r_2$, then (32) converges on the annulus centered at p defined by $r_1 < |z-p| < r_2$, and diverges if $0 \leq |z-p| < r_1$ or if $|z-p| > r_2$.

The radii r_1 and r_2 are referred to as the **inner** and **outer radii** of the Laurent series (32). These radii are determined by the coefficients $\{a_n\}$ in the series (32). We may have $r_1 = 0$ or $r_2 = +\infty$ (or both); for example, if $0 = r_1 < r_2 = +\infty$, the ring-shaped domain of convergence is just the punctured plane obtained by deleting the base point $z = p$.

The sums of (33A) and (33B) are analytic on their respective domains of convergence (why?), so we see that a Laurent series sums to an analytic function on its ring-shaped natural domain of convergence. Let us consider any ring shaped set $R_1 \leq |z-p| \leq R_2$ which lies entirely within the domain of convergence $r_1 < |z-p| < r_2$ (so that $r_1 < R_1 < R_2 < r_2$). We have previously shown that series (33A) converges uniformly on any domain like $|z-p| \geq R_1$ with inner radius greater than r_1, and series (33B) is uniformly convergent on any disc $|z-p| \leq R_2$ with radius less than r_2 (recall Theorem 3.7, Section 3.9 for these standard results). Therefore, the Laurent series (32) must converge *uniformly* to its limit on the smaller annulus $R_1 \leq |z-p| \leq R_2$, although the convergence may fail to be uniform on the full domain of convergence $E = \{z : r_1 < |z-p| < r_2\}$.

Obviously, Laurent series are made to order for describing analytic functions that are defined on a ring shaped domain; the main question is whether *every* analytic function $f(z)$ defined on such a domain can be described by a Laurent series that converges to $f(z)$ throughout this domain. The main result of this section gives an affirmative answer. If the inner radius of the annulus is zero, we obtain the important special case in which $f(z)$ is analytic with an isolated singularity at $z = p$.

Theorem 6.6 (Laurent Series Expansion) *Let $f(z)$ be analytic on the annulus E defined by $r_1 < |z - p| < r_2$. Then there exists a Laurent series*

$$\sum_{n=-\infty}^{+\infty} a_n(z - p)^n \tag{35}$$

which converges to $f(z)$ throughout E. The coefficients are unique, and are given by

$$a_n = \frac{1}{2\pi i}\int_\gamma \frac{f(z)}{(z - p)^{n+1}}\,dz \qquad n = 0, \pm 1, \pm 2, \ldots \tag{36}$$

where $\gamma(t) = p + re^{it}$ $(0 \le t \le 2\pi)$ is any circular contour centered at p whose radius is chosen so that $r_1 < r < r_2$.

Note: If we take two counterclockwise oriented closed circular contours γ' and γ'' about $z = p$ which lie within the annulus E, it is clear that these contours are homologous within E (have the same winding properties with respect to every point outside of E), while the integrand $f(z)(z - p)^{-(n+1)}$ appearing in (36) is analytic on E. Thus, by the general Cauchy Theorem, we get the same value for the contour integral (36) using either of these contours; the value a_n given by formula (36) does not depend on the particular circular contour (or radius r) we use.

PROOF: We shall set up a cycle ϕ for which the Cauchy-type integral $\dfrac{1}{2\pi i}\displaystyle\int_\phi f(w)/(w - z)\,dw$ describes $f(z)$ at points in the annulus. Then we will replace $1/(w - z)$ by its power series expansion (in powers of the integration variable w), much as we did in proving Theorem 5.20; finally, we will bring the summation outside of the integral. Of course, this is only a rough outline of the procedure, but it may help motivate the details that follow.

Let γ' and γ'' be circular contours, with radii r' and r'' respectively, located near the inner and outer boundaries of the annulus, as shown in Figure 6.9. The cycle $\phi = \gamma'' - \gamma'$ formed by these contours is obviously homologous to zero within E; by the general Cauchy integral formula we have

$$I(\phi, z)f(z) = \frac{1}{2\pi i}\int_\phi \frac{f(w)}{w - z}\,dw$$

for all points z in E which do not lie on the circles $|z - p| = r'$ and $|z - p| = r''$. The index $I(\phi, z) = +1$ for points in the smaller annulus E^* bounded by

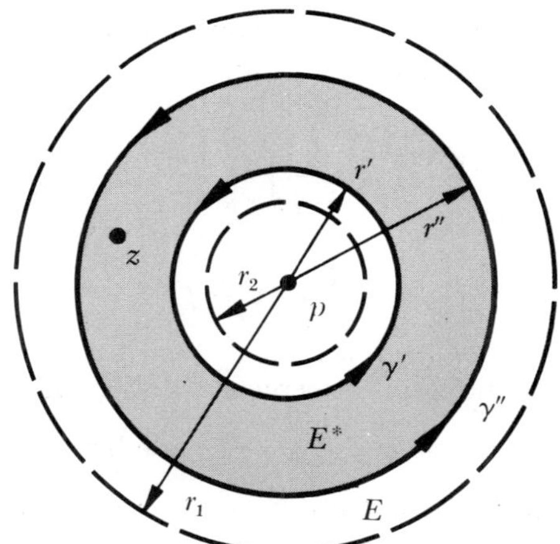

Figure 6.9 The contours γ' and γ'' and the subdomain E^* in the proof of the Laurent series expansion formula.

these trajectory circles, so we have

$$(37)\quad f(z) = \frac{1}{2\pi i}\int_{\phi} \frac{f(w)}{w - z}\, dw = \frac{1}{2\pi i}\int_{\gamma''} \frac{f(w)}{w - z}\, dw - \frac{1}{2\pi i}\int_{\gamma'} \frac{f(w)}{w - z}\, dw$$

throughout the annulus $E^* = \{z : r' < |z - p| < r''\}$. Because $f(w)$ is continuous on the circles $|w - p| = r'$ and $|w - p| = r''$, the integrals

$$(38)\quad \begin{aligned} F_1(z) &= \frac{-1}{2\pi i}\int_{\gamma'} \frac{f(w)}{w - z}\, dw = \frac{1}{2\pi i}\int_{\gamma'} \frac{f(w)}{z - w}\, dw \\ F_2(z) &= \frac{1}{2\pi i}\int_{\gamma''} \frac{f(w)}{w - z}\, dw \end{aligned}$$

produce functions which are analytic throughout the complex plane, off the respective circles; furthermore, (37) shows that $f(z) = F_1(z) + F_2(z)$ in the annulus E^*. (The reader may wish to recall Theorem 5.22 on generating analytic functions via integral formulas like (38).) Now let us consider the function $F_2(z)$ at a typical point z in the disc $|z - p| < r''$ enclosed by γ''. For every w on the boundary circle, we have $|w - p| = r''$ and $\left|\dfrac{z - p}{w - p}\right| = |z - p|/r'' < 1$, so that the geometric series

$$(39)\quad \begin{aligned} \frac{1}{w - z} &= \frac{1}{(w - p) - (z - p)} = \frac{1}{w - p}\left[\frac{1}{1 - \left(\dfrac{z - p}{w - p}\right)}\right] \\ &= \frac{1}{w - p}\sum_{n=0}^{\infty}\left(\frac{z - p}{w - p}\right)^n = \sum_{n=0}^{\infty} \frac{(z - p)^n}{(w - p)^{n+1}} \end{aligned}$$

converges absolutely and uniformly on the circle $|w - p| = r''$ (the variable is w in this series; z is fixed). Thus, substituting (39) into (38) we get

$$F_2(z) = \frac{1}{2\pi i}\int_{\gamma''} \frac{f(w)}{w - z}\, dw = \sum_{n=0}^{\infty} (z - p)^n \left(\frac{1}{2\pi i}\int_{\gamma''} \frac{f(w)}{(w - p)^{n+1}}\, dw\right)$$

$$= \sum_{n=0}^{\infty} a_n (z - p)^n.$$

The uniform convergence of (39) and Theorem 5.10 justify the interchange of $\int_{\gamma''}(\cdots)\, dw$ and $\sum_{n=0}^{\infty}(\cdots)$ made here; the last equality holds by the remarks in the note which precedes this proof. The same formula works for *each* z such that $|z - p| < r''$.

The function $F_1(z)$ is handled in a slightly different way. We consider a typical point z in the domain $|z - p| > r'$. If w is a point on the boundary circle $|w - p| = r'$, we have $\left|\dfrac{w - p}{z - p}\right| = r'/|z - p| < 1$, so that the series

$$\text{(40)} \qquad \frac{1}{z - w} = \frac{1}{z - p}\sum_{n=0}^{\infty}\left(\frac{w - p}{z - p}\right)^n = \sum_{n=0}^{\infty} \frac{(w - p)^n}{(z - p)^{n+1}}$$

converges absolutely and uniformly on this circle (with respect to the variable w). Substituting this series into the integral which defines F_1, we get

$$F_1(z) = \frac{1}{2\pi i}\int_{\gamma'} \frac{f(w)}{z - w}\, dw = \sum_{n=0}^{\infty} \frac{1}{(z - p)^{n+1}} \left(\frac{1}{2\pi i}\int_{\gamma'} f(w)(w - p)^n\, dw\right)$$

$$= \sum_{n=-1}^{-\infty} a_n (z - p)^n.$$

This formula is valid for all z such that $|z - p| > r'$. Combining these results, we see that the Laurent series with coefficients a_n defined as in (36) converges to $f(z) = F_1(z) + F_2(z)$ for all z such that $r' < |z - p| < r''$.

This demonstration that (35) converges to $f(z)$ on E^* works for every choice of the contours γ' and γ''; by choosing these contours appropriately, we can see that (35) converges to $f(z)$ at every point z in the domain $r_1 < |z - p| < r_2$, and our theorem is proved. ▌

The natural domain of convergence of the Laurent series (35) in this theorem may very well extend beyond the domain E on which $f(z)$ was originally defined, so the sum of the series may extend $f(z)$ into a larger annulus. It is important to notice that the Laurent series cannot be used to represent a function which has a *branch singularity* at p; there is no annulus about p on which $f(z)$ is analytic, due to the presence of discontinuities along a cut from p to ∞. Theorem 6.6 simply does not apply in such circumstances. It is also helpful to notice that, because the coefficients are determined by integral formulas (36) involving f, the coefficients in a Laurent expansion of a given analytic function about a base point p are unique (no other choice of coefficients $\{b_n\}$ can lead to a series which represents f in the annulus); we leave the verification as an exercise.

Formula (36) provides one way of calculating Laurent series coefficients, but the integrals involved are often difficult to evaluate. There are other ways of determining the form of the Laurent series.

Example 6.19 If we want to represent $f(z) = \sin(1/z)$ near its isolated singularity at the origin, we can do this by recalling that

$$\sin w = w - \frac{1}{3!}w^3 + \cdots \quad \text{for all } w,$$

so if $z \neq 0$ we may substitute $w = 1/z$ into this series to get

$$\sin\left(\frac{1}{z}\right) = \frac{1}{z} - \frac{1}{3!}\frac{1}{z^3} + \frac{1}{5!}\frac{1}{z^5} - \cdots.$$

This gives a Laurent-type expansion valid for $0 < |w| < +\infty$; by the uniqueness of coefficients in Laurent series (Exercise 1), this is the only series in positive and negative powers of z which can possibly represent $f(z)$ near the origin. Since this series converges uniformly on any circle $|z| = R$ about the origin (recall Theorem 3.7), we may determine the residue at the essential singularity $z = 0$ by integrating the Laurent series term-by-term along the contour $\gamma(t) = Re^{it}$ (defined for $0 < t < 2\pi$):

$$\begin{aligned}\operatorname{Res}(f, 0) &= \frac{1}{2\pi i}\int_\gamma \left(\frac{1}{z} - \frac{1}{3!}\frac{1}{z^3} + \cdots\right) dz \\ &= \frac{1}{2\pi i}\int_\gamma \frac{1}{z}\, dz - \frac{1}{3!}\frac{1}{2\pi i}\int_\gamma \frac{1}{z^3}\, dz + \ldots \\ &= 1 - 0 + 0 - \cdots = +1.\end{aligned}$$

Notice that $\operatorname{Res}(f, 0)$ is just the coefficient a_{-1} on the term $(z - p)^{-1}$ in the Laurent series about $p = 0$ (earlier we saw this was true when there are a finite number of negative power terms). The residue at an essential singularity like this could be quite difficult to evaluate by direct calculation of contour integrals.

The last remark about residues and the coefficient of $(z - p)^{-1}$ in the Laurent series expansion is in fact a general principle. Given the Laurent series of $f(z)$ about an isolated singular point (of any type)

$$f(z) = \sum_{n=-\infty}^{+\infty} a_n(z - p)^n = \sum_{n=1}^{\infty} a_{-n}\frac{1}{(z - p)^n} + \sum_{n=0}^{\infty} a_n(z - p)^n,$$

the separate series of positive and negative powers converge uniformly on small circles $|z - p| = R$ centered at the singular point. Therefore, we may integrate each series term-by-term along the contours $\gamma(t) = p + Re^{it}$, defined for $0 \leq t \leq 2\pi$, to get the residue

$$\begin{aligned}\operatorname{Res}(f, p) &= \frac{1}{2\pi i}\int_\gamma \left(\sum_{n=1}^{\infty} a_{-n} \frac{1}{(z-p)^n}\right) dz + \frac{1}{2\pi i}\int_\gamma \left(\sum_{n=0}^{\infty} a_n (z-p)^n\right) dz \\ &= \sum_{n=1}^{\infty} a_{-n}\left(\frac{1}{2\pi i}\int_\gamma \frac{1}{(z-p)^n}\, dz\right) + 0 = a_{-1}.\end{aligned}$$

(The series of positive powers sums to a function that is analytic at p, so the second integral is zero.) We showed that

$$\frac{1}{2\pi i}\int_\gamma (z-p)^k\, dz = \begin{cases} +1 & \text{if } k = -1 \\ 0 & \text{if } k \neq -1 \end{cases}$$

some time ago, by direct calculations.

It will often be convenient to calculate residues at poles of high order, or at essential singularities, by determining the coefficient $a_{-1} = \operatorname{Res}(f, p)$ in the Laurent series expansion about p.

Example 6.20 Evaluate $\displaystyle\int_\gamma \frac{\sinh z}{z^6}\, dz$ along the counterclockwise oriented circle $|z| = 1$.

This integral is the same as $2\pi i \cdot \operatorname{Res}(f(z), 0)$ where $f(z) = (\sinh z)/z^6$. It is a simple matter to determine the Taylor (= Laurent) series coefficients of the numerator $\sinh z$ about $z = 0$, namely

$$\sinh z = \tfrac{1}{2}(e^z - e^{-z}) = z + \frac{z^3}{3!} + \frac{z^5}{5!} + \cdots;$$

from this we immediately get the Laurent series expansion of $f(z)$,

$$f(z) = z^{-5} + \frac{1}{3!}z^{-3} + \frac{1}{5!}z^{-1} + \left(\frac{1}{7!}z + \frac{1}{9!}z^3 + \cdots\right)$$

for $z \neq 0$. Evidently, $a_{-1} = \operatorname{Res}(f(z), 0) = 1/5!$.

If $f(z)$ is analytic on a punctured disc $0 < |z - p| < r$, and has a Laurent series expansion

$$f(z) = \sum_{n=-\infty}^{+\infty} a_n (z-p)^n \quad \text{for all } z \text{ such that } 0 < |z-p| < r,$$

Table 6.3

Singularity at $z = p$	Behavior of a_n for $n = -1, -2 \ldots$
removable singularity	all $a_n = 0$
pole (of finite order)	finitely many nonzero a_n
essential singularity	infinitely many nonzero a_n

then the sum $\sum_{n=-1}^{-\infty} a_n(z-p)^n$ of the negative power terms in this series is called the **principal part** of $f(z)$ at the isolated singularity $z = p$. The relationship between the nature of the isolated singularity at p and the form of the principal part of the Laurent series about p is shown in Table 6.3.

These relationships should be clear from the definitions. If p is a pole of order m, then $a_{-m} \neq 0$ and $a_k = 0$ for all $k < -m$.

EXERCISES

1. Prove that the coefficients in a Laurent series are uniquely determined. If

$$f(z) = \sum_{n=-\infty}^{\infty} a_n(z-p)^n \quad \text{and} \quad g(z) = \sum_{n=-\infty}^{\infty} b_n(z-p)^n$$

converge and agree on some annulus $A = \{z : r_1 < |z-p| < r_2\}$, then $a_n = b_n$ for $n = 0, \pm 1, \pm 2, \ldots$.

Hint: Take $R = (r_1 + r_2)/2$ and $\gamma(t) = p + Re^{it}$ for $0 < t < 2\pi$; integrate $f(z)/(z-p)^{n+1} = g(z)/(z-p)^{n+1}$ along γ for $n = 0, \pm 1, \pm 2, \ldots$.

2. The Laurent series

$$\sum_{n=0}^{\infty} z^n = \frac{1}{1-z} \quad \text{and} \quad \sum_{n=1}^{\infty} z^{-n} = \frac{(-1)}{1-z}$$

have nontrivial domains of convergence. It would seem that the combined series, involving *all* powers z^n, converges to zero

$$\sum_{n=-\infty}^{\infty} z^n = \frac{1}{1-z} + (-1)\frac{1}{1-z} = 0 \quad \text{for all } z.$$

Why is this conclusion false?

Hint: What is the domain of convergence of $\sum_{n=-\infty}^{\infty} z^n$?

3. Give Laurent series expansions that converge near ∞, about the base points indicated. Determine the inner radii of convergence.

(i) $\dfrac{1}{1-z}$ about $z=0$; about $z=+i$

(ii) $\dfrac{z+1}{z-1}$ about $z=1$; about $z=0$

(iii) $\dfrac{1}{z-a}$ about $z=0$ (a a fixed complex number, $a\neq 0$)

Hint: Refer back to the geometric series, $1/(1-z)=(-1)\sum_{n=1}^{\infty} z^{-n}$ for $|z|>1$; recall Example 3.15.

Answers: (i) $(-1)\sum_{n=1}^{\infty} z^{-n}$ with $r_1=1$; $(-1)\sum_{n=1}^{\infty}(1-i)^{n-1}/(z-i)^n$ (with $r_1=\sqrt{2}$); (ii) $1+2/(z-1)$, a *finite* series; $1-2\sum_{n=1}^{\infty} z^{-n}$ (with $r_1=1$); (iii) $\sum_{n=1}^{\infty} a^{n-1}/z^n$ (with $r_1=|a|$).

4. Obtain the first few terms of the Laurent series for the following expressions.

(i) $\dfrac{1}{1-\cos z}$ about $z=0$

(ii) $\dfrac{1}{1-e^z}$ about $z=0$

(iii) $\tan z$ about $z=0$

(iv) $\dfrac{\tan z}{z-(\pi/2)}$ about $z=\pi/2$

(v) $\dfrac{\sin z}{1-\cos z}$ about $z=0$

(vi) $(x^2+1)\cdot\csc z$ about $z=0$.

5. Give Laurent series expansions for the domains and base points indicated.

(i) $1/(z-1)(z-3)$ for $|z|<1$; about $z=0$

(ii) $1/(z-1)(z-3)$ for $1<|z|<3$; about $z=0$

(iii) $\dfrac{1}{z-a}$ for $|z|<|a|$ and for $|z|>|a|$

($a\neq 0$ fixed); about $z=0$.

(iv) $\dfrac{e^z}{(z-1)^2}$ for $|z-1|>0$; about $z=1$.

Hint: In (i) and (ii) try writing $f(z) = A/(z-1) + B/(z-3)$; then see Exercise 3.

Answers: (i) $\sum_{n=0}^{\infty} \frac{1}{2}[1 - (1/3^{n+1})]z^n = \frac{1}{2}\sum_{n=0}^{\infty} z^n - \frac{1}{6}\sum_{n=0}^{\infty} z^n/3^n$;

(ii) $-\frac{1}{2}\sum_{n=-1}^{-\infty} z^n - \frac{1}{6}\sum_{n=0}^{\infty} z^n/3^n$;

(iii) $-\sum_{n=0}^{\infty} z^n/a^{n+1}$; $\sum_{n=1}^{\infty} a^{n-1}/z^n$;

(iv) $e \cdot \sum_{n=0}^{\infty} (z-1)^{n-2}/n!$.

6. If $P(z)$ and $Q(z)$ are polynomials of degree M and N respectively $(N \geq 1)$, show that the Laurent series expansion of the rational function P/Q

$$\frac{P(z)}{Q(z)} = \sum_{n=-\infty}^{\infty} a_n z^n \quad \text{(about } z = 0)$$

has only a *finite number* of positive power terms. ($a_n = 0$ if $n \geq M - N + 1$.) Thus its domain of convergence has the form $E = \{z : R < |z| < +\infty\}$.

Hint: P/Q has only finitely many singularities, so P/Q is analytic on some domain of the form $R < |z| < +\infty$; furthermore, $|P/Q| \leq K \cdot |z|^p$ near ∞, for suitable integer p. Use formula (36).

7. If f and g are analytic at p and are both *zero* at p, prove that:

(i) $\lim_{z \to p} \left(\frac{f}{g}\right)$ exists if the limit of derivatives $\lim_{z \to p} \left(\frac{f'}{g'}\right)$ exists

(ii) $\lim_{z \to p} \left(\frac{f}{g}\right) = \lim_{z \to p} \left(\frac{f'}{g'}\right)$.

Prove the same result if f and g have poles at p.

8. Write out a Laurent series expansion of

$$f(z) = \frac{1}{z^2 - 2z + 1} = \frac{1}{(z-1)^2}$$

about $z = 0$, valid on the domain $1 < |z| < +\infty$.

Hint: In formula (36), $a_n = 0$ if $n \geq -1$ since $1/z^{n+1}(z-1)^2$ vanishes rapidly at infinity. For $n < -1$, reduce to calculating integrals about $z = 0$ and $z = +1$. Integral around $z = 0$ is zero for $n < -1$ since the integrand is then analytic at $z = 0$. Integral around $z = +1$ is Cauchy formula for a first derivative.

Answer: $f(z) = -\sum_{n=-1}^{-\infty} (n+1)z^n = \sum_{n=1}^{\infty} (n-1)z^{-n}$.

9. Fix $\zeta \neq 0$ and verify that $f(z) = \text{Log}\left(1 - \frac{\zeta}{z}\right)$ is analytic if $|z| > |\zeta|$. Show that f has the following Laurent series expansion near infinity:

$$\text{Log}\left(1 - \frac{\zeta}{z}\right) = \sum_{n=1}^{\infty} \frac{-1}{n}\left(\frac{\zeta}{z}\right)^n; \quad \text{for} \quad |z| > |\zeta|.$$

If $\gamma_R(t) = Re^{it}$, for $0 \leq t \leq 2\pi$ show that the integrals $\frac{-1}{2\pi i}\int_{\gamma_R} f(z)\,dz$ have the common value $+\zeta$ for large R. (This value is referred to as the **residue at infinity** of $f(z)$.)

6.6 PARTIAL FRACTIONS DECOMPOSITION OF RATIONAL FUNCTIONS

Let $f(z) = P(z)/Q(z)$ be a rational function. Such functions are either polynomials, or else are meromorphic functions that have at most a finite number of poles $\zeta_1, \ldots, \zeta_l$ ($l \leq$ degree Q). At any pole ζ_k, the nature of the singularity is determined by the principal part $p_k(z)$ of the Laurent series about ζ_k; let us write this in the form

$$(41) \qquad p_k(z) = \frac{A_{k,m_k}}{(z-\zeta_k)^{m_k}} + \cdots + \frac{A_{k2}}{(z-\zeta_k)^2} + \frac{A_{k1}}{(z-\zeta_k)}$$

where A_{kj} ($1 \leq j \leq m_k =$ order of the pole at ζ_k) are complex constants. The sum of the various principal parts at all the poles,

$$S(z) = \sum_{k=1}^{l} p_k(z),$$

is meromorphic, and $\lim_{z\to\infty} S = 0$ since each function $A_{kj}/(z - \zeta_k)^j$ in this finite sum vanishes at infinity. If we subtract the principal parts, we get a function $f(z) - S(z)$ that has *removable* singularities at $z = \zeta_k$ (Exercise 5) and is analytic elsewhere; thus $f(z) - S(z)$ is an *entire* function.

If degree $P \geq$ degree Q, long division gives us

$$P(z)/Q(z) = R(z) + P^*(z)/Q(z),$$

where R is a polynomial of degree $=$ degree $P -$ degree Q, and the remaining rational function P^*/Q has degree$P^* <$ degree$Q$. Clearly, $\lim_{z\to\infty} P^*/Q = 0$; furthermore, P/Q and P^*/Q have the same poles ζ_k, and the same principal part $p_k(z)$ at each pole, since they differ by a polynomial (analytic function) near each singularity. Thus, $H = (P^*/Q) - S$ is an entire function such that

$$\lim_{z\to\infty} H = \lim_{z\to\infty} P^*/Q - \lim_{z\to\infty} S = 0.$$

This means that H is a bounded (why?), entire function on the plane, and by Liouville's Theorem is constant, $H(z) = c$ everywhere. Since $\lim_{z\to\infty} H = 0$, the constant is zero, so that $H(z) = 0$ and $P^*/Q = \sum_{k=1}^{l} p_k(z)$ for all z; thus,

(42) $$\frac{P(z)}{Q(z)} = R(z) + \sum_{k=1}^{l} p_k(z) = R(z) + (\text{sum of principal parts of } P/Q)$$

for all z. If degree $P <$ degree Q to begin with, the preliminary long division is unnecessary, and in this special case P/Q is *identical to* the sum of its principal parts,

(43) $$\frac{P(z)}{Q(z)} = \sum_{k=1}^{l} p_k(z) \quad \text{if degree } P < \text{degree } Q.$$

These formulas are known as the **partial fractions decomposition** of $f = P/Q$. They express f as the sum of its principal parts, plus a *polynomial.*

This decomposition can be extremely useful because the distinct poles have been sorted out and are represented by separate terms in the sum, instead of being lumped together as they are in an expression of the form P/Q. Calculating the coefficients in a partial fractions decomposition amounts to calculating the principal parts. However, once we know that a partial fractions decomposition exists, we may write the principal parts of f in the form (41), with undetermined coefficients; then we can calculate these coefficients by straightforward linear algebra, as shown in the following examples.

Example 6.21 Let $f(z) = (z^4 + 1)/(z^4 - 1)$. By long division,

$$\frac{z^4+1}{z^4-1} = 1 + \frac{2}{z^4-1} = 1 + \frac{P^*(z)}{Q(z)}.$$

Now $Q(z) = z^4 - 1$ has roots, each of multiplicity one, which are $\zeta_1 = 1$, $\zeta_2 = +i$, $\zeta_3 = -1$, and $\zeta_4 = -i$, so that P^*/Q can be expressed in the form

$$\frac{2}{z^4-1} = \frac{A}{z-1} + \frac{B}{z-i} + \frac{C}{z+1} + \frac{D}{z+i}.$$

Multiplying both sides by $z^4 - 1$ we get the identity

$$\begin{aligned} 2 &= A(z^2+1)(z+1) + B(z^2-1)(z+i) + C(z^2+1)(z-1) \\ &\qquad + D(z^2-1)(z-i) \\ &= (A+B+C+D)z^3 + (A+iB-C-iD)z^2 + (A-B+C-D)z \\ &\qquad + (A-iB-C-iD) \end{aligned}$$

for all z. Comparing coefficients of powers of z on each side, we see that

$$\begin{aligned} A + B + C + D &= 0 \qquad & A &= 1 \\ A + iB - C - iD &= 0 & B &= i \\ A - B + C - D &= 0 & C &= -1 \\ A - iB - C - iD &= 2 & D &= -i \end{aligned}$$

so that

$$\frac{z^4 + 1}{z^4 - 1} = 1 + \frac{1}{z - 1} + \frac{i}{z - i} + \frac{-1}{z + 1} + \frac{-i}{z + i}.$$

Using this decomposition, it is quite easy to calculate the residues of f:

$$\operatorname{Res}(f, 1) = 1, \quad \operatorname{Res}(f, +i) = i, \quad \operatorname{Res}(f, -1) = -1, \quad \operatorname{Res}(f, -i) = -i.$$

Furthermore, it is also easy to obtain the residues of products, such as $(\tan z)\cdot f(z)$. For example, if γ is a small circular contour about $+i$, we get

$$\begin{aligned} \operatorname{Res}\left(\frac{z^4 + 1}{z^4 - 1} \cdot \tan z, +i\right) &= \frac{1}{2\pi i}\int_\gamma \tan z \, dz + \frac{1}{2\pi i}\int_\gamma \frac{\tan z}{z - 1}\, dz \\ &\quad + \frac{i}{2\pi i}\int_\gamma \frac{\tan z}{z - i}\, dz + \cdots + \frac{-i}{2\pi i}\int_\gamma \frac{\tan z}{z + i}\, dz \\ &= 0 + 1 \cdot 0 + i \cdot \tan(i) - 1 \cdot 0 - i \cdot 0 \\ &= i \tan(i) = -\tanh(+1), \end{aligned}$$

since $(\tan z)/(z - i)$ is the only integrand that is singular at $+i$.

Example 6.22 Partial fraction decompositions can be used to determine the form of antiderivatives of a rational function $f(z)$. In fact, the method of partial fractions is widely used in calculus, although most calculus texts judiciously avoid the question of proving the validity of partial fractions expansions (which is difficult unless one applies complex variable methods). Consider the decomposition

$$\begin{aligned} f(z) = \frac{2z^2 - z}{(z + 1)(z^2 - 2z + 1)} &= \frac{2z^2 - z}{(z + 1)(z - 1)^2} \\ &= \frac{1}{z + 1} + \frac{1}{z - 1} + \frac{1}{(z - 1)^2}. \end{aligned}$$

Its antiderivative is given by

$$\begin{aligned} F(z) = \int_0^z f(z)\, dz &= \int_0^z \frac{1}{z + 1}\, dz + \int_0^z \frac{1}{z - 1}\, dz + \int_0^z \frac{1}{(z - 1)^2}\, dz \\ &= \log(z + 1) + \log(z - 1) - \left(\frac{1}{z - 1}\right) + c \\ &= \log(z^2 - 1) - \left(\frac{1}{z - 1}\right) + c; \end{aligned}$$

here $\int_0^z (\cdots)\,dz$ indicates integration along any contour from 0 to z, and c is an arbitrary complex constant. On any domain which admits analytic determinations of the logarithms involved, $F(z)$ is an antiderivative of $f(z)$. For example, on a simply connected domain that avoids $+1$ and -1, there are analytic determinations of $\log(z + 1)$ and $\log(z - 1)$, and $\log(z + 1) + \log(z - 1)$ is an analytic determination of $\log(z^2 - 1)$. An antiderivative defined on most of the plane can be defined by introducing suitable cuts originating at $+1$ and -1.

The theory of partial fractions can be developed further in many ways. If $f(z)$ is a general meromorphic function that is not necessarily rational, such as $f(z) = \tan z$, we might inquire whether the sum $\sum_{k=1}^{\infty} p_k(z)$ of the principal parts at the poles $\zeta_1, \zeta_2, \ldots$ converges and represents $f(z)$. If there are infinitely many poles, the convergence of $S(z) = \sum_{k=1}^{\infty} p_k(z)$ can present non-trivial difficulties. We shall give two examples, suppressing most computational details.

Example 6.23 Suppose $f(z)$ has *finitely many* poles. Then $f(z) - \sum_{k=1}^{l} p_k(z)$ is an entire function (Exercise 5). The representation

$$f(z) = (\text{entire function}) + (\text{sum of principal parts})$$

is about the best we can expect; however, in many applications, such as calculating residues, the nature of the entire function is irrelevant.

Example 6.24 The function $f(z) = \text{ctn}\, z$ has simple poles at $\zeta_n = \pi n + i0$ for $n = 0, \pm 1, \pm 2, \ldots$. The residues are all the same, due to periodicity,

$$\text{Res}(\text{ctn}\, z, \zeta_n) = \left[\frac{\cos z}{\dfrac{d}{dz}(\sin z)} \Bigg|_{z=\zeta_n} \right] = +1,$$

and the sum of all the principal parts is

$$\text{(44)} \quad S(z) = \sum_{n=-\infty}^{+\infty} \frac{+1}{(z - \zeta_n)} = +\frac{1}{z} + \sum_{n=1}^{\infty} \frac{2z}{z^2 - \zeta_n^{\,2}} = +\frac{1}{z} + \sum_{n=1}^{\infty} \frac{2z}{z^2 - \pi^2 n^2}.$$

Using residue techniques, it is not difficult to show that the series $S(z)$ converges (to an analytic function) on the domain $E = \{z : z \neq n\pi,\ n \text{ an integer}\}$, and that

$$\text{(45)} \qquad \text{ctn}\, z = +\frac{1}{z} + \sum_{n=1}^{\infty} \frac{2z}{z^2 - \pi^2 n^2} \quad \text{for} \quad z \text{ in E}.$$

This representation of ctn z as the sum of its principal parts at the poles ζ_n can be generalized to other meromorphic functions (for some functions a suitable entire function must be added to the sum of principal parts). For these developments we refer the reader to Hille [9], Sections 8.5 and 9.3, or Nevanlinna and Paatero [18], Section 10.6 and Chapter 13. For details on the particular formula (45), see Exercise 7.

EXERCISES

1. If P/Q has only poles $\zeta_1, \ldots, \zeta_n$ *of order 1*, prove that the coefficients in the partial fractions decomposition

$$\frac{P(z)}{Q(z)} = \frac{A_1}{(z - \zeta_1)} + \cdots + \frac{A_n}{(z - \zeta_n)}$$

are exactly the residues, $A_k = \text{Res}(P/Q, \zeta_k)$. This observation is sometimes very useful in calculating decompositions.

2. Give partial fractions decompositions for

(i) $\dfrac{z+1}{z-1}$

(ii) $\dfrac{1}{z^2-1}$

(iii) $\dfrac{z^4}{z^3-1}$

(iv) $\dfrac{1}{z(z^2+2z+1)(z+2)^3}$

(v) $\dfrac{z^4+1}{z^2-1}$.

3. Give the Laurent series expansion of $f(z) = 1/(z-1)(z-3)$ on $E_1 = \{z:|z| < 1\}$. Do the same for $E_2 = \{z:|z| > 3\}$.

Hint: Write f in partial fractions; then calculate Laurent series for $1/(z-1)$ and $1/(z-3)$ (geometric series).

Answer: See Exercise 5, Section 6.5.

4. Determine the Laurent series expansion of $z^6/(1-z)^3$ about $z = +1$. What is the principal part?

Answer: The series is *finite.*

5. Suppose $f(z)$ is meromorphic, with a *finite* number of poles $\zeta_1, \ldots, \zeta_n$. Let $p_k(z)$ be the principal part of the Laurent series about ζ_k. Show that $f(z) - \sum_{k=1}^{n} p_k(z)$ has only removable singularities in the plane, and hence is an *entire* function. (There is no need to assume that f is a rational function.)

6. Using partial fractions, calculate the Laurent series

$$f(z) = \frac{1}{z(z-1)(z-2)} = \sum_{n=-\infty}^{\infty} a_n z^n$$

for the respective domains (i) $0 < |z| < 1$, (ii) $1 < |z| < 2$, and (iii) $2 < |z| < +\infty$. How would you evaluate the coefficients $\{a_n\}$ for these domains from formula (36) *without* using partial fractions? Try it.

Answers: (i) $a_n = 0$ if $n < -1$, $a_{-1} = \frac{1}{2}$, $a_n = \left(1 - \frac{1}{2^{n+2}}\right)$ if $n \geq 0$, for $0 < |z| < 1$; (ii) $a_n = 1$ if $n \leq -2$, $a_n = 1/2^{n+2}$ if $n \geq -1$ for $1 < |z| < 2$; (iii) $a_n = 1/2^{n+2}$ if $n \leq -2$, $a_n = 0$ if $n \geq -1$ for $2 < |z| < +\infty$.

7. For ζ in the domain $E = \{z : z \neq n\pi,\ n = 0, \pm 1, \pm 2, \ldots\}$ estimate the integrals

$$S_N = \frac{1}{2\pi i} \int_{\gamma_N} \frac{\operatorname{ctn} z}{z - \zeta}\, dz \quad \text{large } N = 1, 2, \ldots$$

along the rectangular contour γ_N with vertices at $\pm(N\pi + (\pi/2)) \pm iN$; prove that $\lim_{N\to\infty} S_N = 0$. By keeping track of residues, prove that

$$\operatorname{ctn} \zeta = +\frac{1}{\zeta} + \sum_{n=1}^{\infty} \frac{2\zeta}{\zeta^2 - \pi^2 n^2} \quad \text{for all } \zeta \text{ in } E.$$

Hint: Write

$$\frac{\operatorname{ctn} z}{z - \zeta} = \operatorname{ctn}(z) \cdot \left[\frac{1}{z} - \frac{1}{z} + \frac{1}{z - \zeta}\right] = \frac{\operatorname{ctn} z}{z} + \frac{\zeta}{z(z - \zeta)} \operatorname{ctn} z.$$

The integral of the first term is *zero* for $N = 1, 2, \ldots$; the residue at $z = 0$ is zero, and the residues at $+\pi k$, $-\pi k$ cancel (direct calculations). Meanwhile, for large N, we have $|\operatorname{ctn}(z)| \leq 2$ on the trajectory of γ_N (direct calculations, or use ideas from Exercise 3, Section 6.7); thus, the integral of the second term approaches zero as $N \to \infty$ because the integrand is bounded by K/N^2 on γ_N.

8. In Exercise 7, $\frac{1}{2\zeta}\left[\operatorname{ctn} \zeta - \frac{1}{\zeta}\right]$ has a removable singularity at $\zeta = 0$ whose value may be determined from low order terms of the Laurent series of $\operatorname{ctn} \zeta$ at $\zeta = 0$. Use this observation to verify that $\sum_{n=1}^{\infty} 1/n^2 = \pi^2/6$.

9. A function $f(z)$ that is defined "near infinity" (that is, for all large values of $|z|$), is said to have a **pole at infinity** if the function $g(z) = f(1/z)$ has a pole at $z = 0$. Similar terminology applies to **removable singularities,** or **essential singularities** at infinity. Suppose that $f(z)$ is meromorphic with a finite number of poles $\zeta_1, \ldots, \zeta_m$ in the plane. Assume that $\zeta_0 = \infty$ is also a *pole* or removable singularity (so that f is meromorphic on $\mathbf{C}^*$). Prove that f must be a *rational function* of z.

Note: Conversely, it is obvious that every rational function is meromorphic on $\mathbf{C}^*$.

Hint: Subtract off the principal parts at $\zeta_1, \ldots, \zeta_m$ and show (from the condition at $\zeta_0 = \infty$) that $f(z) - \sum_{k=1}^{\infty} p_k(z)$ has polynomial growth at ∞. Apply Exercise 5 of Section 5.13.

10. If $f(z)$ is analytic at infinity (that is, for $R_0 < |z| < +\infty$), the nature of the singularity at infinity is classified as in Exercise 9. If $\gamma_R(t) = Re^{it}$ $(0 \le t \le 2\pi)$, the integrals along γ_R are the same for all large $R > 0$; explain why this is so as a consequence of the general Cauchy Theorem. Thus, we define the **residue at infinity** to be

$$\operatorname{Res}(f, \infty) = \frac{-1}{2\pi i} \int_{\gamma_R} f(z)\, dz \quad \text{(all large } R\text{).}$$

Classify the singularity at ∞, and calculate $\operatorname{Res}(f, \infty)$, for

(i) e^z

(ii) $e^{1/z}$

(iii) $z^6 + 1$

(iv) $z/(z^6 + 1)$

(v) $z + \dfrac{1}{z} + \dfrac{1}{z^2}$

(vi) $1/(1 + z^2)$

(vii) $\dfrac{e^z}{z(z-1)^2}$.

Hint: Use series expansions (positive or negative powers of z) where appropriate, keeping in mind Theorems 5.9 and 5.10. Sometimes Section 5.4 will be helpful.

Answers: (i) ess. sing.; Res $= 0$; (ii) rem. sing.; Res $= -1$; (iii) pole, order $m = 6$; Res $= 0$; (iv) zero, order $m = 5$; Res $= 0$; (v) pole, order $m = 1$; Res $= -1$; (vi) removable sing. $\left(\lim_{z\to\infty} f = +1\right)$; Res $= 0$; (vii) ess. sing.; Res $= -e$.

11. There is a unique determination of $f(z) = \sqrt{1 - z^2}$ that is analytic on $E = \mathbf{C} \sim [-1, +1]$, with limit value $+\sqrt{1 - x^2}$ on the top edge of the cut (Exercise 23, Section 4.10). Show that f has a pole of order $m = 1$ at infinity. Then calculate

$$\operatorname{Res}(f, \infty) = \lim_{R\to\infty} \frac{-1}{2\pi i} \int_{\gamma_R} f(z)\, dz = -i.$$

Hints: Transform $f(z)$ and the contour $z = \gamma_R(t)$ via $z = 1/w$ (so that $w = 1/z$). Then $h(w) = f(1/w)$ is analytic for $0 < |w| < 1$, and the transformed contours $w = \eta_R(t) = 1/\gamma_R(t)$ approach $w = 0$ as $R \to \infty$. Let $\zeta^{1/2} = (re^{i\theta})^{1/2} = r^{1/2}e^{i\theta/2}$, for $0 \le \theta < 2\pi$, so the discontinuity is along $[0, +\infty)$; then $+(w^2 - 1)^{1/2}$ and $-(w^2 - 1)^{1/2}$ are both analytic for $|w| < 1$. From the definition of f, show that $h(w)$ and $-(w^2 - 1)^{1/2}/w$ agree for $w = iy$ $(y > 0)$; by analytic continuation, they agree for $0 < |w| < 1$. Now the limit of the integrals $\int_{\gamma_R} f(z)\, dz = \int_{\eta_R} h(w) \dfrac{dz}{dw}\, dw = \int_{\eta_R} -h(w)/w^2\, dw$ can be calculated in a routine way.

12. Using a contour like the one in Figure 6.8, calculate

$$\int_{-1}^{+1} \frac{\sqrt{1 - x^2}}{1 + x^2}\, dx = \pi(\sqrt{2} - 1).$$

The integral along $|z| = R$ does *not* go to zero as $R \to +\infty$; you must calculate its value (a residue at infinity) using ideas developed in Exercise 11. (Compare with Exercise 17, Section 6.4; in those more elementary problems, the integrals along $|z| = R$ went to zero, and did not require detailed consideration.)

*6.7 LOGARITHMIC RESIDUES; COUNTING ZEROS INSIDE A CONTOUR

If $f(z)$ is analytic on some domain D in which its only singularities are poles, we are frequently interested in locating its zeros and poles. Even when $f(z)$ is a polynomial, the task of locating its zeros often cannot be carried out exactly; determining their location to some prescribed degree of accuracy then becomes a problem of great practical importance. To deal with problems like this we introduce a residue-like contour which counts the number of zeros and poles enclosed by the contour.

The quotient $f'(z)/f(z)$ is defined and analytic on the domain D, except where $f(z)$ has a pole or zero. This quotient is called the **logarithmic derivative** of $f(z)$, because whenever it is possible to define a single valued determination of $\log(f(z))$ on D, the derivative of this function is $f'(z)/f(z)$. However, this quotient makes sense whether or not it is possible to define a logarithm of $f(z)$ on D. Our main result is the following theorem.

Theorem 6.7 *Let $f(z)$ be analytic on the domain D, except at a finite number of poles $p_1, \ldots, p_m$ in D. Let γ be a positively oriented simple closed contour which, together with its set $E_{\text{int}}(\gamma)$ of interior points, lies within D. Assume that the trajectory of γ avoids the zeros and poles of $f(z)$. Then, if $f(z)$ is not identically zero, we have*

$$\frac{1}{2\pi i} \int_\gamma \frac{f'(z)}{f(z)}\, dz = Z_f - P_f, \tag{46}$$

where P_f is the number of poles and Z_f the number of zeros enclosed by γ (each pole and zero is counted as many times as its order).

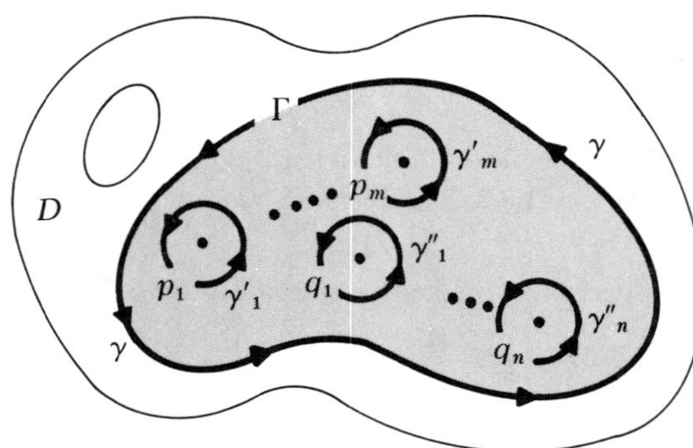

Figure 6.10 $E_{\text{int}}(\gamma)$ is shaded, γ is a simple closed contour, and $\Gamma \cup E_{\text{int}}(\gamma)$ lies within D.

PROOF: Unless $f(z)$ is identically zero on D, there can be only a finite number of zeros enclosed by γ, so that Z_f is well defined.

Otherwise, there would be infinitely many distinct zeros of f in the closed bounded set E consisting of $E_{\text{int}}(\gamma)$ and the trajectory Γ which is its boundary. These zeros cannot approach a pole of f, since $\lim_{z \to p} f = \infty$ at a pole.

Given any *closed bounded* set in the plane, and any infinite set of points in this set, one can extract a sequence of these points that converges to a point within the set. This result for plane sets (which will not be proved here) is the Bolzano-Weierstrass Theorem. Applying it to E, we get a sequence $\{z_n\}$ of zeros of $f(z)$ that converges to a point z^* in E.

Since the z_n cannot approach a pole, f must be analytic at z^*; by the analytic continuation principle (Theorem 3.19), we would then have $f(z) = 0$ near z^*, and throughout D. Thus, we conclude that γ can enclose only a finite number of zeros.

Let $p_1, \ldots, p_m$ be the poles and $q_1, \ldots, q_n$ the zeros enclosed by γ. Form small counterclockwise oriented circular contours $\gamma_1', \ldots, \gamma_m'; \gamma_1'', \ldots, \gamma_n''$ centered at these points, as in Figure 6.10. If their radii are chosen sufficiently small, the discs bounded by these contours will lie within the set $E_{\text{int}}(\gamma)$ of points enclosed by γ and the cycles γ and $\phi = \gamma_1' + \cdots + \gamma_m' + \gamma_1'' + \cdots + \gamma_n''$ will have the same winding properties with respect to all points in the complement of D (where the index is zero) and at the points $p_1, \ldots, p_m; q_1, \ldots, q_n$ (where the index is $+1$). Contour integrals along γ and ϕ will then have the same value for any integrand which is analytic on the domain $D^* = D \sim \{p_1, \ldots, p_m; q_1, \ldots, q_n\}$ obtained by deleting the points $\{p_k\}$ and $\{q_k\}$ from D. Applying this to $f'(z)/f(z)$, we see that

$$\frac{1}{2\pi i}\int_\gamma \frac{f'(z)}{f(z)}\,dz = \sum_{k=1}^{m} \frac{1}{2\pi i}\int_{\gamma_k'} \frac{f'(z)}{f(z)}\,dz + \sum_{k=1}^{n} \frac{1}{2\pi i}\int_{\gamma_k''} \frac{f'(z)}{f(z)}\,dz$$

$$= \sum_{k=1}^{m} \operatorname{Res}\left(\frac{f'(z)}{f(z)}, p_k\right) + \sum_{k=1}^{n} \operatorname{Res}\left(\frac{f'(z)}{f(z)}, q_k\right).$$

At a typical zero $z = q$, say of order N, we have

$$\frac{f'(z)}{f(z)} = \frac{[Na_N(z-q)^{N-1} + \cdots]}{[a_N(z-q)^N + \cdots]} = \frac{N}{z-q} + c_0 + c_1(z-q) + \cdots$$

for $z \neq q$, so that $\text{Res}\left(\frac{f'(z)}{f(z)}, q\right) = N =$ order of q as a zero of $f(z)$. Likewise if $z = p$ is one of the poles, with order N, we have

$$\frac{f'(z)}{f(z)} = \frac{\left[\frac{-Na_{-N}}{(z-p)^{N+1}} + \cdots\right]}{\left[\frac{a_{-N}}{(z-p)^N} + \cdots\right]} = \frac{-N}{z-p} + d_0 + d_1(z-p) + \cdots$$

for $z \neq p$, and $\text{Res}\left(\frac{f'(z)}{f(z)}, p\right) = -N = (-1)\cdot$(order of p as a pole). Obviously,

$$\frac{1}{2\pi i}\int_\gamma \frac{f'(z)}{f(z)}\,dz = Z_f - P_f$$

as desired. ▮

Formula (46) is interesting in that it allows us to count zeros and poles enclosed by a contour without actually locating them. However, the integral on the left side may be quite difficult to compute as it stands. What is remarkable is that this integral has another simple interpretation. The mapping $w = f(z)$ transforms γ (defined on some interval $[a, b]$) into an entirely new contour $\eta = f \circ \gamma$ defined by $\eta(t) = f(\gamma(t))$ for $a \leq t \leq b$. The transformation of γ into η is illustrated in Figure 6.11. The image contour η will be closed, but might not be simple any longer; also, it avoids the point $w = 0$ since we assume that $f(z)$ is never zero on the trajectory of γ.

Notice that the integral in (46) is precisely the winding number of the

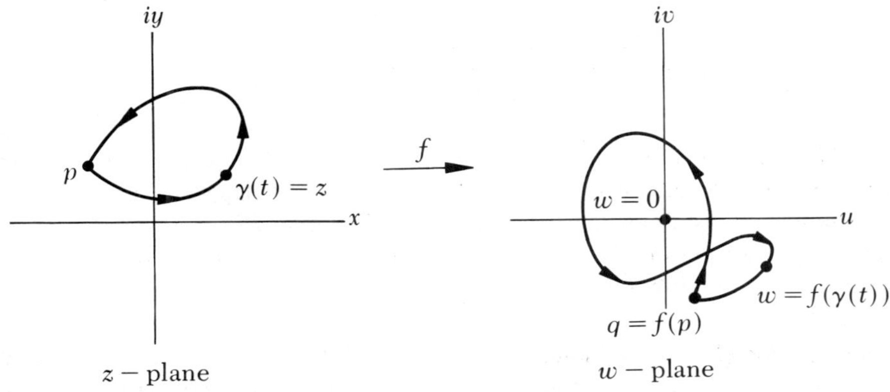

Figure 6.11 The transformation of the contour γ to the new contour $\eta = f \circ \gamma$ by the mapping $w = f(z)$.

transformed contour η with respect to the origin $w = 0$. This is seen by straightforward computations:

$$\begin{aligned}\int_\gamma \frac{f'(z)}{f(z)}\,dz &= \int_a^b \frac{f'(\gamma(t))}{f(\gamma(t))}\cdot\frac{d\gamma}{dt}\,dt\\ &= \int_a^b \frac{1}{f(\gamma(t))}\,\frac{df}{dz}(\gamma(t))\,\frac{d\gamma}{dt}\,dt\\ &= \int_a^b \frac{1}{f(\gamma(t))}\,\frac{d}{dt}(f\circ\gamma)(t)\,dt\\ &= \int_\eta \frac{1}{w}\,dw = \int_\eta \frac{1}{w-0}\,dw\\ &= 2\pi i\cdot I(f\circ\gamma, 0);\end{aligned}$$

recall that $d\eta/dt = f'(\gamma(t))\cdot(d\gamma/dt)$. This winding number measures the angle increment $\Delta_{f\circ\gamma}[\arg w]$ as $w = f(\gamma(t))$ moves along the trajectory of $\eta = f\circ\gamma$. This is precisely the increment $\Delta_\gamma[\arg f(z)]$ in $\arg(z)$ as $z = \gamma(t)$ moves along the original contour γ. Thus we have an important relation between these angle increments and the count of zeros and poles enclosed by γ:

$$Z_f - P_f = \frac{1}{2\pi}\Delta_\eta[\arg w] = \frac{1}{2\pi}\Delta_\gamma[\arg f(z)]. \tag{47}$$

This formula is known as the **principle of the argument.** Although formula (47) can sometimes be used to evaluate contour integrals, as illustrated below, this application is not the most important one we have in mind.

Example 6.25 We may evaluate the integral $\displaystyle\int_\gamma \frac{z}{z^2+1}\,dz$ along the counterclockwise oriented circle $|z| = 2$ by calculating residues at the singular points $z = +i, -i$. But we can also use formula (47); the integrand has the form $\dfrac{f'(z)}{f(z)}$ if we take $f(z) = z^2 + 1$, and $f(z)$ has zeros at $z = +i, -i$ so that

$$\int_\gamma \frac{z}{z^2+1}\,dz = \frac{1}{2}\int_\gamma \frac{2z}{z^2+1}\,dz = i\pi\cdot Z_f = 2\pi i.$$

The principle of the argument can be used to estimate the positions of the zeros of a polynomial, or other analytic function, and to determine a number of qualitative results about the behavior of a polynomial $P(z) = a_N z^N + \cdots + a_1 z + a_0$ as we perturb its coefficients. Our main tool is the following result, derived from the principle of the argument.

Theorem 6.8 (Rouche's Theorem) *Let γ be a positively oriented simple closed contour in the plane and assume that $f(z)$ and $g(z)$ are analytic on the trajectory Γ and the set of points enclosed by γ. Suppose that f is never zero on Γ and that $|f(z)| > |g(z)|$ at* *all points on the trajectory* Γ. *Then the functions* $f(z)$ *and* $h(z) = f(z) + g(z)$ *have the same number of zeros enclosed by* γ *(counting each according to its multiplicity).*

Thus, if f "dominates" g on the boundary Γ, we may "perturb" f by adding on g without changing the number of zeros lying interior to γ; however, it is very likely that we will change the actual location of these zeros within γ by adding on g.

PROOF: Let Z_f and Z_h be the number of zeros of $f(z)$ and $h(z)$ enclosed by γ. According to the principle of the argument, these are given by angle increments

$$Z_f = \frac{1}{2\pi}\Delta_\gamma[\arg f(z)] = \frac{1}{2\pi i}\int_\gamma \frac{f'(z)}{f(z)}\,dz$$

$$Z_h = \frac{1}{2\pi}\Delta_\gamma[\arg h(z)] = \frac{1}{2\pi i}\int_\gamma \frac{h'(z)}{h(z)}\,dz,$$

since there are no poles. Thus,

$$Z_f - Z_h = \frac{1}{2\pi i}\int_\gamma \left(\frac{f'}{f}\right) - \left(\frac{f'+g'}{f+g}\right) dz = \frac{1}{2\pi i}\int_\gamma \frac{f'g - fg'}{f^2 + fg}\,dz. \tag{48}$$

Since $|f(z)| > |g(z)|$ and $f(z) \neq 0$ for z on the trajectory Γ, $h(z) = f(z) + g(z)$ must also be nonzero on Γ, because $|f(z) + g(z)| \geq |f(z)| - |g(z)| > 0$. Thus, for any z on Γ,

$$f(z) - g(z) = f(z)\cdot\left[1 - \frac{g(z)}{f(z)}\right] = f(z)\cdot F(z),$$

and the estimate $|g(z)/f(z)| < 1$ on Γ implies that $w = F(z) = 1 - g(z)/f(z)$ lies in the disc $|w - 1| < 1$ centered at $w = +1$. Now apply Theorem 6.7 to $F(z)$. By the principle of the argument,

$$\frac{1}{2\pi i}\int_\gamma \frac{F'(z)}{F(z)}\,dz = \text{index of } w = 0 \text{ with respect to the transformed contour } \eta = F\cdot\gamma. \tag{49}$$

The values $w = \eta(t) = F(\gamma(t))$ all lie in the *simply connected* domain $D = \{w: |w - 1| < 1\}$, and $w = 0$ lies outside of D. Any closed contour in D, such as $\eta = F\circ\gamma$, is homologous to zero within D, so that $I(F\circ\gamma, 0) = 0$. On the other hand, direct calculations show that $F'/F = (f'g - fg')/(f^2 + fg)$, so that

$$Z_f - Z_h = \frac{1}{2\pi i}\int_\gamma \frac{f'g - fg'}{f^2 + fg}\,dz = \frac{1}{2\pi i}\int_\gamma \frac{F'(z)}{F(z)}\,dz = I(F\circ\gamma, 0) = 0. \quad ∎$$

We will now illustrate the applications of Rouché's theorem (and, indirectly, the principle of the argument).

Example 6.26 (Locating zeros of a polynomial) Consider the cubic polynomial $P(z) = z^3 + z + 1$. Since $|z^3| \geq |z| + 1 \geq |z + 1|$ if $|z| \geq \frac{4}{3}$, we conclude that the functions

$$f(z) = z^3 \quad \text{and} \quad h(z) = f(z) + g(z) = z^3 + z + 1 \text{ (where } g(z) = z + 1\text{)}$$

have the same number of zeros inside the circle $|z| = \frac{4}{3}$. In particular, all three roots of $P(z)$ must lie within this circle, since $f(z)$ has a zero of order three at the origin. On the other hand, notice that $|z + 1| \geq 1 - |z| > |z|^3$ for all z such that $0 \leq |z| \leq \frac{2}{3}$; that is, $f(z) = z + 1$ dominates $g(z) = z^3$ on the circle $|z| = \frac{2}{3}$. Thus $f(z) = z + 1$ and $P(z)$ both have no zeros at all inside this circle.

Taken together, these observations show that all roots of $P(z)$ lie in the annulus $\frac{2}{3} \leq |z| \leq \frac{4}{3}$. Further estimates can be made by trial and error, or by numerical analysis. Since the coefficients of $P(z)$ are all real, one of the roots (say z_1) must be real, and the other two must be complex conjugates of one another, $z_2 = \bar{z}_3$ (recall Exercise 2, Section 5.14).

Example 6.27 (A transcendental equation) Let $\lambda > 0$ be a real parameter. The roots of the equation $h(z) = \tan z - \lambda z = 0$ are important in certain problems related to heat propagation. Since $h(-z) = -h(z)$, the roots are located symmetrically with respect to the origin; furthermore, it is not difficult to locate the *real* roots by graphical methods. If we plot the graphs of $\tan x$ and λx (for real variable) as in Figure 6.12, the roots will correspond to the values of x for the points where the graphs intersect. In the interval $(-\pi/2, +\pi/2)$ there are three roots when $1 < \lambda < +\infty$, and one root when $0 < \lambda \leq 1$. Whatever the value of λ, we get one positive root in each of the

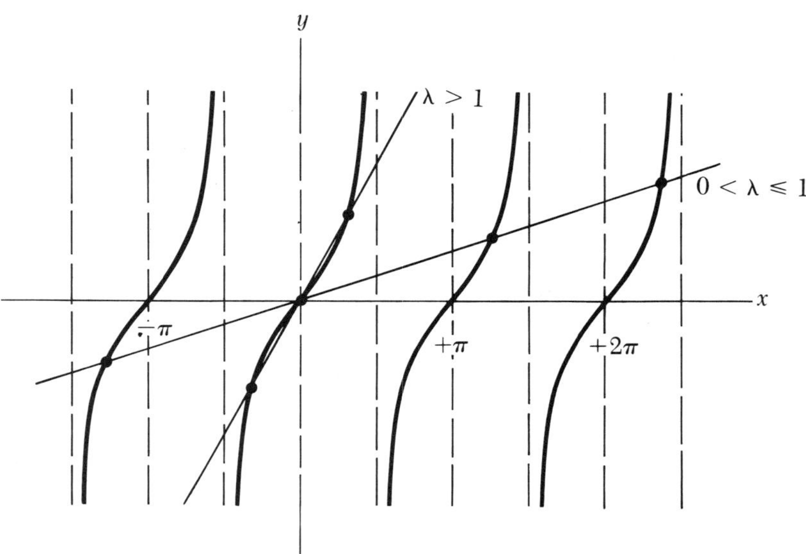

Figure 6.12 Graphical solution for the real roots of the equation $\tan x - \lambda x = 0$, shown for $0 < \lambda \leq 1$ and for $1 < \lambda < +\infty$.

intervals $(n\pi, n\pi + (\pi/2))$, and one negative root in each of the intervals $(-n\pi - (\pi/2), -n\pi)$ for $n = 1, 2, \ldots$.

In locating the *complex* roots, we face a difficult task if we attempt brute force calculation. Fortunately, we can simplify the task by using Rouché's theorem. Let us write out the real and imaginary parts of tan z:

$$\tan(x+iy) = \frac{1}{i}\left(\frac{e^{iz}-e^{-iz}}{e^{iz}+e^{-iz}}\right) = -i\left(\frac{e^{2iz}-1}{e^{2iz}+1}\right)$$

$$= \frac{(2\sin 2x) + i(e^{2y}-e^{-2y})}{e^{-2y}+2\cos 2x + e^{2y}}$$

(50)
$$= \left[\frac{\sin 2x}{\cosh 2x + \cos 2x}\right] + i\left[\frac{\sinh 2y}{\cosh 2y + \cos 2x}\right]$$

Now consider the magnitude of $|\tan z|$ for z on the boundary of the square S_N with vertices at $\pm N\pi \pm iN\pi$; using formula (50) it is not hard to show that $|\tan z| \leq 2$ on the boundary of S_N for all large N (we leave the details as Exercise 3). On the other hand, if $f(z) = -\lambda z$, then $|f(z)| = \lambda\,|z| \geq \lambda N$ on the boundary of S_N, because S_N includes the disc $|z| \leq N$. So, for all sufficiently large N we have (i) $\lambda N > 2$, and (ii) $|\tan z| \leq 2$ on the boundary of S_N, so that

$$|f(z)| \geq \lambda N > 2 > |\tan z| \quad \text{for all } z \text{ in bdry}(S_N).$$

Thus $h(z) = \tan z - \lambda z$ is a small perturbation of $f(z)$ on the boundary of S_N and these functions have the same number of poles and zeros within S_N, by Rouché's theorem:

$$Z_h - P_h = Z_f - P_f = 1 - 0 = 1.$$

But $h(z)$ has poles only at the poles of tan z, that is, at the points $z_k = (\pi/2) + \pi k$; there are $2N$ such poles within S_N; therefore, $Z_h = 2N + 1$.

When $1 < \lambda < +\infty$, there are $2N + 1$ real roots of $h(z)$, three in the interval $(-\pi/2, +\pi/2)$, within S_N, so there are no other roots in this domain. Since this is true for all large N, we see that the real roots are the only ones in this case. When $0 < \lambda \leq 1$, there are only $2N - 1$ real roots within S_N, so there must be two extra complex roots in this square. Since this is true for all large N, the transcendental equation has just two complex roots in addition to the infinite number of real roots.

The location of the extra roots, when $0 < \lambda \leq 1$, can be determined by examining the behavior of $h(z)$ along the imaginary axis. If $z = 0 + iy$ (y real), we get $\tan(iy) = i \tanh y$. A comparison of the graphs of the functions of real variable y,

$$\tanh y \quad \text{and} \quad \lambda y,$$

indicates that there are two imaginary roots, which can be located by graphical or numerical methods, when $0 < \lambda < 1$. When $\lambda = 1$, $\tan z - z$ has a zero of *third order* at the origin; since zeros are counted according to their orders, this accounts for the three zeros. Rouché's theorem does not guarantee that the zeros will be *distinct*, unless they are of order one.

EXERCISES

1. By comparing $f(z) = a_n z^n$ with $h(z) = f(z) + (a_{n-1}z^{n-1} + \cdots + a_0)$ on circles $|z| = R$ of large radius, prove the Fundamental Theorem of Algebra using Rouché's Theorem.

2. Suppose $f(z)$ is *holomorphic* on the trajectory of γ and on $E_{\text{int}}(\gamma)$, in Theorem 6.7. What is the significance of the integrals

$$\text{(i)}\ \frac{1}{2\pi i}\int_\gamma \frac{f'(z)}{f(z)}\,dz \quad \text{and} \quad \text{(ii)}\ \frac{1}{2\pi i}\int_\gamma \frac{f'(z)}{f(z) - \alpha}\,dz$$

(α a fixed complex number)?

3. On the boundary of the square determined by $\pm N\pi \pm iN\pi$ show that $|\tan z| \leq 2$ for all large $N = 1, 2, \ldots$. On the vertical segment $\text{Re}(z) = 0$ this is obvious; it follows for $\text{Re}(z) = \pm N\pi$ by periodicity. On horizontal segments use formula (50).

4. If $\lambda > 1$ show that $f(z) = z + e^{-z}$ takes the value λ at exactly one point in the right half plane; then decide whether the solution is real. Are there any solutions on the boundary line $\text{Re}(z) = 0$?

Hint: Consider semicircles of radius $R > 0$.

7 HARMONIC FUNCTIONS AND BOUNDARY VALUE PROBLEMS

A **harmonic function** of two real variables $H(x, y)$ is any solution of Laplace's equation

(1) $$\nabla^2 H = \frac{\partial^2 H}{\partial x^2} + \frac{\partial^2 H}{\partial y^2} = 0.$$

To avoid getting ourselves embroiled in technical questions about just how much differentiability is required of H, we will always include the following requirement as part of the definition of a harmonic function.

(2) H is continuous and has continuous first and second partial derivatives.

There is a similar definition of harmonic functions of several real variables $H(x_1, \ldots, x_n)$, except that Laplace's equation takes the higher dimensional form

$$\nabla^2 H = \frac{\partial^2 H}{\partial {x_1}^2} + \cdots + \frac{\partial^2 H}{\partial {x_n}^2} = 0.$$

Laplace's equation, and the harmonic functions that appear as its solutions, play a central role in an astonishing number of physical problems. In Chapter 8 we will show how this equation enters into problems concerning electromagnetic fields, fluid flow, and heat conduction. In two dimensional situations, complex variable methods can be used to great advantage in solving these problems. This possibility arises because harmonic functions $H(x, y)$ are intimately associated with analytic functions of the complex variable $z = x + iy$; we have already indicated some of these associations in Chapter 2, and will explore others in the present chapter.

A word of caution is in order concerning the use of complex variable methods in studying physical problems. Many physical problems are intrinsically three dimensional; but harmonic functions of many variables are not related in any simple way to analytic functions of one complex variable, so the methods of complex analysis can not be applied directly to these higher dimensional problems. If we wish to find exact solutions to an intrinsically three dimensional problem, we would have to develop completely different methods to solve it, if it can be solved at all in closed form. Problems of this kind are one of the main topics in the theory of partial differential equations, and will not be pursued here. In spite of this warning, two dimensional models of physical situations (and the complex variable techniques which are so helpful in understanding them) have played, and continue to play, an important role in the development of the physical sciences. There are two reasons for this. First, many three dimensional problems have a sufficient degree of symmetry that we may consider them as essentially two dimensional problems. For example, in determining the electromagnetic field surrounding a long conductor or the flow of air around a wing whose length is large in comparison to its cross section, we are faced with essentially two dimensional problems. Second, there are many realistic problems which are so difficult to solve that there is little hope of working out an exact solution. Sometimes there is an instructive two dimensional analog of the real problem, and by solving this two dimensional problem we can get useful insights into what may be expected in the full scale problem.

7.1 BOUNDARY VALUE PROBLEMS FOR LAPLACE'S EQUATION

We will consider the problem of finding solutions of Laplace's equation in various domains in the plane. For our purposes it will be sufficient to restrict our attention to domains that have a reasonably smooth boundary, one which is made up of smooth or piecewise smooth curves. Once a domain D has been specified, there will be many different functions that are harmonic throughout D. In the problems that arise from physics, we want to find a function $H(x, y)$ that satisfies Laplace's equation within D, but we also want $H(x, y)$ to behave in a particular way as we approach the boundary of the domain. In physical problems, the boundary behavior we may expect is known from the start; by insisting that $H(x, y)$ be harmonic in D and have the desired boundary behavior, we place a sufficient number of restraints on the solution that there is essentially only one harmonic function defined on D that satisfies these conditions.

Problems of this kind are called **boundary value problems,** and may be posed for partial differential equations other than Laplace's equation. We will consider Laplace's equation together with two basic types of boundary conditions; these lead to the **Dirichlet problem** and the **Neumann problem** for harmonic functions.

Any function $H(x,y)$ of two real variables may be regarded as a function of the complex variable $z = x + iy$, and we will find it convenient to use these two points of view interchangeably, without further comment. Let D be a domain whose boundary $\Gamma = \text{bdry}(D)$ is made up of piecewise smooth contours; that is, we assume that we can introduce parametrizations so that bdry(D) becomes the trajectory of finitely many contours. (For a multiply connected domain, such as the annulus $1 < |z| < 2$, the boundary may consist of more than one contour.) Let $h(z)$ be a continuous real valued function that is defined on the boundary set Γ.

The Dirichlet Problem. *We want to find a function $H(z) = H(x,y)$ that is defined and continuous throughout the closure $\bar{D} = D \cup \text{bdry}(D)$ of our domain, and is harmonic within D, and satisfies the boundary condition,*

$$H(z_0) = h(z_0) \quad \textit{for all points } z_0 \textit{ on } \text{bdry}\,(D).$$

In other words, we ask that the values of $H(z)$ within D match up with the prescribed values $h(z)$ on the boundary.

There is a complementary family of problems in which we specify the values of the inward normal derivative $\partial H/\partial n$, the directional derivative of H along the unit vector $\mathbf{n}$ that is perpendicular to the boundary and directed into the domain, rather than the values of $H(z)$ on the boundary. The normal derivative $\partial H/\partial n$ may be computed at a boundary point p in the following way. We first compute the gradient of H, a vector attached to p that is defined by

$$\nabla H = \textbf{grad}\ H = \frac{\partial H}{\partial x}(p)\,\mathbf{i} + \frac{\partial H}{\partial y}(p)\,\mathbf{j};$$

here $\mathbf{i}$ and $\mathbf{j}$ are the unit vectors in the x and y directions, respectively (we will always use boldface type to indicate vectors). Then we take a unit vector $\mathbf{n} = n_1\mathbf{i} + n_2\mathbf{j}$ that is perpendicular to the boundary at p and is directed into the domain D, and form the usual inner product (or dot product) of the vectors ∇H and $\mathbf{n}$. This gives us the inward normal derivative at p,

$$\frac{\partial H}{\partial n} = \nabla H \cdot \mathbf{n} = n_1 \frac{\partial H}{\partial x}(p) + n_2 \frac{\partial H}{\partial y}(p).$$

Recall that the inner product $\mathbf{a} \cdot \mathbf{b}$ of a pair of vectors attached to the same point in the plane is a scalar; if $\mathbf{a} = a_1\mathbf{i} + a_2\mathbf{j}$ and $\mathbf{b} = b_1\mathbf{i} + b_2\mathbf{j}$, then $\mathbf{a} \cdot \mathbf{b} = a_1b_1 + a_2b_2$. The length $\|\mathbf{a}\|$ of a vector $\mathbf{a} = a_1\mathbf{i} + a_2\mathbf{j}$ is given by $\|\mathbf{a}\| = \sqrt{a_1^2 + a_2^2} = \sqrt{\mathbf{a} \cdot \mathbf{a}}$. In terms of the lengths $\|\mathbf{a}\|$ and $\|\mathbf{b}\|$, the inner product $\mathbf{a} \cdot \mathbf{b}$ takes the form

$$\mathbf{a} \cdot \mathbf{b} = \|\mathbf{a}\| \cdot \|\mathbf{b}\| \cos\theta,$$

where θ is the angle between the vectors $\mathbf{a}$ and $\mathbf{b}$. Thus, $\partial H/\partial n = \nabla H \cdot \mathbf{n}$ measures the component of ∇H in the direction of the inward normal vector $\mathbf{n}$ associated with the boundary point p, as shown in Figure 7.1. In particular, if

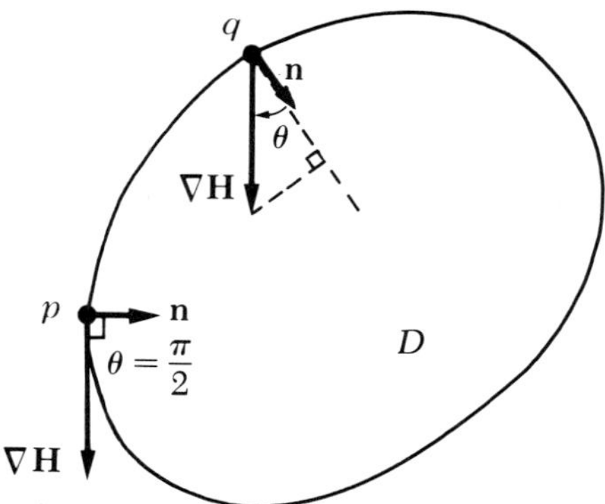

Figure 7.1 $\partial H/\partial n = \nabla H \cdot \mathbf{n}$ is given by projecting ∇H orthogonally onto the ray determined by the unit normal vector $\mathbf{n}$. If $\theta = \pm\pi/2$, then $\partial H/\partial n = 0$.

$\partial H/\partial n = 0$ at p, then ∇H is *tangent to* the boundary curve at p, since $\cos\theta = 0$ if and only if ∇H is perpendicular to the normal vector $\mathbf{n}$.

The Neumann Problem. *We want to find a function $H(z) = H(x, y)$ that is continuous on the closure $\bar{D} = D \cup \mathrm{bdry}(D)$ and harmonic on D, and satisfies the boundary condition*

$$\frac{\partial H}{\partial n}(z_0) = h(z_0) \quad \textit{for every boundary point } z_0.$$

Obviously, $H + \alpha$ (α a real constant) solves the problem if H is a solution, so that solutions of the Neumann problem are undetermined up to an added constant.

The reader should understand that many other types of boundary conditions arise in practice, but these should be treated as part of a general discussion of partial differential equations. We will show how to solve the Dirichlet and Neumann problems for simple domains, such as the unit disc $D = \{z : |z| < 1\}$ and the upper half plane $D = \{z : \mathrm{Im}(z) > 0\}$; the Dirichlet and Neumann problems for more intricate domains can often be reduced to corresponding problems for these simple domains by using the conformal mapping principle, which will be explained in Section 7.5.

We conclude these introductory remarks by indicating a few variants of the Dirichlet and Neumann problems that will be allowed in our discussion. Boundary values $h(z)$ that are bounded and have a finite number of discontinuities are commonly encountered in physical problems. It makes no sense to specify Neumann or Dirichlet boundary conditions at boundary points where $h(z)$ is discontinuous. Thus, it is customary to exclude these points when we discuss boundary conditions of this kind. However, if we leave the boundary behavior completely unspecified at one or more boundary points, this allows new families of solutions of the boundary value problem to exist, as is shown by the next example. In order to exclude the appearance of these extraneous solutions, we must impose *some* condition on the behavior of $H(z)$ near these points; most commonly, one requires that the solution $H(z)$ be *bounded* as z approaches (from within D) a boundary point where the value was not specified.

Example 7.1 Consider the disc $D = \{z : |z| < 1\}$ and try to solve the boundary value problem

$$H(z) \text{ harmonic within } D$$

$$H(z) = 0 \text{ for all boundary points } |z| = 1,$$

except that the boundary behavior is left unspecified at the single boundary point $z = 1 + i0$. The function $H_1(z) = 0$ on the closed disc $\bar{D}$ is evidently one solution to this problem. On the other hand, consider

$$H_2(z) = \text{Re}\left(\frac{z+1}{z-1}\right) = \frac{|z|^2 - 1}{|z-1|^2} \quad (\text{defined for } z \neq 1).$$

This function is evidently harmonic, being the real part of an analytic function, and if z is any point on the boundary circle with $z \neq 1 + i0$, then $H(z) = 0/|z-1|^2 = 0$. However, $H_2(z)$ is clearly non-vanishing within the disc D, and, as a matter of fact, $|H_2(z)| \to +\infty$ as $z \to 1 + i0$ along any ray that is not tangential to the circle. Both harmonic functions H_1 and H_2 appear to satisfy the boundary conditions we have posed, since the behavior at $z = 1 + i0$ was left completely unspecified. If we now impose the additional boundedness condition

$$|H(z)| \text{ is bounded as } z \to 1 + i0 \text{ from within } D,$$

the extra solution H_2 is excluded. Actually, it can be shown that H_1 is the *only* solution which satisfies the full set of boundary conditions, although we will not try to prove this here.

We will also allow the domain in our problem to be unbounded if this is required by the application we have in mind. Sometimes, with an unbounded domain, we must impose an additional condition specifying the behavior of the solution $H(z)$ "at infinity" (actually, as $|z| \to +\infty$, with z in D), in order to completely determine the solution. In the next example the domain is unbounded, and the boundary values are also discontinuous.

Example 7.2 Let D be the upper half plane (defined by $\text{Im}(z) > 0$), and impose the boundary conditions

$$H(x + i0) = +1 \quad \text{for } x < 0$$

$$H(x + i0) = 0 \quad \text{for } x > 0$$

$$H(z) \text{ is bounded as } z \to 0 \text{ from within } D$$

$$H(z) \text{ is bounded as } z \to \infty \text{ from within } D.$$

In this example we can almost guess the solution. The analytic function $\text{Log}(z)$ is well defined on D, and its imaginary part $\text{Arg}(z)$ is harmonic. Now $\text{Arg}(z)$ is constant on rays extending from the origin; to get the desired behavior on the two rays that make up the real axis, we need only multiply by $1/\pi$.

$$H(x, y) = \frac{1}{\pi}\operatorname{Arg}(x + y) = \frac{1}{\pi}\arccos\left(\frac{x}{\sqrt{x^2 + y^2}}\right).$$

We have left the boundary value at 0 unspecified because there is a jump discontinuity in the boundary values $h(x + i0)$ when $x = 0$. If we had not imposed the boundedness condition on$H(z)$ as $z \to 0$, there would be extraneous solutions which are unbounded as $z \to 0$ (see Exercise 8). The same comment applies to the boundedness condition "at infinity."

Example 7.3 (The principle of superposition of solutions) Since the Laplace operator $\nabla^2 = \partial^2/\partial x^2 + \partial^2/\partial y^2$ is linear, it is obvious that any linear combination $H = \sum_{k=1}^{N} r_k H_k$ of harmonic functions is again harmonic; furthermore, the boundary conditions satisfied by H are given by the same linear combination of the boundary conditions satisfied by the individual functions H_k. Now let us take the results of the last example and use this "principle of superposition" to solve much more sophisticated Dirichlet problems in the upper half plane. If $p = x_0 + i0$ is a fixed point on the real axis, it is evident that

$$H(z) = \frac{1}{\pi}\operatorname{Arg}(z - p)$$

has boundary values $H(x + i0)$ on the real axis equal to $+1$ to the left of $x = x_0$ and equal to zero to the right of $x = x_0$. By taking the difference of two such functions (still a harmonic function on the upper half plane) for points p and q $(p > q)$ on the real axis, we get a function

$$H(z) = \frac{1}{\pi}[\operatorname{Arg}(z - p) - \operatorname{Arg}(z - q)] = \frac{1}{\pi}\operatorname{Arg}\left(\frac{z - p}{z - q}\right) \tag{3}$$

whose boundary values are equal to $+1$ on the interval (p, q) and are zero elsewhere,

$$H(x + i0) = \begin{cases} 0 & \text{if} \quad x < q \quad \text{or if} \quad x > p \\ +1 & \text{if} \quad q < x < p. \end{cases}$$

The geometric significance of the function (3) is indicated in Figure 7.2; it is constant on every circular arc joining p and q. We may even take $q = -\infty$ or $p = +\infty$ in formula (3), if we assign the values $\operatorname{Arg}(z + \infty) = 0$ and $\operatorname{Arg}(z - \infty) = +\pi$ for all z.

By taking finite linear combinations of these functions we can easily solve problems in which the boundary values $h(x + i0)$ on the real axis are given by a step function with a finite number of steps; that is, a function obtained by dividing the real axis into a finite number of intervals $I_0 = (-\infty, p_1), \ldots,$

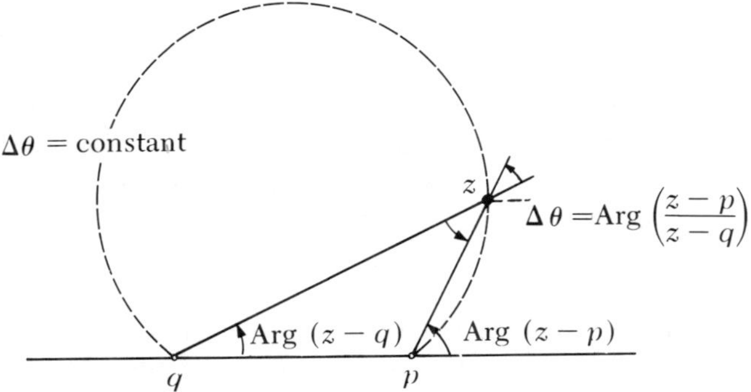

Figure 7.2 In formula (3), $H(z)$ may be interpreted as the angle $\Delta\theta$. This angle is measured from the segment $[q, z]$ to the segment $[p, z]$; thus it is positive throughout the upper half plane.

$I_N = (p_N, +\infty)$ and defining $h(x + i0) = r_k$ (some real number) throughout I_k. The solution is obtained by superposition of functions (3),

$$(4)\quad H(z) = \frac{r_0}{\pi}\operatorname{Arg}(z - p_1) + \frac{1}{\pi}\sum_{k=1}^{N-1} r_k \operatorname{Arg}\left(\frac{z - p_{k+1}}{z - p_k}\right) + \frac{r_N}{\pi}[\pi - \operatorname{Arg}(z - p_N)];$$

$H(z)$ is even bounded at infinity and at the exceptional points $p_1, \ldots, p_N$ where the boundary values are discontinuous.

Notice that the values of $H(z)$ within the upper half plane are fully determined by the behavior of $h(x + i0)$ on the boundary, and the coordinates of the point z. It cannot be emphasized too strongly that the boundary conditions completely determine the solution of the problem. That is, if we have a fixed domain with smooth boundary in mind and alter the boundary conditions, we may end up with a radically different solution of Laplace's equation, even though the domain, and the partial differential equation to be solved, remain the same. Here is an example which illustrates the primacy of the boundary conditions in determining the solution.

Example 7.4 In the vertical half strip, shown in Figure 7.3, we consider various boundary conditions as indicated in the diagram. (Problem (II) has

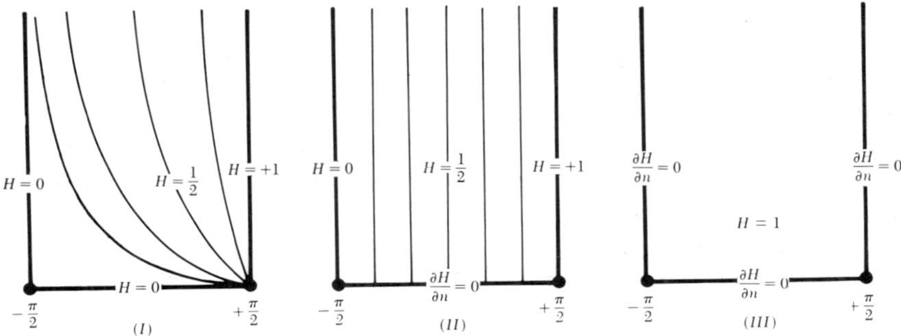

Figure 7.3 Level loci H = constant for the boundary value problems in Example 7.4.

mixed Dirichlet and Neumann boundary conditions.) In each problem we ask that $H(z)$ be bounded at infinity and at the points where the boundary values change abruptly, or (with the Neumann condition) where the boundary has a corner. Problem (III) has the trivial solutions $H(z) = \textit{constant}$, and these are the only solutions which satisfy the boundedness condition; the value of the constant is not determined by our conditions.

There *are* other solutions of (III) that are sometimes of physical significance, even though they are *unbounded* at infinity. In Figure 4.27, we sketched the level lines of $U(z) = \text{Re}(\sin z)$ in this strip, and from that diagram it is clear that the level curves of U are perpendicular to the boundary, so that $\partial U/\partial n = 0$ (except at the corners, where the normal derivative $\partial U/\partial n$ is undefined). This could also be verified by directly calculating the normal derivative. It is also clear from Figure 4.27 (or by direct calculation based on formula (39) of Section 4.10) that

$$\lim_{z\to\infty} |U(z)| = +\infty \qquad \lim_{z\to+\pi/2} U(z) = +1 \qquad \lim_{z\to-\pi/2} U(z) = -1.$$

Likewise, we may guess the solution to problem (II):

$$H(x+iy) = \frac{x}{\pi} + \frac{1}{2} \left(\text{so that } H(z) = \text{Re}\left(\frac{z}{\pi} + \frac{1}{2}\right) \text{ on the strip}\right).$$

Obviously, the level curves $H(z) = \textit{constant}$ are vertical lines perpendicular to the real axis, so that $\partial H/\partial n = \partial H/\partial y = 0$ on the bottom of the strip.

The solution of problem (I) is more difficult; however, this is due to a momentary lack of the proper tools. Once we have discussed the "conformal mapping principle" for boundary value problems (Sections 7.5 and 7.6), it will be clear that the desired solution can be obtained by solving a related boundary value problem in the upper half plane,

$$\text{(5)} \qquad H(w) = \begin{cases} 0 & \text{if } w = u + i0 \text{ with } -\infty < u < +1 \\ +1 & \text{if } w = u + i0 \text{ with } +1 < u < +\infty \end{cases}$$

and then substituting $w = \sin z$. Roughly speaking, we may pass over to the more tractable problem in the upper half plane because the familiar mapping $w = \sin z$ transforms the strip conformally onto the upper half plane (as we showed in Section 4.10), and at the same time transforms the boundary of the strip to the real axis in a way that makes boundary values for the strip match up with the boundary values for the upper half plane specified in (5).

As we have seen, the problem in the upper half plane has the solution

$$H(w) = \frac{1}{\pi}[\pi - \text{Arg}(w-1)] \quad \text{for } w \text{ in the upper half plane,}$$

so that the original problem has the solution

$$H(z) = \left[H(w) \Big|_{w=\sin z} \right] = \frac{1}{\pi} [\pi - \operatorname{Arg}(\sin(z) - 1)]. \tag{6}$$

Level curves for this function can be obtained by sketching the level curves $H(w) = \textit{constant}$ in the upper half plane (radial lines originating at $w = 1$), and then transforming them into the strip by the inverse mapping $z = \operatorname{Arcsin} w$; the end result is indicated in Figure 7.3 (I).

Another point should be kept in mind as we study boundary value problems. Although the desired solution $H(z)$ may be irregular at certain boundary points, it must be harmonic at every point inside the domain, *without exceptions.* If we allow the solution to be singular, or undefined, at even one interior point, this allows entirely new solutions to appear. Allowing a singularity at a point p in D amounts to changing the shape of the domain in a fundamental way; the point p should now be considered as part of the boundary, and not as a point in our domain. The "punctured domain" obtained by removing p is multiply connected, and it is this change in geometry which leads to new types of solutions. This is illustrated in the next example.

Example 7.5 Let D be the unit disc $|z| < 1$. If we impose the Dirichlet condition $H(z) = 0$ on the boundary circle $|z| = 1$, the constant function $H_1(z) = 0$ is certainly one solution of our problem. Now let us relax our conditions, allowing the solution function $H(z)$ to be undefined or otherwise singular at the single point $z = 0$. Suddenly we are confronted with a new family of solutions which are harmonic in the punctured disc, namely

$$H_\alpha(x, y) = \alpha \cdot \log |x + iy| = \log |x + iy|^\alpha = \left(\frac{\alpha}{2}\right) \cdot \log(x^2 + y^2).$$

These functions are harmonic and, although they are quite singular at the origin, they all satisfy the boundary condition $H(z) = 0$ for $|z| = 1$. The new point $z = 0$ is a very significant part of the boundary of the new domain $E^* = \{z : 0 < |z| < 1\}$, and the solution to our problem is not fully determined until we impose conditions on its behavior at this new point.

EXERCISES

1. Show that the Laplace operator $\nabla^2 = \dfrac{\partial^2}{\partial x^2} + \dfrac{\partial^2}{\partial y^2}$ acts *linearly* on functions $H(x, y)$ with continuous partial derivatives of first and second order; if α and β are real constants, then

$$\nabla^2(\alpha F + \beta G) = \alpha \nabla^2 F + \beta \nabla^2 G.$$

Show that any linear combination $H = \sum_{j=1}^{m} \alpha_j H_j$ of harmonic functions is harmonic.

2. Solve the Dirichlet problems in the unit disc $E = \{z: |z| < 1\}$ with the boundary values indicated.

$$\text{(i)}\ h(e^{i\theta}) = \cos n\theta \quad \text{for} \quad 0 \leq \theta \leq 2\pi$$
$$\text{(ii)}\ h(e^{i\theta}) = \sin n\theta \quad \text{for} \quad 0 \leq \theta \leq 2\pi.$$

Hint: Are there obvious analytic functions $f(z)$ whose real parts do this?

3. Use the superposition principle to determine an analytic function $f(z)$ on the disc $|z| < 1$ whose real part $H(z) = \text{Re}(f(z))$ has as its boundary values the trigonometric polynomial

$$h(e^{i\theta}) = a_0 + \sum_{k=1}^{N} a_k \sin k\theta + \sum_{k=1}^{N} b_k \cos k\theta$$

(a_0, a_k, b_k real) for $0 \leq \theta \leq 2\pi$.

Answer: $f(z) = a_0 + \sum_{k=1}^{N} (b_k - ia_k) z^k$.

4. Find harmonic functions which solve the boundary value problems indicated below:

(i) On the quadrant $E = \{z:\text{Re}(z) > 0 \text{ and } \text{Im}(z) > 0\}$,

$$\frac{\partial H}{\partial n}(0 + iy) = 0 \quad (y > 0) \quad \text{and} \quad H(x + i0) = 1 \quad (x > 0).$$

(ii) On $E = \{z:1 < |z| < 2\}$,

$$H(z) = 1 \text{ on } |z| = 1 \text{ and } H(z) = 3 \text{ on } |z| = 2.$$

(iii) On $E = \{z : 0 < \text{Im}(z) < 1\}$,

$$\frac{\partial H}{\partial n} = 0 \quad \text{on both boundary segments}$$

(and H not identically zero on E.)

Answers: (i) $1 + y$; (ii) $\left(\dfrac{2}{\log 2}\right) \cdot \log |z| + 1$; (iii) $H(x, y) = x + \text{constant}$.

5. Calculate the normal derivatives of the following harmonic functions, defined on $E = \{z:\text{Im}(z) > 0\}$, on the parts of the real axis indicated. Take $\mathbf{n}$ directed into the domain.

(i) $H(z) = \log |z|$ on $(-\infty, 0)$ and on $(0, +\infty)$
(ii) $H(z) = \text{Arg}(z)$ on $(-\infty, 0)$ and on $(0, +\infty)$
(iii) $H(z) = \text{Re}(e^{-z^2})$ on $(-\infty, +\infty)$
(iv) $H(z) = \text{Im}(e^{-z^2})$ on $(-\infty, +\infty)$.

6. Check that $w = \sin z$ maps the sides of the semi-infinite strip $E = \{z : \text{Im}(z) > 0 \text{ and } -\pi/2 < \text{Re}(z) < +\pi/2\}$ one-to-one onto the segments $(-\infty, -1)$, $(-1, +1)$, $(1, +\infty)$ in the real axis, and that the domain E is mapped *conformally* onto the upper half plane $\text{Im}(w) > 0$.

7. The exceptional solution $H_2(z) = \dfrac{|z|^2 - 1}{|z - 1|^2}$ in Example 7.1 has unusual behavior as $z \to 1$ from within the disc $|z| < 1$. Produce sequences $\{z_n'\}$ and $\{z_n''\}$ of points that approach $1 + i0$ from within the disc such that

(i) $\lim_{n\to\infty} H_2(z_n') = +\infty$ (ii) $\lim_{n\to\infty} H_2(z_n'') = 0$

Hint: Let $z_n'' \to 1$ along a curve tangent to the circle; z_n' may approach along any non-tangential line.

8. Write out $H(x, y) = \text{Im}(1/z)$. Show that $H(z) = 0$ along the real axis (except at $z = 0$ where it is undefined). Is $H(z)$ bounded as $z \to \infty$ or $z \to 0$ in the upper half plane? These observations extend the results of Example 7.1 to apply to the half plane $E = \{z : \text{Im}(z) > 0\}$.

9. Use the function given in Exercise 8 to demonstrate that multiple solutions may arise in Example 7.2 unless we impose some restrictions on the behavior of the solution at the boundary point $z = 0 + i0$ where the boundary values are discontinuous.

10. Let $f(z)$ be analytic near $z = p$ and let $\gamma(t)$ be a smooth curve passing through p when $t = t_0$, with non-zero tangent vector $\dfrac{d\gamma}{dt}(t_0)$. If $\mathbf{n}$ is the unit vector obtained by rotating the unit tangent vector $\dfrac{\gamma'(t_0)}{|\gamma'(t_0)|}$ by $+\pi/2$ radians, show that $U = \text{Re}(f(z))$ has normal derivative

$$\frac{\partial U}{\partial n}(p) = \nabla U \cdot \mathbf{n} = \frac{\text{Re}\left\{\dfrac{df}{dz}(p)\,\dfrac{d\gamma}{dt}(t_0)\right\}}{\left|\dfrac{d\gamma}{dt}(t_0)\right|}.$$

Notice that there are *two* vectors normal to γ at p, namely $+\mathbf{n}$ and $-\mathbf{n}$.

11. Verify that $\text{Arg}(z - p) - \text{Arg}(z - q) = \text{Arg}\left(\dfrac{z - p}{z - q}\right)$ (an equality, not a congruence), if $p > q$ on the real axis and if z is in the upper half plane.

12. If $p > q$ on the real axis, prove that $\text{Arg}\left(\dfrac{z - p}{z - q}\right)$ may be interpreted as the angle $\Delta\theta$ shown in Figure 7.2 (measured from segment

$[z, q]$ to segment $[z, p]$). Show that this function is constant on any circular arc from p to q. On which arc is its value $+\pi/2$?

13. If $p > q$ on the real axis, consider $H(z) = \text{Arg}\left(\frac{z-p}{z-q}\right)$ defined for *all* $z \neq p, q$. Show that H is harmonic, except along the segment $[q, p]$ where it has a jump discontinuity. Is the geometric interpretation of $H(z)$ (Exercise 12) still valid if $\text{Im}(z) < 0$? Show that $H(z) \to +\pi$ as z approaches a point on $[q, p]$ from the upper half plane, and that $H(z) \to -\pi$ as z approaches such points from the lower half plane.

14. The harmonic function on the upper half plane,

$$H(z) = \frac{1}{\pi}\,\text{Arg}\left(\frac{z-a}{z+a}\right) \quad (a > 0 \quad \text{real}),$$

has boundary values $H(x + i0) = 0$ if $|x| > a$ and $H(x + i0) = 1$ if $-a < x < +a$. Show that $H(z) = H(x, y)$ may be expressed in the form

$$H(x, y) = \frac{1}{\pi}\,\text{arctan}^*\left(\frac{2ay}{x^2 + y^2 - a^2}\right)$$

for $z = x + iy$ in the upper half plane, where $\text{arctan}^*(t)$ is the determination of arctangent whose values lie in the interval $[0, \pi]$ for $-\infty < t < +\infty$.

15. Determine an analytic function $f(z)$ on the half plane $E = \{z : \text{Im}(z) > 0\}$ whose *imaginary* part $H(z) = \text{Im}(f(z))$ satisfies the boundary conditions

(i) $H(x + i0) = +1$ for $x > 0$ and $H(x + i0) = -1$ for $x < 0$

(ii) $|H(z)|$ is bounded as $z \to 0$ and as $z \to \infty$, from within E.

Find a function $g(z)$ whose *real* part has these properties.

Answers: $f(z) = \frac{2}{\pi}\,[i\pi/2 - \text{Log}\, z]$; $H(z) = \frac{2}{\pi}\left[\frac{\pi}{2} - \text{Arg}\, z\right]$; $g(z) = -if(z)$.

16. Use the ideas of Example 7.3 to solve the following boundary value problems in the upper half plane $\text{Im}(z) > 0$. All solutions are to be bounded at ∞, and bounded near points of discontinuity for the boundary values $h(z)$.

$$\text{(i)}\quad h(x + i0) = \begin{cases} +1 & \text{for} \quad -\infty < x < -1 \\ & \qquad \text{and} \quad +1 < x < +\infty \\ 0 & \text{for} \quad -1 < x < +1 \end{cases}$$

$$\text{(ii)}\quad h(x + i0) = \begin{cases} +1 & \text{for} \quad -1 < x < 0 \\ -1 & \text{for} \quad 0 < x < 5 \\ 0 & \text{otherwise} \end{cases}$$

17. Write out the solution to the boundary value problem in the upper half plane in which

(i) $H(x + i0) = 0 \qquad -\infty < x < -2$
(ii) $H(x + i0) = 10 \qquad -2 < x < +\infty$
(iii) $H(z)$ bounded as $z \to -2$ or as $z \to \infty$.

Sketch the level lines $H(z) = constant$ in the upper half plane.
Answer: $10 - (10/\pi)\text{Arg}(z + 2)$.

18. Consider $f(z) = \dfrac{1}{2}\left(z + \dfrac{1}{z}\right) = H(z) + iN(z)$ on the exterior domain $E = \{z: |z| > 1\}$. Write out explicit formulas for $H(x, y)$ and $N(x, y)$ as functions of x and y. Prove that H and N have the following boundary behavior.

(i) $N = 0$ on the circle $|z| = 1$, and also on the segments $(-\infty, -1)$ and $(1, +\infty)$ on the real axis

(ii) $\dfrac{\partial H}{\partial n} = 0$ on the circle $|z| = 1$ and also along the segments $(-\infty, -1)$ and $(1, +\infty)$ on the real axis

(iii) $\dfrac{\partial N}{\partial n} = 0$ on the segments $(+i, +i\infty)$ and $(-i, -i\infty)$ in the imaginary axis; $H = 0$ on these segments.

Hint: In (ii) and (iii) use polar coordinates, to study $\partial\phi/\partial n = +\partial\phi/\partial r$ on the circle.

19. Consider $\text{Arcsin}(w) = X(u, v) + iY(u, v)$, defined on the upper half plane $\text{Im}(w) > 0$. From the fact that $w = \sin(X + iY)$ we get

$$u = \sin X \cosh Y \quad \text{and} \quad v = \cos X \sinh Y,$$

so that

$$\frac{u^2}{\sin^2 X} - \frac{v^2}{\cos^2 X} = 1 \quad \text{and} \quad \frac{u^2}{\cosh^2 Y} + \frac{v^2}{\sinh^2 Y} = 1.$$

From these, derive the following explicit formulas for $X(u, v) = \text{Re}[\text{Arcsin}(w)]$ and $Y(u, v) = \text{Im}[\text{Arcsin}(w)]$ in the upper half plane.

$$\begin{aligned} X(u, v) &= \arcsin \tfrac{1}{2}[|w + 1| - |w - 1|] \\ &= \arcsin \tfrac{1}{2}[\sqrt{(u + 1)^2 + v^2} - \sqrt{(u - 1)^2 + v^2}] \end{aligned}$$

$$\begin{aligned} Y(u, v) &= \cosh^{-1} \tfrac{1}{2}[|w + 1| + |w - 1|] \\ &= \cosh^{-1} \tfrac{1}{2}[\sqrt{(u + 1)^2 + v^2} + \sqrt{(u - 1)^2 + v^2}]. \end{aligned}$$

Here, take the usual determination of $\arcsin t$, for $-1 \le t \le 1$; $s = \cosh^{-1}(t)$ is the inverse of $t = \cosh s$ (for $0 \le s < +\infty$), and may be calculated explicitly as $\cosh^{-1}(t) = \log\left[t + \sqrt{t^2 - 1}\right]$ for $1 \le t < +\infty$.
Hint: Recall Exercise 10 (Section 4.11) for more details.

20. Solve the following boundary value problem in the upper half plane:

$$H(u + i0) = -1 \quad \text{for} \quad -\infty < u < -1$$
$$H(u + i0) = +1 \quad \text{for} \quad +1 < u < +\infty$$
$$\frac{\partial H}{\partial n}(u + i0) = 0 \text{ for } -1 < u < +1.$$
$|H(w)|$ bounded near discontinuities, and at infinity.

Use Exercise 19 to show that the solution can be expressed as

$$H(w) = \frac{2}{\pi}\,\text{Re}[\text{Arcsin}\, w] = \frac{2}{\pi}\,\text{Arg}\big[iw + \sqrt{1 - w^2}\big],$$

or as

$$H(u, v) = \frac{2}{\pi} \arcsin \tfrac{1}{2}\big[\sqrt{(u + 1)^2 + v^2} - \sqrt{(u - 1)^2 + v^2}\big].$$

Note: Without Exercise 19 it would be difficult to express the solution as an explicit function of u and v.

7.2 CONJUGATE HARMONIC FUNCTIONS

Any analytic function of a complex variable $f(x + iy) = U(x, y) + iV(x, y)$ gives us a pair of solutions of Laplace's equation, $U = \text{Re}(f)$ and $V = \text{Im}(f)$, which are related through the Cauchy-Riemann equations

$$\frac{\partial U}{\partial x} = \frac{\partial V}{\partial y} \qquad \frac{\partial U}{\partial y} = -\frac{\partial V}{\partial x}.$$

This connection between analytic functions and harmonic functions works *both* ways. In fact, with one qualification, every harmonic function turns out to be the real part of some analytic function of a complex variable, and this very strong connection lets us bring all the machinery of complex analysis to bear on boundary value problems associated with Laplace's equation.

Let $U(z) = U(x, y)$ be harmonic on an open set D in the complex plane. Our task is to find, if possible, a second harmonic function $V(z)$ which, when pieced together with $U(z)$, gives us a complex differentiable (hence analytic) function $f(z) = U(z) + iV(z)$. Any harmonic function $V(z)$ that combines with $U(z)$ in this way will be referred to as a **conjugate harmonic function** of $U(z)$. Since we can look for conjugate functions separately in each connected subset of D, we may as well assume that D is a (connected) domain. First notice that

(i) On a domain D the conjugate function is unique up to an added real constant, if such a conjugate function exists at all.

This is proved in Exercises 5 and 6, using the Cauchy-Riemann equations. Whether or not there is a conjugate harmonic function defined on D depends on the shape of D and the nature of the harmonic function $U(z)$ we start with, as will be shown in the examples that follow. Nevertheless, we can always define a suitable conjugate function $V(z)$ *locally*, and this is all we really need for most applications.

(ii) If $U(z)$ is harmonic on a domain D, and p is any point in D, consider any disc D_p, centered at p, that lies entirely within D. Then there is a conjugate harmonic function $V(z)$ for $U(z)$ defined on D_p.

The proof of this fact is suggested by working backwards, and noticing what would happen if we could find such a conjugate function. Then $f = U + iV$ would be analytic, and its derivative could be expressed entirely in terms of the (known) partial derivatives of U;

$$\frac{df}{dz} = \frac{\partial U}{\partial x} - i\frac{\partial U}{\partial y}.$$

Thus, V would be obtained as $\mathrm{Im}(f)$, where f is the function obtained by antidifferentiating the *known* function $f' = (\partial U/\partial x) - i(\partial U/\partial y)$;

$$f(z) = c + \int_p^z \left[\frac{\partial U}{\partial x}(\zeta) - i\frac{\partial U}{\partial y}(\zeta)\right] d\zeta. \tag{7}$$

We include the details below, but the reader may omit them if he wishes. Besides, in practical examples, there are simpler methods of determining conjugate harmonic functions; these methods will be illustrated in the examples below.

Proof of (ii). We start with U and must construct a conjugate harmonic function V on the disc D_p. Define $G(z) = \partial U/\partial x(z) - i\,\partial U/\partial y(z)$. This is certainly continuous and well defined throughout the original domain D. It is also holomorphic on D because its real and imaginary parts $\mathrm{Re}(G) = \partial U/\partial x$ and $\mathrm{Im}(G) = -\partial U/\partial y$ have continuous first partial derivatives that satisfy the Cauchy-Riemann equations

$$\frac{\partial}{\partial x}(\mathrm{Re}(G)) - \frac{\partial}{\partial y}(\mathrm{Im}(G)) = \frac{\partial^2 U}{\partial x^2} + \frac{\partial^2 U}{\partial y^2} = 0$$

$$\frac{\partial}{\partial y}(\mathrm{Re}(G)) + \frac{\partial}{\partial x}(\mathrm{Im}(G)) = \frac{\partial^2 U}{\partial y\,\partial x} + \left(-\frac{\partial^2 U}{\partial x\,\partial y}\right) = 0.$$

Analytic functions such as G have well defined antiderivatives on any simply connected domain, such as D_p; these antiderivatives are defined up to an added constant by

$$F(z) = \int_p^z G(\zeta)\,d\zeta \quad \text{for } z \text{ in } D_p$$

(the integral indicates line integration along any contour, contained in the disc, that connects p to z). Obviously,

$$\frac{dF}{dz} = \frac{\partial}{\partial x}(\mathrm{Re}(F)) - i\frac{\partial}{\partial y}(\mathrm{Re}(F)) = G = \frac{\partial U}{\partial x} - i\frac{\partial U}{\partial y},$$

so that $\text{Re}(F)$ and U have the same first partial derivatives throughout the disc D_p. This means that the first partial derivatives of $(U - \text{Re}(F))$ are identically zero on D_p, so that $U = \text{Re}(F) + c$, where c is some real constant. Clearly, the analytic function $f(z) = F(z) + c$ gives $\text{Re}(f) = U$ on the disc, so its imaginary part $V = \text{Im}(f) = \text{Im}(F)$ is a conjugate harmonic function for U on the disc. ▮

If the domain D in (ii) is simply connected, we can use the whole domain D in place of the smaller disc D_p in the preceding proof, without altering the argument. We obtain a very useful theorem about the existence of conjugate harmonic functions on *simply connected* domains.

Theorem 7.1 *If $U(z) = U(x, y)$ is harmonic on a simply connected domain D, there is a conjugate harmonic function $V(z)$ that is well defined throughout D.*

Here are some methods for determining conjugate harmonic functions in practical situations; other methods are discussed in Section 7.3 and in Chapter 8.

Example 7.6 $U(x, y) = x^2 - y^2 + 1$ is easily seen to be harmonic on the plane. This function is just the real part of $f(z) = z^2 + 1$. Since $f(x + iy) = (x^2 - y^2 + 1) + i(2xy)$, $V(x, y) = 2xy$ is a harmonic function conjugate to $U(x, y)$; obviously, $V^*(x, y) = 2xy + c$ (c a real constant) works equally well. How could we have systematically arrived at a suitable conjugate function if we didn't remember that $U(x, y) = \text{Re}(f(x + iy))$? It is crucial to notice that the function V cannot be chosen freely, since the complex differentiability of $f(z)$ requires that U and V be related as in the Cauchy-Riemann equations. We can arrive at the desired conjugate functions by working backward from these equations:

$$\frac{\partial V}{\partial y} = \frac{\partial U}{\partial x} = 2x \qquad \frac{\partial V}{\partial x} = -\frac{\partial U}{\partial y} = 2y.$$

By partial integration (taking indefinite integral with respect to one of the variables x or y, while the other variable is treated as a constant), these equations imply that V has the form

$$V(x, y) = 2xy + \phi(y) \qquad V(x, y) = 2xy + \psi(x),$$

where $\phi(y)$ and $\psi(x)$ are arbitrary functions of x and y. Comparing these requirements, we see that V must have the form $V(x, y) = 2xy + c$. A simple computation shows that we have, indeed, obtained a conjugate harmonic function for $U(x, y)$.

The method of integrating the Cauchy-Riemann equations to determine conjugate functions works in many situations. The next example shows that it is not always possible to define a conjugate harmonic function throughout a multiply connected domain.

Example 7.7 Consider $U(z) = \log|z| = \log\sqrt{x^2 + y^2}$ on the punctured plane $D = \{z: z \neq 0\}$. Obviously, $U(z)$ has continuous partial derivatives of all orders on D, and is harmonic on D. Now $\log|z|$ is trying to be the real part of an analytic determination of the function $\log z$. Suppose that we *could* find a conjugate harmonic function $V(z)$ that is well defined throughout D; then $f(z) = U(z) + iV(z)$ would be analytic. We could then compare $f(z)$ with the function $\operatorname{Log} z$ defined on the smaller domain D^* obtained by removing the negative real axis from the plane. Both functions would be analytic on D^*, and both would have the same real part $\operatorname{Re}(f(z)) = \log|z| = \operatorname{Re}(\operatorname{Log} z)$. Thus, $f(z) - \operatorname{Log}(z)$ would be analytic and would have purely imaginary values on the domain D^*. This would imply that the function is constant (see Section 2.10), so that $f(z)$ would equal $\operatorname{Log} z + ic$ (c real) on D^*. But $\operatorname{Log} z + ic$ has a discontinuity at every point on the negative real axis, while $f(z)$ is supposed to be continuous (analytic!) there. Obviously, we arrive at this discrepancy if we assume that there is a conjugate harmonic function defined throughout D; we must therefore conclude that $\log|z|$ does not admit such a globally defined conjugate function on D.

On the other hand, conjugate functions for $\log|z|$ may be defined *near* any point p in the domain D. Take any holomorphic determination of $\log z$ defined near p; then $V(z) = \operatorname{Im}(\log z)$, a determination of $\arg z$, is a conjugate harmonic function for $U(z) = \log|z|$ near p.

In the last example it is interesting to notice that a "conjugate function" does exist throughout D, but it is one that is *multiple valued*, namely $V(z) = \arg z$. We cannot define it as a single valued function on a domain like the punctured plane. We have already shown (Theorem 7.1) that a smooth single valued conjugated harmonic function always exists on a domain that is simply connected.

EXERCISES

1. If $H(z)$ is harmonic near $z = p$, we have shown that there is an analytic function $f(z)$, defined near p, such that $H(z) = \operatorname{Re}(f(z))$. Show that there is also an analytic function $g(z)$ such that $H(z) = \operatorname{Im}(g(z))$ near $z = p$.

2. Verify that $H(x, y)$ is harmonic and find a conjugate harmonic function.

(i) x

(ii) $ax + by + c$ (a, b, c real constants)

(iii) $2y + e^{-x}\cos y$

(iv) $\log(x^2 + y^2)^{3/2}$ (for $z = x + iy \neq 0$)

(v) $(-y/(x^2 + y^2)) + x$ (for $z = x + iy \neq 0$)

3. Identify the harmonic functions $H(x, y)$ below as real parts of elementary analytic functions $f(z)$.

(i) $x^3 - 2xy^2$

(ii) $\sin x \cosh y$

(iii) $e^{-y} \sin x + x$

(iv) $\dfrac{1 - (x^2 + y^2)}{(1 - x)^2 + y^2}$

Answers: (i) z^3; (ii) $\sin z$; (iii) $z - ie^{iz}$; (iv) $(1 + z)(1 - z)$.

4. Write out analytic functions $g_1(z)$ and $g_2(z)$ such that $H = \text{Re}(g_1)$ and $H = \text{Im}(g_2)$, respectively, for the harmonic functions

(i) $H(z) = \text{Arg}(z - 1) - \text{Arg}(z + 1) + (\pi/2)$

(ii) $(1/\pi)[\text{Arg}(z - p) - \text{Arg}(z - q)] = \dfrac{1}{\pi} \text{Arg}\left(\dfrac{z - p}{z - q}\right)$

defined on the upper half plane.

Answers: $g_1(z) = -ig_2(z)$. (i) $g_2(z) = \text{Log}(z - 1) - \text{Log}(z + 1) + (i\pi/2) = \text{Log}\left(\dfrac{z - 1}{z + 1}\right) + \dfrac{i\pi}{2}$; (ii) $g_2(z) = \dfrac{1}{\pi} \text{Log}\left(\dfrac{z - p}{z - q}\right)$.

5. Use the Cauchy-Riemann equations (and the comments of Section 2.10) to prove that an analytic function $f(z)$ is constant on E if (i) E is a *domain*, and (ii) $f(z)$ has purely imaginary values on E.

6. If $G_1(z)$ and $G_2(z)$ are conjugate harmonic functions for a harmonic function $H(z)$ on a *domain* (connected open set), show that there is a real constant c such that $G_2(z) = G_1(z) + c$ for all z.

Hint: $g_k = H + iG_k (k = 1, 2)$ are analytic; $f(z) = g_1(z) - g_2(z)$ has *purely imaginary* values.

7.3 SOME APPLICATIONS OF CONJUGATE HARMONIC FUNCTIONS

The previous theorems on the existence of conjugate harmonic functions can be used to demonstrate a number of properties of harmonic functions on the plane. Here are a few examples.

1. The Mean Value Property. If $U(z) = U(x, y)$ is harmonic, then the average value of U taken over any small circle about a point p is always equal to the value $U(p)$ at the center of the circle. Thus, for all small radii $r > 0$,

$$\frac{1}{2\pi} \int_0^{2\pi} U(p + re^{i\theta})\, d\theta = U(p) \tag{8}$$

(see remarks in Section 5.12 on defining average values). In fact, if D is any disc centered at p, on which $U(z)$ is defined, then D is simply connected and U is the real part of some analytic function $f(z) = U(z) + iV(z)$ on D. Since $f(z)$ has the mean value property on circles $|z - p| = r$ within this disc, we get

$$U(p) + iV(p) = f(p) = \frac{1}{2\pi}\int_0^{2\pi} f(p + re^{i\theta})\, d\theta$$
$$= \frac{1}{2\pi}\int_0^{2\pi} U(p + re^{i\theta})\, d\theta + i\frac{1}{2\pi}\int_0^{2\pi} V(p + re^{i\theta})\, d\theta.$$

We get formula (8) by comparing real parts in this expression.

The mean value property is actually characteristic of harmonic functions; one can prove the following result.

Theorem: *Any function $U(z)$ that is defined and continuous on a domain D, and has the mean value property* (8), *must be harmonic on D. In particular, U must have continuous first and second partial derivatives.*

We will not prove this here. The mean value property is also valid for harmonic functions in higher dimensions; for example, in the three dimensional case, the average values are taken by integrating U over the surfaces of small spheres centered at p. Thus the mean value property is still a useful tool in studying general harmonic functions, when complex variable methods are no longer available.

2. *Harmonic Functions Are Infinitely Differentiable.* In defining the notion of a harmonic function for two variables, we assume only that $U(x, y)$ and its partial derivatives of first and second orders exist and are continuous. Actually, a function that is harmonic will automatically have continuous partial derivatives of *all* orders. If we consider a harmonic function $U(x, y)$ on a small disc about a typical point p, there is a well defined harmonic conjugate function $V(x, y)$ such that $f(z) = U(z) + iV(z)$ is complex differentiable, and hence analytic. Thus, $f(z)$ has continuous complex derivatives $f^{(n)}(z)$ of all orders. Let us now recall how the complex derivative of $f = U + iV$ can be computed in terms of partial derivatives of U and V,

$$\frac{df}{dz} = \frac{\partial U}{\partial x} + i\frac{\partial V}{\partial x} = \frac{\partial V}{\partial y} - i\frac{\partial U}{\partial y}.$$

The existence and continuity of df/dz insures the existence and continuity of the first partial derivatives of the component functions U and V. We can differentiate $f'(z)$ to get

$$\frac{d^2f}{dz^2} = \frac{\partial^2 U}{\partial x^2} + i\frac{\partial^2 V}{\partial x^2} = \frac{\partial^2 V}{\partial y\,\partial x} - i\frac{\partial^2 U}{\partial y\,\partial x}$$
$$= \frac{\partial^2 V}{\partial x\,\partial y} - i\frac{\partial^2 U}{\partial x\,\partial y} = -\frac{\partial^2 U}{\partial y^2} - i\frac{\partial^2 V}{\partial y^2};$$

existence and continuity of f'' insures the existence and continuity of all these second partial derivatives. In the next step we would differentiate $f''(z)$ to find that all third partial derivatives of U and V exist and are continuous. This process can be carried out inductively for higher order derivatives.

3. The Maximum (and Minimum) Property for Harmonic Functions.

If $H(z)$ is a real valued harmonic function on a domain E, then H has the following **local maximum property.**

> Let z^* be any point in E. Then $H(z)$ cannot achieve a relative maximum at z^* unless $H(z)$ is identically constant on some disc about z^*.†

To see this, suppose that $H(z)$ achieves a relative maximum at z^*. Let D be any disc about z^* on which $H(z) \leq H(z^*)$. On D, H is the real part of some analytic function $f(z) = H(z) + iH^*(z)$. Consider $g(z) = e^{f(z)}$, which is also analytic on D. Since g is analytic, its absolute value has the local maximum property (Section 3.8), so that $|g(z)| = |e^{f(z)}| = e^{\text{Re}(f(z))} = e^{H(z)}$ cannot achieve a relative maximum at z^* without being constant on D. But $y = e^x$ is a *strictly increasing* function of the real variable x, so that

(i) $H(z)$ has a relative maximum at z^* if and only if $e^{H(z)}$ does;
(ii) $H(z)$ is constant near z^* if and only if $e^{H(z)}$ is constant near z^*.

Thus, $H(z)$ must be constant on some disc about z^* if it has a relative maximum at z^*.

The local maximum property for harmonic functions leads directly to the following *global* maximum principle.

Theorem 7.2 *Let E be a bounded domain. Assume that $H(z)$ is a real valued function that is*

(i) *continuous on the closure $\bar{E} = E \cup bdry(E)$*
(ii) *harmonic within E.*

Let M be the maximum value achieved by $H(z)$ on the closed bounded set E. Then, unless $H(z)$ is identically constant throughout E, we have

(iii) *$H(z) < M$ for all z in E.*

In particular, the maximum M can be achieved only on the boundary set $bdry(E)$, unless $H(z)$ is constant.

Proof: A continuous function on a closed bounded set has a maximum value that is achieved at some point in the set (this is why we must assume that E is *bounded*). Naturally, it will be necessary to invoke the *connectedness* of E (a domain is connected by definition) to get our result. Suppose there is some point z^* in E for which $H(z^*) = M$; we will be finished if we can show that this premise leads to the conclusion that $H(z) = M$ throughout E.

Let X be the set of points for which $H(z) = M$, and let Y be the complement $Y = E \sim X = \{z : z \text{ is in } E \text{ and } H(z) < M\}$. Then X and Y are disjoint sets whose union is all of E, and X is non-empty since it includes at least the point z^*. Now Y is an *open* set because $H(z)$ is continuous; if $H(z_0) < M - \delta$,

† Here *relative maximum* means that $H(z) \leq H(z^*)$ for all points in some disc about z^*.

then for all z sufficiently close to z_0 we get $H(z) < M - (\delta/2) < M$, so that all of the nearby points are in Y, too. Furthermore, X is an open set, because if $H(z_0) = M$, we must have $H(z) = M$ on some disc about z_0, by the local maximum property, and this disc lies within X. If both of the open sets X and Y are non-empty, then E cannot be connected; thus, the only possibility is that Y is empty and $X = E$, so that $H(z) = M$ on E. ▮

There is also a **minimum principle** for harmonic functions. Just replace the word "maximum" with the word "minimum" in Theorem 7.2, and the inequality "$H(z) < M$" with "$H(z) > M$," in statement (iii). There is no need to repeat the proofs; proving a "minimum" statement for a harmonic function $H(z)$ is equivalent to proving a "maximum" statement for the harmonic function $-H(z)$. Hereafter, we will use the phrase "maximum principle" to refer to either the maximum *or* minimum principle.

We can use the maximum principle to demonstrate the uniqueness of a harmonic function on a bounded domain, whose boundary values are specified.

Example 7.8 Let E be a bounded domain and assume that $H_1(z)$ and $H_2(z)$ are functions that are defined and continuous on the closure $\bar{E}$ of E, and are harmonic within E. Assume that H_1 and H_2 have the same values on the boundary, so $H_1(z) = H_2(z)$ for every z in bdry(E). Then apply Theorem 7.2 to the function $H(z) = H_1(z) - H_2(z)$ to show that $H_1(z) = H_2(z)$ throughout the domain E. In other words, the boundary values uniquely determine the values of a harmonic function within the domain E, if they are continuous on bdry(E).

EXERCISES

1. Consider the harmonic function $H(x, y) = y$ on the *unbounded* domain $E = \{z : \text{Im}(z) > 0\}$. Show that the maximum principle is not true for this domain. The *local* maximum property must be valid in E; explain why it does not lead to the global property discussed in Theorem 7.2.

2. By examining $H(x, y) = x = \text{Re}(z)$, show that a harmonic function defined on the plane can be zero at an infinite sequence of points z_n that converge to a point z_0 (where H is harmonic), without being identically zero near z_0. Compare this with the behavior of an analytic function (Theorem 3.17).

3. Prove that harmonic functions have the following "continuation" property.

Theorem: Suppose $H(z)$ is harmonic on a domain E, with $H(z) = 0$ on a small disc contained in E. Then $H(z) = 0$ throughout E.

Thus, $G = H$ everywhere if G and H are harmonic on a domain, and $G = H$ near a point p in the domain (throughout a small disc about p). This continuation theorem is not as strong as Theorem 3.19 (for analytic functions); nevertheless, it is very useful in potential theory. It is valid for harmonic functions of many variables ($n \geq 3$).

Hint: The connectedness of E must be used. Examine the subsets $X = \{z : z \text{ in } E \text{ and } H(z') = 0 \text{ for } z' \text{ near } z\}$, and $Y = E \sim X$. Prove that Y is empty as in proof of Theorem 7.2.

4. Suppose $H(z)$ is harmonic on $E = \{z : 1 < |z| < \infty\}$ and continuous on the closure $\bar{E} = \{z : 1 \leq |z| < \infty\}$, with

(i) $\lim_{z \to \infty} |H(z)| = 0$

(ii) $H(z) = 0$ on the boundary circle $|z| = 1$.

Use the maximum principle to prove that $H(z) = 0$ is the *only* harmonic function that satisfies these boundary conditions.

Hint: Consider subdomains of the form $1 < |z| < R$ and let $R \to +\infty$.

5. (Reflection principle for harmonic functions) Suppose that $H(z)$ is harmonic on a domain E, and let $J(z) = \bar{z}$ be the conjugation map. By "reflection" we may define a new function $H^*(z) = -H(\bar{z})$ on $E^* = J(E)$. Prove that H^* is harmonic on the reflected domain E^*.

7.4 TRANSFORMATION OF FUNCTIONS BY AN ANALYTIC CHANGE OF VARIABLE

The next result has great significance in the study of Laplace's equation in the plane, and clearly illustrates the use of complex analysis in these problems. Our result expresses the fact that Laplace's equation is "invariant" under analytic changes of variable. This is also known as the **conformal mapping principle,** due to the close connection between analytic mappings and conformal mappings of the plane (see Section 4.4).

The situation we want to consider is shown in Figure 7.4; we are given two domains D and E and an analytic mapping $f : D \to E$ which maps D onto E. Let us write the variable in D as $z = x + iy$ and the variable in E as $w = u + iv$, to distinguish what is going on in these domains. We can think of f as effecting an analytic change of variable $w = f(z)$; this allows us to express the u,v-coordinates in the w-plane in terms of the x,y-coordinates of the z-plane:

$$u = U(x, y) = \text{Re}(f(x + iy))$$
$$v = V(x, y) = \text{Im}(f(x + iy)).$$

If we are given a function $H(w) = H(u, v)$ of two real variables defined on E, then the change of variable $w = f(z)$ naturally leads to a function of the variables x and y defined on D, namely $\tilde{H}(x, y) = \tilde{H}(z) = H(f(z))$. The relationship between the functions $H(w)$ and $\tilde{H}(z)$, defined on domains E and

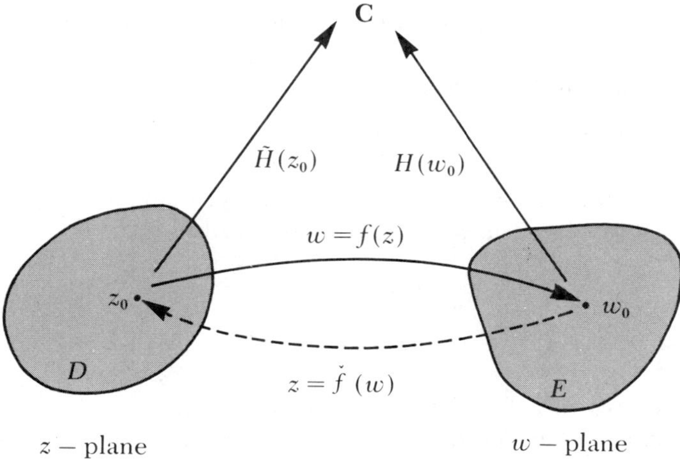

Figure 7.4 Function $H(w)$ on E transformed to a function $\tilde{H}(z)$ on D.

D respectively, is indicated in Figure 7.4. If w is the image of z, so that $w = f(z)$, then $H(w)$ and $\tilde{H}(z)$ have the same value, $H(w) = \tilde{H}(z)$.

Now let us suppose that the mapping $f: D \to E$ is regular (nonvanishing derivative) and invertible, so that the mapping f and the inverse mapping $\check{f}: E \to D$ are both analytic. Then the changes of variable $w = f(z)$ and $z = \check{f}(w)$ are analytic in each direction. This means that a function $H(w) = H(u, v)$ that is defined on E can be transformed into a function $\tilde{H}(z) = \tilde{H}(x, y)$ defined on D by the change of variable $w = f(z)$, while a function $H(z)$ defined on D can be transformed in the reverse direction to get a function $\tilde{H}(w)$, by making the inverse change of variable $z = \check{f}(w)$.

Let us illustrate this transformation of functions for the explicit change of variable $w = f(z) = e^z$. To get an analytic mapping that is invertible, we shall restrict z to the strip $D = \{z: -\pi < \text{Im}(z) < +\pi\}$, shown in Figure 7.5. Then, f maps D invertibly onto the cut plane E obtained by deleting the negative real axis $(-\infty, 0]$ from the w-plane. The inverse mapping $z = \check{f}(w)$ is just the principal determination $z = \text{Log } w$. Substituting $w = f(z)$ in a function

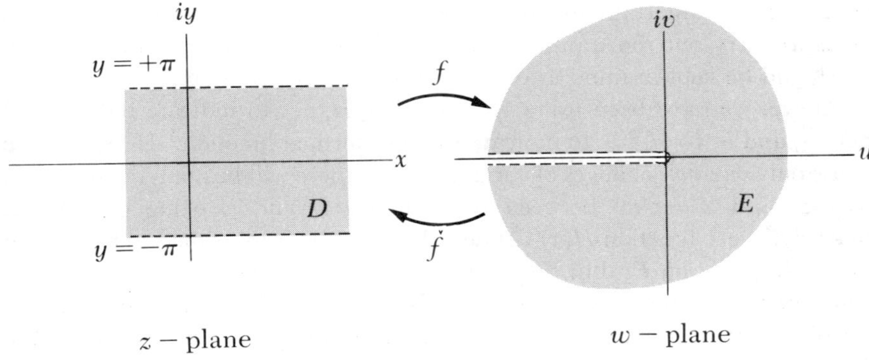

Figure 7.5 The change of variable $w = e^z$.

$H(w)$ defined on E means that we are to replace the real variables u, v with

$$u = \text{Re}(w) = \text{Re}(f(z)) = \text{Re}(e^{x+iy}) = e^x \cos y$$
$$v = \text{Im}(w) = \text{Im}(f(z)) = \text{Im}(e^{x+iy}) = e^x \sin y$$

to get the transformed function $\tilde{H}(z) = \tilde{H}(x, y)$. For example, the polynomial $H(u, v) = u^2 + uv + v^2$, defined on E, is transformed to

$$\begin{aligned}\tilde{H}(x, y) &= e^{2x} \cos^2 y + e^{2x} \sin y \cos y + e^{2x} \sin^2 y \\ &= e^{2x} + e^{2x} \sin y \cos y = e^{2x}(1 + \sin y \cos y).\end{aligned}$$

In order to transform a function $H(z) = H(x, y)$ defined on D in the other direction to obtain a function $\tilde{H}(w) = \tilde{H}(u, v)$ defined on E, we must make the inverse substitutions:

$$x = \text{Re}(z) = \text{Re}(\check{f}(w)) = \text{Re}(\text{Log}\, w) = \log |w| = \log(\sqrt{u^2 + v^2})$$
$$y = \text{Im}(z) = \text{Im}(\check{f}(w)) = \text{Im}(\text{Log}\, w) = \text{Arg}(u + iv) = A(u, v).$$

If we want a more explicit formula for y, we can express $\text{Arg}(u + iv)$ in terms of the inverse functions $\arcsin t$ and $\arccos t$ (with real variable t) in different parts of E, as in Section 2.6. A polynomial such as $H(x, y) = x^2 - y + 1$, defined on the strip D, is then transformed to the following function on the cut plane E by the change of variable $z = \check{f}(w)$:

$$\tilde{H}(u, v) = (\log \sqrt{u^2 + v^2})^2 + 1 - \text{Arg}(u + iv).$$

On the right half plane, where $\text{Arg}(u + iv) = \arcsin(v/\sqrt{u^2 + v^2})$, we could write $\tilde{H}(u, v)$ in a corresponding explicit form; slightly different versions of this can be used in other parts of E.

Other examples of transforming functions by making a change of variable are given in Exercises 1 to 3. The reader should take particular notice of the following fact. Transforming a function by means of a change of variable also transforms its domain of definition, and in some uses of this transformation process it is as important to keep track of what is happening to the domains as it is to carry out the transformation of the functions we are interested in. This should be kept in mind as other examples are worked out.

Above, we have been using symbols F, G, H, . . . to indicate the original function, and $\tilde{F}$, $\tilde{G}$, $\tilde{H}$, . . . to indicate the transformed function. Hereafter, we will use only *invertible* changes of variable; thus, there will be analytic mappings $w = f(z)$ and $z = \check{f}(w)$ between the domains D and E being considered. Obviously, every function $H(z)$ defined on D can be transformed to a function $\tilde{H}(w) = H(\check{f}(w))$ on E; but the converse is also true! Since functions can be transformed in either way, distinctions between the "original" and "transformed" functions are a matter of convention in any particular situation. We will continue to use paired symbols, such as H and $\tilde{H}$, to indicate functions that are transforms of one another by the change of variable under discussion. No special significance will be attached to which function carries the tilde, nor is it important which domain is assigned the variable z, and which the variable w.

EXERCISES

1. Transform $H(x, y)$ (defined on the upper half plane) to $\tilde{H}(u, v)$ defined on the first quadrant, using the transformation of domains $w = z^{1/2}$.

(i) $H(x, y) = y + 1$ (iii) $H(x, y) = x$

(ii) $H(x, y) = 1/y$ (iv) $H(x, y) = \sin x \cdot \sin y$

Hint: The inverse is $z = w^2$.

Answers: (i) $2uv + 1$; (ii) $1/2uv$; (iii) $u^2 - v^2$; (iv) $\sin(u^2 - v^2) \cdot \sin(2uv)$.

2. The exponential $w = e^z$ maps the strip $E = \{z: -\pi < \text{Im}(z) < +\pi\}$ to the cut plane $F = \mathbf{C} \sim (-\infty, 0]$. Transform functions $H(x, y)$ on the strip,

(i) $H(x, y) = \cos(y/2)$
(ii) $H(x, y) = \cos(y/2) \cdot \sin x$,

to functions $\tilde{H}(u, v)$ on F. Express the transformed function in polar coordinates $\tilde{H}(r, \theta)$ for $r > 0$; $-\pi < \theta < +\pi$.

Answers: (i) $\tilde{H}(r, \theta) = \cos(\theta/2)$; (ii) $\tilde{H}(r, \theta) = \sin(\log r) \cdot \cos(\theta/2)$.

3. The fractional linear transformation $w = f(z) = i\dfrac{z - i}{z + i}$ $\left(\text{whose inverse is } z = \check{f}(w) = \dfrac{1}{i}\dfrac{w + i}{w - i}\right)$ transforms the upper half plane $E = \{z: \text{Im}(z) > 0\}$ onto the unit disc $D = \{w: |w| < 1\}$. Determine the function $\tilde{H}(u, v)$ on the disc we get by transforming the following harmonic functions $H(x, y)$ defined on E.

(i) $H(x, y) = y = \text{Re}(-iz)$

(ii) $H(x, y) = \dfrac{-y}{x^2 + y^2} = \text{Re}(1/z)$.

Answers:

$$\text{(i)}\quad \tilde{H}(u, v) = \frac{1 - (u^2 + v^2)}{u^2 + (v - 1)^2} = \frac{1 - |w|^2}{|w - i|^2};$$

$$\text{(ii)}\quad \tilde{H}(u, v) = \frac{2u}{u^2 + (v + 1)^2} = \frac{2u}{|w + i|^2}$$

4. Suppose we use a mapping $w = \phi(z)$, such that $\phi : D \to E$, to transform functions $H(w)$ defined in E to functions $\tilde{H}(z) = [H(w)|_{w=\phi(z)}]$ on D. Show that the following relations are valid:

(i) $(f + g)^{\sim}(z) = \tilde{f}(z) + \tilde{g}(z)$
(ii) $(\alpha f)^{\sim}(z) = \alpha \cdot \tilde{f}(z)$ (α a complex scalar)
(iii) $(f \cdot g)^{\sim}(z) = \tilde{f}(z) \cdot \tilde{g}(z)$.

7.5 THE CONFORMAL MAPPING PRINCIPLE FOR HARMONIC FUNCTIONS

Our main result says that a harmonic function remains harmonic after any analytic change of variable.

Theorem 7.3 *Let $f: D \to E$ be an analytic mapping which carries domain D into domain E. If $H(w) = H(u, v)$ is any solution of Laplace's equation in E:*

(9A)
$$\frac{\partial^2 H}{\partial u^2} + \frac{\partial^2 H}{\partial v^2} = 0 \qquad \textit{everywhere on } E,$$

then the transformed function $\tilde{H}(z) = \tilde{H}(x, y)$ on D, obtained by making the change of variable $u + iv = f(x + iy)$, is a solution of Laplace's equation on D:

(9B)
$$\frac{\partial^2 \tilde{H}}{\partial x^2} + \frac{\partial^2 \tilde{H}}{\partial y^2} = 0 \qquad \textit{everywhere on } D.$$

Proof: (Refer to Figure 7.4). We must show that (9B) is satisfied at every point z_0 in D. Let $w_0 = f(z_0)$. Since $H(u, v)$ is harmonic at w_0, there is a conjugate harmonic function $G(u, v)$, defined near w_0, by Theorem 7.1. Thus $h(w) = h(u + iv) = H(u, v) + iG(u, v)$ is analytic at and near w_0. Now define the composite function $\tilde{h}(z) = (h \circ f)(z) = h(f(z))$; this function is well defined for z near z_0 and it is analytic too, being a composite of analytic functions.

We obtain the transform $\tilde{H}(z)$ of the function $H(w)$ by making the substitution $w = f(z)$, so that $\tilde{H}(z) = H(f(z))$ for z near z_0; similarly, $G(w) = G(u, v)$ is transformed to the function $\tilde{G}(x, y) = \tilde{G}(z) = G(f(z))$. Obviously,

$$\tilde{h}(z) = h(f(z)) = H(f(z)) + iG(f(z)) = \tilde{H}(z) + i\tilde{G}(z)$$

for z near z_0. Thus, $\tilde{H}(z) = \text{Re}(\tilde{h}(z))$, and $\tilde{h}(z)$ is analytic at z_0. We conclude that $\tilde{H}(z)$ is harmonic at z_0 (a typical point in D). ▮

In this theorem it is not necessary that the mapping f be invertible. However, if the analytic mapping $f: D \to E$ has an analytic inverse $\check{f}: E \to D$, our theorem applies to the change of variable in each direction. Therefore, these mappings transform harmonic functions on E to harmonic functions on D, and *vice versa*. This is summed up in the following theorem.

Theorem 7.4 (Conformal mapping principle) *Let $w = f(z)$ be an analytic mapping between two domains D and E that is invertible, and has analytic inverse $z = \check{f}(w)$. Then a function $H(w) = H(u, v)$, defined on E, is a solution of Laplace's equation throughout E if and only if the transformed function $\tilde{H}(z) = \tilde{H}(x, y)$ is a solution of Laplace's equation throughout D.*

Suppose we wish to solve Laplace's equation in some geometrically complicated domain E in the complex plane. We can reduce our task to one of solving Laplace's equation in a simpler domain, such as a half plane or an open disc, if we can recognize when the given region E can be mapped to a convenient standard domain D; we must also be able to calculate the mapping f or $\check{f}$ which accomplishes this task. As we go along, we will gradually develop a catalog of mappings between specific domains. At present we shall give only a few elementary examples.

Example 7.9 Consider the harmonic function $H(w) = e^u \cos v = \mathrm{Re}(e^w)$ defined on the w-plane. The analytic mapping $w = f(z) = z^3$ maps the z-plane onto the w-plane, and the substitution $w = z^3 = (x + iy)^3 = (x^3 - 3xy^2) + i(3x^2y - y^3)$ transforms $H(w) = H(u, v)$ into a corresponding harmonic function in the z-plane $\tilde{H}(z) = \tilde{H}(x, y) = \exp(x^3 - 3xy^2) \cdot \cos(3x^2y - y^3)$. Since we have applied only Theorem 7.3 to get this result, the mapping $w = f(z)$ *need not be invertible*, and in fact $w = z^3$ is not a one-to-one mapping. We would need an invertible mapping if we wished to transform functions in the other direction, converting a function $H(z)$ into a function $\tilde{H}(w)$.

Example 7.10 For an example that illustrates the conformal mapping of domains, let us take D to be the first quadrant in the z-plane, as shown in Figure 7.6. If we recall that $h(z) = z^2 = (x^2 - y^2) + i(2xy)$, we see that $N(x, y) = x^2 - y^2$ and $H(x, y) = 2xy$ are harmonic on D. They satisfy the Neumann and Dirichlet conditions $\partial N/\partial n = 0$ and $H = 0$, respectively, on the boundary of D. This can be verified by direct calculations, or by recalling that the vector $\nabla N =$ **grad** N at a point p is always perpendicular to the level curve $N =$ *constant* that passes through p. The level curves $N =$ *constant* and $H =$ *constant* are shown in Figure 7.6; it is obvious that the level curves of N

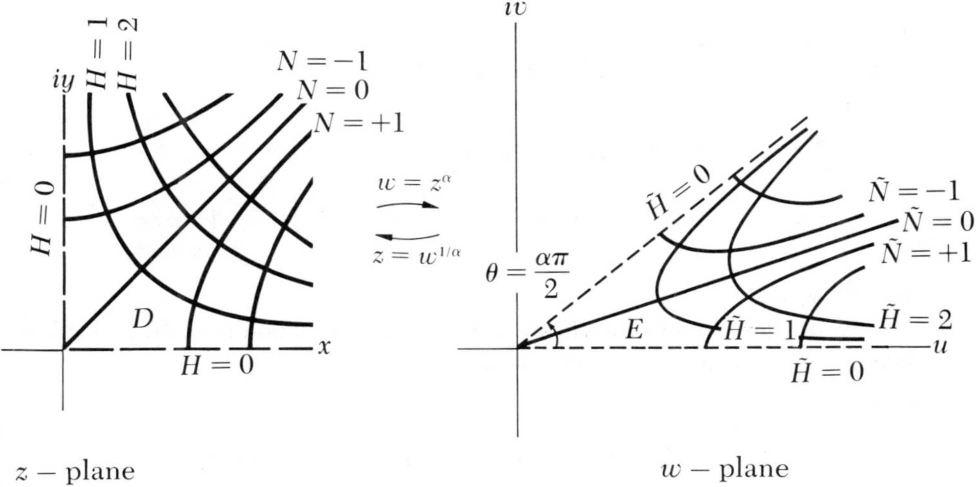

Figure 7.6 The conformal mappings $w = f(z) = z^\alpha$ and $z = \check{f}(w) = w^{1/\alpha}$, and their actions on the functions $H(x, y) = 2xy$ and $N(x, y) = x^2 - y^2$.

are perpendicular to the boundary of D, so that $\nabla N \cdot \mathbf{n} = \partial N/\partial n = 0$ (except at the origin, where there is no well defined normal direction $\mathbf{n}$ and $\partial N/\partial n$ is undefined).

Suppose we wish to solve the corresponding problems for the wedge shaped domain with an opening of $\pi\alpha/2$ radians $(0 < \alpha < 1)$, as in Figure 7.6. There is an obvious analytic invertible mapping of D onto E, namely

$$w = f(z) = z^\alpha = e^{\alpha \cdot \operatorname{Log} z}.$$

The inverse mapping is given by

$$\begin{aligned} z = \check{f}(w) = w^{1/\alpha} = e^{(1/\alpha)\cdot \operatorname{Log} w} &= \exp\left(\frac{\log |w|}{\alpha} + i\frac{\operatorname{Arg} w}{\alpha}\right) \\ &= |w|^{1/\alpha} \cdot \exp\left(\frac{i}{\alpha}\operatorname{Arg} w\right). \end{aligned}$$

The function $h(z) = z^2$, defined on D, is transformed to a new analytic function

$$\tilde{h}(w) = h(\check{f}(w)) = (w^{1/\alpha})^2 = w^{2/\alpha} = \tilde{N}(u, v) + i\tilde{H}(u, v),$$

defined for $w = u + iv$ in the wedge E. In the special case when $\alpha = \frac{1}{2}$ (so that the angle in the wedge is $\theta = \pi/4$), $h(z) = z^2$ is transformed to $\tilde{h}(w) = (w^{1/\alpha})^2 = w^4$; the real and imaginary parts are the harmonic functions

$$\text{(10)} \qquad \begin{aligned} \tilde{N}(u, v) &= u^4 - 6u^2v^2 + \\ \tilde{H}(u, v) &= 4uv(u^2 - v^2) \end{aligned} \qquad \text{for } w = u + iv \text{ in } E.$$

If the exponent α is not a simple fraction (for example, if $\alpha = \sqrt{2}/2$), the expressions for the real and imaginary parts are more complicated:

$$\begin{aligned} \tilde{N}(u, v) &= \operatorname{Re}(w^{2/\alpha}) = (\sqrt{u^2 + v^2})^{2/\alpha} = (u^2 + v^2)^{1/\alpha} \\ \tilde{H}(u, v) &= \operatorname{Im}(w^{2/\alpha}) = (2/\alpha)\operatorname{Arg}(u + iv) = (2/\alpha)\arccos\left(\frac{u}{\sqrt{u^2 + v^2}}\right) \end{aligned}$$

(see Exercise 1).

In the next section we will see that the transformation from H, N to $\tilde{H}$, $\tilde{N}$ leaves the boundary conditions unaltered, so that

$$\frac{\partial \tilde{N}}{\partial n}(w) = 0 \qquad \text{and} \qquad \tilde{H}(w) = 0 \quad \text{on bdry}(E);$$

this can also be verified by direct calculations (Exercise 2).

Suppose that z in D and w in E correspond under the transformations f and $\check{f}$; then $w = f(z)$ and $z = \check{f}(w)$, and the values

$$N(z) + iH(z) = h(z) = c + id$$

$$\tilde{N}(w) + i\tilde{H}(w) = \tilde{h}(w) = h(\check{f}(w)) = h(z) = c + id$$

are equal. By comparing real and imaginary parts, we see that the value of $\left\{\begin{matrix} h \\ N = \mathrm{Re}(h) \\ H = \mathrm{Im}(h) \end{matrix}\right\}$ at z is the same as the value of $\left\{\begin{matrix} \tilde{h} \\ \tilde{N} = \mathrm{Re}(\tilde{h}) \\ \tilde{H} = \mathrm{Im}(\tilde{h}) \end{matrix}\right\}$ at w if z and w correspond under the given transformations between D and E. Thus,

(11) The locus $N(z) = c$ is mapped by $w = f(z)$ to the locus $\tilde{N}(w) = c$

(c a real constant), and similar results hold for the loci $H(z) = d$ and $\tilde{H}(w) = d$ (d a real constant). The inverse mapping $z = \check{f}(w)$ maps loci in the opposite direction. The reader should examine Figure 7.6 carefully, comparing the patterns of loci $H = c$, $N = d$ and $\tilde{H} = c$, $\tilde{N} = d$ in the respective domains D and E.

In the last example it is easy to guess solutions of the Dirichlet and Neumann problems for the standard domain D; it would be much more difficult to guess the solutions (10) to the corresponding problems for the wedge E. We have used the conformal mapping principle to transform the problem in E to a solvable problem for the standard domain D.

In Chapter 9 we will discuss the *Riemann Mapping Theorem*, which shows that every simply connected domain, except for the special (and trivial) domain $E = \mathbf{C}$ (the whole complex plane), can be mapped to the open unit disc $D = \{z\colon |z| < 1\}$ by an invertible analytic mapping. Therefore, every question about harmonic functions on a simply connected domain can be converted, by means of the mapping principle, into a corresponding problem for the unit disc.

EXERCISES

1. Work out the formulas for real and imaginary parts of $w^{2/\alpha}$,

$$\tilde{H}(w) = \mathrm{Re}(w^{2/\alpha}) = (u^2 + v^2)^{1/\alpha}$$

$$\tilde{N}(w) = \mathrm{Im}(w^{2/\alpha}) = \left(\frac{2}{\alpha}\right)\arccos\frac{u}{\sqrt{u^2 + v^2}}$$

for $0 < \alpha < 1$ and $w = u + iv$ in the wedge-shaped domain discussed in Example 7.10. Reconcile these formulas with the elementary formulas (10) obtained when $\alpha = 1/2$.

2. In Example 7.10 verify, by direct calculation, that

$$\tilde{H}(w) = 0 \qquad \text{and} \qquad \frac{\partial \tilde{N}}{\partial n} = 0$$

on the boundary of the domain.

7.6 TRANSFORMATION OF BOUNDARY CONDITIONS

If we want to solve a boundary value problem for a certain domain E, using the conformal mapping principle, we must know how the boundary behavior of a harmonic function is transformed when we make an analytic change of variable. Otherwise, we would not know which boundary conditions to impose on bdry(D) in order that solutions of the new boundary value problem, in the standard domain D, correspond to solutions of the original boundary value problem, posed for E. We will find that boundary conditions of the Dirichlet type, and the commonly encountered *homogeneous* Neumann condition $\partial H/\partial n = 0$, both transform to boundary conditions of the same kind. Other boundary conditions, including the non-homogeneous Neumann condition $\partial H/\partial n = f(z)$, transform in more complicated ways. A detailed and mathematically rigorous account of these matters would involve us in some very technical questions about boundary behavior of harmonic functions, particularly in studying Neumann problems. We will not attempt such a highly detailed account in this book. This section is intended as a brief introduction, adequate for dealing with the concrete examples we will encounter later on; technical fine points will be deliberately overlooked. For an account that squarely faces some of the technical difficulties, the interested reader might consult De Pree and Oehring [5], Sections 63 to 65, or Nevanlinna and Paatero [18], Chapters 11 and 17.

Let D and E be domains in the z-plane and w-plane, respectively, and let $w = f(z)$ be an invertible analytic mapping from D onto E. Let us also assume that $w = f(z)$ and $z = \check{f}(w)$ are defined and analytic on the boundaries. Then it makes sense to say that points z and w "correspond to each other" under the given mappings whenever $w = f(z)$, or equivalently, $z = \check{f}(w)$. In the Dirichlet problem for D we are given piecewise continuous boundary values $h(z)$, defined on bdry(D), and we want a function $H(z)$ which is continuous on the closure $\bar{D}$ and harmonic within D, such that

$$H(z) = h(z) \quad \text{for all } z \text{ on bdry}(D). \tag{12}$$

What values should a function $\tilde{H}(w)$, defined on E, have so that the transformed function $H(z) = \tilde{H}(w) = \tilde{H}(f(z))$ will be a solution of the original problem (12) in D? Since $H(z)$ and $\tilde{H}(w)$ have the same value at points that correspond under the mappings f and $\check{f}$, it should be clear that the boundary values for $\tilde{H}(w)$, appropriate to E, are given by

$$\tilde{h}(w) = \left[h(z) \Big|_{z=\check{f}(w)} \right] = h(\check{f}(w)) \quad \text{for } w \text{ on bdry}(E).$$

That is, $h(z) = \tilde{h}(w)$ if z and w are corresponding boundary points of D and E, respectively. In particular, if we start with an elementary boundary condition of the form

(13) $$H(z) = c \quad (c \text{ some real constant})$$

on part of bdry(D), we must impose the same condition $\tilde{H}(w) = c$ on the corresponding part of bdry(E). Such a transformation of Dirichlet boundary conditions has already been illustrated, in Example 7.10 of Section 7.5 (notice how level curves $H = \textit{constant}$ are transformed by the mappings $w = f(z)$ and $z = \check{f}(w)$ in that example). Here is a more complicated example, typical of the mapping problems that appear in physical applications. It illustrates the solution of a nontrivial mapping problem, together with the transformation of Dirichlet boundary conditions.

Example 7.11 The fractional linear transformation

$$w = f(z) = \frac{1}{i}\left(\frac{z+i}{z-i}\right)$$

maps the unit disc $D = \{z: |z| < 1\}$ conformally onto the upper half plane $E = \{w: \mathrm{Im}(w) > 0\}$ as we indicated in Example 4.16 of Section 4.8. Straightforward algebraic calculations show that $z = \check{f}(w) = i\left(\frac{w-i}{w+i}\right)$. The mappings f and $\check{f}$ are clearly analytic. Let us solve the following boundary value problem in D,

$$H(z) = +1 \quad \text{on the semicircular boundary arc } C_2$$

$$H(z) = 0 \quad \text{on the semicircular boundary arc } C_1,$$

as indicated in Figure 7.7. The boundary values are discontinuous at $z = +i$ and $z = -i$; let us insist that the solution $H(z)$ be bounded as $z \to +i$ or

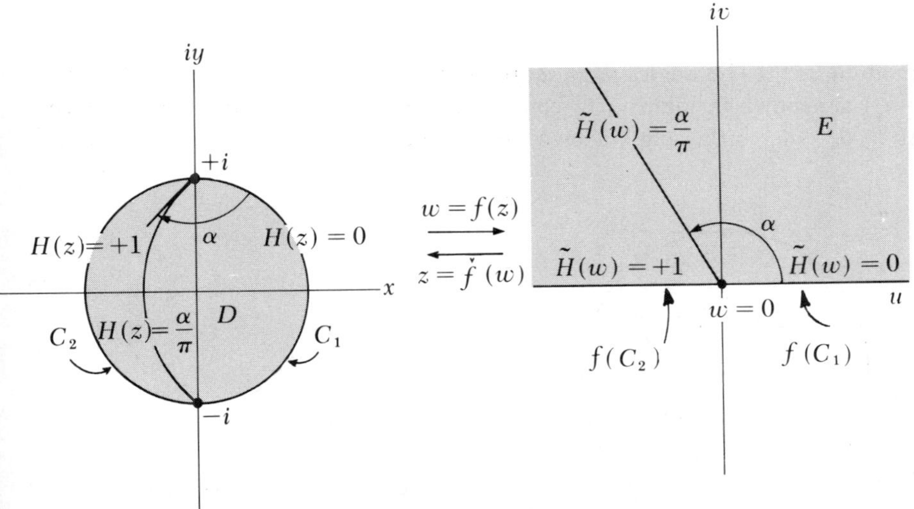

Figure 7.7 The domains and mappings in Example 7.11.

$z \to -i$ from within the disc. In the domain E we should impose the corresponding boundary conditions; since f maps the arc C_2 to the negative real axis $(-\infty, 0)$, and the arc C_1 to the positive real axis $(0, +\infty)$, in the w-plane, we should require that the solution function $\tilde{H}(w)$ on the domain E satisfy the conditions

$$\tilde{H}(u + i0) = +1 \quad \text{for} \quad u < 0$$

$$\tilde{H}(u + i0) = 0 \quad \text{for} \quad u > 0.$$

We should also have $|\tilde{H}(w)|$ bounded as $w \to 0 = f(-i)$ and as $w \to \infty = f(+i)$, from within E, if the transformed harmonic function $H(z) = \tilde{H}(f(z))$ is to have the correct behavior at the exceptional points $z = +i$ and $z = -i$ on the boundary of the disc. The solution $\tilde{H}(w)$ we give does fulfill these additional boundedness requirements. We have already solved the problem for the half plane (see Section 7.1); just take

$$\tilde{H}(w) = \frac{1}{\pi} \operatorname{Arg}(u + iv) = \frac{1}{\pi} \arccos\left(\frac{u}{\sqrt{u^2 + v^2}}\right)$$

throughout the upper half plane. We transform this into a solution $H(z)$ of the original problem posed for the disc, by substituting $w = f(z) = -i\left(\dfrac{z + i}{z - i}\right)$ in $\tilde{H}(w)$ to get

$$H(z) = \frac{1}{\pi} \operatorname{Arg}\left[-i\left(\frac{z + i}{z - i}\right)\right] \quad \text{for} \quad |z| < 1.$$

Notice that

$$\arg\left[-i\left(\frac{z + i}{z - i}\right)\right] \equiv \arg(-i) + \arg(z + i) - \arg(z - i)$$

$$\equiv -\frac{\pi}{2} + \arg(z + i) - \arg(z - i)$$

(modulo 2π). The angles $\theta_1 = \operatorname{Arg}(z - i)$ and $\theta_2 = \operatorname{Arg}(z + i) = \operatorname{Arg}(z - (-i))$ are shown in Figure 7.8 (counterclockwise angles are positive); clearly, $\Delta\theta = \theta_2 - \theta_1$ is the angle shown, for a typical point z in the disc, and $\Delta\theta$

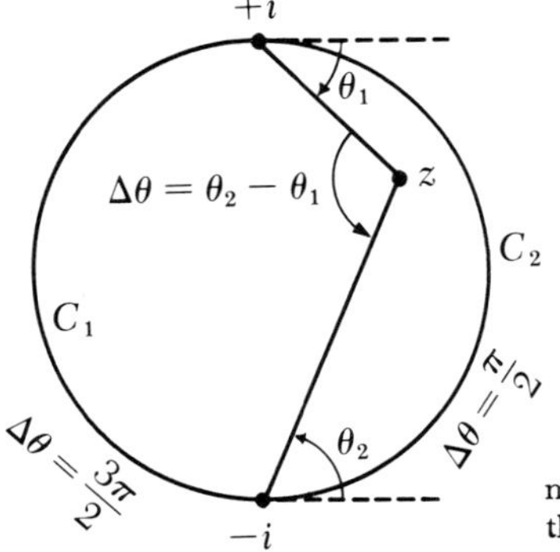

Figure 7.8 The angle $\Delta\theta$ is measured positively, moving counterclockwise from the segment $[z, +i]$ to the segment $[z, -i]$.

varies between $+\pi/2$ (on the right hand boundary arc) and $+3\pi/2$ (on the left hand boundary arc). Therefore, we have the *equality* (rather than congruence)

$$\text{(14)}\qquad H(z) = \frac{1}{\pi}\left[-\frac{\pi}{2} + \operatorname{Arg}(z+i) - \operatorname{Arg}(z-i)\right] = \frac{1}{\pi}\left[\Delta\theta - \frac{\pi}{2}\right];$$

there is no need to adjust the expression in brackets by adding multiples of 2π to get it to agree with $\operatorname{Arg}\left[-i\left(\frac{z+i}{z-i}\right)\right]$ for $|z| < 1$. Obviously, $H(z) = +1$ and $H(z) = 0$ on the appropriate parts of the boundary circle $|z| = 1$, as shown in Figure 7.8.

It would be difficult to arrive at this result by guesswork. This example illustrates the value of being able to transform the original problem for the disc into one for the half plane. One should notice that $\tilde{H}(w)$ is constant on radial lines in the half plane E, and that the ray $\arg(w) = \alpha$ is transformed into a circular arc as shown in Figure 7.7; the function $H(z)$ on the disc must have the same value on this image arc as $\tilde{H}(w)$ had on the original ray, namely $H(z) = \alpha/\pi$. Since $H(z)$ is constant on each circular arc passing through the points $z = +i$ and $z = -i$, it is easy to understand how the harmonic function $H(z)$ behaves within the disc $|z| < 1$. In particular, on the straight line segment $[+i, -i]$ we get $H(z) = +\frac{1}{2}$.

For another example, the reader should re-examine the boundary value problem (I) in Example 7.4 of Section 7.1, where we used $w = \sin z$ to transform a problem in a half strip in the z-plane to a familiar problem for the upper half of the w-plane.

Now let us examine Neumann type boundary conditions. If U and V are a conjugate pair of harmonic functions defined near p, then ∇U and ∇V are perpendicular at p (if they are both nonzero). In fact, the Cauchy-Riemann equations show that the inner product is zero:

$$\text{(15)}\qquad \nabla U \cdot \nabla V = \frac{\partial U}{\partial x}\cdot\frac{\partial V}{\partial x} + \frac{\partial U}{\partial y}\cdot\frac{\partial V}{\partial y} = \frac{\partial U}{\partial x}\left(-\frac{\partial U}{\partial y}\right) + \frac{\partial U}{\partial y}\cdot\frac{\partial U}{\partial x} = 0;$$

therefore, the vectors ∇U and ∇V (attached to p) are perpendicular, as shown in Figure 7.9. This observation about conjugate pairs of harmonic functions has interesting consequences in boundary value problems.

Application 1 Suppose that $F(z) = U(z) + iV(z)$ is analytic on a domain that includes a smooth arc segment Γ. Its real part U satisfies the elementary Dirichlet condition $U =$ *constant* on Γ if and only if its imaginary part V satisfies the homogeneous Neumann condition $\partial V/\partial n = 0$ on Γ.

This pairing of Dirichlet and Neumann problems will be discussed below, but first let us show why the conditions

(i) $U =$ *constant* on Γ

(ii) $\dfrac{\partial V}{\partial n} = 0$ on Γ

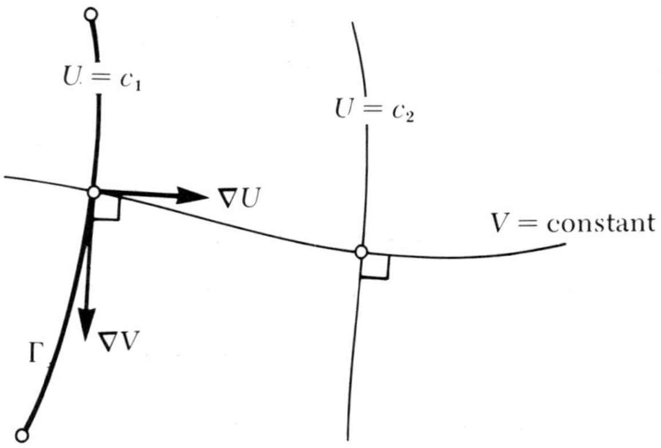

Figure 7.9

are equivalent. Since $\partial V/\partial n = \nabla V \cdot \mathbf{n}$, by definition, statement (ii) is equivalent to showing that ∇V is *tangent to* Γ at every point on this arc. Since the gradients ∇V and ∇U are perpendicular, this in turn is equivalent to the statement that ∇U is *perpendicular to* Γ at every point on this arc. But if $\phi(x, y)$ is *any* smooth function, its gradient vector $\nabla\phi$ is perpendicular to the level curve $\phi(z) = \textit{constant} = \phi(p)$ that passes through p. Applying this to $\phi = U$, we see that ∇U is perpendicular to Γ if $U = \textit{constant}$ on Γ (so that Γ is part of the locus $U = \textit{constant} = U(p)$). Thus, (i) implies (ii). We leave the converse implication, (ii) implies (i), as Exercise 18.

By combining U and V in a slightly different way, to form the analytic function $G = V - iU$, we could equally well prove that

$$\frac{\partial U}{\partial n} = 0 \text{ on } \Gamma \text{ if and only if } V = \textit{constant} \text{ on } \Gamma,$$

so that the relationship between the pair of functions U and V is symmetrical.

Application 2 Suppose that $w = f(z)$ is an invertible analytic mapping, with analytic inverse, that transforms a domain D to a domain E. Let $\tilde{\Gamma}$ be a smooth arc in D and $\Gamma = f(\tilde{\Gamma})$ its image in E under the mapping $f: D \to E$. Let $\tilde{H}(z)$ and $H(w)$ be harmonic functions defined on D and E, respectively, which are transforms of each other under the analytic changes of variable $w = f(z)$ and $z = \check{f}(w)$. Then $\tilde{H}(z)$ satisfies the homogeneous Neumann condition $\partial\tilde{H}/\partial n(z) = 0$ on the arc $\tilde{\Gamma}$ in D, if and only if $H(w)$ satisfies the same condition, $\partial H/\partial n(w) = 0$, on the corresponding arc Γ in E.

PROOF: Consider corresponding points $z = p$ and $w = q = f(p)$ on $\tilde{\Gamma}$ and Γ respectively. There are conjugate harmonic functions $\tilde{U}(z)$ and $U(w)$, defined near p and q respectively, that are related via the change of variable $w = f(z)$; thus, $\tilde{F}(z) = \tilde{H} + i\tilde{U}$ and $F(w) = H + iU$ are analytic. The conditions $\partial\tilde{H}/\partial n(z) = 0$ and $\partial H/\partial n(w) = 0$ are equivalent to the conditions $\tilde{U}(z) = \textit{constant}$ and $U(w) = \textit{constant}$ on the respective arcs $\tilde{\Gamma}$ and Γ, as indicated in

Application 1. But $\tilde{U}(z) = constant$ on $\tilde{\Gamma}$ if and only if $U(w) = constant$ on Γ. Therefore, the Neumann condition $\partial\tilde{H}/\partial n(z) = 0$ corresponds to the Neumann condition $\partial H/\partial n(w) = 0$ under an invertible analytic mapping of domains. ▌

Nonhomogeneous Neumann conditions, of the form $\partial H/\partial n(w) = h(w)$ on an arc Γ, transform in a more complicated way. We cannot simply substitute $w = f(z)$ into $h(w)$ to get the correct boundary values of $\partial\tilde{H}/\partial n(z)$ on the arc $\tilde{\Gamma}$ in the z-plane. Instead, if $H(w)$ satisfies the condition

(16) $$\frac{\partial H}{\partial n}(w) = h(w) \qquad \text{on the arc } \Gamma \text{ in the } w\text{-plane,}$$

then the transformed function $\tilde{H}(z) = \left[H(w)\Big|_{w=f(z)}\right]$ will satisfy a rather different condition

(17) $$\frac{\partial \tilde{H}}{\partial n}(z) = \left|\frac{df}{dz}(z)\right| \cdot \tilde{h}(z) \qquad \text{for } z \text{ on the arc } \tilde{\Gamma},$$

where we take $\tilde{h}(z) = [h(w)|_{w=f(z)}] = h(f(z))$. We leave the verification of (17) as Exercise 21, and will only deal with homogeneous Neumann problems in the main part of this text.

These observations are nicely illustrated by reconsidering Example 7.11 above. On the half plane E we have mutually conjugate harmonic functions $H(w)$ and $N(w)$, associated with $(-i/\pi)\text{Log } w = H(w) + iN(w)$. Here

$$H(w) = \frac{1}{\pi}\text{Arg}(w) \qquad \text{and} \qquad N(w) = -\frac{1}{\pi}\log|w|,$$

and since $H = 0$ or $H = 1$ on the boundary (the real axis), the function $N(w)$ must satisfy the homogeneous Neumann condition $\partial N/\partial n(w) = 0$ on the real axis (except at $w = 0$, where N fails to have a well defined normal derivative). The families of lines $H(w) = constant$ and $N(w) = constant$ are shown in Figure 7.10. These functions in the half plane are transformed by the change

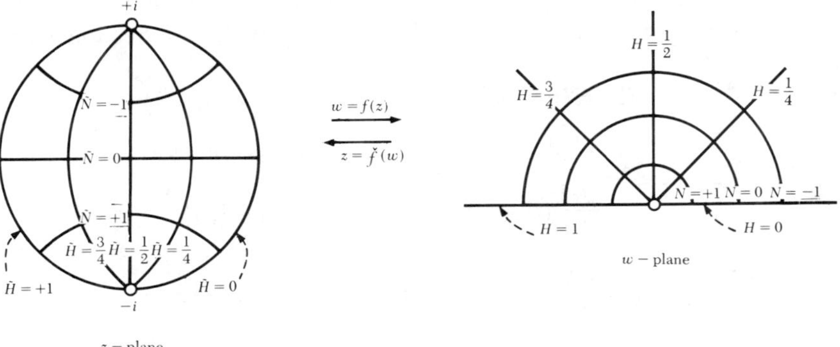

Figure 7.10 Here $z = +i$ goes to $w = \infty$, and $z = -i$ to $w = 0$; boundary arcs of the unit circle are mapped to appropriate halves of the real axis.

of variable $w = -i\left(\dfrac{z+i}{z-i}\right)$ into the harmonic functions defined in the disc $|z| < 1$:

$$\tilde{H}(z) = \frac{1}{\pi}\left[-\frac{\pi}{2} + \operatorname{Arg}(z+i) - \operatorname{Arg}(z-i)\right]$$

$$\tilde{N}(z) = -\frac{1}{\pi}\left[\log|z+i| - \log|z-i|\right] = -\frac{1}{\pi}\log\left|\frac{z+i}{z-i}\right|$$

which make up the real and imaginary parts of the transformed analytic function $(-i/\pi)\operatorname{Log}(f(z)) = (-i/\pi)\operatorname{Log}\left[-i\left(\dfrac{z+i}{z-i}\right)\right]$. Since $\tilde{H}$ is constant ($\tilde{H}(z) = 1$ or $\tilde{H}(z) = 0$) on the boundary arcs that connect $+i$ and $-i$, $\tilde{N}(z)$ satisfies the Neumann condition $\partial\tilde{N}/\partial n(z) = 0$ on these arcs (and $\partial\tilde{N}/\partial n$ is undefined at $+i$ and $-i$). The mapping $w = -i(z+i)/(z-i)$ maps the loci $H(w) = c_1$ and $N(w) = c_2$ to the corresponding curves $\tilde{H}(z) = c_1$ and $\tilde{N}(z) = c_2$ in the disc, as shown in Figure 7.10.

These transformation principles will be illustrated repeatedly as we consider additional examples in this and the following chapters.

EXERCISES

1. Consider $f(z) = (z+1)/(z-1)$, which is singular at $z = +1$.

(i) Verify that its real part is given by $U(z) = \dfrac{|z|^2 - 1}{|z-1|^2}$.

(ii) Write $\operatorname{Im}(f(z)) = V(x, y)$ as a function of the variables x and y.

(iii) Show that $\partial V/\partial n = 0$ on the unit circle $|z| = 1$, except at $z = +1$, where $V(z)$ is undefined.

Hint: The methods of Application 1 are useful.

2. Show that the real and imaginary parts of $\operatorname{Arcsin}(w) = H(w) + iN(w)$, defined in the upper half plane, satisfy the following boundary conditions along the real axis. Refer to Figure 4.34 of Section 4.10.

(i) $H(u + i0) = -\pi/2$ on $(-\infty, -1)$

$H(u + i0) = +\pi/2$ on $(1, +\infty)$

$\dfrac{\partial H}{\partial n}(u + i0) = 0$ on $(-1, 1)$

(ii) $\dfrac{\partial N}{\partial n}(u + i0) = 0$ on $(-\infty, -1)$ and $(1, +\infty)$

$N(u + i0) = 0$ on $(-1, +1)$.

Hint: Use methods from Application 1.

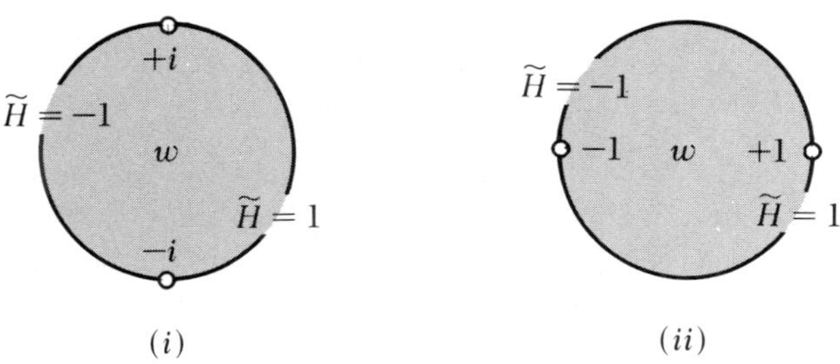

Figure 7.11

3. $\operatorname{Arctan}(w) = H(w) + iN(w)$ is harmonic on the half plane $E = \{w : \operatorname{Re}(w) > 0\}$. Calculate the inward normal derivatives

$$\frac{\partial H}{\partial n}(0 + iv) \quad \text{and} \quad \frac{\partial N}{\partial n}(0 + iv)$$

of the real and imaginary parts as functions of v on the boundary segments $-\infty < v < -1$; $-1 < v < +1$; $1 < v < +\infty$ in the imaginary axis.

Hint: Refer to Figure 4.33, Section 4.10.

4. Show that the imaginary part $H(z) = \operatorname{Im}(f(z))$ of $f(z) = +i - (2/\pi)\operatorname{Log} z$ on the upper half plane has boundary values

$$H(x + i0) = +1 \quad \text{for} \quad x > 0; \qquad H(x + i0) = -1 \quad \text{for} \quad x < 0.$$

Use simple variants of the fractional linear transformations discussed in Example 4.16 of Section 4.8 to transform $H(z)$ to functions $\tilde{H}(w)$ on the unit disc $D = \{w = |w| < 1\}$ which satisfy the boundary conditions indicated in Figure 7.11. ($\tilde{H}$ is bounded on the disc.)

Answers: (i) $1 - (2/\pi)\operatorname{Arg}\left(\frac{1}{i}\frac{w + i}{w - i}\right)$; (ii) $1 - (2/\pi)\operatorname{Arg}\left(\frac{1}{i}\frac{w + 1}{w - 1}\right)$.

5. Determine a fractional linear transformation that maps the disc $D = \{w\colon |w| < 1\}$ onto the upper half plane $E = \{z : \operatorname{Im}(z) > 0\}$ so that $w = 1$ and $w = +i$ are mapped to $z = 0$ and $z = \infty$ respectively. Then extend the methods of Exercise 4 to find a solution of the boundary value problem indicated in Figure 7.12.

Answer: $\tilde{H}(w) = 1 - (1/\pi)\operatorname{Arg}\left[-(1 + i)\left(\frac{w - 1}{w - i}\right)\right]$.

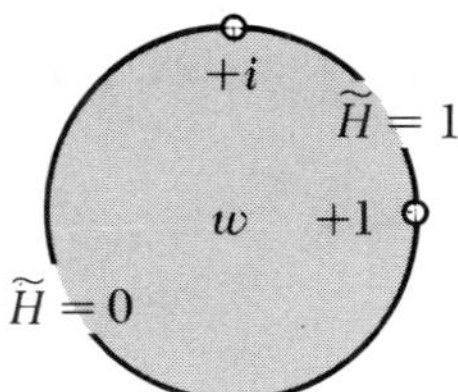

Figure 7.12

6. By noticing that the fractional linear transformation $z = f(w) = -(i+1)\left(\dfrac{w-1}{w-i}\right)$ maps circular arcs between $w = 1$ and $w = +i$ to rays from $z = 0$ to $z = \infty$, sketch the loci $\tilde{H}(w) = \textit{constant}$ in the unit disc for the function $\tilde{H}(w) = 1 - (1/\pi)\text{Arg}(f(w))$ in Exercise 5. The loci $\tilde{H} = 0$ and $\tilde{H} = 1$ form the boundary circle $|w| = 1$. Show that $0 < \tilde{H}(w) < 1$ on the disc, and determine the loci $\tilde{H} = 3/4$ and $\tilde{H} = 1/2$ explicitly.

Hint: To decide which arcs map to which rays, note that $z = f(w)$ is conformal at $w = 1$ (angles preserved).

Answer: $\tilde{H} = 3/4$ on the line segment $[1, +i]$; $\tilde{H} = 1/2$ on the arc through $1, +i$ (within the disc) that is *perpendicular* to the circle $|w| = 1$.

7. From formula (15), deduce the following important result.

Theorem: If $H(z)$ and $N(z)$ are conjugate harmonic functions near p, then the loci through p,

$$H(z) = c = H(p) \qquad N(z) = d = N(p),$$

are perpendicular.

8. If $\tilde{H}(w)$ is the solution of the boundary value problem in Exercise 5, it is constant on each circular arc from 1 to $+i$ (Exercise 6, above). Let $\tilde{G}(w)$ be a conjugate harmonic function; $\tilde{G}$ exists because the disc is a simply connected domain. How are the loci $\tilde{G}(w) = \textit{constant}$ related to the loci $\tilde{H}(w) = \textit{constant}$? Make a rough sketch showing both sets of curves in the disc. The families are *orthogonal,* by Exercise 7.

9. Solve the boundary value problem for the disc $|z| < 1$ indicated in Figure 7.13(i) by transforming it to the problem in the upper half plane in Figure 7.13(ii). Roughly sketch the level loci $\tilde{H}(z)$ and $H(w)$.

Hint: Use a fractional linear transformation.

Answer: $H(w) = -1 + (1/\pi)\text{Arg}(w-1) + (1/\pi)\text{Arg}(w+1)$; substitute $w = f(z) = -i(z+1)/(z-1)$ so that $f(+i) = -1$; $f(-i) = +1$; $f(1) = \infty$.

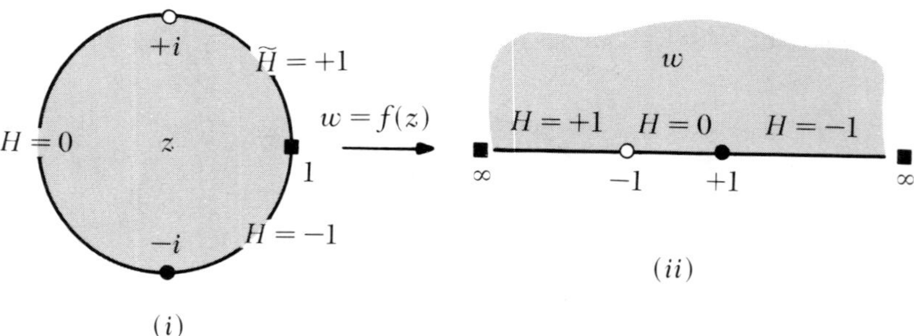

Figure 7.13

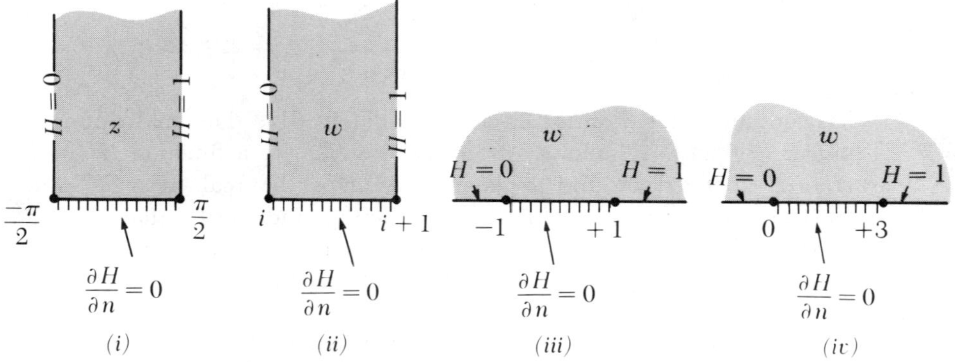

Figure 7.14

10. Solve the boundary value problem shown in (i) in Figure 7.14. Then use suitably chosen mappings of the half strip in (i) to solve the other problems shown.

Hint: Use $w = \sin z$ to get a half plane; use linear mappings to get the shifted domains (ii) and (iv).

Answers: (i) $H(z) = (2x + \pi)/2\pi = \text{Re}((2z + \pi)/2\pi)$; (ii) $\tilde{H}(u + iv) = u$; (iii) substitute $z = \text{Arcsin}(w) = -i\text{Log}(iw + \sqrt{1 - w^2})$ (principle determination of square root), so $\tilde{H}(w) = (\frac{1}{2}) + \text{Re}[(1/\pi)\text{Arcsin}\, w]$; (iv) replace w by $(2w/3) - 1$, so $\tilde{H}(w) = \frac{1}{2} + \text{Re}\left[\frac{1}{\pi}\text{Arcsin}\left(\frac{2w - 3}{3}\right)\right]$.

11. Solve the boundary value problem shown in Figure 7.15(i); then use the mapping $w = z^{1/2}$ to solve the problem in (ii).

Hint: In (i), translate so right hand corner is at origin; then apply $w = -1/z$ and another translation to get a simple problem in the first quadrant.

Answers: (i) $H(z) = \frac{2}{\pi}\text{Arg}\left(-\frac{1}{2}\frac{z+1}{z-1}\right)$; (ii) substitute $z = w^2$ to get $\tilde{H}(w) = \frac{2}{\pi}\text{Arg}\left(-\frac{1}{2}\frac{w^2+1}{w^2-1}\right)$ for w in the quarter disc.

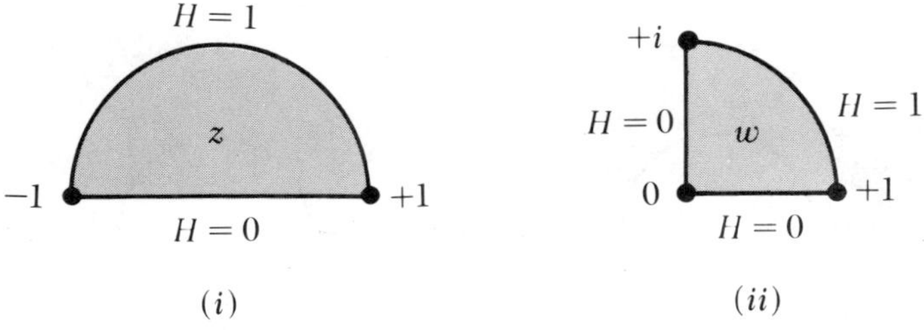

Figure 7.15

12. The harmonic function

$$H(x, y) = 3 - y + \frac{y}{x^2 + y^2} \qquad (x + iy \neq 0)$$

has constant value 3 on the circle $\Gamma = \{z: |z| = 1\}$. The analytic mapping $z = f(w) = e^{iw}$ allows us to transform $H(z)$ to a function $\tilde{H}(w) = H(f(w))$ defined on the w-plane, and maps the real axis Γ' onto Γ. Calculate $\tilde{H}(u, v) = \tilde{H}(u + iv)$ explicitly, and verify that $\tilde{H} = 3$ on the real axis. This amounts to a direct verification of the transformation law (12) for Dirichlet conditions on Γ' and Γ.

13. Calculate the inward normal derivative $\partial H/\partial n(z)$ on the unit circle $|z| = 1$, for $H(x, y)$ in Exercise 12. Calculate $\partial \tilde{H}/\partial n(w)$ along the real axis in the w-plane ($\mathbf{n}$ directed into upper half plane, which corresponds to the disc $|z| < 1$ when $z = e^{iw}$). Show that $\partial H/\partial n(z)$ and $\partial \tilde{H}/\partial n(w)$ *do not* agree at corresponding points z and w. Thus, *nonhomogeneous* boundary values in a Neumann problem are not transformed by assigning the same value at corresponding boundary points.

Answer: In polar coordinates $\partial H/\partial n(r, \theta) = -\partial H/\partial r(r, \theta) = \sin\theta = y/\sqrt{x^2 + y^2}$; $\partial \tilde{H}/\partial n(u, v) = \partial \tilde{H}/\partial v = 2 \sin u \sinh v$.

14. In Exercise 13, show that the Neumann conditions on Γ and Γ' are correctly transformed by the general formula (17).

15. The transformation $w = T(z) = (\frac{1}{2})\left(z + \frac{1}{z}\right)$ maps the domain shown in Figure 7.16 onto the upper half plane, carrying the circular arc to the segment $[-1, +1]$. Accepting the validity of this observation, solve the boundary value problem (i).

Answer: $H(z) = (2/\pi)\text{Arg}\left(\frac{z - 1}{z + 1}\right) = (2/\pi)\arctan^*\left(\frac{2y}{x^2 + y^2 - 1}\right)$, where $0 < \arctan^*(t) < \pi$ for all t.

16. Apply the comments of Exercise 15 to solve the boundary value problem (ii) shown in Figure 7.16.

Answer: $H(z) = \text{Re}(f(z))$, where $f(z) = \frac{2}{\pi}\,\text{Arcsin}\left[\frac{1}{2}\left(z + \frac{1}{z}\right)\right]$ on the half plane.

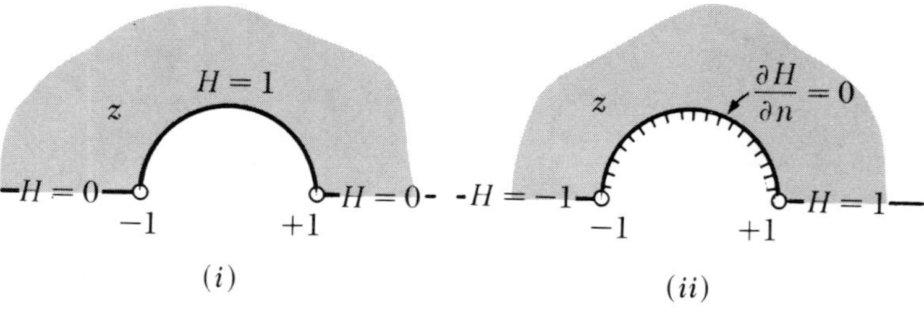

Figure 7.16

17. In a sketch, indicate the boundary conditions satisfied by the conjugate harmonic functions $H^*(z)$ associated with the solutions of Exercises 15 and 16.

18. Let Γ be the trajectory of a smooth parametrized curve $\gamma(t)$, defined for $a \leq t \leq b$, and let $H(x, y)$ be a smooth real valued function defined on an open set which includes Γ. Assume that $d\gamma/dt \neq 0$ everywhere (so there is a definite tangent at each point), and assume that ∇H is perpendicular to Γ. Prove that $\phi = \textit{constant}$ on Γ; that is, $\phi(\gamma(t_1)) = \phi(\gamma(t_2))$ for t_1, t_2 in $[a, b]$.

Hint: Perpendicularity means $\nabla H \cdot \dfrac{d\gamma}{dt} = 0$; relate this to the derivative of the composite function $(H \circ \gamma)(t) = H(\gamma(t))$ and integrate with respect to t.

19. Explain why the result in Exercise 18 does not depend on the particular parametrization we used.

20. Prove the converse of Exercise 18. Let $H(x, y)$ be a smooth real-valued function defined on the trajectory Γ of a smooth curve $\gamma(t)$, defined for $a \leq t \leq b$. If $H(x, y) = c$ (a constant) everywhere on Γ, prove that ∇H is perpendicular to $d\gamma/dt$ for each t.

21. Let $\tilde{\Gamma}$ be a smooth arc in the z-plane which is mapped to a smooth arc Γ in the w-plane under an invertible analytic mapping $w = f(z)$. Let $\tilde{H}(z)$ and $H(w)$ be harmonic functions, defined on $\tilde{\Gamma}$ and Γ respectively, which correspond to one another under analytic, invertible transformations $w = f(z)$ and $z = \check{f}(w)$. If $H(w)$ satisfies the condition $\dfrac{\partial H}{\partial n}(w) = h(w)$ for w on Γ, prove that $\tilde{H}(z)$ satisfies the condition $\dfrac{\partial \tilde{H}}{\partial n}(z) = \left|\dfrac{df}{dz}\right| \cdot \tilde{h}(z)$ for z on $\tilde{\Gamma}$, where $\tilde{h}(z) = \left[h(w)\Big|_{w=f(z)} \right] = h(f(z))$.

Hint: Introduct a parametrization $\gamma(t)$ of $\tilde{\Gamma}$, and use the corresponding parametrization $\eta(t) = f(\gamma(t))$ of Γ.

7.7 THE POISSON FORMULA FOR THE DISC

We have indicated that boundary value problems for a large variety of domains can be transformed into boundary value problems for the unit disc. In this section we derive certain integral formulas that solve the Dirichlet and Neumann problems in the unit disc $D = \{z: |z| < 1\}$. Let $f(z)$ be analytic on the closed disc $|z| \leq 1$ (that is, on some domain that includes this disc). We have already seen how the Cauchy Integral Formula takes the values of f on the boundary circle $|z| = 1$ and reproduces $f(\zeta)$ for points $|\zeta| < 1$ within the disc. We have also seen that the real part $U = \text{Re}(f)$ determines f up to an added constant through the Cauchy-Riemann equations (see Exercises 5 and 6, Section 7.2), so we suspect that there should be an integral formula that takes the boundary values $U(e^{i\theta})$ and directly determines from them the harmonic function $U(\zeta)$ for $|\zeta| < 1$. In the next paragraph we will use these ideas to deduce the correct integral formula; then we will show that this formula, although deduced in rather special circumstances, can be used to solve Dirichlet problems with bounded, piecewise continuous boundary conditions.

Consider the following situation. We are given a function $U(z)$ that is harmonic on the disc $|z| < 1$ and on the boundary circle $|z| = 1$; this means that $U(z)$ is defined and harmonic on some slightly larger disc $D' = \{z: |z| < 1 + \delta\}$. Since D' is simply connected, there is an analytic function f defined on D' such that $\mathrm{Re}(f) = U$ (Theorem 7.1). We may adjust f by adding an imaginary constant so that $U(0) = f(0)$. Now, f has a series expansion

$$f(z) = \sum_{n=0}^{\infty} a_n z^n \qquad \text{(convergent for } |z| < 1 + \delta) \tag{18}$$

with $a_0 = f(0) = U(0)$; this series is uniformly convergent on the circle $\Gamma = \{z: |z| = 1\}$ (indeed, on the closed disc $|z| \leq 1$). The conjugated series $\sum_{n=0}^{\infty} \overline{a_n}(z^n)^- = \sum_{n=0}^{\infty} \overline{a_n}(\bar{z})^n$ converges uniformly also, with sum $\overline{f(z)} = (\sum_{n=0}^{\infty} a_n z^n)^-$ (see Exercise 2), so we can add the conjugated series to the original series and express $U(z)$ as

$$\begin{aligned} U(z) = \mathrm{Re}(f(z)) &= \tfrac{1}{2}(f(z) + \overline{f(z)}) = \tfrac{1}{2}\sum_{n=0}^{\infty}(a_n z^n + \overline{a_n}(\bar{z})^n) \\ &= a_0 + \tfrac{1}{2}\sum_{n=1}^{\infty}(a_n z^n + \overline{a_n}(\bar{z})^n). \end{aligned} \tag{19}$$

Clearly, this series is uniformly convergent on the circle $|z| = 1$. Notice that $\bar{z} = 1/z$ when $|z| = 1$, so that the series in (19) may be written in the form

$$U(z) = a_0 + \tfrac{1}{2}\left[\sum_{n=0}^{\infty}(a_n z^n + \overline{a_n} z^{-n})\right] \qquad \text{for } z \text{ such that } |z| = 1. \tag{20}$$

We want to obtain $f(z)$, starting from $U(z)$. We can accomplish this if we can determine the coefficients a_n in the series (18) starting with only the boundary values of $U(z)$ on the circle $|z| = 1$. To this end, we multiply $U(z)$ by various powers z^k and integrate along the contour $\gamma(\theta) = e^{i\theta}$ defined for $0 \leq \theta \leq 2\pi$; since (20) converges uniformly on the trajectory, we get

$$\begin{aligned} \frac{1}{2\pi i}\int_\gamma U(z) z^k\,dz &= \frac{a_0}{2\pi i}\int_\gamma z^k\,dz + \frac{1}{2}\cdot\frac{1}{2\pi i}\int_\gamma\left(\sum_{n=1}^{\infty} a_n z^n + \overline{a_n} z^{-n}\right) z^k\,dz \\ &= \frac{a_0}{2\pi i}\int_\gamma z^k\,dz + \tfrac{1}{2}\sum_{n=1}^{\infty}\left[\frac{a_n}{2\pi i}\int_\gamma z^{n+k}\,dz + \frac{\overline{a_n}}{2\pi i}\int_\gamma z^{k-n}\,dz\right] \end{aligned}$$

for $k = 0, \pm 1, \pm 2, \ldots$. Recall that $\int_\gamma z^p\,dz = 0$ unless $p = -1$. It follows that, for each choice of exponent k, there is only one nonzero integral in the whole right-hand sum, and from this we easily conclude that

$$a_0 = \frac{1}{2\pi i}\int_\gamma U(z)\frac{1}{z}\,dz = \frac{1}{2\pi}\int_0^{2\pi} U(e^{i\theta})\,d\theta \qquad (k = -1)$$

$$\tfrac{1}{2}a_n = \frac{1}{2\pi i}\int_\gamma U(z)\frac{1}{z^{n+1}}\,dz = \frac{1}{2\pi}\int_0^{2\pi} U(e^{i\theta})e^{-in\theta}\,d\theta \qquad (k = -n-1).$$

Now that we have determined the coefficients a_n, let us substitute their values into (18) and examine the series at a typical point ζ ($|\zeta| < 1$). We get

$$\begin{aligned} f(\zeta) = \sum_{n=0}^{\infty} a_n\zeta^n &= \frac{1}{2\pi i}\int_\gamma U(z)z^{-1}\,dz + \sum_{n=1}^{\infty}\left(2\cdot\frac{1}{2\pi i}\int_\gamma U(z)z^{-n-1}\,dz\right)\zeta^n \\ &= \frac{1}{2\pi i}\int_\gamma \left[1 + 2\cdot\sum_{n=1}^{\infty}\frac{\zeta^n}{z^n}\right]\frac{U(z)}{z}\,dz \\ &= \frac{1}{2\pi i}\int_\gamma U(z)\frac{1}{z}\left(\frac{z+\zeta}{z-\zeta}\right)dz. \end{aligned}$$

The interchange of $\int_\gamma (\cdots)\,dz$ with $\sum_{n=1}^{\infty}(\cdots)$ is justified because ζ is fixed, and so the series $\left[1 + 2\cdot\sum_{n=1}^{\infty}\frac{\zeta^n}{z^n}\right] = \frac{z+\zeta}{z-\zeta}$ converges uniformly on the circle $|z| = 1$. We conclude that $f(\zeta)$ can be written as

$$(21) \qquad f(\zeta) = \frac{1}{2\pi i}\int_\gamma U(z)\frac{1}{z}\left(\frac{z+\zeta}{z-\zeta}\right)dz = \frac{1}{2\pi}\int_0^{2\pi} U(e^{i\theta})\frac{e^{i\theta}+\zeta}{e^{i\theta}-\zeta}\,d\theta$$

for all points ζ such that $|\zeta| < 1$. We finally have an integral formula that determines $f(\zeta)$ for ζ in the unit disc from the boundary values of $U = \mathrm{Re}(f)$. The original harmonic function $U(\zeta)$ is the real part of $f(\zeta)$, and is obtained by taking the real part of this integral. For the Riemann integral on the right, this amounts to taking the real part of the integrand. Since U is already real valued we get :

$$(22) \qquad \begin{aligned} U(\zeta) &= \frac{1}{2\pi}\int_0^{2\pi} U(e^{i\theta})\mathrm{Re}\left(\frac{e^{i\theta}+\zeta}{e^{i\theta}-\zeta}\right)d\theta \\ &= \frac{1}{2\pi}\int_0^{2\pi} U(e^{i\theta})\frac{1-\rho^2}{1-2\rho\cos(\theta-\phi)+\rho^2}\,d\theta \end{aligned}$$

if we write $\zeta = \rho e^{i\phi}$ with $\rho = |\zeta| < 1$. This important formula is called the **Poisson integral formula.**

Note: For Riemann integrals with complex valued integrands it is true that $\mathrm{Re}\{\int_a^b f(t)\,dt\} = \int_a^b \mathrm{Re}\{f(t)\}\,dt$, but this is definitely not so for line integrals! In such integrals $\int_\gamma f(z)\,dz$, the variable of integration dz is complex and contributes to both the real and imaginary parts of the contour integral, since $\int_\gamma f(z)\,dz = (\int_\gamma U\,dx - V\,dy) + i(\int_\gamma V\,dx + U\,dy)$.

Now that we have seen what sort of integral can be expected to reproduce a harmonic function from its boundary values, we shall try to use this integral to solve the Dirichlet problem. Suppose we are given arbitrary continuous boundary values $h(z)$ defined on the unit circle $|z| = 1$. It is not at all clear that the Dirichlet problem with boundary values $h(z)$ is solvable; after all, we do not assume that $h(z)$ has anything to do with a harmonic function on the disc $|z| < 1$. The heart of the Dirichlet problem lies in deciding *which* functions continuous on the circle $|z| = 1$ can be extended to functions $H(z)$ that are defined and continuous for $|z| \leq 1$ and harmonic for $|z| < 1$, with $H(z_0) = h(z_0)$ for every point z_0 on the boundary circle. The answer is that *every* continuous function $h(z)$ gives a solvable Dirichlet problem, and furthermore, the desired harmonic function in the disc $|z| < 1$ is obtained by inserting $h(e^{i\theta})$ in place of $U(e^{i\theta})$ in the Poisson integral formula (22). In fact, we may allow $h(e^{i\theta})$ to be discontinuous but bounded at a finite number of places on the circle, and the formula will still work; the harmonic function $H(\zeta)$ will then be bounded at the points z_0 where h is discontinuous.

Theorem 7.5 *Let $h(z)$ be any piecewise continuous*† *bounded function on the circle $|z| = 1$ which has real values. Then the function*

$$H(\zeta) = \frac{1}{2\pi}\int_0^{2\pi} h(e^{i\theta})\,\mathrm{Re}\left(\frac{e^{i\theta}+\zeta}{e^{i\theta}-\zeta}\right) d\theta$$

is defined and harmonic for $|\zeta| < 1$; furthermore,

$$\lim_{\zeta\to z_0} H(\zeta) = h(z_0) \text{ at points } |z_0| = 1 \text{ where } h(z) \text{ is continuous}$$

and

$$|H(\zeta)| \text{ is bounded as } \zeta \to z_0 \quad (\text{keeping } |\zeta| < 1)$$

at points where $h(z)$ is discontinuous.

Note: If we define $H(z)$ for $|z| < 1$ as in (22), and then define $H(z) = h(z)$ at points on the circle $|z| = 1$, the limit relation $\lim_{z\to z_0} H(z) = h(z_0) = H(z_0)$, valid for boundary points z_0 where $h(z)$ is continuous, insures that the extended function on the closed disc solves the Dirichlet problem as we originally described it.

Proof: The integrand is a bounded and piecewise continuous function for $0 \leq \theta \leq 2\pi$ and for any choice of $|\zeta| < 1$. To see that $H(\zeta)$ is harmonic, we shall actually produce the analytic function $f(\zeta)$ of which $H(\zeta)$ is the real part. Notice that $H(\zeta)$ is just the real part of a line integral along the contour

† That is, $h(z)$ is allowed to have discontinuities at a *finite number* of points on the circle. Aside from the assumption that $|h(z)|$ is bounded for $|z| = 1$, it is not necessary to specify the behavior at the discontinuities.

$\gamma(t) = e^{it}$ $(0 \leq t \leq 2\pi)$; in fact, $H = \text{Re}(f)$ where

(23)
$$f(\zeta) = \frac{1}{2\pi}\int_0^{2\pi} h(e^{i\theta})\left(\frac{e^{i\theta}+\zeta}{e^{i\theta}-\zeta}\right)\frac{ie^{i\theta}}{ie^{i\theta}}\,d\theta$$
$$= \frac{1}{2\pi i}\int_\gamma \frac{h(z)}{z}\left(\frac{z+\zeta}{z-\zeta}\right)dz$$
$$= \frac{1}{2\pi i}\int_\gamma h(z)\left[\frac{2}{z-\zeta}-\frac{1}{z}\right]dz$$
$$= \frac{1}{2\pi i}\int_\gamma \frac{2h(z)}{z-\zeta}\,dz - \frac{1}{2\pi i}\int_\gamma \frac{h(z)}{z}\,dz$$
$$= \frac{1}{2\pi i}\int_\gamma \frac{2h(z)}{z-\zeta}\,dz + \textit{constant}.$$

Since $2 \cdot h(z)$ is bounded and piecewise continuous on the trajectory of γ, the Cauchy-type integral involving ζ determines a function $f(\zeta)$ that is analytic off the trajectory of γ, as explained in Theorem 5.22. Now $H(\zeta)$ is the real part of this analytic function, and hence is harmonic for $|\zeta| < 1$.

To complete the proof we must demonstrate that

(24)
$$\lim_{\zeta\to z_0} H(z) \quad \text{exists and is equal to } h(z_0)$$

for each boundary point z_0 at which $h(z)$ is continuous. The proof is based on a few elementary properties of the **Poisson kernel function**

$$K(\zeta, \theta) = \frac{1}{2\pi}\text{Re}\left(\frac{e^{i\theta}+\zeta}{e^{i\theta}-\zeta}\right) = \frac{1}{2\pi}\frac{1-|\zeta|^2}{|e^{i\theta}-\zeta|^2}$$

that multiplies $h(e^{i\theta})$ when we integrate to get $H(\zeta)$.

Certain facts about formula (22) are almost self-evident. If $|\zeta| < 1$, the numerator and denominator in $K(\zeta, \theta)$ are positive (notice that $|e^{i\theta} - \zeta|^2 \geq (1 - |\zeta|)^2 > 0$), so that

(25)
$$K(\zeta, \theta) > 0 \quad \text{for all real } \theta \text{ and for } |\zeta| < 1.$$

By applying formula (22) to the harmonic function $U(z) = 1$ on the plane, we get

(26)
$$\int_0^{2\pi} K(\zeta, \theta)\,d\theta = 1 \qquad \text{for all } \zeta \text{ such that } |\zeta| < 1.$$

Finally, $H(\zeta)$ has the same upper and lower bounds on the disc $|\zeta| < 1$ as

$h(z)$ has on the unit circle; if $A \leq h(z) \leq B$ for all z such that $|z| = 1$, then

$$A = A \cdot \int_0^{2\pi} K(\zeta, \theta)\, d\theta \leq \int_0^{2\pi} h(e^{i\theta})\, K(\zeta, \theta)\, d\theta \leq B \cdot \int_0^{2\pi} K(\zeta, \theta)\, d\theta = B,$$

so that

$$A \leq H(\zeta) \leq B \quad \text{for} \quad |\zeta| < 1. \tag{27}$$

If H is not constant, these inequalities must be *strict*; otherwise, H would have a relative maximum (or minimum) within the disc, in violation of the maximum principle. In our discussion, we may consider $h(e^{i\theta})$ and $K(\zeta, \theta)$ as periodic functions of the real variable θ, so that integrating over any interval $[a, b]$ has the same effect as integrating over $[a + 2\pi, b + 2\pi]$. To reduce notational confusion, we shall write

$$\int_{|\theta-\theta_0|<\delta} (\cdots)\, d\theta \quad \text{for} \quad \int_{\theta_0-\delta}^{\theta_0+\delta} (\cdots)\, d\theta$$

$$\int_{|\theta-\theta_0|>\delta} (\cdots)\, d\theta \quad \text{for} \quad \int_{\theta_0+\delta}^{\theta_0+2\pi-\delta} (\cdots)\, d\theta$$

Let us begin the serious details of verifying (24) by examining a special case. Take an arc $\Gamma = \{z\colon |z| = 1 \text{ and } \alpha \leq \arg z \leq \beta\}$ in the unit circle $(\alpha < \beta < \alpha + 2\pi)$ and let Γ' be the complementary arc, on which $\beta < \arg z < \alpha + 2\pi$; then consider the boundary values

$$h(z) = \begin{cases} 1 & \text{for } z \text{ on } \Gamma \\ 0 & \text{for } z \text{ on } \Gamma'. \end{cases} \tag{28}$$

Our first observation is that Poisson's formula provides the correct solution for boundary values of this kind. Formula (21) gives us an analytic function

$$f(\zeta) = \phi(\zeta; \Gamma) = \frac{1}{2\pi i} \int_\gamma \frac{h(z)}{z} \left(\frac{z + \zeta}{z - \zeta}\right) dz,$$

where $\gamma(\theta) = e^{i\theta}$ for $0 \leq \theta \leq 2\pi$. But $h(e^{i\theta}) = 1$ for $\alpha \leq \theta \leq \beta$, and is zero elsewhere, so we are really integrating along the smaller arc Γ traced out by $\eta(\theta) = e^{i\theta}$ for $\alpha \leq \theta \leq \beta$; thus,

$$\phi(\zeta; \Gamma) = \frac{1}{2\pi i} \int_\eta \frac{h(z)}{z} \left(\frac{z + \zeta}{z - \zeta}\right) dz = \frac{1}{2\pi i} \int_\eta \frac{1}{z} \left(\frac{z + \zeta}{z - \zeta}\right) dz.$$

This function is defined and analytic at *all* points ζ lying off trajectory(η) $= \Gamma$; moreover, we can calculate ϕ explicitly,

$$\begin{aligned}\phi(\zeta;\Gamma) &= \frac{1}{2\pi i}\int_\eta \frac{1}{z}\left(\frac{z+\zeta}{z-\zeta}\right) dz \\ &= \frac{1}{2\pi i}\int_\eta \left[-\frac{1}{z}+\frac{2}{z-\zeta}\right] dz \\ &= \frac{-1}{2\pi i}\int_\eta \frac{1}{z}\, dz + \frac{1}{i\pi}\int_\eta \frac{1}{z-\zeta}\, dz \\ &= -\left(\frac{\beta-\alpha}{2\pi}\right) + \frac{1}{i\pi}\operatorname{Log}\left(\frac{e^{i\beta}-\zeta}{e^{i\alpha}-\zeta}\right)\end{aligned}$$

for ζ in $E = \mathbf{C} \sim \Gamma$. It is not hard to check that the principal determination of logarithm used here is analytic off the arc Γ, so that the last equality is correct on E. The real part of ϕ is the harmonic function

(29) $$\omega(\zeta;\Gamma) = \operatorname{Re}(\phi(\zeta;\Gamma)) = \frac{1}{\pi}\operatorname{Arg}\left(\frac{e^{i\beta}-\zeta}{e^{i\alpha}-\zeta}\right) - \left(\frac{\beta-\alpha}{2\pi}\right).$$

By noticing how the fractional linear transformation $w = (e^{i\beta}-\zeta)/(e^{i\alpha}-\zeta)$ transforms $\mathbf{C}^* \sim \Gamma$ to $\mathbf{C}^* \sim [-\infty, 0]$, it is easy to see that

$$\omega(\zeta;\Gamma) = 0 \quad \text{for } \zeta \text{ on the complementary arc } \Gamma';$$

and if z_0 is a point on Γ that is not one of the end points, then

(30) $$\begin{aligned}&\omega(\zeta;\Gamma) \to +1 \text{ as } \zeta \text{ approaches } z_0 \text{ from within the disc } |\zeta| < 1\\ &\omega(\zeta;\Gamma) \to -1 \text{ as } \zeta \text{ approaches } z_0 \text{ from within the domain } |\zeta| > 1.\end{aligned}$$

Since $0 \le h(z) \le 1$ in this problem, we also have $0 \le \omega(\zeta;\Gamma) \le 1$ on the disc $|\zeta| < 1$, by (27). Obviously $\omega(\zeta;\Gamma)$ is bounded near points where $h(z)$ is discontinuous, so this function is the solution of the problem with boundary conditions (28). It is important to notice that this discussion applied to *any* arc Γ in the unit circle.

Now let us consider general bounded, piecewise continuous boundary values $h(z)$ in (24). We already know that $H(\zeta)$ is bounded on the disc; if $|h(z)| \le M$ for $|z| = 1$, then $|H(\zeta)| \le M$ for $|\zeta| < 1$. Thus, $H(\zeta)$ is certainly bounded near any point of discontinuity on the circle $|z| = 1$, and it is only necessary to examine the behavior of $H(\zeta)$ near a point $z_0 = e^{i\theta_0}$ at which $h(z)$ is continuous. Continuity of h at z_0 means that the values $h(e^{i\theta})$ and $h(e^{i\theta_0})$ are close together if $|\theta - \theta_0|$ is small enough; if ε is a given positive number, we get

(31) $$|h(e^{i\theta}) - h(z_0)| < \varepsilon/3 \qquad \text{for all } \theta \text{ such that } |\theta - \theta_0| < \delta$$

by taking $\delta > 0$ sufficiently small. For this choice of δ, let Γ_δ be the arc $\theta_0 - \delta \le \arg z \le \theta_0 + \delta$, and let Γ'_δ be the complementary arc, in the unit circle. If we write $h(e^{i\theta}) = [h(e^{i\theta}) - h(z_0)] + h(z_0)$, then formula (22) takes the form

$$\begin{aligned} H(\zeta) &= \int_{|\theta-\theta_0|<\delta} h(e^{i\theta})K(\zeta, \theta)\, d\theta + \int_{|\theta-\theta_0|>\delta} h(e^{i\theta})K(\zeta, \theta)\, d\theta \\ &= \int_{|\theta-\theta_0|<\delta} h(z_0)K(\zeta, \theta)\, d\theta + \int_{|\theta-\theta_0|<\delta} [h(e^{i\theta}) - h(z_0)]K(\zeta, \theta)\, d\theta \\ &\qquad + \int_{|\theta-\theta_0|>\delta} h(e^{i\theta})K(\zeta, \theta)\, d\theta \\ &= h(z_0)\cdot \omega(\zeta; \Gamma_\delta) + A(\zeta) + B(\zeta) \qquad \text{for } |\zeta| < 1. \end{aligned}$$

We may estimate the size of the last two terms as follows. For $|\zeta| < 1$, formulas (31) and (26) give

$$\begin{aligned} |A(\zeta)| &\le \int_{|\theta-\theta_0|<\delta} |h(e^{i\theta}) - h(z_0)|\, K(\zeta, \theta)\, d\theta \\ &\le \int_{|\theta-\theta_0|<\delta} (\varepsilon/3)K(\zeta, \theta)\, d\theta \le (\varepsilon/3)\int_0^{2\pi} K(\zeta, \theta)\, d\theta = \varepsilon/3. \end{aligned}$$

Since $|h(z)| \le M$ for all z on the unit circle, we get

$$|B(\zeta)| \le \int_{|\theta-\theta_0|>\delta} |h(e^{i\theta})|\, K(\zeta, \theta)\, d\theta \le M\int_{|\theta-\theta_0|>\delta} K(\zeta, \theta)\, d\theta = M\cdot \omega(\zeta; \Gamma'_\delta).$$

Combining these observations, we see that

$$\begin{aligned} |H(\zeta) - h(z_0)| &\le |h(z_0)|\cdot|1 - \omega(\zeta; \Gamma_\delta)| + \varepsilon/3 + M\cdot\omega(\zeta; \Gamma'_\delta) \\ &\le M\,|1 - \omega(\zeta; \Gamma_\delta)| + M\cdot\omega(\zeta; \Gamma'_\delta) + \varepsilon/3 \end{aligned}$$

for $|\zeta| < 1$. But we already know how the functions $\omega(\zeta;\Gamma_\delta)$ and $\omega(\zeta;\Gamma'_\delta)$ behave as ζ approaches z_0 from within the disc;

$$\lim_{\zeta\to z_0} \omega(\zeta; \Gamma_\delta) = +1 \quad \text{and} \quad \lim_{\zeta\to z_0} \omega(\zeta; \Gamma'_\delta) = 0 \qquad (\text{keeping } |\zeta| < 1).$$

Thus, we may insure that

$$M\cdot|1 - \omega(\zeta; \Gamma_\delta)| < \varepsilon/3 \quad \text{and} \quad M\cdot|\omega(\zeta; \Gamma'_\delta)| < \varepsilon/3$$

for all ζ such that $|\zeta - z_0| < r(\varepsilon)$ and $|\zeta| < 1$, by taking $r(\varepsilon)$ small enough. For these ζ, which occupy the shaded region $D_{r(\varepsilon)}$ shown in Figure 7.17, we get

$$|H(\zeta) - h(z_0)| < \frac{\varepsilon}{3} + \frac{\varepsilon}{3} + \frac{\varepsilon}{3} = \varepsilon \qquad \text{for all } \zeta \text{ in } D_{r(\varepsilon)}.$$

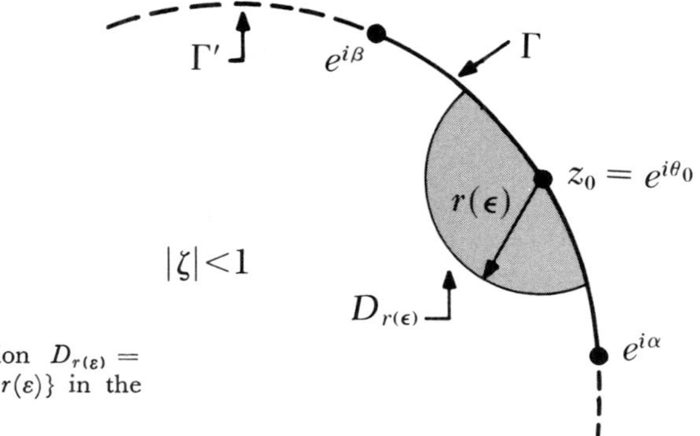

Figure 7.17 The region $D_{r(\varepsilon)} = \{\zeta: |\zeta| < 1 \text{ and } |\zeta - z_0| < r(\varepsilon)\}$ in the proof of Theorem 7.5.

This reasoning works for any choice of $\varepsilon > 0$, so we have proved that

$$H(\zeta) \to h(z_0) \quad \text{as } \zeta \text{ approaches } z_0 \text{ from within the unit disc,}$$

for any point of continuity of the boundary values $h(z)$. This finishes the proof. ∎

The contour integral version of formula (22) can be quite useful in solving Dirichlet problems; it is often easier to deal with formula (21) than with Poisson's formula (22), a Riemann integral, because we may apply the principles of contour integration in evaluating (21). In the preceding discussion, this was done in determining the particular solutions $\omega(\zeta;\Gamma)$ associated with an arc Γ on the unit circle.

For continuous boundary values $h(z)$ the solution of the Dirichlet problem in the disc is *unique*; no solutions are possible other than the one given by the Poisson formula. This follows immediately from our discussion of the maximum/minimum principle for harmonic functions. It is quite a bit harder to show that there is still only one harmonic function $H(\zeta)$ defined for $|\zeta| < 1$ which satisfies the conditions

(i) $\lim_{\zeta \to z_0} H(\zeta) = h(z_0)$ at points z_0 where $h(z)$ is continuous

(ii) $|H(\zeta)|$ is bounded as ζ approaches a point of discontinuity

when we are given *piecewise continuous and bounded* values $h(z)$ on the unit circle. This is the solution given by the Poisson formula (22). We have indicated in Section 7.1 that the solution need not be unique if we drop the boundedness condition (ii). We will not take the time to prove that the bounded solution is unique when we allow bounded, piecewise continuous boundary values.

Slight modifications of the preceding discussion allow us to solve the *exterior* Dirichlet problem for the disc, in which we take the points exterior to the unit circle $|z| = 1$ as our domain. We then want a function $H^*(\zeta)$, defined and harmonic for $|\zeta| > 1$, that satisfies the boundary conditions

(i) $|H^*(\zeta)|$ is bounded on the domain $|\zeta| > 1$

(ii) $H^*(\zeta) \to h(z_0)$ as $\zeta \to z_0$ (keeping $|\zeta| > 1$), at points z_0 where $h(z)$ is continuous.

The solution to the exterior problem is obtained by multiplying the Poisson integral in (22) by -1, and taking $|\zeta| > 1$ instead of $|\zeta| < 1$:

$$(32)\qquad H^*(\zeta) = \frac{(-1)}{2\pi}\int_0^{2\pi} h(e^{i\theta})\operatorname{Re}\left(\frac{e^{i\theta}+\zeta}{e^{i\theta}-\zeta}\right) d\theta \qquad \text{for all } |\zeta| > 1.$$

This function $H^*(\zeta)$ is just the real part of $(-1)\cdot f(\zeta) = f^*(\zeta)$, where

$$(33)\quad f^*(\zeta) = \frac{(-1)}{2\pi i}\int_\gamma h(z)\,\frac{1}{z}\left(\frac{z+\zeta}{z-\zeta}\right) dz = \frac{(-1)}{2\pi}\int_0^{2\pi} h(e^{i\theta})\operatorname{Re}\left(\frac{e^{i\theta}+\zeta}{e^{i\theta}-\zeta}\right) d\theta.$$

The contour integral gives a function that is analytic off the circle $|z| = 1$ (both inside and outside of the circle!), so that $H^*(\zeta)$ is harmonic for $|\zeta| > 1$. For $|\zeta| > 1$, the modified Poisson kernel function appearing in (33),

$$K^*(\zeta, \theta) = (-1)\cdot K(\zeta, \theta) = \frac{1}{2\pi}\frac{|\zeta|^2 - 1}{|e^{i\theta} - \zeta|^2} \qquad \text{for } |\zeta| > 1,$$

satisfies the properties (25), (26), and (27) in the proof of Theorem 7.5. We must introduce the minus sign to get (25) for $|\zeta| > 1$; thus, if $|\zeta| > 1$, we get $(-1)(1 - |\zeta|^2) = |\zeta|^2 - 1 > 0$, and we still have $|e^{i\theta} - \zeta|^2 \geq (|\zeta| - 1)^2 > 0$. We must also use the elementary solutions $\omega^*(\zeta;\ \Gamma) = -\omega(\zeta;\ \Gamma)$, defined for $|\zeta| > 1$, corresponding to an arc Γ in the unit circle. The rest of the proof can be repeated almost word for word to prove that $H^*(\zeta)$ satisfies the stated boundary conditions in the domain $|\zeta| > 1$.

Example 7.12 Let us examine more carefully the solutions to the interior and exterior Dirichlet problems corresponding to the boundary conditions

$h(e^{i\theta}) = +1$ on the arc $\Gamma = \{z : |z| = 1 \text{ and } \alpha \leq \arg z \leq \beta\}$

$h(e^{i\theta}) = 0$ elsewhere on the unit circle (i.e., on Γ')

$|H(\zeta)|$ is bounded on the domain being considered.

In the proof of Theorem 7.5 we evaluated the solution in the disc $|\zeta| < 1$; $H = \operatorname{Re}(f)$, where

$$f(\zeta) = \frac{1}{2\pi i}\int_\gamma \frac{h(z)}{z}\left(\frac{z+\zeta}{z-\zeta}\right) dz = \frac{1}{i\pi}\operatorname{Log}\left(\frac{e^{i\beta}-\zeta}{e^{i\alpha}-\zeta}\right) - \left(\frac{\beta-\alpha}{2\pi}\right)$$

Thus,

$$H(\zeta) = \omega(\zeta;\ \Gamma) = \frac{1}{\pi}\operatorname{Arg}\left(\frac{e^{i\beta}-\zeta}{e^{i\alpha}-\zeta}\right) - \left(\frac{\beta-\alpha}{2\pi}\right).$$

Since the fractional linear transformation $w = T(\zeta) = (e^{i\beta} - \zeta)/(e^{i\alpha} - \zeta)$ maps Γ onto the extended negative real axis $[-\infty, 0] = \{\infty\} \cup (-\infty, 0]$ in $\mathbf{C}^*$, $H(\zeta)$ is actually defined and harmonic everywhere off Γ. Since $T(e^{i\beta}) = 0$ and $T(e)^{i\alpha} = \infty$, it follows that T maps each circular arc through $e^{i\beta}$ and $e^{i\alpha}$ to a radial line between 0 and ∞ in the w-plane. It is a simple matter (Exercise

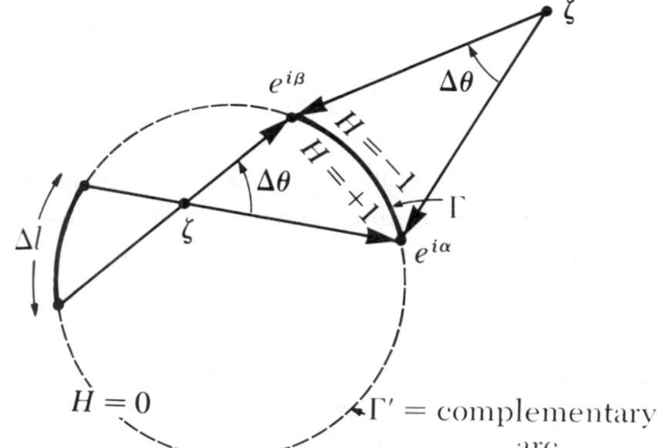

Figure 7.18 Interpretation of the solution $H(\zeta) = \omega(\zeta; \Gamma)$ of the interior Dirichlet problem.

4) to see that $H(\zeta)$ is constant on each circular arc from $e^{i\beta}$ to $e^{i\alpha}$ (there is a discontinuity along the particular arc Γ).

Geometrically, $\text{Arg}[(e^{i\beta} - \zeta)/(e^{i\alpha} - \zeta)]$ is just the angle $\Delta\theta$, shown in Figure 7.18 for typical points ζ in the domains $|\zeta| < 1$ and $|\zeta| > 1$; $\Delta\theta$ is measured from the segment $[\zeta, e^{i\alpha}]$ to the segment $[\zeta, e^{i\beta}]$ in either case, and is positive (counterclockwise) for $|\zeta| < 1$. Thus,

$$\omega(\zeta; \Gamma) = \frac{\Delta\theta}{\pi} - \left(\frac{\beta - \alpha}{2\pi}\right) \qquad \text{for } \zeta \text{ in } \mathbf{C} \sim \Gamma.$$

Notice that $(\beta - \alpha)/2 < \Delta\theta < \pi + (\beta - \alpha)/2$, so that $0 < \omega(\zeta;\Gamma) < 1$, for $|\zeta| < 1$.

If $|\zeta| < 1$, there is another geometric interpretation of $\omega(\zeta;\Gamma)$, one in which we hold ζ fixed and consider various arcs Γ in the unit circle. If we extend lines from the end points $e^{i\beta}$, $e^{i\alpha}$, through ζ to the opposite side of the unit circle, then the arc subtended by these extensions has length

$$\Delta l = [2\Delta\theta - (\beta - \alpha)] = 2\pi \cdot \omega(\zeta;\ \Gamma).$$

This new arc is obtained by "projecting" Γ through ζ; $\omega(\zeta;\Gamma)$ measures the fraction $\Delta l/2\pi$ of the unit circle covered by this projected arc. The value $\omega(\zeta;\ \Gamma)$ is generally referred to as the **harmonic measure** of the arc Γ with respect to ζ. For a more detailed development of this concept, we refer to Nevanlinna and Paatero [18], Sections 11.14 to 11.21.

In the exterior domain, $H(\zeta) = (\Delta\theta/\pi) - (\beta - \alpha)/2\pi$ approaches -1 as ζ approaches any point on Γ, other than one of the end points, keeping $|\zeta| > 1$. Since $H(\zeta)$ is already equal to zero on the complementary arc Γ',

$$H^*(\zeta) = (-1) \cdot \omega(\zeta; \Gamma) = \frac{-1}{\pi} \text{Arg}\left(\frac{e^{i\beta} - \zeta}{e^{i\alpha} - \zeta}\right) + \left(\frac{\beta - \alpha}{2\pi}\right) = -\frac{\Delta\theta}{\pi} + \left(\frac{\beta - \alpha}{2\pi}\right)$$

is the solution to the boundary value problem in the exterior domain, as we expect.

The Poisson formula can be used to get explicit solutions to Dirichlet problems if the boundary conditions lead to an integral (21) or (22) simple enough to be evaluated; however, there are many situations in which such an evaluation is out of the question. Even then, integral formulas of the Poisson type are important because they lend themselves easily to approximate *numerical* calculations of the solution from the given boundary values. The reason is that error estimates are quite easy to make in a numerical calculation based on integration processes (as opposed to differentiation processes), and in deriving the Poisson formula we have obtained an explicit integration process for solving Laplace's equation with boundary conditions of the Dirichlet type. Poisson's formula is also important as a tool in theoretical investigations of how harmonic functions behave; for example (see Exercise 18), by its use we can show that the limit of a *uniformly convergent* sequence of harmonic functions is harmonic also. This is a result which could be very difficult to prove without the use of integral formulas such as (22).

EXERCISES

1. Verify the formula

$$\mathrm{Re}\left(\frac{e^{i\theta}+\zeta}{e^{i\theta}-\zeta}\right)=\frac{1-\rho^2}{1-2\rho\cos(\theta-\phi)+\rho^2}\qquad\text{for all real } \theta,$$

where $\zeta = \rho e^{i\phi} (\rho < 1)$.

2. If a complex series $a = \sum_{n=0}^{\infty} a^n$ converges, prove that the complex conjugate $\bar{a} = \left(\sum_{n=0}^{\infty} a_n\right)^{-}$ is given by $\sum_{n=0}^{\infty} \overline{a_n}$. Show that the conjugated series is absolutely convergent if and only if the original series is absolutely convergent.

3. Verify that the series $1 + 2\sum_{n=1}^{\infty} w^n$ has radius of convergence $r = 1$. Evaluate its sum, using what you know about the geometric series. If $w = \zeta/z$ with z on the unit circle $|z| = 1$, and if ζ is a fixed point such that $|\zeta| < 1$, show that the sum $1 + \frac{1}{2}\sum_{n=1}^{\infty}\left(\frac{\zeta^n}{z^n}\right)$ converges uniformly on $|z| = 1$ to the sum $\left(\frac{z+\zeta}{z-\zeta}\right)$ as a function of z (holding ζ fixed).

Hint: $\sum_{n=0}^{\infty} w^n$ converges uniformly to its limit on the circle $|z| = 1$, since $|\zeta| < 1$.

4. If $p = e^{i\beta}$ and $q = e^{i\alpha}$, examine the mapping properties of the fractional linear transformation

$$w = f(\zeta) = \frac{e^{i\beta} - \zeta}{e^{i\alpha} - \zeta} \qquad (\zeta \neq e^{i\alpha} \text{ or } e^{i\beta})$$

and demonstrate that

$$\operatorname{Log} f(\zeta) = \operatorname{Log}\left(\frac{e^{i\beta} - \zeta}{e^{i\alpha} - \zeta}\right)$$

is constant on circular arcs from $e^{i\alpha}$ to $e^{i\beta}$, and is harmonic except on the arc $\Gamma = \{z\colon |z| = 1 \text{ and } \alpha \leq \arg z \leq \beta\}$ on which there is a jump discontinuity. (Recall Section 4.8.)

5. Extend the discussion of Exercise 4 to verify that the geometric interpretation

$$\operatorname{Arg}\left(\frac{e^{i\beta} - \zeta}{e^{i\alpha} - \zeta}\right) = \Delta\theta,$$

illustrated in Figure 7.18, is correct.

6. Verify the boundary behavior of $\omega(\zeta;\, \Gamma)$ set forth in formula (30). (Also, see Exercise 4.)

7. If $\alpha < \phi_0 < \beta$, consider $\omega(\zeta;\Gamma)$ near the point $z_0 = e^{i\phi_0}$ on the arc Γ. If $\varepsilon > 0$ is given, prove that there exists an $r > 0$ such that

$$|\omega(\zeta;\Gamma) - 1| < \varepsilon \quad \text{for all } \zeta \text{ such that } |\zeta| < 1 \text{ and } |\zeta - z_0| < r$$

as indicated in Figure 7.17. This proves that

$$\omega(\zeta;\Gamma) \to 1 \quad \text{as} \quad \zeta \to z_0 \text{ from within the disc } |\zeta| < 1,$$

as required in the proof of Poisson's formula.

8. Verify the geometric interpretation of $\omega(\zeta;\Gamma)$ as the *harmonic measure* of Γ projected through ζ, as indicated in Figure 7.18.

9. In the Poisson formula, $G(\zeta) = \operatorname{Im}(f(\zeta))$ is a conjugate harmonic function for the solution $H(\zeta) = \operatorname{Re}(f(\zeta))$ of the Dirichlet problem. If $\zeta = \rho e^{i\phi} (\rho < 1)$, calculate $\operatorname{Im}[(e^{i\theta} + \zeta)/(e^{i\theta} - \zeta)]$ and write out an explicit integral formula for the conjugate $G(\zeta)$.

10. In Exercise 4 (Section 7.6) we found that

$$H(z) = 1 - \frac{2}{\pi}\operatorname{Arg}\left(\frac{1}{i}\frac{z+i}{z-i}\right), \qquad \text{defined for } |z| < 1,$$

solves the Dirichlet problem illustrated in Figure 7.11(i). Derive the same result using the Poisson formula. Express $H(x, y)$ in terms of elementary functions of x and y.

Answer: $H(x, y) = 1 - \dfrac{2}{\pi}\arctan^*\left[\dfrac{1 - x^2 - y^2}{2x}\right]$, where $\arctan^* t$ is determined so that its values lie in $[0, \pi]$.

11. If $K^*(\zeta, \theta)$ is the kernel $(-1)\cdot K(\zeta, \theta)$ in the Poisson formula for the exterior domain $E = \{z: |z| > 1\}$, prove that

$$\int_0^{2\pi} K^*(\zeta, \theta)\, d\theta = +1 \qquad \text{for every } \zeta \text{ in } E.$$

12. Prove that the solution $H^*(\zeta)$ to the Dirichlet problem in the exterior domain $E = \{z: |z| > 1\}$ with boundary values $h(z)$ for $|z| = 1$, given by formula (32), has a well defined finite limit at infinity,

$$\lim_{\zeta\to\infty} H = \lim_{\zeta\to\infty}\int_0^{2\pi} K^*(\zeta, \theta)h(e^{i\theta})\, d\theta = \frac{1}{2\pi}\int_0^{2\pi} h(e^{i\theta})\, d\theta.$$

Hint: If $\varepsilon > 0$, then $|(1/2\pi) - K^*(\theta, \zeta)| < \varepsilon$ for $0 \le \theta \le 2\pi$, if $|\zeta|$ is sufficiently large.

13. Prove the Poisson formula for the exterior domain $|\zeta| > 1$ (formula (32)) by adapting the discussion of Poisson's formula for the disc $|\zeta| < 1$.

14. Solve the boundary value problems

(i) $H(z)$ defined on the disc $|z| < 1$, with $H(e^{i\theta}) = \cos n\theta$.
(ii) $H^*(z)$ defined on the exterior domain $|z| > 1$, with

$$H^*(e^{i\theta}) = \cos n\theta \text{ and } \lim_{z\to\infty} H^* \text{ finite}$$

for integers $n = 0, \pm 1, \pm 2, \ldots$.

Hint: Is the Poisson formula really needed here?

Answer: $H(re^{i\theta}) = r^n \cos n\theta$; $H^*(re^{i\theta}) = (1/r^n)\cos n\theta$.

15. Prove the following theorem using Poisson's formula.

Theorem: Let $h(z)$ and $g(z)$ be bounded, piecewise continuous functions on the unit circle $|z| = 1$. If h and g satisfy the condition

$$|h(z) - g(z)| \le \varepsilon \text{ for all } z \text{ on the circle } |z| = 1,$$

then the solutions $H(\zeta)$ and $G(\zeta)$ of the Dirichlet problem in the unit disc satisfy a similar "closeness" condition throughout the disc,

$$|H(\zeta) - G(\zeta)| \leq \varepsilon, \text{ for all } \zeta \text{ in the disc } |\zeta| < 1.$$

Thus, boundary values that are "close together," in the sense that $|h - g| \leq \varepsilon$, yield solutions that are "close together," $|H - G| \leq \varepsilon$ for $|\zeta| < 1$. In other words, solutions H of the Dirichlet problem *depend "continuously"* on the initial values h.

16. Let $h_n(z)$ and $h(z)$ be bounded, piecewise continuous real valued functions on $\Gamma = \{z: |z| = 1\}$, and assume that $h_n \to h$ uniformly on Γ. Prove that the solutions $H_n(\zeta)$ and $H(\zeta)$ of the corresponding Dirichlet problems in the unit disc converge uniformly: $H_n \to H$ uniformly on $D = \{\zeta: |\zeta| < 1\}$.

17. Show that $h(e^{i\theta}) = \sum_{n=1}^{\infty} (\cos n\theta)/n^2$ is uniformly convergent on the interval $[0, 2\pi]$. Use Exercises 14 and 16 to obtain the solution of the Dirichlet problem in the unit disc, expressed as a uniformly convergent series of elementary functions,

$$H(\zeta) = \sum_{n=1}^{\infty} H_n(\rho, \phi) = \sum_{n=1}^{\infty} \frac{1}{n^2} \rho^n \cos n\phi,$$

where $\zeta = \rho e^{i\phi}$ is expressed in polar coordinates.

In general, it is easy to find series representations of solutions if we can determine Fourier series representations of the boundary values $h(e^{i\theta})$.

18. Fill in details in the proof outlined below.

Theorem: If harmonic functions $H_n(z)$ converge uniformly to a limit function $H(z)$ on an open set E, then the limit function $H(z)$ is necessarily harmonic on E.

If p is a typical point in E, consider a disc $|z - p| \leq r$ lying in E. Let $h_n(z)$ and $h(z)$ be the values of H_n and H on the circle $\Gamma = \{z: |z - p| = r\}$ and let $H'(\zeta)$ be the harmonic function on $D = \{z: |z - p| < r\}$ with boundary values $h(z)$, given by the integral formula (35) in Section 7.8. The values $h_n(z)$ give $H_n(\zeta)$, since the H_n are harmonic. Then

(i) $H'(\zeta)$ is harmonic on D (Poisson formulas give harmonic functions)

(ii) $h_n \to h$ uniformly on Γ (since $H_n \to H$ uniformly on E)

(iii) $H_n \to H'$ uniformly on D (Use (ii) plus simple adaptation of Exercise 16.)

(iv) $H_n \to H$ uniformly on D (since $H_n \to H$ uniformly on E).

Now observe that, for ζ in D, we get

$$0 \le |H(\zeta) - H'(\zeta)| \le |H(\zeta) - H_n(\zeta)| + |H_n(\zeta) - H'(\zeta)| \to 0 \text{ as } n \to \infty$$

(uniformly on D). Thus, the given function H is equal to H' (harmonic) near each p in E. We conclude that H is harmonic on E.

7.8 TRANSFORMING THE POISSON FORMULA TO OTHER DOMAINS

If $w = T(z)$ maps a given domain E in the z-plane invertibly onto the unit disc $D = \{w: |w| < 1\}$, boundary value problems in one of these domains may be transformed into corresponding problems for the other domain. Dirichlet problems transform into problems of the same kind, as we have seen in Section 7.6. In the disc D we may apply the Poisson formulas (21) and (22) to solve such problems, and by using the change of variable formula for contour integrals with the transformation T in mind, we may transform the Poisson formula to a new integral formula which operates directly on the boundary values given for E to produce the desired solution in E.

Example 7.13 Let $h(z)$ be a bounded, piecewise continuous function defined on the circle $|z - p| = r$, and consider the Dirichlet problem in the disc $E = \{z: |z - p| < r\}$, with boundary conditions

(i) $\lim_{\zeta \to z_0} H(\zeta) = h(z_0)$ if z_0 is a point of continuity on the circle $|z - p| = r$ (keeping ζ within E)

(ii) $|H(\zeta)|$ is bounded on E.

We shall follow the steps indicated schematically in the following diagram,

(34)

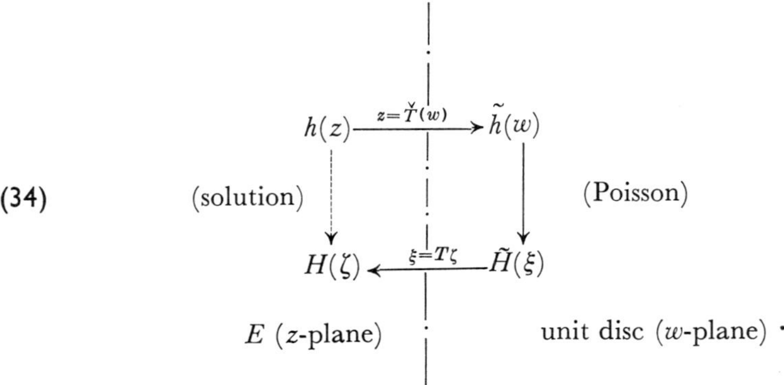

to get the solution $H(\zeta)$ in E from the boundary values $h(z)$, defined on $\text{bdry}(E) = \{z: |z - p| = r\}$.

The mapping $w = T(z) = (z - p)/r$ transforms E conformally onto the unit disc. Substituting $z = \check{T}(w) = rw + p$ transforms the boundary values $h(z)$ on bdry(E) to boundary values $\tilde{h}(w) = h((Tw))$ on the unit circle $|w| = 1$. It is not difficult to check that $h(z)$ is continuous at z_0 on bdry(E) if and only if $\tilde{h}(w)$ is continuous at the corresponding point $w_0 = T(z_0)$ on bdry(D). Now apply Poisson's formula to obtain the solution $\tilde{H}(\xi)$ for Dirichlet's problem in D with boundary values $\tilde{h}(w)$; $\tilde{H}(\xi)$ is the real part of the analytic function

$$\tilde{f}(\xi) = \frac{1}{2\pi i}\int_\gamma \frac{\tilde{h}(w)}{w}\left(\frac{w + \xi}{w - \xi}\right) dw \qquad \text{for } |\xi| < 1,$$

where $\gamma(t) = e^{it}$ for $0 \le t \le 2\pi$. Substituting $\xi = T(\zeta)$, we get an analytic function $f(\zeta) = \tilde{f}(T(\zeta))$ (a composite of analytic functions), and a harmonic function $H(\zeta) = \operatorname{Re}(f(\zeta)) = \tilde{H}(T(\zeta))$, defined for ζ in E:

$$H(\zeta) = \tilde{H}(T(\zeta)) \quad \text{and} \quad f(\zeta) = \frac{1}{2\pi i}\int_\gamma \frac{\tilde{h}(w)}{w}\left(\frac{w + T(\zeta)}{w - T(\zeta)}\right) dw \quad \text{for } \zeta \text{ in } E.$$

Now $H(\zeta)$ is the solution to the original Dirichlet problem, posed for E. In fact, $\tilde{H}(\xi)$ is bounded on the unit disc D, so that $H(\zeta) = \tilde{H}(T(\zeta))$ must be bounded on E. If z_0 is a typical boundary point of E at which $h(z)$ is continuous, then $\tilde{h}(w)$ is continuous at the corresponding point $w_0 = T(z_0)$ on bdry(D). Let us consider any sequence of points ζ_n that approaches z_0 from within E; the image points $\xi_n = T(\zeta_n)$ are in the unit disc D, and approach w_0 because T is continuous. Since $\tilde{H}(\xi)$ is the solution of the Dirichlet problem in the unit disc, we have $\tilde{H}(w_0) = \lim_{n\to\infty} \tilde{H}(\xi_n)$. Thus, for any sequence $\zeta_n \to z_0$ we get

$$H(\zeta_n) = \tilde{H}(\xi_n) \to \tilde{h}(w_0) = h(z_0) \text{ as } n \to \infty.$$

This proves that $\lim_{\zeta\to z_0} H(\zeta) = h(z_0)$ at any point of continuity on bdry(E).

We can get explicit integral formulas for $f(\zeta)$ and $H(\zeta)$ from the change of variable formula. If we parametrize the boundaries of D and E as

$$w = \gamma(t) = e^{it} \quad \text{for} \quad 0 \le t \le 2\pi$$

$$z = \eta(t) = p + re^{it} \quad \text{for} \quad 0 \le t \le 2\pi,$$

respectively, then $\gamma = T \circ \eta$ and, by the change of variable formula,

$$\int_{\gamma = T\circ\eta} g(w)\, dw = \int_\eta g(T(z)) \frac{dT}{dz}\, dz$$

for any function $g(w)$ that is bounded and piecewise continuous on $\Gamma = \{w:$

$|w| = 1\}$. Now $dT/dz = 1/r$, so that

$$\begin{aligned} f(\zeta) &= \frac{1}{2\pi i}\int_{\gamma=T\circ\eta} \frac{\tilde{h}(w)}{w}\left(\frac{w+T(\zeta)}{w-T(\zeta)}\right) dw \\ &= \frac{1}{2\pi i}\int_{\eta} \frac{\tilde{h}(T(z))}{T(z)}\,\frac{T(z)+T(\zeta)}{T(z)-T(\zeta)}\,\frac{1}{r}\,dz \\ &= \frac{1}{2\pi i}\int_{\eta} \frac{h(z)}{z-p}\,\frac{z-2p+\zeta}{z-\zeta}\,dz \\ &= \frac{1}{2\pi}\int_{0}^{2\pi} h(p+re^{i\theta})\left[\frac{re^{i\theta}+(\zeta-p)}{re^{i\theta}-(\zeta-p)}\right] d\theta. \end{aligned}$$

Taking $H(\zeta) = \mathrm{Re}(f(\zeta))$, we get

$$H(\zeta) = \frac{1}{2\pi}\int_{0}^{2\pi} h(p+re^{i\theta})\,\mathrm{Re}\left[\frac{re^{i\theta}+(\zeta-p)}{re^{i\theta}-(\zeta-p)}\right] d\theta \tag{35}$$

for $|\zeta - p| < r$. This is the direct integral formula we sought.

Integration formulas of this kind may be worked out for many other domains E using the scheme of steps (34), the change of variable formula, and an appropriate transformation of E onto the unit disc D. For example, the reader may find it interesting to derive Poisson's formula (32) for the exterior domain $E = \{z\colon |z| > 1\}$ by transforming the usual Poisson formula with the mapping $w = -1/z$ in mind.

An important special case is obtained when we use the fractional linear transformation

$$w = i\,\frac{z-i}{z+i}, \quad \text{with inverse} \quad z = \frac{1}{i}\,\frac{w+i}{w-i},$$

to transform the upper half plane $E = \{z\colon \mathrm{Im}(z) > 0\}$ conformally onto the unit disc. The steps in the scheme (34) may be carried out as before to get the solution $H(\zeta)$ in the half plane such that

(i) $\lim_{\zeta\to x+i0} H(\zeta) = h(x+i0)$ for $-\infty < x < +\infty$

(ii) $|H(\zeta)|$ is bounded for ζ in E,

starting with boundary values $h(x + i0)$ that are bounded and piecewise continuous on the real axis. There are some technical difficulties in applying the change of variable formula; we may use $\gamma(t) = e^{it}$, defined for $-3\pi/2 \le t \le \pi/2$, to parametrize the boundary Γ of the unit disc $|w| < 1$, but the corresponding contour $\eta(t) = \check{T}(\gamma(t))$ is unbounded, and traces out the real axis as t increases within the interval $[-3\pi/2, \pi/2]$, since $\check{T}(+i) = \infty$. If the change of variable formula is applied with due regard for the unboundedness of the transformed contour η, we find that the solution $H(\zeta)$ in the half plane is given by the real part of the analytic function

$$f(\zeta) = \frac{1}{\pi}\int_{-\infty}^{+\infty} h(s+i0)\,\frac{1}{s^2+1}\left[\frac{1}{i}\,\frac{s\zeta+1}{s-\zeta}\right] ds.$$

Thus,

$$H(\zeta) = \frac{1}{\pi}\int_{-\infty}^{+\infty} \frac{h(s+i0)}{s^2+1}\,\mathrm{Re}\left[\frac{1}{i}\,\frac{s\zeta+1}{s-\zeta}\right] ds$$

$$(36) \qquad = \frac{1}{\pi}\int_{-\infty}^{+\infty} h(s+i0)\,\frac{y_0}{(s-x_0)^2+y_0^2}\, ds$$

for $\zeta = x_0 + iy_0$ ($y_0 > 0$) in E. These integrals are improper, but converge (for fixed ζ in E) because the integrands include factors that decrease as $1/s^2$ at infinity. We will not go into the technical details needed to handle unbounded contours. For another approach, which obtains formula (36) as a limiting case of the superposition formula (4), see Kaplan [13], Section 7.4.

In formula (36), notice that the boundary values $h(s + i0)$ are always integrated with the same kernel function

$$M(\zeta, s) = \frac{1}{\pi}\,\frac{\mathrm{Im}(\zeta)}{(s-\mathrm{Re}(\zeta))^2 + (\mathrm{Im}(\zeta))^2} \quad \text{for} \quad -\infty < s < +\infty$$

to produce the solution $H(\zeta)$ at ζ. Since the upper half plane is simply connected, there is a well defined conjugate harmonic function $H^*(\zeta)$; obviously, $H^*(\zeta)$ is just the imaginary part of $f(\zeta) = H(\zeta) + iH^*(\zeta)$, so that

$$H^*(\zeta) = \frac{1}{\pi}\int_{-\infty}^{+\infty} \frac{h(s+i0)}{s^2+1}\,\mathrm{Im}\left[\frac{1}{i}\,\frac{s\zeta+1}{s-\zeta}\right] ds \quad \text{for} \quad \mathrm{Im}(\zeta) > 0.$$

Further comments on integral formulas for general domains can be found in Nehari [16], Section 4.9. Analogous formulas that solve the *Neumann problem* can also be derived; see Churchill [4], Chapter 11.

EXERCISES

1. Transform the Poisson formula (22) for the unit disc, using the mapping $z = 1/w$, to determine the Poisson formula (35) for the solution of Dirichlet's problem with boundary values $h(z)$ in the exterior domain $E = \{z\colon |z| > 1\}$.

2. Consider boundary values along the real axis

$$h(x+i0) = \begin{cases} a_0 & \text{for} \quad -\infty < x < x_0 \\ a_k & \text{for} \quad x_{k-1} < x < x_k \quad (k = 1, 2, \ldots N) \\ a_{N+1} & \text{for} \quad x_N < x < +\infty, \end{cases}$$

for the Dirichlet problem in the upper half plane $\mathrm{Im}(\zeta) > 0$. Show that the integral formula (36) leads to the same solution,

$$H(\zeta) = \frac{a_0}{\pi}\,\mathrm{Arg}(\zeta - x_0) + \sum_{k=1}^{N} \frac{a_k}{\pi}\,\mathrm{Arg}\left(\frac{\zeta - x_k}{\zeta - x_{k-1}}\right)$$

$$+ \, a_{N+1}\left[1 - \frac{1}{\pi}\,\mathrm{Arg}(\zeta - x_N)\right]$$

that was obtained in Section 7.1 by superposition.

3. If $h(x + i0)$ is bounded and piecewise continuous on the real axis, so that $|h(x + i0| \leq M < \infty$ for real x, show that

(i) The improper integral in formula (36) is convergent.

(ii) The solution given by formula (36) is similarly bounded,

$$|H(\zeta)| \leq M \text{ for all } \zeta \text{ such that } \mathrm{Im}(\zeta) > 0.$$

8 PHYSICAL APPLICATIONS OF POTENTIAL THEORY

In this digression into physics we shall list various standard applications of potential theory (theory of harmonic functions). In each application, harmonic functions play a central role in understanding the physical situation. Each application will be illustrated with two-dimensional examples, in which harmonic functions of two real variables, and complex analysis, are useful. Furthermore, a large number of new boundary value problems and conformal mapping problems will be worked out in these examples.

8.1 POTENTIAL THEORY AND THE PREVALENCE OF HARMONIC FUNCTIONS IN PHYSICS

The "vector fields" studied in advanced calculus are almost made to order for describing physical situations in which a force field, or some mathematically equivalent object, is the physically observable quantity. We shall begin this excursion into physics (the "real world," if you will) by recalling the basic terminology associated with vector fields in three-dimensional space; the same ideas apply in two dimensions, as we shall indicate at the end of this section.

Let us adopt the usual notation for vectors based at the origin. These are interpreted as directed line segments from the origin to various points $\mathbf{p} = (x_0, y_0, z_0)$ in three-dimensional Cartesian space $\mathbf{R}^3$. If we write $\mathbf{i}, \mathbf{j}, \mathbf{k}$ for the unit vectors directed along the positive x, y, and z axes, a vector $\mathbf{p}$ is fully described by specifying its components (or Cartesian coordinates) in each of these directions: thus,

(1) $$\mathbf{p} = x_0\mathbf{i} + y_0\mathbf{j} + z_0\mathbf{k}$$

if the coordinates of **p** are given by the set of numbers (x_0, y_0, z_0). A **vector field** is just a "vector valued" function **F** defined on some set D in $\mathbf{R}^3$; it assigns a vector $\mathbf{F}(\mathbf{x}) = \mathbf{F}(x, y, z)$ to each point **x** in D. Geometrically, one can think of the vector $\mathbf{F}(\mathbf{p})$ as being attached to the point **p** in D. Such a vector field is fully described by three scalar valued functions

$$U(\mathbf{x}) = U(x, y, z) \qquad V(\mathbf{x}) = V(x, y, z) \qquad W(\mathbf{x}) = W(x, y, z)$$

which specify the components of the vector $\mathbf{F}(\mathbf{x})$, attached to **x**, as functions of the variables x, y, and z:

$$\mathbf{F}(\mathbf{x}) = \mathbf{F}(x, y, z) = U(x, y, z)\mathbf{i} + V(x, y, z)\mathbf{j} + W(x, y, z)\mathbf{k}.$$

We say that a vector field **F** is **continuous** (**smooth,** etc.) if the component functions U, V, and W are all continuous (smooth, etc.) as scalar valued functions. When we consider two-dimensional vector fields, both the variable **x** and the values $\mathbf{F}(\mathbf{x})$ are vectors in the Cartesian plane $\mathbf{R}^2$. Since a vector $\mathbf{z} = a\mathbf{i} + b\mathbf{j}$ corresponds in a natural way to the complex number $z = a + ib$, a vector field with components $U(x, y)$ and $V(x, y)$

$$\mathbf{F}(x, y) = U(x, y)\mathbf{i} + V(x, y)\mathbf{j}$$

can be identified with an ordinary function of a complex variable, with complex values

$$F(x + iy) = U(x, y) + iV(x, y).$$

Vector fields may be given physically meaningful interpretations in a number of ways. For example, the electrostatic field produced by a distribution of charge in space is described by a vector field, the electric force field, which specifies the strength and direction (a vector quantity!) of the force the field exerts on a positive test charge of unit strength placed at the position **x** in space. In celestial mechanics, gravitational force fields are the objects of primary interest, and these are described by a vector field $\mathbf{F}(\mathbf{x})$ which specifies the gravitational force exerted on a test particle of unit mass placed at **x**. These are examples in which $\mathbf{F}(\mathbf{x})$ is interpreted as a *force field*, specifying magnitude and direction of forces of one kind or another on a test particle located at **x**. A different kind of model arises when we describe the *velocity field* of a fluid flow; here the vector field $\mathbf{F}(\mathbf{x})$ gives the velocity of fluid particles located at various positions **x** in space. Similar applications in which $\mathbf{F}(\mathbf{x})$ describes a more abstract kind of velocity field occur naturally in heat flow problems, in which we want to describe the magnitude and direction of heat flow past various points in space, or in diffusion problems where the flux of diffusing particles (neutrons, dye molecules, or what have you) is described by assigning a vector to each point in space. We will be more specific about these mathematical models in the following sections.

The laws of physics are often expressed as equations involving the relevant vector fields $\mathbf{F}(\mathbf{x})$ and certain vector operators, such as gradient, divergence, and curl. Recall that the **gradient operator** acts on smooth scalar valued functions $\phi(\mathbf{x}) = \phi(x, y, z)$ to give a vector field:

$$\text{(2)} \qquad \mathbf{grad}\ \phi = \nabla\phi = \frac{\partial\phi}{\partial x}\mathbf{i} + \frac{\partial\phi}{\partial y}\mathbf{j} + \frac{\partial\phi}{\partial z}\mathbf{k};$$

the other two operators act on a smooth vector field $\mathbf{F} = U\mathbf{i} + V\mathbf{j} + W\mathbf{k}$ to give us a scalar valued function and a new vector field, respectively:

$$\mathbf{div}\ \mathbf{F} = \nabla \cdot \mathbf{F} = \frac{\partial U}{\partial x} + \frac{\partial V}{\partial y} + \frac{\partial W}{\partial z}$$

$$\mathbf{curl}\ \mathbf{F} = \nabla \times \mathbf{F} = \left(\frac{\partial W}{\partial y} - \frac{\partial V}{\partial z}\right)\mathbf{i} + \left(\frac{\partial U}{\partial z} - \frac{\partial W}{\partial x}\right)\mathbf{j} + \left(\frac{\partial V}{\partial x} - \frac{\partial U}{\partial y}\right)\mathbf{k}$$

These operators have different descriptions in other coordinate systems, such as spherical or cylindrical coordinates in $\mathbf{R}^3$, but their action on vector fields turns out to have a meaning which is independent of the coordinate system used to describe the operators and the vector fields. Thus, these operators have a coordinate-independent geometric meaning, which is the real reason why they are the operators that enter into the equations of physics, rather than other operators formed from combinations of partial differentiations. These ideas can be made precise, but we only mean to suggest the reasons behind the recurrent appearance of these operators in the laws of physics.

Two kinds of laws tend to appear as consequences of the basic conservation laws that apply to the vector field $\mathbf{F}$ that describes a physical situation:

$$\text{(3)} \qquad \begin{aligned} \mathbf{div}\ \mathbf{F} &= \nabla \cdot \mathbf{F} = 0 \\ \mathbf{curl}\ \mathbf{F} &= \nabla \times \mathbf{F} = \mathbf{0} \end{aligned} \qquad \text{in some domain } D.$$

The precise physical significance of these laws is not crucial in understanding why harmonic functions are intimately connected with any vector field which satisfies these laws. For a clear and detailed explanation of the physics behind these "conservation laws" we strongly recommend that the reader examine the Feynman Lectures on Physics [6], particularly Chapters 1–7, 12–14, and 40–41 of v.2.

The law $\nabla \cdot \mathbf{F} = 0$ means that the density of whatever it is that is creating the field (charge, in the case of electrostatic problems; heat sources or sinks, in heat conduction problems; fluid sources in flow problems) is *zero* throughout the region of space where this equation is valid.

The companion equation $\nabla \times \mathbf{F} = \mathbf{0}$ is more subtle. The integral $\int_a^b \mathbf{F}(\mathbf{x}(t)) \cdot \frac{d\mathbf{x}}{dt}(t)\, dt$ (here $(\cdot)$ denotes dot product of vectors) measures the integral of the tangential component of $\mathbf{F}(\mathbf{x})$ along a contour $\mathbf{x}(t)$. If $\nabla \times \mathbf{F} = \mathbf{0}$, these integrals are always *zero* when we integrate along any small closed contour. This fact, which can be viewed as a statement about "local path independence" of line integrals of $\mathbf{F}(\mathbf{x})$ along contours in three dimensional space, has natural physical significance in various situations we have mentioned.

For vector operators certain identities are *always* true, no matter which smooth scalar or vector functions we apply them to. One of these is the equation

$$\textbf{curl}(\textbf{grad}\ \phi) = \nabla \times (\nabla\phi) = \mathbf{0}, \tag{4}$$

which can be verified directly from the definitions (2). This formula suggests that one way to get vector fields that satisfy the law $\nabla \times \mathbf{F} = \mathbf{0}$ is to take $\mathbf{F}$ to be the gradient of some scalar valued function. Indeed, by using standard methods of advanced calculus one can show that this is essentially the *only* way to produce such "curl free" vector fields.

Theorem 8.1 *If* $\mathbf{F}$ *is a smooth vector field with the property* $\nabla \times \mathbf{F} = \mathbf{0}$, *then near any point* $\mathbf{p}$ *we can find a corresponding smooth scalar valued function* $\phi(x, y, z)$ *such that* $\mathbf{F} = -\textbf{grad}\ \phi = -\nabla\phi$ *near* $\mathbf{p}$.

On domains that are simply connected† we get a globally defined "primitive function" $\phi(x, y, z)$ such that $\mathbf{F} = -\textbf{grad}\ \phi$.

Theorem 8.2 *If* $\mathbf{F}$ *is a smooth vector field with* $\nabla \times \mathbf{F} = 0$ *on a simply connected domain* D *in* $\mathbf{R}^3$, *then there is a smooth scalar valued function* $\phi(x, y, z)$ *defined throughout* D *with* $\mathbf{F} = -\textbf{grad}\ \phi$ *everywhere in* D.

The function ϕ for which $\mathbf{F} = -\textbf{grad}\ \phi$, if there is such a function at all, is called a **primitive function** for the vector field $\mathbf{F}$; it plays a role similar to that of an antiderivative of a scalar valued function (think of ∇ as a kind of differentiation operation). Furthermore, one can prove that the primitive function is unique up to an added constant, so that it is almost completely determined by the original vector field. In fact, all information about the vector field $\mathbf{F}$ is implicit within the primitive function ϕ which, being scalar valued, is much easier to deal with than the vector field itself. Besides, this primitive function has its own physical significance in many applications, as we will explain.

If a vector field $\mathbf{F}$ satisfies the conservation law $\nabla \times \mathbf{F} = \mathbf{0}$, then by Theorem 8.1 it must have the form $\mathbf{F} = -\textbf{grad}\ \phi$ on various domains, and we may invoke the second conservation law (3) to see that

$$0 = \textbf{div}\ \mathbf{F} = -\textbf{div}\ (\textbf{grad}\ \phi) = -\nabla\cdot(\nabla\phi).$$

† We have not defined simple connectedness for sets in three-dimensional space, but there is a natural way to define this concept (it is rather different from the definition of simple connectedness for sets in the plane). All convex or star-shaped sets in $\mathbf{R}^3$ are simply connected; so is any spherical shell.

But

$$\nabla \cdot (\nabla \phi) = \nabla^2 \phi = \frac{\partial^2 \phi}{\partial x^2} + \frac{\partial^2 \phi}{\partial y^2} + \frac{\partial^2 \phi}{\partial z^2}$$

is just the familiar Laplacian operator ∇^2 applied to ϕ. Thus, a vector field satisfying the laws (3) arises as the gradient of an essentially unique *harmonic* function. This function will be referred to as the **potential function** associated with $\mathbf{F}$; it has the properties

(5)
$$\mathbf{F} = -\mathbf{grad}\,\phi = -\nabla\phi$$
$$\nabla^2\phi = 0.$$

There is no escape from this connection between harmonic functions and physically significant vector fields as long as conservation laws like (3) are valid; the equations (5) are mathematical consequences of the physical laws (3). Thus, potential theory—the study of harmonic functions of several variables—lies inevitably at the heart of many physical problems.

From what we have said, one might be led to think that all of physics reduces to the same mathematical problem: solving Laplace's equation in various domains. But there is more to the solution of a physical problem based on conservation laws (3) than just solving Laplace's equation in any way we please; there are always boundary conditions which come to us as part of the problem, and this is where the difference between, say, electrostatic and heat propagation problems begins to appear. The boundary conditions natural to each type of application (electrostatics, fluid flow, heat propagation, etc.) are quite different. Thus, a boundary condition such as $\partial\phi/\partial n = 0$, which has a natural interpretation and might be expected to arise in heat conduction problems, is not meaningful (or at least, not natural) in electrostatic problems. Since solutions of boundary value problems for Laplace's equation are very sensitive to changes in the boundary conditions, we face quite different mathematical problems in each kind of application. The boundary conditions natural to each kind of application will be reviewed case-by-case, with illustrating examples, in the rest of this chapter; the physical significance of the potential function ϕ will also be indicated.

Physical problems are intrinsically three-dimensional. Nevertheless, two-dimensional analogs of these problems are often relevant and useful, for reasons which have been outlined at the beginning of Chapter 7. In setting up two-dimensional analogs certain changes must be made, based on physical considerations, but these are not very extensive. For example, the gravitational potential of a point mass located at the origin in three-dimensional space is the well known "$1/r$ potential" given by

$$\phi(\mathbf{x}) = \phi(x, y, z) = \frac{-M}{\sqrt{x^2 + y^2 + z^2}} \qquad (M = \text{mass at origin}),$$

the reciprocal of the distance (in $\mathbf{R}^3$) from $\mathbf{x}$ to the origin $\mathbf{0} = (0, 0, 0)$. There is no two-dimensional space in real life in which we can experimentally determine the potential associated with a point mass at the origin in the plane, but there are natural reasons which prompt us to use the "logarithmic potential"

$$\phi(z) = \phi(x, y) = M \cdot \log \sqrt{x^2 + y^2} \quad (z \neq 0)$$

whenever we want a meaningful two-dimensional analog of the three-dimensional situation. These modeling questions are best left to a physics course. The important point is that in two-dimensional problems the central ideas are similar to those in three-dimensional problems; vector fields are gradients of harmonic functions.

In two-dimensional situations there is also a conjugate harmonic function ϕ^*, and an analytic function of a complex variable $f(z) = \phi(z) + i\phi^*(z)$, associated with the harmonic function ϕ that solves a given boundary value problem for Laplace's equation.† When ϕ is the potential function in a physical problem, the companion conjugate function ϕ^* often has its own special physical significance, and so does the analytic function $f = \phi + i\phi^*$. This (essentially unique) analytic function f is called the **complex potential** associated with the problem. It is especially useful in understanding two-dimensional fluid flow problems.

In concluding this section, let us once again urge the interested reader to spend some time reading the Feynman lectures [6], where the physical aspects of this discussion are beautifully elaborated.

EXERCISES

1. Verify the identities given below for functions $\phi(x, y, z)$ and smooth vector fields $\mathbf{F}(x, y, z)$ in three variables.

$$\text{(i) } \nabla \times (\nabla\phi) = \mathbf{curl}(\mathbf{grad}\ \phi) = \mathbf{0}$$

$$\text{(ii) } \nabla \cdot (\nabla\phi) = \mathbf{div}(\mathbf{grad}\ \phi) = \frac{\partial^2\phi}{\partial x^2} + \frac{\partial^2\phi}{\partial y^2} + \frac{\partial^2\phi}{\partial z^2}$$

$$\text{(iii) } \nabla \cdot (\nabla \times \mathbf{F}) = \mathbf{div}(\mathbf{curl}\ \mathbf{F}) = 0.$$

2. Verify the following properties of the gradient operation applied to smooth functions $f(x)$ and $g(x)$.

$$\text{(i) } \nabla(f \pm g) = \nabla f \pm \nabla g$$

$$\text{(ii) } \nabla(\lambda f) = \lambda\nabla f \quad (\lambda \text{ a real scalar})$$

$$\text{(iii) } \nabla(fg) = f\nabla g + g\nabla f.$$

3. If $\mathbf{F}(\mathbf{x})$ is a given continuous vector field and there is a smooth real valued function $\phi(\mathbf{x})$ such that $\mathbf{F} = -\mathbf{grad}\ \phi$, show that there is also a smooth real valued function $\psi(\mathbf{x})$ such that $\mathbf{F} = +\mathbf{grad}\ \psi$.

---

† At least this is true *locally*—on simply connected subdomains of the domain specified in the problem—as explained in Section 7.2.

4. If a vector field $\mathbf{F}$ is derived from a smooth real valued primitive function $\phi(\mathbf{x})$ via $\mathbf{F} = -\nabla\phi$, prove that $\mathbf{F}$ must satisfy the following conditions:

(i) In three dimensions, $\nabla \times \mathbf{F} = 0$

(ii) In two dimensions, if $\mathbf{F} = U\mathbf{i} + V\mathbf{j}$, the components satisfy the equation

$$\frac{\partial U}{\partial y} - \frac{\partial V}{\partial x} = 0.$$

In (i) write $\mathbf{F} = U\mathbf{i} + V\mathbf{j} + W\mathbf{k}$; what does $\nabla \times \mathbf{F} = \mathbf{0}$ mean in terms of the scalar valued component functions? Does this display an analogy between the conditions in (i) and (ii)?

5. Prove that the following vector fields in two dimensions *cannot* be derived from primitive functions ϕ by the formula $\mathbf{F} = -\nabla\phi$.

(i) $\mathbf{F} = x\mathbf{i} - y\mathbf{j}$

(ii) $\mathbf{F} = \dfrac{x}{(x^2 + y^2)}\,\mathbf{i} + y\mathbf{j}.$

6. Let $\mathbf{F}$ be a continuous vector field in two dimensions defined on a *domain.* Assume there is at least one smooth primitive function $\phi(\mathbf{x})$ such that $\mathbf{F} = -\nabla\phi$. If ψ is any other smooth primitive function for $\mathbf{F}$, prove that $\phi - \psi = c$ (real constant).

Hint: Prove that $\phi - \psi$ is locally constant first; prove $\phi - \psi =$ constant on any disc D within the domain of definition of $\mathbf{F}$. Use partial integration.

7. One might guess that the function $\phi(x, y) = 1/|z| = 1/\sqrt{x^2 + y^2}$ is a potential function on the plane, since its three-dimensional analog $\phi(x, y, z) = 1/\sqrt{x^2 + y^2 + z^2}$ is an important harmonic function in many physical problems. Prove that this guess is misguided by showing that $\phi(x, y)$ is *not* harmonic (even if we exclude the obvious singular point $x = 0, y = 0$). Verify that $\phi(x, y, z)$ is harmonic in three dimensions.

Note: We have already seen that the logarithmic potential

$$\phi(z) = \phi(x, y) = -\log\sqrt{x^2 + y^2} = \log(1/\sqrt{x^2 + y^2})$$

is harmonic for $z \neq 0$.

8.2 ELECTROSTATIC FIELDS

Suppose that $\mathbf{F}(\mathbf{x})$ represents the steady electric field produced by a stationary distribution of charge in space; thus, the force exerted by the field on a test charge of strength e located at $\mathbf{x}$ is given by the vector $e \cdot \mathbf{F}(\mathbf{x})$. The charge e may be positive or negative. The law $\mathbf{curl\ F} = \mathbf{0}$ holds throughout space; it can be demonstrated mathematically that this is a consequence of the fact that the electrostatic force between charged particles is *central* (directed along the line segment between the particles). No other physical properties of the force are needed to establish the validity of this law.

More elaborate physical reasoning based on Gauss' divergence theorem leads to the conclusion that $\mathbf{div\ F}$ is everywhere proportional to the charge density $\rho(\mathbf{x})$, which is a scalar function of position,

$$\text{(6)}\quad \mathbf{div\ F} = \nabla \cdot \mathbf{F} = k \cdot \rho(\mathbf{x}) \quad \text{for all } \mathbf{x} \quad (k \text{ is a physical scale constant}).$$

In any *charge free* region, we must have $\mathbf{div\ F} = 0$, so that both of the conservation laws (3) are valid. Thus, there is a potential function $\phi(\mathbf{x})$ that determines $\mathbf{F}$ in such regions. Potential theory cannot be applied so easily unless we restrict our attention to charge free regions, but this situation is encountered commonly enough.

The potential ϕ is determined only up to an added constant; if we consider the difference in potential between points $\Delta\phi = \phi(\mathbf{p}_2) - \phi(\mathbf{p}_1)$ within a charge free region, the added constant cancels out. These differences, rather than ϕ itself, have great physical significance. If we move a particle with charge e from $\mathbf{p}_1$ to $\mathbf{p}_2$, the amount of work that must be done against the electric field is given by

$$\text{(7)}\qquad \Delta W = e\,\Delta\phi = e[\phi(\mathbf{p}_2) - \phi(\mathbf{p}_1)].$$

(We adopt the convention that $\Delta W > 0$ if work is done *against* the field, and $\Delta W < 0$ if work is done *by* the field.) In fact, the work done in moving a particle against a force field, along a curve $\gamma(t)$, is given by (-1) times the integral, with respect to arc length, of the tangential component of the force on the particle:

$$\text{(8)}\qquad \Delta W = -\int_a^b F_{\tan}(t)\,ds = -\int_a^b \left[e\mathbf{F}(\gamma(t)) \cdot \frac{d\gamma}{dt}\right] dt.$$

The minus sign arises from the convention that $\Delta W > 0$ corresponds to work done *against* the field. These work integrals are defined for vector fields and parametrized curves in Euclidean spaces of any dimension, and should not be confused with the line integrals we have been using with functions of a complex variable. (See Kaplan [12], Sections 5.1 to 5.6 for further comment on such work integrals.)

We shall briefly derive formula (8), taking a vector field $\mathbf{F}(\mathbf{x})$ and a parametrized curve $\gamma(t)$ in three-dimensional space. Let us assume that a particle is moved from $\mathbf{p}_1$ to $\mathbf{p}_2$ along the path given by the parametrized curve

$$\gamma(t) = x(t)\mathbf{i} + y(t)\mathbf{j} + z(t)\mathbf{k} \quad \text{for } a \le t \le b$$

(think of t as a time variable). Evidently, $\gamma(a) = \mathbf{p}_1$ and $\gamma(b) = \mathbf{p}_2$; the instantaneous velocity of the particle (a vector quantity) is given by

$$\frac{d\gamma}{dt} = \lim_{\Delta t \to 0} \frac{\gamma(t + \Delta t) - \gamma(t)}{\Delta t} = \frac{dx}{dt}\mathbf{i} + \frac{dy}{dt}\mathbf{j} + \frac{dz}{dt}\mathbf{k}.$$

At time t the particle, located at $\mathbf{p} = \gamma(t)$, is subjected to a force $e\mathbf{F}(\mathbf{p}) = e\mathbf{F}(\gamma(t))$. Since

$$\mathbf{t} = \frac{d\gamma}{dt} \Big/ \left\| \frac{d\gamma}{dt} \right\|$$

represents a tangent vector of unit length associated with $\mathbf{p} = \gamma(t)$, the tangential component of the force $e\mathbf{F}$ at $\mathbf{p}$ is

$$(e\mathbf{F})_{\tan} = \|e\mathbf{F}\| \cos\theta = (e\mathbf{F}) \cdot \mathbf{t} = \left(e\mathbf{F} \cdot \frac{d\gamma}{dt}\right) \Big/ \left\|\frac{d\gamma}{dt}\right\|.$$

The integral $-\int_a^b (e\mathbf{F})_{\tan}\, ds$ with respect to arc length becomes a Riemann integral if we replace ds by $\|d\gamma/dt\|\, dt$; thus,

$$\begin{aligned}\Delta W &= -\int_a^b (e\mathbf{F})_{\tan}\, ds = -\int_a^b (e\mathbf{F})_{\tan} \left\|\frac{d\gamma}{dt}\right\| dt = -\int_a^b \left[e\mathbf{F} \cdot \frac{d\gamma}{dt}\right] dt \\ &= -\int_a^b e\left[U\frac{dx}{dt} + V\frac{dy}{dt} + W\frac{dz}{dt}\right] dt\end{aligned}$$

if $\mathbf{F}$ has components U, V, and W. If $\mathbf{F}$ also has the form $\mathbf{F} = -\mathbf{grad}\,\phi$, then

$$\begin{aligned}\Delta W &= +e\int_a^b \left[\frac{\partial\phi}{\partial x}\frac{dx}{dt} + \frac{\partial\phi}{\partial y}\frac{dy}{dt} + \frac{\partial\phi}{\partial z}\frac{dz}{dt}\right] dt \\ &= +e\int_a^b \frac{d}{dt}[\phi(\gamma(t))]\, dt = +e[\phi(\gamma(b)) - \phi(\gamma(a))] \\ &= +e\Delta\phi.\end{aligned}$$

The boundary conditions associated with electrostatic problems arise naturally from the following phenomenon. If a steady electric field extends through a region that includes a conductor, as indicated by the shaded region in Figure 8.1, then the electric field $\mathbf{F} = -\mathbf{grad}\,\phi$ must be everywhere perpendicular to the boundary of the conductor. The presence of any tangential component of electric force would cause currents within the conductor, and a consequent redistribution of charge; since the charge (electrons) within the conductor contributes to the total field throughout space, $\mathbf{F}$ would vary as a result of these induced currents. This cannot happen in a steady field. Recall that $\mathbf{grad}\,\phi$ is always perpendicular to the level surfaces $\phi = constant$ (except at a few singular points where $\mathbf{grad}\,\phi = \mathbf{0}$ and the locus $\phi = constant$ may fail to be smooth enough to have a well defined normal direction $\mathbf{n}$); thus,

(9) $\phi = constant$ on the boundary of (and within) a conductor

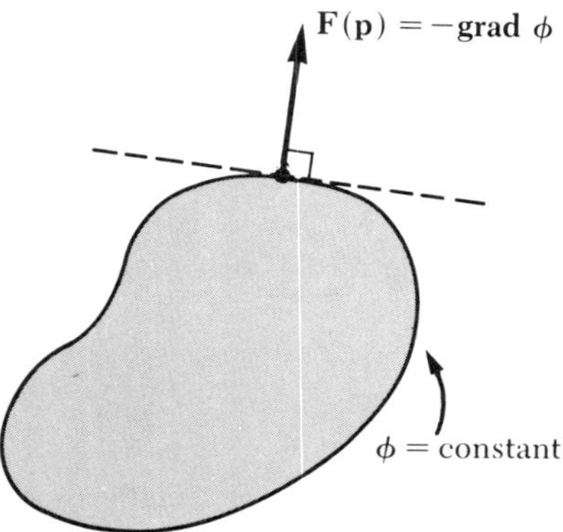

Figure 8.1

in steady state problems. There is a boundary condition of this kind for each isolated conducting surface within the region we are considering. In the rest of the region, we have **equipotential surfaces** $\phi =$ *constant* ; the amount of work required to move a charge from the surface $\phi = V_1$ to the surface $\phi = V_2$ is always $\Delta W = e \cdot (V_2 - V_1)$, regardless of the particular path or initial and final points involved, in view of formula (7).

In practice, one often encounters a charge free region that contains, or is bounded by, various conducting surfaces that are insulated from each other. Fields for which the potential ϕ assumes specified values $V_1, \ldots, V_n$ on the conductors $C_1, \ldots, C_n$ are of great interest; they are easily produced by connecting batteries to the separate conductors. Once the batteries have been connected, charge will flow briefly and distribute itself over the conducting surfaces until a steady-state situation has been achieved, and the potential is constant throughout each conductor. All charges are confined to the conductors, so that the region outside of the conductors remains charge free and the field is given by a potential function ϕ. In the final steady state we must have

$$\phi = V_1 \quad \text{on} \quad C_1; \ldots; \phi = V_n \quad \text{on} \quad C_n.$$

Thus ϕ is the solution of a simple Dirichlet boundary value problem. Notice that the only physical data we start with are the assigned potentials $V_1, \ldots, V_n$ on the conducting surfaces (the boundary values). Here are some two-dimensional versions of these problems.

Example 8.1 Let us derive the potential function within the upper half plane, when the boundary segments $(-\infty, 0)$ and $(0, +\infty)$ are maintained at potentials $V = -10$ and $V = +10$ volts, respectively, by means of a battery, as indicated in Figure 8.2; there is a thin insulator at $z = 0$ to separate these two conducting boundary segments. This problem is easily solved if we recall the discussion of Dirichlet problems in the upper half plane given in Section 7.1:

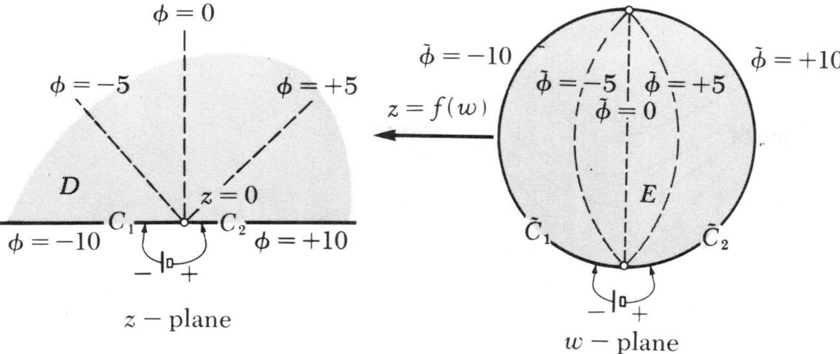

Figure 8.2 Equipotential lines (dashed) and boundary conditions in Example 8.1.

take

$$\phi(z) = 10 - \frac{20}{\pi}\operatorname{Arg}(z)$$

$$= 10 - \frac{20}{\pi}\arccos\left(\frac{x}{\sqrt{x^2 + y^2}}\right) \quad \text{for all } z \text{ such that } \operatorname{Im}(z) > 0.$$

By applying the conformal mapping principle, we may transform this into the solution of related electrostatic problems in more complicated domains. Consider the disc $E = \{w: |w| < 1\}$ and the boundary conditions shown in Figure 8.2. The fractional linear transformation (recall section 4.8) $z = f(w) = -i(w + i)/(w - i)$ maps E conformally onto the upper half plane; at the same time, the circular arcs extending from $-i$ to $+i$ that bound E are mapped onto the corresponding rays from $z = 0$ to infinity, that make up the boundary of the half plane. Therefore, by transforming $\phi(z)$ to a function $\tilde{\phi}(w)$ defined on the unit disc $|w| < 1$

$$\tilde{\phi}(w) = \left[\phi(z)\Big|_{z=f(w)}\right] = 10 - \frac{20}{\pi}\operatorname{Arg}\left(\frac{1}{i}\left(\frac{w+i}{w-i}\right)\right)$$

$$= 10 - \frac{20}{\pi}\arccos\left[\frac{2u}{\sqrt{4u^2 + (u^2 + v^2 - 1)^2}}\right],$$

we get a solution of the boundary value problem in the disc. The equipotential curves in the upper half plane, given by $\phi = \textit{constant}$, are just radial lines extending from the origin to infinity. These are transformed by the inverse mapping $w = \check{f}(z) = i(z - i)/(z + i)$ to the equipotential curves $\tilde{\phi}(w) =$ constant in the unit disc E. One can verify that the latter correspond to circular arcs from $-i$ to $+i$.

Example 8.2 Now let us start with a *given* harmonic function and inquire: What kind of electrostatic problems can it solve? In other words, we are starting with the solution and are looking for the problem(s) it solves. This is done by sketching the equipotential curves and realizing that any such curve

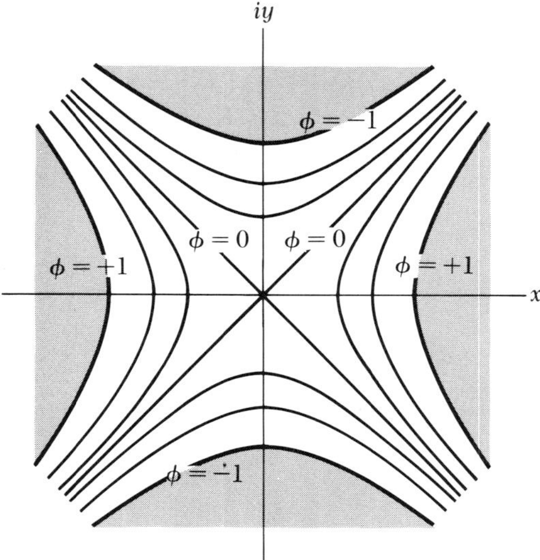

Figure 8.3 Equipotential lines for the quadrupole field (Example 8.2).

$\phi = c_0$ can be taken to be the boundary of a conductor maintained at the potential $V = c_0$.

First consider the function $\phi(z) = \text{Re}(z^2) = x^2 - y^2$; its level curves are indicated in Figure 8.3. If we think of the loci $\phi = +1$ and $\phi = -1$ as conducting "surfaces" maintained at potentials of $V = +1$ and $V = -1$ volts, respectively, then ϕ gives the potential function for the corresponding electric field within the domain E bounded by these curves. The curves in Figure 8.3 are now equipotential curves of ϕ. The field produced by this configuration of conductors is known as a "quadrupole field" and is the sort of field one might want to use in focusing certain kinds of electron beams (a beam oriented perpendicular to the page, and passing through the origin).

For a more complicated example, let us recall our previous discussion of the mapping $w = \sin z$, which transforms the strip $D = \{z: -\pi/2 < \text{Re}(z) < \pi/2\}$ conformally onto the doubly cut domain E obtained by deleting the rays $(-\infty, -1]$ and $[+1, +\infty)$ from the real axis in the w-plane. The inverse mapping $z = \text{Arcsin}\, w$ is analytic on E. We have sketched the lines of constancy of its real and imaginary parts $z = X(w) + iY(w) = \text{Arcsin}\, w$ by methods explained in Section 4.10. The harmonic function $X(w)$ is defined off the cuts and has as its equipotential curves the family of hyperbolas shown in Figure 8.4; the locus $X(w) = c$ is the image of the vertical line $z = c + iy$ under the transformation $w = \sin z$. The function $X(w)$ is the solution of the electrostatic problem in which we have two semi-infinite plates maintained at potentials $+1$ and -1, and wish to find the electrostatic potential between the plates.

EXERCISES

1. Express the solution $\tilde{\phi}(w)$ in Example 8.1 in terms of arctan*(t)

$$\tilde{\phi}(u, v) = \frac{1}{\pi} \arctan^*\left[\frac{1 - u^2 - v^2}{2v}\right],$$

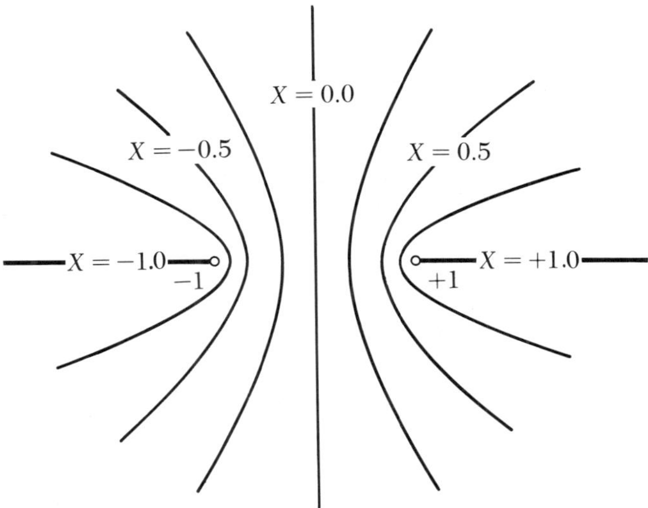

Figure 8.4 The field surrounding two semi-infinite plates.

where $\arctan^*(t)$ is determined so that $0 \leq \arctan^*(t) \leq \pi$. Why is it not legitimate to use the usual determination, for which $-\pi/2 \leq \arctan(t) \leq \pi/2$?

2. If $\alpha \neq 0$ and $\beta > 0$ are real, show that $H(z) = \alpha \cdot \log \beta\,|z|$ is harmonic for $z \neq 0$. Use such functions to determine the electrostatic potential in the region between two coaxial circular conductors, as shown in Figure 8.5.

Answer: (i) $\alpha = 1/\log(r_2/r_1)$ and $\beta = 1/r_1$; (ii) $\alpha = 1/\log(r_1/r_2)$ and $\beta = 1/r_2$.

3. Determine the electric potential within the regions shown in Figure 8.6. These regions are bounded by conductors maintained at the potentials indicated; the potentials $V(z)$ are bounded near the points where thin insulators separate boundary segments, and are bounded at infinity.

Hint: In (iii) transform to the upper half plane using $w = \sin(\pi z/2a)$.

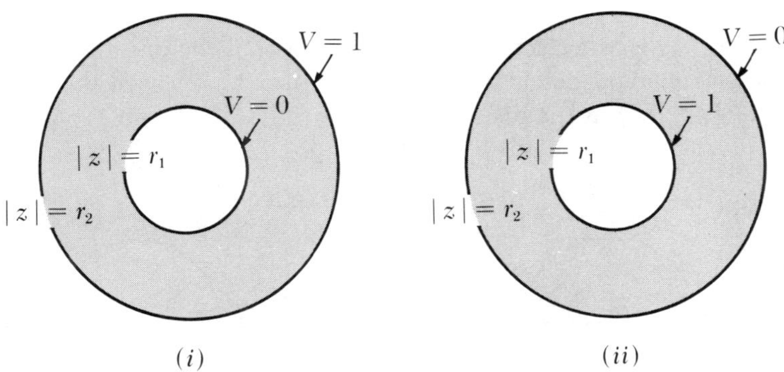

Figure 8.5

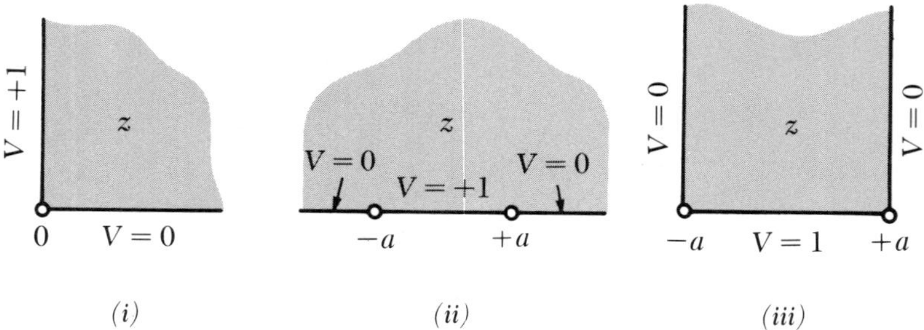

Figure 8.6

Answers: (i) $V(z) = (2/\pi)\text{Arg}(z) = (2/\pi)\arccos[x/\sqrt{x^2 + y^2}]$; (ii) $V(z) = (1/\pi)\text{Arg}\left(\dfrac{z - a}{z + a}\right) = (1/\pi)\arctan^*[2ay/(x^2 + y^2 - a^2)]$ where we take $0 \leq \arctan^*(s) \leq \pi$, since $(z - a)/(z + a)$ is in the upper half plane; (iii) $V(z) = \dfrac{1}{\pi}\text{Arg}\left[\dfrac{\sin(\beta z) - a}{\sin(\beta z) + a}\right]$, where $\beta = \pi/2a$.

4. Solve the electrostatic problem in the semidisc shown in Figure 8.7 (potential $V(z)$ bounded near -1 and $+1$), by mapping the semidisc onto the first quadrant.

Answer: Use $w = (1 + z)/(1 - z)$; $V(z) = (2/\pi)\text{Arg}[(1 + z)/(1 - z)] = (2/\pi)\arctan[2y/(1 - x^2 - y^2)]$

$$= (2/\pi)\arcsin[2y/\sqrt{4y^2 + (1 - x^2 - y^2)^2}]$$

5. Find the electrostatic potential $V(z)$ in the unbounded region shown in Figure 8.8. Assume that $V(z)$ is bounded near the points -1 and $+1$, which separate conducting boundary walls (maintained at different potentials), and assume that $V(z)$ is bounded at infinity. Sketch a few equipotential lines.

Hint: Apply the conformal mapping $w = -1/z$; use Exercise 4.

Note: This result could also be obtained using the observations in Exercise 15, Section 7.6.

6. Find the electrostatic potential within the disc that corresponds to the boundary potentials shown in Figure 8.9. Sketch the loci $V = \frac{1}{4}$, $V = \frac{1}{2}$, $V = \frac{3}{4}$ in the disc.

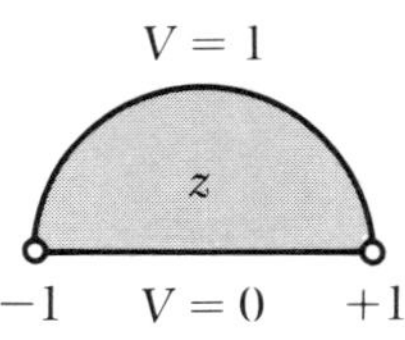

Figure 8.7

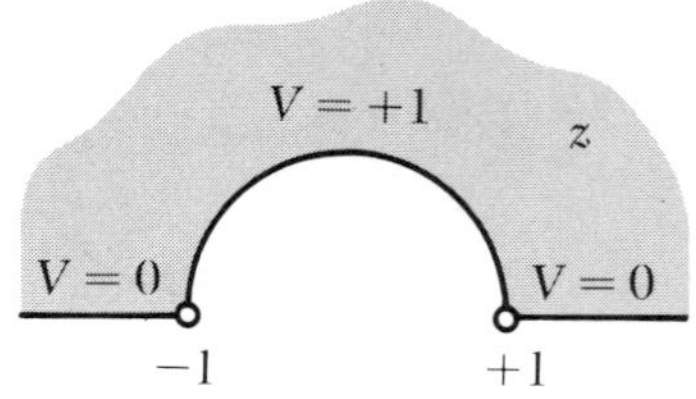

Figure 8.8

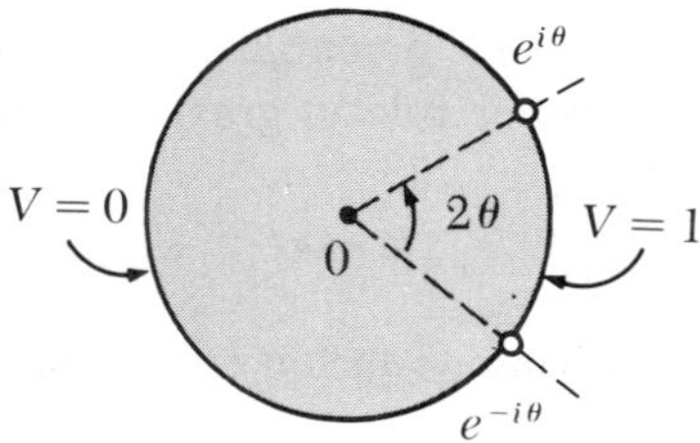

Figure 8.9 Angle θ is fixed.

Hint: Find a fractional linear transformation to get a familiar problem in the upper half plane.

8.3 GRAVITATIONAL FIELDS

Gravitational fields produced by a stationary distribution of mass may be analyzed by techniques similar to those used for electrostatic fields, except that there are no negative masses. Since the force between particles is central, we must have $\mathbf{curl\ F} = \mathbf{0}$. Physical reasoning based on Gauss' Divergence Theorem leads to the conclusion that

$$\mathbf{div\ F} = k \cdot \rho(\mathbf{x}) \quad \text{everywhere} \quad (k \text{ a physical constant of proportionality}),$$

where $\rho(\mathbf{x})$ is the mass density at position $\mathbf{x}$. Therefore, in any mass free region both conservation laws (3) are valid and there is a potential function ϕ that determines the gravitational field, $\mathbf{F} = -\mathbf{grad}\ \phi$. The net force on a particle located at $\mathbf{x}$ is given by the vector $m \cdot \mathbf{F}(\mathbf{x})$, where m is the mass of the particle. Once again, potential *differences* are physically meaningful; $\Delta\phi = \phi(\mathbf{p}_2) - \phi(\mathbf{p}_1)$ is related to the energy required to move a point mass of magnitude m through space from position $\mathbf{p}_1$ to position $\mathbf{p}_2$ by the formula

$$\Delta W = m \cdot \Delta\phi. \tag{10}$$

The reasoning leading to formula (10) is the same as that used to derive formula (7) for electrostatic fields.

When we turn to the question of physically meaningful boundary value problems, we are in for a surprise. In gravitational problems there is no analog of the "conducting region" we encounter so often in electrostatic problems—because there are no "mass carriers," analogous to the charge-carrying electrons present in a conductor, that move freely within a rigid body under the influence of external gravitational fields. Thus, boundary conditions of the form $\phi =$ *constant* do not arise naturally as they do in electrostatics. In fact, further thought indicates that boundary value problems *almost never* arise in a natural way in connection with gravitational problems! The physically natural initial data in such problems are not boundary conditions on the gravitational potential ϕ, such as $\phi =$ *constant* or $\partial\phi/\partial n = 0$, but the mass density distribution $\rho(\mathbf{x})$. We must work from this distribution by other means to determine the gravitational field $\mathbf{F}(\mathbf{x})$, or its potential function $\phi(\mathbf{x})$. Since boundary value problems play a very minor role in gravitational problems, we will not pursue this subject further.

In electrostatic problems in which the potential has specified values on the boundary, the charge distribution on the surface of the conducting walls can be very complicated; it is not known at the beginning of the problem. This should further illustrate the profound differences in the kind of initial data we may expect in electrostatic and gravitational problems.

8.4 HEAT PROPAGATION AND TEMPERATURE DISTRIBUTIONS

Let us consider a region that has a reasonably smooth boundary and is filled with some homogeneous thermally conducting medium. We shall assume that various parts of the boundary are maintained at constant temperatures through contact with external heating or cooling devices, and that the remaining parts of the boundary are perfectly insulated so that no heat can flow across these parts of the boundary. Once the heat sources and sinks are operative, heat will flow within the region until a steady flow is achieved (that is, until the temperature distribution no longer varies with time). The heat flow can be represented by a vector field $\mathbf{F}(\mathbf{x})$; this vector, at a point $\mathbf{p}$ in space, determines the heat flow per unit time, per unit area, across any small rectangle centered at $\mathbf{p}$. If the rectangle has area ΔA and is perpendicular to the unit vector $\mathbf{n} = a\mathbf{i} + b\mathbf{j} + c\mathbf{k}$ at $\mathbf{p}$, then the heat flow (per unit time) across the rectangular surface is given, to first order approximation, by

$$\Delta Q = (\mathbf{F}(\mathbf{p}) \cdot \mathbf{n}) \cdot \Delta A; \tag{11}$$

that is, ΔQ is equal to this expression plus additional terms that become negligible in comparison with (11) as the diameter of the rectangle is decreased (the normal direction $\mathbf{n}$ is kept fixed).

In such problems the vector field $\mathbf{F}$ is not easily measured. Instead, it is the temperature distribution $T(\mathbf{x})$, a scalar valued function, that is directly observed. Heat flow is completely determined once the temperature distribution is known, by means of the equation

$$\mathbf{F} = -k\,\mathbf{grad}\,T = -k\,\nabla T, \tag{12}$$

where $k = k(\mathbf{x})$ is the heat conductivity of the medium at position $\mathbf{x}$. If the medium is homogeneous then $k = \textit{constant}$.

The fact that $\mathbf{F} = -k(\mathbf{x}) \cdot \mathbf{grad}\,T$ follows almost immediately from the operational definition of how one measures the conductivity of the medium. The minus sign appears once we adopt the convention that *positive* heat flow corresponds to heat flow from a high temperature to a lower temperature. This equation is valid even for time-varying heat propagation, and not just the steady state flows which are the focus of our attention. If $\mathbf{F} = \mathbf{F}(\mathbf{x}, t)$ and $T = T(\mathbf{x}, t)$, then

$$\mathbf{F}(\mathbf{x}, t) = -k\left[\frac{\partial T}{\partial x}(\mathbf{x}, t)\mathbf{i} + \frac{\partial T}{\partial y}(\mathbf{x}, t)\mathbf{j} + \frac{\partial T}{\partial z}(\mathbf{x}, t)\mathbf{k}\right]$$

for all $\mathbf{x}$ and all t.

Thus, in heat propagation problems the temperature function $T(\mathbf{x})$ is the potential function associated with the heat flow, and it is this potential function that is measured, rather than the vector field $\mathbf{F} = -\mathbf{grad}\, T$ describing the flow of heat. As a trivial mathematical consequence of equation (4), we see that

$$\text{(13)} \qquad \mathbf{curl}\,\mathbf{F} = -k\,\mathbf{curl}(\mathbf{grad}\ T) = -k\,\nabla \times (\nabla T) = \mathbf{0}$$

in any homogeneous region. On the other hand, physical reasoning based on Gauss' Divergence Theorem leads us to conclude that $\mathbf{div}\,\mathbf{F}$ may be identified with the rate (per unit volume) at which heat is generated or lost at $\mathbf{x}$ for reasons other than simple flow of heat through the medium; that is, $\mathbf{div}\,\mathbf{F} \neq 0$ indicates the presence of some kind of heat source or sink that introduces or removes heat from the system. If there are no heat sources or sinks in the interior of the region E, we must have $\mathbf{div}\,\mathbf{F} = 0$ throughout E. Then $\mathbf{F}$ will satisfy both conservation laws (3), and the temperature distribution T will be a *harmonic* function, since $-k\,\nabla^2 T = \mathbf{div}(-k\,\nabla T) = \mathbf{div}\,\mathbf{F} = 0$.

Let us examine the boundary conditions on $T(\mathbf{x})$ that arise naturally in the situation just described. Obviously the conditions $T = T_1, \ldots, T = T_n$ prevail on the parts of the boundary that are maintained at constant temperatures. At any point $\mathbf{p}$ where the boundary wall is smooth, there is a well defined normal vector $\mathbf{n}$ perpendicular to the bounding wall, and the rate of heat flow *across* the boundary at $\mathbf{p}$ is given by

$$\mathbf{n} \cdot \mathbf{F} = -k(\mathbf{n} \cdot \mathbf{grad}\ T) = -k\frac{\partial T}{\partial n}(\mathbf{p}) \quad \text{(normal derivative)},$$

and this must be identically zero wherever the wall is insulated. Thus $\partial T/\partial n = 0$ along insulated parts of the boundary. Heat propagation problems therefore lead to boundary value problems for Laplace's equation with mixed Dirichlet and Neumann boundary conditions.

Figure 8.10 shows the isothermal curves $T = \textit{constant}$ (isothermal *surfaces* in a three-dimensional problem) as solid curves for a typical two dimensional problem. The reader should notice that these curves are always perpendicular to the insulated portion of the boundary. Indeed, for *any* smooth function $T(x, y)$, the gradient vector ∇T at $\mathbf{p}$ is always perpendicular to the level locus (a surface in three dimensions) $T = \textit{constant} = T(\mathbf{p})$ that passes through $\mathbf{p}$. The condition $0 = \partial T/\partial n = \mathbf{n} \cdot (\mathbf{grad}\ T)$ means that the normal vector $\mathbf{n}$ is perpendicular to $\mathbf{grad}\ T$; therefore, $\mathbf{n}$ is tangent to the locus $T = \textit{constant}$. This is precisely what we mean by saying that the surface $T = \textit{constant}$ is perpendicular to the boundary. It also means that the heat

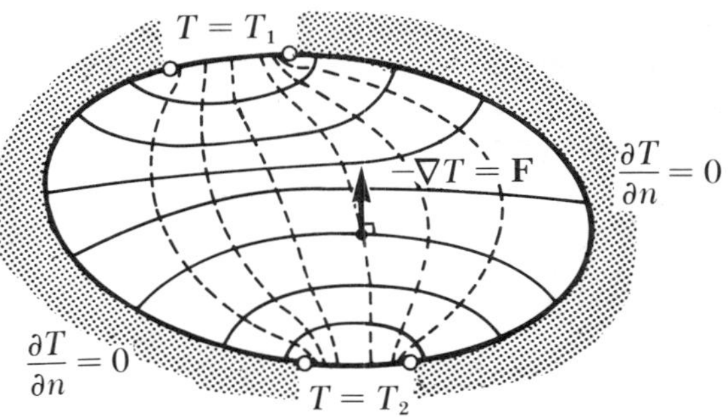

Figure 8.10 Shaded portions of the boundary are insulated ($\partial T/\partial n = 0$). Solid curves are isothermal lines $T =$ constant; dashed lines are the orthogonal curves $T^* =$ constant, which represent lines of heat flux.

flow $\mathbf{F} = -k$ **grad** T has no component perpendicular to the boundary along the insulated sides.

If T is the temperature distribution in a simply connected plane domain, there is a conjugate harmonic function T^* which is determined up to an added constant; this function has its own significance. First recall that the gradient vectors ∇T and ∇T^* are perpendicular at any point (see the discussion of Section 7.6), which means that the loci $T^* =$ *constant* are curves that are perpendicular to the isothermal curves $T =$ *constant*. Consequently, the vectors ∇T and $\mathbf{F} = -k$ **grad** T are always tangent to the loci $T^* =$ *constant*. Since $-$**grad** T indicates the direction of heat flow, one can think of heat as flowing along the curves $T^* =$ *constant*; thus, the loci $T^* =$ *constant* are often referred to as "lines of flux" for the heat flow.

We may take an entirely different view of the conjugate function T^* by recalling that boundary conditions on T of the form $T =$ *constant* or $\partial T/\partial n = 0$ correspond to the *opposite* conditions on T^*, $\partial T^*/\partial n = 0$ and $T^* =$ *constant*, respectively, as explained in Section 7.6. Therefore, for every temperature distribution $T(\mathbf{x})$, there is a conjugate temperature distribution T^*. The latter is a solution of the **conjugate boundary value problem** that is obtained by interchanging boundary conditions, as indicated below.

$$T = \textit{constant} \rightarrow \frac{\partial T^*}{\partial n} = 0$$

$$\frac{\partial T}{\partial n} = 0 \rightarrow T^* = \textit{constant}.$$

Finally, we note that one must sometimes specify the behavior of T at infinity in an unbounded region; usually we encounter conditions of the form $\partial T/\partial n = 0$ (no gain or loss of heat at infinity) or $T =$ *constant* (infinity maintained at a fixed temperature). The first condition is the most common.

Example 8.3 Consider the following boundary value problem for the infinite horizontal strip $E = \{z : 0 < \text{Im}(z) < +\pi\}$;

$$T(x, 0) = +1 \quad \text{if} \quad x < 0 \qquad T(x, 0) = 0 \quad \text{if} \quad x > 0$$

$$T(x, \pi) = +1 \quad \text{if} \quad x < 0 \qquad T(x, \pi) = 0 \quad \text{if} \quad x > 0,$$

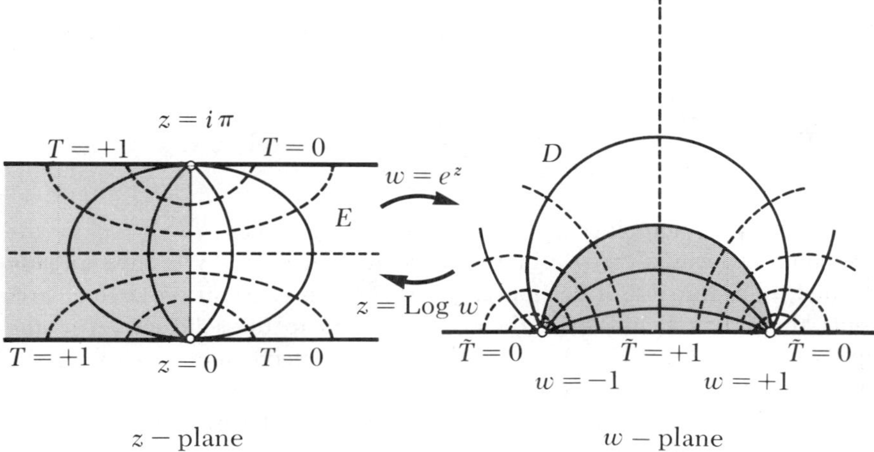

Figure 8.11 In Example 8.3, the function $T(z)$ (and its boundary values) on the strip are transformed to $\tilde{T}(w)$ (with appropriate boundary values) on the half plane.

as indicated on the left in Figure 8.11. The boundary temperatures are discontinuous at $z = 0$ and $z = +i\pi$, which should be thought of as thin insulating strips between isothermal boundary walls. The mapping $w = e^z$ transforms this strip conformally onto a half plane $D = \{w: \text{Im}(w) > 0\}$, so we can convert the given problem into a more tractable problem in D. Our mapping transforms the shaded part of E to the shaded half-disc in D. The given boundary values in E correspond to the following boundary conditions on the real axis (the boundary of D) in the w-plane:

$$\tilde{T}(u, 0) = 0 \quad \text{if} \quad -\infty < u < -1 \quad \text{or} \quad +1 < u < +\infty$$
$$\tilde{T}(u, 0) = +1 \quad \text{if} \quad -1 < u < +1,$$

as indicated in Figure 8.11. The latter Dirichlet problem for the half plane is already familiar from Example 7.3 of Section 7.1, and has as its solution the harmonic function

$$\tilde{T}(w) = \frac{1}{\pi}\text{Arg}\left(\frac{w-1}{w+1}\right) = \text{Re}\left[\frac{-i}{\pi}\text{Log}\left(\frac{w-1}{w+1}\right)\right] \tag{15}$$

for w in D. Now substitute $w = e^z$ to get the desired temperature distribution within the original domain E,

$$T(z) = \frac{1}{\pi}\text{Arg}\left(\frac{e^z-1}{e^z+1}\right) = \frac{1}{\pi}\text{Arg}\left(\frac{e^{z/2}-e^{-z/2}}{e^{z/2}+e^{-z/2}}\right)$$
$$= \frac{1}{\pi}\text{Arg}\left(\tanh\left(\frac{z}{2}\right)\right). \tag{16}$$

Since $\zeta = \tanh(z/2) = \left[\frac{w-1}{w+1}\bigg|_{w=e^z}\right]$ lies in the upper half plane for z in E (and $w = e^z$ in D), we may write $\text{Arg}(\zeta) = \arccos[\text{Re}(\zeta)/|\zeta|]$, and by directly calculating real and imaginary parts of ζ we may express T as a function of the real variables x and y,

$$T(x, y) = \frac{1}{\pi} \arccos\left[\frac{\sinh x}{\sqrt{\sinh^2 x + \sin^2 y}}\right] \quad \text{for } x \text{ real; } 0 < y < \pi. \tag{17}$$

The inverse mapping $z = \operatorname{Log} w$ carries the locus $\tilde{T}(w) = c$ to the corresponding locus $T(z) = c$ in the original domain E. We have already indicated, in Figure 7.2 of Section 7.1, that $\tilde{T}(w)$ is constant on circular arcs passing through -1 and $+1$; these are mapped to the solid curves in the strip shown in Figure 8.11. We leave it to the interested reader to work out the equations of these curves, the isothermal lines in the z-plane. It is not very difficult to start with equation (16) and derive the following formula for the conjugate harmonic function:

$$T^*(z) = \frac{-1}{\pi} \log|\tanh(z/2)| = \frac{1}{\pi} \log\sqrt{\frac{\cosh x + \cos y}{\cosh x - \cos y}}. \tag{18}$$

The level curves $T^* = \textit{constant}$ are indicated by dashed lines in Figure 8.11; these are the lines of heat flux. We leave the derivation of formula (18) and other properties of the conjugate harmonic function T^* as Exercises 3 and 12.

Example 8.4 In Example 7.4 of Section 7.1 we derived solutions to a few Dirichlet and Neumann problems in the semi-infinite strip $E = \{z : -\pi/2 < \operatorname{Re}(z) < +\pi/2 \text{ and } \operatorname{Im}(z) > 0\}$, by noticing that the transformation $w = \sin z$ maps E conformally onto the upper half of the w-plane $D = \{w : \operatorname{Im}(w) > 0\}$, a domain in which the corresponding boundary value problems are far more manageable. The boundary conditions indicated in parts (I), (II), and (III) of Figure 7.3 (in Section 7.1) involve the elementary conditions $T =$ constant and $\partial T/\partial n = 0$. The solutions we obtained were

$$\begin{aligned} H_{\mathrm{I}}(x, y) &= \frac{1}{\pi}\left[\pi - \arccos\left(\frac{\sin x \cosh y - 1}{\cosh y - \sin x}\right)\right] \\ &= \frac{1}{\pi}\left[\pi - \operatorname{Arg}(\sin(x + iy) - 1)\right] \\ H_{\mathrm{II}}(x, y) &= \left(\frac{x}{\pi} + \frac{1}{2}\right) \\ H_{\mathrm{III}}(x, y) &= 1, \end{aligned}$$

respectively. Clearly, these harmonic functions may be interpreted as solutions of two-dimensional heat propagation problems in a homogeneous semi-infinite strip, some of whose sides are maintained at constant temperature, while others are insulated. We interpret the level curves for H_{I}, H_{II}, and H_{III} in Figure 7.3 as isothermal lines, and we note that these curves are perpendicular to the insulated parts of the boundary in each problem.

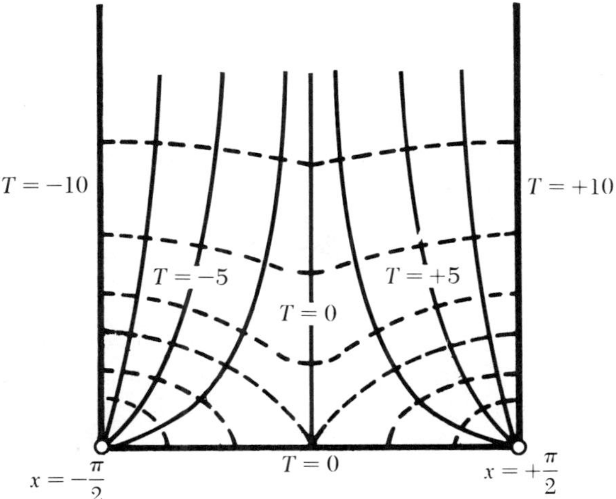

Figure 8.12

We may solve a number of similar problems by using the mapping $w = \sin z$ to transform them to problems in the upper half plane $E = \{w : \mathrm{Im}(w) > 0\}$. For example, we can solve the temperature distribution problem whose boundary conditions are illustrated in Figure 8.12; the appropriate boundary values on the real axis, in the transformed domain E, are

$$\tilde{T}(u + i0) = \begin{cases} +10 & \text{if} \quad -\infty < u < -1 \\ 0 & \text{if} \quad -1 < u < +1 \\ -10 & \text{if} \quad +1 < u < +\infty. \end{cases}$$

The solution $\tilde{T}(w)$ in the half plane is evidently

$$\begin{aligned} \tilde{T}(w) &= 10 - \frac{10}{\pi}\mathrm{Arg}(w - 1) - \frac{10}{\pi}\mathrm{Arg}(w + 1) \\ &= 10 - \frac{10}{\pi}\mathrm{Arg}[(w - 1)(w + 1)] \\ &= 10 - \frac{10}{\pi}\mathrm{Arg}(w^2 - 1). \end{aligned}$$

(This is an equality, rather than a congruence; see Exercise 11.) Substituting $w = \sin z$, we get the solution on the half strip,

$$\begin{aligned} T(z) = \tilde{T}(\sin z) &= 10 - \frac{10}{\pi}\mathrm{Arg}(\sin^2 z - 1) \\ &= 10 - \frac{10}{\pi}\mathrm{Arg}(-\cos^2 z). \end{aligned}$$

But $\text{Arg}(-\cos^2 z) \equiv \text{Arg}(-1) + 2\,\text{Arg}(\cos z) \equiv +\pi + 2\,\text{Arg}(\cos z)$, modulo 2π, and one can show that this congruence is really an equality if z is in the strip (Exercise 11); thus,

$$\begin{aligned} T(z) &= 10 - 10 - \frac{20}{\pi}\text{Arg}(\cos z) \\ &= -\frac{20}{\pi}\text{Arg}(\cos z). \end{aligned}$$

If z is in E, then $w = \cos z$ lies in the *right* half of the w-plane; using the formulas (40) of Section 2.6, we may represent $\text{Arg}(u + iv) = A(u, v)$ explicitly in this half plane, to get

$$T(x, y) = -\frac{20}{\pi}\arcsin\left(\frac{-\sin x \sinh y}{\sqrt{\cos^2 x + \sinh^2 y}}\right) \quad \text{for } z \text{ in } E.$$

Since $T(z)$ is the real part of the analytic function $f(z) = T(z) + iT^*(z) = i(20/\pi)\log(\cos z)$, defined throughout E, the conjugate harmonic function is given by

$$\begin{aligned} T^*(z) &= \text{Im}\, f(z) = (20/\pi)\log|\cos z| = (10/\pi)\log|\cos z|^2 \\ &= (10/\pi)\log(\cos^2 x + \sinh^2 y) \end{aligned}$$

The lines of heat flux, $T^* = \textit{constant}$, are shown as dashed curves in Figure 8.12; naturally, these are orthogonal to the isothermal curves, $T = \textit{constant}$, shown as solid curves.

EXERCISES

1. Solve the temperature distribution problem with boundary conditions indicated in Figure 8.13. Then calculate the conjugate harmonic function T^*, and describe its boundary behavior. What is the physical significance of the conditions satisfied by T and T^* at the boundary?

2. Solve the temperature distribution problem shown in Figure 8.14. Sketch a few isothermal curves $T(z) = \textit{constant}$; notice that, by the symmetry of the problem, $T = 0$ along the y-axis. Verify that $|T(z)|$ is bounded as $z \to 0$, or as $z \to \infty$, from within the half strip.

Answer:

$$\begin{aligned} T(x, y) &= 10 - (20/\pi)\text{Arg}[\sin(\pi z/2)] \\ &= \arccos\left[\frac{\sin(\pi x/2)\cosh(\pi y/2)}{\sqrt{\sin^2(\pi x/2) + \sinh^2(\pi y/2)}}\right]. \end{aligned}$$

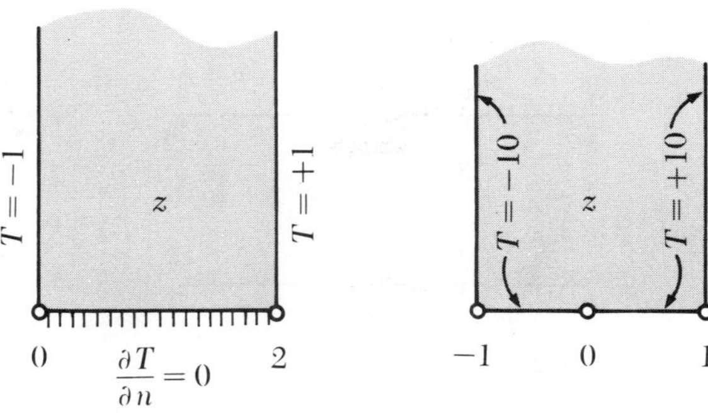

Figure 8.13

Figure 8.14

3. The function $T(z) = (1/\pi)\text{Arg}[\tanh(z/2)]$, defined for $0 < \text{Im}(z) < \pi$, appears as the solution of the temperature distribution problem in Example 8.3. Perform the calculations needed to express T in the form

$$T(x, y) = \frac{1}{\pi}\arccos\left[\frac{\sinh x}{\sqrt{\sinh^2 x + \sin^2 y}}\right]$$

for x real; $0 < y < \pi$. Here $s = \arccos(t)$ is the usual determination, normalized so that $0 \leq s \leq \pi$ for $-1 \leq t \leq +1$.

4. Solve the heat propagation problem illustrated in Figure 8.15 by examining the solution of Example 8.3, and noticing what boundary conditions are satisfied in the half strip $E' = \{z : \text{Re}(z) > 0 \text{ and } 0 < \text{Im}(z) < \pi\}$.

Hint: Locate the $T = +\frac{1}{2}$ isotherm in Example 8.3; use $z = -i\pi(w - 1)$ to map the unshaded half strip in Figure 8.11 onto E'.

Answer:

$$\begin{aligned} T(w) &= (2/\pi)\text{Arg}[\tanh[-i\pi(w-1)/2]] \\ &= \frac{2}{\pi}\arccos\left[\frac{\sinh \pi v}{\sqrt{\sinh^2 \pi v + \sin^2 \pi(1-u)}}\right]. \end{aligned}$$

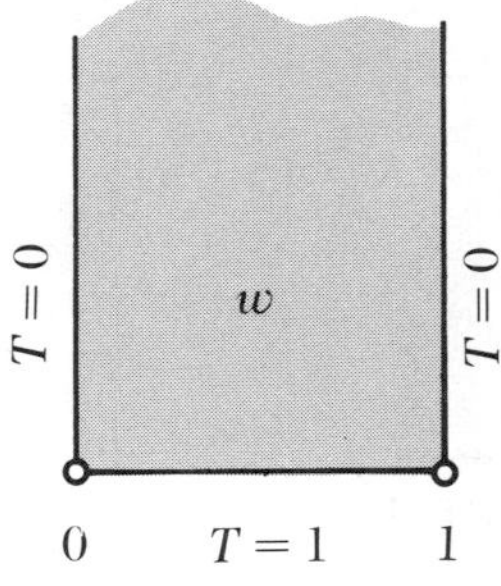

Figure 8.15

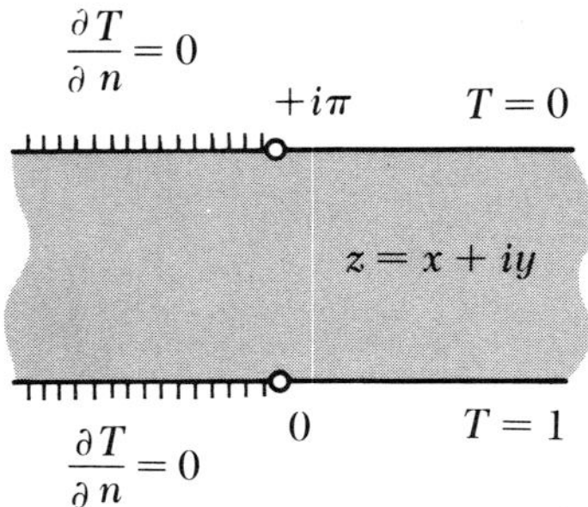

Figure 8.16

5. Solve the problem in Figure 8.15 by another method; map the half strip in Figure 8.15 onto the upper half plane $\text{Im}(\zeta) > 0$ via $\zeta = \phi(w) = \sin[\pi w - (\pi/2)]$. The corresponding problem in the half plane has the familiar solution

$$\tilde{H}(\zeta) = \frac{1}{\pi} \text{Arg}\left(\frac{\zeta - 1}{\zeta + 1}\right),$$

so the solution in the half strip is

$$H(w) = \frac{1}{\pi} \text{Arg}\left[\frac{\phi(w) - 1}{\phi(w) + 1}\right].$$

Reconcile this answer with the one obtained in Exercise 4.

6. Solve the temperature distribution problem in Figure 8.16 using conformal mappings from Example 8.3.
Answer: $T(z) = (1/2) + (1/\pi)\text{Re}[\text{Arcsin}(e^z)]$

$$= \frac{1}{2} + \frac{1}{\pi} \arcsin \tfrac{1}{2}\left[\sqrt{e^{2x} + 2e^x(\cos y) + 1} - \sqrt{e^{2x} - 2e^x(\cos y) + 1}\right].$$

(Recall Exercise 10, Section 4.10 for the last version of the answer.)

7. Show that the temperature distribution problems in Figure 8.17 are equivalent under a suitable conformal mapping. Solve both

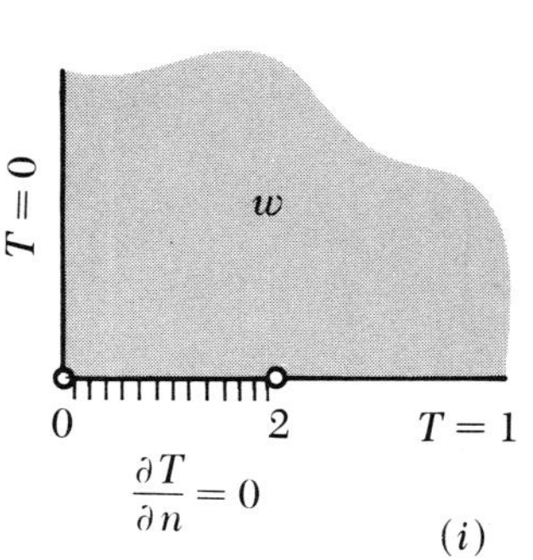

Figure 8.17

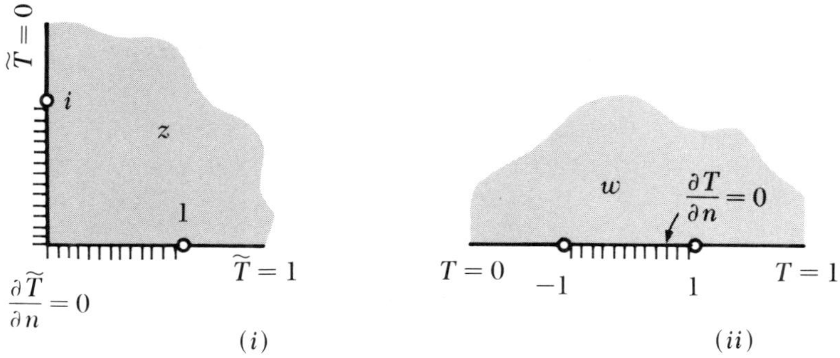

Figure 8.18

problems. Sketch and compare the level curve $T = constant$ in the two domains.

Answer: $T(w) = (2/\pi)\text{Re}[\text{Arcsin}(w/2)]$.

8. Use $w = z^2$ to demonstrate the conformal equivalence of the boundary value problems in Figure 8.18, and find explicit solutions T in each domain.

Answer:

$$T(w) = (1/2) + (1/\pi)\text{Re}[\text{Arcsin } w];$$

$$\tilde{T}(z) = (1/2) + (1/\pi)\text{Re}[\text{Arcsin}(z^2)]$$

$$= \frac{1}{2} + \frac{1}{\pi} \arcsin \tfrac{1}{2}\left[\sqrt{(x^2 - y^2 + 1)^2 + 4x^2y^2} - \sqrt{(x^2 - y^2 - 1)^2 + 4x^2y^2}\right]$$

(Recall Exercise 10, Section 4.10, for the last version of the answer.)

9. Solve the temperature distribution problem on the left in Figure 8.19 by converting it to the problem on the right. Find a fractional linear transformation that effects this equivalence. Sketch a few level curves in each domain (start with the right-hand problem).

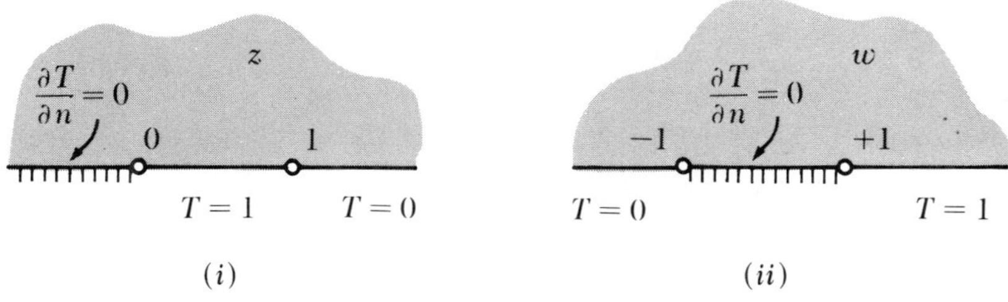

Figure 8.19

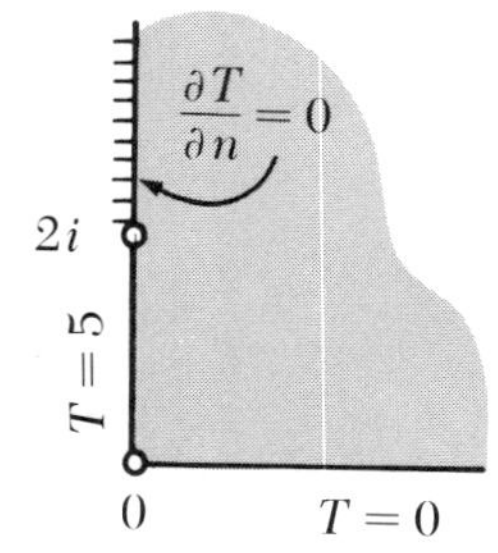

Figure 8.20

Answer: Set $w = (1 + z)/(1 - z)$; $T(w) = (1/2) + (1/\pi)\text{Re}[\text{Arcsin } w]$; $T(z) = \dfrac{1}{2} + \dfrac{1}{\pi}\text{Re}\left[\text{Arcsin}\left(\dfrac{1+z}{1-z}\right)\right]$. To express as functions of (u, v) or (x, y), use Exercise 10, Section 4.10.

10. Using Exercise 9, solve the boundary value problem illustrated in Figure 8.20.

Answer: Use mapping $\zeta = (z^2/4) - 1$; finally,

$$T(z) = \frac{5}{2} + \frac{5}{\pi}\text{Re}\left[\text{Arcsin}\left(\frac{z^2+8}{-z^2}\right)\right].$$

11. If $\text{Im}(w) > 0$, verify that the *equality*

$$\text{(i)}\ \text{Arg}(w^2 - 1) = \text{Arg}(w + 1) + \text{Arg}(w - 1)$$

is valid (rather than a *congruence* modulo 2π). If z is in the half strip $E = \{z : \text{Im}(z) > 0 \text{ and } -\pi/2 < \text{Re}(z) < +\pi/2\}$, show that the following identity is valid:

$$\begin{aligned}\text{(ii)}\ \text{Arg}(-\cos^2 z) &= \text{Arg}(-1) + 2\,\text{Arg}(\cos z)\\ &= \pi + 2\,\text{Arg}(\cos z).\end{aligned}$$

Hint: If z is in E, then $\cos z$ is in the right half of the w-plane.

12. Show that the formula (used in Example 8.3)

$$-\log\left|\tanh\left(\frac{z}{2}\right)\right| = \log\left|\text{ctnh}\left(\frac{z}{2}\right)\right| = \log\sqrt{\frac{\cosh x + \cos y}{\cosh x - \cos y}}$$

is valid in the strip $0 < \text{Im}(z) < +\pi$. Calculate the limits

$$\lim_{z\to 0} T^* \quad\text{and}\quad \lim_{z\to i\pi} T^*$$

of the conjugate temperature distribution in Example 8.3. Also, calculate the limits (if they exist) of $T^*(x + iy)$ as $y \to +\infty$, and as $y \to -\infty$, keeping x fixed $(0 < x < +\pi)$.

Hint: $\tanh(z/2) = (e^z - 1)/(e^z + 1)$.

Answers: (i) $T^* \to -\infty$ at $z = i\pi$; (ii) $T^* \to +\infty$ at $z = 0$; T^* does not approach any definite limit as $z \to \infty$ from within the strip.

8.5 STEADY FLOW OF PERFECT FLUIDS

Under certain conditions one may ignore the effects of viscosity in describing the flow of a fluid; there are also situations in which real fluids do exhibit perfect (that is, frictionless) flow. For example, liquid helium at very low temperatures becomes a "superfluid" due to quantum effects and, in this state, viscosity effects are not observed. When it is reasonable to disregard the effect of viscosity in a fluid flow problem, one can define a vector field $\mathbf{F}(\mathbf{x})$ that specifies the instantaneous velocity of the fluid at $\mathbf{x}$, and by physical reasoning deduce that $\mathbf{F}$ should satisfy the conservation laws (3) in simply connected regions. We will summarize this reasoning below. Thus the theory of perfect fluids is intimately connected with potential theory. It is interesting to note that these conservation laws are only *approximately* true for real fluids, and the description of fluid velocity fields $\mathbf{F}(\mathbf{x})$ by potential functions ϕ with $\mathbf{F} = -\nabla\phi$ rests upon this approximation. In an exact treatment we would have to solve a complicated and non-linear set of partial differential equations, the Navier-Stokes equations, which are formidable even in two-dimensional problems, and potential theory would be much less useful. The other applications of potential theory we have discussed so far do not rest on approximations of this kind; for example, the conservation laws (3) seem to be satisfied *exactly* by electrostatic fields in charge free regions and by temperature distributions in homogeneous, source-free media.

We will consider *steady flows* of a perfect fluid; thus, the velocity field $\mathbf{F}(\mathbf{x})$ is assumed to be time independent. In a full scale hydrodynamic theory we would allow $\mathbf{F}$ to change with time, so that $\mathbf{F} = \mathbf{F}(\mathbf{x}, t)$, but these generalizations are more appropriate as part of a physics course. We will also assume that the fluid is *incompressible*, which implies that the fluid is homogeneous (has constant density) throughout the region in which the flow is being considered. Physical reasoning based on Gauss' Divergence Theorem once again allows us to identify the physical significance of the divergence $\nabla \cdot \mathbf{F}$; this turns out to be the rate at which fluid is created near $\mathbf{x}$. Therefore, if $\mathbf{div}\,\mathbf{F} \neq 0$ at a $\mathbf{p}$, then there is a fluid *source* (if $\nabla \cdot \mathbf{F} > 0$) or a *sink* (when $\nabla \cdot \mathbf{F} < 0$) at $\mathbf{p}$. This source may be concentrated at a point or may be distributed continuously over an area, as when fluid is lost by diffusion through a porous medium. Within the region in which the fluid is flowing (i.e., away from the boundary walls) it is usually true that there are no fluid sources or sinks and there the first of our conservation laws holds,

(19) $$\mathbf{div}\,\mathbf{F} = 0 \quad \text{in any region without sources or sinks.}$$

The reasoning that leads to the companion law $\nabla \times \mathbf{F} = \mathbf{0}$ is more subtle, and is based on an examination of the *circulation* of fluid along various closed curves. The circulation along a closed curve γ is, by definition, the integral with respect to arc length of the component of fluid velocity tangent to the curve; expressing this as a Riemann integral, we get

(20) $$\text{Circulation} = \int_a^b \left[\mathbf{F}(\gamma(t)) \cdot \frac{d\gamma}{dt}\right] dt.$$

There are physical reasons to expect that the circulation must be zero along all small closed curves. We will not go into this reasoning here (see, for example, Feynman [6], Chapter 40 and 41 of v.II). Once we know this much, based on the physics of the situation, we may proceed by mathematical reasoning. There is an intimate relation between these circulation integrals and the curl $\nabla \times \mathbf{F}$ and by judicious use of Stokes' Theorem we may conclude that the curl is zero everywhere, $\nabla \times \mathbf{F} = \mathbf{0}$. Thus, both conservation laws hold, although we have been a little vague about the precise circumstances in which we may expect the curl of the velocity field $\mathbf{F}$ to be zero. Flows in which $\nabla \times \mathbf{F} = \mathbf{0}$ are referred to as **irrotational flows** in many texts.

An adequate discussion of irrotational flows would require at least a brief study of time dependent flows in which the velocity field changes with time. One can prove the following remarkable fact for any time dependent flow of a perfect fluid: if at some instant of time $t = t_0$ we have $\nabla \times \mathbf{F} = \mathbf{0}$, then we must have $\nabla \times \mathbf{F} = \mathbf{0}$ at all later times. In particular, any steady flow of a perfect fluid obtained by starting with a body of fluid *at rest* ($\mathbf{F}(\mathbf{x}) = \mathbf{0}$ everywhere, so that $\nabla \times \mathbf{F} = \mathbf{0}$ everywhere), and letting the flow develop into a steady flow due to the presence of sources and sinks on the boundary must be an irrotational flow.

Next we examine the boundary conditions relevant to fluid flows. An obvious condition is that there can be no transverse flow across a solid (impenetrable) boundary wall. This means that the normal component of fluid velocity vanishes at the wall,

(21) $$\mathbf{F} \cdot \mathbf{n} = -\nabla\phi \cdot \mathbf{n} = -\frac{\partial \phi}{\partial n} = 0,$$

where ϕ is the potential function that determines the velocity field $\mathbf{F} = -\nabla\phi$. Here $\mathbf{n}$ is the unit normal vector, perpendicular to the boundary wall. Dirichlet conditions on the potential function, such as $\phi = \textit{constant}$, do not arise naturally as boundary conditions for a steady fluid flow. There is a rather peculiar situation which can arise for the homogeneous Neumann condition (21), in which the boundary condition $\partial\phi/\partial n = 0$ is imposed everywhere except at a finite number of points $\mathbf{p}_1, \ldots, \mathbf{p}_m$. If we demand that $\partial\phi/\partial n = 0$ *everywhere*, without exceptions, one can show that constant functions are the only possible solutions; these lead to extremely trivial flows, since $-\nabla\phi = \mathbf{0}$ everywhere.

But if we allow exceptional points like $\mathbf{p}_1, \ldots, \mathbf{p}_m$ there may be non-trivial solutions with $\partial\phi/\partial n = 0$ everywhere else on the boundary. These solutions $\phi(z)$ will behave quite singularly as z approaches the exceptional points from within the domain of the flow; in particular, we cannot expect the potential to be bounded near these points. Flows with this kind of behavior arise naturally; the exceptional points may be regarded as one-point fluid sources or sinks located on the boundary. To fully specify the solution $\phi(z)$, additional conditions must be imposed on the behavior of ϕ at such points. Most often physicists use integral conditions which, in physical terms, specify the rate at which fluid enters or is lost at each of the exceptional points. We will give some indication of what these conditions are, at least for two-dimensional problems, in the examples below.

Now let us examine a few special features of *two-dimensional* problems. In view of the conservation laws, there will be a potential function $\phi(z)$ such that $\mathbf{F} = -\nabla\phi$; but in any simply connected domain there is also a conjugate harmonic function $\phi^*(z)$, and an associated **complex potential** for the flow, the analytic function $f(z) = \phi(z) + i\phi^*(z)$. The potential ϕ does not have any direct physical significance (although it does fully determine the velocity field), but the complex potential is very useful. The velocity vector at p,

$$\mathbf{F}(p) = -\nabla\phi(p) = \left(-\frac{\partial\phi}{\partial x}\right)\mathbf{i} + \left(-\frac{\partial\phi}{\partial y}\right)\mathbf{j} = -\left[\left(\frac{\partial\phi}{\partial x}\right)\mathbf{i} + \left(\frac{\partial\phi}{\partial y}\right)\mathbf{j}\right],$$

can be identified with the complex number

$$-\left[\frac{\partial\phi}{\partial x} + i\frac{\partial\phi}{\partial y}\right] = -\left[\frac{\partial\phi}{\partial x} - i\frac{\partial\phi^*}{\partial x}\right] = \overline{\left(-\frac{df}{dz}\right)},$$

the complex conjugate of $-df/dz$. Furthermore, when we identify a vector $\mathbf{x} = a\mathbf{i} + b\mathbf{j}$ with the corresponding complex number $z = a + ib$, the length or magnitude of the vector $\|\mathbf{x}\| = \sqrt{a^2 + b^2}$ is the same as the absolute value of the complex number, so that the speed $\|\mathbf{F}(p)\|$ of the fluid passing the position p is given in terms of the complex potential by

$$\|\mathbf{F}(p)\| = \left|\frac{df}{dz}(p)\right|. \tag{22}$$

There are other properties of the flow which may be determined from the complex potential $f(z)$ and its derivative df/dz. One important observation is that the velocity vector at a point p (where $\mathbf{F} \neq \mathbf{0}$) is always tangent to the curve $\phi^*(z) = \textit{constant} = \phi^*(p)$ which passes through the point p, and is perpendicular to the curve $\phi(z) = \textit{constant} = \phi(p)$. To see this, one should first recall that the loci $\phi = \mathrm{Re}(f) = \textit{constant}$ and $\phi^* = \mathrm{Im}(f) = \textit{constant}$ are orthogonal at any point where $|df/fz| \neq 0$, as we indicated in Chapter 4. It is well known that the gradient $\mathbf{F} = -\nabla\phi$ is always perpendicular to the locus $\phi = \textit{constant}$, whenever this gradient vector is nonzero. Thus, $\mathbf{F}$ is

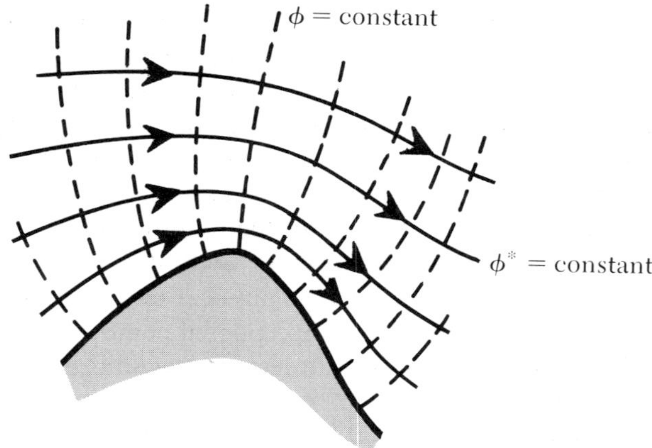

Figure 8.21 Equipotential lines ϕ = constant (dashed) and stream lines ϕ^* = constant (solid) near an impenetrable wall.

perpendicular to the curve ϕ = *constant*, and is therefore parallel (tangent) to the curve ϕ^* = *constant*. Since the fluid velocity is tangent to the curves ϕ^* = *constant*, one can show that a small particle released into the steady flow will move along the appropriate curve ϕ^* = *constant*; thus, fluid molecules move along these curves. For this reason, the curves ϕ^* = *constant* are called the **stream lines** of the flow. The orthogonal family of curves ϕ = *constant* are called **equipotential lines,** but do not have a direct physical significance.

There is an interesting relationship between this pattern of curves and the boundary condition $\partial\phi/\partial n = 0$ that holds along the walls which bound the region of the flow. Since $\partial\phi/\partial n = \nabla\phi \cdot \mathbf{n}$, where $\mathbf{n}$ is the unit normal vector directed into the region, the gradient $\nabla\phi$ must be tangent to the boundary curve. This means that equipotential curves ϕ = *constant* are always perpendicular to the wall at points where $\partial\phi/\partial n = 0$; at the same time, the velocity vector is tangent to the wall along such a segment of the boundary. Thus, this piece of the boundary must coincide with part of a stream line. This is shown in Figure 8.21, and is also illustrated in a number of the examples that follow.

This observation can also be used in a rather different way; if we have a known flow and locate some particular stream line $\phi^* = c_0$ within the body of the fluid, it is possible to replace this stream line with an impenetrable boundary without affecting the flow in the rest of the region. This is allowed since the curves ϕ = *constant* that meet this stream line are perpendicular to it, so that

$$0 = -\nabla\phi \cdot \mathbf{n} = -\partial\phi/\partial n \quad \text{everywhere on it.}$$

Our final comment on two-dimensional problems concerns boundary conditions that specify the rate at which fluid enters or leaves at exceptional boundary points (sources or sinks on the boundary). These conditions are based on the fact that, for any positively oriented simple closed contour γ, we may calculate the net rate at which fluid enters the region enclosed by γ by the

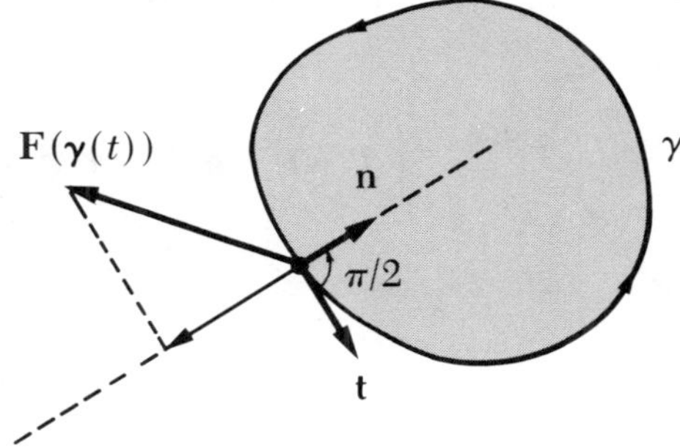

Figure 8.22 The vector **t** and normal vector **n**, used to derive formula (24). The component of **F** parallel to **n** is the component of velocity normal to the boundary curve, into the (shaded) domain enclosed by γ.

following integral,

$$\rho \int_\gamma \mathbf{F} \cdot \mathbf{n}\, ds = \rho \int_a^b \left[\frac{\partial \phi}{\partial x} \frac{dy}{dt} - \frac{\partial \phi}{\partial y} \frac{dx}{dt} \right] dt, \tag{23}$$

where ρ is the density (mass per unit area) of the fluid. (For simplicity, we shall always assume that $\rho = 1$). The normal vector **n** is obtained by taking the normalized *tangent vector* $\mathbf{t} = \dfrac{d\gamma}{dt} \Big/ \left| \dfrac{d\gamma}{dt} \right|$ at each point on the curve; when this tangent vector is rotated counterclockwise by $\pi/2$ radians, we get a unit vector $\mathbf{n} = i \dfrac{d\gamma}{dt} \Big/ \left| \dfrac{d\gamma}{dt} \right|$ that is normal to the curve and directed into the domain enclosed by γ, as indicated in Figure 8.22. The dot product $\mathbf{F} \cdot \mathbf{n} = -\nabla \phi \cdot \mathbf{n}$ is the component of fluid velocity normal to the boundary, measured positively when directed into the domain, so that the net flow of fluid into the domain per unit time is obtained by integrating this dot product with respect to arc length $ds = \left| \dfrac{d\gamma}{dt} \right| dt$,

$$\int_\gamma -\nabla \phi \cdot \mathbf{n}\, ds = \int_a^b (-\nabla \phi(\gamma(t)) \cdot \mathbf{n}) \left| \frac{d\gamma}{dt} \right| dt$$

$$= \int_a^b \left[\frac{\partial \phi}{\partial x} \frac{dy}{dt} - \frac{\partial \phi}{\partial y} \frac{dx}{dt} \right] dt \tag{24}$$

if $\gamma(t) = x(t) + iy(t)$ for $a \le t \le b$. We will use this integral in a few of the examples which follow, but we cannot pretend that these few examples illustrate all of the uses of integrals like this in studying flow problems. In three-dimensional problems there are similar integrals, now surface integrals, for calculating the net rate of fluid loss across the boundary of a region.

Here is another version of formula (24). If vectors **a** and **b** are interpreted as complex numbers, it is easy to see that the value of the dot product is given by $\mathbf{a} \cdot \mathbf{b} = \operatorname{Re}(\bar{a}b)$. If the velocity field $\mathbf{F} = U\mathbf{i} + V\mathbf{j}$ is regarded as a function $F = U + iV$, having complex variable and complex values, the integral of

$\mathbf{F} \cdot \mathbf{n}$ along a contour γ (closed or not) may be expressed as a contour integral,

$$\int_a^b \mathbf{F}(\gamma(t)) \cdot \mathbf{n}\, ds = \int_a^b \mathbf{F}(\gamma(t)) \cdot \frac{i\dfrac{d\gamma}{dt}}{\left|\dfrac{d\gamma}{dt}\right|} \left|\frac{d\gamma}{dt}\right| dt$$

$$= \int_a^b \left[\mathbf{F}(\gamma(t)) \cdot i\frac{d\gamma}{dt}\right] dt$$

$$= \int_a^b \operatorname{Re}\left[i\overline{F(\gamma(t))}\,\frac{d\gamma}{dt}\right] dt$$

$$= \operatorname{Re}\left[i\int_a^b \overline{F(\gamma(t))}\,\frac{d\gamma}{dt}\, dt\right]$$

$$= -\operatorname{Im}\left[\int_\gamma \overline{F(z)}\, dz\right]. \tag{25}$$

If the potential function ϕ admits a conjugate ϕ^* on a region containing γ, then $F = -\overline{(df/dz)}$, where $f = \phi + i\phi^*$. The net flow across the arc γ (parallel to the direction of the normal vector determined by $i\, d\gamma/dt$) is given by a contour integral involving the complex potential f:

$$\Delta Q = -\operatorname{Im}\left[\int_\gamma \overline{F(z)}\, dz\right]$$

$$= +\operatorname{Im}\left[\int_\gamma \frac{df}{dz}\, dz\right]$$

$$= \operatorname{Im}[f(\gamma(b)) - f(\gamma(a))]$$

$$= [\phi^*(\gamma(b)) - \phi^*(\gamma(a))]$$

$$= \Delta_\gamma \phi^* \tag{26}$$

the *net variation in the conjugate function* ϕ^* as we move from one end of the contour to the other. It is important to notice that these formulas are valid even if γ is not a *closed* curve; when γ is closed, ΔQ measures the rate at which fluid is accumulating in the domain enclosed by γ.

Example 8.5 Consider the flow associated with the complex potential $f(z) = z$,

$$\phi(x + iy) = x \quad \text{and} \quad \phi^*(x + iy) = y.$$

The fluid velocity is given by the function $-\overline{df/dz} = -1 + i0$, interpreted as a vector field; there is a constant velocity at each point. The stream lines $\phi^* = \textit{constant}$ are just the horizontal lines $y = \textit{constant}$, since the velocity vector is directed horizontally.

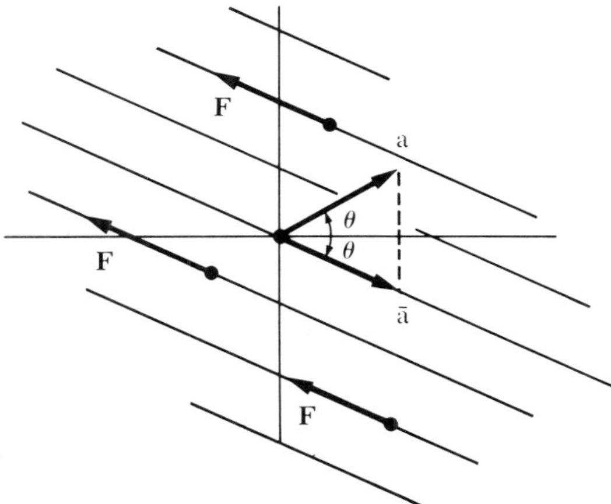

Figure 8.23 Flow associated with the complex potential $f(z) = az + b$; $\mathbf{F} = -\bar{a}$ everywhere.

Similarly, any linear function $f(z) = az + b$ (with $a \neq 0$) is the complex potential of a constant velocity flow whose velocity vector corresponds to the complex number $-\overline{df/dz} = -\bar{a}$; the stream lines are all parallel to $-\bar{a}$, as indicated in Figure 8.23. Obviously, the choice of constant b does not affect the velocity field. The complex potential of *any* flow is determined only up to an added complex constant, which disappears when we differentiate to calculate the velocity.

Example 8.6 A more subtle flow is the one with potential

$$\phi(z) = \log|z| = \tfrac{1}{2}\log(x^2 + y^2) \tag{27}$$

defined on the domain $E = \{z : z \neq 0\}$. The origin is not part of the domain, and will turn out to be a fluid sink. Since the domain is not simply connected, the conjugate harmonic function ϕ^* is given by the multiple valued harmonic function $\phi^*(z) = \arg z$. The complex potential $f = \phi + i\phi^* = \log z$ is also multiple valued, but single valued determinations of $\log z$ can be defined on simply connected subdomains of E. The derivative of the complex potential is *single valued*, although the potential itself is multiple valued; $df/dz = 1/z$ no matter which locally defined determination of $f = \log z$ we start with. This derivative determines the velocity field,

$$\mathbf{F} = -\overline{\frac{df}{dz}} = \overline{(-1/z)} = -1/\bar{z} \quad \text{for} \quad z \neq 0.$$

At every point on the circle $|z| = r$, the magnitude of the velocity is constant, $|df/dz| = |1/z| = 1/r$, and the velocity vector is directed toward the origin. There is an inward flow of fluid from infinity to a sink located at the origin, associated with the potential $\phi(z)$. Stream lines are rays emanating from the origin, and equipotentials are circles centered at the origin. If we parametrize the circle $|z| = r$ in a counterclockwise direction and calculate the net rate of fluid flow across the boundary (toward the origin) using formula (26), we find that

$$\Delta Q = \int_a^b -\nabla\phi \cdot \mathbf{n}\, ds = +\text{Im}\left[\int_\gamma \frac{df}{dz}\, dz\right] = +\text{Im}\left[\int_\gamma \frac{1}{z}\, dz\right]$$
$$= \Delta_\gamma[\phi^*(z)] = \Delta_\gamma[\arg z] = 2\pi.$$

Now **div F** $= 0$ everywhere in E, so there are no sources or sinks anywhere else in E; thus, the origin (or infinity) is the only place where we can have a net loss of fluid that could account for this net flow across the circles $|z| = r$. It is not surprising then, that this integral has the same value for all radii $r > 0$. The origin is a source of strength -2π (a sink); since fluid is *lost*, at the rate of 2π units per second, the strength is assigned a *negative* value. Fluid removed at the origin is replaced by fluid flowing in from infinity. The flow pattern becomes quite singular near the origin; in particular, the velocity becomes infinite.

If we multiply the potential ϕ by a positive or negative constant $k \neq 0$, the new potential $\psi = k \cdot \log |z|$ describes a flow toward a source of strength $-2\pi k$ located at the origin. The pattern of stream lines and equipotential lines for the potential ψ is the same as that for ϕ, although the velocities are all scaled by the factor k.

Example 8.7 Next we shall consider a simple flow in the half plane $\text{Im}(w) > 0$, and use it to solve problems in more elaborate domains by conformal transformation of the potential function $\phi(w)$, and also of the complex potential $f(w)$. We shall also make use of the principle that any stream line $\phi^* = \textit{constant}$ can be replaced by a solid, impenetrable boundary. A similar idea has been used in electrostatics problems (recall Example 8.2).

We start with the flow toward the origin along radial lines associated with the potential $\phi(w) = \log |w|$ in the upper half plane. The two parts of the real axis are regarded as impenetrable boundaries, where $\partial\phi/\partial n = 0$; the potential $\log |w|$ obviously satisfies this requirement. In effect we are replacing two stream lines of the flow in Example 8.6 with solid boundaries. The flow pattern associated with this potential is indicated on the right in Figure 8.24. There is

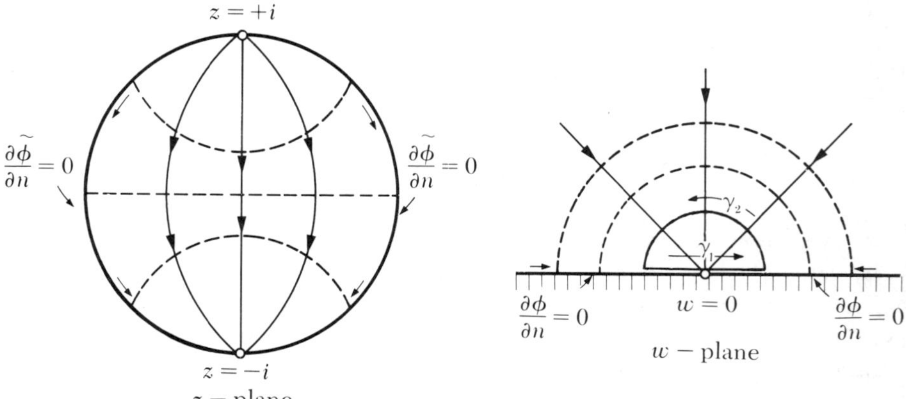

Figure 8.24 Flow lines (solid) and equipotential lines (dashed) in Example 8.7.

a single valued conjugate harmonic function $\phi^*(w) = \operatorname{Arg} w$, determined only up to an added constant, and we may take $f(w) = \operatorname{Log} w$ as the complex potential. There is a source located on the boundary at $w = 0$, and its strength is determined by taking a semicircular closed contour $\gamma = \gamma_1 + \gamma_2$ (diameter included), such as the one shown in Figure 8.24, and calculating

$$\Delta Q = \int_a^b -\nabla\phi \cdot \mathbf{n}\, ds = \operatorname{Im}\left[\int_\gamma \frac{df}{dz}\, dz\right].$$

Consider the integrals along the boundary segment γ_1 and along the circular arc γ_2 separately. Along the segment γ_1 the curve normal coincides with the inward normal $\mathbf{n}$ to the boundary wall, so that $-\nabla\phi \cdot \mathbf{n} = -\partial\phi/\partial n = 0$ and the integral is *zero*. Thus, the amount of fluid flowing across γ (toward the origin) per unit time is

$$\Delta Q = \operatorname{Im}\left[\int_\gamma \frac{df}{dz}\, dz\right] = \operatorname{Im}\left[\int_{\gamma_2} \frac{df}{dz}\, dz\right]$$
$$= \operatorname{Im}[\Delta_{\gamma_2}[f(z)]] = \Delta_{\gamma_2}[\phi^*(z)] = +\pi.$$

Notice that the sink at $w = 0$ has *half* the strength of the sink in Example 8.6 (where the flow covered the whole plane).

Let us transform ϕ, ϕ^*, and $f = \phi + i\phi^*$ using the fractional linear transformation $w = T(z) = -i(z + i)/(z - i)$ to map the unit disc $|z| < 1$ onto the half plane $\operatorname{Im}(w) > 0$. The rays $(-\infty, 0)$ and $(0, +\infty)$ that bound the half plane correspond to the left and right hand arcs extending from $-i$ to $+i$ in the circle $|z| = 1$. The boundary condition $\partial\phi/\partial n = 0$ is transformed without alteration, as we indicated in Section 7.6. The transformed harmonic function

$$\tilde{\phi}(z) = \left[\phi(w)\Big|_{w=T(z)}\right] = \log\left|\frac{z+i}{z-i}\right| \quad \text{for} \quad |z| < 1$$

satisfies the Neumann condition $\partial\tilde{\phi}/\partial n = 0$ on the boundary arcs, and the complex potential is

$$\tilde{f}(z) = \left[f(w)\Big|_{w=T(z)}\right] = \operatorname{Log}\left(\frac{1}{i}\left(\frac{z+i}{z-i}\right)\right)$$
$$= \log\left|\frac{z+i}{z-i}\right| + i\operatorname{Arg}\left(\frac{1}{i}\left(\frac{z+i}{z-i}\right)\right).$$

It is clear that the loci $\phi^*(w) = $ *constant*, the stream lines of the original flow in the w-plane consisting of radial lines from the origin, are transformed by $z = \check{T}(w)$ to the loci $\tilde{\phi}^*(z) = $ *constant* in the disc in the z-plane, so that the stream lines of the transformed flow in the disc $|z| < 1$ are circular arcs from $-i$ to $+i$, as indicated in Figure 8.24. Likewise, the equipotential lines $\phi = $ *constant* for one flow are transformed into the corresponding equipotential lines for the other flow under the mappings $w = T(z)$ and $z = \check{T}(w)$.

In the original flow the point at infinity acted as a source of fluid of strength $+\pi$, to match the sink at $w = 0$. In the transformed flow, $w = \infty$ is mapped to $z = +i$, and it is quite apparent that there is a source on the boundary circle at $z = +i$ to match the sink of equal strength located at $z = -i = \check{T}(0)$. In Exercise 5 we indicate how to verify that the strengths of the transformed sources, at $+i$ and $-i$, are $+\pi$ and $-\pi$ (just as they were in the half plane).

Example 8.8 This example is fundamental to a number of advanced topics in hydrodynamics. In it we examine the potential function

$$\phi = \mathrm{Re}(f) \quad \text{where} \quad f(z) = \frac{1}{2}\left(z + \frac{1}{z}\right),$$

and notice that there are certain natural loci where $\partial\phi/\partial n = 0$. Identifying these loci with boundary walls, we obtain irrotational flow patterns that are quite interesting since they correspond to the perturbation of a constant velocity flow caused by a stationary obstacle. The mapping $w = f(z)$ by itself is extremely useful as a conformal mapping, and with it we will be able to solve a variety of problems.

First consider $f(z)$ as the complex potential of a fluid flow in the domain $E = \mathbf{C}$. When z is very far from the origin, the term $1/2z$ may be regarded as a small perturbation of the basic potential function $f_0(z) = z/2$. If we calculate the velocity fields associated with these potentials, it is clear that the flow

$$\mathbf{F} = \overline{\left(-\frac{df}{dz}\right)} = (-1/2)\overline{\left(1 - \frac{1}{z^2}\right)} \tag{28}$$

differs only slightly from the constant velocity field $\mathbf{F}_0 = (-1/2) + i0$, associated with the unperturbed complex potential f_0. Thus the complex potential $f(z)$ satisfies a new kind of boundary condition, one in which its asymptotic behavior at infinity is prescribed. Conditions like this are very important, especially in problems concerning the flow of fluid around a finite obstacle; in such situations it is just about the only physically natural boundary condition at infinity. Explicitly, the behavior we expect at infinity is that

$$\lim_{z\to\infty} \frac{f(z)}{z} = a$$

should exist. The value of this limit (a complex number) corresponds to a constant flow that has complex potential $f_0(z) = az$.

As a second step in the analysis of the flow (28) we try to determine the curves on which $\partial\phi/\partial n = 0$. We may use the principle that $\nabla\phi$ is always perpendicular to the curves of the form $\phi^* = \textit{constant}$; thus, we will always have $\partial\phi/\partial n = 0$ along such curves. Write out the real and imaginary parts of $f(z)$,

$$f(z) = \phi + i\phi^* = \frac{x}{2}\left(1 + \frac{1}{x^2 + y^2}\right) + i\frac{y}{2}\left(1 - \frac{1}{x^2 + y^2}\right)$$

$$= \frac{x}{2}\left(1 + \frac{1}{|z|^2}\right) + i\frac{y}{2}\left(1 - \frac{1}{|z|^2}\right)$$

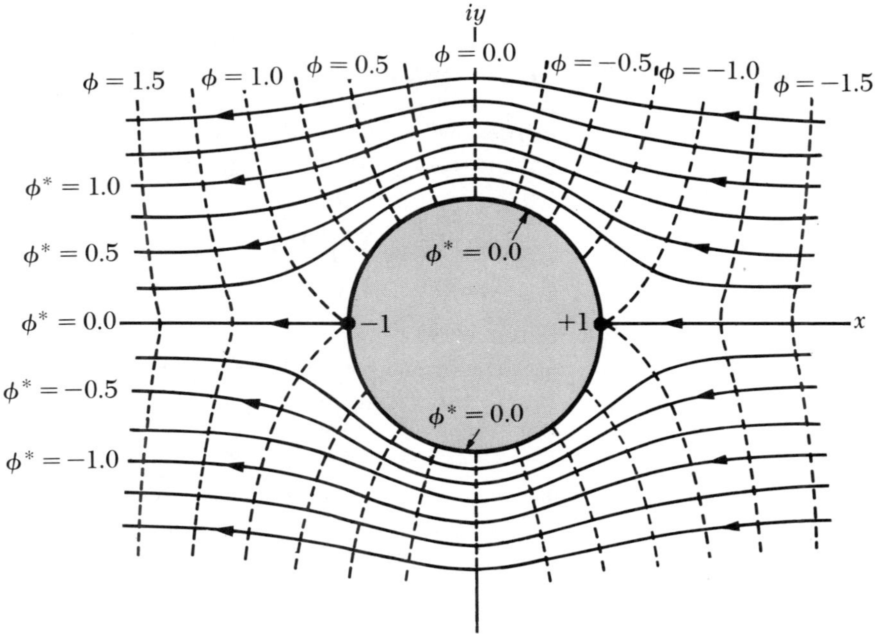

Figure 8.25 The loci $\phi = \text{Re}(f) = constant$ (dashed lines) and $\phi^* = \text{Im}(f) = constant$ (solid lines) for $f(z) = \dfrac{1}{2}\left(z + \dfrac{1}{z}\right)$.

if $z = x + iy$. Plotting the loci $\phi^* = c$ requires tedious calculations, but there is one locus of particular interest whose shape may be determined easily. The locus $\phi^* = 0$ is given by

$$\frac{y}{2}\left(1 - \frac{1}{|z|^2}\right) = 0,$$

and consists of the unit circle $|z| = 1$, together with the whole real axis (where $y = 0$). For our purposes we shall regard the unit circle as a solid obstacle and examine the flow pattern in the exterior domain $|z| > 1$, simply ignoring the domain interior to the unit circle. The equipotential curves $\phi = constant$ must be perpendicular to the locus $\phi^* = 0$ where they meet it,† so we have $\partial\phi/\partial n = 0$ there. The circle $|z| = 1$ may legitimately be thought of as an impenetrable boundary for the flow. The stream lines $\phi^* = c$ with $c \neq 0$ are the smooth curves within the exterior domain $E = \{z\colon |z| > 1\}$ shown in Figure 8.25. Notice how the stream lines tend to become horizontal and equally spaced at large distances from the origin, as we expect from the asymptotic behavior of the complex potential of this flow near infinity. This flow pattern in the exterior domain E describes the flow which results when a circular obstacle is introduced into a constant velocity flow parallel to the real axis.

† Except at points $z = -1$ and $z = +1$; here the locus branches and there is no well defined normal direction to refer to in verifying the boundary condition; the normal derivative cannot be defined at these points.

Example 8.9 Now we shall use $w = T(z) = (\frac{1}{2})(z + (1/z))$ as a conformal mapping, rather than as a complex potential, to determine the flow pattern of a fluid around objects with irregular shapes. For this application we must know the mapping properties of T on the exterior domain $E = \{z: |z| > 1\}$. The mapping is conformal since

$$\left|\frac{dT}{dz}\right| = \frac{1}{2}\left|1 - \frac{1}{z^2}\right| \geq \frac{1}{2}\left(1 - \frac{1}{|z|^2}\right) > 0 \quad \text{for } z \text{ in } E.$$

For our purposes we only need to know that it maps E invertibly onto the slit domain obtained by removing the closed interval $[-1, +1]$ from the real axis in the w-plane (recall Exercises 19 and 20, Section 4.10).

Given the level curves $u = constant$ and $v = constant$ for $w = T(z) = u + iv$, as shown in Figure 8.25, we can deduce all global mapping properties of T from this pattern. For example, we have $v = \phi^*(z) = 0$ and $-1 \leq u \leq +1$ for z on the circle $|z| = 1$; it follows that this circle is mapped twofold onto the segment $[-1, +1]$, which is the boundary of the domain D in the w-plane. One can also verify that T is univalent (one-to-one) from E into D, and that $T(E) = D$, by inspecting Figure 8.25.

The interval $[-1, +1]$ will now be considered as an obstacle for flows in the w-plane. We will use $w = T(z)$ to transform known flows around the circular obstacle in the z-plane to obtain interesting flows around the slit. Let us start with a flow from top to bottom in the z-plane, around the disc $|z| \leq 1$, which behaves like a constant velocity flow near infinity. The desired flow may be obtained by rotating the flow shown in Figure 8.25 counterclockwise by an angle of $\pi/2$ radians. In the new flow the velocity (and complex potential, etc.) at a point $z' = iz$ should be the same as the velocity (complex potential, etc.) of the original flow at the point z, since z' is rotated $\pi/2$ radians counterclockwise from z. Thus the new complex potential $g(z)$ has the property $g(iz) = f(z)$, which is the same as saying that:

(29) $$g(z) = f\left(\frac{1}{i}\,z\right) = \frac{i}{2}\left(\frac{1}{z} - z\right) \quad \text{for} \quad |z| > 1.$$

At great distances from the origin this potential is virtually indistinguishable from the constant velocity potential $g_0(z) = -iz/2 = z/2i$; the velocity of the corresponding flow,

$$-\overline{dg_0/dz} = \overline{(i/2)} = -i/2,$$

is directed downwards, as it should be. Let us transform this flow around the disc to a corresponding flow around the slit. This can be done by using the inverse mapping $z = \check{T}(w)$, which maps the cut plane D onto E. Substituting $z = \check{T}(w)$ into the complex potential $g(z)$, we get the complex potential $\tilde{g}(w)$

of the corresponding flow around the slit,

$$\tilde{g}(w) = \left[g(z) \Big|_{z=\check{T}(w)} \right] = \frac{i}{2}\left[\frac{1}{\check{T}(w)} - \check{T}(w) \right]$$

$$= i\left[\frac{1}{2}\left[\frac{1}{\check{T}(w)} + \check{T}(w) \right] - \check{T}(w) \right]$$

$$= iT(\check{T}(w)) - i\check{T}(w) = iw - i\check{T}(w) \tag{30}$$

(remember that $T(\check{T}(w)) = w$). Putting aside the question of finding explicit formulas for $z = \check{T}(w)$, we know that the curves $\phi(z) = constant$ and $\phi^*(z) = constant$ are transformed by the forward map $w = T(z)$ to the curves $\tilde{\phi}(w) = constant$ and $\tilde{\phi}^*(w) = constant$ in the w-plane, where $\tilde{g}(w) = \tilde{\phi}(w) + i\tilde{\phi}^*(w)$. Since we have at least a rough idea of how $w = T(z)$ deforms the exterior domain E to the slit domain D, we can make a rough sketch of these transformed curves in the w-plane. This is done in Figure 8.26. Notice that the flow associated with complex potential $\tilde{g}(w)$ is substantially perturbed by the presence of the barrier along the slit because the fluid flows broadside to this plate.

Thus we have the solution to a rather non-trivial problem concerning flow around a linear barrier. By similar reasoning we could start with a flow around the disc in the z-plane which behaves asymptotically like a constant velocity flow oriented at any prescribed angle θ to the positive real axis; we have just considered the angles $\theta = -\pi$ and $\theta = -\pi/2$. By transforming this flow to the w-plane (by $z = \check{T}(w)$) we would get a flow around the slit that behaves asymptotically like a constant velocity flow in the w-plane with the same direction θ. We leave it as an exercise for the reader to work out the form of the complex potential $f(z)$ associated with such a flow in the z-plane, and to

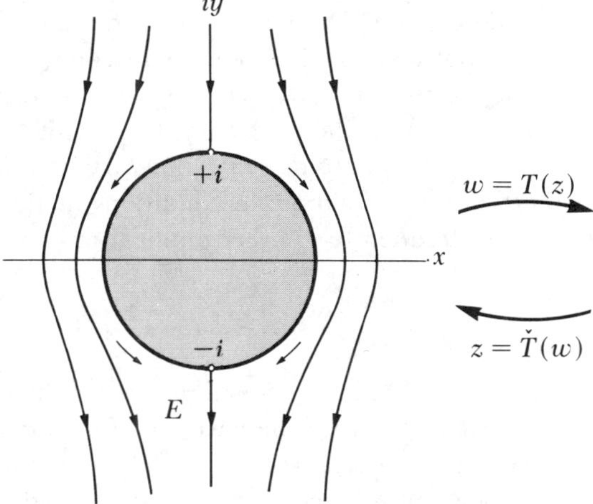

Figure 8.26 Transformation of a known flow in the z-plane via $w = T(z) = \frac{1}{2}\left(z + \frac{1}{z}\right)$ as in Example 8.9.

sketch the level curves of the corresponding flow around the slit in the w-plane (Exercise 14).

As for the problem of writing out an explicit formula for the inverse mapping $z = \check{T}(w)$, we note that all the possible solutions of the equation $w = T(z) = (1/2)(z + (1/z))$ are the same as the solutions z of the quadratic equation $z^2 - 2wz + 1 = 0$ (w fixed), so that

$$z = w \pm \sqrt{w^2 - 1}. \tag{31}$$

Since, by other means, we may demonstrate that the mapping $T{:}E \to D$ is univalent and onto, there is just one choice of the square root that gives the correct value for z, with z in E, for a given point w in the cut plane D. We leave it to the reader (recall Exercises 21 to 25, Section 4.10) to determine how the proper choice of $\sqrt{w^2 - 1}$ should be made for w in various parts of the domain D to give $z = \check{T}(w) = w + \overset{*}{\sqrt{w^2 - 1}}$; the symbol $\overset{*}{\sqrt{\cdots}}$ indicates the analytic function on D so defined. It may disagree with the principal determination of square root, denoted by $\sqrt{\cdots}$, but clearly $\overset{*}{\sqrt{w^2 - 1}} = (\pm 1) \cdot \sqrt{w^2 - 1}$ for each w. Substituting this formula for $\check{T}$ into equation (30), we may express the complex potential $\tilde{g}(w)$ explicitly as

$$\tilde{g}(w) = -i\overset{*}{\sqrt{w^2 - 1}},$$

from which we can obtain the potential function $\tilde{\phi}(w) = \text{Re}(\tilde{g}(w))$ and the velocity field $\mathbf{F} = \overline{(-d\tilde{g}/dw)}$ by direct calculations.

Other fluid flows are considered in the exercises. In the problems just reviewed we generally started with a known (complex) potential and determined the stream lines and other features of the associated flow. In effect, we found, or described, the problem that fits the given potential, or else we made conformal transformations of simple known flows to describe more complicated ones. We generally do not proceed from boundary conditions in a domain to the desired flow that meets these requirements. One reason for this is that there are sources and sinks located on the boundary in most of these problems, for which we have rather troublesome integral boundary conditions that specify their strengths; there are also asymptotic conditions at infinity. Conditions like these do not arise so frequently in other applications.

EXERCISES

1. Calculate the conjugate function ϕ^* and the complex potential $f(z) = \phi + i\phi^*$ associated with the harmonic function $\phi(x, y) = x^2 - y^2$ in the first quadrant. Interpret ϕ as the potential function of a fluid flow. Sketch equipotential curves ($\phi = \textit{constant}$) and flow lines ($\phi^* = \textit{constant}$), and indicate the direction of the flow. Does the fluid velocity go to zero as $z \to \infty$?

2. Transform the potential $\phi(z)$ in the quadrant, in Exercise 1, to a potential $\tilde{\phi}(w)$ in the sector $0 < \arg w < \pi/4$. Calculate $\tilde{\phi}(u, v) = \tilde{\phi}(u + iv)$ explicitly; sketch equipotential lines and flow lines. Calculate the complex potential.

Hint: Use $w = z^{1/2}$.

3. If $a > 0$ and $c > 0$, consider the potential $\phi(z) = c \log |z - a| + c \log |z + a| = c \log |z^2 - a^2|$. Prove that $\partial\phi/\partial n = 0$ on the imaginary axis, so this axis may be regarded as a wall that bounds a flow in the right half plane. Sketch flow lines and a few equipotential lines in this half plane; indicate the direction of flow. Is there a source or a sink at $z = a + i0$?

4. If vectors $\mathbf{a} = x\mathbf{i} + y\mathbf{j}$ and $\mathbf{b} = u\mathbf{i} + v\mathbf{j}$ are interpreted as complex numbers $a = x + iy$, $b = u + iv$, show that the value of the dot product is given by $\mathbf{a} \cdot \mathbf{b} = \mathrm{Re}(\bar{a}b)$.

5. In Example 8.7, consider the arcs γ_2 and $\tilde{\gamma}_2$ (transforms of one another) illustrated in Figure 8.27. Show that the strengths of the sources, at $w = 0$ and $z = -i$ respectively, are equal. Use formula (26) to express the net flow (per unit time) of fluid across each curve, toward the sink, in the forms $\Delta Q = \Delta_{\gamma_2}[\phi^*(w)]$ and $\Delta\tilde{Q} = \Delta_{\tilde{\gamma}_2}[\tilde{\phi}^*(z)] = \Delta_{\tilde{\gamma}_2}[\phi^*(T(z))]$. Since $\phi^*(w)$ and $\tilde{\phi}^*(z) = \phi^*(T(z))$ have the same value at corresponding points w and z, it is easy to see that $\Delta Q = \Delta\tilde{Q}$.

6. Calculate the strength of the source/sink at $z = a + i0$ in Exercise 3.

7. Verify that $w = T(z) = (z - 1)/(z + 1)$ maps the upper half plane $\mathrm{Im}(z) > 0$ to the upper half plane $\mathrm{Im}(w) > 0$, and that

(i) $T(-1) = \infty$, $T(0) = -1$, $T(+1) = 0$

(ii) T maps the arc $|z| = 1$ in the half plane to the positive imaginary axis.

Use T to transform the potential $\phi(w) = (1/\pi)\log |w|$ to the potential $\tilde{\phi}(z) = (1/\pi)\log \left|\dfrac{z-1}{z+1}\right|$. Then,

(iii) calculate the conjugate functions $\phi^*(w)$ and $\tilde{\phi}^*(z)$;

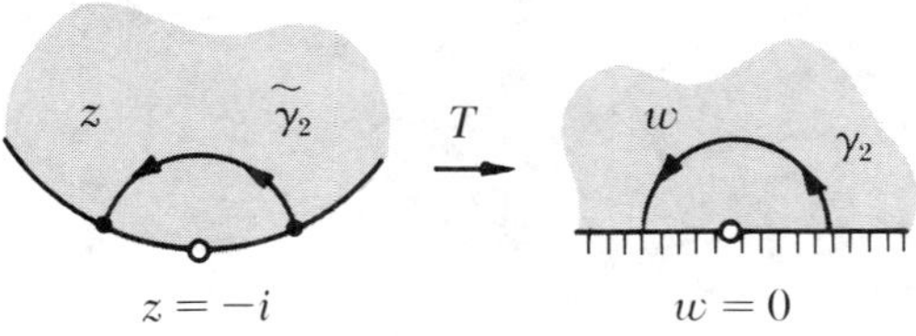

Figure 8.27

(iv) sketch the equipotential lines and flow lines $\phi(w) =$ *constant*, and $\phi^*(w) =$ *constant*; sketch the transformed loci $\tilde{\phi}(z) =$ *constant* and $\tilde{\phi}^*(z) =$ *constant* in the z plane. (Indicate direction of flow in each case.)

Hint: In (ii), remember that the fractional linear transformation T maps lines to circles.

Answer: $\phi^*(w) = (1/\pi)\text{Arg } w$; $\tilde{\phi}^*(w) = (1/\pi)\text{Arg}[(z - 1)/(z + 1)]$ taking principal determinations (why?).

8. In Exercise 7, calculate the strengths of the source at $z = -1$ and the sink at $z = +1$.

Hint: Use formula (26) to avoid calculating contour integrals.
Answer: Strength $= +1$ (source) at $z = -1$; strength $= -1$ (sink) at $z = +1$.

9. In Exercise 7, calculate the fluid velocities in each domain, expressing them as functions of the complex variables w and z. Does the velocity approach zero as $z \to \infty$ in the z-plane? Does the velocity have a limit (improper value ∞ allowed) as $z \to -1$ or $z \to 1$?

10. Consider the complex potential $f(z) = \phi + i\phi^* = (1/\pi)\text{ Log}(z)$ on the domain D obtained by deleting the segment $[+1, +\infty)$ from the half plane $\text{Re}(z) > 0$. Transform it to the complex potential $\tilde{f}(\zeta)$ of a flow in the strip $E = \{\zeta : 0 < \text{Im}(\zeta) < \pi\}$, using the conformal mapping $z = \sin(\zeta/2i)$. Verify that the transformed equipotential curves (dashed) and flow curves (solid) have the form indicated in Figure 8.28.

11. In Exercise 10, show that the sink at $\zeta = 0$ has strength -1 using the methods outlined in Exercise 5. Show that $\zeta = +i\pi$ is neither a source nor a sink (strength $= 0$). Then consider the velocity $F(\zeta) = -\overline{(d\tilde{f}/d\zeta)}$, and show that F has limit velocities at infinity

$$\lim_{s \to +\infty} F(s + it) = -\frac{1}{2\pi} \quad \text{and} \quad \lim_{s \to +\infty} F(s + it) = +\frac{1}{2\pi}$$

for $0 < t < \pi$. Explain the physical meaning of these limits; compare them with the strength of the sink at $\zeta = 0$.

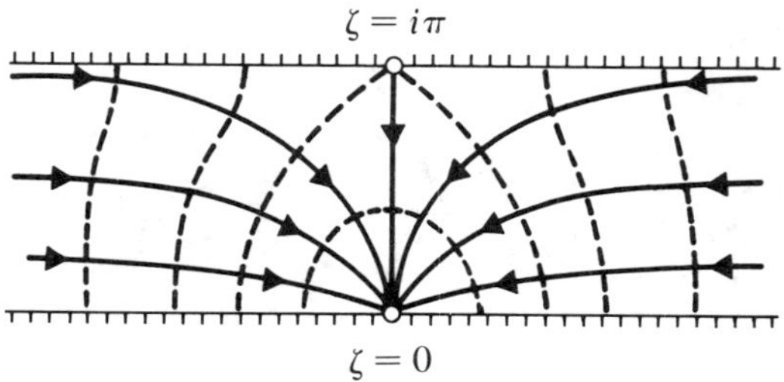

Figure 8.28

12. The flow considered in Example 8.8 has somewhat irregular behavior at -1 and $+1$ on the boundary of the disc (examine the flow lines in Figure 8.25). In particular, the normal derivative $\partial\phi/\partial n(z)$ is not defined at $z = -1$ and $z = +1$, and we should consider the possibility that there are sources or sinks at these points. By applying the method outlined in Exercise 5, and by examining the behavior of the conjugate function $\phi^*(z)$, show that fluid does not enter or leave the exterior domain $|z| > 1$ at these points (i.e., $+1$ and -1 have strength *zero*).

13. Use the results of Exercises 19 and 20 (Section 4.10) to show that $w = T(z) = (1/2)[z + (1/z)]$ maps the exterior domain $D = \{z: |z| > 1\}$ onto the cut plane $E = \mathbf{C} \sim [-1, +1]$. Give an algebraic proof that T is univalent on D. Use the symmetry property $T(z) = T(1/z)$ to determine the image of the punctured disc $D' = \{z: 0 < |z| < 1\}$ in the w-plane.

14. Determine the complex potential $f(z)$ of a flow around the unit disc in the z-plane, whose velocity "at infinity" is described by the unit vector $e^{+i\theta}(0 < \theta < \pi/2)$. Then transform this to a corresponding potential $g(w)$ around the cut $[-1, +1]$ in the w-plane. Verify that the flow lines in the w-plane have the form indicated in Figure 8.29 and show that the velocity of this flow approaches $e^{+i\theta}$ as $w \to \infty$ in the cut plane.

Answer: To get $f(z)$, replace z by $e^{i(\pi-\theta)}z = -e^{-i\theta}z$ in $T(z) = \frac{1}{2}\left(z + \frac{1}{z}\right)$; thus $f(z) = \frac{-\cos\theta}{2}\left[z + \frac{1}{z}\right] + \frac{i\sin\theta}{2}\left[z - \frac{1}{z}\right] = -e^{-i\theta}T(z) - \frac{i\sin\theta}{z}$. Then, substitute $z = \check{T}(w) = -i\sqrt[*]{w^2 - 1}$ to get $g(w)$:

$$g(w) = f(\check{T}(w)) = -e^{-i\theta}w + \sin\theta/\sqrt[*]{w^2 - 1}.$$

15. Sketch the flow lines associated with the complex potential $f(z) = 1/z$, for $z \neq 0$. This is a new kind of flow. There is a source and a

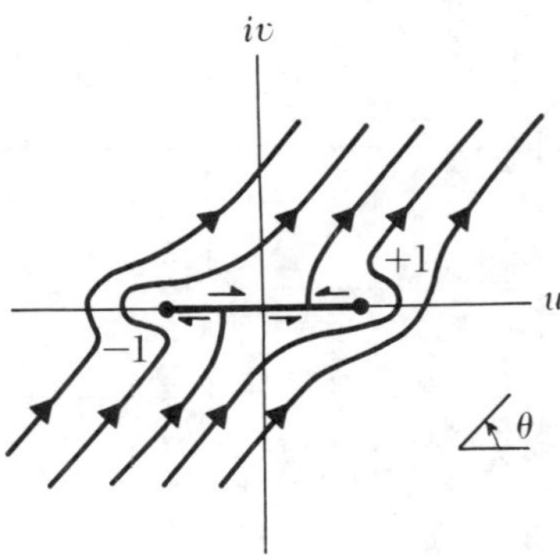

Figure 8.29

Table 8.1 A Summary of the Boundary Conditions in the Physical Applications Discussed in This Chapter.

	Conservation laws valid	*Natural boundary conditions*	*Physical meaning of ϕ or $\Delta\phi$, if any*	*Physical meaning of conjugate function ϕ^*, if any*
Electrostatic fields	static field; charge-free region	$\phi = constant$	$\Delta\phi$ = difference in energy at different positions (per unit charge)	none
Gravitational fields	static field; mass-free region	none	$\Delta\phi$ = difference in energy at different positions (per unit mass)	none
Heat propagation	steady heat flow; homogeneous medium; no sources or sinks within region	$\phi = constant$ or $\frac{\partial\phi}{\partial n} = 0$	ϕ = temperature; $\phi = constant$ gives isothermal lines	$\phi^* = constant$ gives lines of heat flux
Fluid flow	perfect incompressible fluid; steady irrotational flow; no sources or sinks within region	$\frac{\partial\phi}{\partial n} = 0$	none	$\phi^* = constant$ gives stream lines of flow

sink at the origin; they are adjacent, but isolated so that their effects do not cancel. Fluid flows away from the origin in one direction, circulates in the plane, and then returns to the origin from the other side. Such a compound source generates what is known as a "dipole flow."

8.6 ADDITIONAL COMMENTS ON PHYSICAL APPLICATIONS

Table 8.1 summarizes the basic details of each of the applications of potential theory we have examined. Perhaps the most important entries are (i) the boundary conditions natural to each type of problem, and (ii) the physical conditions under which we may expect the vector field $\mathbf{F}$ to satisfy the basic conservation laws $\nabla \times \mathbf{F} = \mathbf{0}$ and $\nabla \cdot \mathbf{F} = 0$. In a more detailed exposition we would encounter situations in which the conservation laws are not always true (as we have stated them), and the problems are not time-independent. Potential theory can still be used to advantage, although other mathematical tools must also be introduced. Whenever potential theory can be applied, complex analysis can be used in the study of important two-dimensional analogs of the actual three-dimensional problems.

Two general principles deserve some final emphasis. One might be called the principle of duality between problems with Dirichlet boundary conditions of the form $\phi = \textit{constant}$ and Neumann problems with boundary condition $\partial\phi/\partial n = 0$. In any simply connected two-dimensional domain, a harmonic function ϕ satisfies one of these conditions along a smooth boundary arc if and only if the conjugate harmonic function ϕ^* satisfies the **conjugate boundary value problem** with the other condition. For example, in a temperature distribution problem with mixed boundary conditions, the conjugate harmonic function T^* may be interpreted as the solution of a conjugate temperature distribution problem, as we indicated in Section 8.3. The general idea is nicely illustrated by the level lines $\phi = \textit{constant}$ (solid lines) and $\phi^* = \textit{constant}$ (dashed lines) in Figure 8.10; $T = \phi$ is the temperature distribution in the original problem, and $T^* = \phi^*$ is the temperature distribution in the conjugate boundary value problem. Other examples may be re-examined from the same viewpoint. The only complication in this pairing of temperature problems is illustrated by the situation in Figure 8.12 accompanying Example 8.4. The original boundary conditions are all of the form $T = \textit{constant}$, with discontinuities at the corners of the domain, and in the conjugate problem the temperature T^* satisfies $\partial T^*/\partial n = 0$, except at these corners; the isotherms of the conjugate temperature distribution are given by the dashed curves in Figure 8.12. In the conjugate problem the walls are insulated, but there are point heat sinks at the corners and a heat source at infinity which keep the solution from being a constant function throughout the domain. Indeed, the reader can see for himself that

$$\lim_{z\to-\pi/2} T^*(z) = \lim_{z\to\pi/2} T^*(z) = -\infty$$

$$\lim_{z\to\infty} T^*(z) = +\infty.$$

On the other hand, there are certain applications in which the conjugate boundary value problem is not at all like the original one. When we start with an electrostatic field produced by potentials

$$\phi = V_1, \ldots, \phi = V_n$$

on the boundary of the domain, the conjugate harmonic function ϕ^* satisfies the condition $\partial\phi^*/\partial n = 0$ on each of the corresponding parts of the boundary; but this kind of boundary condition is not at all natural to electrostatic problems. In order to make it physically meaningful, the conjugate problem should be interpreted as some other kind of problem, possibly a fluid flow or heat propagation problem. In Figure 8.4 we have shown the equipotential lines $\phi =$ *constant* for one of the problems discussed in Example 8.2. Here $\phi + i\phi^* =$ Arcsin w and the curves $\phi^* =$ *constant* for the conjugate harmonic function (see Figure 4.34 in Chapter 4) are semi-ellipses orthogonal to the family of hyperbolas shown in Figure 8.4. The function $\phi^*(w) = \text{Im}(\text{Arcsin } w)$ may be regarded as the potential associated with the flow of fluid through a gap between two semi-infinite barriers in the plane, corresponding to the segments $(-\infty, -1]$ and $[+1, +\infty)$ in the real axis. The complex potential $f(w)$ of the flow should have $\text{Re}(f) = \phi^*$, so that $f(w) = -i \text{ Arcsin } w$. This harmonic function $\phi^*(w)$ may also be regarded as the solution to a rather unrealistic temperature distribution problem in which heat flows between source and sink at $\pm i\infty$, and the segments $(-\infty, -1]$ and $[+1, +\infty)$ represent insulating portions of the real axis.

As another example, the reader might find it interesting to sketch the level curves of the conjugate harmonic function $\phi^*(w)$ in Example 8.1, comparing them with the equipotential curves $\phi =$ *constant* shown in Figure 8.2 (recall Figure 7.10 of Chapter 7), and to consider how the conjugate boundary value problem whose solution is ϕ^* might be interpreted as a heat flow or a fluid flow problem (the latter somewhat unrealistic unless we examine the function $\phi^*(w)$ in the larger domain $E = \mathbf{C} \sim \{+i, -i\}$). Other examples of dual, or conjugate, boundary value problems are mentioned in the exercises which follow.

At the other extreme there are physical problems which seem very different, as physical problems, but lead to the same kind of boundary value problems for Laplace's equation. Once this happens, we realize that the physical problems have mathematically equivalent formulations, and there is a direct correspondence between problems in one physical situation and problems in the other. We might refer to this as the principle of analogy of physical problems, as opposed to the principle of duality (relation through conjugacy) we have just been discussing. For example, electrostatic problems may be identified with a certain class of temperature distribution problems—those in which the boundary condition $\partial\phi/\partial n = 0$ does not appear—but there are temperature problems which do not correspond to any meaningful electrostatic problem. Likewise, certain fluid flow problems are analogous to a special kind of temperature distribution problem. On the other hand, there is no analogy at all between the kinds of electrostatic boundary value problems we have been discussing and fluid flows in an enclosed region; these problems are related only by being conjugate to one another.

EXERCISES

1. The function $\phi(z) = \text{Re}(1/z)$ is discussed in Exercise 15 of Section 8.5 as the potential of a "dipole flow" in the plane. Interpret ϕ as the electrostatic potential obtained when positive and negative charges of equal strength are brought together at the origin. The charge distribution that produces this potential is called an electric dipole; these arise quite often in practical applications.

2. Find a conjugate harmonic function $T^*(z)$ for the solution of the temperature distribution problem in Exercise 2 (Section 8.4). What boundary conditions are satisfied? Explain how T^* can be interpreted as the potential of a fluid flow (with a source or sink at $z = 0$). Sketch the flow lines, and calculate the derivative df/dz of the complex potential. Explain why the point $z = 0$ must be a *source* and *not a sink* (examine the velocity field associated with the potential T^*).

Answer: $T + iT^* = f(z) = 10i - (20/\pi)\text{Log}[\sin(\pi z/2)]$; $df/dz = -10\,\text{ctn}(\pi z/2)$.

3. Exercise 2 can, alternatively, be interpreted as a temperature distribution in which

(i) heat flows from a source at $z = 0$, toward infinity;

(ii) walls are insulated, except at $z = 0$.

Show that $\lim_{z\to 0} T^* = +\infty$ ($z \to 0$ within the half strip).

Note: One can show that the quantity of heat entering at $z = 0$ (per unit time) is fixed; this is the boundary condition satisfied at $z = 0$, rather than any other condition, such as having $|T^*(z)|$ bounded.

4. Show that $w = T(z) = e^z + z$ maps the lines $y = +\pi$ and $y = -\pi$ to the half lines C_1 and C_2, shown in Figure 8.30; show that the strip $D = \{z: -\pi < \text{Im}(z) < +\pi\}$ is mapped one-to-one onto the complementary region $E = \mathbf{C} \sim (C_1 \cup C_2)$.

5. Using Exercise 4, transform the constant velocity horizontal flow in the strip D, associated with the complex potential $f(z) = -z$, to a flow in E. Verify that the flow lines have the general form indicated in Figure 8.30; then draw in a few lines of constant fluid potential.

Note: $\check{f}(w) = -\check{T}(w)$ describes flow from an open channel into the rest of the plane.

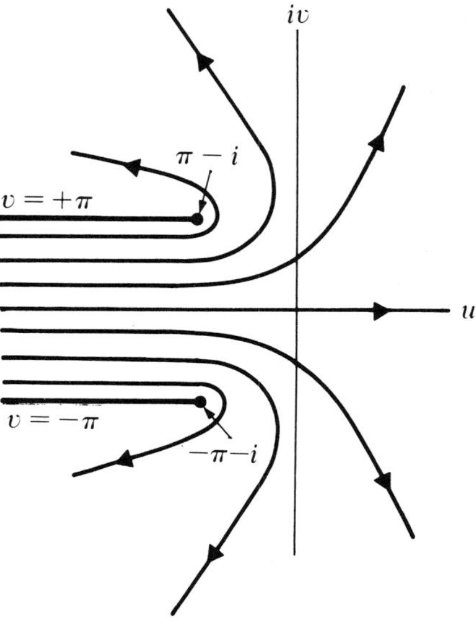

Figure 8.30

6. Interpret the flow lines in Figure 8.30 as the equipotential curves of an electric field between two semi-infinite charged plates. If the inverse function $z = \check{T}(w)$ were known, would the electrostatic potential $\phi(w)$ be given by $\text{Re}[-\check{T}(w)]$ or by $\text{Im}[-\check{T}(w)]$? How would you interpret $\phi(w)$ as the solution of a temperature distribution problem?

9 CONFORMAL MAPPING PROBLEMS

We have seen, in Chapter 7 and elsewhere, that it is extremely useful to be able to map certain domains conformally onto a standard domain such as the unit disc $|z| < 1$ or the upper half plane $\mathrm{Im}(z) > 1$. We have already built up a substantial catalog of particular conformal mappings in the course of working out examples, and we will go a bit further in this direction in this chapter. Special attention will be given to techniques based on the Schwarz reflection principle and on the Schwarz-Christoffel formula. We will also examine new kinds of mappings not previously discussed. Finally, we shall discuss a theoretical result, the Riemann mapping theorem, which states that every simply connected domain in $\mathbf{C}$, except for the special domain $E = \mathbf{C}$, can be mapped conformally onto the unit disc. This theorem guarantees that most boundary value problems for Laplace's equation in plane domains can be converted into corresponding problems for Laplace's equation in the unit disc.

9.1 GEOMETRIC PROPERTIES OF FRACTIONAL LINEAR TRANSFORMATIONS

When fractional linear transformations were introduced in Section 4.8 we were not yet ready to study their global mapping properties. In this section we will concentrate on these properties, so that the reader can use fractional linear transformations more effectively in practical mapping problems. We first show how to express fractional linear transformations in normal form, from which we can obtain geometric information more easily than we could from the usual formula $T(z) = (az + b)/(cz + d)$.

Theorem 9.1 *Let T be a fractional linear transformation and let $\{p, q\}$ be points in the plane with $T(p) \neq \infty$ and $T(q) \neq \infty$. Then there is a complex constant $\mu \neq 0$*

such that the image point $w = T(z)$ is always related to z via the equation:

$$\text{(1)} \qquad \frac{w - T(p)}{w - T(q)} = \mu \frac{z - p}{z - q} \qquad \textit{for all } z \textit{ in } \mathbf{C}^*.$$

Formula (1) is called the **normal form** of T with respect to the base points $\{p, q\}$. The normal form is particularly simple (and interesting) when T has two distinct fixed points; if we use these as the base points then $p = T(p)$ and $q = T(q)$, by definition of a fixed point, and $w = T(z)$ is given by

$$\text{(2)} \qquad \frac{w - p}{w - q} = \mu \frac{z - p}{z - q} \qquad \text{for all } z \text{ in } \mathbf{C}^*.$$

PROOF OF THEOREM 9.1: Set $z_1 = p$, $z_2 = q$, and select any third point z_3 different from z_1 and z_2 such that $T(z_3) \neq \infty$. The points $\{z_1, z_2, z_3\}$ are mapped to points $\{w_1, w_2, w_3\}$, and T is uniquely determined once we know this much about its action. Then T must satisfy the following identity (equation (32) of Section 4.8):

$$\left(\frac{w_3 - w_2}{w_3 - w_1}\right) \frac{w - w_1}{w - w_2} = \left(\frac{z_3 - z_2}{z_3 - z_1}\right) \frac{z - z_1}{z - z_2}$$

On dividing both sides by $(w_3 - w_2)/(w_3 - w_1)$ we obtain

$$\frac{w - w_1}{w - w_2} = \mu \frac{z - z_1}{z - z_2}, \quad \text{where} \quad \mu = \left(\frac{w_3 - w_1}{w_3 - w_2}\right)\left(\frac{z_3 - z_2}{z_3 - z_1}\right). \qquad \blacksquare$$

If we are given the normal form of T, we may easily solve for w in terms of z and obtain the usual formula. Conversely, to determine the normal form once the base points p and q have been chosen, all we need is the value of μ; this can be obtained by inserting the values $w = T(z)$ into (2) for some third point z different from p and q. The choice $z = \infty$ is allowed.

Example 9.1 Express $w = 1/z$ in normal form with respect to the fixed points $p = +1$ and $q = -1$. Since $T(p) = +1$ and $T(q) = -1$, the normal form is given by

$$\frac{w - 1}{w + 1} = \frac{w - 1}{w - (-1)} = \mu \frac{z - 1}{z + 1}.$$

To determine the constant μ we might insert the values $z = +i$ and $w = 1/i = -i$ into this equation; then $\mu = \left(\frac{i + 1}{i - 1}\right)\left(\frac{i + 1}{i - 1}\right) = -1$. This calculation is even easier if we use the special values $z = 0$ and $w = 1/0 = \infty$, so that $1 = \frac{\infty - 1}{\infty + 1} = \mu\left(\frac{-1}{1}\right) = -\mu$.

When we take absolute values of both sides in (1) we get an equation with great geometric significance,

(3) $$\frac{|w - T(p)|}{|w - T(q)|} = |\mu| \frac{|z - p|}{|z - q|} \quad \text{for all } z \text{ in } \mathbf{C}^*.$$

This shows that the ratio of distances $|z - p|$ and $|z - q|$ to the base points in the z-plane is altered by a constant factor $|\mu| > 0$ when we measure from $w = T(z)$ to the new base points $\{T(p), T(q)\}$ in the w-plane. The scale factor $|\mu|$ is fixed once T and the base points $\{p, q\}$ have been given; it does not vary as we examine different points z and $w = T(z)$. This formula becomes especially interesting when we recall the following geometric fact, whose reasonably straightforward proof will be left as an exercise.

Theorem 9.2 *Let p and q be distinct points in the plane, let α be any positive real constant $(0 < \alpha < \infty)$, and let A_α be the set of points z that satisfy the equation*

$$\left| \frac{z - p}{z - q} \right| = \alpha.$$

If $\alpha = 1$, A_1 is a line (the perpendicular bisector of the segment from p to q); if $\alpha \neq 1$ the locus A_α is a circle that separates p from q in the complex sphere.

For fixed base points p and q, the curves A_α form the family $\mathscr{A}\{p, q\}$ of circles (including an extended line A_1) displayed in Figure 9.1. These are called the **circles of Apollonius.** Their centers all lie on the straight line L passing through p and q, on either side of the segment $[p, q]$. It is easily seen that the A_α close down on the point p as $\alpha \to 0$ and close down on q as $\alpha \to +\infty$. Different circles $A_{\alpha'}$ and $A_{\alpha''}$ in this family are disjoint. The exceptional circle A_1 is regarded as a degenerate circle with center at infinity.

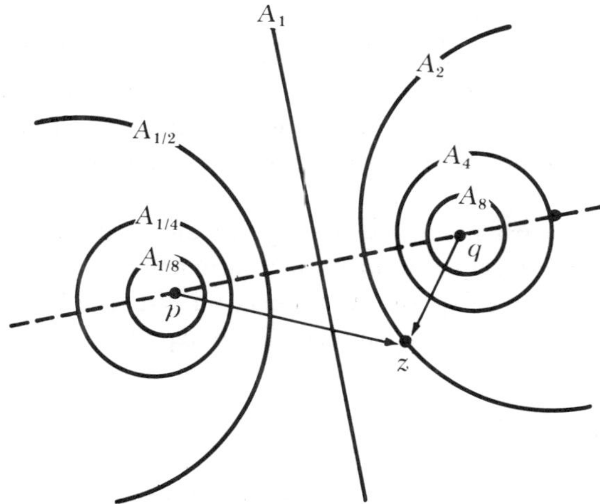

Figure 9.1 Apollonian circles A_α $(0 < \alpha < \infty)$. The arrows from p and q represent the distances $|z - p|$ and $|z - q|$.

A transformation with normal form (1) must map the family $\mathscr{A} = \{A_\alpha: 0 < \alpha < \infty\}$ of Apollonian circles based on points p and q in the z-plane to the family $\tilde{\mathscr{A}} = \{\tilde{A}_\alpha: 0 < \alpha < \infty\}$ of Appolonian circles in the w-plane based on the image points $\{T(p), T(q)\}$. In fact, the circle or extended line (line plus point at infinity)

$$A_\alpha = \left\{z: \left|\frac{z-p}{z-q}\right| = \alpha\right\}$$

is transformed to the circle (or extended line)

$$\tilde{A}_{|\mu|\alpha} = \left\{w: \left|\frac{w-T(p)}{w-T(q)}\right| = |\mu|\,\alpha\right\},$$

for each $0 < \alpha < +\infty$.

Example 9.2 The transformation $w = (3z+1)/(z+3)$ has just two fixed points, $p = +1$ and $q = -1$. The constant μ in the normal form with respect to these base points,

$$\frac{w-1}{w+1} = \mu \frac{z-1}{z+1},$$

may be obtained by substituting $z = 0$ and $w = T(0) = \frac{1}{3}$ into this equation; thus, $\mu = (\frac{1}{2}) + i0$ and $|\mu| = \frac{1}{2}$ in equation (1′). Considering the families $\mathscr{A}$ and $\tilde{\mathscr{A}}$ of Apollonian circles with respect to the base points $+1$ and -1 in both the z-plane and w-plane, we see that T maps A_α one-to-one onto the circle $A_{|\mu|\alpha} = A_{\alpha/2}$ in the w-plane. It may be better to think of T as transforming the z-plane onto itself, so that the circles A_α and their images $T(A_\alpha) = A_{\alpha/2}$ may be compared directly. Each point z on A_α is transformed to a point $T(z)$ on $A_{\alpha/2}$, which is a circle closer to the base point $p = +1$; thus, T moves points away from $q = -1$ toward $p = +1$; Apollonian circles in the family $\mathscr{A}\{p, q\}$ are transformed to different circles in the family. The action of T is illustrated in Figure 9.2. Later in this section we shall demonstrate that points are indeed shifted along the dashed circular arcs that connect p and q, as we have indicated in the figure. (This need not be the case for other transformations).

Before we go on to examples, we must point out that the relation (3) obtained by taking absolute values in the normal form has as its counterpart the equation obtained by comparing the arguments on either side of (1). We obtain a congruence (modulo 2π), instead of an equality, because of the indeterminacy of $\arg z$:

$$\text{(4)}\quad [\arg(w - T(p)) - \arg(w - T(q))] \equiv \arg(\mu) + [\arg(z-p) - \arg(z-q)]$$

This is valid provided that $z \neq p$ and $z \neq q$.

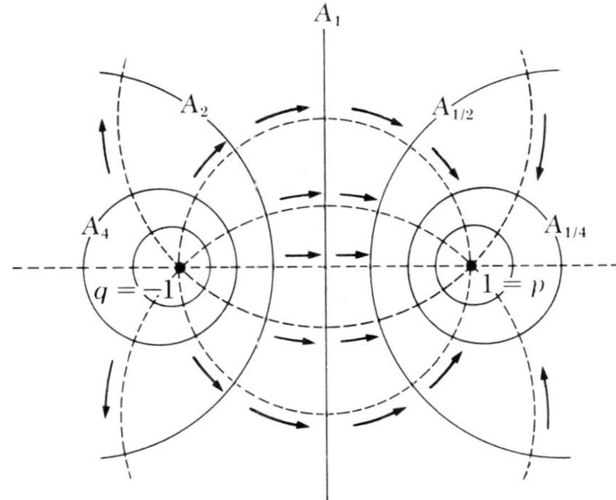

Figure 9.2 The action of $T(z) = \dfrac{3z + 1}{z + 3}$ on Apollonian circles.

The reduced equations (3) and (4) taken together contain all the information present in the original equation (1), and therefore they completely determine the transformation T; the separate equations (3) and (4) display this information in two convenient separate parts. Consideration of the first equation shows how T transforms the Apollonian circles based on $\{p, q\}$; there is a complementary geometric interpretation of the second equation. To make this interpretation we must identify the sets B_θ of all points z in the plane that satisfy the equation

$$\arg(z - p) - \arg(z - q) \equiv \theta \pmod{2\pi}, \tag{5}$$

where θ is a fixed real number. This is a straightforward task in analytic geometry, so we will only summarize the important conclusions as Theorem 9.3, without giving a detailed proof. Since $\arg(0)$ is not defined, even modulo 2π, the left side of (5) is undefined if $z = p$ or $z = q$; let us interpret this to mean that the points p and q never appear in the sets B_θ. It is obvious that θ and $\theta + 2\pi k$ (k any integer) determine the same set of points, so $B_\theta = B_{\theta+2\pi k}$, and all possible curves B_θ are obtained as θ varies through some half open interval such as $[0, 2\pi)$.

Theorem 9.3 *Let $\{p, q\}$ be distinct points in the complex plane and consider the set of points B_θ determined by equation* (5), *where θ is a fixed real number. Then $B_\theta = B_{\theta+2\pi k}$ for any integer k. Furthermore,*

(i) *If $\theta \equiv \pi \pmod{2\pi}$, then B_θ is the straight line segment connecting p to q, with the points $\{p, q\}$ deleted.*

(ii) *If $\theta \equiv 0 \pmod{2\pi}$, then B_θ is the straight line L determined by p and q, with the segment from p to q and the points $\{p, q\}$ deleted.*

If we regard a line as a circular arc in $\mathbf{C}^*$, *these special cases fit nicely into the general situation:*

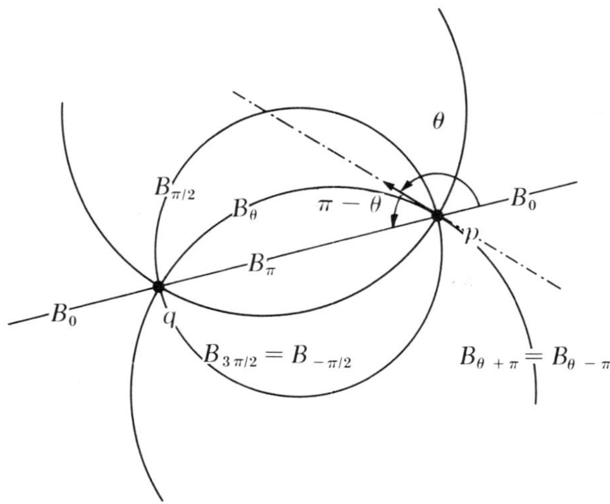

Figure 9.3 The circular arc B_θ. Angle θ is measured as shown (counterclockwise angles are positive).

(iii) *If $\theta \not\equiv 0$ (mod 2π) and $\theta \not\equiv \pi$ (mod 2π), then B_θ is the circular arc from p to q whose tangent line at p makes an angle $(\pi - \theta)$ with the directed line segment from p to q (measured from the arc to the segment).*

The location of B_θ is illustrated in Figure 9.3. Notice that the angle condition in (iii) is the same as saying that the tangent line at p makes an angle θ with the ray from p to ∞ that points in the direction opposite that of the directed line segment from p to q. From this description of the curves B_θ the following important point is clear.

Corollary 9.4 *The curves B_θ and $B_{\theta+\pi} = B_{\theta-\pi}$, together with the points p and q, fit together to form a circle through the points $\{p, q\}$. Furthermore, if $\theta \not\equiv \phi$ (mod 2π), the curves B_θ and B_ϕ never intersect.*

The reader should examine these curves to understand how B_θ varies as θ increases from $\theta = 0$ to $\theta = 2\pi$, keeping Figure 9.3 in mind. He should also try to understand how the exceptional curves B_0 and B_π fit naturally into these families (a look at what is happening on the complex sphere is helpful).

The family $\mathscr{B}\{p, q\}$ of all curves B_θ we obtain this way is complementary to the family of Apollonian circles $\mathscr{A}\{p, q\} = \{A_\alpha : 0 < \alpha < +\infty\}$ based on the points $\{p, q\}$. A circle A_α and circular arc B_θ are orthogonal where they meet, and intersect in exactly one point. Every point z in the domain $E = \{z : z \neq p$ and $z \neq q\}$ is the intersection of exactly one pair of these curves A_α and B_θ; therefore, the families $\mathscr{A}$ and $\mathscr{B}$ are orthonormal and give us an orthogonal curvilinear coordinate system on E. The families $\mathscr{A}$ and $\mathscr{B}$, taken together, are called the **Steiner circles** based on the points $\{p, q\}$.

The points p and q are singular in this coordinate system, much as the points $z = 0$ and $z = \infty$ are singular with respect to polar coordinates. The point ∞ has well defined coordinates in the circular coordinate system just described (it is the intersection of A_1 and B_0), and the coordinates are well behaved near ∞ on the complex sphere.

Let T be a fractional linear transformation and let $\{p, q\}$ be distinct points, neither a singular point for T. The congruence (5) shows us that the circular arc B_θ in $\mathscr{B}\{p, q\}$ is transformed by T to the circular arc $B_{\theta+\arg(\mu)}$ in $\mathscr{B}\{T(p), T(q)\}$, for each θ. Now that we know how T transforms both kinds of curves in the coordinate system of Steiner circles based on $\{p, q\}$ to curves in the system based on $\{T(p), T(q)\}$, we have a complete geometric picture of how T acts once it has been written in the normal form (1). If z is any point in the plane (other than p or q), take the unique curves A_α and B_θ in the coordinate system about $\{p, q\}$ that meet at z, so that $\{z\} = A_\alpha \cap B_\theta$; then T transforms A_α to $T(A_\alpha) = \tilde{A}_{|\mu|\alpha}$ in the system of Apollonian circles $\mathscr{A}\{T(p), T(q)\}$ and transforms B_θ to the circular arc $T(B_\theta) = \tilde{B}_{\theta+\arg(\mu)}$ in the system $\mathscr{B}\{T(p), T(q)\}$. The unique point where the transformed curves meet is just $T(z)$. The description of T is particularly simple when $\{p, q\}$ are *fixed points*, so that $T(p) = p$ and $T(q) = q$.

Example 9.3 We have calculated the normal form of $w = (3z + 1)/(z + 3)$ about the fixed points $p = +1$ and $q = -1$:

$$\frac{w-1}{w+1} = \left(\frac{1}{2}\right)\frac{z-1}{z+1} \qquad \text{for all } z \text{ in } \mathbf{C}^*.$$

Here $\arg(\mu) = 0$ and $|\mu| = \frac{1}{2}$, so a point z lying on the circles A_α and B_θ is mapped to a point $w = T(z)$ lying on a different Apollonian circle $A_{\alpha/2}$, but the image remains within B_θ. The new point has coordinate parameters $(\alpha/2, \theta)$ and is obtained by shifting z along the circular arc B_θ from A_α to $A_{\alpha/2}$, as we have indicated in Figure 9.2. Our geometric picture of how T acts as a "shift" of points toward $p = +1$ and away from $q = -1$ is now completely justified.

Example 9.4 The transformation $w = i(z - i)/(z + i)$ maps the upper half plane $\operatorname{Im}(z) > 0$ conformally onto the unit disc $|w| < 1$; we shall now consider it as a mapping of $\mathbf{C}^*$ onto itself. The fixed points of T are $p = +1$ and $q = -1$ (an easy calculation). To determine μ in the normal form

$$\frac{w-1}{w+1} = \mu\,\frac{z-1}{z+1} \qquad \text{for all } z \text{ in } \mathbf{C}^*,$$

we may substitute $z = +i$ and $w = T(i) = 0$; then $\mu(i - 1)/(i + 1) = -1$, so that $\mu = (1 + i)/(1 - i) = i = e^{i\pi/2}$. Equations (3) and (4) take the form

$$\left|\frac{w-1}{w+1}\right| = 1 \cdot \left|\frac{z-1}{z+1}\right|$$

$$\arg\left(\frac{w-1}{w+1}\right) \equiv \frac{\pi}{2} + \arg\left(\frac{z-1}{z+1}\right) \qquad (\text{mod } 2\pi)$$

if z is not one of the fixed points. Since $|\mu| = 1$, it follows that $T(A_\alpha) = A_\alpha$, so that T maps each Apollonian circle into itself. On the other hand, an arc B_θ is

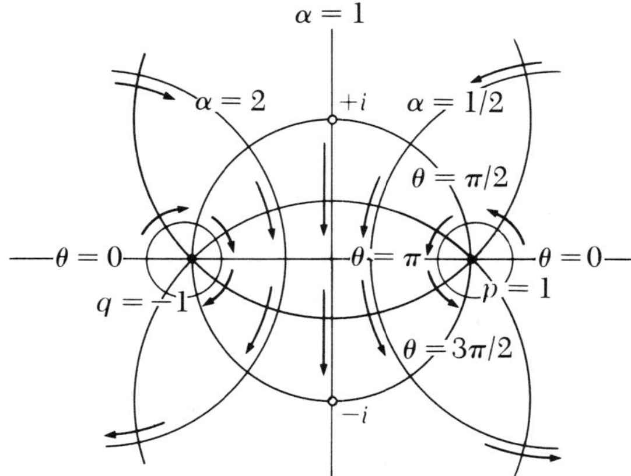

Figure 9.4A Apollonian circles A_α are left fixed, but B_θ is shifted to $B_{\theta+\pi/2}$, as indicated by the arrows.

mapped to the arc $B_{\theta+\pi/2}$. A point with Steiner coordinates (α, θ) is transformed to one with coordinates $(\alpha, \theta + \pi/2)$. Points on an Apollonian circle A_α are moved around within this circle; T causes a rotation-like shift of points along A_α as shown in Figure 9.4A. The reader should try to see, by inspecting this figure, that T really does map the upper half plane onto the unit disc, the extended real axis onto the unit circle, and the lower half plane onto the exterior domain $\{z:|z| > 1\}$.

On the exceptional circle A_1 the action of T can be regarded as a rotation-like shift, which moves points while preserving their relative order, if we examine the action in $A_1^* = A_1 \cup \{\infty\}$, a circle through ∞ in $\mathbf{C}^*$. Thus T moves points *across* the point at infinity and moves ∞ to the point $+i$. This action on A_1^* is illustrated in Figure 9.4B. The exceptional behavior of T

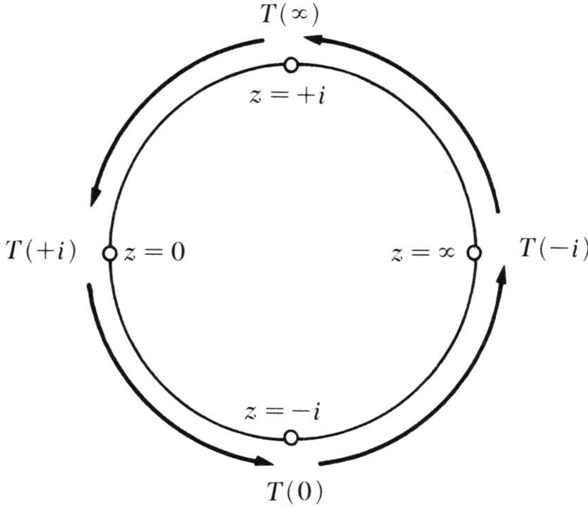

Figure 9.4B The action of $w = i\dfrac{z - i}{z + i}$ on the extended line A_1 (the imaginary axis) in Example 9.4.

on A_1 can also be understood by examining the action of T on nearby circles A_α with α close to 1 (but $\alpha \neq 1$); the action on A_1 is the "limit" of the rotation-like actions on A_α as we let $\alpha \to 1$.

Finally, we examine a family of transformations which will be useful in later sections.

Example 9.5 Let $0 < R < 1$ be fixed and consider the fractional linear transformation $w = T_R(z) = \dfrac{R - z}{Rz - 1}$. The extended real axis $\mathbf{R}^*$ is clearly mapped onto itself, and

$$T(1) = 1; T(-1) = -1; T(R) = 0; T(0) = -R; T(-R) = \frac{-2R}{1 + R^2}.$$

The fixed points are $p = +1$ and $q = -1$, and the normal form is

$$\frac{w - 1}{w + 1} = \mu \frac{z - 1}{z + 1} \quad \text{where} \quad \mu = \frac{1 + R}{1 - R}.$$

Here $\mu = |\mu| > 0$ and $\arg \mu \equiv 0$, since $0 < R < 1$. Points are shifted along the circles B_θ, toward -1 and away from $+1$. The arcs $B_{\pi/2}$ and $B_{3\pi/2}$ are the two halves of the boundary of the unit disc $D = \{z : |z| < 1\}$, so T_R maps the disc *conformally onto itself* and also maps the circle $|z| = 1$ onto itself. The general features of the shift of points within D are illustrated in Figure 9.5, where we show how various parts of the unit disc are transformed. The circle Γ through $R + i0$ and orthogonal to the unit circle is just the Apollonian circle A_α with $\alpha = \dfrac{1 - R}{1 + R}$; it separates D into two pieces which are mapped by T_R to half discs. Notice that $T_R(\Gamma) = A_1$, the imaginary axis.

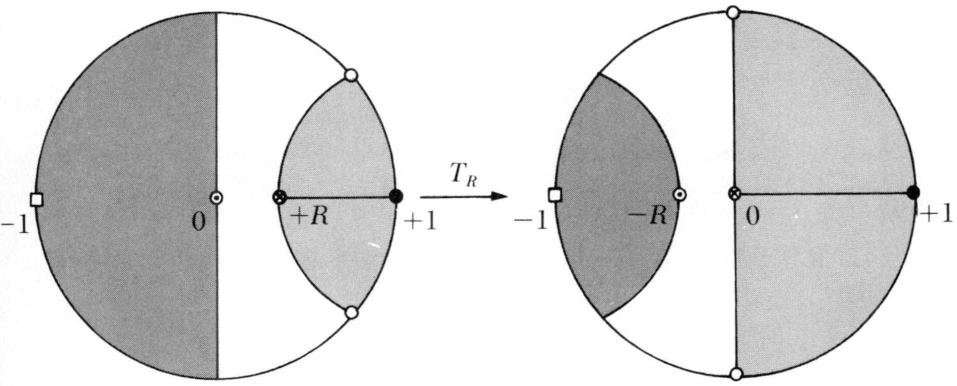

Figure 9.5 The transformation $w = \dfrac{R - z}{Rz - 1}$ $(0 < R < 1)$ maps points and subregions in the manner shown.

These examples illustrate the particularly simple interpretation of T when the constant μ in the normal form has arg $\mu = 0$ or $|\mu| = 1$ respectively, and when $\{p, q\}$ are fixed points for T. Other examples appear in the exercises.

Sometimes we seek geometric information that is best revealed by taking base points other than the fixed points of T; furthermore, there are transformations with just one fixed point, and there are not enough fixed points to set up a normal form whose base points are fixed by T. For a more detailed account of normal forms, and the geometry of fractional linear transformations, see Ford [7].

EXERCISES

1. Let A_α and $A_{\alpha'}$ be Apollonian circles based on $\{p, q\}$; show that A_α and $A_{\alpha'}$ are disjoint if $\alpha \neq \alpha'$. Explain why A_α closes down on p as $\alpha \to 0$ and closes down on q as $\alpha \to +\infty$.

2. If p and q are fixed points in $\mathbf{C}$ for a fractional linear transformation $w = T(z)$, show that the constant μ in the normal form

$$\frac{w - p}{w - q} = \mu \frac{z - p}{z - q}$$

may be calculated as

$$\mu = \left[\frac{dw}{dz}\bigg|_{z=p}\right] \quad \text{or} \quad \frac{1}{\mu} = \left[\frac{dw}{dz}\bigg|_{z=q}\right].$$

3. If $0 < \alpha < +\infty$ (with $\alpha \neq 1$), and if $p \neq q$ are points in $\mathbf{C}$, prove:

Theorem: The locus $\left|\dfrac{z - p}{z - q}\right| = \alpha$ is a circle of radius $r_\alpha = \dfrac{\alpha\ |p - q|}{(1 - \alpha^2)}$ centered at the point $(p - \alpha^2 q)/(1 - \alpha^2)$.

Verify that reciprocal parameters α and $1/\alpha$ give circles of the same radius. Verify that the center lies on the line determined by the points p and q, and lies off the segment $[p, q]$.

Hint: Set $p = a + ib$ and $q = r + is$, and use straightforward analytic geometry to convert $|z - p| = \alpha\ |z - q|$ to the form $(x - A)^2 + (y - B)^2 = R^2$.

4. Take $p = -1$ and $q = +1$, and explicitly determine the loci

$$A_\alpha = \left\{z: \left|\frac{z - p}{z - q}\right| = \alpha\right\}$$

for $\alpha = \frac{1}{10}$, $\alpha = \frac{1}{2}$, $\alpha = 1$, $\alpha = 2$, $\alpha = 100$. Sketch their locations.

5. If base points $p \neq q$ are given, show that $w = T(z) = (z - p)/(z - q)$ transforms Steiner circles in the following way:

(i) $T(p) = 0$ and $T(q) = \infty$

(ii) $\mathscr{B}\{p, q\}$ is transformed to the family of all *rays* extending from 0 to ∞.

(iii) $\mathscr{A}\{p, q\}$ transforms to the family of all *circles* centered at the origin.

Hint: Circles map to circles in $\mathbf{C}^*$; orthogonality relations are preserved since T is conformal. There is a simple reason why $T(B_\theta)$ equals the ray $\arg w \equiv \theta$. Statement (iii) follows easily from (ii).

6. Use Exercise 5 to verify that the loci $B_\theta = \{z : \arg(z - p) - \arg(z - q) \equiv \theta \pmod{2\pi}\}$ are indeed the circular arcs shown in Figure 9.3.

Hint: T maps the locus B_θ to the ray $R_\theta = \{w : \arg(w) \equiv \theta\}$; thus $B_\theta = \check{T}(R_\theta)$ is the desired circle.

7. Take $p = 1$ and $q = +i$; sketch the loci B_θ for $\theta = 0, \pm\pi/4, \pm\pi/2, \pm 3\pi/4$, and $+\pi/6$.

8. Take the Steiner coordinates (α, θ) based on the points $p = +1$ and $q = -1$, and locate the points in $\mathbf{C}^*$ with the following coordinates. Give their Cartesian coordinates.

(i) $\alpha = 1, \theta = 0$ (iv) $\alpha = \frac{1}{2}, \theta = \pi$

(ii) $\alpha = 1, \theta = \pi$ (v) $\alpha = \frac{1}{2}, \theta = \pi/2$

(iii) $\alpha = 1, \theta = +\pi/2$ (vi) $\alpha = 2, \theta = \pi/2$

Hint: Use the results in Exercise 3.

Answers: (i) ∞; (ii) 0; (iii) $+i$; (iv) $(\frac{1}{3}) + i0$; (v) $(3 + 4i)/5$; (vi) $(-3 + 4i)/5$.

9. Express the following transformations in normal form with respect to their fixed points:

$$w = \frac{z}{2z - 1} \qquad w = \frac{2z}{3z - 1} \qquad w = \frac{3z - 4}{z - 3} \qquad w = \frac{z}{2 - z}.$$

10. Verify that $H(z) = \log \left| \dfrac{z - p}{z - q} \right|$ is harmonic on the domain $E = \mathbf{C} \sim \{p, q\}$. Show that $H(z) = 0$ on the perpendicular bisector of the segment $[p, q]$, and that

$$\lim_{z \to p} H = -\infty \qquad \lim_{z \to q} H = +\infty.$$

Is $|H(z)|$ bounded as $z \to \infty$? Evaluate $\lim_{z \to \infty} H$, if this limit exists.

11. Let Γ_1 and Γ_2 be the circles $|z - 1| = \frac{1}{2}$ and $|z + 1| = \frac{1}{2}$, respectively. Find a harmonic function $H(z)$, defined for points z that are exterior to both circles, such that

$$\begin{aligned} &H(z) = -1 \quad \text{for } z \text{ on } \Gamma_1 \\ &H(z) = +1 \quad \text{for } z \text{ on } \Gamma_2 \\ &|H(z)| \text{ bounded at infinity} \end{aligned}$$

This is an extremely important boundary value problem in electrostatics and the theory of heat propagation.

Hint: Use Exercise 3 to locate points p and q $(q = -p)$ on the real axis, such that Γ_1 and Γ_2 become loci A_α and $A_{1/\alpha}$.

Answer: (i) $r = 1/2$ implies $p = (1 - \alpha^2)/4\alpha$; (ii) center $p = 1$ implies $p = (1 - \alpha^2)/(1 + \alpha^2)$. Thus $\alpha = 2 - \sqrt{3}$, $p = \sqrt{3}/2$, and

$$H(z) = A \cdot \log\left|\frac{2z - \sqrt{3}}{2z + \sqrt{3}}\right| = \frac{A}{2}\log\left[\frac{(2x - \sqrt{3})^2 + 4y^2}{(2x + \sqrt{3})^2 + 4y^2}\right]$$

where $A = 1/\log(2 - \sqrt{3}) = 1/\log \alpha$.

12. Use Exercise 11 to find harmonic functions that satisfy the boundary conditions indicated below.

(i) $H(z) = 0$ on the imaginary axis, $H(z) = -10$ on the circle $|z - 5| = 2$, $H(z)$ bounded on the domain between these curves.

(ii) $H(z) = 0$ on the circle $|z + 1| = \frac{1}{2}$, $H(z) = +10$ on the circle $|z - 1| = \frac{1}{2}$, $H(z)$ bounded on the domain between these circles.

13. We may consider the boundary value problem in Exercise 11 as either an electrostatic or a temperature distribution problem. Sketch the level curves $H(z) = c(-1 < c < +1)$. Sketch the lines of constancy $H^*(z) = \textit{constant}$ for the conjugate harmonic function (the lines of heat flux, in a temperature distribution problem). How are these related to the families $\mathscr{A}\{-1, +1\}$ and $\mathscr{B}\{-1, +1\}$ of Steiner circles?

14. Map the domain bounded by $|z| = 1$ and $|z - (\frac{1}{4})| = \frac{1}{4}$ onto an annulus $1 > |w| > R$. How large is the outer radius R?

Answer: $R = 2 + \sqrt{3}$.

15. Given two *disjoint* circles Γ_1 and Γ_2, prove that it is always possible to choose $p \neq q$ in $\mathbf{C}$, and parameters $\alpha \neq \beta$ in the interval $(0, +\infty)$, such that Γ_1 and Γ_2 are the Apollonian circles A_α and A_β, respectively.

Hint: By rotating the plane we may reduce to the case where the centers of Γ_1 and Γ_2 are on the real axis. It may help to distinguish cases: (i) each circle lies exterior to the other; (ii) one lies in the disc enclosed by the other.

16. Make a diagram that illustrates how $w = T_R(z) = (R - z)/(Rz - 1)$ acts on the Steiner coordinates $(\alpha, 0)$ with respect to the points $p = +1$ and $q = -1$, the fixed points of T_R. Use Figure 9.4 as a model.

17. Let R be fixed $(0 < R < 1)$. Prove that $w = T_R(z) = (R - z)/(Rz - 1)$ maps the unit disc conformally onto itself and verify that T_R maps points, arcs, and regions as indicated in Figure 9.5. Then,

(i) Show that the inverse has the same form:

$$z = \check{T}_R(w) = (R - w)/(Rw - 1).$$

(ii) Calculate the fixed points.

(iii) Write out the normal form of T_R with respect to the fixed points.

Hint: Use the results of Exercise 16.

18. Explain why the circle that passes through $R + i0$ and is perpendicular to the unit circle $|z| = 1$ (as shown in Figure 9.5) is actually part of an Apollonian circle A_α, with respect to the base points $p = +1$ and $q = -1$. Show that $\alpha = (1 - R)/(1 + R)$.

Hint: $\check{T}_R$ and $\check{T}_R$ map Apollonian circles in $\mathscr{A}(p, q)$ to circles of the same kind. What is $\check{T}_R(A_1)$?

19. Make a diagram similar to Figure 9.5 that shows how the *negative* of T_R (the mapping $N_R(z) = (-1) \cdot T_R(z)$) transforms the regions and arcs on the left-hand side of Figure 9.5. This negative is the mapping we will actually use in applications.

Hint: Use the facts worked out in Exercise 17. Think of N_R as the transformation $\zeta = T_R(z)$, followed by the elementary transformation $w = -\zeta$.

Note: We will use N_R later on, but T_R has a simpler normal form. Thus, it is convenient to examine geometric properties of T_R and then multiply by (-1) to determine the behavior of N_R.

20. Consider the transformations T whose normal forms are

(i) $\dfrac{w-1}{w+1} = \left(\dfrac{1}{2}\right)\dfrac{z-1}{z+1}$ (ii) $\dfrac{w-1}{w+1} = (e^{i\pi/4})\dfrac{z-1}{z+1}$

(iii) $\dfrac{w-1}{w+1} = \left(\dfrac{e^{i\pi/4}}{2}\right)\dfrac{z-1}{z+1} = \left(\dfrac{1+i}{2\sqrt{2}}\right)\dfrac{z-1}{z+1}$

Compare their actions by setting up individual diagrams, similar to the one in Figure 9.4, which show how the families $\mathscr{A}\{+1, -1\}$ and $\mathscr{B}\{-1, +1\}$ of Steiner circles are transformed.

21. Use the results of Exercise 20 to determine the images of $z = 0$ under repeated applications of T and $\check{T}$. Do this for each of the mappings (i), (ii), and (iii). Do the points

$$T^n(0) = \underbrace{T(\cdots(T(0))\cdots)}_{n \text{ times}}$$

$$T^{-n}(0) = \underbrace{\check{T}(\cdots(\check{T}(0))\cdots)}_{n \text{ times}}$$

approach limits in $\mathbf{C}^*$ as $n \to \infty$?

9.2 SCHWARZ' LEMMA

An entire function is constant if it is bounded on $\mathbf{C}$, and is a polynomial if it satisfies a growth condition of the form $|f(z)| \leq A\,|z|^N$ for sufficiently large $|z|$.† There are many results like this in complex analysis; a holomorphic function that satisfies certain extra conditions may be so restricted that it is almost completely determined. Schwarz' lemma is closely related to this family of results, and is quite useful in resolving some emminently practical mapping problems. As one application we demonstrate that every invertible conformal mapping $f: D \to D$ of the unit disc onto itself must be a *fractional linear transformation*, and is more or less determined once we specify the image $f(0)$ of the origin.

***Theorem 9.5* (*Schwarz' lemma*)** *Let $w = f(z)$ be a holomorphic function defined on the unit disc $D = \{z : |z| < 1\}$, and assume that*

(i) *$|f(z)| \leq 1$ for $|z| < 1$ (thus f maps D into the closed disc $|z| \leq 1$).*
(ii) *$f(0) = 0$.*

Then $|f(z)| \leq |z|$ for all z in D. Furthermore, f satisfies the stronger inequality

(6) $$|f(z)| < |z| \quad \text{for all } z \text{ in the punctured disc } 0 < |z| < 1$$

unless f is a rotation, $f(z) = e^{i\phi} \cdot z$ (ϕ a real constant).

PROOF: There is a convergent series expansion $f(z) = a_1 z + a_2 z^2 + \cdots$ throughout D, whose constant term is zero since $f(0) = 0$. Accordingly, $g(z) = f(z)/z$ has a removable singularity at $z = 0$, and series expansion $g(z) = a_1 + a_2 z + \cdots$. On a circle $|z| = R$ with radius $R < 1$, we have $|g(z)| = |f(z)/z| = |f(z)|/R \leq 1/R$; now apply the maximum principle to $g(z)$ to see that $|g(z)| \leq 1/R$ throughout the disc $|z| \leq R$. This is true for every $R < 1$; discs of radius $R < 1$ fill D, and $1/R \to 1$ as R increases to 1, so that $|g(z)| \leq 1$, and thus $|f(z)| \leq |z|$, throughout D.

Suppose that there is a point z^* such that $0 < |z^*| < 1$ and $|f(z^*)| = |z^*|$, so that the last inequality fails to be strict; write $f(z^*)/z^* = e^{i\phi}$. Now

† This is referred to as "polynomial growth" at infinity; see Exercise 4, Section 5.13.

$|g(z^*)| = 1$ and g achieves its maximum modulus at an *interior* point of the disc; by the maximum principle, $g(z) = \text{constant} = g(z^*) = e^{i\phi}$ throughout D. Therefore $f(z) = e^{i\phi}z$ for all z and f is a rotation, unless the strict inequality (6) holds for every point such that $0 < |z| < 1$. ▮

Here are some applications of the Schwarz lemma, arranged in order of increasing difficulty.

Example 9.6 If $w = f(z)$ is an *invertible* conformal mapping of the unit disc onto itself, and if f leaves the origin fixed, then f is a rotation about the origin, $f(z) = e^{i\phi}z$.

In this situation the inverse $z = \check{f}(w)$ is also a holomorphic map $\check{f}: D \to D$, and leaves the origin fixed. Applying Schwarz' lemma to both mappings, we see that $|f(z)| \leq |z|$ and $|\check{f}(w)| \leq |w|$. Thus, $|z| = |\check{f}(f(z))| \leq |f(z)| \leq |z|$, which implies that $|f(z)| = |z|$ for all z. Obviously, the strict inequality $|f(z)| < |z|$ cannot hold for $z \neq 0$, so that f must be a rotation. If the origin is not left fixed, f need not be a rotation; there are, in fact, numerous other fractional linear transformations of the disc D onto itself.

Example 9.7 Determine all fractional linear transformations of the unit disc onto itself.

Theorem 9.6 *If S and T are fractional linear transformations of D onto D, and if $S(0) = T(0)$, then S and T can differ only by a rotation $R_\lambda = \lambda \cdot z = e^{i\phi} \cdot z$ (where $\lambda = e^{i\phi}$ has absolute value 1), in the sense that $S(z) = T(\lambda \cdot z) = (T \circ R_\lambda)(z)$. If the derivatives at the origin have the same values, or even the same argument,* $\arg dS/dz(0) \equiv \arg dT/dz\,(0)$ (modulo 2π), *then $S = T$.*

PROOF: The composite $\check{T} \circ S$ maps the disc invertibly onto itself and the origin to the origin. It must then be a rotation, $\check{T}(S(z)) = \lambda \cdot z$ for some λ such that $|\lambda| = 1$, and so $S(z) = T(\lambda \cdot z)$ for all z. The scalar λ is determined by the arguments of the derivatives at the origin, since $dS/dz(0) = \lambda \cdot dT/dz(0)$. ▮

To produce a transformation $T: D \to D$ that maps the origin to a different point $w_0 = Re^{i\theta}$ in the disc, we may transform the origin to $R + i0$ using the negative $w = (-1) \cdot T_R(z) = (z - R)/(Rz - 1)$ of the mapping defined in Example 9.5. If this is followed by the rotation $w = e^{i\theta}z$, we obtain a particular mapping $w = e^{i\theta}(z - R)/(zR - 1)$ of the disc such that $T(0) = Re^{i\theta}$. In view of Theorem 9.6, all other fractional linear transformations with this behavior are of the form

(7)
$$S(z) = T(e^{i\phi}z) = e^{i\theta}\,\frac{e^{i\phi}z - R}{Re^{i\phi}z - 1} = \frac{e^{i(\phi+\theta)}z - Re^{i\theta}}{Re^{i\phi}z - 1},$$

where ϕ is any real number. Notice that there is one adjustable (real) parameter ϕ left, once $T(0)$ is specified. There is a different family of mappings if we wish to map the origin to some other point in D; replace (R, θ) with the new polar coordinates (R', θ') of $T(0)$. If we let all three parameters, R, θ, and ϕ, vary, keeping $0 < R < 1$ and θ and ϕ real, we obtain *all possible* fractional linear transformations of the unit disc onto itself. This family may also be described as the set of transformations having the special form

$$(8) \qquad T(z) = \frac{az + b}{\bar{b}z + \bar{a}} \qquad (a \text{ and } b \text{ complex, such that } |a| > |b|)$$

See Exercise 7 for the details needed to identify the families (7) and (8).

Example 9.8 Determine all invertible conformal mappings of the unit disc D onto itself.

Using Schwarz' lemma we will show that such a mapping T must be *fractional linear*, one of the mappings determined in the last example. In fact, if $T(0) = w_0$ there is a fractional linear transformation $S: D \to D$ which carries w_0 back to the origin, so $(S \circ T)(0) = 0$. Obviously, $S \circ T$ is an invertible conformal map of D onto D, so the remarks of Example 9.6 apply: $(S \circ T)(z) = e^{i\phi} \cdot z$ and $T(z) = \check{S}(e^{i\phi}z) = (\check{S} \circ R_\lambda)(z)$, where $\lambda = e^{i\phi}$. Clearly, $\check{S}$ and the composite $\check{S} \circ R_\lambda$ are fractional linear, and so is T.

The preceding remarks may be varied in a routine way to identify all invertible conformal mappings from one disc $|z - p| < r'$ to another disc $|z - q| < r''$, and to prove that they are all fractional linear transformations. More interesting variants, discussed below, allow us to determine all conformal mappings between more complicated domains. For example, suppose we wish to determine all conformal mappings of the upper half plane $E = \{z : \text{Im}(z) > 0\}$ onto itself. Certain obvious transformations come to mind: *real* translations $w = z + b$ (with b real), and scaling operations $w = \lambda z$ where $\lambda > 0$; successive composites of these map E onto E and are sufficient in number to shift the point $z = +i$ to any prescribed new position w_0—simply take $w = (\text{Re}(w_0)) + (\text{Im}(w_0)) \cdot z$.

How shall we determine *all* invertible conformal mappings $T: E \to E$ which map $+i$ to w_0? One way is to set up an analogous problem for the unit disc D so we may apply the results of Examples 9.6 to 9.8. This is done using the familiar conformal map $f: E \to D$ and its inverse:

$$w = f(z) = i\frac{z - i}{z + i} \qquad z = \check{f}(w) = \frac{1}{i}\frac{w + i}{w - i}.$$

If T maps E onto E, we get a corresponding map $\tilde{T}: D \to D$ by transforming *both the variable and the values* in $w = T(z)$ via f and $\check{f}$. Starting from D, apply the mappings $w = \check{f}(z)$, $w = T(z)$, and $w = f(z)$ in succession. As indicated in Figure 9.6, we obtain a mapping $\tilde{T} = f \circ T \circ \check{f}: D \to D$, and if $T(+i) = w_0$ it is clear that $\tilde{T}(0) = (f \circ T)(+i) = f(w_0)$, since $\check{f}(0) = +i$. Let us label $f(w_0) = w_0^*$ so that $\tilde{T}(0) = w_0^*$. We know that $\tilde{T}$ is a fractional linear transformation, and have actually calculated all possible transformations $\tilde{T}: D \to D$ such that $\tilde{T}(0) = w_0^*$. The transformation T we really want is easily recovered from $\tilde{T}$ since†

$$T = (\check{f} \circ f) \circ T \circ (\check{f} \circ f) = \check{f} \circ (f \circ T \circ \check{f}) \circ f = \check{f} \circ \tilde{T} \circ f$$

† Remember that $(f \circ \check{f})(z) = z = (\check{f} \circ f)(z)$, since f and $\check{f}$ are inverses of one another.

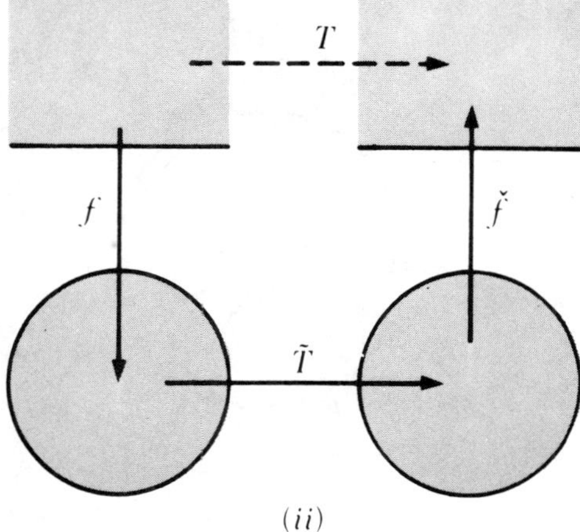

Figure 9.6

Obviously $T = \check{f} \circ \tilde{T} \circ f$ is fractional linear, being a composite of such transformations; $\tilde{T}$ can be any transformation of D onto itself such that $\tilde{T}(0) = w_0^*$.

Example 9.9 Determine all invertible conformal mappings $T: E \to E$ that leave $z = +i$ fixed.

The only conformal mappings of the disc onto itself that leave the origin fixed are the rotations $R_\lambda(z) = \lambda \cdot z$ $(|\lambda| = 1)$. Thus, if $\lambda = e^{i\phi}$ is any scalar with absolute value one, there is a corresponding mapping of the half plane $T_\lambda = \check{f} \circ R_\lambda \circ f$; thus,

(9) $$w = T_\lambda(z) = \check{f}\left(e^{i\phi}\frac{iz+1}{z+i}\right) = i\left[\frac{z(1+\lambda) + i(1-\lambda)}{z(1-\lambda) + i(1+\lambda)}\right]$$

maps E onto E so that $T_\lambda(i) = i$. We can see that the transformations T_λ retain many characteristics of rotations by examining their normal form with respect to the fixed points $p = +i$ and $q = -i$. Substitute $z = 0$ and $w = T_\lambda(0) = i\dfrac{1-\lambda}{1+\lambda}$ into the normal form to determine μ:

(10) $$\frac{w-i}{w+i} = \mu\,\frac{z-i}{z+i} \qquad \mu = \lambda = e^{i\phi}.$$

Since $|\mu| = 1$, Apollonian circles A_α about $\{+i, -i\}$ are mapped onto themselves, and in particular points on the extended real axis $\mathbf{R}^* = \mathbf{R} \cup \{\infty\} = A_1$ are shifted around within $\mathbf{R}^*$. The Steiner circles B_θ are transformed into $B_{\theta + \arg\mu} = B_{\theta+\phi}$, and the net action of T_λ is illustrated in Figure 9.7. When this action is viewed on the complex sphere, it really *is* a rotation of $\mathbf{C}^*$ (about the axis through the equatorial points $+i$ and $-i$); see Exercise 14. The action of T_λ on $\mathbf{C}^*$ should be compared with the action of the original rotation of the disc $w = \lambda z$ from which it was derived.

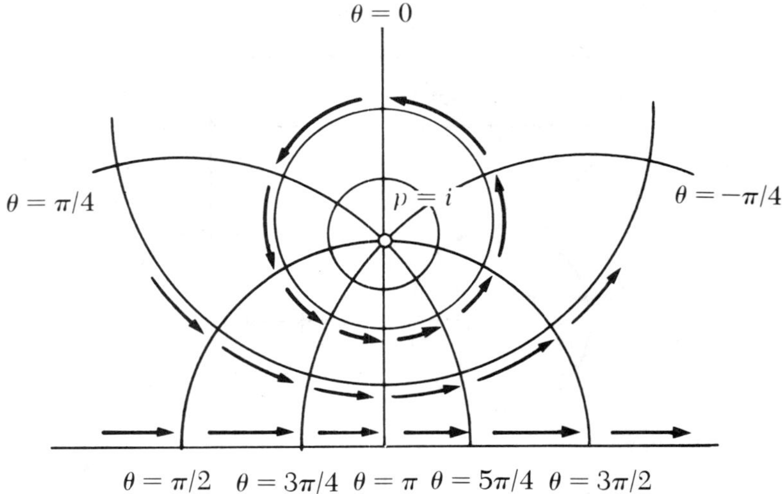

Figure 9.7 Rotation-like action of T_λ on the upper half plane, illustrated for $\lambda = e^{i\phi} = (i+1)/\sqrt{2}$, so that $\phi = +\pi/4$. Thus, T_λ maps B_θ to $B_{\theta+\pi/4}$.

Similar reasoning allows us to devise mappings of the half plane onto itself for special purposes. We might ask that $+i$ be mapped to some other point w_0, or that $+i$ be fixed, while certain points on the boundary are shifted in a particular way.

There is an interesting characterization of the transformations which map E onto E. In any fractional linear transformation $w = T(z) = \dfrac{az+b}{cz+d}$ we may always adjust the coefficients by multiplying all of them by a common complex scalar so that $\det \begin{bmatrix} a & b \\ c & d \end{bmatrix} = ad - bc = +1$; this adjustment does not change the mapping T. If this normalization is made, the transformations of the half plane are precisely the ones with *real* coefficients a, b, c, d.

In general, the transformations $T_\lambda : E \to E$ in (9) do not have real coefficients, but this happens because their determinants, $ad - bc = -4\lambda$, have not been normalized. Multiplying each coefficient by $\alpha = i/(2\sqrt{\lambda}) = (i/2)\, e^{-i\phi/2}$ yields normalized determinants equal to $\alpha^2(-4\lambda) = +1$, and makes each coefficient in T_λ real.

EXERCISES

1. Prove that the coefficients a, b, c, d in a fractional linear transformation may be multiplied by a common complex factor μ, so that the new coefficients $a' = \mu a, \ldots, d' = \mu d$ have determinant $+1$. (This is not always possible if we restrict the multiplier μ to be *real*; give an example.)

2. If S and T are invertible conformal maps of the unit disc D onto itself that leave fixed a point p in D, prove that $S = T$ if we also have $\arg S'(p) \equiv \arg T'(p)$. If $T: D \to D$, and if

(i) $T(p) = p$ (ii) $\arg T'(p) \equiv 0$,

show that T must be the *identity map*, so that $T(z) = z$ for all z in D.

3. The fractional linear transformation $w = T(z) = -i(z + i)/(z - i)$ maps the unit disc D to the upper half plane H. Write out the fractional linear transformations S obtained by the following constructions.

(i) S equals a rotation $R(z) = \lambda z$ ($|\lambda| = 1$) followed by T.

(ii) S equals T followed by a translation $P(w) = w + b$ (b real).

In either case, we get new mappings $S: D \to H$.

4. If p and q are points in the unit disc D, show that there is a fractional linear transformation T of D onto D such that $T(p) = q$. Is this mapping unique? Write out a formula for such a mapping when $p = \frac{1}{2}$ and $q = 2i/3$.

Hint: Map p to 0; then map 0 to q. Map is not unique (try case $p = q = 0$).

5. The general mapping $S: D \to D$ of the unit disc onto itself that carries the origin to $w_0 = Re^{i\theta}$ $(0 \le R < 1)$ has been determined in (7). The inverse $z = \check{S}(w)$ maps D onto D, carrying w_0 to the origin; show that $\check{S}$ has the simple form

$$\check{S}(w) = e^{i\alpha}\frac{w - w_0}{\bar{w}_0 w - 1} \qquad (\alpha \text{ real})$$

Note: In fact $\alpha = -(\phi + \theta)$ where ϕ is the angle appearing in (7).

6. Exercise 5 shows that every mapping of the unit disc D onto itself that carries w_0 to the origin ($|w_0| < 1$) can be put into the form

$$S(w) = e^{i\alpha}\frac{w - w_0}{\bar{w}_0 w - 1} \qquad (\alpha \text{ real}) \tag{11}$$

for suitably chosen real α. Give a direct proof of the converse statement: given $|w_0| < 1$ and α real, every mapping of this form transforms D onto D, carrying w_0 to the origin. Thus a fractional linear transformation maps D onto D, carrying w_0 to the origin, *if and only if* it has this form.

7. Show that a fractional linear transformation $w = T(z)$ maps the unit disc D onto itself if and only if it has the form

$$T(z) = \frac{az + b}{\bar{b}z + \bar{a}} \qquad (a, b \text{ complex, such that } |a| > |b|) \tag{12}$$

Two steps are required in a careful proof; you must show that:

(i) If T has the form (12), then it does map D onto D.

(ii) If T maps D onto D, it can be expressed in the form (12).

Hints: For (i), divide numerator and denominator by $|a|$ and write $a/|a| = e^{i\alpha/2}$, $\bar{a}/|a| = e^{-i\alpha/2}$ (α real); then cast T into the form (11), which is known to map D onto D. For (ii), T maps some points z_0 to the origin, so T can be written in the form (11) for suitable real α. Then write $e^{i\alpha} = \mu/\lambda$ where $\mu = e^{i\alpha/2}$ and $\lambda = e^{-i\alpha/2}$; after multiplying numerator by μ and denominator by λ, T is cast in the desired form (12).

8. Try to obtain a simple formula for the most general invertible conformal map of the unit disc onto itself that leaves the point $z_0 = Re^{i\theta_0}$ fixed.

9. Suppose that

$$T(z) = \frac{az + b}{cz + d}$$

and suppose that the coefficients a, b, c, d have been *normalized*, so that $ad - bc = +1$. Show that T maps the half plane $\text{Im}(z) > 0$ onto itself if and only if a, b, c, d are all *real*. If the determinant $ad - bc$ is not equal to $+1$ (-1 will not do), the transformation may have real coefficients but fail to map the half plane onto itself; try $a = 0$, $b = 1$, $c = 1$, $d = 0$. (See also the comments at the end of Section 9.2.)

10. Let points $\{p_1, p_2, p_3\}$ on the extended real axis $\mathbf{R}^*$ be labeled so that the upper half planes lies to the *left* as we move through p_1, p_2, p_3. Prove that there exists a fractional linear transformation $w = T(z)$ such that:

(i) T maps the half plane $\text{Im}(z) > 0$ onto itself

(ii) $T(p_1) = 0$, $T(p_2) = +1$, $T(p_3) = \infty$.

Hint: Recall Example 4.16, Section 4.8.

11. Extend Exercise 10 to prove that: given two such triples, $\{p_1, p_2, p_3\}$ and $\{q_1, q_2, q_3\}$, on $\mathbf{R}^*$ there is a *unique* fractional linear transformation $w = T(z)$ that maps the half plane onto itself and $T(p_k) = q_k$ for $k = 1, 2, 3$.

Hint: Uniqueness follows from Theorem 4.10.

12. State and prove a version of the result in Exercise 11 valid for triples of points labeled in counterclockwise order around the circumference of the unit disc $|z| < 1$.

13. Explain why there is a one-to-one correspondence between invertible analytic mappings of the unit disc D onto itself, and invertible analytic mappings of the half plane $H = \{z : \text{Im}(z) > 0\}$ onto itself. Use this correspondence to prove that:

Theorem: Every invertible analytic mapping of H onto itself is a fractional linear transformation.

14. If $|\lambda| = 1$ and $\phi = \arg \lambda$, show that the transformation

$$T_\lambda(z) = i\left[\frac{(1+\lambda)z + i(1-\lambda)}{(1-\lambda)z + i(1+\lambda)}\right],$$

discussed in Example 9.9, acts as a *rotation* of the complex sphere $\mathbf{C}^*$ by an angle ϕ, around the axis through $+i$ and $-i$.

15. If b is real, the translation $T_b(z) = z + b$ of the upper half plane H gives a mapping $S_b = \phi \circ T_b \circ \bar{\phi}$ of the unit disc D onto itself, if we use $w = \phi(z) = i(z-i)/(z+i)$ to map H onto D and to "lift" mappings of H to mappings of D. Write an explicit formula for $S_b(w)$. Which points in D are the image $S_b(0)$ of the origin as b varies through R?

Answer: Images $S_b(0)$ fill the circle $|w - (i/2)| = \frac{1}{2}$ except for the boundary point $+i$ $(-\infty < b < +\infty)$.

9.3 THE RIEMANN MAPPING THEOREM

Our commentary on the Riemann mapping theorem will deal with the main ideas of the proof, some of which are useful in other conformal mapping problems. In order to solve boundary value problems in the plane it is obviously important to know whether two domains D_1 and D_2 are **conformally equivalent,** in the sense that there is an invertible analytic mapping f, with nonvanishing derivative, which maps one domain onto the other. The simplest version of this question is the following: given a standard domain, such as the unit disc $D = \{z:|z| < 1\}$, which domains are conformally equivalent to D? Boundary value problems for this category of domains can be transformed into equivalent problems on the disc by the conformal mapping principle.

Theorem 9.7 (Riemann mapping theorem) *Let E be a domain in the complex plane, $E \subseteq \mathbf{C}$. Then E is conformally equivalent to the unit disc $D = \{z:|z| < 1\}$ if*

(i) *E is simply connected, and*

(ii) *E is not equal to the whole complex plane* $\mathbf{C}$.

Since there are fractional linear transformations between the disc $|z| < 1$ and the upper half plane $\mathrm{Im}(z) > 0$, any domain that can be mapped conformally onto the unit disc can also be mapped onto the upper half plane.

A few preliminary concepts are needed. The exceptional domain $E = \mathbf{C}$ *cannot* be mapped conformally onto the disc. If such a mapping $f:\mathbf{C} \to D$ existed, $w = f(z)$ would be analytic throughout the plane, and would be bounded since $|w| = |f(z)| < 1$ for every z; by Liouville's theorem, constants are the only functions meeting these requirements, and evidently these are not conformal mappings onto the disc. One can also prove that every domain conformally equivalent to the disc $|w| < 1$ must, like the disc itself, be *simply connected;* thus, there is no hope of mapping multiply connected domains onto the disc.

Theorem 9.7 is proved in two steps. First, E is mapped onto a simply connected domain E' that lies within the unit disc D; this can be done with familiar conformal transformations. Then one must prove that any simply connected domain in the disc, such as E', can be mapped conformally *onto* the disc. The conformal mappings $T:E \to E'$ and $S:E' \to D$ obtained in each step may be composed to give a mapping $w = S(T(z)) = (S \circ T)(z)$ which is a solution of the Riemann mapping problem for E. We will prove Step 1 and will indicate the general idea behind Step 2. A proof of Step 2 would require confident use of some advanced techniques from analysis, in particular the concept of "normal families" of analytic functions. These details are not important for the rest of the discussion in this book, nor are they necessary for a full understanding of the significance of the mapping theorem. Thus we have chosen to omit them, giving suitable references for further reading.

Proof of Step 1. Since E is not all of $\mathbf{C}$, there is a point p outside of E. If there is actually a disc of positive radius, such as $|z - p| < \delta$, outside of E we can transform E into a subset of the unit disc via the invertible conformal mapping $w = \delta/(z - p)$. However, there might not be any disc like this, as when E is a cut plane or is obtained by removing other curve segments from the plane. To circumvent this difficulty, the reader should recall how the square root function $w = z^{1/2}$ "opens up" a cut plane to give a half plane; because the image set covers only "half" of the plane, the transformed domain has a substantial complementary set. We are assuming that E is simply connected, so there must be well defined analytic determinations of $\log(z - p)$ on E, and also determinations of the square root function $(z - p)^{1/2}$ given by $f(z) = \exp((\frac{1}{2}) \log(z - p))$; recall Section 5.19. For each z in E the number $f(z)$ is one of the two possible square roots $\pm\sqrt{z - p}$, and since $f(z)^2 = (\pm 1)^2(z - p) = z - p$, we see that

$$\frac{df}{dz} = \frac{1}{2}\frac{1}{f(z)} \quad \text{and} \quad \left|\frac{df}{dz}\right| = \left|\frac{1}{2}\frac{1}{f(z)}\right| = \frac{1}{2}|z - p|^{-1/2} \tag{13}$$

for all z in E.

The mapping $w = f(z)$ is one-to-one since $f(z_1) = f(z_2)$ implies that $z_1 - p = f(z_1)^2 = f(z_2)^2 = z_2 - p$, so $z_1 = z_2$. The derivative never vanishes on E since $z = p$ is not in E, so f is an invertible conformal mapping of E onto a new simply connected domain E'.† Our most important observation is that the negative $-z$ of a point z in E' must lie outside of E'.

Suppose that z and $-z$ are both in E'. Consider the points z' and z'' in E such that $f(z') = z$ and $f(z'') = -z$. Then $z' - p = f(z')^2 = z^2 = (-z)^2 = f(z'')^2 = z'' - p$, so that $z' = z''$ and $z = -z$. This can

† The image set $E' = f(E)$ is an open set, by the open mapping theorem (Theorem 2.21). For simple connectedness of E', recall Exercise 10, Section 5.19.

only happen if $z = 0$. But a determination of $\sqrt{z - p}$ cannot take the value zero on E (this value can only arise at $z = p$, which is outside of E). It follows that z and $-z$ cannot both be in E' for any z.

Now take any point q^* in E'; since E' is an open set, there is also a disc D^* of positive radius δ and center q^* that is contained within E'. The negatives of points in D^* compose a disc $D = -D^*$ about $q = -q^*$, with radius δ, and D lies outside of E'. Now we may transform E' into a simply connected domain E'' lying inside the unit disc, by applying $g(z) = \delta/(z - q)$. The composite mapping $h(z) = (g \circ f)(z) = \delta \cdot (f(z) - q)^{-1}$ transforms E into E'', which lies in the unit disc $|w| < 1$. If E'' does not include the origin, we may perform an additional fractional linear transformation of the disc onto itself that shifts E'' to a domain that does include the origin (see Example 9.7, Section 9.2).

PROOF OF STEP 2 (OUTLINE). Let E be any simply connected subdomain of the unit disc that contains $z = 0$. We seek a procedure for constructing conformal mappings T that map E into D, leaving the origin fixed, so that the image $E_1 = T(E)$ fills up *more* of the disc. As a measure of how much of D is filled up by a domain containing the origin, we define the *inner radius* $r_{in}(E)$ to be the distance from $z = 0$ to the nearest point in bdry(E); thus, the disc $|z| < r_{in}$ lies within E and is the largest open disc centered at the origin that does so.†

We seek a transformation defined on E that has the following properties.

(14)
(i) $|T(z)| < 1$ for each z in E (so that T maps E *into* D)
(ii) $T(0) = 0$
(iii) T is an invertible conformal map of E onto a domain E_1 which satisfies $r_{\text{in}}(E_1) > r_{\text{in}}(E)$.

The domain E is simply connected, but this is all we know about it and it could have a very irregular boundary. As a prototype of the problem, we might try to transform the cut disc E (segment $[R, 1]$ removed) shown on the left in Figure 9.8 to a new domain in D with larger inner radius. This special case can

† E is an open set, so bdry(E) is disjoint from E and is a closed bounded set in the plane. All boundary points lie within the closed disc $|z| \leq 1$.

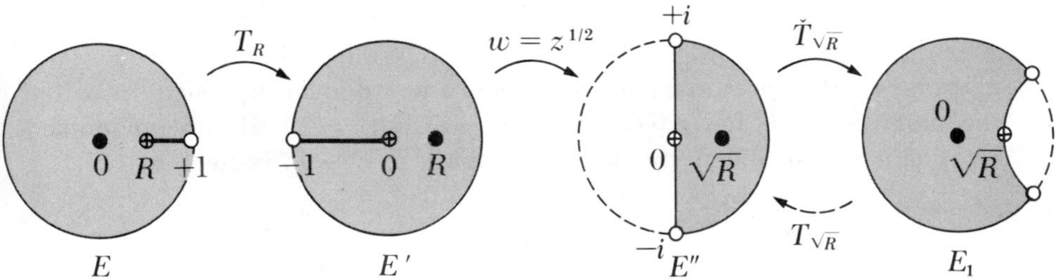

Figure 9.8 Transformation of a slit disc with $r_{\text{in}} = R < 1$ to a "notched disc" with larger inner radius $r'_{\text{in}} = \sqrt{R}$.

be solved using transformations only slightly more complicated than fractional linear ones.† More important, a similar construction can be devised for *any* simply connected domain E, so the prototype provides the key to the general problem.

The fractional linear transformations $w = T_\alpha(z) = \dfrac{z-\alpha}{\alpha z - 1}$ map D conformally onto itself and have inverses $z = \check{T}_\alpha(w) = \dfrac{w-\alpha}{\alpha w - 1}$, for each choice of $0 < \alpha < 1$, as explained in Example 9.5. In Figure 9.8 we have $r_{in}(E) = R$; T_R maps D onto itself and the segment $[R, 1]$ to $[-1, 0]$. Now apply $w = z^{1/2}$ (principal determination) to transform $E' = T_R(E)$ onto a half disc E'', followed by the inverse $\check{T}_{\sqrt{R}}$ which maps E'' to the "notched disc" E_1 on the right side of the figure. Obviously, $r_{\text{in}}(E_1) = \sqrt{R} > R = r_{\text{in}}(E)$, so that the composite mapping

$$w = T(z) = \check{T}_{\sqrt{R}}\left[\sqrt{T_R(z)}\right] \quad \text{for } z \text{ in } E$$

has each of the properties (14). Notice that T is obtained by solving the rational quadratic equation

$$\text{(15)} \qquad \left(\frac{w-\sqrt{R}}{\sqrt{R}w - 1}\right)^2 = \frac{z-R}{Rz-1} \qquad \text{(solve for } w \text{ as a function of } z\text{)},$$

because $(T_{\sqrt{R}}(w))^2 = ((T_{\sqrt{R}} \circ \check{T}_{\sqrt{R}})(\cdots))^2 = (T_R(z)^{1/2})^2 = T_R(z)$. The solution $w = T(z)$ is *multiple valued*, since a square root is involved; the construction indicated in Figure 9.8 defines a single valued determination of T on E.

Now let E be an arbitrary simply connected domain that includes the origin, and let $R = r_{\text{in}}(E)$. There is a point z^* on bdry(E) such that $|z^*| = R$; a suitable rotation $w = e^{i\phi} \cdot z$ moves z^* to $R + i0$ on the positive real axis. By using the results of Section 5.19, one can show that, for a suitably chosen determination of square root $\sqrt[*]{z}$, the transformation

$$w = f(z) = \check{T}_{\sqrt{R}}\left[\sqrt[*]{T_R(z)}\right] = \check{T}_{\sqrt{R}}\left[\sqrt[*]{\frac{z-R}{Rz-1}}\right]$$

maps the rotated domain E_0 invertibly onto a new domain E_1 which lies within the unit disc, and has $r_{\text{in}}(E_1) > r_{\text{in}}(E_0) = r_{\text{in}}(E) = R$. If the preliminary rotation is now included, the desired mapping $T: E \to E_1$ is given by

$$\text{(16)} \qquad w = T(z) = \check{T}_{\sqrt{R}}\left[\sqrt[*]{\frac{e^{i\phi}z - R}{Re^{i\phi}z - 1}}\right] \qquad \text{for all } z \text{ in } E.$$

---

† No fractional linear transformation can have all of the properties (14). Why not?

Once we know how to transform any simply connected domain in the unit disc which includes $z = 0$ to another domain of this kind with strictly larger inner radius, we can repeat the whole procedure, now starting with E_1 instead of E. Repeated applications yield a sequence of successive domains $E, E_1, E_2, \ldots$ which contain the origin and fill up increasingly larger portions of the unit disc.† At the same time we obtain invertible conformal mappings from one domain to the next.

$$E \xrightarrow{T_1} E_1 \xrightarrow{T_2} E_2 \cdots E_{n-1} \xrightarrow{T_n} E_n \to \cdots; \qquad F_n = T_n \circ \cdots \circ T_1 \quad (17)$$

each map $T_k: E_{k-1} \to E_k$ has the properties (14). The composite mappings $F_n = T_n \circ \cdots \circ T_1$ are also invertible and conformal, and map E onto E_n. The inner radii of the image domains $E_n = F_n(E)$ are strictly increasing (with $R_n \leq 1$); the construction of each T_k is designed to accomplish this. If after finitely many steps we get $R_n = r_{\text{in}}(E_n) = 1$, then $E_n = D$ and $F_n: E \to E_n = D$ is the solution to the Riemann mapping problem. If this does not occur, one can still prove that $R_n \to 1$ as $n \to \infty$. Thus, the domains E_n come closer and closer to filling the disc, and the mappings $F_n: E \to E_n$ may be thought of as more and more refined approximate solutions to the mapping problem we started with.

We could actually prove that

(i) There is a well defined limit $F(z) = \lim_{n\to\infty} F_n(z)$ for each z in E, and

(ii) The mapping $w = F(z)$ is an invertible analytic mapping of E onto the unit disc.

The main difficulty lies in proving that the limit $F(z)$ exists, and in verifying that it is both analytic (with non-vanishing derivative) and univalent on E. Full accounts of the Riemann mapping theorem compatible with the outline we have just given may be found in Nevanlinna and Paatero [18], Sections 17.1 to 17.12, or De Pree and Oehring [5], Sections 59 and 60.

This completes our discussion of the proof of Theorem 9.7. ▌

The Riemann mapping theorem is essentially an *existence* theorem. If we are given any simply connected domain $E \neq \mathbf{C}$, there exists at least one invertible analytic mapping of E onto the unit disc $D = \{z: |z| < 1\}$. But in the end, after a few preliminary transformations of E, the desired map $w = F(z)$ is obtained as the limit of approximating maps $w = F_n(z)$ which do not themselves accomplish our goal. It is usually very difficult to give an explicit formula for $F(z)$, and the mapping theorem is not often helpful in determining the specific map (or mappings) that solve a given conformal mapping problem, although it does assure us that the problem can be solved *in principle*.

Most often we will continue to solve conformal mapping problems as we have in the past, by referring to our ever-expanding list of conformal mappings

† The domains need *not* be successively larger in the sense that E_{n+1} *contains* E_n; we only know that the inner radii R_n are strictly increasing.

and making educated guesses about how these mappings should be put together to transform given domains onto each other, or onto the unit disc.

In any conformal mapping problem we seek an invertible analytic mapping $F: E \to D$ between given domains D and E. The reader should now observe that the mapping F which satisfies this requirement is *not unique*. We can always follow F by any conformal mapping $T: D \to D$ to get a different solution $F^* = T \circ F$, as indicated in diagram (i) below. The mappings $T: D \to D$

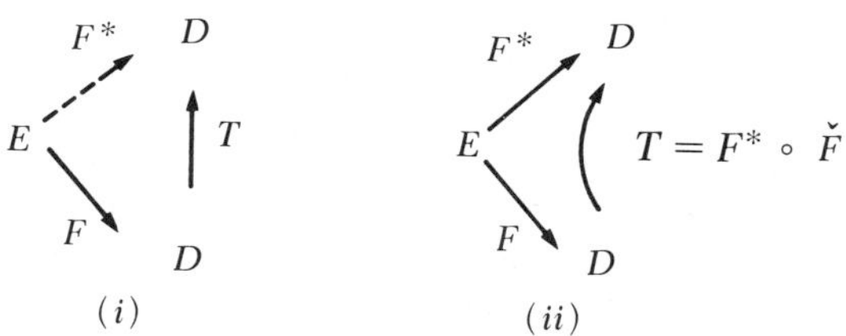

have been identified in Section 9.2, and are all fractional linear. Conversely, if F and F^* are two solutions of the mapping problem, then $T = F^* \circ \check{F}$ maps D onto D, because $F^*(\check{F}(D)) = F^*(E) = D$, and it is one of the fractional linear mappings of the disc onto itself. Clearly, $F^* = T \circ F$, as in diagram (ii). The solution $F: E \to D$ is uniquely determined once we specify the values $F(p)$ and $\arg(F'(p))$ at some point p in E (see Exercise 2). In many of the mapping problems we examined in earlier chapters, this uncertainty about which mapping $F: E \to D$ should be used is eliminated because F must also satisfy side conditions on the behavior of $F(z)$ as z approaches the boundary of E. When these side conditions are taken into account, the class of solution mappings is often reduced to the point that the solution is unique; at the very least, its ambiguity is reduced so that different choices of the mapping lead to the same solution of the boundary value problem being considered.

We should also note that Theorem 9.7 has nothing to say about the behavior of $F(z)$ as z approaches points on the boundary of E; the mapping F is only known to be analytic and conformal within the domain E, and may in fact become quite singular at the boundary, especially if the boundary set is very irregular. However, if the boundary is smooth enough one may extend $F: E \to D$ so it maps $\text{bdry}(E)$ continuously one-to-one onto $\text{bdry}(D) = \{w : |w| = 1\}$.

Theorem 9.8 *Let E be the interior domain of a simple closed contour γ, and let f be any invertible conformal mapping of E onto the unit disc D. Then there is a mapping F of the closure $\bar{E} = E \cup \text{bdry}(E)$ onto the closure $\bar{D} = \{w : |w| \leq 1\}$ which has the following properties:*

(i) *F maps E one-to-one onto D, and $\text{bdry}(E)$ one-to-one onto $\text{bdry}(D) = \{w : |w| = 1\}$.*

(ii) *F, and its inverse $\check{F}: \bar{D} \to \bar{E}$, are both continuous.*

(iii) *F agrees with f on E, and likewise $\check{F}$ agrees with $\check{f}$ on D.*

We will not attempt to prove this result. These details may be pursued in the more advanced accounts found in Nevanlinna and Paatero [18], Sections 17.13 to 17.20, or De Pree and Oehring [5], Section 60.

Only simply connected domains are considered in Theorem 9.7; multiply connected domains are much more difficult to deal with. One problem lies in the choice of a suitable "standard domain." The disc, being simply connected, will not do, but the plane with a number of isolated cuts in it might work. For an introduction to these matters we recommend Ahlfors [1], Sections 6.5.1 to 6.5.3.

EXERCISES

1. Apply the ideas of Step 1 in the proof of the Riemann mapping theorem to devise invertible, conformal mappings of the following simply connected domains to domains that lie within the unit disc $|w| < 1$ and include the point $w = 0$.

(i) The strip $0 < \text{Re}(z) < 100$

(ii) The doubly cut plane obtained by deleting the rays $(-\infty, -1]$ and $[1, +\infty)$ from $\mathbf{C}$.

2. Suppose that E is a simply connected domain such that $E \neq \mathbf{C}$, so that there is at least one invertible conformal mapping of E onto the unit disc D. Let p in E be fixed. If T_1 and T_2 map E onto D, and if

$$T_1(p) = T_2(p) \quad \text{and} \quad \arg T_1'(p) \equiv \arg T_2'(p),$$

prove that $T_1 = T_2$. Thus, the mapping $T: E \to D$ is uniquely determined once $T(p)$ and $\arg T'(p)$ are specified at some point in E.

Hint: $S = T_2 \circ \check{T}_1$ maps D onto D, leaving the image of p fixed. Use Exercise 2, Section 9.2.

3. In Step 2 of the proof of Riemann's theorem we have indicated (see Figure 9.8) how the cut disc E, with segment $[R, 1]$ removed, is transformed to a notched disc E_1 that has somewhat larger inner radius, so that $r_{\text{in}}(E) < r_{\text{in}}(E_1)$. This mapping $T_1: E \to E_1$ is the first of a sequence of transformations that increase r_{in}; determine the general shape of the domain E_2 obtained from E_1 in the second step of this iteration process. Refer to the diagram (17) and related discussion.

4. Annular domains $E(r_1, r_2) = \{z : r_1 < |z - p| < r_2\}$ are multiply connected; therefore the Riemann mapping theorem cannot be applied to them. In fact, two annuli (even ones with the same base point) need *not* be conformally equivalent. Prove the following positive result: if $r_1/r_2 = r_1'/r_2'$, then $E(r_1, r_2)$ is conformally equivalent to $E(r_1', r_2')$, even if they have different base points.

5. Suppose Γ_1 and Γ_2 are disjoint circles such that Γ_2 lies in the disc enclosed by Γ_1. Let E be the (multiply connected) bounded domain between Γ_1 and Γ_2. Prove the E is conformally equivalent to *some* annulus $r_1 < |z| < r_2$, for suitably chosen r_1, r_2.

Hint: Identify Γ_1 and Γ_2 as members of a single family of Apollonian circles $\mathscr{A}\{p, q\}$ (Exercise 15, Section 9.1). Then apply Exercise 5, Section 9.1.

9.4 THE SCHWARZ REFLECTION PRINCIPLE

If f is analytic on E, then $g(z) = \overline{f(\bar{z})}$ is analytic on the domain E^* consisting of the conjugates $\bar{z}$ of points in E; E^* is obtained by reflecting E across the real axis (see Exercise 11, Section 2.9). Similarly, if $U(z) = U(x, y)$ is harmonic on E, we can show that $-U(\bar{z}) = -U(x, -y)$ is harmonic on E^* by identifying U with the imaginary part of some analytic function. When the domain E has a segment of the real axis as part of its boundary, and $U = 0$ on this boundary segment, these facts can be used to produce continuations of harmonic functions across the real axis (Exercise 5).

Definition 9.1 *Let E be a domain and suppose that a segment $I = (p, q)$, between points p and q in* $\mathbf{C}$, *is part of* bdry(E). *If z is on I, every small disc about z is divided into two open semidiscs if we delete points from I. We say that I is a* **free side** *of E if, for each z on I, there is a disc centered at z such that one half lies within E and the other half is disjoint from E.*

The size of the disc about z may depend on the position of z. Figure 9.9 shows that an interval can be part of bdry(E) without being a free side. Similarly, we could define what it means for a circular arc to be a *free arc* in the boundary of a domain E.

Theorem 9.9 Schwarz reflection principle *Let E be a domain in the upper half plane whose boundary includes a segment (a, b) in the real axis as a free side. Let E^* be the image domain $E^* = J(E)$ under the conjugation map $J(z) = \bar{z}$, as in Figure* 9.10. *Consider any function $w = f(z)$ defined on the union $E \cup (a, b)$ such that:*

(i) *f is continuous on $E \cup (a, b)$*

(ii) *f is analytic on the domain E*

(iii) *the values $f(x + i0)$ are all real for $a < x < b$.*

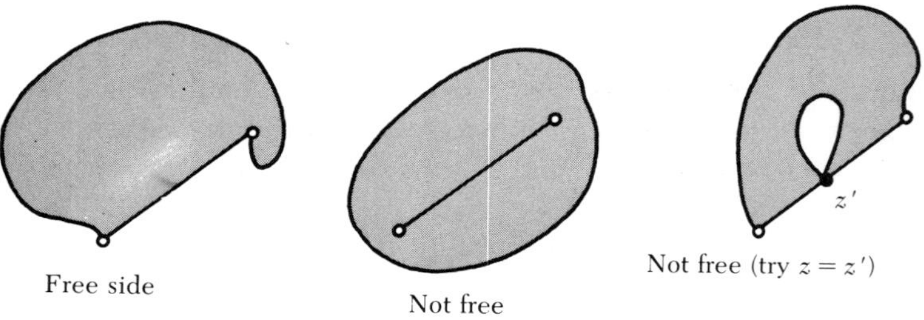

Figure 9.9 Domains with a segment (p, q) as part of their boundaries.

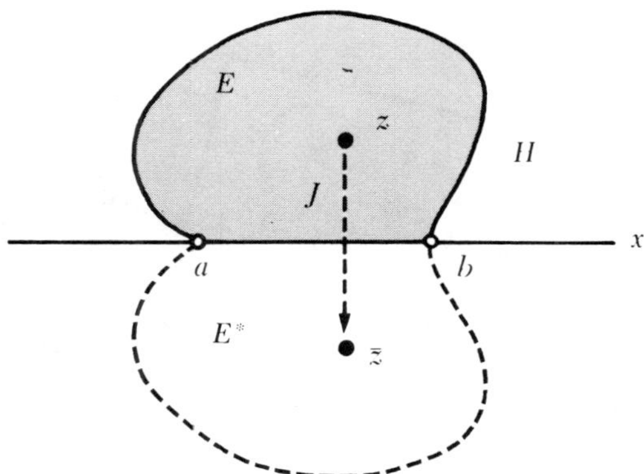

Figure 9.10 The situation in Theorem 9.9.

Then the function

(18)
$$F(z) = \begin{cases} f(z) & \textit{for } z \textit{ in } E \cup (a, b) \\ \overline{f(\bar{z})} & \textit{for } z \textit{ in } E^* \end{cases}$$

is analytic on the domain $D = E \cup (a, b) \cup E^*$ and agrees with f on the original domain E.

In effect, the values of f are "reflected" across the interval (a, b), and we get an analytic continuation of f into the lower half plane.

Proof: If $f(z) = u(z) + iv(z)$ on E, the real and imaginary parts of F are

$$F(x + iy) = U(x, y) + iV(x, y) = \begin{cases} u(x, y) + iv(x, y) & \text{for } z = x + iy \text{ in } E \\ u(x, y) + i0 \text{ for } y = 0 \text{ and } a < x < b \\ u(x, -y) - iv(x, -y) & \text{for } z = x + iy \text{ in } E^*. \end{cases}$$

We know that $F = f$ is holomorphic on E; on E^* the components $U = \text{Re}(F)$ and $V = \text{Im}(F)$ obviously have continuous partial derivatives and satisfy the Cauchy-Riemann equations

$$\frac{\partial U}{\partial x}(z) = \frac{\partial u}{\partial x}(\bar{z}) = \frac{\partial v}{\partial y}(\bar{z}) = \frac{\partial V}{\partial y}(z); \; \frac{\partial U}{\partial y}(z) = \frac{\partial u}{\partial y}(\bar{z}) = -\frac{\partial v}{\partial x}(\bar{z}) = -\frac{\partial V}{\partial x}(z).$$

Finally, F is *continuous* at each point on the interval (a, b) which separates E from E^*, by hypothesis (i), but it is not at all clear that dF/dz, or even the partial derivative $\partial u/\partial y$, exists at these points. Direct calculations with difference quotients do not resolve this differentiability question; we need a more powerful result, based on integration methods, to prove that F is differentiable at each point on (a, b). Once Theorem 9.10 (below) has been proved, it will be clear that F has all the desired properties. ▌

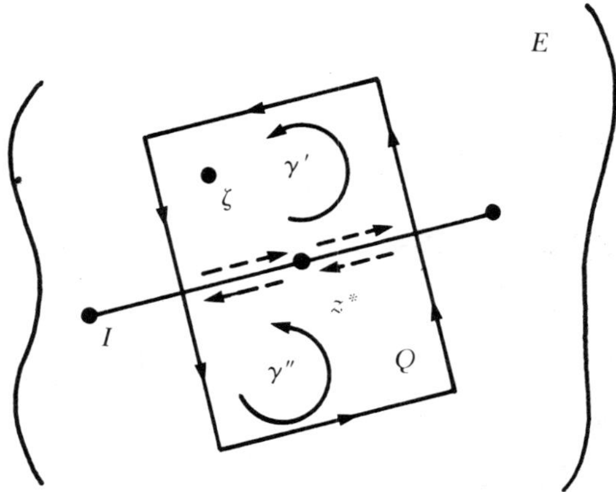

Figure 9.11 The situation near z^* in Theorem 9.10.

Theorem 9.10 *Let E be a domain that includes a line segment I. Suppose that f is defined and continuous on E, and that f is known to be analytic at all points in the complement $E \sim I$. Then f must actually be analytic at each point on the segment I.*

PROOF: Let z^* be a typical point on I. We will produce a function g, defined at and near z^*, that agrees with f at points off I and, by its very definition, is analytic at z^*.

Points close to z^* lie in E (E is open), so there is a small open rectangle Q about z^* whose boundary and interior are within E, as in Figure 9.11. The Cauchy-type integral around the boundary of Q,

$$g(\zeta) = \frac{1}{2\pi i}\int_\gamma \frac{f(z)}{z-\zeta}\,dz \quad \zeta \text{ in } Q,$$

is obviously analytic on Q (recall Section 5.10), since f is continuous on the trajectory of γ. To show that $g(\zeta) = f(\zeta)$ for ζ in $Q \sim I$, we form the closed contours γ' and γ'' shown. Integrals in opposite directions along I cancel, so that

$$\frac{1}{2\pi i}\int_\gamma \frac{f(z)}{z-\zeta}\,dz = \frac{1}{2\pi i}\int_{\gamma'} \frac{f(z)}{z-\zeta}\,dz + \frac{1}{2\pi i}\int_{\gamma''} \frac{f(z)}{z-\zeta}\,dz.$$

Furthermore, f is analytic on the rectangular domains enclosed by γ' and γ'', and is continuous on their closures, so the Cauchy Integral Formula (together with Exercise 17, Section 5.3) insures that the integral along the contour that encloses ζ has the value $f(\zeta)$, while the integral along the other contour is zero. Thus,

$$g(\zeta) = \frac{1}{2\pi i}\int_\gamma \frac{f(z)}{z-\zeta}\,dz = f(\zeta) \quad \text{for } \zeta \text{ in } Q \sim I.$$

Each point in $Q \cap I$ may be approached by sequences of points lying in $Q \sim I$ (where $f = g$), so we must have $f = g$ at *every* point in Q, by the continuity of f and g on Q. Since g is analytic on Q, and $f = g$, f is analytic at z^*. ▮

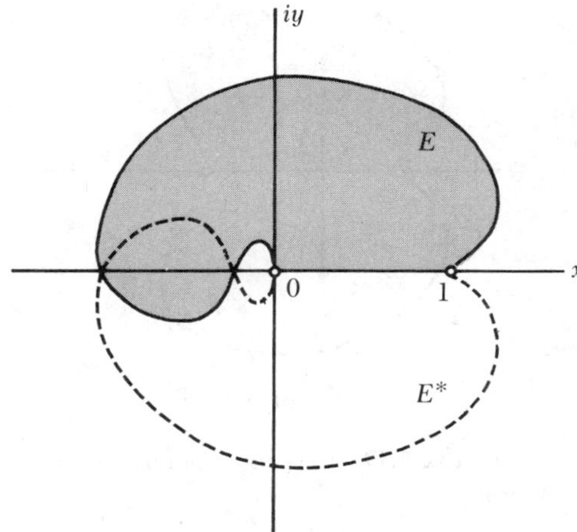

Figure 9.12 A reflection problem in which E and E^* overlap.

This result is a powerful generalization of Riemann's theorem on removability of isolated singularities. The result can be generalized further in many ways (use a curve instead of a segment I, etc.).

The Schwarz reflection principle may also be applied in certain cases when E does not lie entirely within the upper half plane, provided that (a, b) is a free side; but there *are* cases when the principle is not valid. For example, consider the domain in Figure 9.12, and let $f(z)$ be the determination of square root defined by taking $f(re^{i\theta}) = r^{1/2}e^{i\theta/2}$, normalizing θ so that $-\pi/2 < \theta < 3\pi/2$. The function f on E, and its "reflection" $\tilde{f}(z) = \overline{f(\bar{z})}$ on E^*, agree along the segment $(a, b) = (0, 1)$, but $\tilde{f}$ is the other determination of square root on the intersection $E \cap E^*$.

Another variant of the basic reflection principle allows us to continue functions across the real axis in certain situations when the values on the segment (a, b) are no longer real. Without *some* restrictions on these boundary values (continuity is not enough) there might not be any way of continuing f analytically across (a, b); however, the reflection principle will work if the values taken by f on (a, b) all lie along some *line* L (real values correspond to $L = \mathbf{R}$).

Theorem 9.11 *Let E be a domain in the upper half plane whose boundary includes a segment (a, b) in the real axis as a free side. Let E^* be the image $J(E)$ under the conjugation map $J(z) = \bar{z}$. Suppose that f satisfies the boundary conditions specified in Theorem 9.9, except that its values on (a, b) are no longer real; instead, assume that the values $w = f(x + i0)$ all lie on some line L, for $a < x < b$. If J_L is the reflection across L, then the function*

$$F(z) = \begin{cases} f(z) & \text{for } z \text{ in } E \cup (a, b) \\ J_L(f(\bar{z})) & \text{for } z \text{ in } E^* \end{cases} \tag{19}$$

is analytic on the domain $D = E \cup (a, b) \cup E^*$, and agrees with f on E.

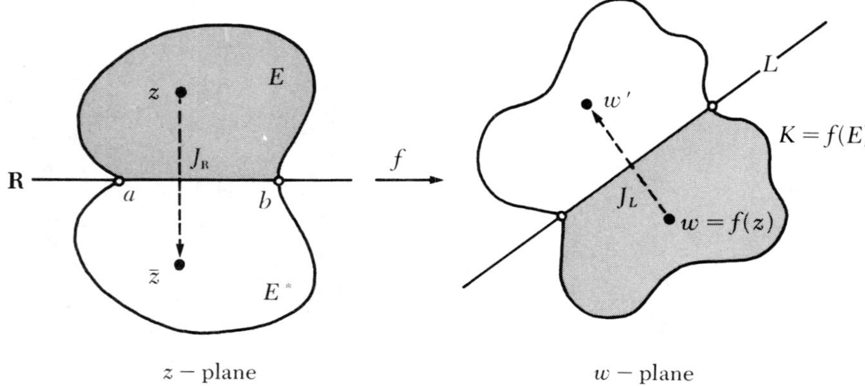

Figure 9.13 Reflection/continuation across the real axis when the values $f(x + i0)$ lie on a line L.

In other words, the value $w' = F(\bar{z})$ of the continuation F at a typical point $\bar{z}$ (in E^*) is obtained by reflecting the value $w = f(z)$ at z (in E) across the line L in the w-plane. Thus,

$$F(\bar{z}) = w' = J_L(w) = J_L(f(z)) \quad \text{for all } z \text{ in } E, \tag{20}$$

as indicated in the following diagram:

$$\begin{array}{ccc} z & \xrightarrow{f} & w \\ J\downarrow & & \downarrow J_L \\ \bar{z} & \xrightarrow{F} & w' \end{array}$$

In Figure 9.13, f maps the shaded region E onto the shaded region $K = f(E)$; in view of (19), the continuation F must map the (unshaded) reflected region E^* onto the (unshaded) reflected region $K^* = J_L(K)$.

If L were the real axis, then f would have real values (as before) and $J_L = J$; thus, formula (19) would reduce to the one obtained in Theorem 9.9. To prove Theorem 9.11 we follow the map $\zeta = f(z)$ by any invertible conformal mapping $w = \phi(\zeta)$ that transforms L to the real axis; a rotation will do. Then $g(z) = (\phi \circ f)(z) = \phi(f(z))$ has real values on (a, b), and Theorem 9.9 may be applied to continue g across (a, b) to an analytic function $G(z)$ defined on $E \cup (a, b) \cup E^*$. Then apply $\check{\phi}$ to the values of G; we get a function $F(z) = \check{\phi}(G(z))$ that is analytic on $E \cup (a, b) \cup E^*$ and agrees with f on E. (For z in E, $F(z) = \check{\phi}(G(z)) = \check{\phi}(\phi(f(z)) = f(z)$, since $(\check{\phi} \circ \phi)(w) = w$.) We leave the reader to check that $F(z)$ is indeed given by formula (19) for points z in E^*.

EXERCISES

1. If Γ is a circle in the complex sphere $\mathbf{C}^*$, and if $J_\Gamma : \mathbf{C}^* \to \mathbf{C}^*$ is the reflection map, explain why

$$J_\Gamma(J_\Gamma(z)) = z \quad \text{for all } z.$$

The fact that J_Γ applied twice in succession equals the identity map $I(z) = z$ is characteristic of "reflections."

2. Let Γ be each of the following lines in **C**. Give explicit formulas for the reflection mapping $w = J_\Gamma(z)$ in each case.

(i) $\Gamma =$ imaginary axis
(ii) $\Gamma = \{z : \text{Im}(z) = -2\}$
(iii) $\Gamma = \{z : \arg z \equiv \pi/4 \text{ or } -3\pi/4\}$
(iv) $\Gamma =$ the line through $+i$ and $+1$.

Calculate the images of the points $z = 0, +i, +1, (2 + i), -5$, and ∞.

3. Carry out the calculations that were left to the reader in the discussion of Theorem 9.11.

4. Generalize the work in Theorem 9.11 to prove the following result.

Theorem: Let E be a domain that lies to one side of circle Γ and has an arc $(p, q) \subseteq \Gamma$ as a free arc. Let J_Γ be the reflection through Γ and let $E^* = J_\Gamma(E)$, the reflected image of E. If $f(z)$ is holomorphic on E and continuous on (p, q), and the values $w = f(z)$ all lie on some circle (or extended line) $\tilde{\Gamma}$ in $\mathbf{C}^*$, then f may be continued analytically across (p, q) to be defined on $D = E \cup (p, q) \cup E^*$ by taking

$$F(z) = \begin{cases} f(z) & \text{for } z \text{ in } E \cup (p, q) \\ J_{\tilde{\Gamma}}[f(J_\Gamma(z)] & \text{for } z \text{ in } E^*. \end{cases}$$

If $\Gamma = \mathbf{R}^*$ and $\tilde{\Gamma} = L$ (some line), verify that we get the result of Theorem 9.11. If $\Gamma = \tilde{\Gamma} = \mathbf{R}^*$, verify that the formula for $F(z)$ at a point in E^* reduces to the Schwarz reflection formula $F(z) = \overline{f(\bar{z})}$.

5. Prove the following result, using the ideas outlined below.

Theorem: (Reflection principle for harmonic functions). Let E be a domain in the upper half plane which has $I = (a, b)$ as a free side. Let $u(z)$ be real valued and continuous on $E \cup (a, b)$, harmonic on E, and assume that $u(x + i0) = 0$ for $a < x < b$. Then

$$U(z) = \begin{cases} u(z) & \text{for } z \text{ in } E \cup (a, b) \\ -u(\bar{z}) & \text{for } z \text{ in } E^* = J_{\mathbf{R}}(E) \end{cases}$$

is harmonic on the larger domain $E \cup (a, b) \cup E^*$.

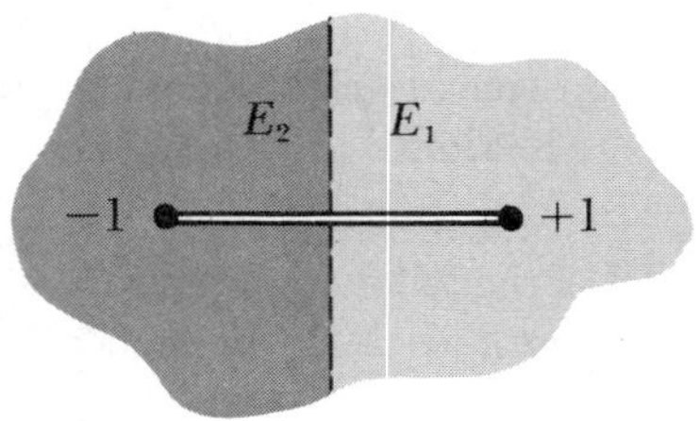

Figure 9.14

Particular attention is needed only at points p on (a, b). If $\gamma(\theta) = p + re^{i\theta}$ $(0 \le \theta \le 2\pi)$ is a small circular contour around p, and we take boundary values $h(z) = U(z)$ in Poisson's formula (24) for the disc $|z - p| < r$ (Section 7.7), we get a *harmonic* function $V(z)$ on the disc, with boundary values $U(z)$. By definition of $U(z)$, the integral formula gives $V(z) = 0$ on the real axis. Compare $U(z)$ and $V(z)$ in the upper half disc; both U and V are *harmonic* on the half disc (why?), and their boundary values are equal (on the circular arc *and* the real axis). Likewise for the lower half disc. Thus $U = V$ on disc; V is manifestly harmonic on the disc (even on the segment (a, b)).

6. Let E be the cut plane $\mathbf{C} \sim [-1, +1]$; let $\sqrt{\cdots}$ be the principal determination of square root. Divide E into domains E_1 and E_2, as indicated in Figure 9.14. It is fairly obvious that the functions $\pm\sqrt{z^2 - 1}$ are holomorphic within E_1 and E_2, and are continuous on the dashed boundary segments. Starting with $f_1(z) = +\sqrt{z^2 - 1}$ on E_1, show that the sign ± 1 can be chosen so that $f_2(z) = \pm\sqrt{z^2 - 1}$ equals $f_1(z)$ on the dashed rays where the boundaries of E_1 and E_2 meet. This straightforward construction gives a determination $f(z) = \overset{*}{\sqrt{z^2 - 1}}$ that is holomorphic on each domain E_k, but only *continuous* on the dashed rays.

Use Theorem 9.10 to verify that $f(z)$ is automatically holomorphic on the rays; thus $f(z)$ is an analytic determination of $\sqrt{z^2 - 1}$ on the cut plane E.

Hint: How are E_1 and E_2 transformed by $\zeta = z^2 - 1$ and $w = \sqrt{\zeta}$?

Note: Compare with Exercise 25, Section 4.10. This use of Theorem 9.10 to justify "pasting together" determinations of a multiple valued function has many generalizations. How would you verify differentiability (rather than just continuity) at points on the rays without using Theorem 9.10?

9.5 THE SCHWARZ-CHRISTOFFEL FORMULA

The simplest kind of domain we might wish to transform onto the disc, or the upper half plane $H = \{z : \mathrm{Im}(z) > 0\}$, is one bounded by a closed polygonal curve. These include rectangles, triangles, and other convex polygons; but we will not restrict our attention to polygons that are convex, and sometimes we will allow the point at infinity to be a vertex, so that domains such as a half strip may be though of as unbounded polygonal domains. For some of the simplest polygonal domains, the mapping that transforms E to a half plane cannot be expressed in terms of elementary functions. For example, if we try to map a square onto the upper half plane (or equivalently, onto the unit disc), our attention is inevitably directed toward the family of *elliptic functions*, which are not elementary and can only be described through series expansions, integral formulas, or other limit processes.

Given a polygonal domain E, we may discuss either the mapping $f:H \to E$ or its inverse $\check{f}:E \to H$, since each implicitly determines the other. The map from the half plane H onto E has properties which allow us to find nearly explicit formulas for f. Although f itself is usually non-elementary, its derivative df/dz turns out to be an elementary function on the half plane, which can be expressed in closed form; f is then determined as the antiderivative of a known function. The Schwarz-Christoffel formula, explained below, expresses df/dz in terms of the parameters that characterize the domain E.

A polygonal domain may be described by listing its successive vertices $w_1, \ldots, w_n$ in **positive order;** that is, the vertices are to be listed so that the domain is always to the *left* of the segments $[w_k, w_{k+1}]$ as the boundary is traversed. The boundary may also be described by listing the angles $\alpha_k = \arg(w_{k+1} - w_k)$, measured from the positive real axis to the segment $[w_k, w_{k+1}]$, together with the lengths of the successive sides; specifying all these values determines E up to a translation. We shall also refer to the exterior angles $\beta_1 = \alpha_1 - \alpha_n$ and $\beta_k = \alpha_k - \alpha_{k-1}$ for $k = 1, 2, \ldots, n$. These are measured from the incoming side to the outgoing side at the vertex w_k, as indicated in Figure 9.15. Obviously, we may determine these exterior angles so that $-\pi \leq \beta_k \leq +\pi$; their sum is $2\pi = \sum_{k=1}^{n} \beta_k$ for any bounded polygonal domain. We definitely mean to allow the possibility that $\beta_k = \pm\pi$; this corresponds to a *cut* whose sides are regarded as different sides of a degenerate polygonal domain. Such cases will be considered at the end of this section.

Since E and H have piecewise smooth boundaries, the mapping f may be extended so that it is continuous on their boundaries as well (it will be continuous at ∞ too, if we view the situation in the complex sphere). Let $x_1 + i0, \ldots, x_n + i0$ be the points on the (extended) real axis that correspond to the vertices of the polygon. We may alter f by composing it with a preliminary fractional linear transformation of the half plane so that any three of these points (listed in

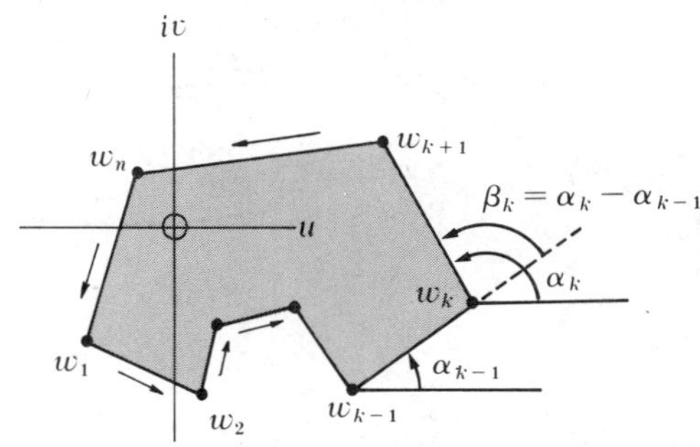

Figure 9.15 The parameters α_k and β_k. Vertices of E are labeled in *positive order.*

positive order), say x_{k_1}, x_{k_2}, and x_{k_3}, are moved to prescribed positions in $\mathbf{R}^*$ (the new positions also listed in positive order with respect to the half plane). Once this is done, the mapping $f:H \to E$ is fully determined, and the locations of the remaining points are implicitly determined along with the mapping f. As x increases in the interval (x_k, x_{k+1}), the values $w = f(x + i0)$ lie on the line L_k determined by the side (w_k, w_{k+1}) of the polygon.† By Schwarz' reflection principle, f can be continued analytically across (x_k, x_{k+1}); the continuation $F_k(z)$ agrees with f on H and on (x_k, x_{k+1}), and is analytic throughout the cut plane $Q_k = H \cup (x_k, x_{k+1}) \cup H^*$. As in Theorem 9.11, the continuation is defined on $H^* = \{z:\text{Im}(z) < 0\}$ by

$$F_k(z) = \begin{cases} f(z) & \text{for } z \text{ in } H \cup (x_k, x_{k+1}) \\ J_k(f(\bar{z})) & \text{for } z \text{ in } H^*, \end{cases}$$

where $w' = J_k(w)$ is the reflection through the line L_k in the w-plane. Obviously the values $w = f(z)$ lie in the polygon E, while the values $w' = J_k(w) = F_k(\bar{z})$ lie in the reflected polygon $E_k = J_k(E)$ for each point z in H (recall z is in H and $\bar{z}$ is a typical point in H^*). Now F_k is univalent near any point in the combined domains $H \cup (x_k, x_{k+1}) \cup H^*$ shown in Figure 9.16. Since it is analytic on Q_k, its derivative must be nonvanishing throughout this domain, even on the interval (x_k, x_{k+1}); F_k cannot be univalent near a critical point.

To carry out the preceding continuation across the exceptional "interval" $(x_n, +\infty) \cup \{\infty\} \cup (-\infty, x_1)$, which corresponds to the side (w_n, w_1) of E, we should view the situation in $\mathbf{C}^*$. The continuation F_k will be analytic

† We will let (w_n, w_{n+1}) stand for (w_n, w_1) to avoid introducing a special notation for the last side.

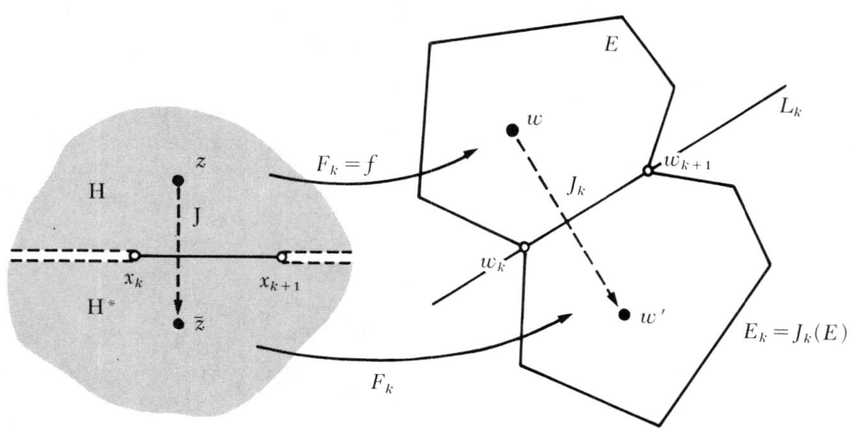

Figure 9.16 Mapping properties of the continuation F_k of f to the domain $H \cup (x_k, x_{k+1}) \cup H^*$

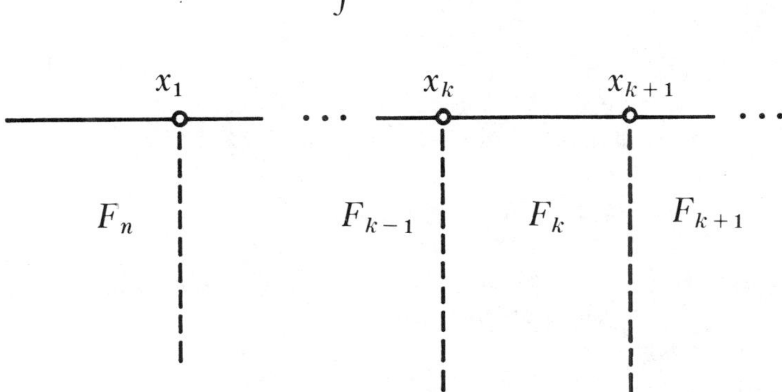

Figure 9.17 The domain Q with cuts from x_k to ∞. In the kth strip, f is continued as F_k.

near infinity and bounded, so that ∞ is a removable singularity. Exceptional intervals could be avoided entirely if we were to shift one of the points, say x_n, to ∞ at the outset.

Continuations of f across different intervals (x_k, x_{k+1}) usually will not agree in the lower half plane; f has multiple valued continuations. To avoid this we define the domain Q by introducing cuts from x_k to infinity in the lower half plane, as indicated in Figure 9.17, and we continue f into each strip separately. Now f may be discontinuous along the cuts, but it will be analytic otherwise. Notice that df/dz is defined and nonvanishing on the intervals (x_k, x_{k+1}). If $x_k < x < x_{k+1}$, we know something about $f'(x + i0)$ because $f(x + i0)$ is known to lie on the side (w_k, w_{k+1}) of E. In particular, difference quotients $[f(x + \Delta t) - f(x)]/\Delta t$ are directed parallel to this segment, and must have argument $\alpha_k = \arg(w_{k+1} - w_k)$. The same is true of the limit;† thus,

$$\text{(21)} \qquad \arg f'(x + i0) \equiv \alpha_k \quad \text{for } x_k < x < x_{k+1} \ (k = 1, 2, \ldots, n).$$

Thus, $\arg f'(x + i0)$ is a step function with discontinuities at $x_1, \ldots, x_n$.

It is not hard to produce analytic functions $G(z)$ on Q such that $\arg G(z)$ takes on these values on the extended real line. Let $\text{Log}^*(z)$ be the determination of $\log z$ that agrees with $\log x$ on the positive real axis, and has its discontinuities confined to the negative imaginary axis, which we shall denote by $[0, -i\infty)$. Clearly the functions

$$\text{Log}^*(z - x_{k+1}) - \text{Log}^*(z - x_k) \qquad 1 \leq k \leq n$$

are analytic on Q; as before, we identify $x_{n+1} = x_1$. Then let

$$h(z) = i\alpha_n + \sum_{k=1}^{n} \frac{\alpha_k}{\pi} [\text{Log}^*(z - x_{k+1}) - \text{Log}^*(z - x_k)];$$

† We need the fact that $f'(x + i0) \neq 0$ to be sure that the argument of the derivative makes any sense at all.

it follows that

$$\operatorname{Im}(h(z)) = \alpha_n + \sum_{k=1}^{n} \frac{\alpha_k}{\pi} [\operatorname{Arg}^*(z - x_{k+1}) - \operatorname{Arg}^*(z - x_k)],$$

and by straightforward calculations we see that

$$\operatorname{Im}(h(z)) = \alpha_k \quad \text{on } (x_k, x_{k+1}), \text{ for } 1 \le k \le n.$$

In particular, $\operatorname{Im}(h(z)) = \alpha_n$ on $(x_n, x_{n+1}) = (x_n, +\infty) \cup \{\infty\} \cup (-\infty, x_1)$. Now $h(z)$ may be rewritten in the form

$$\begin{aligned} h(z) &= i\alpha_n - \left[\sum_{k=1}^{n} \frac{\alpha_k}{\pi} \operatorname{Log}^*(z - x_k) - \sum_{k=1}^{n} \frac{\alpha_k}{\pi} \operatorname{Log}^*(z - x_{k+1})\right] \\ &= i\alpha_n - \left[\sum_{k=2}^{n} \frac{\alpha_k - \alpha_{k-1}}{\pi} \operatorname{Log}^*(z - x_k) + \frac{\alpha_1}{\pi} \operatorname{Log}^*(z - x_1) - \frac{\alpha_n}{\pi} \operatorname{Log}^*(z - x_1)\right] \\ &= i\alpha_n - \sum_{k=1}^{n} \frac{\beta_k}{\pi} \operatorname{Log}^*(z - x_k). \end{aligned}$$

The imaginary part of h is the *argument* of the exponential $g(z) = \exp(h(z))$. Let us simplify the notation for products by writing

$$\prod_{k=1}^{n} s_k \text{ for the product } s_1 \cdot \ldots \cdot s_n \qquad (s_k \text{ complex numbers});$$

then

$$g(z) = e^{h(z)} = A \cdot \prod_{k=1}^{n} \exp\left(-\frac{\beta_k}{\pi} \operatorname{Log}^*(z - x_k)\right),$$

where $A = \exp(i\alpha_n)$. The function $w = \exp((-\beta_k/\pi)\operatorname{Log}^*(z))$ is a determination of $w = z^{-\beta_k/\pi}$ that is analytic on the cut plane $\mathbf{C} \sim [0, -i\infty)$; if we interpret powers in this manner ($-\beta_k/\pi$ is real), the product can be written in the form

$$g(z) = A \cdot \prod_{k=1}^{n} (z - x_k)^{-\beta_k/\pi} \qquad \text{for } z \text{ in } Q.$$

Since Q is simply connected, g has an antiderivative on Q,

$$\text{(22)} \quad G(z) = \int_p^z g(\zeta)\, d\zeta + B = A\int_p^z (\zeta - x_1)^{-\beta_1/\pi} \cdot \ldots \cdot (\zeta - x_n)^{-\beta_n/\pi}\, d\zeta + B,$$

whose derivative $G'(z) = g(z)$ has the same argument on the intervals (x_k, x_{k+1}) as the derivative of the mapping $f(z)$ we wish to determine.

It seems reasonable to expect that G will be closely related to f. After all, the condition $\arg G' \equiv \alpha_k$ on (x_k, x_{k+1}) means that $G(x + i0)$ traces out a segment (w_k^*, w_{k+1}^*), inclined at an angle α_k radians, as x increases within (x_k, x_{k+1}).

To see this, write $w_k^* = G(x_k + i0)$ and express the displacement $G(x) - G(x_k)$ as an integral:

$$G(x) - G(x_k) = \int_{x_k}^{x} G'(s + i0)\, ds$$
$$= \int_{x_k}^{x} e^{i \arg G'(s)}\, |G'(s)|\, ds$$
$$= \int_{x_k}^{x} e^{i\alpha_k}\, |G'(s)|\, ds$$
$$= e^{i\alpha_k} \int_{x_k}^{x} |G'(s)|\, ds.$$

Thus,

$$\arg[G(x) - G(x_k)] \equiv \alpha_k \pmod{2\pi}$$
$$|G(x) - G(x_k)| = \int_{x_k}^{x} |G'(s)|\, ds;$$

the absolute value $|G(x) - G(x_k)| = |w(x) - w_k^*|$ is continuous and strictly increasing as x increases within (x_k, x_{k+1}), so it is clear that $G(x)$ traces out a segment (w_k^*, w_{k+1}^*) with the proper orientation.

Therefore, $G(x + i0)$ traces out a closed polygonal arc whose successive segments $(w_1^*, w_2^*), \ldots, (w_n^*, w_1^*)$ have the same orientation to the positive real axis, (α_k radians) and the same exterior angles ($\beta_k = \alpha_k - \alpha_{k-1}$), as the sides of the given polygon E. This is not quite enough to prove that $G(x + i0)$ traces out the boundary of a polygonal domain congruent to E; for that we would have to know that the side lengths $|w_{k+1}^* - w_k^*|$ and $|w_{k+1} - w_k|$ match, as well as the angles. By carrying out a more detailed analysis of f and its continuations $F_1, \ldots, F_n$ we could prove that f is indeed given by

$$\textbf{(23)} \qquad f(z) = A\int_p^z (\zeta - x_1)^{-\beta_1/\pi} \cdot \ldots \cdot (\zeta - x_n)^{-\beta_n/\pi}\, d\zeta + B$$

if the scalars A and B are suitably chosen. This is the **Schwarz-Christoffel formula.** We shall go on to examine the practical uses of this formula, leaving the remainder of its derivation to more advanced courses; for further reading we recommend Nehari [16], Section 5.5, or Nevanlinna and Paatero [18], Sections 17.25 to 17.28.

Note 1: If we adjust the mapping function $f: H \to E$ so that $x_n = \infty$, the discussion leading to (23) may be repeated to obtain a formula

$$\textbf{(24)} \qquad f(z) = A\int_p^z \prod_{k=1}^{n-1} (\zeta - x_k)^{-\beta_k/\pi}\, d\zeta + B.$$

The only difference between this and (23) is the absence of the factor involving $x_n = \infty$ in the integrand. This formula is a little simpler to use than (23);

however, once we set $x_n = \infty$ we may adjust the position of only *two* other points x_k by fractional linear transformation of the upper half plane.

Note 2: The whole discussion could have been carried out for a mapping of the *unit disc* onto E. Alternatively, we can substitute the well known transformation $z = T(w) = -i(w+i)/(w-i)$ (of the disc onto the half plane) into the Schwarz-Christoffel formula to get a formula for the mapping $h = f \circ T: D \to E$ which transforms the disc onto E; we get

$$h(w) = A' \int_0^w \prod_{k=1}^{n} (w - \zeta_k)^{-\beta_k/\pi}\, dw + B' \quad \text{for } |w| \le 1. \tag{25}$$

Here A' and B' are constants to be determined (not the same as the ones in (23)), and $\zeta_k = \check{T}(x_k)$ are the points on the boundary of the disc that correspond to the vertices of E. Determinations of $(w - \zeta_k)^\mu$ (for μ real) are to be chosen so they are analytic on the disc $|w| < 1$; this can be done by determining $\log(w - \zeta_k)$ so it is analytic on the half plane (containing the disc) that is bounded by the tangent line passing through $w = \zeta_k$.

If γ is a curve from $w = 0$ to w in D, then $\eta = T \circ \gamma$ is a curve from $z = +i$ to a typical point $z = T(w)$ in H. Applying the change of variable formula for contour integrals, we get

$$\begin{aligned} h(w) = f(T(w)) &= A \int_\gamma \prod_{k=1}^{n} (z - x_k)^{-\beta_k/\pi}\, dz + B \\ &= A \int_{\eta = T\circ\gamma} \prod_{k=1}^{n} (T(w) - T(\zeta_k))^{-\beta_k/\pi} \cdot \frac{dz}{dw}\, dw + B \\ &= A \int_0^w \prod_{k=1}^{n} (T(w) - T(\zeta_k))^{-\beta_k/\pi} \frac{-2}{(w-i)^2}\, dw + B. \end{aligned}$$

Direct calculations show that

$$T(w) - T(\zeta_k) = \frac{-2(w - \zeta_k)}{(w-i)(\zeta_k - i)}, \quad \text{for } |w| < 1.$$

Now $z = T(w)$ is analytic and $T: D \to H$, so that the composite of analytic functions

$$\phi(w) = \prod_{k=1}^{n} (T(w) - T(\zeta_k))^{-\beta_k/\pi} = \prod_{k=1}^{n} (T(w) - x_k)^{-\beta_k/\pi}$$

is analytic for $|w| < 1$. Also, the function

$$\psi(w) = \prod_{k=1}^{n} (w - \zeta_k)^{-\beta_k/\pi}$$

is analytic for $|w| < 1$, if we use determinations of $(w - \zeta_k)^{-\beta_k/\pi}$ as indicated above.

Finally, we note that

$$(a \cdot b)^{-\beta_k/\pi} = c_k \cdot (a^{-\beta_k/\pi}) \cdot (b^{-\beta_k/\pi}) \quad (c_k = e^{-2im_k\beta_k};\ m_k \text{ some integer})$$

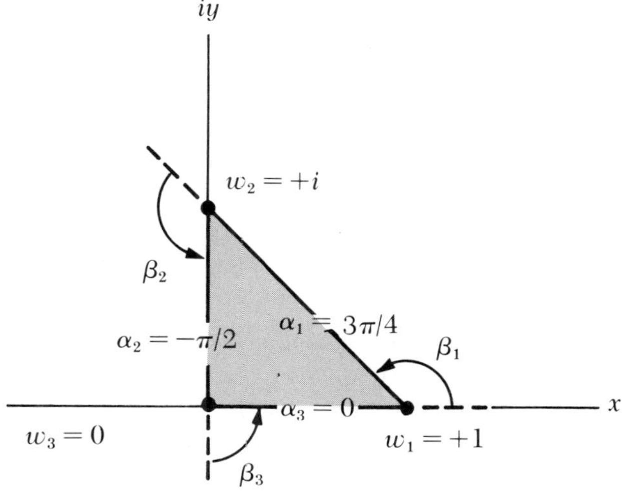

Figure 9.18 The triangular domain in Example 9.10.

and

$$\sum_{k=1}^{n} -\beta_k/\pi = +2.$$

Using these facts, it is not hard to see that

$$\frac{-2}{(w - i)^2} \cdot \phi(w) = A'' \cdot \psi(w)$$

for w near the origin, where A'' is some fixed constant. By analytic continuation, we must have $-2\phi(w)/(w - i)^2 = A''\psi(w)$ everywhere on the disc. We get the desired formula by taking $A' = A \cdot A''$.

Example 9.10 Determine the mapping f that transforms the upper half plane to the triangle shown in Figure 9.18. The points x_k that correspond (under f) to the vertices w_k of E may be assigned freely, as long as they are arranged in positive order along the real axis (the boundary of the half plane); the vertices w_k of E must also be labelled in positive order, as we explained earlier. We may assume that f has been adjusted so that

$$x_1 = -1; \; x_2 = +1; \; x_3 = \infty.$$

Clearly,

$$\left.\begin{array}{l} \alpha_1 \equiv 3\pi/4; \; \alpha_2 \equiv -\pi/2; \; \alpha_3 \equiv 0 \\ \beta_1 \equiv 3\pi/4; \; \beta_2 \equiv 3\pi/4; \; \beta_3 \equiv \pi/2 \end{array}\right. \quad (\text{modulo } 2\pi).$$

The Schwarz-Christoffel formula assures us that there is a unique choice of constants A, B for which

$$f(z) = B + A\int_0^z (\zeta + 1)^{-\beta_1/\pi}(\zeta - 1)^{-\beta_2/\pi}\, d\zeta$$

$$= B + A\int_0^z (\zeta + 1)^{-3/4}(\zeta - 1)^{-3/4}\, d\zeta$$

is the desired mapping function $f: H \to E$ such that $f(x_k + i0) = w_k$. Remember that $w = z^\mu$ (μ real) is always to be defined so that its discontinuities lie along the cut $[0, -i\infty)$ and $(x + i0)^\mu = x^\mu > 0$ for positive real x. A simple comparison shows that $(z+1)^{-3/4}(z-1)^{-3/4} = (z^2-1)^{-3/4} = 1/(z^2-1)^{3/4}$ on the upper half plane and on the segments (x_k, x_{k+1}), so that f can be written

$$f(z) = B + A\int_0^z \frac{1}{(\zeta^2 - 1)^{3/4}}\,d\zeta \qquad \text{for } z \text{ in } H.$$

Care is needed in multiplying together fractional powers; it is always true that $1/z^\mu = z^{-\mu}$, but the validity of the equation $z^\mu \cdot w^\mu = (z \cdot w)^\mu$ depends upon the relative position of z and w (see Exercise 4).

Our first step in picking A and B is to arrange that $\arg f'(x + i0) \equiv \alpha_k \pmod{2\pi}$ for $x_k < x < x_{k+1}$ $(1 \le k \le n)$. We have

$$\begin{aligned}\arg f'(x + i0) &\equiv \arg\left[A \cdot \prod_{k=1}^{n-1} (x - x_k)^{-\beta_k/\pi}\right] \\ &\equiv \arg A + \sum_{k=1}^{n-1} -\frac{\beta_k}{\pi}\arg(x - x_k)\end{aligned} \qquad \pmod{2\pi}$$

for real x. The value of the sum on the right-hand side above differs by a fixed constant from the desired value α_k on each interval (x_k, x_{k+1}); choosing $\arg A$ to compensate for this, we get $\arg f'(x + i0) \equiv \alpha_k$, as desired. The values arising in our example are indicated below.

$$\infty \xrightarrow{\;0\;} -1 \xrightarrow{\;3\pi/4\;} +1 \xrightarrow{\;-\pi/2\;} \infty \qquad \text{values of } \alpha_k$$

$$\infty \xrightarrow{\;\pi/2\;} -1 \xrightarrow{\;-3\pi/4\;} +1 \xrightarrow{\;0\;} \infty \qquad \arg\left[\prod_{k=1}^{n-1} (x - x_k)^{-\beta_k/\pi}\right].$$

To make them match, we choose $\arg A = -\pi/2$; thus, $A = e^{-i\pi/2}\,|A| = -i\,|A|$.

Since $f'(x + i0) \equiv \alpha_k$ for $x_k < x < x_{k+1}$, $w = f(x + i0)$ traces out a segment (w_k^*, w_{k+1}^*) parallel to the side (w_k, w_{k+1}) of E. By choosing $|A|$ properly, we will make the side lengths match, so that $|w_{k+1}^* - w_k^*| = |w_{k+1} - w_k|$ for $1 \le k \le n$; then f will map $\mathbf{R}^*$ onto the boundary of a polygonal domain which differs from E by a simple translation. Choosing B suitably eliminates the need for a translation.

The requirement $|w_2^* - w_1^*| = |w_2 - w_1| = |i - 1| = \sqrt{2}$ is enough to determine $|A|$. The definite integral

$$k = \int_{-1}^{1} \frac{1}{(1 - x^2)^{3/4}}\,dx \qquad \text{(integrand is \textit{positive})}$$

is convergent, though improper at the end points $x = \pm 1$. It is a non-elementary integral, but is value might be calculated numerically, or through the use of residue methods. Whatever its value is, we should have

$$\begin{aligned}
\sqrt{2} = |w_2 - w_1| &= \left|\int_{-1}^{+1} f'(x + i0)\, dx\right| \\
&= \left|A\int_{-1}^{+1} (x^2 - 1)^{-3/4}\, dx\right| = \left|A\int_{-1}^{+1} [(-1)\cdot(1 - x^2)]^{-3/4}\, dx\right| \\
&= \left|e^{-i3\pi/4} A\int_{1}^{+1} (1 - x^2)^{-3/4}\, dx\right| \\
&= k\,|A|,
\end{aligned} \tag{26}$$

if f is to be the desired mapping $f: H \to E$. Thus $A = (-i\sqrt{2})/k$.

The constant B can be determined by the requirement that $f(x_1) = f(-1) = w_1 = 1$. Obviously,

$$\begin{aligned}
1 = f(-1) &= A\int_0^{-1} (x + 1)^{-3/4}(x - 1)^{-3/4}\, dx + B \\
&= A \cdot e^{-i3\pi/4}\int_0^{-1} (1 + x)^{-3/4}(1 - x)^{-3/4}\, dx + B \\
&= -A \cdot e^{-i3\pi/4}\int_{-1}^{0} (1 - x^2)^{-3/4}\, dx + B \\
&= \frac{Ae^{i\pi/4}}{2}\int_{-1}^{+1} (1 - x^2)^{-3/4}\, dx + B \\
&= \frac{kAe^{i\pi/4}}{2} + B = \frac{\sqrt{2}\, e^{i\pi/4}}{2i} + B.
\end{aligned}$$

Thus

$$B = \tfrac{1}{2}(2 + i\sqrt{2}\, e^{i\pi/4}) = \tfrac{1}{2}(2 + i(-1 + i)) = (1 - i)/2,$$

and the desired mapping function is

$$f(z) = \frac{\sqrt{2}}{ik}\int_0^z (\zeta^2 - 1)^{-3/4}\, d\zeta + \frac{(1 - i)}{2}.$$

Calculating the lengths of other sides of E, we get other conditions on $|A|$:

$$\begin{aligned}
1 = |w_3 - w_2| &= |A|\int_1^{\infty} (x^2 - 1)^{-3/4}\, dx \\
1 = |w_1 - w_3| &= |A|\int_{-\infty}^{-1} (x^2 - 1)^{-3/4}\, dx.
\end{aligned} \tag{27}$$

Clearly, these two conditions are equivalent, since the definite integrals have the same value in each, but they seem to be different from the third condition (26) we used to determine $|A|$. However, the derivation of the

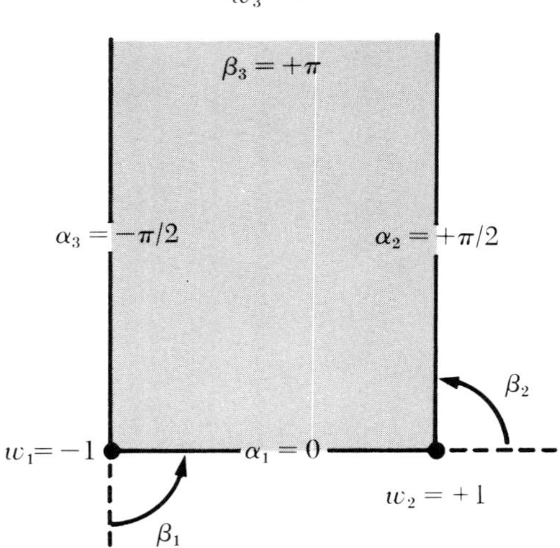

Figure 9.19 The semi-infinite strip in Example 9.11.

Schwarz-Christoffel formula assures us that there is just one map $f: H \to E$ such that $f(z_k) = w_k$. Once the constants A and B are determined by comparing some of the lengths $|w_{k+1}^* - w_k^*|$ and $|w_{k+1} - w_k|$, the map F must satisfy all the remaining requirements. One could also show by direct (residue) calculations that (26) and (27) lead to the same value of $|A|$.

When ∞ is allowed as a vertex of E, formula (23) remains valid. For example, E might be a semi-infinite strip as in the next example. Then the change in angle at $w_k = \infty$ is $\beta_k = \pm\pi$ (examining the polygon in $\mathbf{C}^*$ should make this clear). Rather than re-examine the reasoning behind (23), allowing one or more infinite vertices, we shall verify directly that the function (23) has the desired mapping properties.

Example 9.11 Consider the semi-infinite strip shown in Figure 9.19. There is a conformal mapping $f: H \to E$ since E is simply connected, and f may be adjusted so that the points $x_1 = -1$, $x_2 = +1$, and $x_3 = \infty$ in $\mathbf{R}^*$ correspond to the vertices $w_1 = -1$, $w_2 = +1$, and $w_3 = \infty$. We expect that f will be given by a suitable choice of A and B in†

$$(28)\qquad \begin{aligned} f(z) &= B + A\int_0^z (\zeta+1)^{-\beta_1/\pi}(\zeta-1)^{-\beta_2/\pi}\, d\zeta \\ &= B + A\int_0^z (\zeta+1)^{-1/2}(\zeta-1)^{-1/2}\, d\zeta. \end{aligned}$$

† For the exponents in this example we have $(x^2 - 1)^{-1/2} = (-1)(x+1)^{-1/2}(x-1)^{1/2}$ if $-\infty < x < -1$, so these functions do not agree and we may not use the function $(\zeta^2 - 1)^{-1/2} = 1/(\zeta^2 - 1)^{1/2}$ in place of the intergrand in (28).

To determine A, notice that

$$\frac{df}{dz} = A\,\frac{1}{(z+1)^{1/2}}\,\frac{1}{(z-1)^{1/2}},$$

so that

$$\arg f'(z) \equiv \arg A - \tfrac{1}{2}\arg(z+1) - \tfrac{1}{2}\arg(z-1).$$

Substituting $z = x + i0$, and letting $x > 1$, we want to have $\arg f'(x + i0) \equiv \alpha_2 = +\pi/2$ since the segment $(1, \infty)$ corresponds to the side (w_2, w_3). Since the terms other than $\arg A$ are both zero for $1 < x < +\infty$, we should take $\arg A = +\pi/2$, so that $A = i\,|A|$.

If we are to have $|w_2 - w_1|$ equal to $|w_2^* - w_1^*|$, we must set $|A| = 2/\pi$, so that $A = 2i/\pi$, since

$$\begin{aligned} 2 = |w_2 - w_1| = |w_2^* - w_1^*| &= \left| A\int_{x_1}^{x_2} (x+1)^{-1/2}(x-1)^{-1/2}\,dx \right| \\ &= \left| \frac{A}{i}\int_{-1}^{+1} \frac{1}{\sqrt{1-x^2}}\,dx \right| = |A| \int_{-1}^{+1} \frac{1}{\sqrt{1-x^2}}\,dx \\ &= |A| \left[\arcsin x \right]_{x=-1}^{x=+1} = \pi\,|A|. \end{aligned}$$

We used the fact that

$$\begin{aligned} (x+1)^{-1/2}(x-1)^{-1/2} = (x^2-1)^{-1/2} &= ((-1)(1-x^2))^{-1/2} \\ &= e^{-i\pi/2}(1-x^2)^{-1/2} \end{aligned}$$

on the interval $-1 < x < +1$. Further calculations show that $B = 0$; indeed, $F(1) = B + 1$, while we should have $F(1) = w_2 = 1$. Thus

$$f(z) = \frac{2i}{\pi}\int_0^z (\zeta+1)^{-1/2}(\zeta-1)^{-1/2}\,d\zeta \quad \text{if} \quad \operatorname{Im}(z) \geq 0.$$

This result could also be obtained by directly antidifferentiating $f'(z) = A(z+1)^{-1/2}(z-1)^{-1/2}$. In certain parts of the upper half plane (the first quadrant, for example) this may be written $A(z^2-1)^{-1/2} = -iA(1-z^2)^{-1/2} = A/i\sqrt{1-z^2}$, where $\sqrt{z}$ is the principal determination of square root. The antiderivative of this function is already familiar; thus the Schwarz-Christoffel formula leads us to suspect that the mapping function f is some determination of $w = A \cdot \arcsin z$. A little thought reveals that the principal determination serves our purposes, and that $f(z) = (2/\pi)\operatorname{Arcsin} z$. Conversely, the single-valued inverse function $z = \sin(\pi w/2)$ maps the strip onto the half plane (recall Example 4.11).

Although we could have guessed the answer to this mapping problem, the Schwarz-Christoffel formula is often invaluable in directing one's guesswork when the mapping function is elementary, but the domain is not as familiar as the half strip (more on this in Example 9.13 below).

The locations of the points $\{x_k\}$ are interdependent; they are implicitly determined by the mapping function f once three of them have been specified. Thus, when E has more than three vertices we face the serious difficulty that some of the parameters x_k in formula (23) depend on the function f we are trying to determine. In certain special cases, especially if E is highly symmetric, the location of the points x_k may be determined; but in general, the problem of finding f and the points $x_4, \ldots, x_n$ simultaneously can be very difficult.

One crude approach is to *guess* a distribution of points. By numerical calculations one can determine the lengths of the segments (w_k^*, w_{k+1}^*) to which the intervals (x_k, x_{k+1}) are mapped by F. The angles of these segments can easily be made to agree with those of the sides (w_k, w_{k+1}) of E by making a suitable choice of arg A. If the side lengths match, we have solved the mapping problem; if they don't, we try another distribution of points $x_4, \ldots, x_n$ and successively approach the correct distribution of these parameters.

Example 9.12 Consider the rectangle E shown in Figure 9.20. The bilateral symmetry of this polygon leads us to expect that we should be able to map H onto E so that the symmetrically located points $x_1 = -a$, $x_2 = -1$, $x_3 = +1$, and $x_4 = +a$ ($a > 1$ is a parameter to be determined) are mapped to the vertices w_k. The point at infinity on the extended real axis does not correspond to a vertex, so the mapping $f: H \to E$ should be determined by the Schwarz-Christoffel formula as presented in (23); the exceptional interval (x_4, x_1) should be thought of as the arc $(a, +\infty) \cup \{\infty\} \cup (-\infty, -a)$ in the complex sphere. Since $\beta_k = \pi/2$ at each vertex, f should be determined by a suitable choice of a, A, and B in

$$F(z) = B + A\int_0^z (\zeta + a)^{-1/2}(\zeta + 1)^{-1/2}(\zeta - 1)^{-1/2}(\zeta - a)^{-1/2}\, d\zeta.$$

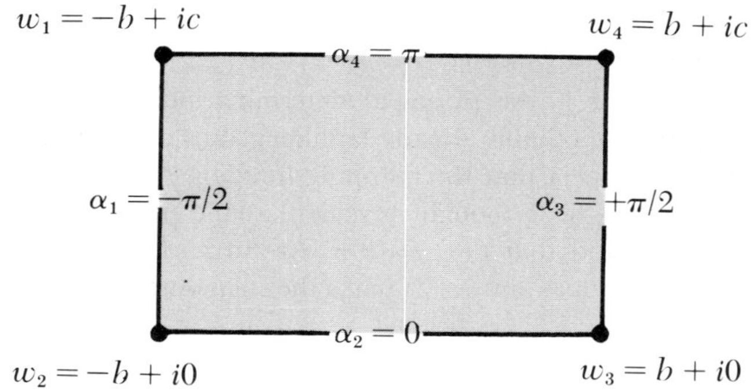

$w_1 = -b + ic$ $\qquad$ $w_4 = b + ic$

$\alpha_4 = \pi$

$\alpha_1 = -\pi/2$ $\qquad$ $\alpha_3 = +\pi/2$

$\alpha_2 = 0$

$w_2 = -b + i0$ $\qquad$ $w_3 = b + i0$

Figure 9.20 The rectangular domain in Example 9.12.

Table 9.1

Interval	$\|\Delta w\|$	$\|\Delta w^*\|$
$I_1 = (x_1, x_2)$	c	$\|A\| \int_{-a}^{-1} \frac{1}{\sqrt{a^2 - x^2}\sqrt{x^2 - 1}}\, dx$
$I_2 = (x_2, x_3)$	$2b$	$\|A\| \int_{-1}^{+1} \frac{1}{\sqrt{a^2 - x^2}\sqrt{1 - x^2}}\, dx$
$I_3 = (x_3, x_4)$	c	$\|A\| \int_{1}^{a} \frac{1}{\sqrt{a^2 - x^2}\sqrt{x^2 - 1}}\, dx$
$I_4 = (x_4, x_1)$	$2b$	$\|A\| \left[\int_{-\infty}^{-a} \frac{1}{\sqrt{x^2 - a^2}\sqrt{x^2 - 1}}\, dx + \int_{a}^{+\infty} \cdots dx\right]$

It is easy to verify that the integrand agrees with $[(\zeta^2 - a^2)^{1/2}(\zeta^2 - 1)^{1/2}]^{-1}$ if $\text{Im}(\zeta) \geq 0$, so that

(29) $$f(z) = B + A\int_0^z \frac{1}{(\zeta^2 - a^2)^{1/2}(\zeta^2 - 1)^{1/2}}\, d\zeta.$$

These integrals give *elliptic functions*, which are not elementary. The integrand is constructed so that its argument and $\arg f'(x + i0)$ differ only by a fixed constant on the intervals (x_k, x_{k+1}); a simple comparison of arguments on the particular interval $(x_4, +\infty)$ indicates that we must choose $\arg A = \pi$ to get $\arg f'(x + i0) \equiv \alpha_k$ on each of the intervals. Now let us compare the lengths of the sides Δw of E and the lengths of the corresponding segments Δw^* to which f maps the intervals (x_k, x_{k+1}). These comparisons are listed in Table 9.1 (we have extracted factors of modulus one from the integrals in each case, so that each integrand will be positive valued).

The constants $|A|$ and $a > 1$ must be chosen so these lengths agree for each I_k. A comparison of lengths corresponding to I_2 and I_3 gives us two equations which implicitly determine $|A|$ and a—only one choice of these constants can satisfy both equations:

(30) $$2b = |A|\int_{-1}^{+1} \frac{1}{\sqrt{a^2 - x^2}\sqrt{1 - x^2}}\, dx$$
$$c = |A|\int_1^a \frac{1}{\sqrt{a^2 - x^2}\sqrt{x^2 - 1}}\, dx$$

An explicit solution of this system will not be attempted here; it would require numerical methods or a (numerically calculated) table of elliptic functions. The equation corresponding to I_1 is redundant because I_1 and I_3 lead to the same conditions on $|A|$ and a. The condition associated with I_4 is also redundant, although this is not so obvious; one can use residue calculations, or the change of variable formula, to show that the definite integrals for I_2 and I_4 are

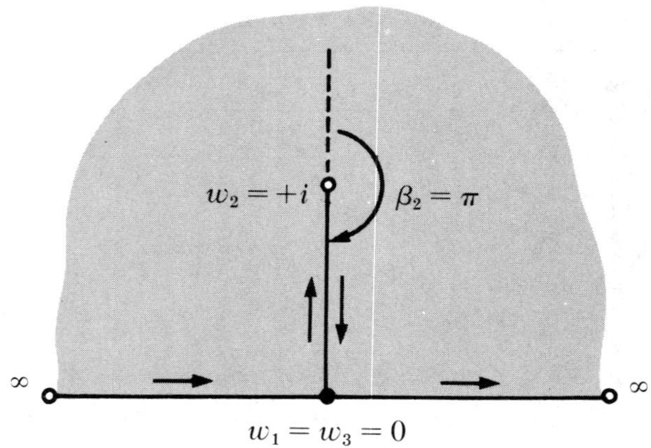

Figure 9.21 The re-entrant polygon in Example 9.13.

equal, so that they lead to the same values for $|A|$ and a. For the residue calculation one would take a determination of $(z^2 - a^2)^{-1/2}(z^2 - 1)^{-1/2}$ whose singularities are located on the cuts $(-1, +1) = I_2$ and $(-\infty, -a) \cup (a, +\infty) = I_4$. To get matching of vertices, $w_k^* = F(x_k) = w_k$, it is not hard to see that we should take $B = 0$, due to the choice of base point $p = 0$ in (29).

The Schwarz-Christoffel formula applies to polygons with highly irregular boundaries. In these complicated situations its main use is to determine the form of f'.

Example 9.13 Let E be the re-entrant polygon shown in Figure 9.21. It might not be clear whether ∞ should be regarded as a vertex; no harm would be done in carrying it along as a (possibly superfluous) vertex w_4, but Table 9.2 of values for α_k and β_k indicates that it is redundant to regard ∞ as a vertex. (Since $\beta_4 = 0$, this "vertex" would not contribute a factor to the integrand in (23), and so may be omitted.) There *is* a mapping $f: H \to E$, since E is a simply connected domain, and by adjusting f we may insure that the points $x_1 = -1$, $x_2 = 0$, $x_3 = +1$ map to $w_1 = 0$, $w_2 = +i$, $w_3 = 0$ respectively. This uniquely determines f. We expect that f is given by

$$f(z) = B + A\int_0^z (\zeta + 1)^{-1/2}(\zeta - 0)^1(\zeta - 1)^{-1/2}\, d\zeta.$$

Table 9.2

	$k = 1$	$k = 2$	$k = 3$	$k = 4$
w_k	0	$+i$	0	∞
α_k	$\pi/2$	$-\pi/2$	0	0
β_k	$\pi/2$	$-\pi$	$\pi/2$	0
μ_k	$-\frac{1}{2}$	$+1$	$-\frac{1}{2}$	0

The product $(\zeta + 1)^{-1/2}(\zeta - 1)^{-1/2}$ gives us a determination of $1/\sqrt{\zeta^2 - 1}$ which is analytic on the cut plane Q, but it cannot be obtained throughout Q, or even on the upper half plane, by taking the principal determination of $\sqrt{\zeta^2 - 1}$. Let us use $\overset{*}{\sqrt{\zeta^2 - 1}}$ to indicate the correct determination of this function on Q; then we get

$$f(z) = B + A\int_0^z \frac{\zeta}{\overset{*}{\sqrt{\zeta^2 - 1}}}\, d\zeta \qquad \text{if } z \text{ is in } Q.$$

The antiderivative can be evaluated. It is the elementary function $\overset{*}{\sqrt{z^2 - 1}}$, and we only have to determine the constants which make $f(z) = A\overset{*}{\sqrt{z^2 - 1}} + B$ map H onto E so that $f(x_k + i0) = w_k$. It is clear that $B = 0$ and, with due regard for the way $\overset{*}{\sqrt{z^2 - 1}}$ is defined, we find that

$$+i = A\overset{*}{\sqrt{(-1)}} = +iA \quad \text{and} \quad A = +1,$$

since $x_2 = 0$ is mapped to $w_2 = +i$. Thus,

$$f(z) = \overset{*}{\sqrt{z^2 - 1}} = (z + 1)^{1/2}(z - 1)^{1/2}$$

is the desired mapping. The inverse mapping $\check{f}: E \to H$ is given by a suitable determination of $z = \check{f}(w)$ in the equation $w^2 + 1 = z^2$.

This example has numerous uses, illustrated in Exercise 13. It would have been fairly difficult to guess that the solution would be a function of the form $f(z) = A\overset{*}{\sqrt{z^2 - 1}} + B$; the Schwarz-Christoffel formula was instrumental in providing us with this clue.

EXERCISES

1. If we label the vertices of a bounded polygonal domain E so that the boundary is traversed in a positive sense (E to the left of each segment), we get a list of exterior angles $\beta_1, \ldots, \beta_n$ such that $\sum_{k=1}^{n} \beta_k = 2\pi$. Show that this list of angles *alone* is not sufficient to determine the shape of E. Construct two domains E_1 and E_2 that are not even similar (congruent under scaling, rotation, and translation), but have the same set of exterior angles $\{\beta_1, \ldots, \beta_n\}$.

Hint: You must take $n > 3$ ($n = 3$ gives a triangle). Why?

2. If the exterior angles $\beta_1, \ldots, \beta_n$ at successive vertices, and the lengths l_k = length $[w_k, w_{k+1}]$ of successive sides, are specified for a polygonal domain E, to what extent is the shape, position, orientation, etc., of E determined?

Answer: Two such figures can differ only by a rotation and translation, a rigid motion (they are *congruent*).

3. Using Exercise 11, Section 9.2, explain why the location of *three* of the points $x_1, \ldots, x_n$ in the Schwarz-Christoffel formula can be adjusted at will as long as the order in which the points occur is preserved.

4. If z^μ is determined (for real μ) so its discontinuities lie along the ray $[0, -i\infty)$, verify that

$$(z+1)^{-3/4}(z-1)^{-3/4} = (z^2-1)^{-3/4} = \frac{1}{(z^2-1)^{3/4}}$$

on the upper half plane and on the segments $(-\infty, -1)$, $(-1, +1)$, and $(+1, +\infty)$. Start by calculating the arguments of

$$w = (x+1)^{-3/4} \cdot (x-1)^{-3/4} \quad \text{and} \quad w = (x^2-1)^{-3/4},$$

for real x on these segments.

5. Verify, by direct change of variable, that

$$\int_1^\infty \frac{1}{\sqrt{x}\sqrt{x^2-1}}\,dx = \int_0^1 \frac{1}{\sqrt{x}\sqrt{1-x^2}}\,dx.$$

Use this result to prove that the mapping

$$w = f(z) = \int_0^z \frac{1}{\zeta^{1/2}(\zeta-1)^{1/2}(\zeta+1)^{1/2}}\,d\zeta$$

maps the half plane $\text{Im}(z) > 0$ onto a *square* in the w-plane (not a rectangle); take principal determinations of square root. Show that the sides of the square are *parallel* to the real and imaginary axes by examining $f'(x + i0)$. The side lengths are given by the definite integrals above; show that the length is also given by

$$\tfrac{1}{2}\int_0^1 t^{-3/4}(1-t)^{-1/2}\,dt$$

(which is related to the Γ-function).

6. Show that

$$w = f(z) = \int_0^z \frac{1}{(\zeta-1)^{3/4}(\zeta+1)^{3/4}}\,d\zeta \quad \text{(principal determinations of } \tfrac{3}{4} \text{ powers)}$$

maps the upper half plane onto a triangle. Demonstrate that the triangle has position and side lengths as shown in Figure 9.22, where

$$a = \int_0^1 \frac{1}{(1-x^2)^{3/4}}\,dx.$$

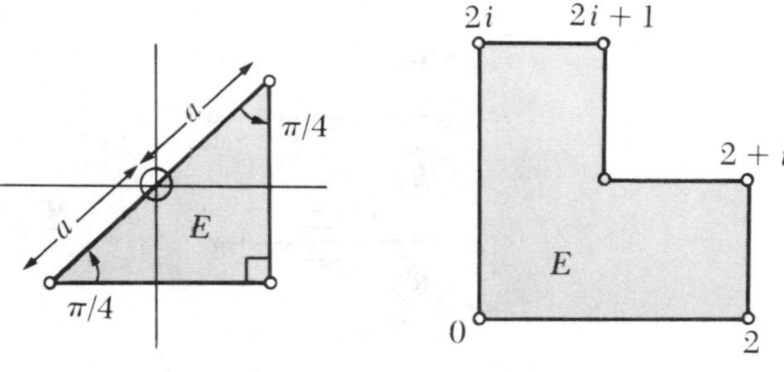

Figure 9.22 **Figure 9.23**

7. Devise a mapping that transforms the half plane $\text{Im}(z) > 0$ onto the polygonal domain shown in Figure 9.23.

Hint: Take $x_1 = -1$, $x_2 = 0$, $x_3 = +1$. By symmetry it seems reasonable to try $x_4 = k$, $x_5 = \infty$, $x_6 = -k$ (where $k > 1$ is to be determined).

8. Show that

$$w = \int_0^z (1 - \zeta^n)^{-2/n}\, d\zeta$$

maps the disc $|z| < 1$ onto the interior of a *regular* polygon of order n. Express the side length as a definite integral.

9. Apply the Schwarz-Christoffel formula to the degenerate polygonal domain E (shown in Figure 9.24) which has vertices

$$w_1 = \infty, \quad w_2 = +i\pi, \quad w_3 = \infty, \quad w_4 = \infty$$
$$\beta_1 = +\pi, \quad \beta_2 = -\pi, \quad \beta_3 = +\pi, \quad \beta_4 = +\pi.$$

Show that $w = \log^*(z^2 - 1)$ maps the half plane conformally onto E. (Take $\log^*(re^{i\theta}) = \log r + i\theta$ for $0 < \theta < 2\pi$.)

Hint: Take $x_1 = -1$, $x_2 = 0$, $x_3 = +1$; by symmetry, x_4 should equal ∞. Integrate from $p = 0$ to z; the integral (24) can then be expressed in closed form. Verify the mappings properties of $\log^*(z^2 - 1)$ by direct calculations.

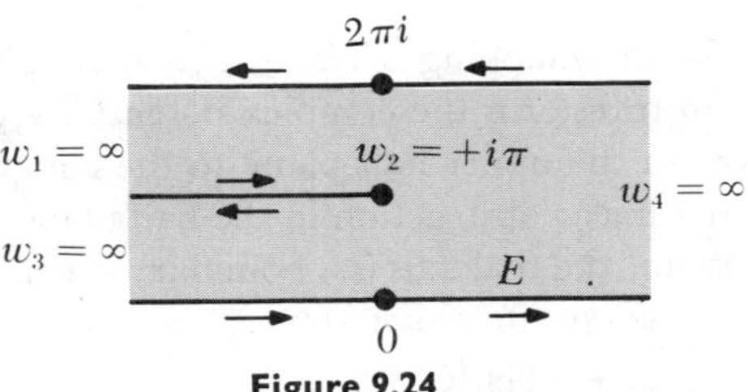

Figure 9.24

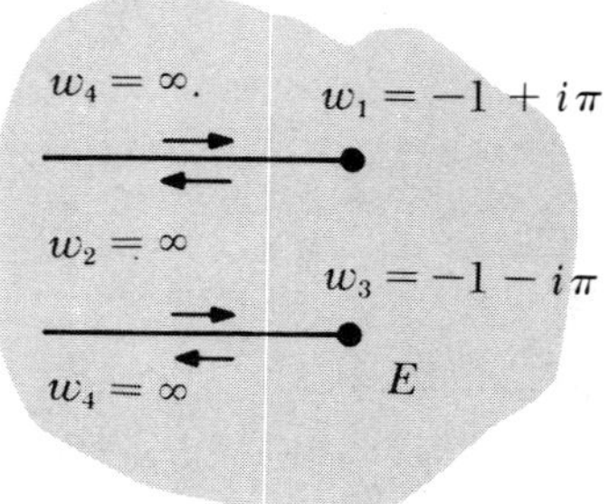

Figure 9.25

10. Consider the degenerate polygonal domain E shown in Figure 9.25, which has

$$\begin{array}{llll} w_1 = -1 + i\pi & w_2 = \infty & w_3 = -1 - i\pi & w_4 = \infty \\ \beta_1 = -\pi & \beta_2 = +\pi & \beta_3 = -\pi & \beta_4 = -\pi. \end{array}$$

In (23), take $x_1 = -1$, $x_2 = 0$, $x_3 = +1$; by symmetry, it is reasonable to expect that $x_4 = \infty$. Show that the antiderivative in (23) can be evaluated in closed form:

$$F(z) = \int_1^z (\zeta - 1)(\zeta + 1)\zeta^{-1}\, d\zeta = \int_1^z \frac{(\zeta^2 - 1)}{\zeta}\, d\zeta = \tfrac{1}{2}[\zeta^2 - \log^*(\zeta^2)]$$

taking $\log^*(re^{i\theta}) = \log r + i\theta$ $(0 < \theta < 2\pi)$. Show that F maps the upper half plane to a doubly cut plane E' similar to E by locating the points $w_k = F(x_k + i0)$, and by determining $\arg(w_{k+1} - w_k)$ and $|w_{k+1} - w_k|$. How should A and B be chosen so that $f(z) = A \cdot F(z) + B$ maps the half plane onto E?

Hint: Obviously $w_3 = 0$ (due to base point $p = 1$ in integral). We cannot evaluate $w_3 - w_1$ by integrating along real interval $(-1, +1)$ because we get $+\infty - \infty$; instead, integrate from $+1$ to -1 along the arc $|z| = 1$, $\text{Im}(z) > 0$ to prove that $w_3 - w_1 = +2\pi i$.

Answer: $E' = \mathbf{C}$ with horizontal segments to the *right* of $w_3 = 0$ and $w_1 = 2\pi i$ deleted; $A = -1$, $B = (\pi i - 1)$.

11. Use the exponential mapping, and the mapping $w = f(z)$ in Exercise 9, to show that $w = e^z + z$ maps the strip $-\pi/2 < \text{Im}(z) < \pi/2$ onto the doubly cut plane in Figure 9.25. (We used this fact to solve heat flow and electrostatic problems in Exercise 4, Section 8.6.)

12. Use the mapping $T: H \to E$ in Example 9.13, and its inverse $z = \check{T}(w)$, to transform the complex potential $g(z) = -z$ of a constant velocity flow in the upper half plane to the complex potential $\tilde{g}(w)$ of a flow around a linear obstruction in the half plane (the segment $[0, +i]$ is the obstruction; the real axis is a boundary wall). Determine the inverse mapping $z = \check{T}(w)$ in closed form; use it to calculate the velocity function $-d\tilde{g}/dw$ explicitly.

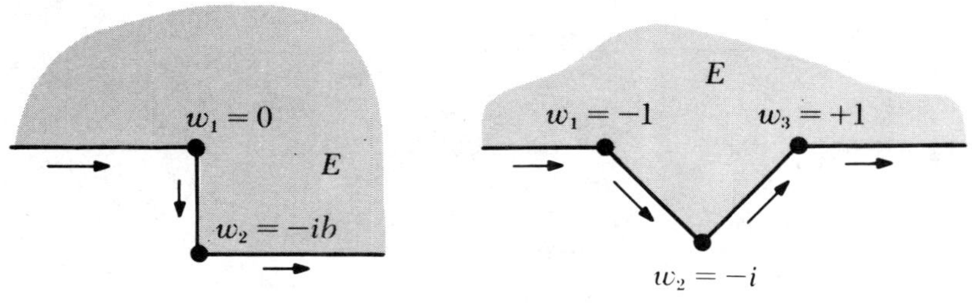

Figure 9.26 **Figure 9.27**

13. Use formula (23) to obtain a mapping of H onto the domain shown in Figure 9.26. Take $x_1 = -1$, $x_2 = +1$; take base point $p = 0$ in the integral. Note that $w_3 = \infty$ is not really a vertex, since $\beta_3 = 0$.

Answer: $f(z) = A \int_0^z \frac{\sqrt{\zeta + 1}}{\sqrt{\zeta - 1}}\, d\zeta + B$ (principal determination of square roots); $A = b/k > 0$, where $k = \int_{-1}^{+1} \sqrt{\frac{1+x}{1-x}}\, dx$; $B = -ib/2$.

14. Consider the constant velocity flow with complex potential $g(z) = -z$, in H. Using the map f described in Exercise 13, we get a "perturbed" flow with potential $\tilde{g}(w) = -\check{f}(w)$ in E. Although we cannot solve explicitly for $w = f(z)$, or its inverse $z = \check{f}(w)$, we may nevertheless calculate derivatives; using implicit differentiation we get

$$\frac{d\check{f}}{dw} = \left[\frac{1}{\left(\dfrac{df}{dz}\right)} \Bigg|_{z=\check{f}(w)} \right].$$

Use these ideas to demonstrate the following properties of the velocity field $\tilde{G}(w) = -\overline{d\tilde{g}/dw}$ of the flow in E:

(i) $\lim_{w \to w_1} |\tilde{G}(w)| = +\infty$

(ii) $\lim_{w \to w_2} |\tilde{G}(w)| = 0$

(iii) $\lim_{w \to \infty} \tilde{G}(w) = k/b = 1/A$ (A from Exercise 13).

Sketch the general shape of flow lines in E. Does the behavior near w_1 and w_2 agree (intuitively) with (i) and (ii)?

Hint: If $w \to w_k$, then $z = \check{f}(w) \to \check{f}(w_k) = x_k$; if $w \to \infty$ in E, then $z = \check{f}(w) \to \infty$ in H.

15. Use formula (23) to obtain a mapping of the deformed half plane (Figure 9.27) onto the upper half plane H. Take $x_1 = -1$, $x_2 = 0$,

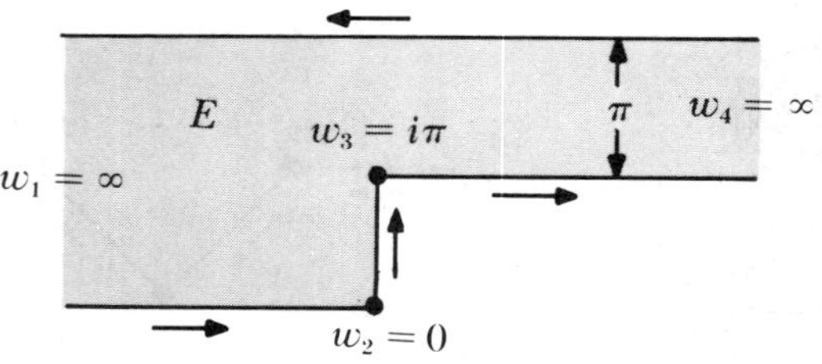

Figure 9.28

$x_3 = +1$ (and $p = +1$ in the integral).

$$Answer:\ f(z) = A\int_1^z \frac{(\zeta+1)^{1/4}(\zeta-1)^{1/4}}{\sqrt{\zeta}}\,d\zeta + B;\ \ B = +1;$$

$$A = k/\sqrt{2} \text{ where } k = \int_0^1 \frac{(1-x^2)^{1/4}}{\sqrt{x}}\,dx.$$

16. Transform the flow with potential $g(z) = -z$ in H to a flow in the domain E in Exercise 14. Show that the fluid velocity has limit *zero* as $w \to w_2 = -i$ from within E. Sketch a few flow lines in E.

17. Find the conformal map that carries the upper half plane to the "offset channel" shown in Figure 9.28. Use $x_1 = 0$, $x_2 = 1$, and $x_4 = \infty$; then $x_3 = k > 1$ (some constant to be determined along with A and B). Determining the constants A and k is not so easy. Obviously, $\arg(A) = 0$. One condition on A and k is $|w_3 - w_2| = \pi$. The other condition is the more subtle requirement that the imaginary parts of points on (w_4, w_1) and (w_1, w_2) must differ by 2π. These sides correspond to segments $(-\infty, 0)$ and $(0, 1)$ in the real axis. If $z(\theta) = \gamma_r(\theta) = re^{i\theta}$ $(0 \le \theta \le \pi)$, and $z' = r + i0$ and $z'' = -r + i0$ are the end points of this arc, we should have

$$2\pi = \operatorname{Im}[f(z'') - f(z')] = \operatorname{Im}\left[A\int_{z'}^{z''} \prod_{k=1}^{n} (\zeta - x_k)^{-\beta_k/\pi}\,d\zeta\right]$$

for all $r > 0$. Let $r \to 0$ to get the relation $A = \sqrt{k}/2$. (Notice that the integral has a pole at $x_1 = 0$.)

Answer: B depends on base point p in (24); relations $A = \sqrt{k}/2$ and

$$\pi = |w_3 - w_3| = |A|\int_1^k \frac{1}{x}\sqrt{\frac{k-x}{x-1}}\,dx$$

determine A and k.

REFERENCES

[1] L. Ahlfors, Complex Analysis (2nd ed.), McGraw-Hill Book Co., 1966.

[2] R. C. Buck, Advanced Calculus (2nd ed.), McGraw-Hill Book Co., 1965.

[3] H. Cartan, Elementary Theory of Analytic Functions of One or Several Complex Variables, Addison-Wesley Publishing Co., 1963.

[4] R. Churchill, Complex Variables and Applications (2nd ed.), McGraw-Hill Book Co., 1960.

[5] J. De Pree and C. Oehring, Elements of Complex Analysis, Addison-Wesley Publishing Co., 1969.

[6] R. Feynman, R. Leighton, and M. Sands, The Feynman Lectures on Physics (v. 1–v. 3), Addison-Wesley Publishing Co., 1964.

[7] L. Ford, Automorphic Functions (corrected ed.), Chelsea Publishing Co., 1951.

[8] W. Fulks, Advanced Calculus (2nd ed.), John Wiley and Sons, Inc., 1969.

[9] E. Hille, Analytic Function Theory (v. 1), Ginn and Co., 1959.

[10] E. Hille, Analytic Function Theory (v. 2), Ginn and Co., 1962.

[11] R. Johnson and F. Kiokemeister, Calculus (4th ed.), Allyn and Bacon, Inc., 1969.

[12] W. Kaplan, Advanced Calculus, Addison-Wesley Publishing Co., 1952.

[13] W. Kaplan, Introduction to Analytic Functions, Addison-Wesley Publishing Co., 1966.

[14] H. Kober, Dictionary of Conformal Representations, Dover Publications, 1957.

[15] N. Levinson and R. Redheffer, Complex Variables, Holden-Day, Inc., 1970.

[16] Z. Nehari, Intorduction to Complex Analysis (revised ed.), Allyn and Bacon, Inc., 1968.

[17] Z. Nehari, Conformal Mapping, McGraw-Hill Book Co., 1952.

[18] R. Nevanlinna and V. Paatero, Introduction to Complex Analysis, Addison-Wesley Publishing Co., 1969.

[19] L. Pennisi, Elements of Complex Variables, Holt, Rinehart and Winston, Inc., 1963.

[20] W. Rudin, Principles of Mathematical Analysis (2nd ed.), McGraw-Hill Book Co., 1964.

SYMBOL INDEX

INDEX

Page numbers in *italics* indicate illustrations.